ENDLESS FORMS

ENDLESS FORMS

Species and Speciation

EDITED BY
Daniel J. Howard
Stewart H. Berlocher

New York   Oxford

Oxford University Press

1998

Oxford University Press

Oxford   New York
Athens   Auckland   Bangkok   Bogotá   Bombay   Buenos Aires   Calcutta
Cape Town   Chennai   Dar es Salaam   Delhi   Florence   Hong Kong   Istanbul
Karachi   Kuala Lumpur   Madrid   Melbourne   Mexico City   Mumbai
Nairobi   Paris   São Paulo   Singapore   Taipei   Tokyo   Toronto   Warsaw

and associated companies in
Berlin   Ibadan

Library of Congress Cataloging-in-Publication Data
Endless forms : species and speciation / edited by Daniel J. Howard
and Stewart H. Berlocher.
p.   cm.
Includes bibliographical references and index.
ISBN 0-19-510900-7; 0-19-510901-5 (pbk.)
1. Species.   I. Howard, Daniel J.   II. Berlocher, Stewart H.
QH380.E54   1998
576.8'6—dc21   97-31461

9 8 7 6 5

Printed in the United States of America
on acid-free paper

Preface

One of the triumphs of the Modern Synthesis, as trumpeted by Mayr in his imposing 1963 book, *Animal Species and Evolution*, was the construction of an essentially universal theory of species formation in animals—the theory of geographic speciation. One might have thought that this achievement would suppress thinking on other modes of speciation. Yet even as Mayr applied the finishing touches to an edifice with roots stretching back to Wagner in the 1860s, the challenges were underway. The most influential assault on the idea that all animal speciation involved geographic isolation was led by Guy Bush, who, while a firm adherent to the theory of geographic speciation, nonetheless came to question its universality. Bush promulgated the idea that sympatric speciation via a shift in host plant or habitat not only was possible, but was also the most likely mode of speciation for phytophagous insects and other habitat specialists. Bush's thinking was influenced by Thoday and Gibson's provocative but frustratingly unrepeatable 1962 demonstration that disruptive selection in the laboratory could lead to the evolution of reproductive isolation, and by John Maynard Smith's 1966 theoretical model indicating that sympatric speciation was possible, albeit under fairly restrictive conditions. However, the real driving force in Bush's thinking was his extensive work on the phytophagous fruit flies in the genus *Rhagoletis*.

Bush noted that species groups within the genus displayed two patterns of host-plant usage, distributional relationships and morphological change. In some taxa, such as the *R. suavis* species group, the species infest the same genus of host plants, are morphologically distinct, and are parapatric or allopatric in distribution. Bush argued that speciation in these flies occurred, in accordance with Mayr's views, via geographic isolation. In other taxa, however, such as the *R. pomonella* complex, the species are morphologically similar, are broadly sympatric, and infest distantly related host plants. Bush argued that these species were likely to have evolved sympatri-

cally as a result of alterations in genes associated with host-plant selection. He was particularly swayed by historical records indicating that a host race (a partially isolated population specializing on a particular host) had originated on apples in the mid-nineteenth century in the Hudson River Valley of New York, well within the range of its source, the native hawthorn-infesting population of *R. pomonella*.

The work of Bush and colleagues on *Rhagoletis* stimulated a great controversy on modes of speciation that persists to this day, but perhaps more important, it aroused renewed interest in speciation and catalyzed a great increase in process-oriented studies of speciation. Systematists continued to play an important role in the field, but the influence of other disciplines grew, most noticeably evolutionary genetics. With new tools to plumb the genome, the genetics of speciation was a natural target for evolutionary geneticists, and the 1970s witnessed a wealth of investigations devoted to measuring allozymic differences between populations at various stages of evolutionary divergence.

Another target of evolutionary geneticists was hybrid zones, and during the 1980s studies of hybrid zones flourished. This work was driven by a new appreciation that hybrid zones are a common phenomenon, even among animals, and that they provide a window on evolutionary process. The stability of hybrid zones was underscored during this period, and controversy grew over the factors that account for the maintenance of zones. Traditional questions about the nature and evolution of reproductive isolation in hybrid zones persisted, but they were joined by new questions about the genetic architecture of reproductive isolation and the properties of the species boundary.

Long-term work on incipient speciation combining genetic, ecological, and behavioral studies began to bear fruit in the 1980s. Among the notable examples were the efforts of Carson, Kaneshiro, Templeton, and their students on Hawaiian *Drosophila*, the work of Patton and his students on small-mammal populations, and the work of Bush and his colleagues on *Rhagoletis*.

The pace of new developments has not abated in the 1990s. A flush of hybridization studies devoted to understanding the genetic basis of reproductive isolation between closely related species, particularly closely related *Drosophila*, have recently appeared. Perhaps more important, the development of new techniques to detect molecular variation, most notably restriction fragment length polymorphisms (RFLPs) and random amplification of polymorphic DNA (RAPDs), has made it possible to quickly uncover genetic markers in virtually any group of organisms. These markers combined with the development of new statistical methods have considerably simplified and sped up the identification and mapping of quantitative trait loci. These new methods promise to revolutionize our understanding of the genetic basis of evolutionarily important traits, such as those directly responsible for reproductive isolation. Evolutionists can finally move beyond the small group of organisms that have been well characterized genetically via morphological markers and tap into the wealth of information stored in the genomes of organisms that are well understood from other perspectives.

The rapid pace of developments in the field and the length of time that has elapsed since the publication of the last book devoted to a broad overview of speciation, *Speciation and Its Consequences* edited by Otte and Endler in 1989, compelled us to invite and organize the essays that comprise the present volume. The essays are dedicated to Guy L. Bush in recognition of his contributions to the field over the course of 30 years.

We want to take this opportunity to thank the contributors to this volume for their thought-provoking essays and the care with which they were prepared. You made our jobs as editors much easier. At the same time, our families are acutely aware of the many late nights and early mornings that went into putting this volume together. Many thanks for your patience and support Jeni, Brittany, and Allison, and Jeanine and Austin.

Las Cruces, New Mexico   D. J. H.
Urbana, Illinois   S. H. B.
July 1998

Contents

Contributors

Michael L. Arnold
Department of Genetics
University of Georgia
Athens, Ga. 30602
U.S.A.

Stewart H. Berlocher
Department of Entomology
University of Illinois at Urbana-Champaign
Urbana, Ill. 61801
U.S.A.

Guy L. Bush
Department of Zoology
Michigan State University
East Lansing, Mich. 48824-1115
U.S.A.

Roger Butlin
Department of Biology
The University of Leeds
Leeds LS2 9JT
U.K.

Michael L. Cain
Department of Biology
New Mexico State University
Las Cruces, N. Mex. 88003
U.S.A.

Jiming Chu
Department of Biology
New Mexico State University
Las Cruces, N. Mex. 88003
U.S.A.

Maria Nazareth F. da Silva
Coordenacão de Entomologia
Instituto Nacional de Pesquisas da Amazônia
CP 478, 69.083
Manaus, AM
Brazil

Kevin de Queiroz
Department of Vertebrate Zoology
National Museum of Natural History
Smithsonian Institution
Washington, D.C. 20560
U.S.A.

Simon K. Emms
Mail #4293
University of Saint Thomas
2115 Summit Ave.
St. Paul, MN 55105

Jeffrey L. Feder
Department of Biological Sciences
University of Notre Dame
Notre Dame, Ind. 46556
U.S.A.

Mike Gardner
Galton Laboratory
Department of Biology
University College London
4 Stephenson Way
London NW1 2HE
England

B. Rosemary Grant
Department of Ecology and Evolutionary Biology
Princeton University
Princeton, N.J. 08544-1003
U.S.A.

Peter R. Grant
Department of Ecology and Evolutionary Biology
Princeton University
Princeton, N.J. 08544-1003
U.S.A.

Pamela G. Gregory
Department of Biology
269/2 Maung District
Mahasarakham University
Maha Sarakham 44000
Thailand

Urban Gullberg
Department of Plant Breeding Research
Swedish University of Agricultural Sciences
S-750 07 Uppsala
Sweden

Richard G. Harrison
Section of Ecology and Systematics
Corson Hall
Cornell University
Ithaca, N.Y. 14853
U.S.A.

David J. Hawthorne
Department of Entomology
University of Maryland
College Park, Md. 20742
U.S.A.

Charles S. Henry
Department of Ecology and Evolutionary Biology
University of Connecticut
U-43, 75 North Eagleville Road
Storrs, Conn. 06269
U.S.A.

Gregory D. Hocutt
Department of Biology
Arizona State University
Tempe, Ariz. 85287
U.S.A.

Hope Hollocher
Department of Ecology and Evolutionary Biology
Princeton University
Princeton, N.J. 08544
U.S.A.

Daniel J. Howard
Department of Biology
New Mexico State University
Las Cruces, N. Mex. 88003-8001
U.S.A.

Chris D. Jiggins
Galton Laboratory
Department of Biology
University College London
4 Stephenson Way
London NW1 2HE
England

Paul A. Johnson
Department of Plant Biology
Swedish University of Agricultural Sciences
Box 7080
S-750 07 Uppsala
Sweden

Alexey S. Kondrashov
Section of Ecology and Systematics
Corson Hall
Cornell University
Ithaca, N.Y. 14853
U.S.A.

H. A. Lessios
Smithsonian Tropical Research Institute
Box 2072
Balboa
Panama

Nathan R. Lovejoy
Section of Ecology and Systematics
Corson Hall
Cornell University
Ithaca, N.Y. 14853
U.S.A.

Mark R. Macnair
Department of Biological Sciences
University of Exeter
Prince of Wales Rd
Exeter EX4 6EZ
U.K.

James Mallet
Galton Laboratory
Department of Biology
University College London
4 Stephenson Way
London NW1 2HE
England

Therese Ann Markow
Department of Biology
Arizona State University
Tempe, Ariz. 85287
U.S.A.

Xulio Rodríguez Maside
Departamento de Biologia Celular y Molecular
Facultad de Ciencias
Universidad de La Coruña
La Coruña
Spain

Amy R. McCune
Section of Ecology and Systematics
Corson Hall
Cornell University
Ithaca, N.Y. 14853
U.S.A.

W. Owen McMillan
Departamento de Biología
Universidad de Puerto Rico
Río Piedras
P.O. Box 23360
San Juan, PR 00931-3360

Steph B. J. Menken
Institute for Systematics and Population Biology
University of Amsterdam
P.O. Box 94766
1090 GT Amsterdam
The Netherlands

Horacio Fachal Naveira
Departamento de Biologia Celular y Molecular
Facultad de Ciencias
Universidad de La Coruña
La Coruña
Spain

Stephen R. Palumbi
Department of Organismic and Evolutionary Biology
Harvard University
Cambridge, Mass. 02138
U.S.A.

James L. Patton
Museum of Vertebrate Zoology
University of California
Berkeley, Calif. 94720
U.S.A.

Stephen D. F. Phillips
Biological and Medical Sciences
University of St Andrews
St Andrews
Scotland KY16 9TS

Dorothy Pashley Prowell
Department of Entomology
Louisiana State University
Baton Rouge, La. 70803
U.S.A.

Marta Reece
Department of Biology
New Mexico State University
Las Cruces, N. Mex. 88003
U.S.A.

William R. Rice
Department of Biology
University of California
Santa Cruz, Calif. 95064
U.S.A.

Michael G. Ritchie
Biological and Medical Sciences
University of St Andrews
St Andrews
Scotland KY16 9TS

Peter Roessingh
Institute for Systematics and Population Biology
University of Amsterdam
P.O. Box 94766
1090 GT Amsterdam
The Netherlands

Dolph Schluter
Department of Zoology and Center for Biodiversity Research
University of British Columbia
Vancouver, BC
V6T 1Z4 Canada

Svetlana A. Shabalina
Section of Ecology and Systematics
Corson Hall
Cornell University
Ithaca, N.Y. 14853
U.S.A.

Kerry L. Shaw
Department of Organismic and Evolutionary Biology
Harvard University
Cambridge, Mass. 02138
U.S.A.

Franco Spirito
Department of Genetics and Molecular Biology
'La Sapienza' University
Rome
Italy

Alan R. Templeton
Department of Biology
Washington University
St. Louis, Mo. 63130-4899
U.S.A.

Sara Via
Department of Zoology and Department of Entomology
University of Maryland
College Park, Md. 20742
U.S.A.

Marta Martínez Wells
Department of Ecology and Evolutionary Biology
University of Connecticut
U-43, 75 North Eagleville Road
Storrs, Conn. 06269
U.S.A.

John H. Werren
Biology Department
University of Rochester
Rochester, N.Y. 14627
U.S.A.

Chung-I Wu
Department of Ecology and Evolution
University of Chicago
Chicago, Ill. 60637
U.S.A.

Lev Yu. Yampolsky
Section of Ecology and Systematics
Corson Hall
Cornell University
Ithaca, N.Y. 14853
U.S.A.

Part I

History

1

Origins

A Brief History of Research on Speciation

Stewart H. Berlocher

People have thought about biological origins for millennia, and the modern Western scientific literature on speciation can be traced well into the nineteenth century. But given recent advances in understanding speciation—advances evinced by the chapters in this book—the newcomer, whether a beginning student or a visitor from some other field, may well ask why one should read the older literature. There are several answers to this question. Acknowledgment of intellectual debt is certainly one of them, but much more is involved. None of the threads of speciation research exemplified by the following chapters can be completely understood without knowing the context of the arguments and issues in each thread. Moreover, the best of the older speciation literature can still provide us today with ideas, and with pointers toward intriguing, unstudied taxa ideally suited for testing current hypotheses. The history of evolutionary biology is more like the history of mathematics—a legacy of incompletely answered questions that have taxed the imagination of a long sequence of original thinkers—than that of, say, molecular biology, where literature older than five years is largely irrelevant.

In this chapter I present a brief history of speciation research from the viewpoint of a biologist. My goal is primarily to provide a framework for the chapters in this volume, and to suggest readings to those new to speciation. Various aspects of speciation have been discussed in works dealing with the history of evolutionary biology in general (Mayr, 1982; Mayr and Provine, 1980; Provine, 1971, 1979, 1986), but much is lost in this piecemeal approach. Speciation needs a historical study of its own (and one much more extensive than I have space for here). As an organizational tool I have rather arbitrarily broken the continuous flow of history into eras, artificial but useful.

Grains of Salt: Problems with the Older Literature on Speciation

For all its value, I must note certain difficulties posed by the older (pre-1937) speciation literature. First, no word

was consistently used for the process of species formation before 1939, when a symposium with the word "speciation" in the title (Cole, 1940) finally established the term. The American entomologist Orator F. Cook had coined the term decades earlier, but it had seen only sporadic usage. In Cook's words, "Speciation, to give the . . . process a name, is the origination or multiplication of species by subdivision, usually, if not always, as a result of environmental incidents" (Cook, 1906, p. 506; he meant primarily geographic isolation by the term "environmental incidents"). In the German literature the term "Artbildung" had gained wide usage, but in the English-speaking world prior to 1939 a variety of terms such as species formation, species fission, the origin of species, specification, and in part "the species problem" confound efforts to follow the historical development of ideas on speciation.

The 34-year lag between proposal of the term "speciation" and its broad acceptance illustrates a general feature of the older literature: newer terms and ideas coexist with older ones for lengthy periods. Overlap between old and new occurs in all science, but seems to reach a maximum in the evolutionary literature. For example, Nihlson-Ehle demonstrated experimentally that so-called "blending" inheritance could be explained in terms of polygenic (or multifactorial) Mendelian inheritance in 1909 (see Provine, 1971, pp. 115–118), but 30 years later Carl Hubbs, the premier American ichthyologist of his time, was seemingly reluctant to abandon the "generally discarded Galtonian system" to explain the inheritance of population differences in vertebral counts in fish, as "we would need to argue that a very high number of multiple factors is operating—if multiple factors are involved" (Hubbs, 1940, p. 206).

This coexistence of the old and the new leads the reader through surges of excitement and disappointment when reading the older literature: viewed through modern eyes, it is a mosaic of brilliant insight and befuddlement (or worse). A writer with a keen and modern-sounding understanding of, say, the role of geographic isolation in speciation may be revealed on the next page to favor neo-Lamarkism (inheritance of acquired characters), orthogenesis (evolution constrained to a linear trend), or vitalism (the existence of a life force inexplicable by chemistry).

A further great impediment in reading some older literature is that the science is mixed with social commentary of a kind most late twentieth-century workers view as disconcerting, and in some cases repellent. A browsing expedition to the evolution shelves of one's local university library should be required of all new students of evolutionary biology. Some journal articles, and many of the texts and books for general readership between 1900 and 1930, include misleading and sometimes appalling discussions of human racial differences, blind acceptance of the idea of evolutionary progress, naive ideas on eugenics, and speculative projections of future human evolution. Even R. A. Fisher's (1930) renowned *The Genetical Theory of Natural Selection* has several chapters on the genetics of class differences and the like. One cannot learn from the old literature without making allowance for historical context. At the same time, there is more than hard science to be learned from the older literature. (For a realistic contemporary view of eugenics, I suggest Haldane [1949].)

The Nineteenth Century: Species Concepts, Geography, Chance, and Selection

The mid-nineteenth century independent discovery of natural selection by Darwin and Wallace (1858) tends to dwarf other evolutionary insights of the time. Yet many current debates, over species concepts, allopatric versus sympatric speciation, the role of chance in producing evolutionary change, and many others, took form in the last century.

"What is a species?" is perhaps the oldest question in speciation research (for current answers, see chapters by Shaw, de Queiroz, Templeton, and Harrison, this volume). The nineteenth century was dominated by a simple morphological species concept—tigers look like tigers and lions look like lions. (The morphological concept remains the most useful concept for taxonomists working on groups with many undescribed species, because it is efficient and effective in the majority of cases.) At the same time, however, "modern" species concepts, like small mammals in the time of dinosaurs, subsisted with the morphological species concept (also referred to as the "typological" concept [Mayr, 1959]). Consider the "biological species concept," which defines a species in terms of reproductive isolation from other species. Dobzhansky (1935) is often considered to have originated the biological species concept (e.g., Galiana et al., 1995, p. 281), but the idea that species do not interbreed with other species is much older (Mayr [1963, pp. 14–15] traces the germ

of the idea to John Ray in 1692). The American entomologist B. D. Walsh (Smith and Sheppard, 1990) clearly understood that lack of interbreeding could discriminate species when morphology could not: "two supposed species . . . do not now in general mix sexually together, or if geographically separated they would not do so supposing them to be placed in juxtaposition" (Walsh, 1863, p. 220). Walsh recognized morphologically identical insects specialized on different host plants as what we would now call sibling or morphologically cryptic species.

The idea of defining species based on reproductive isolation was well enough known in the nineteenth century that both Darwin and Wallace discussed it. However, they both rejected it—a positively "postmodern" attitude. For example, A. R. Wallace states in a monograph on papilionid butterflies of the Malay archipelago that species could be defined as follows: "Species are merely those strongly marked races or local forms which when in contact do not intermix, and when inhabiting distinct areas are generally believed to have had a separate origin, and to be incapable of producing fertile offspring" (Wallace, 1864, p. 32). Wallace chose not to use this definition because, among other things, "the test of nonmixture is useless, except in those rare cases where the most closely allied species are found inhabiting the same region" (p. 31). (Wallace was working with many insular, allopatric populations.) Interestingly enough, Wallace did believe that a defining feature of species was "the constant transmission of some characteristic peculiarity of organization" (p. 13), a phrase that hints at the idea of autapomorphy and the "phylogenetic" species concepts that have arisen in recent decades.

One of the clearest pre-Synthesis statements of the evolutionary implications of reproductive isolation (or rather its converse, gene flow within species) was made by the British entomologist Poulton in 1904. In a famous after-dinner speech to the Entomological Society of London, Poulton not only stressed the importance of reproductive isolation, indicated that it often arises in geographic isolation, and outlined a rudimentary model for the evolution of reproductive isolation, but clearly recognized the evolutionary consequences of interbreeding: "By inter-breeding the favorable variations arising in one direction are combined with others arising in different directions; by the kaleidoscopic changes produced by inter-breeding more varied results are presented for selection, and the beneficial qualities arising in one part of the mass may quickly become the heritage of the whole" (Poulton, 1904, p. cxvi). I have placed Poulton's statement with nineteenth-century definitions because in 1904 he had yet to give the 1900 rediscovery of Mendel much attention (Poulton, 1908, p. xiii).

The question of whether the first stage of speciation absolutely requires geographic isolation also emerged in the nineteenth century, and remains vigorously debated today. No fewer than six chapters in this volume, by

Berlocher, Feder, Menken and Roessingh, Kondrashov et al., Johnson, and Schluter, discuss sympatric speciation, while geographic speciation is the focus of Patton and da Silva, and Lessios. Belief in sympatric speciation, speciation in which geographic isolation is unnecessary, actually predates the idea of geographic speciation (Mayr, 1949, p. 294). Mayr has consistently argued that many of the older sympatric ideas were untenable, and in the case of instantaneous speciation by single mutations of large effect (de Vries, 1901–1903) he is certainly correct. However, Mayr has consistently ignored viable models of sympatric speciation. For example, Benjamin Walsh outlined a theory in which "phytophagic varieties" (what we would now call host races of phytophagous insects) evolve in sympatry through series of stages to become "phytophagic species" (which we now call sibling species specialized on different hosts) (Walsh, 1864). Although his ideas were flawed by invoking the inheritance of acquired characters along with natural selection, Walsh deserves great credit for designing the first reasonably explicit model of sympatric speciation. Walsh's legacy of research on sympatric speciation in phytophagous insects has been well tended by Guy Bush and colleagues (see chapters by Bush, Menken and Roessingh, Feder, and Berlocher, this volume).

Allopatric speciation, speciation requiring geographic isolation in the first stages, originated with von Buch (1825; cited in Mayr, 1963, p. 483) and was developed by Moritz Wagner (1868; cited in Mayr, 1963, pp. 484–485). Both von Buch and Wagner observed that closely related species often occurred in adjacent geographic areas. (I must note that Wagner, unlike Walsh, rejected natural selection [Provine, 1986, pp. 214–215].) The late nineteenth-century development of the geographical race or subspecies concept, though its value is debated today (Wiens, 1982), was invaluable in accommodating biological variation into species concepts, and in focusing attention on geographic variation as a clue to the origin of species.

A third critical issue of nineteenth-century speciation research was the relative importance of chance and natural selection. Although a thorough understanding of genetic drift and founder effect would have to wait until well after the rediscovery of Mendelian genetics, the basic ideas are quite old. Gulick (1872), working on endemic Hawaiian snails of the genus *Achatinella*, in which strikingly different species occupy different valleys, argued that random variation after geographic colonization by a small number of founders could produce different evolutionary outcomes, even in identical environments. Darwin and Wallace (Provine, 1986, pp. 219–220), as could be expected, argued that environments were never identical, and that natural selection would always play a part in the outcome. Provine (1986, pp. 217–220) discusses the long shadow cast by Gulick's work; argument over the relative importance of natural selection and drift in speciation continues unabated today.

Fits and Starts: 1900–1925

In an ideal world the discovery of Mendel's work at the beginning of the twentieth century would have produced an immediate giant step forward in understanding speciation. Instead, the period from 1900 to about 1925 proved to be very complex, with almost as many steps backward (or at least sideways) as forward. A significant source of confusion was that Mendelism, and Mendelian terms like mutation, became linked with de Vries's (1901–1903) model of evolution by mutational saltations—evolutionary "jumps" claimed to instantly produce new species (Mayr, 1980). De Vries's ideas produced an extraordinarily unproductive polarization in early twentieth-century evolutionary biology.

Yet the more realistic views of naturalists survived. David Starr Jordan's well-known 1905 paper "The Origin of Species through Isolation" provides an overview of speciation research at the beginning of our now fading century. This short work is not only clearly written but, because Jordan quotes correspondence from other naturalists, presents more of a consensus view on speciation than can be gained from most papers of the period.

Jordan begins by stating that the importance of geographic isolation for species formation "is accepted as almost self-evident by every competent student of species" (p. 545). (In an aside that I assume is directed in part at early Mendelians, Jordan laments that not all contemporary biologists understood the role of geography, "this showing itself often in a semi-contemptuous attitude of morphologists and physiologists towards species mongers and towards outdoor students of nature generally" [p. 550]; a similar attitude is not difficult to find in today's molecular biology departments.) The interrelation of gene flow and geographic barriers is clearly understood: "Whenever the free movement of a species is possible, this involving the free interbreeding of its members, the characters of a species remain substantially uniform. Whenever free movement and interbreeding is checked, the character of the species itself is altered." (p. 546). The idea of founder events is stated in a somewhat clearer form than in Gulick (1872), with examples from livestock breeding. Jordan is very explicit about the biogeographic evidence for allopatric speciation: sister species occupying adjacent areas separated by a geographic barrier (Berlocher, this volume). Jordan claimed no originality for this idea, citing Wagner, but he discusses in detail many examples. Later, Jordan (1908) listed 34 pairs of what he called "geminate species" of fish on either side of the Isthmus of Panama (Lessios, this volume). Jordan does not discuss species concepts, but the colleagues he queried do. Grinnell states, "I am about ready to deny the value of trinomials in nomenclature. . . . Call all distinguishable groups species" (qtd. in Jordan, 1905, p. 553), an opinion that is operationally little different from the current phylogenetic species concept of, say, Nixon and Wheeler (1990).

Jordan also, with the assistance of his correspondents, discusses potential problems with an allopatric model of speciation. Stejneger notes the problem posed by with what we would now call a ring species, a pair of non-interbreeding populations connected by a geographic "ring" of interconnecting, interbreeding populations (Jordan, 1905, p. 552). Stejneger also points out that crossbills (birds) may represent a possible exception to the allopatry of sister species (p. 552); intriguingly, the broad sympatry and high mobility of the possibly eight very close species in the Red Crossbill complex still present "stumbling blocks" to an allopatric model (Groth, 1993). Stejneger also seems aware of the possibility of what we would now call allochronic isolation, lack of interbreeding due to different mating times. Jordan (1905, pp. 559–560) discusses a classic violation of his geminate species rule, the observation of species flocks in lake fish (McCune and Lovejoy, this volume), such as the Mexican atherinid genus *Chirostoma*. The *Chirostoma* of Lake Chapala do appear to be monophyletic (Barbour and Chernoff, 1984), so sympatric speciation is not ruled out.

Of course, the paper has errors when viewed from our vantage point. Jordan states in several places that the characters that distinguish closely related species are the result of isolation à la Gulick, not natural selection, and also disparages explanations of such characters by sexual selection (p. 559). Jordan essentially argues that natural selection is responsible for the big trends in evolution, but not for speciation (pp. 550, 559); current views could not be more different (see chapters by Schluter, Feder, Macnair and Gardner, and Mallet et al., this volume).

The most glaring omission from Jordan's paper is, of course, Mendelian genetics. However, other early twentieth-century biologists quickly realized that a connection between the older traditions of systematics and the new Mendelian genetics could in fact be made. By crossing interfertile species, the genetic basis of species differences could be analyzed. Despite ongoing confusion over mutation and its meaning, a beginning was made in the first two decades of this century on understanding the genetics of morphological species differences in a variety of organisms (see brief review in Dobzhansky, 1937, pp. 63–68).

Remarkably, only nine years after the rediscovery of Mendel, the prominent English Mendelian Bateson published a theory for the genetics of interspecific sterility that remains the best explanation today (see Orr, 1996). If the genotype $AaBb$ is sterile, but $AAbb$ and $aaBB$ are not (in other words, sterility is epistatic), then sterility *between* two populations can evolve (one evolving from $aabb$ to $AAbb$, the other from $aabb$ to $aaBB$) without either population ever encountering a sterility barrier itself (see Wu and Hollocher, Macnair and Gardner, and Naviera and Maside, this volume). Unfortunately, the paper was forgotten, seemingly even by Bateson; the idea was independently developed much later by Dobzhansky and Muller (Orr, 1996).

Other early work on the genetics of species differences did not suffer the fate of Bateson's 1909 paper. An important generalization to emerge from the early work was "Haldane's Rule" (Haldane, 1922), the observation that postzygotic reproductive isolation was strongest in the heterogametic sex; Prowell (this volume) discusses Haldane's Rule in the female-heterogametic Lepidoptera, and Ritchie and Phillips (this volume) assess whether prezygotic isolation follows any such rules. Another important landmark of this period is Sturtevant's 1920 paper on the discovery and potential for interspecific hybridization of *Drosophila simulans*. Sturtevant was deeply disappointed to find that the F_1 of the *D. melanogaster* $\times$ *D. simulans* cross was sterile, but the work inspired later workers, especially Dobzhansky, to search for *Drosophila* species that could be crossed successfully. (East [1921] successfully observed segregation of sterility in interspecific crosses in *Nicotiana*, but tobaccos did not offer the genetic insights of *Drosophila*.) Diverse current work such as Shaw (1996) and chapters in this volume by Howard et al., Via and Hawthorne, Wu and Hollocher, Naviera and Maside, and Markow and Hocutt can all be seen as stemming from these early attempts to locate "speciation genes" by interspecific hybridization. Overall, though, the period 1900–1925 must be judged as failing to combine the different but equally valid insights of systematics and genetics into speciation.

The Modern Synthesis Phase I: 1926–1936

From the standpoint of speciation, the Modern Synthesis (from J. Huxley's 1942 *Evolution, The Modern Synthesis*) began with the publications of Dobzhansky (1937) and Mayr (1942). But the decade before the publication of these works saw a tremendous advance in our understanding of evolution *within* species (Provine, 1971, chapter 5), without which little progress on speciation would have occurred. I divide the Synthesis period into an initial phase, which saw the blossoming of population genetics, and a second phase, in which speciation became a central focus. The period has received attention from historians, and Mayr and Provine (1980) is essential reading for anyone who wants to understand the forces that produced the synthesis.

Five works changed the face of evolutionary biology during the period 1926–1936. The Russian biologist Chetverikov (1926) had a very clear grasp of how natural selection could act on preexisting genetic variation, but he published in a time of great social turmoil in Russia, and the impact of his work was felt only indirectly through its effect on Th. Dobzhansky (Dobzhansky, 1980; Adams, 1980). The work of F. B. Sumner was quickly recognized as important by population geneticists like Wright but has received the historical attention it deserves only recently (Provine, 1979). Working with *Peromyscus* mice, Sumner in the late 1920s (summarized in Sumner, 1932) combined his systematic observations on size and coat color with the idea of polygenic Mendelian control of continuous variation originating from Nilhson-Ehle (1909) and East (1910), and with the concept of natural selection producing local differentiation, to achieve a very satisfying explanation for differentiation of local populations and subspecies. Sumner saw speciation as an extension of subspecies differentiation (Sumner, 1932, p. 59).

The remaining three works are the great trinity that produced the core of mathematical population genetics: Fisher (1930), Wright (1931), and Haldane (1932). As with Chetverikov and Sumner, the focus is on evolution within species. However, Fisher (1930, pp. 139–146) did briefly discuss speciation. He recognized that geographic isolation could set the stage for speciation, but also discusses the ideas of allochronic isolation, habitat isolation, and parapatric speciation (geographically varying selection breaking a single species into two or more species with abutting ranges). Had Fisher developed any of his ideas mathematically the history of speciation might have been quite different. The monumental paper of Wright (1931) reveals an awareness of speciation (e.g., Wagner is cited on geographical isolation, p. 99), but it would be another decade until Wright wrote explicitly about speciation.

The Modern Synthesis Phase II: 1937–1947

The 10-year period from 1936–1947 (as delimited by Mayr, 1980, p. 1) saw a series of major works that successfully brought together the widely disparate intellectual strands of not only systematics and population genetics, but also embryology and paleontology. Dobzhansky's (1937) *Genetics and the Origin of Species* and Mayr's (1942) *Systematics and the Origin of Species* ensured that speciation would henceforth occupy a prominent place on the stage of evolutionary biology. (Stebbins's [1950] *Variation and Evolution in Plants*, which devoted much more attention than the preceding works to genetic systems such as apomixis and polyploidy, is also of great significance to speciation, but its impact was to some extent lessened by being the last work of the Synthesis.)

A major factor contributing to the wide readership of Dobzhansky (1937) and Mayr (1942) is that both writers were successful "bridge-builders" (Mayr, 1980, pp. 40–42), biologists who attained in themselves a synthesis of different disciplines. This is especially true of Dobzhansky, who was a practicing systematist (especially coccinellid beetles) who then totally "retooled" himself as an experimental *Drosophila* geneticist. But more than just personal synthesis was needed. Previous workers who themselves bridged the gap between systematics and genetics, such as Sumner and Sturtevant, did not succeed in building wide bridges for the masses to cross. The ability to communicate effectively across disciplines was critical.

Dobzhansky (1937) includes four chapters devoted exclusively to speciation and the genetics of speciation, the first such review ever. Dobzhansky defined species as a "step of the evolutionary process" at which "forms . . . become incapable of interbreeding" (p. 312), a definition that reflects Dobzhansky the evolutionist interested in mechanisms, not Dobzhansky the systematist. Perhaps Dobzhansky's most important contribution was in classifying the characteristics that prevent interbreeding of species, and discussing what was then known of the genetics of those characteristics. Dobzhansky's term "isolating mechanisms" has received much recent criticism (e.g., Paterson, 1981) on the grounds that an isolation characteristic is a "mechanism" only if it has been directly selected for, as for example in reinforcement (Howard, 1993). Such criticism, while valid, overlooks how Dobzhansky's term helped focus a complex area and stimulate future work (such as chapters by Wells and Henry, Palumbi, McNair and Gardner, Ritchie and Phillips, and Markow and Hocutt, this volume). Deserving of special mention is the summary of Dobzhansky's series of papers on the genetics of hybrid sterility between *D. pseudoobscura* and *D persimilis* (starting with Dobzhanksy, 1933). This work remained the most sophisticated genetic analysis of hybrid sterility, if not of any isolating mechanism, in any organism, for some 40 years (Lewontin, 1981, p. 98); it represents the beginning of a line of powerful genetic analysis of postzygotic isolation in *Drosophila* that continues today (see Wu and Hollocher, Naviera and Maside, this volume). Yet another milestone was the estimation of chromosome inversion phylogenies in the *D. pseudoobscura* subgroup (Dobzhansky, 1937, p. 93). Forty years before the term was invented (Avise et al., 1987), Dobzhansky was combining intraspecific phylogeny with geography to study *phylogeography*. His contrast of *D. pseudoobscura* with *D. azteca*, in which "phylogenetically closely related chromosome structures tend to occur in . . . geographically close regions" (Dobzhansky, 1941, p. 124), indicates that Dobzhansky was on the road to the kind of phylogeographic insights that have been made recently, although he did not travel very far down it.

Mayr (1942) focused on the systematic side (and biogeographic) of the Synthesis. Unlike Dobzhansky, Mayr never left systematics (of birds), although he did carry out some experimental *Drosophila* work on reproductive isolation (e.g., Dobzhansky and Mayr, 1944). Mayr's contribution was to bring together a very wide range of insights on geographic variation and complex taxa from the systematics literature, and reinterpret them in terms of modern genetics; his strength was in being a widely read and effective synthesizer. His concise version of the biological species concept ("Species are groups of actually or potentially interbreeding natural populations, which are reproductively isolated from other such groups" [Mayr, 1942, p. 120]) was to become the standard species definition for the next several decades. Mayr's organization of allopatric speciation into stages (1942, pp. 159–162, Fig. 16) was to set the stage for much subsequent work.

The history of an era can never be known from its major books alone. To give a small sampling of the diversity of research, major and minor, stimulatory and not so, during the Synthesis, I briefly review two symposium proceedings. Symposia propitiously held in 1939 and 1947 form almost perfect temporal "bookends" for phase II of the Modern Synthesis. The 1939 symposium was actually two meetings published in the same volume of *American Naturalist* (Cole, 1940). Hubbs (1940) carries over the most from the past, in the form of his willingness to consider Galtonian inheritance. This aside, however, his stress on adaptation as a driving force in speciation is in agreement with much recent work (e.g., Schluter, this volume). Dice (1940a, 1940b), building on the work of Sumner in *Peromyscus*, argues for the role of natural selection in adaptation to local habitats (e.g., coat color matches local soil color) and in speciation. Dice (1940b, p. 293) also describes a textbook example of a ring species in *P. maniculatus*. Wright's (1940a) paper on breeding structure relating to speciation is his first to deal explicitly with speciation (see also Wright 1940b) and, together with his earlier work, had a profound impact on both Dobzhansky and Mayr. Mayr's (1940) analysis of speciation in birds previews his 1942 book, but the treatment of species concepts is less polished; after proposing a definition for birds, Mayr states uncharacteristically, "It remains to be seen how useful it is when applied to other groups" (p. 256). Spencer's (1940) paper will probably be surprising to many, as it reveals that he and his group at Wooster College in Ohio, away from the great metropolitan centers of genetics research, had nonetheless made significant progress studying speciation in several native North American *Drosophila* taxa. In fact, given that the natural larval habitat of Dobzhansky's beloved *D. pseudoobscura* has never been found, more progress ultimately might have been made using one of Spencer's taxa. Dobzhansky (1940) covers much of the same ground as his 1937 book. Finally, Irwin and Cumley (1940) may represent the first attempt to employ molecular data, blood group frequencies and primitive immunological analysis, to study speciation. Overall, the 1939 symposium reveals that with few exceptions, the synthesis of population genetics and systematics remained mostly conceptual.

The 1947 Princeton symposium (Jepsen et al., 1949), much better known than the 1939 meetings (and embracing the entirety of evolution, not just speciation), demonstrates the rapidity with which the Synthesis occurred (Mayr, 1980, pp. 42–43). Mayr's (1949) paper, a continuing part of his long argument for the universality of allopatric speciation, concludes with questions for the paleontologist, geneticist, and systematist. Two of the questions posed for geneticists—how many genes differentiate species that have just barely speciated, and how much gene flow is permissible between incipient species—remain critical today. Lack's (1949) paper on the significance of ecological isolation raises issues that remain contentious (see Schluter, this volume), while also providing an example of the powerful effect Mayr had on the field. Lack writes, "To account for the habitat differences between closely related species, various writers, formerly including Lack . . . suggested that where a population is spread over several habitats ecological isolation might lead to subspecific differentiation, and so eventually to speciation" (p. 300), but then goes on to state that arguments in Mayr (1942) caused him to eschew sympatric speciation. Ford's (1949) conclusion that evolution in small isolated populations was due not to drift but to "selection adjusting them to the varied environments to which they are exposed" (p. 314) was very different from that of Gulick, Jordan, and their colleagues—but unlike earlier workers he had experimental data to support his claim. Moore (1949) concluded from his review of the ecology of the frog genus *Rana* that "adaptation to different environmental conditions can in itself lead to genetic isolation" (p. 337), a view in agreement with much of the present volume, but one must bear in mind that Moore's concept of a widespread, highly variable *R. pipiens* has now been revealed to be incorrect, as "*R. pipiens*" is actually a parapatrically distributed complex of some 15 species (Hillis, 1988). Hovanitz (1949) demonstrated that, in butterflies, interspecific hybridization could clearly increase the variability of a population, but he did not develop the evolutionary implications of this observation. Mason's (1949) short paper on "genetic submergence" of *Pinus remorata* via introgressive hybridization has implications for current concerns about both species concepts and conservation biology. Interspecific gene flow is today recognized as an important source of variation for evolution within species (see Grant and Grant, and Arnold and Emms, this volume). In summary, the Princeton conference reveals a substantially broader experimental base than in 1939, and the final demise of non-Mendelian inheritance, but also revealed a population genetics still in search of data. The year 1947 was also of note because the Society for the Study of Evolution (initially, of Speciation), and its journal *Evolution* began that year.

Consolidation: 1947–1966

The years following the Modern Synthesis are probably best characterized by its most important publication, Mayr's 1963 *Animal Species and Evolution*—a masterful weaving of the strands of the Modern Synthesis into an apparently seamless whole. Among the most remarkable things about the book is its dual nature, being both an effective summary of contemporary ideas and a history of the field. The impact of the book on the next generation of speciation workers was enormous, even (or especially) on those who disagreed with it. Mayr's version of the biological species concept continued to be almost universally accepted, and allopatric speciation, in particular via founder event, was very widely accepted.

Another strong force during this period was Dobzhansky. Fitting Dobzhansky's work into any time frame is difficult, as his publications extend from 1918 to 1976 (Lewontin et al., 1981). However, I believe that the many publications by Dobzhansky and his students during the 1947–1966 period can generally be viewed as consolidating the gains of the Synthesis. From the standpoint of speciation, several achievements stand out. Prezygotic isolation was shown to vary intraspecifically, and to respond to artificial selection (e.g., Kessler, 1966). The observation of greater prezygotic isolation of *D. paulistorum* semispecies in sympatry than allopatry (Ehrman, 1965) remains one of the relatively few demonstrations of reinforcement of prezygotic isolation (Howard, 1993). This period also saw the beginning of the most comprehensive study of speciation in a taxon up to that time, the long-term study of Dobzhansky and South American colleagues on the *Drosophila willistoni* complex of sibling species, semispecies, and races showing partial reproductive isolation (summarized in Ehrman and Powell, 1982; see also relevant chapters in Levine, 1995).

The role of chromosomal change in speciation became the chief focus of British entomologist M. J. D. White (1945 and subsequent editions; see White, 1978). Impelled by his training with the cytologist Sally Hughes-Schrader and his friendship with Dobzhanksy, White argued for a central role for chromosomal rearrangements as a cause of speciation (review of recent work in Spirito, this volume).

An important and productive line of research on the spectacular endemic *Drosophila* and other organisms of Hawaii was initiated by the challenge of Zimmerman (1958). Wagner and Funk (1995) summarize recent progress.

One important discovery of the 1950s that cannot be seen as simply consolidation or extension of previous knowledge was the finding by Laven (1959) in *Culex* mosquitoes and Ehrman in the *D. paulistorum* complex (see Ehrman et al., 1995) that endosymbiotic bacteria can produce interspecific sterility. Werren (this volume) discusses current understanding of this widespread phenomenon.

The Gates Are Open Wide: 1966–1978

The year 1966 initiated a decade of enormous changes in speciation research. The period saw an explosion of enzyme electrophoresis, the return of sympatric speciation, the birth of cladistics, the final acceptance of continental drift and the growth of vicariance biogeography, numerical simulation via computer, the genesis of the neutral theory of molecular evolution, and a renewed interest in speciation in the fossil record.

The impact of enzyme electrophoresis on speciation research, initiated primarily by Hubby and Lewontin (1966), cannot be overemphasized (see Powell, 1994, for a history of molecular population genetics). For the first time it was possible to quantify the genetic divergence between species. While the genes under study had nothing to do with reproductive isolation, one could at least answer Mayr's (1947) question about the *number* of genetic differences between species. The range of species in which it was possible to study evolutionary genetics expanded from a select few—some *Drosophila* species, some butterflies, some snails—to the entirety of life. Species-level phylogenies began to be estimated wholesale (albeit with imperfect analytical tools at first). Intraspecific population structure and its relation to speciation, the focus of Wright's work of the 1930s and 1940s, began to be studied in great detail. (Ironically, many of the early electrophoretic papers did not use Wright's statistical tools, F statistics, but rather genetic distances and cluster analysis, reducing what could be learned from the new data.) The electrophoretic study of the *Drosophila willistoni* complex (Ayala et al., 1974), building on the work of Dobzhansky, is a classic example of the use of electrophoresis to study the stages of speciation hypothesized by Mayr. Hybrid zones, known before 1966 from a restricted number of cases, were discovered seemingly everywhere using electrophoresis (Butlin, this volume; reviewed in Harrison, 1993). Interspecific gene flow became easily measurable for the first time, and is now known to be both common and important (Spence and Gooding, 1990; Smith, 1992; see also Grant and Grant, and Arnold and Emms, this volume).

Of perhaps the greatest importance, enzyme electrophoresis changed the set of biologists working on speciation, in two ways. First, it "egalitarianized" the field, drawing in many new researchers who thought of themselves as "evolutionary biologists," or even just as naturalists, but not as population geneticists (too much math) or systematists (too close to taxonomy and all that nomenclature). Second, the common tool of electrophoresis, and the common set of allele frequency data it produced, aided in building experimental, not just conceptual, bridges between population genetics and systematics. At a minimum, systematists doing electrophoresis felt compelled to carry out tests for random mating and the like—to actually *be* population geneticists, if only temporarily. More substantial synthesis came later with the realization that alleles had phylogenies (or genealogies).

The advent of cladistic systematics (sparked by the German entomologist Willi Hennig's [1966] *Phylogenetic Systematics*) initially had the opposite effect on the unity of speciation research. The explicitly philosophical approach and the seeming requirement for doctrinal conformity was unattractive to both population geneticists, with their experimental approach, and the "electrophoretic naturalists," who (at least in America) had generally received no training whatsoever in philosophy. Ultimately, however, the Hennigian controversies of the 1970s brought a much needed clarity to the estimation of phylogeny. Today phylogenies, whether estimated with the maximum parsimony approach that grew out of Hennig (1966), or more sophisticated maximum likelihood approaches (Swofford et al., 1996), are critical to understanding species and speciation (see chapters by Shaw, Templeton, Harrison, de Queiroz, Berlocher, Patton and da Silva, Schluter, and Wells and Henry, this volume).

Several developments generated a resurgence of interest in allopatric speciation via vicariance (geographic subdivision of an ancestral species range). Continental drift, a major discovery of the 1960s (Hurley, 1968) helped legitimize the idea that geography might change more rapidly than the geographic range of a species (although few speciations are driven by actual between-continent vicariances). However, the strongest impetus for vicariance biogeography (Nelson and Platnick, 1981, Brundin, 1966, and Rosen, 1974, are important progenitors) was cladistic systematics, because the (ideally) fully resolved, bifurcating phylogenies it produced, much more so than earlier "impressionistic" phylogenies, invited comparison with geological events. Further, electrophoretic data provided a means for detecting "cryptic" vicariance events in the past through their effect on population structure (e.g., Avise and Smith, 1974). Patton and da Silva (this volume) provide a fascinating study of vicariance speciation of South American mammals.

At the same time, sympatric speciation, which Mayr had believed banished from the realm (Mayr, 1963, pp. 449–480), underwent a resurgence of both theoretical and empirical work, both ironically started as reactions to Mayr's stance. John Maynard Smith's (1966) theoretical paper on sympatric speciation via habitat shift demonstrated that sympatric speciation was theoretically possible (see Johnson and Gullberg, and Kondrashov et al., this volume). Bush's empirical work with host races and species of *Rhagoletis* flies (Bush, 1966, 1969), a story in which Mayr was involved at the inception (Bush, this volume), led to several decades of work demonstrating the reality of host race formation (Feder, this volume) and the likelihood of speciation via host race formation (Berlocher, this volume). In general, the period increased the number of modes of speciation that were taken seriously (Bush, 1975).

The 1966–1978 era also saw a resurgence of interest in studying speciation in the fossil record. Eldridge and Gould's hypothesis of punctuated equilibrium immediately comes to mind, but it must be noted that their hypothesis was largely a consumer, not a producer of ideas about speciation. In its initial formulation (Eldridge and Gould, 1972), punctuated equilibrium relied heavily on

a belief in founder-effect speciation as envisioned by Mayr (1954), a belief that has been seriously questioned recently (e.g., Galiana et al., 1995).

The neutral theory of molecular evolution (Kimura, 1968) spawned not only an acrimonious debate about selection and drift (Powell, 1994), but also raised great hopes that speciation events could be dated using the "molecular clock." A cautious reading of the current state of the "clock" is that it *is* useful—if its limitations are kept in mind (Hillis et al., 1996). A good example of the use of the clock idea (along with electrophoretic data) in speciation research is Coyne and Orr's (1997) identification of patterns in *Drosophila* speciation. Their demonstration that total reproductive isolation increases with divergence time, for example, would be impossible without at least a "sloppy" clock.

Finally, the contribution of computers to speciation research should not be overlooked. The electrophoretic, Hennigian, and later DNA revolutions would have had been very different without mainframe, and later personal computer, programs for computing genetic distance (the venerable BIOSYS-1; Swofford and Selander, 1981), estimating phylogenetic trees (HENNIG, J. S. Ferris; PHYLIP, J. Felsenstein and colleagues; and PAUP, D. S. Swofford), and manipulating trees (MacClade, W. P. and D. R. Maddison; for information on recent versions of all these programs, see Swofford et al., 1996). Another area in which computers had and continue to have great impact was on simulations of speciation (e.g., Dickinson and Antonovics, 1973).

DNA! DNA! DNA! 1979–1996

Writing history about the very recent past is very risky. Moreover, the literature on speciation in the last decade has grown enormously (an excellent earlier summary is provided by Otte and Endler, 1989, the volume resulting from the 1987 Philadelphia speciation meetings), and reviewing it will seriously tax future historians of science. I make no pretense of a comprehensive review here. However, one development has had such an unambiguous and striking effect that one can legitimately say that it has ushered in a new era: the ability to analyze DNA directly. From the standpoint of speciation, the watershed paper is the work of Avise, Lansman, and Shade (1979) on DNA variation in *Peromyscus* species (some of the same species studied by Sumner). Their combination of RFLP (restriction fragment length polymorphism) analysis and mtDNA (mitochondrial DNA) proved to be both powerful and technically suited for the analysis of large sample sizes; for awhile in the 1980s, RFLP/mtDNA became the "new enzyme electrophoresis."

The next logical step in the analysis of speciation using DNA data was quickly shown by the pathbreaking population genetics work of Kreitman (1983) on sequence analysis of alleles of *Adh* (alcohol dehydrogenase) in *D. melanogaster*. Although not itself concerned with speciation, the paper demonstrated that it was feasible to study variation—both within and between species, in the manner of an electrophoretic study—all the way down to the level of individual base pairs. However, studies did not become common until after the development of the polymerase chain reaction (PCR) with Taq polymerase in the late 1980s. PCR sped up the process of producing enough of a given sequence for analysis, and facilitated the application of DNA analysis to a wide range of life forms, but the number of sequence studies is still very small compared to allozyme works. See Avise (1994) for an excellent review of recent work.

What has been learned at the DNA level? At least four areas have been deeply invigorated by the new capacity. First, interest in species concepts is at an all-time high, driven primarily by the discovery (Avise et al., 1983) that many species may be "paraphyletic." That is, if one sequences, say, 10 alleles of a gene from each of two closely related species, and then estimates an allele phylogeny, the alleles from one of the species will form a small branch emerging from deeper branches constituting the other species. No tabulation has been done, but reciprocally monophyletic sister species pairs (where one finds a single fork at the base of the phylogeny, one branch leading to all the alleles of one species and the other leading to all the alleles of the other species) may be in the minority. This leads to a problem when one attempts to apply strict Hennigian logic, with its requirement that all taxa be monophyletic, to any species that has a paraphyletic gene tree (see "Fear of Paraphyly" in Harrison, this volume). In retrospect, paraphyletic species should have been anticipated; *D. pseudoobscura* is paraphyletic in Dobzhansky's (1937, p. 93) first extensive chromosome inversion tree (assuming, as Dobzhansky was inexplicably reluctant to do, that *D. miranda* is the outgroup), and a few cases of paraphyly can be found in the electrophoretic literature.

Second, the DNA data has expanded the field of population structure, adding a new facet termed "phylogeography" (Avise et al., 1987; for similar ideas applied to speciation, see Harrison, 1991). By studying the geographic distribution of alleles or haplotypes together with their phylogeny and frequency distributions, deeper inferences can be made about the history of population structure than is possible from frequency data (e.g., from enzyme electrophoresis) alone. Ironically, phylogeography sees paraphyletic gene trees as highly informative (the paraphyletic species is ancestral), rather than as fearsome. Although phylogeography can potentially make use of complete sequence information on any gene, in practice RFLP analysis of mtDNA has constituted the bulk of the data. As an example of what can be learned, the phylogeographic approach has revealed that Gulf of Mexico–Atlantic Ocean vicariances (Avise, 1994) may have affected a very large number of species; it is conceivable that it could be almost as impor-

tant as the long-known Isthmus of Panama marine vicariance (Jordan, 1908; Lessios, this volume).

Third, DNA markers enable classical gene mapping techniques to be applied rapidly to any organism. The resulting development of new statistical mapping procedures such as QTL (quantitative trait locus) analysis has just begun to move into the field of speciation (Via and Hawthorne, this volume), but will ultimately allow the kind of powerful genetic studies now possible only in *Drosophila* (Wu and Hollocher, this volume) to be carried out on any form of life. Bradshaw and colleagues (1995) were among the first to start on this path.

But without doubt the most significant development engendered by DNA techniques is the ability to sequence genes *directly* involved in reproductive isolation–the Holy Grail of speciation research. As this chapter is written, the situation is unlike 1966–when one could read Hubby and Lewontin's paper and begin enzyme electrophoresis on one's chosen organism the next month—but rather like 1983, when Kreitman had demonstrated allele sequencing was possible, but few could follow immediately. But the path to the Grail is now clear, if steep. Palumbi (this volume) has sequenced and carried out phylogenetic analysis in sea urchins of the gene for bindin, which mediates sperm attachment/fusion to eggs. Sequences of genes involved in postzygotic reproductive isolation in *Drosophila* are also within sight (see Wu and Hollocher, this volume).

However, I would be remiss if I did not mention two important recent findings that do not involve DNA sequences, at least directly. Sperm–egg interactions of the kind studied by Palumbi, long recognized as important in marine organisms with external fertilization, have recently been shown to be important in internally fertilized terrestrial organisms as well. Both Rice's (this volume) selection experiments on sperm–egg interactions in *Drosophila* and Howard et al.'s (this volume) work on postinsemination barriers in ground crickets open new doors on the study of reproductive isolation. Finally, the "rescue mutation" approach of selecting for reduced reproductive isolation has, in the hands of Ashburner and colleagues, produced a remarkable result: the ability to obtain fertile offspring from the *D. melanogaster* × *D. simulans* cross (Davis et al., 1996). Sturtevant, thwarted by his inability to make this cross in 1920, would be more than pleased.

Whither the Future?

When we now contemplate a tangled bank, with its cloth of varied host plants, its flitting phytophagous insects, and its clones of crawling worms, we see not only the grandeur of the Darwinian framework of evolution, but many of the fine details and specific mechanisms as well. Yet no small number of mysteries remain. Perhaps the greatest of these is the genetics of prezygotic isolation.

We still possess precious few genetic insights into that imposing array of prezygotic isolating traits that involve changes in behavior or ecology, or both. The reason is simply that the phenotypes of behavior and ecology are more complex than the postzygotic phenotypes of sterility or inviability.

One of the clearest messages that emerges from the conference that gave rise to this book—and from the work of Guy Bush—is that the initial events of many speciations are in fact changes in ecology or behavior. Many of the first steps that need to be taken to understand the evolution of ecology and behavior are clear. We need to better document and determine the causes of rapid ecological shifts in ecological and historical time (see chapters by Feder, and Schluter, this volume), we need to understand the historical evolution of ecology and behavior from phylogenetic studies (Wells and Henry, this volume), we need to map loci that control ecology or behavior (e.g., Shaw, 1996; Via and Hawthorne, this volume), we need to identify and understand patterns in the evolution of such loci (Ritchie and Phillips, this volume), and we need, ultimately, to understand how such loci function, vary, and evolve at the base level. I anticipate that a history of speciation research written in the not too distant future will find the promise of these first steps to have been fulfilled.

Acknowledgments I thank G. L. Bush, E. Mayr, D. J. Howard, E. G. MacLeod, J. Mallet, S. Lyons, C. Sheppard, and J. L. Feder for information, discussions, and reviewing, but any errors are solely my responsibility. This chapter is dedicated to the memory of Milton Huettel: on that final collecting trip, may there always be good beer, good music, and good talk at the end of the day.

References

Adams, M. B. 1980. Sergei Chetverikov, the Kol'stov Institute, and the evolutionary synthesis. In E. Mayr and W. B. Provine (eds.), The evolutionary synthesis. Cambridge, Mass.: Harvard University Press, pp. 242–278.

Avise, J. C. 1994. Molecular markers, natural history, and evolution. New York: Chapman and Hall.

Avise, J. C., and M. H. Smith. 1974. Biochemical genetics of sunfish. I. Geographical subdivision and subspecific intergradation in the bluegill. Evolution 28:42–56.

Avise, J. C., R. A. Lansman, and R. O. Shade. 1979. The use of restriction endonucleases to measure mitochondrial DNA sequence relatedness in natural populations. I. Population structure and evolution in the genus *Peromyscus*. Genetics 92:279–295.

Avise, J. C., J. Arnold, R. M. Ball, E. Bermingham, T. Lamb, J. E. Neigel, C. A. Reeb, and N. C. Saunders. 1987. Intraspecific phylogeography: the mitochondrial DNA bridge between population genetics and systematics. Annual Review of Ecology and Systematics 18:489–522.

Avise, J. C., J. F. Shapira, S. W. Daniel, C. F. Aquadro, and R. A. Lansman. 1983. Mitochondrial DNA differentiation

during the speciation process in *Peromyscus*. Molecular Biology and Evolution 1:38–56.

Ayala, F. J., M. L. Tracy, D. Hedgecock, and R. C. Richmond. 1974. Genetic differentiation during the speciation process in *Drosophila*. Evolution 28:576–592.

Barbour, C. D., and B. Chernoff. 1984. Comparative morphology and morphometrics of the pescados blancos (genus *Chirostoma*) from Lake Chapala, Mexico. In A. A. Echelle and I. Kornfield (eds.), Evolution of fish species flocks. Orono, Mass.: University of Maine, pp. 111–127.

Bradshaw, H. D., S. M. Wilbert, K. G. Otto, and D. W. Schemske. 1995. Genetic mapping of floral traits associated with reproductive isolation in monkeyflowers (*Mimulus*). Nature 376:762–765.

Brundin, L. 1966. Transantarctic relationships and their significance, as evidenced by chironomid midges. Kungl. Svenska Vetenskapsakademiens Handlingar. Fjärde series, 11:1–472.

Bush, G. L. 1966. The taxonomy, cytology, and evolution of the genus *Rhagoletis* in North America (Diptera, Tephritidae). Bulletin of the Museum of Comparative Zoology 134:431–562.

Bush, G. L. 1969. Sympatric host race formation and speciation in frugivorous flies of the genus *Rhagoletis* (Diptera, Tephritidae). Evolution 23:237–251.

Bush, G. L. 1975. Modes of animal speciation. Annual Review of Ecology and Systematics 6:339–364.

Chetverikov, S. S. 1926. On certain aspects of the evolutionary process from the standpoint of modern genetics (in Russian). Zhurnal Eksperimental 'noi Biologii A2:3–54. (English translation, 1961, Proceedings of the American Philosophical Society 105:167–195)

Cole, L. J. 1940. The relation of genetics to geographic distribution and speciation; speciation I. Introduction. American Naturalist 74:193–197.

Cook, O. F. 1906. Factors of species-formation. Science 23:506–507.

Coyne, J. A., and H. A. Orr. 1997. "Patterns of speciation in *Drosophila*" revisited. Evolution 51:295–303.

Darwin, C., and A. R. Wallace. 1858. On the tendencies of species to form varieties; and on the perpetuation of varieties and species by natural means of selection. Journal of the Proceedings of the Linnaean Society of London 3:45–62.

Davis, A. W., J. Roote, T. Morely, K. Sawamura, S. Herrmann, and M. Ashburner. 1996. Rescue of hybrid sterility in hybrids between *D. melanogaster* and *D. simulans*. Nature 380:157–159.

de Vries, H. 1901–1903. Die Mutationstheorie, Vols. I and II. Leipzig: Verlag von Veit and Company.

Dice, L. R. 1940a. Ecological and genetic variability within species of *Peromyscus*. American Naturalist 74:212–221.

Dice, L. R. 1940b. Speciation in *Peromyscus*. American Naturalist 74:289–298.

Dickinson, H. and J. Antonovics. 1973. Theoretical considerations of sympatric divergence. American Naturalist 107:256–274.

Dobzhansky, Th. 1933. On the sterility of the interracial hybrids in *Drosophila pseudoobscura* hybrids. Proceedings of the National Academy of Sciences USA 19:397–403.

Dobzhansky, Th. 1935. A critique of the species concept in biology. Philosophy of Science 2:344–355.

Dobzhansky, Th. 1937. Genetics and the origin of species. New York: Columbia University Press.

Dobzhansky, Th. 1940. Speciation as a stage in evolutionary divergence. American Naturalist 74:312–321.

Dobzhansky, Th. 1941. Genetics and the origin of species (2nd ed.). New York: Columbia University Press.

Dobzhansky, Th. 1980. The birth of the genetic theory of evolution in the Soviet Union in the 1920s. In E. Mayr and W. B. Provine (eds.), The evolutionary synthesis. Cambridge, Mass.: Harvard University Press, pp. 229–242.

Dobzhansky, Th., and E. Mayr. 1944. Experiments on sexual isolation in Drosophila. I. Geographic strains of Drosophila willistoni. Proceedings of the National Academy of Sciences USA 30:238–244.

East, E. M. 1910. A Mendelian interpretation of variation that is apparently continuous. American Naturalist 44:65–82.

East E. M. 1921. A study of partial sterility in certain hybrids. Genetics 6:311–365.

Ehrman, L. 1965. Direct observation of sexual isolation between allopatric and between sympatric strains of different Drosophila paulistorum races. Evolution 19:459–464.

Ehrman, L., and J. R. Powell. 1982. The Drosophila willistoni species group. In M. Ashburner, H. L. Carson, and J. N. Thompson, Jr. (eds.), The genetics and biology of Drosophila, vol. 3b. New York: Academic Press, pp. 193–225.

Ehrman, L., I. Perelle, and J. R. Factor. 1995. Endosymbiont infectivity in Drosophila paulistorum semispecies. In L. Levine (ed.), Genetics of natural populations. The continuing importance of Theodosius Dobzhansky. New York: Columbia University Press, pp. 241–261.

Eldridge, N., and S. J. Gould. 1972. Punctuated equilibrium: an alternative to phyletic gradualism. In T. J. M. Schopf (ed.), Models in paleontology. San Francisco: Freeman, Cooper, pp. 82–115.

Fisher, R. A. 1930. The genetical theory of natural selection. Oxford: Clarendon Press.

Ford, E. B. 1949. Early stages in allopatric speciation. In G. L. Jepsen, E. Mayr, and G. G. Simpson (eds.), Genetics, paleontology, and evolution. Princeton, N.J.: Princeton University Press, pp. 339–314.

Galiana, A., A. Moya, and F. J. Ayala. 1995. The founder effect in speciation: Drosophila pseudoobscura as a model case. In L. Levine (ed.), Genetics of natural populations: The continuing influence of Theodosius Dobzhansky. New York: Columbia University Press, pp. 281–297.

Groth, J. G. 1993. Evolutionary differences in morphology, vocalization, and allozymes among nomadic sibling species in the North American Red Crossbill (Loxia curvirostris) complex. University of California Publications in Zoology 127:1–143.

Gulick, J. T. 1872. On diversity of evolution under one set of external conditions. Journal of the Linnaean Society of London (Zoology) 11:496–505.

Haldane, J. B. S. 1922. Sex ratio and unisexual sterility in animals. Journal of Genetics 12:101–109.

Haldane, J. B. S. 1932. The causes of evolution. London: Longmans, Green.

Haldane, J. B. S. 1949. Human evolution: past and future. In G. L. Jepsen. E. Mayr, and G. G. Simpson (eds.), Genetics, paleontology, and evolution. Princeton, N.J.: Princeton University Press, pp. 407–418.

Harrison, R. G. 1991. Molecular changes at speciation. Annual Review of Ecology and Systematics 22:281–308.

Harrison, R. G. (ed.). 1993. Hybrid zones and the evolutionary process. New York: Oxford University Press.

Hennig, W. 1966. Phylogenetic systematics. Urbana, Ill.: University of Illinois Press.

Hillis, D. M. 1988. Systematics of the Rana pipiens complex: puzzle and paradigm. Annual Review of Ecology and Systematics 19:39–63.

Hillis, D. M., B. K. Mabel, and C. Moritz. 1996. Applications of molecular systematics. In D. M. Hillis, C. Moritz, and B. K. Mable (eds.), Molecular systematics (2nd ed.). Sunderland, Mass.: Sinauer. pp. 407–514.

Hovanitz, W. 1949. Increased variability in populations following natural hybridization. In G. L. Jepsen, E. Mayr, and G. G. Simpson (eds.), Genetics, paleontology, and evolution. Princeton, N.J.: Princeton University Press, pp. 339–355.

Howard, D. J. 1993. Reinforcement: the origin, dynamics, and fate of an evolutionary hypothesis. In R. G. Harrison (ed.), Hybrid zones and the evolutionary process. New York: Oxford University Press, pp. 46–69.

Hubbs, C. L. 1940. Speciation of fishes. American Naturalist 74:198–211.

Hubby, J. L., and R. C. Lewontin. 1966. A molecular approach to the study of genic heterozygosity in natural populations I. The number of alleles at different loci in Drosophila pseudoobscura. Genetics 54:577–594.

Hurley, P. M. 1968. The confirmation of continental drift. Scientific American 218:52–64.

Huxley, J. 1942. Evolution, the modern synthesis. London: Allen and Unwin.

Irwin, M. R., and R. W. Cumley. 1940. Speciation from the point of view of genetics. American Naturalist 74:222–231.

Jepsen, G. L., E. Mayr, and G. G. Simpson (eds.). 1949. Genetics, paleontology, and evolution. Princeton, N.J.: Princeton University Press.

Jordan, D. S. 1905. The origin of species through isolation. Science 22:545–562.

Jordan, D. S. 1908. The law of geminate species. American Naturalist 42:73–80.

Kessler, S. 1966. Selection for and against ethological isolation between Drosophila pseudoobscura and Drosophila persimilis. Evolution 20:634–645.

Kimura, M. 1968. Evolutionary rate at the molecular level. Nature 217:624–626.

Kreitman, M. 1983. Nucleotide polymorphism at the alcohol dehydrogenase locus of Drosophila melanogaster. Nature 304:412–417.

Lack, D. 1949. The significance of ecological isolation. In G. L. Jepsen, E. Mayr, and G. G. Simpson (eds.), Genetics, paleontology, and evolution. Princeton, N.J.: Princeton University Press, pp. 299–308.

Laven, H. 1959. Speciation by cytoplasmic isolation in the Culex pipiens complex. Cold Spring Harbor Symposia on Quantitative Biology 24:166–173.

Levine, L. (ed.). 1995. Genetics of natural populations. The continuing importance of Theodosius Dobzhansky. New York: Columbia University Press.

Lewontin, R. C. 1981. The scientific work of Th. Dobzhansky. In R. C. Lewontin, J. A. Moore, W. B. Provine, and B. Wallace (eds.), Dobzhansky's genetics of natural populations I-XLII. New York: Columbia University Press, pp. 93–115.

Lewontin, R. C., J. A. Moore, W. B. Provine, and B. Wallace (eds.). 1981. Dobzhansky's genetics of natural populations I-XLII. New York: Columbia University Press.

Mason, H. L. 1949. Evidence for the genetic submergence of *Pinus remorata*. In G. L. Jepsen, E. Mayr, and G. G. Simpson (eds.), Genetics, paleontology, and evolution. Princeton, N.J.: Princeton University Press, pp. 356–362.

Maynard Smith, J. 1966. Sympatric speciation. American Naturalist 100:637–650.

Mayr, E. 1940. Speciation phenomena in birds. American Naturalist 74:249–278.

Mayr, E. 1942. Systematics and the origin of species. New York: Columbia University Press.

Mayr, E. 1947. Ecological factors in evolution. Evolution 1:263–288.

Mayr, E. 1949. Speciation and systematics. In G. L. Jepsen, E. Mayr, and G. G. Simpson (eds.), Genetics, paleontology, and evolution. Princeton, N.J.: Princeton University Press, pp. 281–298.

Mayr, E. 1954. Change of genetic environment and evolution. In J. S. Huxley, A. C. Hardy, and E. B. Ford (eds.), Evolution as a process. London: Allen and Unwin, pp. 157–180.

Mayr, E. 1959. Darwin and the evolutionary theory in biology. In Evolution and anthropology: a centennial appraisal. Washington, DC: The Anthropological Society of Washington, pp. 409–412.

Mayr, E. 1963. Animal species and evolution. Cambridge, Mass.: Belknap Press.

Mayr, E. 1980. Prologue: some thoughts on the history of the evolutionary synthesis. In E. Mayr and W. B. Provine (eds.), The evolutionary synthesis. Cambridge, Mass.: Harvard University Press, pp. 1–48.

Mayr, E. 1982. The growth of biological thought. Cambridge, Mass.: Harvard University Press.

Mayr, E., and W. B. Provine (eds.). 1980. The evolutionary synthesis. Cambridge, Mass.: Harvard University Press.

Moore, J. A. 1949. Patterns of evolution in the genus *Rana*. In G. L. Jepsen, E. Mayr, and G. G. Simpson (eds.), Genetics, paleontology, and evolution. Princeton, N.J.: Princeton University Press, pp. 315–338.

Nelson, G., and N. I. Platnick. 1981. Systematics and biogeography: Cladistics and vicariance. New York: Columbia University Press.

Nilsson-Ehle, H. 1909. Kreuzungsuntersuchen an Hafer und Weizen. Lunds Universitets Arsskrift, n.s., ser. 2.5, no. 2.

Nixon, K. C., and Q. D. Wheeler. 1990. An amplification of the phylogenetic species concept. Cladistics 6:211–223.

Orr, H. A. 1996. Dobzhansky, Bateson, and the genetics of speciation. Genetics 144:1331–1335.

Otte, D., and J. A. Endler. 1989. Speciation and its consequences. Sunderland, Mass.: Sinauer.

Paterson, H. E. H. 1981. The continuing search for the unknown and the unknowable: a critique of contemporary ideas on speciation. South African Journal of Science 77:113–119.

Poulton, E. B. 1904. What is a species? Proceedings of the Entomological Society of London 1903:lxvii-cxvi.

Poulton. E. B. 1908. Essays on evolution. Oxford: Clarendon Press.

Powell, J. R. 1994. Molecular techniques in population genetics: a brief history. In B. Schierwater, B. Streit, G. P. Wagner, and R. Desalle (eds.), Molecular ecology and evolution: approaches and applications. Basel: Birkhäuser Verlag, pp. 131–156.

Provine, W. B. 1971. The origins of theoretical population genetics. Chicago: University of Chicago Press.

Provine, W. B. 1979. Francis B. Sumner and the evolutionary synthesis. Studies in the History of Biology 3:211–240.

Provine, W. B. 1986. Sewall Wright and evolutionary biology. Chicago: University of Chicago Press.

Rosen, D. E. 1974. Phylogeny and zoogeography of salmoniform fishes and relationships of *Lepidogalaxias salmandroides*. Bulletin of the American Museum of Natural History 153:265–326.

Shaw, K. L. 1996. Polygenic inheritance of a behavioral phenotype: interspecific genetics of song in the Hawaiian cricket genus *Laupala*. Evolution 50:256–266.

Smith, E. B., and C. A. Sheppard. 1990. A heritage of distinctive journalism: a look at the paper trail. American Entomologist 36:6–17.

Smith, G. K. 1992. Introgression in fishes: significance for paleontology, cladistics, and evolutionary rates. Systematic Biology 41:41–57.

Spence, J. R., and R. H. Gooding. 1990. Introduction: evolutionary significance of hybridization and introgression in insects. Canadian Journal of Zoology 68:1699–1805.

Spencer, W. P. 1940. Levels of divergence in *Drosophila* speciation. American Naturalist 74:299–311.

Stebbins, G. L. 1950. Variation and evolution in plants. New York: Columbia University Press.

Sturtevant, A. H. 1920. Genetic studies on *Drosophila simulans*. I. Introduction. Hybrids with *Drosophila melanogaster*. Genetics 5:488–500.

Sumner, F. B. 1932. Genetic, distributional, and evolutionary studies of deer mice (*Peromyscus*). Bibliographia Genetica 9:1–106.

Swofford, D. S., and R. B. Selander. 1981. BIOSYS-1: A FORTRAN program for the comprehensive analysis of electrophoretic data in population genetics and systematics. Journal of Heredity 72:281–283.

Swofford, D. L., G. J. Olsen, P. J. Wadell, and D. M. Hillis. 1996. Phylogenetic inference. In D. M. Hillis, C. Moritz, and B. K. Mable (eds.), Molecular systematics (2nd ed.). Sunderland, Mass.: Sinauer Press, pp. 407–514.

Wagner, W. L., and V. A. Funk. 1995. Hawaiian biogeography: evolution over a hot spot. Washington, DC: Smithsonian Institution Press.

Wallace, A. R. 1864. On the phenomena of varieties and geographic distribution, as illustrated by Papilionidae of the Malayan region. Transactions of the Linnaean Society of London 25:1–71.

Walsh, B. D. 1863. Observations on certain North American Neuroptera, by H. Hager. M. D, of Koenigsberg, Prussia; translated from the original French MS, and published by permission of the author, with notes and descriptions of about 20 new North American species of Pseudoneuroptera. Additional notes by Walsh, pp. 182–267. Proceedings of the Entomological Society of Philadelphia 3:167–267.

Walsh, B. D. 1864. On phytophagic varieties and phytophagic species. Proceedings of the Entomological Society of Philadelphia 3:403–430.

White, M. J. D. 1945. Animal cytology and evolution. Cambridge: Cambridge University Press.

White, M. J. D. 1978. Modes of speciation. San Francisco: Freeman.

Wiens, J. A. 1982. Forum: avian subspecies in the 1980s. Auk 99:593–615.

Wright, S. 1931. Evolution in Mendelian populations. Genetics 16:97–159.

Wright, S. 1940a. Breeding structure of populations in relation to speciation. American Naturalist 74:212–221.

Wright, S. 1940b. The statistical consequences of Mendelian heredity in relation to speciation. In J. S. Huxley (ed.), The new systematics. Oxford: Oxford University Press, pp. 161–183.

Zimmerman, E. C. 1958. Three hundred species of *Drosophila* in Hawaii? A challenge to geneticists and evolutionists. Evolution 12:557–558.

Part II

Species Concepts

2

Linking Evolutionary Pattern and Process

The Relevance of Species Concepts for the Study of Speciation

Richard G. Harrison

Species are fundamental units of natural diversity—of obvious interest to systematists, evolutionary biologists, ecologists, and conservation biologists. However, both within and among these disciplines, there is considerable disagreement about the way in which species should be defined. As a result, the last two decades have witnessed a proliferation of new species concepts and a rapidly expanding literature, in which the merits of the various definitions are vigorously debated. Much of the debate has been carried out within the community of systematic biologists, who see the definition of species as the province of their discipline. The emergence of phylogenetic systematics resulted in profound unhappiness with the Biological Species Concept (BSC), which until recently remained the default framework for thinking about species and speciation. One faction among the phylogenetic systematists has argued for the primacy of pattern over process in defining species. Wheeler and Nixon (1990) articulated the views of such pattern cladists: "The appropriately militant view that systematists need to embrace is that the responsibility for species concepts lies *solely* with systematists. If we continue to bow to the study of process over pattern, then our endeavors to elucidate pattern become irrelevant" (p. 79).

Speciation is the process by which new species arise, and evolutionary biologists who study this process should be concerned with how species are defined. However, with only a few exceptions, students of speciation have not engaged systematists in a debate about species concepts. Unlike phylogenetic systematists, many evolutionary geneticists appear quite satisfied with the BSC of Mayr and Dobzhansky, which focuses attention on reproductive isolation or barriers to gene exchange. They see no need for any change. Coyne (1994) summarized this point of view: "It is a testament to the power of the BSC that virtually everyone studying the origin of species concentrates on reproductive isolating mechanisms" (p. 22). Despite accusations that the BSC is both conceptually and operationally inadequate, Coyne (1994) wrote that he has "no idea why the BSC . . . seems to ignite so much controversy" (p. 22).

A second reason that evolutionary geneticists have generally absented themselves from debates about species concepts is that many are willing to acknowledge that there cannot be a single concept that serves the needs of both evolutionary geneticists and systematists. The two groups have very different perspectives on evolutionary pattern and process. Systematists are primarily interested in fixed character state differences between species (or higher taxa) and inferences about phylogenetic relationships among defined terminal taxa (nonreticulating lineages). In contrast, population biologists study variation within "species," patterns of reticulation (mating), and the processes (mechanisms) by which one lineage splits into two. The interface between these two disciplines has only recently emerged as a subject of intensive research, a transition catalyzed by the increasing availability of DNA sequence data and the recognition that there is a direct connection between ancestor–descendant relationships within populations (genealogy) and phylogeny (Avise et al. 1987; Harrison 1991; Templeton 1994; Hey 1994). Analysis of gene genealogies using coalescent theory (e.g., Hudson 1990) has focused attention on that connection.

Finally, discussions of the nature of species seem to arouse passion and self-righteousness not found in most "scientific" debates. Each side tends to caricature the logic of its opponents, emphasizing the obvious weaknesses and not mentioning the strengths. Furthermore, some of the issues appear to be philosophical, not scientific. For these reasons, it is hardly surprising that the response of many evolutionary geneticists has been to ignore (if not disparage) these discussions. Again, Coyne (1992) expresses what may be a common response among population geneticists and process-oriented evolutionary biologists: "It is clear that the arguments [about species concepts] will persist for years to come, but equally clear that, like barnacles on a whale, their main effect is to retard slightly the progress of the field" (p. 290).

I am not so pessimistic. Recent contributions to the literature on species concepts offer a variety of new and useful perspectives, although no single concept has emerged that will satisfy all parties. Defining relationships among these concepts will help to integrate the disciplines of systematic and evolutionary biology. My goal in this chapter is simple: to discuss a sample of species concepts or definitions and to consider how they might provide a context for research on speciation. I do not propose any new concepts—but rather, try to explicate, critique, and organize those already in the literature. I suggest that each species (or lineage; see de Queiroz, this volume) has a distinctive life history, which includes a series of stages that correspond to some of the named species concepts discussed below.

Species Concepts

Table 2.1 summarizes seven different species concepts or definitions, using quotations taken directly from the literature. The list is by no means comprehensive, and a number of frequently cited alternatives (e.g., the phenetic species concept of Sokal and Crovello [1970], the ecological species concept of Van Valen [1976], and Simpson's [1961] original version of the evolutionary species concept) are not included.

Taxonomy of Species Concepts

It is useful to start by attempting a taxonomy of species concepts. A number of potential contrasts have been proposed as organizing principles for characterizing species concepts. For example, definitions of species may be motivated by an interest in pattern alone or by attempts to understand the speciation process. Concepts may be retrospective or prospective (O'Hara 1993), mechanistic or historical (Luckow 1995), character based or history based (Baum and Donoghue 1995). Unfortunately, none of these contrasts allows unambiguous characterization of all species concepts. Moreover, many species concepts do not easily satisfy one or more of the established dichotomies; that is, the dichotomous categories are not necessarily mutually exclusive alternatives. Nonetheless, it is useful to consider how they might apply to the species definitions listed in table 2.1.

Prospective species concepts explicitly invoke criteria (e.g., genetic cohesion or isolation) that have implications for the future status of populations. In contrast, retrospective concepts view species as "end-products" of evolution (Luckow 1995), as units that have evolved rather than lineages that are evolving (Frost and Hillis 1990). O'Hara (1993) suggests that "our judgement as to what individuals belong to a particular population or reproductive community will always depend to some extent upon our expectation of the future behavior of those individuals and their descendants" (p. 242). Therefore, he argues that all species concepts "depend to some extent upon prospective narration" (p. 242).

According to Luckow (1995), mechanistic species concepts "begin with a theory of how speciation (evolution) works" (p. 590). They rely on knowledge of the genetics and ecology of natural populations and use that knowledge in defining species, which are viewed as "participants" in the evolutionary process. In contrast, historical species concepts focus exclusively on the outcome of evolution. In Luckow's (1995) terminology, both the "character-based" and "history-based" concepts of Baum and Donoghue (1995) are historical species concepts.

The contrast between character-based and history-based species concepts reflects a fundamental division within phylogenetic systematics. Proponents of these two sorts of phylogenetic species concepts have engaged in a lengthy debate in the systematic biology literature (e.g., de Queiroz and Donghue 1988, 1990; Wheeler and Nixon 1990; Nixon and Wheeler 1990; Baum 1992; Baum and Donoghue 1995; Luckow 1995). The essence of the disagreement is whether species definitions should rely simply on characters (with no necessary prior inference of historical relationships) or whether species should be defined using the inferred historical relationships (genealogy) among component organisms. In the rest of this chapter, I refer to the character-based definition (Cracraft 1983; Nixon and Wheeler 1990) as the phylogenetic species concept and use the genealogical species concept (Baum and Shaw 1995) as an example of the historical approach. In fact, both sorts of definitions are phylogenetic (see de Queiroz [this volume] for a clear discussion of this issue).

The several dichotomies useful for characterizing species concepts are not necessarily independent. Historical concepts are, of course, retrospective and tend to focus on patterns of variation; mechanistic concepts are prospective and usually derive from a fundamental interest in the evolutionary process. Furthermore, patterns of variation are used to infer past processes and thereby shed light on how speciation has occurred.

The first three concepts in table 2.1 (BSC, recognition, cohesion) are clearly motivated by an interest in the process of speciation. They are prospective, in the sense that isolation and cohesion are viewed as important because they allow us to predict future patterns of variation (e.g., whether an advantageous mutation will spread, whether two distinct entities can persist in sympatry, whether a lineage will become an exclusive group). These concepts would be termed mechanistic by the pattern cladists, although I will argue below that this characterization is inaccurate. Contrary to the claims of their critics, prospective species concepts do not necessarily imply the action of particular evolutionary mechanisms in species formation.

Table 2.1. Seven species concepts or definitions from the systematic biology and evolutionary biology literature.

1. Biological Species Concept (Isolation Concept) (BSC)

"[G]roups of actually or potentially interbreeding natural populations which are reproductively isolated from other such groups" (Mayr 1963, p. 19).

"[S]ystems of populations, the gene exchange between these systems is limited or prevented in nature by a reproductive isolating mechanism or by a combination of such mechanisms" (Dobzhansky 1970, p. 357).

2. Recognition Species Concept

"[T]he most inclusive population of individual biparental organisms which share a common fertilization system [specific mate recognition system]" (Paterson 1985, p. 15).

3. Cohesion Species Concept

"[T]he most inclusive population of individuals having the potential for phenotypic cohesion through intrinsic cohesion mechanisms [genetic and/or demographic exchangeability]" (Templeton 1989, p. 12).

4. Phylogenetic Species Concept (Character-Based)

"[A]n irreducible (basal) cluster of organisms, diagnosably distinct from other such clusters, and within which there is a parental pattern of ancestry and descent" (Cracraft 1989, p. 34).

"[T]he smallest aggregation of populations (sexual) or lineages (asexual) diagnosable by a unique combination of character states in comparable individuals" (Nixon and Wheeler 1990, p. 218).

5. Genealogical Species Concept

"'[E]xclusive' groups of organisms, where an exclusive group is one whose members are all more closely related to each other than to any organisms outside the group. . . . [B]asal taxa . . . , that is taxa that contain no included taxa" (Baum and Shaw 1995, p. 290).

6. Evolutionary Species Concept

"[A] single lineage of ancestor-descendant populations which maintains its identity from other such lineages and which has its own evolutionary tendencies and historical fate" (Wiley 1978, p. 18).

7. Genotypic Species Cluster Definition

"[D]istinguishable groups of individuals that have few or no intermediates when in contact. . . .

". . . [C]lusters are recognized by a deficit of intermediates, both at single loci (heterozygote deficits) and at multiple loci (strong correlations or disequilibria between loci that are divergent between clusters)" (Mallet 1995, p. 296).

The phylogenetic species concept and genealogical species concept are clearly retrospective and emphasize pattern rather than process (although consideration of process is not excluded). The phylogenetic species concept and the genotypic clusters definition are explicitly character based, in contrast to the genealogical species concept, which is history based (in the sense of Baum and Donoghue 1995).

Criteria for Recognizing and Defining Species

De Queiroz (this volume) clearly distinguishes between species concepts and species criteria. The distinction is an important one because different species concepts may share a common criterion for species delimiatation. The seven species concepts or definitions in table 2.1 suggest a number of possible criteria for evaluating whether groups of individuals are distinct species: (1) species are characterized by genetic isolation or the absence of cohesion (i.e., there are intrinsic barriers to gene exchange); (2) species are demographically nonexchangeable (ecologically distinct); (3) species are diagnosable (characterized by fixed character state differences); (4) species are exclusive groups; (5) species have a separate identity and independent evolutionary tendencies; (6) species are recognized as distinct genotypic clusters.

All species delimitations (even those based on isolation or cohesion concepts) ultimately depend on inferences from patterns of variation and character state distributions. The need to infer process from pattern argues that the pattern/process dichotomy discussed in the previous section will often break down. For similar reasons, implementing isolation or cohesion concepts (which I have termed prospective) may require an explicitly retrospective approach (e.g., see Templeton 1994; see also Templeton, this volume).

The criteria of diagnosability and separate genotypic clusters would appear to have the advantage of being operational, because species are defined directly in terms of the distribution of character states rather than in terms of descent relationships or interbreeding inferred from such distributions. The genotypic clusters definition, however, can only be applied directly to entities that are sympatric or parapatric. Furthermore, patterns revealed by character-state distributions are not always easy to interpret and often require either knowledge of or assumptions about current evolutionary process or mechanism (see critique by Baum and Donoghue 1995). For example, if two distinct morphs co-occur at a single site, do we conclude that this pattern is evidence of a single polymorphic population or of two sympatric species? Clearly, additional information about patterns of genetic exchange is needed.

Populations are diagnosable (in theory and practice) when they exhibit one or more fixed character-state differences. If taxa are exclusive groups only when all gene genealogies become concordant, with coalescence of genes within each group occurring more recently than any coalescence of genes between groups, then exclusivity is a far more stringent requirement for species status. For example, humans and chimpanzees are not exclusive groups for all parts of the genome, because some human major histocompatability complex (MHC) alleles are more closely related to chimp alleles than to other alleles in humans. Baum and Shaw (1995) do not say that all gene genealogies must be concordant if two populations are to be considered genealogical species, but the threshold beyond which lineages become genealogical species remains obscure. How to apply the criterion of separate identities and independent evolutionary tendencies is also not clear, especially since Wiley (1981) explicitly requires that species be lineages in the sense of sharing a common history (being monophyletic or exclusive).

Isolation and Cohesion as Criteria for Defining Species

Both cohesion and isolation are clearly important components of the BSC. Mayr characterizes species as groups of interbreeding populations, and Dobzhansky describes species as "systems of populations," emphasizing that single species are cohesive. At the same time, both versions of the BSC view reproductive isolation or "isolating mechanisms" as defining the boundaries between species in the natural world. Thus, Mayr (1963) wrote: "The mechanisms that isolate one species reproductively from others are perhaps the most important set of attributes a species has, because they are, by definition, the species criteria" (p. 89).

In contrast, the recognition and cohesion concepts attempt to be "nonrelational" and to define species as inclusive groups. Genetic cohesion is at the heart of the recognition concept. Paterson (1985) describes a species as a common "field for recombination"—or as a group of organisms sharing a common fertilization system. However, he is inconsistent in application of the first criterion, since postzygotic barriers are not viewed as reason to recognize two entities as different species (despite the fact that such entities no longer constitute a single field for recombination). Templeton (1989) recognizes the importance of genetic exchangeability (gene flow) in defining species boundaries, but argues that it is not the only sort of cohesion mechanism that needs to be considered. He views the isolation and recognition concepts as "exclusively concerned with genetic relatedness promoted through the exchange of genes via sexual reproduction" (p. 14).

In addition to genetic cohesion, Templeton (1989, this volume) adds the criterion of demographic exchangeability to emphasize ecological (rather than genetic) interactions. Groups of organisms that are demographically exchangeable are ecologically equivalent, occupying the same niche. This criterion for determining species status is particularly useful when applied to sympatric, asexual lineages, which are considered conspecific (even in the absence of gene exchange) when they are demographically exchangeable. At the other extreme, hybridizing populations may be considered distinct species (despite some gene flow) if they are demographically nonexchangeable.

The notion that a nonrelational species concept (e.g., the recognition concept) is superior has been argued by Paterson and his supporters (Paterson 1985; Masters et al. 1987; Lambert et al. 1987). However, description of a mate recognition system or of an array of cohesion mechanisms that operate within a single species or lineage will not directly provide insights into how new species arise. Relational concepts may be a more appropriate framework for studying speciation; it is essential to obtain comparative data from sister species or sister populations and to infer the changes (apomorphies) that have resulted in (or at least co-occurred with) fission. Species may be held together by multiple cohesion mechanisms, but the breakdown of only one of those may lead to lineage splitting and speciation (figure 2.1). Only by comparing sister taxa will it become evident which cohesion mechanisms are labile (hence likely to be important in speciation events) and which are conservative.

Decisions as to species status depend on information about patterns of genetic exchange, regardless of whether one attempts to delineate species in terms of the boundaries between them ("intrinsic barriers to gene flow") or the limits of common cohesion mechanisms or common fertilization systems. Thus, the distinction between dichotomous states in the taxonomy of species concepts (in

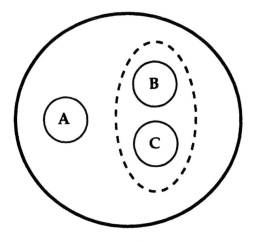

—— calling song, gamete recognition

- - life cycle, courtship display

—— host plant association

Figure 2.1. Example of three species of insects (A–C) with overlapping arrays of "cohesion mechanisms." The cohesion mechanisms operating in these imaginary insects are host plant association, life cycle (phenology), calling song, courtship display, and gamete recognition. Lines enclose groups of individuals that share one or more cohesion mechanisms. Differences in any one of these may result in descendant lineages becoming separate "fields for recombination." Characterizing the array of possible cohesion mechanisms is only a first step in the study of speciation. Understanding what makes A, B, and C distinct species involves identifying which cohesion mechanisms have changed over time. In the example shown here, host association differs among all three species, life cycle and courtship display distinguish A from B+C, but calling song and gamete recognition are cohesion mechanisms still shared by all three species.

this case, relational versus nonrelational) again becomes fuzzy when different concepts are used to generate criteria that actually delimit species.

Proponents of historical and character-based species concepts also recognize the importance of interbreeding, reproductive isolation, or barriers to genetic exchange, although these are not the criteria on which they determine species status.

Species must be reproductively isolated from each other to the extent that this is required for maintaining their separate identities, tendencies and fates. (Wiley 1981, p. 27)

Interbreeding is relevant not only for population biologists, but also to phylogenetic systematists. (de Queiroz and Donoghue 1990, p. 89)

[P]hylogenetic species are the least inclusive populations or set of populations among which there is character-based evidence in the form of fixed differences that gene exchange does not occur. (Davis and Nixon 1992, p. 428)

Wiley (1981) and Davis and Nixon (1992) presumably invoke absence of genetic exchange because diagnosable differences (separate identities) between populations most often arise and persist in that context. However, boundaries between populations can be semipermeable (Harrison 1986, 1990); that is, the extent of introgression can vary across the genome. As a consequence, fixed allelic differences can be maintained at some loci, whereas gene exchange leads to homogenization of allele frequencies at other loci. Therefore, it is possible for populations to be diagnosably distinct or maintain separate identities without being completely genetically or reproductively isolated. Mallet's (1995) genotypic cluster definition seems consistent with this view, because it allows some gene exchange to occur as long as the distribution of multilocus genotypes or phenotypes remains bimodal where the distinct entities come into contact. The genealogical species concept is potentially far more stringent, if a group must exhibit exclusivity at a large proportion of loci in order to be considered a genealogical species.

The role of isolation or barriers to gene exchange, however, is fundamentally different in historical or character-based species concepts than in the BSC. For purposes of species definition, proponents of the BSC are clearly interested only in intrinsic barriers to gene exchange (those that are due to biological differences between species and not simply due to geography). In contrast, the phylogenetic, evolutionary, and genealogical species concepts do not discriminate between intrinsic and extrinsic barriers. Thus, Davis and Nixon (1992) argue that intrinsic barriers hold "no special position [in defining species] except in suggesting stability of the observed situation" (p. 429). But extrinsic barriers are ephemeral, whereas distinct cohesion mechanisms or intrinsic barriers to gene exchange often represent permanent genetic changes in the evolutionary fabric that can be used as predictors of future patterns of evolution. It is these sorts of changes that have traditionally attracted the attention of students of the speciation process.

Do Species Concepts Constrain or Bias Our Views about How Speciation Occurs?

"What is a species? This fundamental question must be answered before the process of species formation can be investigated" (Templeton 1989, p. 3). Most systematists and evolutionary biologists would agree with Templeton that without a clear definition of species, we cannot begin to study the process of speciation. However, some propose that we first define speciation (usually as the reduction and eventual elimination of gene flow) and then ask what will be the properties of the "species" that result (Bush 1994). Whichever path we follow, an obvious danger is that our definition of species or speciation will make assumptions that constrain our views about evolutionary process.

Speciation from the Perspective of the Phylogenetic and Genealogical Species Concepts

If species are simply diagnosably distinct populations, then a new species can appear as the result of the fixation of a single new mutation within a local population or from random sorting of an ancestral polymorphism. With this perspective there is no longer a clear distinction between the genetics of speciation and the genetics of species differences (Templeton 1981), because all fixed differences are cause for recognizing distinct species. Speciation and divergence become synonymous and the study of speciation becomes the study of the relative importance of migration, genetic drift, and natural selection in the evolution of fixed differences between local populations. This is an important problem in population genetics, but it is certainly not what most evolutionary biologists have traditionally viewed as the speciation process!

The genealogical species concept is concerned with the evolution of exclusive groups. Like the phylogenetic species concept, it assigns no special status to the origin of intrinsic barriers to genetic exchange (although such barriers obviously promote the evolution of exclusivity). Baum and Shaw (1995) focus attention "on mechanisms by which divergent (phylogenetic) patterns of relationship emerge out of reticulating patterns" (p. 301). Because coalescent times vary among genes (due to history, chance, and differences in selection), diverging lineages gradually become exclusive groups for an increasing fraction of the genome, and species necessarily have "fuzzy boundaries." Maddison (1995) articulates the same view of species boundaries, suggesting that a species phylogeny is not a "single, simple entity, but rather appears more like a statistical distribution" (p. 285).

Implications of "Isolating Mechanisms" and "Cohesion Mechanisms"

The BSC defines speciation in terms of the origin of barriers to gene exchange or of reproductive isolation. However, both Dobzhansky and Mayr used the term "isolating mechanisms," which implies that biological differences that limit or prevent gene exchange are indeed mechanisms to isolate, that they arose for that pur-

pose, that isolation is a function rather than an effect (in the sense of Williams 1966). Dobzhansky was convinced of the importance of the process of reinforcement, in which prezygotic barriers arise as a result of selection against hybridization in areas of secondary contact. In this scenario intrinsic barriers to gene exchange are indeed isolating mechanisms. Mayr was not a strong supporter of the reinforcement model and more often viewed barriers to gene exchange as incidental by-products of divergence in allopatry. Nonetheless, his language was not always consistent with that view: "It is a function of the isolating mechanisms to prevent such a [hybrid] breakdown and to protect the integrity of the genetic system of species" (Mayr 1963, p. 109). Advocates of the recognition concept (Paterson 1985; Masters et al. 1987; Lambert et al. 1987) have vigorously campaigned against use of the term "isolating mechanism," principally because they deny the possibility of reinforcement as a mechanism for speciation. However, others (Templeton 1989; Mallet 1995) have also been quite critical.

I agree that the term "isolating mechanism" is misleading and that use of the term should generally be avoided. Without sacrificing clarity it is possible to substitute more neutral language (e.g., "barriers to gene exchange") that does not imply a particular origin (process). Only when differences between species have evolved as a result of selection against hybrids (=reinforcement) should the term "isolating mechanism" be used. Because recent reviews do not suggest that reinforcement is a common mode of speciation (Butlin 1989; Howard 1993; but see Coyne and Orr 1989), it is certainly important to avoid the implication that all prezygotic barriers have arisen in this way.

Mallet (1995) also criticizes use of the term "isolating mechanisms" because it includes "under a single label" an incredible diversity of possible biological differences between taxa (e.g., from chromosomal differences to behavioral differences to presence/absence of reproductive parasites). One could easily extend the argument to reject terms like reproductive isolation or barriers to gene exchange. But it is the common effect of all these differences (limiting or preventing gene exchange) that provides the rationale for grouping them. I see no reason not to adopt a single term (e.g., "barriers to gene exchange") to refer to the set of differences that have this very important effect.

The terms "fertilization mechanisms" (used by Paterson 1985) and "cohesion mechanisms" (used by Templeton 1989), although appropriate in some contexts, may be as misleading as the term "isolating mechanisms." Many (perhaps most) biological properties of organisms that confer "cohesion" did not arise for that purpose. They are also effects not functions! Thus, life cycles that result in adults appearing at the same season, or habitat/resource associations which lead to aggregation of individuals in particular places, facilitate fertilization or lead to genetic and/or demographic cohesion. But in most cases, life cycles and habitat associations have not been molded by selection for the purpose of "cohesion." Gamete recognition systems and behavioral components of mate recognition systems are more likely candidates for true cohesion mechanisms.

Pattern, Process, and Mechanism in Species Concepts

Proponents of the phylogenetic species concept are adamant that pattern not process should form the basis of any species definition. According to Cracraft (1983), "a species concept is best formulated from the perspective of the results of evolution rather than from one emphasizing the processes thought to produce those results" (p. 169). I think that most evolutionary biologists would agree. But Cracraft (1983) and other proponents of the phylogenetic species concept (e.g., Luckow 1995) go one step further and imply that the BSC defines species in terms of particular processes or mechanisms. For example, Luckow (1995) states that the BSC "recognizes the inability to interbreed as being the most important causal factor (mechanism) in speciation" (p. 590). Mallet (1995) also suggests that mechanism should be excluded from species definitions. His argument is that "since theories of speciation involve a reduction in ability or tendency to interbreed, species cannot themselves be defined by interbreeding without confusing cause and effect" (p. 295).

These arguments represent a mistinterpretation of the BSC. Reproductive isolation (or a barrier to gene exchange) is a result, not a process (as is a new cohesion mechanism). The inability to interbreed is not the "cause" or "mechanism" of speciation; it is the signal that speciation is complete. I see no confusion of cause and effect if species are defined in terms of barriers to gene exchange. The "cause" of speciation is divergence of populations due to drift or selection (with or without the presence of *extrinsic* barriers to gene exchange) and the "effect" of the resulting differences is the appearance of *intrinsic* barriers to gene exchange. These barriers may be either pre- or postzygotic, thereby including factors that determine the probability of hybrid zygote formation and factors that affect the relative fitnesses of zygotes of mixed ancestry. The BSC (stripped of the term "isolating mechanism") does not, by itself, prejudice us with respect to mechanisms of speciation (or its geographic context). It is essential that we distinguish the implications of particular concepts and the arguments/biases of those who invoke them. I am afraid that many critics of the BSC are guilty of this confusion.

Cracraft (1983, 1989) touts the phylogenetic species concept as the most appropriate framework for what he calls "speciation analysis" (which might be thought to be the study of speciation). But speciation analysis focuses primarily on the description of historical patterns (e.g., areas of endemism, scenarios for vicariance). There seems to be little concern for what an evolutionary geneticist would call mechanism, no apparent interest in knowing what evolutionary forces have acted. A phylogenetic species concept may be an appropriate framework for systematists and biogeographers, but a quite different outlook will be necessary if we hope to unravel what goes on at the boundary between genealogy and phylogeny.

Must Species Be Consistent with "Recovered Evolutionary History"?

The logic of phylogenetic systematics leads to complete rejection of the BSC and cohesion species concepts, because interbreeding or other mechanisms for cohesion almost always represent a shared ancestral condition (plesiomorphy) and thus cannot provide reason for grouping. De Queiroz and Donoghue (1988) identified interbreeding and common descent as the two "processes" through which organisms are related. They clearly characterized the tension between these two processes in providing a basis for species definition. If species are defined solely on the basis of current interbreeding or cohesion, they frequently will represent "paraphyletic assemblages of populations" (Mishler and Donoghue 1982; see also Bremer and Wanntorp 1979); that is, they will include some but not all descendants of a common ancestor (figure 2.2).

Fear of Paraphyly

Such paraphyletic assemblages are anathema to those phylogenetic systematists who argue that species should be monophyletic or exclusive groups. If species B is paraphyletic with respect to species A, then some members of species B will be more closely related to members of species A than they are to other conspecifics (figure 2.2). Obviously, if recency of common ancestry is to

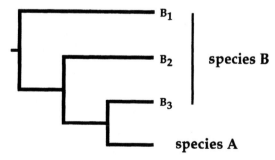

Figure 2.2. Example of a species (B) that appears to be a paraphyletic assemblage of populations. In this gene genealogy, B_1, B_2, and B_3 represent distinct genotypes within species B (perhaps derived from three discrete localities).

provide the criterion for species definition, then such "paraphyletic species" are not allowed. Indeed, many phylogenetic systematists view them as positively misleading. Thus, Frost and Hillis (1990) not only suggest that species concepts must be consistent with "recovered evolutionary history," they also suggest that the BSC "often hinders attempts to recover the history of evolution" (p. 88). Similarly, Cracraft (1989) argues that "non-monophyletic species imply that history has been misrepresented" (p. 39). These claims are exaggerated. If we accept that species are defined by isolation and/or cohesion and do not start with the assumptions that they must be exclusive groups and the units of phylogeny, then including paraphyletic assemblages as species does not mispresent history.

There is no doubt that a robust phylogeny is a prerequisite for studying speciation, because inferences about change over time depend upon knowing the relationships of the populations or species being compared. The observation that one species comprises a paraphyletic assemblage of lineages and a second species is a lineage embedded within that assemblage can provide valuable information about the history of divergence/speciation. In the example shown in figure 2.2, we would want to know whether, in a set of independent gene genealogies, genotypes of species A consistently appear as sister to genotypes from the same population (e.g., B$_3$) or whether the tree topology depends on which gene we choose. In the former case, we might conclude that speciation has involved local divergence, either in sympatry or as a result of a local vicariance or founder event. In the latter case, gene genealogies may simply represent random lineage sorting from a polymorphic ancestor.

Both isolation and cohesion species are often paraphyletic. If A and B (e.g., figure 2.2) are reproductively isolated (due to changes along the branch leading to A), then they are isolation species. But they are not comparable entities in terms of evolutionary history, because A is a monophlyletic group and B is the paraphyletic assemblage of lineages remaining after divergence of A. In order to distinguish these two distinct entities, alternative names have been proposed for such paraphyletic assemblages of populations ("metaspecies" [Baum and Shaw 1995], a modification of the original definition of that term by Donoghue [1985]; or "ferespecies" [Graybeal 1995]).

Molecular phylogenies of groups of closely related species have revealed a number of clear examples of currently recognized species that appear to be paraphyletic assemblages of populations. Figure 2.3 shows two examples within the prodoxid moth genus *Greya*, in which a mitochondrial DNA (mtDNA) phylogeny indicates that a widely distributed species is paraphyletic with respect to a close relative that has a limited geographic range and has shifted to a new host plant (see Brown et al. 1994). In one case, the mtDNA haplotype found in the narrowly distributed species shares a most recent common ances-

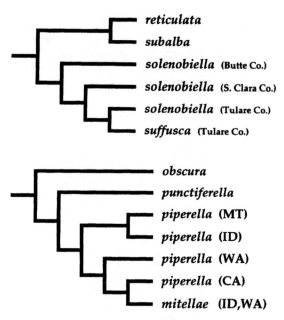

Figure 2.3. Two species of moths in the genus *Greya* (Prodoxidae) appear as paraphyletic assemblages of populations in mtDNA phylogenies. (a) The *Greya solenobiella* group. *Greya solenobiella* is a widespread species in California and Oregon found on *Yabea microcarpa* (Apiaceae). *G. suffusca* is restricted to Tulare Co., California, and is found on *Osmorhiza brachypoda* (Apiaceae). (b) The *Greya punctiferella* group. *Greya piperella* is widespread throughout the western United States and Canada, feeding on plants in the genus *Heuchera* (Saxifragaceae). *Greya mitellae* is found only in northern Idaho and southeastern Washington on *Mitellae stauropetala* (Saxifragaceae). The mitochondrial DNA data are from Brown et al. (1994).

tor with a haplotype of the widespread "paraphyletic" species from the only locality where the two species occur together. A plausible model for divergence involves a host shift "within" one region (either sympatrically or through divergence of a small isolated population), giving rise to a narrowly distributed daughter species on a new host plant. Funk et al. (1995) provide additional examples of insect host shifts resulting in two daughter species, one paraphyletic with respect to the other. In fact, divergence of peripheral isolates or island populations may commonly lead to the same sort of pattern, with a widespread "ancestral" species giving rise to local derivatives (figure 2.4). Data from mice (Avise et al. 1983), pocket gophers (Patton and Smith 1994), macaques (Melnick et al. 1993), and *Drosophila* (Powell 1991; Hey and Kliman 1993) provide likely examples of such situations. Rieseberg and Brouillet (1994) argue that paraphyletic species will be common in plants, given the prevalence of "local speciation" (Levin 1993).

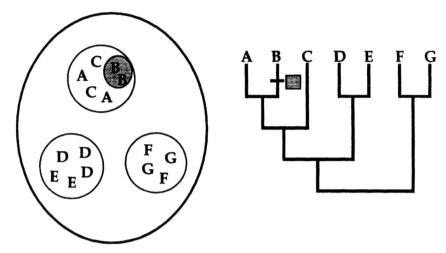

Figure 2.4. Origin of a new species "within" one of several populations of a widely distributed ancestral species. Regardless of the local geography of speciation, the resulting gene genealogy shows that the widely distributed daughter species is paraphyletic with respect to the locally derived species. A–G are distinct genotypes (DNA sequences). Only genotype B is found in the narrowly distributed daughter species. In the example shown here, the three populations of the widely distributed species exhibit fixed differences and would be considered three distinct phylogenetic species according to the criteria of Nixon and Wheeler (1990).

Gene Genealogies, Species, and Speciation

Analyses of gene genealogies in recently subdivided populations suggest that paraphyly (and perhaps polyphyly) will be common and that the phylogenetic status of populations changes over time (Tajima 1983; Neigel and Avise 1986; Hey 1994). Consider a single panmictic population that is subdivided into two daughter populations with complete interruption of gene flow at time t (figure 2.5). The daughter populations each contain N diploid individuals, and selection and recombination are assumed not to occur. If two gene copies are sampled from each of the daughter populations, relationships must conform to one of four possible tree topologies (figure 2.5). These four gene genealogies correspond to situations in which the daughter populations are polyphyletic, one is paraphyletic with respect to the other, or the two populations are both monophyletic. Tajima (1983) calculated the probabilities of each genealogy as a function of the time since interruption of gene flow (table 2.2). With this sampling scheme, the most likely gene genealogy immediately after divergence is that both populations appear polyphyletic, but the probability of polyphyly declines quickly and is less than 10% after $2N$ generations. The probability of paraphyly increases to nearly 40% after N generations and then gradually declines. The initial probability of reciprocal monophyly is small, but this probability continues to increase over time. However, even $4N$ generations after interruption of gene flow, only about 83% of gene genealogies will show this pattern.

Using computer simulations, Neigel and Avise (1986) reached very similar conclusions, arguing for a progression from polyphyly through paraphyly to reciprocal monophyly. They extended the analysis by examining the influence of "mode of speciation"—varying the way in which the original population was partitioned and the subsequent demography of the daughter populations. Probabilities of initial polyphyly, paraphyly or monophyly depend on the numbers of founders and whether

Table 2.2. Probabilities of tree topologies (a)–(d) in figure 2.5, as a function of the number of generations since interruption of gene flow between populations A and B.

Generations	Topology			
	(a)	(b)	(c)	(d)
$N/10$	0.134	0.263	0.201	0.402
$N/2$	0.231	0.364	0.135	0.270
N	0.354	0.400	0.082	0.164
$2N$	0.570	0.340	0.030	0.060
$4N$	0.828	0.160	0.004	0.008

Data were calculated using the formulae given by Tajima (1983). Topology (a) corresponds to reciprocal monophyly, topology (b) to paraphyly, and topologies (c) and (d) to polyphyly. N is the size of the two populations (assumed to be constant).

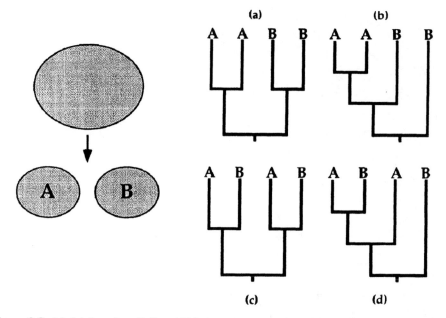

Figure 2.5. Model (based on Tajima 1983) in which a single ancestral population is subdivided into daughter populations A and B. Gene genealogies (a)–(d) represent the four possible tree topologies if two gene copies are sampled from each of populations A and B at some time after interruption of gene flow.

these individuals are chosen at random or from a restricted portion of the ancestral population. When one of the daughter populations derives from a small number of founders, the initial probability of paraphyly is very high. Hey (1994) showed that the probability of one or both daughter populations forming exclusive groups depends on the number of genes sampled from the daughter populations, decreasing as the sample size of genes increases.

Obviously, time to reciprocal monophyly (both daughter populations exclusive groups) depends on the effective population size (Tajima 1983; Neigel and Avise 1986). The time also depends on the nature of natural selection (assumed to be absent in all of the above models). Divergent selection (different alleles favored in the two daughter populations) will speed the approach to exclusivity. In contrast, balancing selection will slow it down (and can, in some cases like human and chimp MHC, prevent populations that have diverged for a very long time from becoming exclusive groups). Finally, all of the above models invoke complete interruption of gene flow. In a steppingstone or isolation-by-distance model, probabilities of the alternative tree topologies depend on migration rate and geographic distance (Barton and Wilson 1995).

Life History of a Species

I propose that species (in a very broad sense) have a life history (figure 2.6). The details of the life history will depend on the geography of speciation, demographic events during and after population subdivision, the impact of natural selection, and constraints on the evolution of cohesion (isolation). First, consider the case when speciation is allopatric. In this case, populations pass through a series of life stages following interruption of gene flow. The trajectory displayed in figure 2.6a is probably a common one—especially in organisms like many insects that have labile mate recognition systems, life cycles, and host associations. Soon after gene flow interruption, daughter populations A and B become phylogenetic species. This stage is completed either when (1) detectable mutations occur and are fixed in a single population or (2) random sorting of ancestral polymorphism results in fixed character state differences between populations. The consequence of either sort of event is a pair of populations that are diagnosably distinct. The probability of observing fixed character state differences depends on the effort expended in looking for such differences. Therefore, the timing of this initial "speciation" event will be a function of the level of resolution achieved in defining character state differences. It will obviously also depend on effective population sizes (small size promotes sorting of ancestral polymorphisms) and selection regimes (divergent selection will more likely lead to alternative fixation).

The second stage in the trajectory shown in figure 2.6a is the evolution of barriers to gene exchange or new cohesion mechanisms. These events are not apparent from

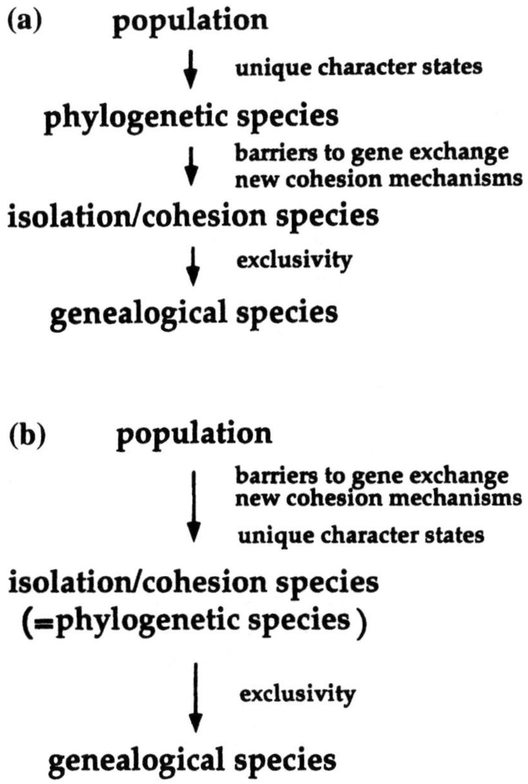

(a)   **population**

$\downarrow$ **unique character states**

phylogenetic species

$\downarrow$ **barriers to gene exchange**
new cohesion mechanisms

isolation/cohesion species

$\downarrow$ **exclusivity**

genealogical species

(b)   **population**

$\downarrow$ **barriers to gene exchange**
new cohesion mechanisms

unique character states

isolation/cohesion species
(=phylogenetic species)

$\downarrow$ **exclusivity**

genealogical species

Figure 2.6. Species' "life histories" that trace change over time in the status of a pair of populations. In (a) speciation occurs in allopatry and the populations first become phylogenetic species, then isolation/cohesion species, and finally genealogical species. In (b) speciation is sympatric and populations become isolation/cohesion species at the same time as they become phylogenetic species.

a gene genealogy and are, to some degree, unpredictable. In many cases they may simply occur as a result of further divergence between populations (i.e., as a by-product of divergence in allopatry). The rate at which this stage is completed is no doubt taxon dependent; in some groups mate recognition systems and habitat/resource associations change quickly, whereas in others they are likely to be very conservative.

To be genealogical species, the two daughter populations must be exclusive groups. This third stage takes a very long time, unless N is small or we relax the requirement that exclusivity apply to all segments of the genome. Avise and Ball (1990) suggested "genealogical concordance" as a criterion for recognizing taxa (subspecies). This is a somewhat arbitrary criterion, representing the stage in the ontogeny of a species in which some but not all gene genealogies show concordant patterns of reciprocal monophyly.

In some circumstances, diverging populations may become genealogical species before they become isolation/cohesion species. This will be most likely when a polymorphic ancestral population becomes subdivided and sustained population bottlenecks result in pairs of exclusive groups in the absence of the evolution of any intrinsic barriers to gene exchange. Graybeal (1995) suggests that "exclusivity marks the point where the connections [between populations] can be considered lost" (p. 249). I agree that exclusivity can be viewed as a genealogical endpoint—but it is not one that is by any means irreversible (e.g., exclusive taxa may be able to hybridize, and introgression will destroy exclusivity).

When speciation is sympatric, populations often become isolation/cohesion species at the same time as they become phylogenetic species (figure 2.6b), because fixed character state differences first arise at the strongly selected loci that are responsible for the elimination of gene flow. With parapatric divergence, models of isolation by distance apply and fixed differences are also more likely to be due to selection.

Evolutionary geneticists have generally regarded the transition to isolation/cohesion species as the essence of the speciation process. Systematists have clearly been more concerned with the evolution of diagnosable and exclusive groups. Both of these views are legitimate. The origin of isolation and cohesion is of particular interest not because the underlying evolutionary processes are unique but because of the nature of the changes that occur (defined by their subsequent effect). As evolutionary biologists interested not only in what has been but also in what will be, most students of the speciation process remain convinced that it is the evolution of new barriers to gene exchange or new cohesion mechanisms (i.e., the "second stage" in figure 2.6a or the "first stage" in figure 2.6b) that should be the focus of attention for those who claim to be studying speciation. For organizing the diversity of life and naming taxa, other stages in the life history of species may prove more useful.

Acknowledgments   I thank Kevin de Queiroz, Alan Templeton, Stewart Berlocher, and an anonymous reviewer for comments on an earlier version of this chapter. The chapter was written while I was on sabbatical leave in the Department of Evolution and Ecology at University of California, Davis. A number of people in that department were helpful in this endeavor; thanks especially to Rick Grosberg and the members of his lab and to David Begun.

References

Avise, J. C., and Ball, R. M. 1990. Principles of genealogical concordance in species concepts and biological taxonomy. Oxford Surveys in Evolutionary Biology 7:45–67.

Avise, J. C., Shapira, J. F., Daniel, S. W., Aquadro, C. F., and Lansman, R. A. 1983. Mitochondrial DNA differentiation during the speciation process in *Peromyscus*. Mol. Biol. Evol. 1:38–56.

Avise, J. C., Arnold, J., Ball, R. M., Bermningham, E., Lamb, T., Neigel, J. E., Reed, C. A., and Saunders, N. C. 1987. The mitochondrial bridge between population genetics and systematics. Annu. Rev. Ecol. Syst. 18:489–522.

Barton, N. H., and Wilson, I. 1995. Genealogies and geography. Phil. Trans. R. Soc. Lond. B 347:49–59.

Baum, D. A. 1992. Phylogenetic species concepts. Trends Ecol. Evol. 7:1–2.

Baum, D. A., and Donoghue, M. J. 1995. Choosing among alternative "phylogenetic" species concepts. Syst. Bot. 20:560–573.

Baum, D. A., and Shaw, K. L. 1995. Genealogical perspectives on the species problem. In P. C. Hoch and A. G. Stevenson, eds. Experimental and Molecular Approaches to Plant Biosystematics. St. Louis, Mo.: Missouri Botanical Garden, pp. 289–303.

Bremer, K., and Wanntorp, H.-E. 1979. Geographic populations or biological species in phylogeny reconstruction. Syst. Zool. 28:220–224.

Brown, J. M., Pellmyr, O., Thompson, J. N., and Harrison, R. G. 1994. Phylogeny of *Greya* (Lepidoptera: Prodoxidae), based on nucleotide sequence variation in mitochondrial cytochrome oxidase I and II: congruence with morphological data. Mol. Biol. Evol. 11:128–141.

Bush, G. L. 1994. Sympatric speciation in animals: new wine in old bottles. Trends Ecol. Evol. 9:285–288.

Butlin, R. K. 1989. Reinforcement of premating isolation. In D. Otte and J. A. Endler, eds. Speciation and Its Consequences. Sunderland, Mass.: Sinauer, pp. 158–179.

Coyne, J. A. 1992. Much ado about species. Nature 357:289–290.

Coyne, J. A. 1994. Ernst Mayr and the origin of species. Evolution 48:19–30.

Coyne, J. A., and Orr, H. A. 1989. Patterns of speciation in *Drosophila*. Evolution 43:362–381.

Cracraft, J. 1983. Species concepts and speciation analysis. Curr. Ornithol. 1:159–187.

Cracraft, J. 1989. Speciation and its ontology: the empirical consequences of alternative species concepts for understanding patterns and processes of differentiation. In D. Otte and J. A. Endler, eds. Speciation and Its Consequences. Sunderland, Mass.: Sinauer, pp. 28–59.

Davis, J. I., and Nixon, K. C. 1992. Populations, genetic variation, and the delimitation of phylogenetic species. Syst. Biol. 41:421–435.

de Queiroz, K., and Donoghue, M. J. 1988. Phylogenetic systematics and the species problem. Cladistics 4:317–338.

de Queiroz, K., and Donoghue, M. J. 1990. Phylogenetic systematics and species revisited. Cladistics 6:83–90.

Dobzhansky, T. 1970. Genetics of the Evolutionary Process. New York: Columbia University Press.

Donoghue, M. J. 1985. A critique of the biological species concept and recommendations for a phylogenetic alternative. Bryologist 88:172–181.

Frost, D. R., and Hillis, D. M. 1990. Species in concept and practice: herpetological applications. Herpetologica 46:87–104.

Funk, D. J., Futuyma, D. J., Orti, G., and Meyer, A. 1995. A history of host associations and evolutionary diversification for *Ophraella* (Coleoptera: Chrysomelidae): new evidence from mitochondrial DNA. Evolution 49:1008–1017.

Graybeal, A. 1995. Naming species. Syst. Biol. 44:237–250.

Harrison, R. G. 1986. Pattern and process in a narrow hybrid zone. Heredity 56:337–349.

Harrison, R. G. 1990. Hybrid zones: windows on evolutionary process. Oxford Surv. Evol. Biol. 7:69–128.

Harrison, R. G. 1991. Molecular changes at speciation. Annu. Rev. Ecol. Evol. 22:281–308.

Hey, J. 1994. Bridging phylogenetics and population genetics with gene tree models. In B. Schierwater, B. Streit, G. P. Wagner, and R. DeSalle, eds. Molecular Ecology and Evolution: Approaches and Applications. Basel, Switzerland: Birkhauser, pp. 435–449.

Hey, J., and Kliman, R. M. 1993. Population genetics and phylogenetics of DNA sequence variation at multiple loci within the *Drosophila melanogaster* species complex. Mol. Biol. Evol. 10:804–822.

Howard, D. J. 1993. Reinforcement: origin, dynamics, and fate of an evolutionary hypothesis. In R. G. Harrison, ed. Hybrid Zones and the Evolutionary Process. New York: Oxford University Press, pp. 46–69.

Hudson, R. R. 1990. Gene genealogies and the coalescent process. Oxford Surv. Evol. Biol. 7:1–44.

Lambert, D. M., Michaux, B., and White, C. S. 1987. Are species self-defining? Syst. Zool. 36:196–205.

Levin, D. A. 1993. Local speciation in plants: the rule not the exception. Syst. Bot. 18:197–208.

Luckow, M. 1995. Species concepts: assumptions, methods, and applications. Syst. Bot. 20:589–605.

Maddison, W. P. 1995. Phylogenetic histories within and among species. In P. C. Hoch and A. G. Stevenson, eds. Experimental and Molecular Approaches to Plant Biosystematics. St. Louis, Mo.: Missouri Botanical Garden, pp. 273–287.

Mallet, J. 1995. A species definition for the modern synthesis. Trends Ecol. Evol. 10:294–299.

Masters, J. C., Rayner, R. J., McKay, I. J., Potts, A. D., Nails, D., Ferguson, J. W., Weissenbacher, B. K., Allsopp, M., and Anderson, M. L. 1987. The concept of species: recognition versus isolation. S. African J. Sci. 83:534–537.

Mayr, E. 1963. Animal Species and Evolution. Cambridge, Mass.: Belknap Press.

Melnick, D. J., Hoelzer, G. A., Absher, R., and Ashley, M. V. 1993. mtDNA diversity in rhesus monkeys reveals overestimates of divergence time and paraphyly with neighboring species. Mol. Biol. Evol. 10:282–295.

Mishler, B. D., and Donoghue, M. J. 1982. Species concepts: a case for pluralism. Syst. Zool. 31:491–503.

Neigel, J. E., and Avise, J. C. 1986. Phylogenetic relationships of mitochondrial DNA under various demographic models of speciation. In S. Karlin and E. Nevo, eds. Evolutionary Processes and Theory. New York: Academic Press, pp. 515–534.

Nixon, K. C., and Wheeler, Q. D. 1990. An amplification of the phylogenetic species concept. Cladistics 6:211–223.

O'Hara, R. J. 1993. Systematic generalization, historical fate and the species problem. Syst. Biol. 42:231–246.

Paterson, H. E. H. 1985. The recognition concept of species. In E. S. Vrba, ed. Species and Speciation. Pretoria: Transvaal Museum, pp. 21–29.

Patton, J. L., and Smith, M. F. 1994. Paraphyly, polyphyly, and the nature of species boundaries in pocket gophers (genus *Thomomys*). Syst. Biol. 43:11–26.

Powell, J. R. 1991. Monophyly/paraphyly/polyphyly and gene/species trees: an example from *Drosophila*. Mol. Biol. Evol. 8:892–896.

Rieseberg, L. H., and Brouillet, L. 1994. Are many plant species paraphyletic? Taxon 43:21–32.

Simpson, G. G. 1961. Principles of Animal Taxonomy. New York: Columbia University Press.

Sokal, R. R., and Crovello, T. J. 1970. The biological species concept: a critical evaluation. Amer. Natur. 104:127–153.

Tajima, F. 1983. Evolutionary relationship of DNA sequences in finite populations. Genetics 105:437–460.

Templeton, A. R. 1981. Mechanisms of speciation—a population genetic approach. Annu. Rev. Ecol. Syst. 12:23–48.

Templeton, A. R. 1989. The meaning of species and speciation: a genetic perspective. In D. Otte and J. A. Endler, eds. Speciation and Its Consequences. Sunderland, Mass.: Sinauer, pp. 3–27.

Templeton, A. R. 1994. The role of molecular genetics in speciation studies. In B. Schierwater, B. Streit, G. P. Wagner, and R. DeSalle, eds. Molecular Ecology and Evolution: Approaches and Applications. Basel, Switzerland: Birkhauser, pp. 455–477.

Van Valen, L. 1976. Ecological species, multispecies and oaks. Taxon 25:233–239.

Wheeler, Q. D., and Nixon, K. C. 1990. Another way of *looking at* the species problem: a reply to de Queiroz and Donoghue. Cladistics 6:77–81.

Wiley, E. O. 1978. The evolutionary species concept reconsidered. Syst. Zool. 27:17–26.

Wiley, E. O. 1981. Phylogenetics. New York: Wiley.

Williams, G. C. 1966. Adaptation and Natural Selection. Princeton, N.J.: Princeton University Press.

3

Species and Speciation

Geography, Population Structure, Ecology, and Gene Trees

Alan R. Templeton

In 1969 Guy Bush published a paper on *Ragoletis* advocating "sympatric speciation." Bush (1969) argued that an ancestral population living in an area containing two or more habitats could split into two species, despite initial and ongoing gene flow during the process of splitting. Such habitat divergence models of speciation have proven to be very controversial (Templeton 1981), as the reception by zoologists to much of Guy Bush's work has demonstrated (Futuyma and Mayer 1980; Paterson 1981). Interestingly, in 1969 another controversial paper in the speciation literature was published by Ehrlich and Raven (1969). They questioned the central role given to gene flow by one of the dominant definitions of species at that time, the Biological Species Concept (BSC). The BSC defines a species as a "reproductive community" (Mayr 1970) whose boundaries are defined by the lack of potential for gene flow with other such communities, that is, by isolating mechanisms. Ehrlich and Raven (1969) argued that other evolutionary forces can play an important and direct role in defining what is or is not a species. It is the balance of gene flow with other evolutionary factors (such as natural selection) that must be considered. Under some circumstances, this balance will be in favor of gene flow, but in other circumstances it will not. In particular, they argued that in many cases the effects of selection can override the effects of gene flow. Thus, for speciation to occur, it is not necessary for selection to cause the cessation of gene flow, merely to override it. Under this view, speciation can occur even when gene flow continues between the speciating populations, and indeed speciation mechanisms similar to Bush's (1969) are commonly invoked by botanists (e.g., Raven and Raven 1976).

Just as Bush's (1969) work provoked controversy among zoologists, so did the Ehrlich and Raven (1969) paper. For example, Shapiro (1970), an entomologist, calls their paper "provocative" and claimed that they underestimated the evolutionary significance of intermit-tent gene flow. In contrast, Baker (1970), a botanist, regarded their paper as mostly "unexceptionable" and simply a restatement of similar views that already existed in the botanical literature. One contributing factor to this different reception is that the BSC is the dominant species concept among zoologists but not botanists (Mayr 1992). The BSC formally defines species only in terms of the potential for gene flow (Mayr 1970), and hence many advocates of the BSC, such as Shapiro (1970), feel that even small amounts of gene flow will prevent speciation. This is a recurrent theme in many of the criticisms of Bush's work as well (e.g., Paterson 1981). Many botanists, such as Baker (1970) or Raven and Raven (1976), do not give gene flow such a central role in defining species and are therefore less reluctant to embrace speciation in the presence of gene flow. Thus, the definition of species that one uses has a major impact upon how one interprets the significance for speciation of basic evolutionary mechanisms such as gene flow or selection. Accordingly, species must be defined before the issue of speciation can be studied. Indeed, once the definition of species is clear, much of the controversy over speciation mechanisms becomes resolved. This chapter therefore focuses on the meaning of species.

Species Definitions

For a species definition to have some claim for biological generality, it must define species in terms of some biological "universal" that extends beyond the particular cases of individual species. Two commonly used universals are the universal of a species being an evolutionary lineage, such as the evolutionary species concept (Wiley 1981) or the phylogenetic species concept (Cracraft 1989), and the universal of a species being a "reproductive community" (Mayr 1970), such as the biological species concept (Mayr 1970) or the recognition species concept (Paterson 1985). These two universals are intimately interrelated: an evolutionary lineage requires a reproducing population, and the act of reproduction creates lineages. Hence, these two universals need not be regarded as antagonistic, but rather as complementary ways of looking at the same evolutionary phenomenon (see also de Queiroz, this issue). The cohesion species concept (Templeton 1989) explicitly acknowledges this intimate relationship between lineage and reproductive community. The cohesion species concept defines species as evolutionary lineages, with the lineage boundaries that define species arising from the forces that create reproductive communities (i.e., cohesion mechanisms). The cohesion species concept has the benefit of making clear that processes (cohesion mechanisms) generate patterns (the evolutionary branches that define lineages), so both process and pattern are part of the cohesion species concept. By integrating these two universals, the cohesion species concept takes advantage of the strengths of both classes of species concepts while avoiding the limitations and difficulties that each universal has when regarded by itself.

For example, one of the great strengths of the BSC is that by defining species in terms of isolating mechanisms, it provides solid guidance to experimental, observational, and theoretical studies of the speciation process. The speciation literature has been greatly strengthened by the BSC focusing on various measures of reproductive isolation as candidate traits for understanding the process of speciation (Templeton 1981, 1989). Because the classical isolating mechanisms are a proper logical subset of the cohesion mechanisms (Templeton 1989), all of the studies performed under the BSC that focus on isolation measures as candidate traits for speciation have equal validity and applicability under the cohesion species concept. The cohesion concept differs from the BSC not only in its primary "universal" (evolutionary lineage versus reproductive community), but also by recognizing a broader set of candidate traits as being relevant for speciation. For example, "fertilization mechanisms" also define a proper logical subset of the cohesion mechanisms (Templeton 1989), so the candidate traits identified by advocates of the recognition concept are also included as legitimate foci for studies of species and speciation.

One major difference between the cohesion species concept and the BSC/recognition concept is in how the phrase "reproductive community" (Mayr 1970) is interpreted. As stated above, lineages are created by acts of reproduction. One aspect of an act of reproduction is the passing on of genetic material from one generation to the next. The BSC defines reproductive communities solely in terms of this genetic aspect of reproduction, often equating the concept of a reproductive community to a shared gene pool (Dobzhansky 1950; Mayr 1970). The BSC further limits the type of passage of genetic material to "a reproductive community of sexual and crossfertilizing individuals" (Dobzhansky 1950). However, lineages can be generated genetically by modes of reproduction other than sexual outcrossing (as is evident by the widespread use of mitochondrial DNA, an asexual genome, to define evolutionary lineages), so the BSC does indeed create a real discrepancy between the universals of evolutionary lineages and reproductive communities by this limitation. Indeed, as limited by the BSC and the recognition concept, the "universal" of a reproductive community is not a biological universal at all, but rather one that explicitly excludes many of the major clades of life on this planet (Mayr 1970; Vrba 1985). In contrast, the cohesion species concept does not limit the reproductive acts that define lineages to sexual outcrossing, but rather recognizes that any mode of reproduction that allows the transfer of genetic material from one generation to the next has the potential for creating lineages.

The cohesion species concept also acknowledges that reproduction requires living parental organisms (although they need not be alive at the exact time of reproduction) and involves the transfer of energy and materials across generations. That is, reproduction also requires that individuals live in an environment to which they are sufficiently adapted so as to secure and utilize the resources needed for survival and reproduction. The range of resources that are essential for survival and reproduction is called the fundamental niche and is determined by adaptations, developmental constraints, and other factors (Templeton 1989). This ecological aspect of reproduction is ignored by the BSC species definition (although Mayr [1970] does argue for the BSC species having a strong ecological significance, he excludes the ecological components from the definition of species; see Templeton [1989] for more discussion). In contrast, the cohesion species concept explicitly acknowledges that reproduction has both an ecological and a genetic context and that both contexts are necessary for actual reproduction and the generation of an evolutionary lineage. Therefore, cohesion mechanisms broaden the spectrum of candidate traits for species status to include the adaptations and other factors that define the fundamental niche. Hence, ecological adaptations play a direct role in the speciation process under the cohesion species concept. A reproductive community is simultaneously both a genetic and an ecological entity under the cohesion concept, and this greatly expands the guidance that is provided for work on speciation and provides true compatibility between the universals of evolutionary lineage and reproductive community.

Although the BSC does provide guidance for speciation studies, the BSC is difficult to apply to natural populations when no supplementary experimental data are available. This difficulty stems from the fact that the BSC defines species in terms of the *lack of potential* for gene flow. Proving a lack of potential is difficult without controlled, experimental contrasts. As a consequence, there are few instances of the BSC being rigorously applied to natural populations in the absence of experimental data (Sokal and Crovello 1970). This is not to say that there have been frequent claims of successful implementation. For example, Mayr (1992) claimed that the BSC worked well for 93.6% of the native vascular plant species near Concord, New Hampshire. The 6.4% of the species that were excluded were done so primarily because they did not reproduce sexually, but the remaining species were regarded as good biological species solely on the grounds of morphological integrity (Whittemore 1993). There was not a single instance in which there was even an attempt to prove the lack of potential for gene flow; indeed, many of the plant species considered were known to interbred by Mayr (1992), but these facts were deemed unimportant because morphological distinctiveness was being maintained. However, the documented ability of plant "species" to maintain their morphological and ecological integrity in the face of recurrent gene flow or introgression (e.g., Anderson 1953; Whittemore and Schaal 1991; Templeton 1994) is one of the major reasons why many botanists have failed to embrace the BSC (see Templeton 1989 for a more extended discussion). Hence, all 93.6% of the species were identified exclusively under a morphological species concept (Whittemore 1993).

By embracing the universal of an evolutionary lineage and broadening the range of candidate traits for species, the cohesion species concept can be implemented and be more readily applied to natural populations. For practical implementation, the cohesion species concept is rephrased as a set of testable null hypotheses (Templeton 1994). The inference of a species requires that the relevant null hypotheses be rejected on the basis of data subjected to objective statistical analysis. As the details of defining species as testable null hypotheses are given elsewhere (Templeton 1994), only a brief summary is given in the following section.

Cohesion Species Inferred from Testable Null Hypotheses

Identifying Lineages

The first null hypothesis to be tested is that the population under study constitutes a single evolutionary lineage. If this null hypothesis cannot be rejected, there is no significant evidence for more than one species in the sample. This may be due to a lack of statistical power, or to the fact that the sample constitutes a single species or a part of a single species. Given that there is no way to discriminate among these alternatives with the current sample, the inference chain ends in this inconclusive state until additional data are obtained.

Molecular biology provides a powerful means for testing the null hypothesis of a single evolutionary lineage. As mentioned in the previous section, an evolutionary lineage is created by acts of reproduction that pass genetic material from one generation to the next. Modern molecular techniques allow one to survey genetic variation in a sample and to estimate the passage of the genetic material across the generations in a backward sense (see, e.g., Shaw, this volume). That is, starting with the initial sample of genetic variation one can estimate backward through time the pattern of mutational and replication events that generated the current sample from a single, ancestral piece of DNA. Tracing the sample of current variation at a locus backward to a common ancestral gene is called the coalescent process, and the pathways interconnecting the current sample through the common ancestor constitute a gene tree. In general, it is impossible to estimate the exact pattern of coalescent events that occur with a subset of the sampled genes that share the same allelic state, so the only portion of the gene tree that is generally estimable is that portion marked by the mutational events that create allelic change. This lower resolution tree is called an allele or haplotype tree.

There is nothing about the theory of coalescence that limits gene or haplotype trees to within species (however defined). Such trees can have both an inter- and an intraspecific component. It is this property that makes such trees so important in defining species and studying speciation: haplotype trees can straddle the interface between intraspecific populations and species (Templeton 1994). In particular, these trees provide an evolutionary historical framework for extending the comparative method— a powerful, nonexperimental analytical technique used traditionally at the interspecific level—to the tokogenetic/ phylogenetic interface and to within species. There are many practical uses for doing comparative studies at the intra- and combined intra/interspecific levels (Templeton 1996), but of particular relevance to the hypothesis at hand is the ability to use haplotype trees to define a nested statistical design that allows one to separate the effects of historical events (such as fragmentation events that start new evolutionary lineages) from the effects of recurrent evolutionary forces that operate within lineages (such as recurrent flow, even if it is restricted). The statistical details of how this is done are given elsewhere (Templeton et al. 1995), and additional worked examples of testing the null hypothesis of a single lineage for both allopatric and sympatric populations, and for plants and animals, are all given in Templeton (1994) and Templeton and Georgiadis (1996). Consequently, only one brief, new example will be given here to illustrate the technique.

The example is based on data on Hawaiian *Drosophila* given in DeSalle (1984), only some of which has been published in nonthesis form. DeSalle (1984) mapped the mitochondrial DNA (mtDNA) for 163 restriction site markers in several species from the planitibia subgroup. The current analysis deals only with the subset of DeSalle's data that deals with 31 individuals that were surveyed for mtDNA variation that fell within a single recognized species, *D. silvestris*. All populations are found on the Island of Hawaii; figure 3.1 shows the loca-

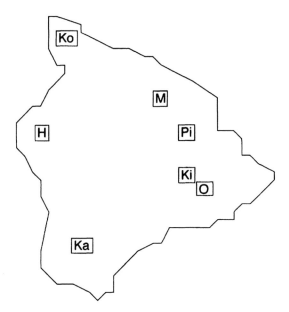

Figure 3.1. An outline map of the Island of Hawaii showing the collection locations for *Drosophila silvestris* used by DeSalle (1984). The collection locations are abbreviated as follows in the figure: Ko is Kohala, H is Hualalai, Ka is Kahuku, M is Maulua, Pi is Piihonua, Ki is Kilauea, and O is Olaa.

tions of the collecting sites. In addition, one individual of *D. planitibia* from the island of Maui was included as an outgroup. The restriction site data for these 32 individuals were used to estimate the haplotype tree shown in figure 3.2 with the methodology described in Templeton et al. (1992). This algorithm does not estimate just a single tree, but rather includes all trees such that the cumulative probability of the connections among haplotypes is equal to or greater than 0.95. This is called the 95% plausible set of trees. The 95% plausible set includes the unique maximum parsimony tree connecting these haplotypes, which is shown in figure 3.2. However, in this case, the 95% plausible set includes some nonparsimonious connections. In particular, three of the branches were likely to deviate by one additional mutation above parsimony. These probable deviations from parsimony create some ambiguity in topology. In figure 3.2, two circles are drawn around the maximum parsimony nodes in the maximum parsimony tree of radius of one mutation. The actual connections between the branches could occur at any mutational transition within these circles in addition to the maximum parsimony node. There is also some ambiguity in the 95% plausible set for the exact connection of the outgroup, *D. planitibia*, to the remainder of the *D. silvestris* haplotype tree. In this case, the probable deviations from parsimony go up to two additional steps; the ambiguity of the outgroup connection to the

D. silvestris haplotype tree is indicated by the brackets in figure 3.2. As shown by this example, the algorithm of Templeton et al. (1992) not only provides an estimate (or set of estimates) of the haplotype tree, but also provides an assessment of statistical confidence in that estimate.

The next step in the analysis is to convert the 95% plausible set of trees into a nested statistical design using the rules given in Templeton et al. (1987) and Templeton and Sing (1993). Basically, these nesting rules start at the tips of the haplotype network and move one mutational step into the interior, uniting all haplotypes that are connecting by this procedure into a "1-step clade." After pruning off the initial 1–step clades from the tips, this procedure is then repeated on the more interior portions of the haplotype network if needed until all haplotypes have been placed into 1–step clades. The next level of nesting uses the 1-step clades as its units, rather than individual haplotypes. The nesting rules are the same but result in "2-step clades" this time. This nesting procedure is repeated until a nesting level is reached such that the next higher nesting level would result in only a single category spanning the entire original haplotype network. The resulting nested clades are designated "C-N" where "C" is the nesting level of the clade and "N" is the number of a particular clade at a given nesting level. Some special nesting rules are needed to deal with symmetries and ambiguities in the estimated haplotype network (Templeton and Sing 1993). For the *D. silvestris* example, the ambiguities shown in figure 3.2 had no impact on the nested design. The content and designations of all nested clades are given in figure 3.3.

After the nested design is determined, the geographical data are overlaid upon the nested design and analyzed as described in Templeton et al. (1995). Basically, the geographical data are quantified in two main fashions: the clade distance, D_c, which measures the geographical range of a particular clade, and the nested clade distance, D_n, which measures how a particular clade is geographically distributed relative to its closest evolutionary sister clades (i.e., clades in the same higher level nesting category). These distances are also given in figure 3.3. Contrasts in these distance measures between tip clades and the clades immediately interior to them in the cladogram are important in discriminating the potential causes of geographical structuring of the genetic variation (Templeton et al. 1995). Given that the outgroup lies closer to the left cluster of haplotypes (and indeed, the 95% plausible set does not exclude the possibility that the outgroup connects within the left cluster), clade 4-1 is regarded as an interior clade and clade 4-2 as a tip clade in the terminology of Templeton et al. (1995), in the contrast of clade 4-1 versus 4-2 that occurs at the highest nesting level. The statistical significance of the different distance measures and the interior-tip contrasts are determined by random permutation testing that simulates the null hypothesis of a random geographical distribution for all clades within

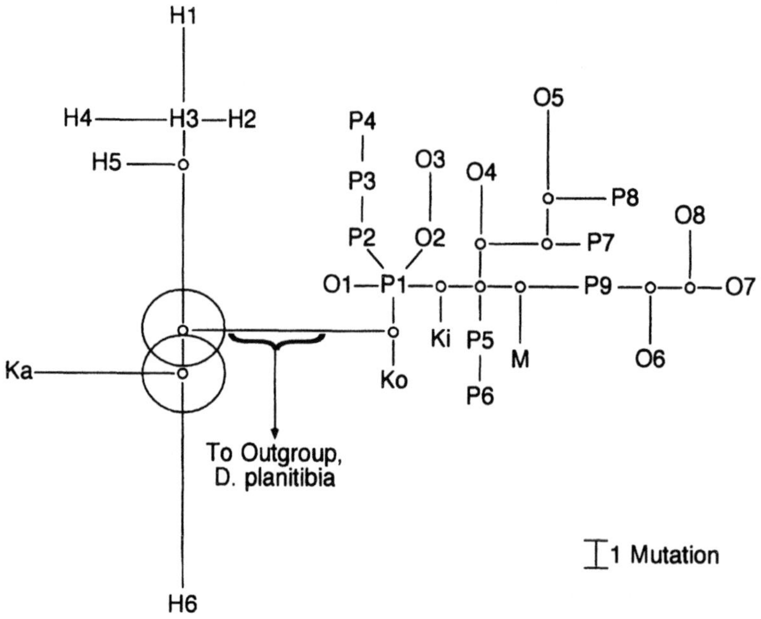

Figure 3.2. Haplotype tree of 31 *D. silvestris* individuals and one *D. planitibia* based upon restriction site mapping data of mitochondrial DNA. Haplotypes are designated by a letter indicating sample location (using the abbreviations in figure 3.1) followed by a number if more than one haplotype was found at that location. Haplotype H3 was found in four individuals from Hualalai, and P3 was found in two individuals from Piihonua; all other haplotypes were found in only one individual. The tree was estimated using the algorithm of Templeton, Crandall, and Sing (1992). Inferred nodes not represented in the sampled individuals are indicated by small circles. Branches are drawn proportional to the number of mutations (restriction site changes) that were estimated to have occurred. All branches between haplotypes or nodes of length 3 or less are in the limits of parsimony, and the topology shown in the figure is the unique maximum parsimony tree. Four branches within *D. silvestris* exceed this length, but have a probability of greater than 0.95 of exceeding the parsimonious length by no more than one additional mutation. This creates some topological ambiguity, with circles indicating the range of alternative connections possible for these longer branches. The outgroup branch length may deviate by up to two mutational changes from the maximum parsimony reconstruction, and brackets indicate the 95% confidence range for how the outgroup connects to the remainder of the haplotype tree.

a nesting category given the marginal clade frequencies and sample sizes per locality. Figure 3.3 shows the results obtained from this analysis. Only one statistically significant result occurred, and this was at the highest level of nesting (4-1 vs. 4-2).

Templeton et al. (1995) also make the biological interpretation of statistically significant results explicit by providing an inference key. The purpose of this key is to use the patterns of statistically significant results to discriminate among different biological processes that can create a phylogeographic association, such as restricted gene flow, range expansion and colonization events, and fragmentation or splitting events. When this key is applied to the results shown in figure 3.3, the inference chain is 1-2-3-4-9(No) (all numbers refer to questions in the inference key of Templeton et al. 1995). This inference chain leads to the conclusion of significant fragmenta-

tion (i.e., historical lineages). Therefore, the null hypothesis of only one lineage is rejected in this case, and two different lineages of *D. silvestris* have been identified. As can be seen from figures 3.1 and 3.2, the two lineages identified as statistically significant are found on the two different sides of the island, and hereafter will be called the Kona-side (clade 4-1, on the left in figures 3.1 and 3.2) and Hilo-side (clade 4-2, on the right in figures 3.1 and 3.2) lineages. Note that *within* each lineage, the haplotype tree clearly shows dispersal among all geographical sites (i.e., no sample site represents a topologically contiguous subset of haplotypes in the haplotype tree within either lineage). Hence, there is not only no evidence for sublineages within the Hilo- and Kona-side lineages, but there is also evidence for movement within each lineage over their respective geographical ranges.

Clade	D_c	D_n	Clade	D_c	D_n	Clade	D_c	D_n	Clade	D_c	D_n	Clade	D_c	D_n
H1	------------		1-1	na	na									
H2	na	na												
H3	na	na	1-2	na	na	2-1	na	na						
H4	------------		1-3	na	na				3-1	2.2	33.9			
H5	------------		1-4	------------		2-2	na	na						
Ka	------------		1-11	------------		2-5	------------		3-2	0	33.9	4-1	33.9	55.0[L]
H6	------------		1-15	------------		2-7	------------		3-3	2.2	33.9			
									I-T	-2.2	0			
Ko	------------		1-19	------------		2-9	------------		3-4	0	53.9			
P1	0	11.6												
P2	0	11.6	1-20	14.3	14.6									
O1	0	18.7												
I-T	0	-3.5												
P3	na	na				2-10	15.0	13.8						
P4	na	na	1-21	0	13.0									
O2	na	na												
O3	na	na	1-22	0	17.2									
			I-T	14.3	-0.5				3-5	19.5	22.6			
Ki	------------		1-23	0	23.7									
P5	na	na										4-2	29.7[S]	36.3
P6	na	na	1-25	0	3.8	2-11	21.8	21.9				I-T	4.2	18.7[L]
M	------------		1-26	0	23.6	I-T	-6.9	-8.1						
			I-T	0	10.0									
O4	------------		1-28	0	13.6									
P7	------------		1-29	0	16.7	2-12	15.0	15.0						
			I-T	0	-3.2				3-6	15.0	24.0			
O5	------------		1-31	0	13.6									
P8	------------		1-32	0	16.7	2-13	15.0	15.0						
						I-T	0	0						
P9	------------		1-33	0	16.7									
O6	------------		1-34	0	13.6	2-14	15.0	14.1						
			I-T	0	3.2				3-7	10.0	29.5			
O7	------------		1-35	na	na				I-T	-2.7	11.5			
O8	------------		1-36	na	na	2-15	0	6.2						
						I-T	15.0	7.9						

Figure 3.3. Nested cladistic analysis of the biogeographical distribution of haplotypes found in *D. silvestris*. The haplotype tree given in figure 3.2 was converted into a nested statistical design using the rules given in Templeton et al. (1987) and Templeton and Sing (1993), with haplotypes (the lowest level) nested into 1-step clades, 1-step clades nested into 2-step clades, and so on, until all the observations are contained within two 4-step clades (4-1 and 4-2) that provide mutually exclusive and exhaustive coverage of the entire haplotype tree. Clades at one level that are covered by an unbroken straight line to their right are nested together, with the nesting clade name being to the right of these vertical lines. Not all clade numbers are shown in the figure because many are null sets (i.e., they only include haplotypes that are inferred intermediates but not present in the actual sample). Whenever there was more than one nested clade within a nesting clade with observations from two or more sampling locations, the average clade distance and nested clade distances were calculated and tested for significant deviations from the null hypothesis of no association between nested clades and geographical location within a nesting clade. Details of this testing procedure are given in Templeton et al. (1995). A superscript "L" by a distance measure indicates a significantly large distance (probability < 0.05 under the null hypothesis), and a superscript "S" indicates a significantly small distance. In addition, clades were classified as either "tips" or "interiors" using the terminology found in Templeton, Routman, and Phillips (1995), and the average distances between interior minus tip clades (I-T) within a nesting clade were calculated and tested.

Deciding Which Lineages or Sets of Lineages Define Cohesion Species

Given that the null hypothesis of a single evolutionary lineage has been rejected, there is now the possibility that two or more of the lineages inferred to be in the sample constitute two or more species. Basically, the pattern of past evolution has been revealed by the inference of lineages through the rejection of the first null hypothesis. However, the cohesion species concept seeks to integrate pattern and process. As a consequence, additional null hypotheses concerning cohesion mechanisms must be rejected before elevating a lineage or set of lineages to the status of species (Templeton 1994). Just as the BSC has helped guide the efforts of researchers of speciation by identifying measures of reproductive isolation as candidate traits for species status, so does the cohesion concept identify candidate traits. The two major classes of cohesion mechanisms are genetic exchangeability (applicable primarily to sexual outcrossers) and ecological interchangeability (demographic exchangeability) (Templeton 1989). For sexual organisms, hypotheses about candidate traits for both genetic exchangeability and ecological interchangeability should ideally be tested, although the rejection of either type of null hypotheses is sufficient to infer a cohesion species. For nonsexual or nonoutcrossing organisms, only the null hypothesis of ecological interchangeability is relevant for inferring species status. However, it is not enough to show that variation in candidate traits has a nonrandom overlay on the haplotype network; rather, the rejection of either the null hypothesis of genetic exchangeability or of ecological interchangeability must be made *in a manner congruent with* the previously defined lineages to infer a cohesion concept species. Note that all evaluations of traits related to genetic exchangeability and ecological interchangeability are done in the context of two or more populations that have already been inferred to be statistically distinct historical lineages. Hence, effective de facto reproductive and/or demographic isolation among lineages and cohesiveness within has already been established to the extent needed to detect significantly distinct lineages as outlined above. However, de facto isolation is necessary but *not sufficient* for the rigorous inference of species under the BSC (Mayr 1970), and the same is true for de facto isolation or cohesiveness under the cohesion species concept. A second stage of the inference scheme must now be performed that constitutes a statistical evaluation of the lineage associations of reproductive and ecological variation that are reasonable candidates for contributing to genetic exchangeability and/or ecological interchangeability.

The null hypothesis at this second stage of testing is that all lineages belong to one species and hence should show no significant differentiation across lineages with respect to traits that are candidates for contributing to genetic exchangeability and/or ecological interchangeability. In practice, this inference approach will be primarily applied to close lineages, such as the Hilo-side and Kona-side lineages of *D. silvestris*. Accordingly, many traits would be expected to be shared across the lineages. Such traits do not contribute to the inference of a species under a statistical null hypothesis approach. With such traits, the null hypothesis cannot be rejected, but that does not mean that the null hypothesis is true. The failure to reject this second-stage null hypothesis means that either all lineages are one species, insufficient samples have been obtained to provide adequate statistical power, or the candidate traits being examined are not the ones that actually define the species in this case. It is impossible to discriminate among these alternatives, so the inference procedure ends in this inconclusive state when the second-stage null hypothesis is not rejected and remains in that state until further data can be collected.

The strong inference of two or more species among the lineages previously identified is possible only when one or more candidate traits allow the rejection of the second-stage null hypothesis at an appropriate level of significance (when multiple traits are tested, the significance level may need to be readjusted). Worked examples of this inference procedure for both plants and animals are given in Templeton (1994). A continuation of the *D. silvestris* example is given below.

Testing Genetic Exchangeability

Genetic exchangeability can be defined either in terms of isolating mechanisms or fertilization mechanisms (Templeton 1989). Either perspective can provide candidate traits for experimental or observational study. When experimental data exist, such data can be used in conjunction with the haplotype tree to test the null hypothesis of genetic exchangeability. For example, experimental data on the extent and genetic basis of premating isolation exists among these *D. silvestris* populations (Kaneshiro and Kurihara 1981; Templeton and Ahearn 1989), and the experimental data are completely concordant with the lineages defined in the previous subsection. However, the purpose of this chapter is to emphasize the use of nonexperimental data, so details of how experimental data can be rigorously integrated into a haplotype tree analysis will be given elsewhere. Here, it is only important to note that experimental data, when it exists, can be used to infer species under the cohesion concept in a manner that satisfies the criteria for species status under the BSC.

Experimental data on the candidate trait of reproductive isolation will not be available in many, probably most, cases. In this situation, the null hypothesis of genetic exchangeability can best be tested through fertilization mechanisms. Natural history observations can often help identify and even quantify many of the factors that allow or make fertilization possible in sexual outcrossers. For example, in plants, one can measure flowering time or identify pollinators. In animals, observations can be made on mating season, behavioral displays, and secondary sexual characteristics used in courtship. Only rarely is the impact of these traits on reproductive isolation directly measured or even relevant in a natural context (e.g., the Kona-side and Hilo-side lineages of *D. silvestris* are allopatric, so studies of reproductive isolation are strictly a laboratory phenomenon that have no direct applicability to cohesion in nature). However, many of these traits can be documented as influencing the probability of fertilization. When such documentation is possible for a trait, it is a legitimate candidate trait under the fertilization mechanism perspective, which has equal validity to the isolation mechanism perspective under the cohesion concept.

As an example, consider the male secondary sexual trait of bristle rows on the front tibia in *Drosophila silvestris* (Carson et al. 1982; Carson and Lande 1984). The last part of the male courtship in this species just prior to copulation involves the male vibrating his wings while behind the female and simultaneously vibrating his front tibia bristles over the female's abdomen (Boake and Hoikkala 1995). The female then either walks away, thereby ending the courtship, or remains motionless, thereby enabling the male to attempt to copulate (Hoikkala and Welbergen 1995). Analysis of videotapes of male courtships reveal that the duration of this final vibration stage strongly discriminates between successful versus unsuccessful courtships, whereas most other components of the male courtship do not show a significant difference (Boake and Hoikkala 1995; Hoikkala and Welbergen 1995). These observations indicate that this secondary sexual trait is a reasonable candidate for contributing to the mate recognition system and the probability of fertilization. In this case, there is also experimental evidence confirming this conclusion: sexual selection occurs in laboratory populations displaying genetic variation for the number of bristles, and this selection can be either for increased or decreased bristle number in different laboratory populations (Carson et al. 1994).

In nature, males of *D. silvestris* have been found to fall into two non-overlapping categories with respect to bristle phenotype: males with two bristle rows versus males with three bristle rows (Carson et al. 1982). The bristle trait status could be assigned to the flies studied by DeSalle (1984) because even when a female was used for the molecular studies, she either came from or was used to establish an isofemale line that produced both males and females. The same nested design that was used to analyze geographical data in order to infer lineages can now be used to look for significant evolutionary transitions in the categorical phenotypes of 2- versus 3-row flies using the nested categorical statistical analysis described in Templeton and Sing (1993). Only one of the nested categories in this case has any phenotypic variability, so only that test is relevant. Hence, the nested categorical analysis collapses into a two-by-two contingency analysis of the contrast of clade 4-1 versus 4-2 with respect to bristle-row number. The results of such an analysis are shown in table 3.1, which localizes a highly significant transition in bristle phenotype that is completely congruent with the Kona-side (2–row flies) and Hilo-side (3–row flies) lineages that were previously defined. Hence, using the criterion of species inference given in Templeton (1994) and restated at the beginning of this section, Kona-side and Hilo-side *D. silvestris* represent two different cohesion species.

Testing Ecological (Demographic) Interchangeability

Given that the null hypothesis of a single lineage has already been rejected, the other major null hypothesis to be tested is the null hypothesis of ecological interchangeability. Ecological interchangeability arises from the shared fundamental adaptations that define a population's fundamental niche or selective regime (in the sense of Baum and Larson 1991). Hence, the null hypothesis of ecological interchangeability (demographic exchangeability) can be tested by overlaying measurements (either quantitative or categorical) of adaptations or selective regimes on the same haplotype trees used to identify the lineages. The null hypothesis of no association between such adaptations or selective regimes with the haplotype tree can be tested using the statistical procedures given by Templeton et al. (1987) or Templeton and Sing (1993). Examples of such testing for both plants and animals and for both allopatric and sympatric populations are given in Templeton (1994). When the null hypothesis of no association is rejected, the statistical analyses of Templeton et al. (1987) or Templeton and Sing (1993) also localize where the significant transitions occurred in the haplotype tree. Hence, congruence with the previously defined lineages is also determined by these tests. If the null hypothesis of ecological interchangeability is rejected in a fashion that is congruent with the previously

Table 3.1. Nested contingency analysis of bristle row character state using the haplotype tree shown in Figure 3.2.

Clade	Character State	
	2-Row	3-Row
4-1	10	0
4-2	0	21

Because there is only one nested category with variation in the character state, the nested analysis reduces to a Fisher's Exact Test for the highest level of nesting (see figure 3.3). The Fisher *p* value is less than 0.0001 under the null hypothesis of no association between clade and character state.

identified lineages, the relevant lineages are regarded as cohesion species. Adaptations that display no significant haplotype tree associations or show incongruent associations with the previously defined lineages are regarded as intraspecific polymorphisms and not species-defining traits. Thus, this inference procedure provides an explicit basis for discriminating between phylogenetic versus tokogenetic traits, as is shown by worked examples given in Templeton (1994).

Advantages of Inferring Species as Testable Null Hypotheses

Inferring species from testable null hypotheses has many advantages. The most obvious advantage is that the criteria, data, methods of analysis, and degree of support for the inference are all made completely explicit and objective. All too often, species criteria are subjective or, as illustrated by Mayr (1992) in light of Whittemore's (1993) criticism, the stated criteria are different from the actual criteria used (see also Sokal and Crovello 1970). This leads to confusion, inconsistency, and artificial controversies.

Another advantage of inferring two or more species only when null hypotheses can be rejected is that the inference of speciation is ensured to be a data-rich one that automatically leads to much insight into likely speciation mechanisms. Thus, the investigator must have information about biogeography, reproductive communities, and/or adaptive transitions that are significantly associated with the inferred speciation events. Such information delimits or even defines the types of processes that were involved during speciation, as shown by examples given in Templeton (1994). As discussed in Templeton (1994), only rarely will an investigator have all these data available, and the dominant species criterion in practice has been and will remain morphological distinctiveness. However, the practical dominance of the morphological species concept does not obviate the need for outlining an ideal species inference procedure. Science has frequently been driven and even redirected by a few well-worked examples, so even an occasional implementation of the data-rich inference procedure outlined in this chapter could have a major impact on thinking about species. Moreover, there are cases in which the species status of a particular group has sufficient importance (either as a model system for basic research, as an endangered or threatened species group, or for practical purposes in agricultural or pharmaceutical research) to merit the type of effort required under the inference procedure given in this chapter.

Although this inference procedure places strong demands upon the investigator in terms of data gathering, it does make this task somewhat easier by identifying a broader range of candidate traits than those under the BSC or recognition concepts. Both experimental data and natural history data can be used, and the natural history data include both the genetics and the ecology of reproduction. Moreover, the inference procedure outlined here serves as an excellent springboard for additional studies on the mechanisms of speciation and gives detailed guidance to what further data need to be gathered when some of the critical null hypotheses could not be rejected. Thus, the cohesion concept provides for both breadth and specificity of data used in inferring species and speciation.

Although it should seem obvious that a scientist should base inference on data, the data-rich inference procedure outlined here and in Templeton (1994) is more the exception than the rule in much of the literature on species and speciation. For example, no data on reproductive isolation were gathered or presented on most of the plant species included in Mayr's (1992) survey of the flora of Concord. Indeed, the only use of reproductive data to reject a hypothesis was the rejection of 6.4% of the morphologically distinct populations as biological species because of data indicating that they did not reproduce sexually (Mayr 1992). For the remainder of the plant species included in Mayr's (1992) survey, the de facto null hypothesis was that a morphological distinct population of sexually reproducing organisms is a biological species. Hence, there was no need to present supporting data for the 93.6% of the morphologically distinct populations that were simply *accepted* as biological species. Indeed, even when data did exist on reproductive communities for some of these plants (e.g., knowledge of selfing or of hybridization), it was actually ignored in species inference (Whittemore 1993). This could be done because there was no formal hypothesis testing framework given by Mayr (1992), thereby allowing a subjective and implicit assessment to accept the null hypothesis of a biological species as being true regardless of data on selfing and hybridization.

However, far worse than merely the absence of data in species inference is the use of imaginary data sets. Imaginary scenarios are a particularly common type of argument when dealing with antagonistic balances between selection and gene flow with respect to speciation. For example, in his response to Ehrlich and Raven's (1969) paper, Shapiro (1970) did not attack the data that Ehrlich and Raven (1969) presented on the lack of gene flow in *Euyphyrdryas editha*, but rather simply dismissed the relevance of these observations because these data imply nothing about "the future" (p. 1636). Similarly, Futuyma (1989) dismissed the evolutionary significance of divergent adaptations in some phytophagous insects because these adaptations "would not persist if their hosts became broadly sympatric" and that "the distinctive features . . . will be lost, swamped by gene flow" (p. 563). Arguments based upon imagined future scenarios have been given a degree of credibility under the biological species concept because the BSC emphasizes the importance of *potential* to interbred. There is nothing wrong with invoking potentials in science. The real issue is in how potentials are detected and described within a framework of scientific inference. Invoking imaginary scenarios is useful and valid for suggesting testable hypotheses, but such scenarios should never be used to *reject* hypotheses. For example, one counterargument to Futuyma (1989) would be that the divergent adaptations observed in some phytophagous insects would persist if their hosts became broadly sympatric because selection would overwhelm gene flow. Obviously, both the argument and the counterargument are just statements of theoretical plausibility (or implausibility) of imagined outcomes. Neither argument constitutes a valid scientific test, although both suggest testable hypotheses. In the inference procedure outlined in this chapter, cohesion mechanisms are hypothesized to have the potential to define evolutionary lineages. This potential predicts an observable pattern of association between candidate cohesion traits with evolutionary lineages. Because the existence or more than one evolutionary lineage is the *first* hypothesis to be tested, the potential of cohesion mechanisms is evaluated only in terms of what has happened (which is scientifically observable), not what might happen (which lies outside the bounds of scientific hypothesis testing unless coupled with explicit tests of predictions).

A final advantage of inferring species by rejecting null hypotheses under the cohesion species concept is the explicit incorporation of both pattern and process into species inference. As indicated in the above paragraph, processes are often tested in science by the patterns that they generate. However, defining species only in terms of pattern leads to biological absurdities. For example, certain authors argue that a pattern of strict monophyly (i.e., one that excludes paraphyly as a special type of monophyly) is necessary for species status (Bonde 1981; Ghiselin 1984; Mishler and Donoghue 1982). To see how such a pattern rule leads to absurd conclusions from an evolutionary mechanism perspective, consider an example from the work of Hedin (1995) on spiders in the genus *Nesticus* in the Appalachian Mountains. *N. mimus* is a surface form found in North Carolina. During the Pleistocene, the ecological conditions needed by these spiders were found at lower altitudes. With post-Pleistocene warming, the *N. mimus* populations followed their habitats up the mountain slopes, leading to the current fragmented distribution of high-altitude populations separated by lower elevations with no suitable habitat. Although this fragmentation is relatively recent, modern molecular techniques allow the detection of two mountain top populations in North Carolina as different lineages. There is no evidence for morphological or ecological diversification among those lineages or changes in their mate recognition system, or evidence of reproductive isolation. Hence, *N. mimus* represents a single, subdivided species under the cohesion concept, the recognition concept, the biological species concept, or the morphological species concept. Moreover, these subdivided surface lineages

coalesce into a strictly monophyletic basal group relative to all other surface species of *Nesticus*. However, a cave exists on the lower slopes of one of the mountains inhabited by *N. mimus*, and that cave is inhabited by another *Nesticus* population. However, this population is a troglobytic form that has been greatly altered morphologically and ecologically and lives in a new, derived adaptive zone. It also defines a unique molecular lineage and therefore can be regarded as a cohesion species. This population is also a recognized taxonomic (morphological) species, *N. carolinensis*. As discussed in Hedin (1995), the troglobytic population probably evolved from a *N. mimus* population that was left stranded in the cave (the cave environment mimics many of the habitat requirements of the surface form) as the remainder of the surface population tracked the surface environment to higher elevations. Not surprisingly, the sister lineage to *N. carolinensis* is the lineage of *N. mimus* found on the same mountain. This means that *N. mimus* is now paraphyletic. In circumstances such as this, Mishler and Donoghue (1982) have argued that a paraphyletic ancestor must be avoided by elevating the lineages in the ancestral group to species status, even when they did have species status prior to the recognition of the derived species (*N. carolinensis* in this case). Note that no evolutionary changes have occurred in the *N. mimus* populations that result in this elevation to species status; rather, the pattern rule given in Mishler and Donoghue (1982) that would result in *N. mimus* being split into two or more species is due exclusively to evolutionary changes that occurred in the cave population of *N. carolinensis*. This creates a new speciation mechanism never before considered: speciation by remote control. Under the remote-control mechanism, speciation occurs between two populations (the two mountaintop populations of *N. mimus* in this case) because of evolutionary change in a third population (the cave population of *N. carolenensis* in this case). Obviously, speciation via remote control makes no sense in terms of evolution as a process, yet the biogeographic/climatic change situation described by Hedin (1995) for *Nesticus* is not an unusual one in the speciation literature. Indeed, most speciation mechanisms that have been invoked can and often will yield paraphyletic ancestors. Hence, the strict monophyletic pattern rule of Mishler and Donoghue (1982) is fundamentally incompatible with virtually all known and hypothesized speciation mechanisms. One should never forget that process generates pattern, not the other way around. Hence, when there is a perceived incompatibility between process and pattern, process should have priority. To do otherwise would be to create a species concept that is an active impediment to understanding the process of speciation. By explicitly using both pattern and process, the cohesion species concept avoids this difficulty and can serve as a powerful vehicle for studying the process of speciation in a manner that arises naturally out of the definition of species (Templeton 1994).

The cohesion species concept explicitly acknowledges that processes (cohesion mechanisms) create patterns (evolutionary lineages). By recasting the cohesion species concept as a set of testable null hypotheses, patterns can be used to make the cohesion concept testable using explicit and statistically rigorous criteria. Among the patterns that *must be tested* under this implementation of species inference are patterns hypothesized to be generated from processes related to cohesion mechanisms. As the cohesion species concept shows, both pattern and process play critical but complementary roles in our understanding of species and speciation. Moreover, the observed patterns are produced by both genetic and ecological factors acting upon reproducing communities, and hence both genetics and ecology have important roles to play in the study of species and speciation.

Acknowledgments   Guy Bush has long been an advocate of the need to integrate genetics and ecology in the study of species and speciation. Guy's influence on my thinking in this regard has been strong and deep. Over the years, Guy has frequently challenged my views on species and speciation in a constructive and rigorous—but always friendly—fashion. My ideas have certainly been improved and brought into a sharper focus by these friendly challenges. I thank Guy and Dori Bush for all the wonderful conversations that we have had over the years on matters both scientific and nonscientific.

I also thank Stewart Berlocher, Kevin de Queiroz, and Richard Harrison for their excellent suggestions for improving an earlier version of this chapter, and Stewart Berlocher and Dan Howard for the opportunity to contribute this chapter to honor an outstanding scientist and a good friend, Guy Bush. Finally, I thank my graduate students, both past and present, for their invaluable input and questioning.

References

Anderson, E. 1953. Introgressive hybridization. Biol. Reviews 28:280–307.

Baker, H. G. 1970. Differentiation of populations [Letter to the Editor]. Science 167:1637.

Baum, D. A., and Larson, A. 1991. Adaptation reviewed: a phylogenetic methodology for studying character macroevolution. Syst. Zool 40:1–18.

Boake, C. R. B., and Hoikkala, A. 1995. Courtship behavior and mating success of wild-caught *Drosophila silvestris* males. Anim. Behav. 49:1303–1313.

Bonde, N. 1981. Problems of species concepts in paleontology. In J. Martinell (ed.). International Symposium on "Concept and Method in Paleontology." Barcelona: Universitat de Barcelona, pp. 19–34.

Bush, G. L. 1969. Sympatric host race formation and speciation in frugivorous flies of the genus *Rhagoletis* (Diptera, Tephritidae). Evolution 23:237–251.

Carson, H. L., and Lande, R. 1984. Inheritance of a secondary sexual character in *Drosophila silvestris*. Proc. Natl. Acad. Sci. USA 81:6904–6907.

Carson, H. L., Val, F. C., Simon, C. M., and Archie, J. W. 1982. Morphometric evidence for incipient speciation in *Drosophila silvestris* from the island of Hawaii. Evolution 36:132–140.

Carson, H. L., Val, F. C., and Templeton, A. R. 1994. Change in male secondary sexual characters in artificial interspecific hybrid populations. Proc. Natl. Acad. Sci. USA 91:6315–6318.

Cracraft, J. 1989. Speciation and its ontology: the empirical consequences of alternative species concepts for understanding pattern and process of differentiation. In D. Otte and J.A. Endler (eds.). Speciation and Its Consequences. Sunderland, Mass.: Sinauer, pp. 28–59.

DeSalle, R. 1984. Mitochondrial DNA evolution and phylogeny in the *planitibia* subgroup of Hawaiian *Drosophila*. Ph.D. thesis, Washington University.

Dobzhansky, T. 1950. Mendelian populations and their evolution. Amer. Nat. 84:401–418.

Ehrlich, P. R., and Raven, P. H. 1969. Differentiation of populations. Science 165:1228–1232.

Futuyma, D. J. 1989. Macroevolutionary consequences of speciation: inferences from phytophagous insects. In D. Otte and J. A. Endler (eds.). Speciation and Its Consequences. Sunderland, Mass.: Sinauer, pp. 557–578.

Futuyma, D. J., and Mayer, G. C. 1980. Non-allopatric speciation in animals. Syst. Zool. 29:254–271.

Ghiselin, M. T. 1984. "Definition," "character," and other equivocal terms. Syst. Zool. 33:104–110.

Hedin, M. C. 1995. Speciation and morphological evolution in cave spiders (Araneae: Nesticidae: *Nesticus*) of the Southern Appalachians. Ph.D. thesis, Washington University.

Hoikkala, A., and Welbergen, P. 1995. Signals and responses of females and males in successful and unsuccessful courtships of three Hawaiian lek-mating *Drosophila* species. Anim. Behav. 50:177–190.

Kaneshiro, K. Y., and Kurihara, J. S. 1981. Sequential differentiation of sexual behavior among populations of *Drosophila silvestris*. Pacific Sci. 35:177–183.

Mayr, E. 1970. Populations, Species, and Evolution. Cambridge, Mass.: Belknap Press.

Mayr, E. 1992. A local flora and the biological species concept. Amer. J. Bot. 79:222–238.

Mishler, B. D., and Donoghue, M. J. 1982. Species concepts: a case for pluralism. Syst. Zool. 31:491–503.

Paterson, H. 1981. The continuing search for the unknown and the unknowable: a critique of contemporary ideas on speciation. S. Afr. J. Sci. 77:119–133.

Paterson, H. 1985. The recognition concept of species. In E. Vrba (ed.). Species and Speciation. Pretoria: Transvaal Museum, Monograph No. 4, pp. 21–29.

Raven, P. R., and Raven, T. E. 1976. The genus *Epilobium* in Australia. Christchurch, New Zealand: New Zealand Dept. Sci. Indust. Res. Bull. 216.

Shapiro, A. M. 1970. Differentiation of populations [Letter to the Editor]. Science 167:1636–1637.

Sokal, R. R., and Crovello, J. T. 1970. The biological species concept: a critical evaluation. Amer. Nat. 104:127–153.

Templeton, A.R. 1981. Mechanisms of speciation—a population genetic approach. Annu. Rev. Ecol. Syst. 12:23–48.

Templeton, A. R. 1989. The meaning of species and speciation: A genetic perspective. In D. Otte and J. A. Endler (eds.). Speciation and Its Consequences. Sunderland, Mass.: Sinauer, pp. 3–27.

Templeton, A. R. 1994. The role of molecular genetics in speciation studies. In B. Schierwater, B. Streit, G. P. Wagner, and R. DeSalle (eds.). Molecular Ecology and Evolution: Approaches and Applications. Basel: Birkhäuser-Verlag, pp. 455–477.

Templeton, A. R. 1996. Cladistic approaches to identifying determinants of variability in multifactorial phenotypes and the evolutionary significance of variation in the human genome. In G. Cardew (ed.). Variation in the Human Genome. Chichester: Wiley, Ciba Foundation Symposium 197, pp. 259–283.

Templeton, A. R., and Ahearn, J. N. 1989. Interspecific hybrids of Drosophila heteroneura and D. silvestris. I. Courtship success. Evolution 43:347–361.

Templeton, A. R., and Georgiadis, N. J. 1996. A landscape approach to conservation genetics: conserving evolutionary processes in the African Bovidae. In J. C. Avise, and J. L. Hamrick (eds.). Conservation Genetics: Case Histories From Nature. New York: Chapman and Hall, pp. 398–430.

Templeton, A. R., and Sing, C. F. 1993. A cladistic analysis of phenotypic associations with haplotypes inferred from restriction endonuclease mapping. IV. Nested analyses with cladogram uncertainty and recombination. Genetics 134:659–669.

Templeton, A. R., Boerwinkle, E., and Sing, C. F. 1987. A cladistic analysis of phenotypic associations with haplotypes inferred from restriction endonuclease mapping. I. Basic theory and an analysis of alcohol dehydrogenase activity in Drosophila. Genetics 117:343–351.

Templeton, A. R., Crandall, K. A., and Sing, C. F. 1992. A cladistic analysis of phenotypic associations with haplotypes inferred from restriction endonuclease mapping and DNA sequence data. III. Cladogram estimation. Genetics 132:619–633.

Templeton, A. R., Routman, E., and Phillips, C. 1995. Separating population structure from population history: a cladistic analysis of the geographical distribution of mitochondrial DNA haplotypes in the tiger salamander, Ambystoma tigrinum. Genetics 140:767–782.

Vrba, E. S. 1985. Introductory comments on species and speciation. In E. Vrba (ed.). Species and Speciation. Pretoria, South Africa: Transvaal Museum, Monograph No. 4, pp. ix–xviii.

Whittemore, A. T. 1993. Species concepts: a reply to Ernst Mayr. Taxon 42:573–583.

Whittemore, A. T., and Schaal, B. A. 1991. Interspecific gene flow in sympatric oaks. Proc. Natl. Acad. Sci. USA 88:2540–2544.

Wiley, E. O. 1981. Phylogenetics: The Theory and Practice of Phylogenetic Systematics. New York: Wiley.

4

Species and the Diversity of Natural Groups

Kerry L. Shaw

The robustness of Darwin's (1859) theory of evolution provides sound justification that species, whatever their nature, are products of evolution. Nonetheless, a nebulous understanding of the nature of species remains, despite wide acceptance of Darwin's theory. The successful study of the origins of species depends critically on a firm concept of the nature of species. Furthermore, speciation studies rely on many areas of active research which depend heavily on some concept of species. In this chapter, I argue that genealogical groups (Baum and Shaw 1995) that arise directly from basic processes of heredity and genetic drift provide the most powerful species concept for the study of early lineage divergence and speciation.

One consequence of taking a genealogical view of species, and hence, species formation, is an expanded consideration of the relevant evolutionary forces that cause "speciation." In order to appreciate how this might come about, I first extend a discussion by Mayr (1942) of the two major difficulties that hinder the development of an acceptable definition of species. I begin by discussing the multifarious demands on the species concept and how this has slowed progress in the understanding of the nature of species. Specifically, I argue that several distinct natural groups have been subsumed under the single term "species." While this position has been appreciated for some time (e.g., Mayr 1942; Cain 1954), it is critical because it highlights the plurality of the evolutionary process and its outcomes. To illustrate this plurality, I consider possible concepts that refer to "natural" groups, or "individuals" (Holsinger 1984; see below), because such groups correspond to species entities with objective existence caused by an organizing force (i.e., evolution). A major point, however, is that the existence of individuals above the level of the organism (i.e., a system of organisms) is not sufficient to designate the species category because multiple natural groups exist at more or less the same level of biological hierarchy. Thus, species may be natural groups, but not all natural groups are species. I

conclude that, of the various natural entities that exist, the view of species as genealogical groups provides the most cohesive structure for theories of speciation in evolutionary biology.

The Role of Species Concepts

Different constraints on species concepts, professed or assumed by various subdisciplines in evolutionary biology, present obstacles to the study of speciation. The disparate views on species inhibit attempts to integrate advances from each of these different subdisciplines into a consistent theory on the origins of species. One of the goals of systematics, for example, is to produce a systematic reference system, wherein the species taxon is named as the least inclusive group in the biological hierarchy (Mayr 1942; Cain 1954; Simpson 1961; Hennig 1966). Such an endeavor requires (1) the grouping of organisms by some criterion, (2) applying an appropriate categorical rank (or somehow formally recognizing a level of inclusiveness), and (3) the naming of such recognized categories (i.e., species). The resulting reference system is of central consequence to all of the biological sciences, and it formally facilitates the study of speciation by recognizing the species entity. Despite extensive disagreement among systematists as to what concept the species category should represent (e.g., de Queiroz and Donoghue 1988; Baum and Donoghue 1995; vs. Nixon and Wheeler 1990; Davis 1995, 1996), most agree that it should refer to a real entity in nature (Mayr 1942; Cain 1954; Stebbins 1950; Simpson 1961; Grant 1971; Hennig 1966; Cracraft 1987; Kluge 1990; but see Raven 1986; Veron 1995). Furthermore, a current goal of at least some systematists (Mishler and Donoghue 1982; Cracraft 1983, 1987; Baum and Shaw 1995) is that species be defined and delimited by the same principles as higher taxa.

A precisely defined and consistently applied definition of species is vital for ecological studies of diversity and phylogenetic investigations of diversification, which offer valuable insights for the study of speciation. Species concepts dictate the criteria used to delimit individual species taxa in any given case. Comparative biologists frequently estimate species numbers in biodiversity assays in order to test hypotheses about rates of diversification under different historical or environmental conditions (e.g., Farrell et al. 1991; Nee et al. 1992; Slowinski and Guyer 1993; Sanderson and Donoghue 1994; Mitra et al. 1996; McCune and Lovejoy, this volume) and thereby assume comparable units of measure in the species category. If species within a comparative study reflect different kinds of groups, delimitations of these species will be made by different criteria and potentially bias conclusions regarding rates of speciation (see McCune and Lovejoy, this volume). A valid claim of cause (conditions that propel diversification) and effect (e.g., an accelerated rate of speciation) requires equivalent units of comparison in the species category.

Population biologists also place demands on the species concept, but in a different way from systematists. For those interested in evolutionary processes of diversification, the species is often conceived to be a *causal* unit, with a permanent integrity relative to closely related species (White 1978; Paterson 1985; Lambert et al. 1987; Mayr 1987; Coyne et al. 1988; Howard 1988; Templeton 1989; Andersson 1990; Harrison 1993; Levin 1993). Species' boundaries and organization contribute to the course of evolution. In particular, the extent and limits of gene flow, the selection intensities on ecologically or reproductively functional phenotypes across the species' range, and the genetic architecture and coadapted-adapted coordination of such phenotypes are indispensable pieces of information for predicting the course of early lineage divergence and the origins of new species.

Demanding that the species category refer to a natural group that plays a causal evolutionary role may be the most challenging and difficult concept to explicate. Yet, it may be of primary interest to those who actively study the process of speciation (e.g., see Otte and Endler 1989 and references therein; Howard 1993; Mallet 1993; Endler and Houde 1995). An appropriate definition of a species concept that meets the requirements of populations biologists requires grouping criteria consistent with the evolutionary role the entity is perceived to play; such criteria must simultaneously offer an explanation for the existence of the evolutionary unit. According to Andersson (1990) for example, the species concept should provide a *universal* causal explanation for the widespread observation of phenotypic compartmentalization in "ecospace." Thus, Andersson (1990), in support of Van Valen (1976), argues that distinct ecomorphs are coextensive with causal evolutionary units. In contrast, others argue that distinct mating systems explain (and are therefore coextensive with) "species" as causal evolutionary units (e.g., Paterson 1985; Lambert et al. 1987). Clearly, there are diverse opinions on what species con-

cepts are required to explain, but there is general consensus among population biologists that a causal evolutionary role is inherent in a statement of species status.

Thus, the species taxon must simultaneously hold a place in the systematic reference system, comprise a group of organisms of comparable nature in different lineages, and constitute a causal unit in the evolutionary process. A vast literature on the species "problem" (e.g., see Ereshefsky 1992; Davis 1995) testifies that reconciliation of these varied demands is not trivial. The study of speciation stands to gain substantially from integrating systematics, comparative approaches, and population biology, and thus the lack of agreement on an appropriate definition of species prevents the development of a unified effort in speciation research. Before attempting to resolve this conflict, let us consider the second difficulty facing our understanding of the nature of species and the study of speciation.

Natural Groups and Species

Several authors in the recent literature (Holsinger 1984; Kitcher 1984; Mishler and Brandon 1987; de Queiroz and Donoghue 1988, 1990; Ereshefsky 1992; Baum and Shaw 1995; see also Cain 1954) have reinforced the pluralistic position of Mayr (1942): "A second difficulty which confronts us in our attempt at a species definition is that there is, in nature, a great diversity of different kinds of species" (p. 114). An implication of this view is that species, whatever their nature, are products of evolution. As a "natural" group of organisms, each kind of species exists by the action of one of several evolutionary processes (Holsinger 1984). Moreover, because evolution can be caused by several different processes, several different kinds of natural groups exist, which may or may not be composed of coextensive groups of organisms. We therefore might choose any one of them to elevate to the rank of "species." I will deal with these two implications in turn, limiting my focus to three prominent outcomes of evolution in order to demonstrate that at least one species concept has been developed with each of these outcomes as its central focus.

First, what is meant by a "natural" group of organisms? It is the endogenous *interactions* (within generations) or *connections* (between generations) among organisms that philosophers of biology emphasize when stressing that natural groups are individuals (e.g., Ghiselin 1974; Hull 1976; Holsinger 1984; Mayr 1987; de Queiroz and Donoghue 1988; Kluge 1990). As Holsinger (1984) points out, it is important to realize that natural groups (i.e., individuals) are "not defined in terms of properties they have, but by the interactions that its members have with one another and by their cohesiveness in certain processes" (p. 297). Group membership based on the possession of attributes by each member of a group is the hallmark of a class (as opposed to an individual; e.g., see

Ghiselin 1974). For example, interactions among organisms may be reproductive in nature, or ecological (usually concerned with survivorship). Alternatively, connections among organisms may be genealogical in nature, with organisms united by exclusive common ancestry. Thus, organisms may be parts of a whole (the individual) through contemporary interactions or historical connections.

The decay of these interactions or connections among organisms of more broadly inclusive groups gives rise to the group's spatiotemporal boundaries. Therefore, the same process that gives internal structure to natural groups (e.g., interbreeding relations) also leads to spatial or temporal boundaries of such groups through diminished interaction at the margins between groups. In studies of species origins, we may hypothesize that the limits of natural groups are directly related to, indeed are caused by, evolutionary processes such as (1) differential mating success (sexual selection, in the sense of Arnold 1994), (2) differential survivorship (natural selection), or (3) stochastic gene-lineage extinction (genetic drift) resulting in gaps between natural groups. How and why these interactions or connections diminish (and natural groups divide), and how interactions of one natural group affect the evolution of other natural groups are of direct concern to the study of speciation.

Consider, for example, an interbreeding community, one kind of natural group (Mayr 1942; Ghiselin 1974; Paterson 1985). The defining feature of an interbreeding group is the presence of reproductive interactions between mates and the resulting fertilization. Thus, members of an interbreeding group are more likely to mate with each other than with any organisms outside the group. Likewise, it is the decayed probability of reproductive interaction and shared descendants (whatever the cause) in more broadly inclusive groups of organisms that give rise to the boundaries of an interbreeding group. We may hypothesize that strong sexual selection corresponds to the boundaries between groups, but other forces or circumstances may result in decayed probabilities of reproductive interactions at the group boundaries as well (e.g., geographical separation). Furthermore, we may focus on the possession of courtship behaviors and morphological structures to generate predictions about what sexual events will occur over time, but it is the reproductive interaction, itself a property of the group, that delimits interbreeding entities. Sexually reproducing organisms and their attributes constitute the units, but interactions comprise the system; in this way, organisms are parts of a whole: an individual. It should be obvious that while many lineages have the conditions necessary for interbreeding groups to form, such groups do not exist in lineages composed of strictly asexual organisms.

The interbreeding group is represented by the most widely accepted and well-known species definition, the Biological Species Concept (BSC; e.g., Mayr 1940, 1942, 1963, 1982). At least part of the original formulation of the BSC was based on the single criterion of reproductive cohesion: "Species are groups of actually or potentially interbreeding natural populations, which are reproductively isolated from other such groups" (Mayr 1942, p. 120). Clearly, this concept refers to a natural group that arises due to reproductive interactions. Paterson (1985) made an important contribution to our understanding of interbreeding groups by pointing out that such groups do not depend on the possession of "isolating mechanisms." Paterson's (1985) definition of the recognition concept of species, "that most inclusive population of individual bi-parental organisms which share a common fertilization system" (p. 120), shifted emphasis from interactions between interbreeding groups to interactions within the fertilization system of an interbreeding group. In doing so, Paterson (1985) exposed the dependency of the BSC on isolating mechanisms (the clause in Mayr's definition). According to Mayr (1963, p. 20; 1996) isolating mechanisms prevent interbreeding between highly coadapted gene pools. Thus, isolating mechanisms draw attention to interactions between *adaptive* groups (the nature of which I do not attempt to discuss here.)

In some lineages, ecological interactions give rise to a second kind of natural group. For example, if the interactions among organisms enhance viability or survivorship (e.g., through heightened predator avoidance or foraging effort), then such organisms also function as parts of a whole, but with a fundamentally different organizing principle than in the interbreeding group. Aposematic coloration provides an illustration of how such natural groups might be structured. The organisms of some groups exhibiting warning coloration (such as the monarch butterfly *Danaus plexippus* [Brower 1988] and the brightly banded coral snakes of the genus *Micrurus* [Brodie 1993]) are thought to interact through a concerted contribution to the learned or innate aversion responses of potential predators (Brower 1988; Brodie 1993; Guilford 1988; Pough 1988; Mallet et al., this volume). Thus, the organisms of an aposematic group benefit interactively (and thereby function as an individual) by causing avoidance responses in potential predators. When the coordination among organisms in more broadly inclusive groups (corresponding to gaps between groups) breaks down, the boundaries of such ecological individuals arise. Organisms of such groups are therefore more likely to form ecological aggregations (e.g., for survivorship) with each other than with any organisms outside the group. The function of the interaction in such cases is survivorship, as opposed to sexual reproduction considered above. Analogous examples where density-dependent benefits to survival are accrued through aggregation include fish schools, bird flocks, mammalian herds (for defense), and schooling pods (for foraging). As was the case for interbreeding individuals, many organisms do not participate in natural groups (individuals) that are ecological in nature. In a later section, I elaborate on the important observation that organismal participation in ecological individuals (such as mimicry rings or multispecies flocks) can result in boundaries that show a departure from the traditional limits of species (often based on reproductive criteria).

The Ecological Species Concept (ESC; Van Valen 1976; endorsed by Andersson 1990) focuses on shared ecology among organisms of a group. Below I examine how the ESC, with an emphasis on "adaptive zone" (in the sense of Van Valen 1971), might serve to illustrate a single-criterion species concept that could represent the ecological individual. An *ecological species* is a "lineage (or closely related set of lineages) which occupies an adaptive zone minimally different from that of any other lineage in its range and which evolves separately from all lineages outside its range" (Van Valen 1976, p. 233), where "an adaptive zone is some part of the resource space together with whatever predation and parasitism occurs on the group considered" (p. 234). Thus, the ESC clearly emphasizes that "species" are primarily ecologically differentiated groups of organisms. As in the case of interbreeding groups, we may hypothesize that strong selection (in this case, survivorship selection) corresponds to the boundaries between ecological groups; other forces or circumstances may result in decayed probabilities of ecological interactions at the group boundaries as well.

Is it necessarily the case, however, that all ecologically differentiated organisms participate in ecological groups? For ecological species to be individuals it is necessary for the organisms of such a group to form a whole, cohesive via ecological interaction. Therefore, the ecological properties emphasized must be characteristics attributable to group individuation, not organismal individuation. If such attributes are organismal features, then a collection of such organisms would form an ecological class, not an individual. (The features in common may, however, indicate common descent and thus participation in an historical individual; see below.) Indeed, Van Valen (1976) effectively proposes that species are classes, whose member organisms each possess an ecological space in nature. Moreover, Van Valen (1976) argues that "adaptive zones" may come to be occupied again by a different group. While such a view gives us insight into organismal attributes and organism's relations to its environment, we gain little understanding of the nature of the species entity as a natural group. Surely if two "species" can come to occupy the same adaptive zone (at different times), there must be something other than the adaptive zone that distinguishes them. If species are individuals, the only sustainable view of ecological species are examples like those discussed above, where ecological interactions among organisms constitute the organization of a natural group. This view of an ecological individual has little overlap with the ecological species definition of Van Valen (1976), except that it suggests the same causal process, survival selection, as its most prominent mechanistic theme.

A third kind of natural group is one united on the basis of common history (see discussions by Hennig 1966; Cracraft 1983, 1987; Mishler and Donoghue 1982; Donoghue 1985; de Queiroz and Donoghue 1988; de Queiroz 1992, 1994; Kluge 1990; Baum and Donoghue 1995; Baum and Shaw 1995). At any particular point in time, the organisms of an historical group comprise a whole if all members of the group share exclusive common ancestry. In a genealogical group, for example, the parts (organisms) are connected through common historical relationship, such that members of the group are more closely related to one another than to any organisms outside the group (Baum and Shaw 1995). Genealogical relationship in this context refers to how far back in time members of a group of organisms must trace their paths of ancestry before they are connected by progenitors in common. In strictly asexual systems, the relationship of organisms relative to one another is simple and hierarchic (figure 4.1), reflecting monophyletic ancestry (Hennig 1966). In addition, all parts of an asexual genome have ancestries that match the organismal pedigree. Boundaries between genealogical groups can be accentuated as a simple consequence of finite intergenerational sample error (i.e., the differential perpetuation of asexual lineages due to stochastic forces); they may also be enhanced by other causal processes such as directional selection from one generation to the next. Emphasis must be placed on the connections comprising the individual (i.e., exclusive genealogical relationship) and not the putative causes that produced such relationship.

Hereditary patterns in bisexual organisms are more complex. However, one form of historical relationship shows similar structure to the asexual system discussed above. A genealogical group (Baum and Shaw 1995) in a bisexual lineage is also a group of organisms whose members are all more closely related to one another than to any organisms outside the group. Likewise, genealogical relationship in this context is a measure of how far back in time members of the group must trace their paths of ancestry before they are connected by ancestors in common. However, in the bisexual case, the connections and boundaries of such genealogical groups are only manifest at the level of gene lineages because each organism derives a mixture of genes from two parents. Thus, while the pedigrees of bisexual organisms expand as organismal ancestry is traced back in time (Hudson 1990), a sample of homologous, nonrecombining, genic lineages shows a coalescent pattern of ancestry within the organisms of such a pedigree (Avise and Ball 1990; Hudson 1990; Baum and Shaw 1995). As in the analogous case of asexual pedigrees, at the level of genes, relationships are simple and hierarchic (figure 4.1). Likewise, if all genes (at a given locus) share an exclusive common ancestor, the pattern of gene-genealogy mirrors the concept of monophyly (as defined by de Queiroz and Donoghue 1988; Baum and Shaw 1995).

Figure 4.1. The genealogies of two divergent groups, A and B, that each have exclusive, treelike histories and hierarchic structure. A and B may be considered groups of asexual organisms, or genes sampled from organisms of reticulate, bisexual groups.

The action of genetic drift alone will cause boundaries to arise around gene lineages, producing exclusive relations within groups of descendent genes (within organisms), as a consequence of stochastic lineage extinction (due to finite sampling of gene lineages between generations; Avise and Ball 1990; Hudson 1990; Slatkin 1991). Exclusive relationship has been observed in many groups to date (e.g., Gill et al. 1993; Avise et al. 1994; Brown et al. 1994; Jacobs et al. 1995; Moore 1995; Brower 1996) where gene phylogenies have been estimated to infer taxic relationships.

Patterns of exclusive genealogy will arise at other loci as well, if a finite, sampling of genes occurs between generations (either by random or some deterministically governed processes). However, due to recombination and the probabilistic contribution of genes from parents to offspring, the historical paths to common ancestry at other loci will be different from the first. Furthermore, the time to a common ancestor for a sample of genes may vary from locus to locus due simply to the stochastic nature of gene-lineage extinction (Avise and Ball 1990; Wu 1991; Hudson 1992; Baum and Shaw 1995; Maddison 1995). Consequently, the patterns of relationship among sexual organisms within a genealogical group, as measured by genomic relationship, will be reticulate (as was recognized by Hennig 1966). Nonetheless, such reticulation gives way to visible hierarchic relations at the level of the nonrecombining unit (operationally, the gene). This, in turn, can serve as a basis for estimating relationship, and the existence of genealogical groups. The boundaries of genealogical groups occur where reticulate relationships among organisms of the group give way to

divergent relationships between groups (figure 4.2). Such groups of organisms will be bounded in space and time as a simple consequence of gene-lineage attrition (Avise and Ball 1990; Baum and Shaw 1995).

The Genealogical Species Concept (GSC; Baum and Shaw 1995) emphasizes that organismal relationship within a reticulating group should be thought of at the level of the recombining genome. A *genealogical species* is a basal, exclusive group of organisms, whose members are all more closely related to each other than they are to any organisms outside the group, and that contains no exclusive groups within it. Thus, the structure of exclusive relationship can be seen only through an account of recency of *genomic* coalescence. The genealogical species is a third kind of single-criterion natural group, based solely on genealogical relationships within a lineage. The boundaries of genealogical species lie at the transition between reticulate and divergent genealogy, where divergent genealogy might be envisioned as the early splitting of lineages. Just as the boundaries of interbreeding groups (and ecological groups) are manifest as reduced probabilities of interaction between groups, the boundaries of each genealogical group are manifest as a reduced degree of relationship between groups. An explicit mathematical treatment of the boundaries of genealogical groups (indeed, of *any* natural group) is badly needed. Such an endeavor is nontrivial. For a possible statistical treatment, see Templeton (1994).

If we accept that species, whatever their ontological status, are individuals resulting from an evolutionary process, these three examples illustrate that evolution has the potential to produce several kinds of "species." We

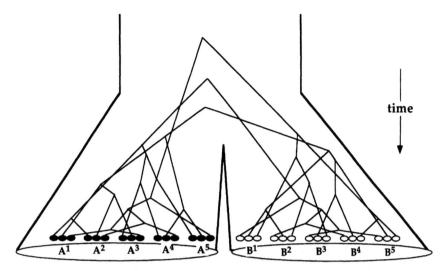

time

Figure 4.2. The genealogies of two bisexual groups, A and B, that each have exclusive genealogi-cal relationship. Superscript numbers refer to single bisexual organisms within A or within B. The genealogies from three genes (each from different loci) sampled from the same organisms (organisms 1–5 from both A and B) are represented. Exclusive genomic relationship can be generalized across the organisms sampled (within either A or B), despite the reticulate relationships evident within A and within B.

have, in the context of evolutionary theory, forces that effect the evolution of interbreeding groups, ecological groups, and historical groups. It could be argued that these groups will be coextensive in nature and are therefore the same. To counter this, I show in the next section that multiple evolutionary forces can act to produce a number of kinds of individuals with conflicting boundaries that are particularly apparent when (1) the same organisms are parts of multiple individuals (such as in gender-specific mimicry), and (2) different evolutionary forces affect daughter lineages to a greater or lesser degree or at different rates during cladogenesis, leading to variation in the presence or frequency of different kinds of natural groups among lineages.

Conflicting Boundaries of Natural Groups

Gender-limited mimicry is an example where multiple entities arise from a pluralistic evolutionary process. In *Papilio glaucus*, the tiger swallowtail, the yellow and black color patterns of male butterflies are thought to resemble the ancestral condition, possibly maintained by stabilizing sexual selection (Turner 1978; but see Scriber et al. 1996). The color patterns of female *P. glaucus* are dimorphic, with one lightly colored morph resembling the male, and a second darkly colored morph mimicking the color pattern of an unpalatable relative, *Battus philenor*. Burns (1966) proposed that the color dimorphism in females is maintained by a counterbalance between sexual selection and natural selection. Dark females of

P. glaucus are favored by natural selection through protective mimicry of *B. philenor*, whereas light morphs of *P. glaucus* experience greater mating success. Thus, the dark mimetic females participate in two natural groups: an ecological mimicry complex with *B. philenor*, and an interbreeding group with other members of *P. glaucus*.

Dual participation in natural groups is switched from females to males in *Chrysobothris humilis*, a beetle of the family Buprestidae, whose males mimic models of the beetle subfamily Clytrinae (Hespenheide 1975). While both males (black with red patches on the elytra) and females (metallic green) of *C. humilis* participate in the same interbreeding community, only males participate in the mimicry complex. This gender-specific dimorphism is thought to coincide with differences in spatial use of the habitat, with males more active in conspicuous locations, thereby experiencing selective pressures that females avoid.

These two systems are exemplars of the intriguing polymorphisms that exist in mimicry systems, indicating the presence of several evolutionary forces acting within a lineage (natural and sexual selection). Mimicry continues to challenge taxonomists. Most (if not all) evolutionary biologists and experts dealing with mimicry rings view species from the point of view of interbreeding individuals. From this perspective, "misidentification" has on occasion resulted in mimics being identified as part of the local species. Recently, a new species, *Heliconius tristero* (Brower 1996), was described, despite over 100 years of intensive study on the taxonomy, evolution, and behavior of this butterfly mimicry complex. Brower

(1996) attributes this oversight to the exceptional resemblance of *H. tristero* to its comimics, *H. melpomene* and *H. cydno*.

The complexities of mimicry rings illustrate that single organisms can participate in more than one natural group (individual). Each natural group results from some aspect of the evolutionary process, leading to the conclusion that multiple, partial, or nonoverlapping natural groups can exist at more or less the same level of biological organization. I do not mean to suggest that, for example, mimetic *P. glaucus* females ought to be considered part of the species *B. philenor*, but rather to point out that it is not enough to say that species are individuals (see de Queiroz and Donoghue [1988] for a different argument with a similar conclusion). The ecological individual illustrated by the monarch butterfly, *Danaus plexippus*, does not present a problem to our traditional notion of species boundaries because the grouping based on ecological interaction appears to be coextensive with interbreeding group boundaries. However, when confronted with gender-specific ecological aggregations and mimicry rings, extended ecological interactions give rise to boundaries beyond traditional species boundaries, creating conflict among different natural groups.

The Trinidad guppy, *Poecilia reticulata*, illustrates a situation where multiple evolutionary forces act to different degrees in daughter populations during cladogenesis. Intensive research on a variety of populations of *P. reticulata* (Breden and Stoner 1987; Houde and Endler 1990; Endler and Houde 1995; Magurran et al. 1996; for a recent review, see Endler 1995) has produced evidence for the existence of interbreeding, ecological, and genealogical groups, dependent on a variety of environmental conditions. In *P. reticulata*, populations have diversified in parallel (Fajen and Breden 1992; Carvalho et al. 1991) due to ecological, reproductive and geographical circumstances.

Trinidad *P. reticulata* occur in a series of isolated drainages, within which populations vary from high to low exposure to specialized predators (such as predacious cichlids). Guppies in high-predation localities have accentuated schooling tendencies (Seghers 1974; Breden et al. 1987; Magurran and Seghers 1994) and associated antipredator characteristics such as smaller, more fusiform bodies, cryptic coloration, and higher degrees of predator vigilance (Endler 1995). In these populations natural selection for predator avoidance appears to be strong enough to generate diversity in ecological traits and aggregation behaviors that are characteristic of ecological individuals. In low predation-risk localities, natural selection for predator avoidance plays a subordinate role in population differentiation. In such cases, sexual selection acts as the predominant force affecting phenotypic divergence, influencing both male courtship traits and female preferences for those traits (Breden and Stoner 1987; Breden et al. 1987; Houde and Endler 1990; Endler and Houde 1995). Thus, both ecological and interbreed-

ing natural groups arise among different populations within the same streams.

Historical boundaries have also formed within the taxonomic species of *P. reticulata*. In some cases, extreme genetic differentiation among populations accompanies the ecological and reproductive diversity found at various sampling scales. For example, large differences were found among populations between east (Oropuche) and west (Caroni) drainages with allozyme studies (see Carvalho et al. 1991). Monophyletic relationships inferred from mitochondrial DNA (mtDNA) variation among samples from these drainages corroborate the allozyme findings (Fajen and Breden 1992). However, a low level of differentiation was found between rivers of the Oropuche drainage. These conditions suggest that different locations within drainages and different populations between drainages represent increasing scales of lineage divergence (Carvalho et al. 1991; Fajen and Breden 1992) and genealogical group formation.

At the most phylogenetically divergent extremes (to the extent known with the available allozyme and sequence data, keeping in mind that mtDNA is but a fraction of the genome), genealogical groups are coextensive with interbreeding groups. Houde and Endler (1995) found that females from different populations show a net preference for males from their own populations. In addition, geographic isolation contributes to reproductive group boundaries. However, at the very lowest level of divergence discernible with the available data, interbreeding and ecological groups have formed despite a currently reticulate genealogy in the Oropuche drainage (see Fajen and Breden 1992). Below, I elaborate on this latter finding, because it illustrates what we expect for lineages that are in the early stages of splitting. (In this argument, I refer to the actual patterns of interbreeding, as opposed to the potential patterns of interbreeding. Magurran et al. [1996] found that, in the laboratory environment, reproductive compatibility exists across the most extreme levels of genetic differentiation. However, as these different populations may not have the opportunity to interbreed, and show some degree of nonrandom mating tendencies in the laboratory, I refer to these populations as different reproductive groups.)

As one particular focus, let us consider the existence of two interbreeding groups that have recently split from a single interbreeding group. For comparison, consider the most clear-cut case, when all migration between two newly separated interbreeding groups has ceased; hence, there is no ambiguity about the fact that two interbreeding groups exist, whatever the nature of the barrier. Emphasizing the outcome of the reproductive process, each interbreeding group comprises organisms that are all more likely to share common descendants with each other than with any other organism outside the group (O'Hara 1993). Contrasting this prospective view of the interbreeding group with a retrospective view of genealogical groups provides another (probably common) example of conflicting boundaries among individuals produced by the evolutionary process. Initially, the genes within a locus within the two interbreeding groups will share many ancestors in common. Over time, due to severed gene exchange and genetic drift, the genes within each interbreeding group will evolve exclusive relations, from a state of polyphyly to reciprocal monophyly (figure 4.3; Neigel and Avise 1986; Avise and Ball 1990; Baum and Shaw 1995). Other hereditary units in the genome will also evolve through a polyphyly–monophyly transition. However, the rate of this transition will vary for neutral genes by the stochastic nature of the evolutionary sampling process and as a function of the inbreeding effective size of the locus considered (Hudson 1990; Moore 1995; see Harrison, this volume, for expected probabilities of monophyly after a variable number of generations). When lineages are actively splitting, an organism may simultaneously be part of one or more natural groups; in particular, an organism may be a member of one of two newly formed interbreeding groups, but also a more inclusive genealogical group that includes the organisms from both interbreeding groups. Thus, we expect that a retrospective (genealogical) and a prospective (interbreeding) view will at some point circumscribe different groups of organisms. Both kinds of groups are real, the first (two) existing by virtue of reproductive relations, the second existing by connections of genealogy. Between northern and western drainages, enough lineage attrition has occurred among *P. reticulata* populations that reciprocal monophyly of mitochondrial DNA has evolved (Fajen and Breden 1992). The situation appears to be different in the eastern drainages: although several interbreeding and ecological groups exist, reticulate relationships have

not yet given way to exclusive genealogical groups between descendent populations.

As speciation biologists, we should pay special attention to discrepancies between prospective and retrospective natural groups. Such cases may offer opportunities to observe and measure evolutionary forces that cause lineage splitting (e.g., see discussion by Endler and Houde 1995). It seems inevitable that the formation of different natural groups will occur at different rates (see also Harrison, this volume). How is the species problem to be addressed in such circumstances? My point in describing the excellent work done with *P. reticulata* is not to argue whether the differentiated populations should be considered one species or many, but to show that such a decision will depend upon which natural group we choose to represent the species category. Whatever is decided will not invalidate the existence of the remaining natural groups.

Because species concepts in the literature place different degrees of emphasis on different natural groups, the importance of a diversity of observations on the nature of *the* species has been debated, resulting in confusion about definitions. In addition, important forces that may affect group formation (such as predation regimes) are underappreciated in speciation studies because they are not directly relevant to reproductive isolation, the most prevalent focus of speciation research (Coyne 1992). Certainly the forces that cause these groups to form have the potential to simultaneously affect the perpetuation of a lineage. However, study systems that allow us to see clear distinctions among multiple evolutionary forces, such as *P. glaucus* or *P. reticulata*, provide the best opportunity to study the causal forces involved in early lineage divergence and speciation.

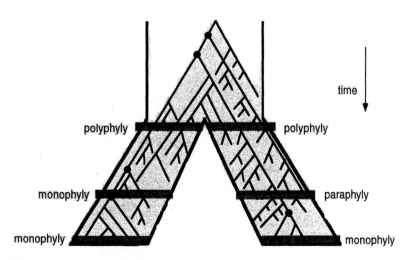

Figure 4.3. An illustration of the transition from polyphyly to monophyly (Avise and Ball 1990) expected to occur among genes sampled at a locus, when a lineage of organisms is split into two. Extinctions of genes at any particular point in time will occur due to stochastic sampling error.

Monistic Species Concepts

Despite the potential for confusion, most published species concepts implicitly refer to several interactions simultaneously, leading to species entities that are characterized by a multitude of biological properties resulting from those interactions. The implication of this monistic view is that there is a single kind of species entity with many biological properties stemming from a multitude of interactions. In contrast, a pluralistic view holds that there may be several kinds of species entities, each owing its existence to a singular kind of interaction. The views of Simpson (1961), Wiley (1978), and Templeton (1989) serve to illustrate a monistic position. Appreciating that ecology and reproduction can be important evolutionary targets, and that common history or future fate will result from ecological and reproductive processes, the species concepts of these authors incorporate all these facets into one conceptual definition. Consequently, any outcomes that are predicted by such composite theory are taken as symptomatic of "species-ness" (see also de Queiroz, this volume).

Indeed, later versions of the BSC, at odds with Mayr's earlier pluralistic position, incorporate ecological considerations into the definition: "A species is a reproductive community of populations (reproductively isolated from others) that occupies a specific niche in nature" (Mayr 1982, p. 273). Likewise, Paterson (1985, pp. 25–26; see also Paterson 1981) argues that adaptive shifts to new habitats will likely cause the evolution of new fertilization systems. In arguing thus, Paterson equates the evolution of new fertilization systems, and hence the very nature of unique species, with new ecologies. To maintain consistency with Paterson's view, one is forced to conclude that if ecological differentiation is absent between two reproductively distinct groups, only one species exists, by definition. Despite a purported focus on an interbreeding group by the recognition concept, reproduction is subordinate to ecological differentiation. In addition, while the ESC of Van Valen (1976) highlights ecological differentiation as the primary feature of distinct species, ancestor–descendent relationships and the potential for gene exchange among populations comprising the lineage are also formulated into the concept.

Finally, most phylogenetic approaches to the species problem also draw on multiple criteria to recognize species taxa. They thereby support a monistic position by subsuming multiple natural groups into a single species definition. Most definitions and discussions of "phylogenetic species" (e.g., Cracraft 1983, 1987; de Queiroz and Donoghue 1988; Nixon and Wheeler 1990; Davis and Manos 1991; Davis and Nixon 1992; Davis 1996) rely on some sort of reproductive cohesion criterion for delimiting a group of organisms prior to the investigation of character variation indicative of "a parental pattern of ancestry and descent" (in the sense of Cracraft 1983). This additional criterion of reproductive cohesion is necessary because many levels of biological inclusiveness can be organized by patterns of ancestry and descent (Baum and Shaw 1995). The result is the added assumption of coextension between two individuals: a reproductively cohesive group and an historical group.

For the study of speciation, we need a clear understanding of what species are. Monistic approaches to the species problem fail to explicate different natural groups, which leads to a weak conceptual framework for asking what causes species to form. If it becomes evident that there are multiple mechanisms incorporated into one species definition, while each mechanism alone can cause boundaries to form around natural groups, then the definition is bound to result in confusion and controversy for three reasons. First, it is likely that different evolutionary processes leading to group formation will operate at different rates, leading to conflicting natural groupings based on different criteria (like interbreeding cohesion and genealogical history). Second, while we often may find that multiple causes act simultaneously to produce what appear to be unitary natural groupings, what can we claim about the nature of species so identified? We end up with multiple explanations for the existence of a single group of organisms! Third, there will undoubtedly be, in the diversity of life, cases where all criteria cannot be meaningfully applied within some lineages. Such cases are well known in asexual organisms, where interbreeding communities don't exist. What do we conclude about the "species" entities in such cases?

Such considerations have undoubtedly influenced many to propose a conservative approach to the species problem. For example, strictly phenetic definitions (e.g., "genotypic clusters" of Mallet [1995a] and the phenetic definition of Sokal and Crovello [1970]) delimit species on the basis of covarying character distributions. Thus, this approach focuses entirely on boundaries to spatial distributions at a particular time, the object of focus being nothing more than the congruence of characters. Spatial boundaries alone are sought for "species" delimitation, and are not considered to be an indication of any real sort of entity or taxon (Mallet 1995b). Therefore, such a view considers the existence of natural groups to be irrelevant in the formulation of a species concept.

The "ease" of applying these approaches is heralded as a solution to the species problem. However, as one finds more difficult patterns of group formation and a lack of coextension among groups (such as the case discussed above for interbreeding groups and genealogical groups), such allegedly "process-free" ideas will become more difficult to apply, not less. We need, at some level, organizing principles able to cope with the complex and sometimes conflicting patterns of variation, especially for the study of speciation (Shaw 1996).

Species Concepts and the Study of Speciation

What is it about the origin of species that compels us to argue for decades over the problem's important foci? The answer to this question should dictate the best frame of reference to mark the species category, because it is this category that formalizes the study of speciation.

Perhaps the two most remarkable foci were pointed out by Darwin: descent and modification. An increase in the number of natural groups that result from the process of descent is remarkable in itself, and an observation we feel compelled to explain. The modification and elaboration that accompanies such an increase in the number of entities is all the more inspiring, for we look to the repeated patterns of such modification for clues about the mechanistic origins of that diversity. The species category has traditionally defined the unit for investigating this phenomenon. Viewing species as individuals is critical to studies of species origins because it focuses our attention on interactions (or connections), what drives those interactions, and what causes those interactions to decay. These interactions are what make the species entity a unique level of biological organization. Without this focus, we do little more than study the origins of novel organisms.

In this chapter, I have argued that species concepts are, first and foremost, concepts about natural groups of organisms, by discussing three prominent alternatives: the interbreeding group, the ecological group, and the genealogical group. These different natural groups can be noncoextensive in some cases, and in other cases may arise at different rates within and across lineages. A multifaceted process of evolution inevitably leads to many different kinds of natural groups.

The clearest way to explicate the process of speciation is to have the species category represent a single-criterion natural group. To do so while remaining consistent with modern evolutionary theory requires that we maintain a pluralistic position: that there is more than one kind of natural group, whose ontological status may be referred to by a single criterion. There are two good reasons for the species category to be represented by a single-criterion natural group. First, single-criterion entities are succinctly definable, and therefore can be explained by recourse to the action of one evolutionary process. As a result, the species definition continues to function as a model of a real entity in nature (Shaw 1996). Such a model will lend itself far more easily to implementation in the real world. Second, ambiguities about the nature of the species taxon are minimized because of this distinct definition. As a result, comparable entities across the lineages of life may be identified so as to conduct meaningful studies in a comparative framework.

Given that we choose among the single-criterion alternatives discussed above, it is clear that the traditions of the last five decades have favored the interbreeding group as the most appropriate natural group to represent the species category. This has been justified by the argument that species are individuals (Ghiselin 1974; Mayr 1987; see also discussion by Buss 1987). However, that species are individuals is not a sufficient condition for a choice of a species definition, because evolution can produce several kinds of individuals (natural groups) above

the organismal level (also see discussion in de Queiroz and Donoghue 1988, 1990).

The fact that several kinds of natural groups exist raises the question of whether one group or another is applicable to a broader diversity of life. One reason for supporting a history-based, genealogical approach is that it has a broad applicability in the biological world. Although the prevalence of ecological individuals appropriate for the species category remains unknown, several biologists have pointed out that many lineages do not produce interbreeding groups (Ehrlich 1961; Mishler and Donoghue 1982; Donoghue 1985; Cracraft 1987). It is known, however, that all of life has genealogical history and that boundaries form around entities of close relationship (corresponding to gaps).

The genealogical species concept proposes a criterion (i.e., exclusive relationship) suitable for the definition of higher taxa. Hence, the demand that species taxa be similar kinds of natural group as higher taxa is satisfied. Such an achievement would result in a clear genealogical relationship among increasingly more inclusive taxa (as has been argued by de Queiroz and Donoghue 1990). Taxa delimited by the same criterion (in this case, exclusive genealogical relationship) can be considered equivalent kinds.

A further demand on the species concept discussed above was the insistence that the species entity be a causal player in evolution. I propose that we turn the argument around and, rather than demand a causal role for the species unit, find out what causes affect the evolution of the unit. Such a shift will not belittle the importance of any process. However, it would relieve the species concept from the duty of explaining all forms of "compartmentalization" (Andersson 1990) that present themselves in the world.

Focusing on the genealogical group will integrate our understanding of historical population genetics directly into the understanding of how species form. So, although genealogical species are an outcome of evolution and say little about future permanence of the entity, they can be directly understood by the same processes that speciation biologists regard as important. A focus on genealogical groups puts us in the position to pose questions such as, how does sexual selection of male secondary sexual characteristics, or natural selection on ecological novelty leading to niche shifts, accelerate the evolution of exclusive genealogical relationship? Furthermore, demographic and population genetic structures become directly relevant to speciation questions. For example, do highly subdivided lineages with small population sizes generate species (as genealogical groups) at a faster rate than large panmictic lineages? Such questions have direct and clear relevance to the study of speciation as the origin of genealogical species.

Summary

Species are natural groups of organisms. Acknowledgment of a pluralistic evolutionary process leads to the realization that there are several different kinds of natural groups (i.e., individuals) that reside at similar, but not identical, levels of biological organization. Examples discussed include biological groups, ecological groups, and genealogical groups. Boundaries that demarcate the natural groups in one natural system may conflict with the boundaries arising in other natural systems. The species debate provides evidence of the struggle among biologists to deal with the pluralistic outcomes of the evolutionary process and the struggle for the primacy of one group over others in the definition of species. Not all natural groups are suitably equipped for the role of species, for the diverse aims in evolutionary biology, and in particular for the study of the origins of "species." The genealogical group, composed of organisms connected by exclusive historical relationship, is (1) united by a grouping principle that is also suitable for higher level systematics, (2) an equivalent group in a broad array of taxa, and (3) a direct outcome of microevolutionary processes. Thus, the Genealogical Species Concept provides a well-suited reference system for investigations of the roles of different processes in the diversification of life.

Acknowledgments   For useful discussions and/or comments on species and earlier drafts of this chapter, I thank Damhnait McHugh, Michael Donoghue, David Baum, Stewart Berlocher, Alan Templeton, and Kevin de Queiroz. I am grateful to Dan Howard and Stewart Berlocher for their efforts in arranging this symposium and this publication and the opportunity to share my views.

References

Andersson, L. 1990. The driving force: species concepts and Ecology. Taxon 39:375–382.

Arnold, S. J. 1994. Is there a unifying concept of sexual selection that applies to both plants and animals? Am. Nat. 144:S1–S12.

Avise, J. C., and Ball, R. M. 1990. Principles of genealogical concordance in species concepts and biological taxonomy. Oxford Surv. Evol. Biol. 7:45–67.

Avise, J. C., Nelson, W. S., and Sugita, H. 1994. A speciational history of "living fossils": molecular evolutionary patterns in horseshoe crabs. Evolution 48:1986–2001.

Baum, D. A., and Donoghue, M. J. 1995. Choosing among alternative "phylogenetic" species concepts. Syst. Bot. 20:560–573.

Baum, D. A., and Shaw, K. L. 1995. Genealogical perspectives on the species problem. In P. C. Hoch and A. G. Stevenson, eds. Experimental and Molecular Approaches to Plant Biosystematics. St. Louis, Mo.: Missouri Botanical Garden, pp. 289–303.

Breden, F., and Stoner, G. 1987. Male predation risk determines female preference in the Trinidad guppy. Nature 329:831–835.

Breden, F., Scott, M., and Michel, E. 1987. Genetic differentiation for anti-predator behaviour in the Trinidad guppy, *Poecilia reticulata*. Anim. Behav. 35:618–620.

Brodie, E. D. 1993. Differential avoidance of coral snake banded patterns by free-ranging avian predators in Costa Rica. Evolution 47:227–235.

Brower, A. V. Z. 1996. A new mimetic species of *Heliconius* (Lepidoptera: Nymphalidae), from southeastern Colombia, revealed by cladistic analysis of mitochondrial DNA sequences. Zool. J. Linn. Soc. 116:317–332.

Brower, L. P. 1988. Avian predation on the monarch butterfly and its implications for mimicry theory. Am. Nat. 131: S4–S6.

Brown, J. M., Pellmyr, O., Thompson, J. N., and Harrison, R. G. 1994. Phylogeny of *Greya* (Lepidoptera: Prodoxidae), based on nucleotide sequence variation in mitochondrial cytochrome oxidase I and II: congruence with morphological data. Mol. Biol. Evol. 11:128–141.

Burns, J. M. 1966. Preferential mating versus mimicry: disruptive selection and sex-limited dimorphism in *Papilio glaucus*. Science 153:551–553.

Buss, L. W. 1987. The Evolution of Individuality. Princeton, N.J.: Princeton University Press.

Cain, A. J. 1954. Animal Species and Their Evolution. Princeton, N.J.: Princeton University Press. (Reprinted in 1996)

Carvalho, G. R., Shaw, P. W., Magurran, A. E., and Seghers, B. H. 1991. Marked genetic divergence revealed by allozymes among populations of the guppy, *Poecilia reticulata* (Poeciliidae) in Trinidad. Biol. J. Linn. Soc. 42:389–405.

Coyne, J. A. 1992. Much ado about species. Nature 357:289–290.

Coyne, J. A., Orr, H. A., and Futuyma, D. J. 1988. Do we need a new species concept? Syst. Zool. 37:190–200.

Cracraft, J. 1983. Species concepts and speciation analysis. Curr. Ornithol. 1:159–187.

Cracraft, J. 1987. Species concepts and the ontology of evolution. Biol. Philos. 2:329–346.

Darwin, C. 1859. On the Origin of Species (1st ed.). London: John Murray. (Reprinted 1964, Cambridge, Mass.: Harvard University Press)

Davis, J. I. 1995. Species concepts and phylogenetic analysis–introduction. Syst. Bot. 204:555–559.

Davis, J. I. 1996. Phylogenetics, molecular variation, and species concepts. BioScience 46:502–511.

Davis, J. I., and Manos, P. S. 1991. Isozyme variation and species delimitation in the *Puccinellia nuttalliana* complex Poaceae: an application of the phylogenetic species concept. Syst. Bot. 16:431–445.

Davis, J. I., and Nixon, K. C. 1992. Populations, genetic variation, and the delimitation of phylogenetic species. Syst. Bot. 41:421–435.

de Queiroz, K. 1992. Phylogenetic definitions and taxonomic philosophy. Biol. Philos. 7:295–313.

de Queiroz, K. 1994. Replacement of an essentialistic perspective on taxonomic definitions as exemplified by the definition of "Mammalia." Syst. Biol. 43:497–510.

de Queiroz, K., and Donoghue, M. J. 1988. Phylogenetic systematics and the species problem. Cladistics 4:317–338.

de Queiroz, K., and Donoghue, M. J. 1990. Phylogenetic systematics and species revisited. Cladistics 6:83–90.

Donoghue, M. J. 1985. A critique of the biological species concept and recommendations for a phylogenetic alternative. Bryologist 88:172–181.

Ehrlich, P. R. 1961. Has the biological species concept outlived its usefulness? Syst. Zool. 10:167–176.

Endler, J. A. 1995. Multiple-trait coevolution and environmental gradients in guppies. Trends Ecol. Evol. 10:22–29.

Endler, J. A., and Houde, A. E. 1995. Geographic variation in female preferences for male traits in *Poecilia reticulata*. Evolution 49:456–468.

Ereshefsky, M. 1992. Species, higher taxa, and the units of evolution. In M. Ereshefsky, ed. The Units of Evolution: Essays on the Nature of Species. Cambridge, Mass.: MIT Press, pp. 381–398.

Fajen, A., and Breden, F. 1992. Mitochondrial DNA sequence variation among natural populations of the Trinidad guppy, *Poecilia reticulata*. Evolution 46:1457–1465.

Farrell, B. D., Dussourd, D. E., and Mitter, C. 1991. Escalation of plant defense: do latex and resin canals spur plant diversification? Am. Nat. 138:881–900.

Ghiselin, M. T. 1974. A radical solution to the species problem. Syst. Zool. 23:536–544.

Gill, F. B., Mostrom, A. M., and Mack, A. L. 1993. Speciation in North American chickadees: I. Patterns of mtDNA genetic divergence. Evolution 47:195–212.

Grant, V. 1971. Plant Speciation. New York: Columbia University Press.

Guilford, T. 1988. The evolution of conspicuous coloration. Am. Nat. 131:S7–S21.

Harrison, R. G., ed. 1993. Hybrid Zones and the Evolutionary Process. New York: Oxford University Press.

Hennig, W. 1966. Phylogenetic Systematics. Chicago: University of Illinois Press.

Hespenheide, H. A. 1975. Reversed sex-limited mimicry in a beetle. Evolution 29:780–783.

Holsinger, K. E. 1984. The nature of biological species. Philos. Sci. 51:293–307.

Houde, A. E., and Endler, J. A. 1990. Correlated evolution of female mating preferences and male color patterns in the guppy, *Poecilia reticulata*. Science 248:1405–1408.

Howard, D. J. 1988. The species problem. Evolution 42:1111–1112.

Howard, D. J. 1993. Reinforcement: origin, dynamics, and fate of an evolutionary hypothesis. In R. G. Harrison, ed. Hybrid Zones and the Evolutionary Process. New York: Oxford University Press, pp. 46–69.

Hudson, R. R. 1990. Gene genealogies and the coalescent process. Oxford Surv. Evol. Biol. 7:1–44.

Hudson, R. R. 1992. Gene trees, species trees and the segregation of ancestral alleles. Genetics 131:509–512.

Hull, D. L. 1976. Are species really individuals? Syst. Zool. 25:174–191.

Jacobs, S. C., Larson, A., and Cheverud, J. M. 1995. Phylogenetic relationships and orthogenetic evolution of coat color among tamarins (genus *Saguinus*). Syst. Biol. 44:515–532.

Kitcher, P. 1984. Species. Philos. Sci. 51:308–333.

Kluge, A. G. 1990. Species as historical individuals. Biol. Philos. 5:417–431.

Lambert, D. M., Michaux, B., and White, C. S. 1987. Are species self-defining? Syst. Zool. 36:196–205.

Levin, D. A. 1993. Local speciation in plants: the rule not the exception. Syst. Bot. 18:197–208.

Maddison, W. 1995. Phylogenetic histories within and among species. In P. C. Hoch and A. G. Stevenson, eds. Experimental and Molecular Approaches to Plant Biosystematics. St. Louis, Mo.: Missouri Botanical Garden, pp. 273–289.

Magurran, A. E., and Seghers, B. H. 1994. Predator inspection behaviour covaries with schooling tendency amongst wild guppy, *Poecilia reticulata*, populations in Trinidad. Behaviour 128:121–134.

Magurran, A. E., Paxton, C. G. M., Seghers, B. H., Shaw, P. W., and Carvalho, G. R. 1996. Genetic divergence, female choice and male mating success in Trinidadian guppies. Behaviour 133:503–517.

Mallet, J. 1993. Speciation, raciation, and color pattern evolution in *Heliconius* butterflies: evidence from hybrid zones. In R. G. Harrison, ed. Hybrid Zones and the Evolutionary Process. New York: Oxford University Press, pp. 226–260.

Mallet, J. 1995a. A species definition for the modern synthesis. Trends Ecol. Evol. 10:294–299.

Mallet, J. 1995b. Reply from Mallet. Trends Ecol. Evol. 10:490–491.

Mayr, E. 1940. Speciation phenomena in birds. Am. Nat. 74:249–278.

Mayr, E. 1942. Systematics and the Origin of Species. New York: Columbia University Press.

Mayr, E. 1963. Animal Species and Evolution. Cambridge, Mass.: Harvard University Press.

Mayr, E. 1982. The Growth of Biological Thought. Cambridge, Mass.: Belknap Press.

Mayr, E. 1987. The ontological status of species: scientific progress and philosophical terminology. Biol. Philos. 2:145–166.

Mayr, E. 1996. What is a species, and what is not? Philos. Sci. 63:261–276.

Mishler, B. D., and Brandon, R. N. 1987. Individuality, pluralism, and the phylogenetic species concept. Biol. Philos. 2:397–414.

Mishler, B. D., and Donoghue, M. J. 1982. Species concepts: a case for pluralism. Syst. Zool. 31:491–503.

Mitra, S., Landel, H., and Pruett-Jones, S. 1996. Species richness covaries with mating system in birds. Auk 113:544–551.

Moore, W. S. 1995. Inferring phylogenies from mtDNA variation: mitochondrial-gene trees versus nuclear-gene trees. Evolution 49:718–726.

Nee, S., Mooers, A. O., and Harvey, P. H. 1992. Tempo and mode of evolution revealed from molecular phylogenies. Proc. Natl. Acad. Sci. USA 89:8322–8326.

Neigel, J. E., and Avise, J. C. 1986. Phylogenetic relationships of mitochondrial DNA under various demographic models of speciation. In S. Karlin and E. Nevo, eds. Evolutionary Processes and Theory, New York: Academic Press, pp. 515–534.

Nixon, K. C., and Wheeler, Q. D. 1990. An amplification of the phylogenetic species concept. Cladistics 6:211–223.

O'Hara, R. J. 1993. Systematic generalization, historical fate and the species problem. Syst. Biol. 42:231–246.

Otte, D., and Endler, J. A., eds. 1989. Speciation and Its Consequences. Sunderland, Mass.: Sinauer.

Paterson, H. E. H. 1981. The continuing search for the unknown and unknowable: a critique of contemporary ideas on speciation. S. Afr. J. Sci. 77:113–119.

Paterson, H. E. H. 1985. The recognition concept of species. In E. S. Vrba, ed. Species and Speciation. Pretoria: Transvaal Museum, pp. 21–29.

Pough, F. H. 1988. Mimicry of vertebrates: are the rules different? Am. Nat. 131:S67–S102.

Raven, P. H. 1986. Modern aspects of biological species in plants. In K. Iwatsuki, P. H. Raven, and W. J. Bock, eds. Modern Aspects of Species. Tokyo: University of Tokyo Press, pp. 11–29.

Sanderson, M. J., and Donoghue, M. J. 1994. Shifts in diversification rate with the origin of angiosperms. Science 264: 1590–1593.

Scriber, J. M., Hagen, R. H., and Lederhouse, R. C. 1996. Genetics of mimicry in the tiger swallowtail butterflies, *Papilio glaucus* and *P. canadensis* (Lepidoptera: Papilionidae). Evolution 50:222–236.

Seghers, B. H. 1974. Schooling behaviour in the guppy *Poecilia reticulata*: an evolutionary response to predation. Evolution 28:486–489.

Shaw, K. L. 1996. What are good species? Reply to Mallet. Trends Ecol. Evol. 11:174.

Simpson, G. G. 1961. Principles of Animal Taxonomy. New York: Columbia University Press.

Slatkin, M. 1991. Inbreeding coefficients and coalescence times. Genet. Res. Camb. 58:167–175.

Slowinski, J. B., and Guyer, C. 1993. Testing whether certain traits have caused amplified diversification: an improved method based on a model of random speciation and extinction. Am. Nat. 142:1019–1024.

Sokal, R. R., and Crovello, T. J. 1970. The biological species concept: a critical evaluation. Am. Nat. 104:127–153.

Stebbins, G. L. 1950. Variation and Evolution in Plants. New York: Columbia University Press.

Templeton, A. R. 1989. The meaning of species and speciation: a genetic perspective. In D. Otte and J. A. Endler, eds. Speciation and Its Consequences. Sunderland, Mass.: Sinauer, pp. 3–27.

Templeton, A. R. 1994. The role of molecular genetics in speciation studies. In B. Schierwater, B. Streit, G. P. Wagner, and R. DeSalle, eds. Molecular Ecology and Evolution: Approaches and Applications. Basel, Switzerland: Birkhauser-Verlag, pp. 3–27.

Turner, J. R. G. 1978. Why male butterflies are non-mimetic: natural selection, sexual selection, group selection, modification and sieving. Biol. J. Linn. Soc. 10:385–432.

Van Valen, L. 1971. Adaptive zones and the orders of mammals. Evolution 25:420–428.

Van Valen, L. 1976. Ecological species, multispecies, and oaks. Taxon 25:233–239.

Veron, J. E. N. 1995. Corals in Space and Time: The Biogeography and Evolution of the Scleractinia. Sydney: University of New South Wales Press.

White, M. J. D. 1978. Modes of Speciation. San Francisco: Freeman

Wiley, E. O. 1978. The evolutionary species concept reconsidered. Syst. Zool. 27:17–26.

Wu, C. I. 1991. Inferences of species phylogeny in relation to segregation of ancient polymorphism. Genetics 127:429–435.

5

The General Lineage Concept of Species, Species Criteria, and the Process of Speciation

A Conceptual Unification and Terminological Recommendations

Kevin de Queiroz

Speciation, the process through which new species come into being, is one of the central topics of evolutionary biology. It links the great fields of micro- and macroevolutionary biology and intersects a wide variety of related biological disciplines, including behavioral biology, ecology, genetics, morphology, paleontology, physiology, reproductive biology, and systematics. For this reason, a persistent controversy regarding the definition of the term *species* may seem disconcerting. The continual proposal of new species definitions—commonly characterized as alternative species concepts—seems to suggest that there is no general agreement about what species are, and if this is the case, then the possibility of understanding how species come into being also seems unlikely. At the very least, there seems to be considerable potential for misinterpretation and confusion about what different biologists mean when they talk about species and speciation.

But the situation is not really as troublesome as it may appear. Although real differences underlie alternative species definitions, there is really less disagreement about species concepts than the existence of so many alternative definitions seems to suggest. Each species definition has a different emphasis, but the various phenomena that they emphasize are all aspects or properties of a single kind of entity. In other words, almost all modern biologists have the same general concept of species. Differences among the many versions of this general concept are at least partly attributable to the complex and temporally extended nature of species and the process or processes through which they come into existence. In many respects, considering speciation as a temporally extended process is the key to understanding the diversity of species definitions.

In this chapter, I provide a general theoretical context that accounts for both the unity and the diversity of ideas represented by alternative species definitions. First, I review the major categories of species definitions adopted

by contemporary biologists. Next, I present evidence that all modern species definitions describe variants of a single general concept of species. I then discuss how the time-extended nature of species and the diversity of events that occur during the process of speciation provide the basis for the diversity of alternative species definitions. Based on a distinction between species concepts and species criteria, I propose a revised and conceptually unified terminology for the ideas described by contemporary species definitions, and I discuss the significance and limitations of different classes of species definitions. Finally, I examine an interpretation of species criteria that places alternative criteria in direct conflict and thus contains the key to resolving the species problem.

Alternative Species Definitions

Over the last half century, biologists have established a minor industry devoted to the production of new definitions for the term *species*. In this section I present a summary of those definitions, using a terminology proposed by the authors of those definitions and others commenting on their work (see also Haffer, 1986; Häuser, 1987; Panchen, 1992; King, 1993; Ridley, 1993; Smith, 1994; Vrba, 1995; Hull, 1996; Shaw, this volume; Harrison, this volume). My purpose is not to catalog modern species definitions exhaustively but rather to represent their diversity, and my use of an existing terminology is not intended to endorse that terminology–indeed, I propose what I believe to be a more useful one later in this chapter—but rather to reflect current views on the historical and conceptual relationships among alternative species definitions. References in the headings are for the terms themselves (as opposed to their definitions); emphasis has been removed from quoted passages.

Biological Species Concept (e.g., Mayr, 1942, 1963). This term has been applied to definitions that emphasize interbreeding, specifically, the idea that species are populations of interbreeding organisms. Although such definitions have ancient roots (e.g., the writings of Buffon, discussed by Mayr, 1982) and were more or less clearly articulated by authors at the beginning of the twentieth century (e.g., Poulton, 1903; Jordan, 1905; see Mayr, 1955), they are most commonly associated with the New Systematics (e.g., Huxley, 1940) of the Evolutionary Synthesis (reviewed by Mayr and Provine, 1980). Some examples are as follows: "groups within which all subdivisions interbreed sufficiently freely to form intergrading populations wherever they come in contact, but between which there is so little interbreeding that such populations are not found" (Wright, 1940:162); "the largest and most inclusive Mendelian population . . . [a] Mendelian population is a reproductive community of sexual and cross-fertilizing individuals which share in a common gene pool" (Dobzhansky, 1950:405); "groups of actually or potentially interbreeding natural populations, which are reproductively isolated from other such groups" (Mayr, 1942:120; see also Mayr, 1963, 1970). Other names given to this class of definitions are "interbreeding-population concept" (Mayr, 1942), "genetical concept of species" (Simpson, 1951), "reproductive species concept" (Van Valen, 1976), and "species concepts based on interbreeding" (de Queiroz and Donoghue, 1988). Some authors (e.g., Paterson, 1981, 1985, 1986; Lambert and Paterson, 1982; Masters et al., 1987) see an important distinction between two subcategories of "biological" or "genetical" species concepts (see below); others (e.g., Mayr, 1988) do not.

Isolation [Species] Concept (e.g., Paterson, 1985). This term has been applied, by advocates of the alternative "recognition concept" (see below), to the views on species articulated by Dobzhansky and Mayr, which emphasize reproductive isolation between organisms of different species. Mayr's (1942, 1963, 1970) definition, quoted above, falls into this category, as does the following definition of Dobzhansky: "Species are . . . systems of populations; the gene exchange between these systems is limited or prevented in nature by a reproductive isolating mechanism or perhaps by a combination of several such mechanisms" (Dobzhansky, 1970:357).

Recognition [Species] Concept (e.g., Paterson, 1985). This term has been applied to the views on species articulated by Paterson (e.g., 1980, 1981, 1985, 1986, 1993a; see also Lambert and Paterson, 1982, 1984; Masters et al., 1987; Lambert and Spencer, 1995) as an alternative to the "isolation concept." Species definitions associated with the "recognition concept" emphasize the unification of species rather than their separation from one another. More specifically, they emphasize the common fertilization and specific mate recognition systems

shared by conspecific organisms, rather than the reproductive isolation between heterospecific organisms. For example, "members of a species share a common specific mate recognition system" (Paterson 1978:369); "a species [is] that most inclusive population of individual biparental organisms which share a common fertilization system" (Paterson 1985:25).

Evolutionary Species Concept (e.g., Wiley, 1978, 1981). This term has been used for definitions emphasizing the extension of species through time and attempting to accommodate both the observation that some populations appear to maintain their distinctness despite interbreeding with other populations and the idea that asexual organisms form species. "An evolutionary species is a lineage (an ancestral–descendant sequence of populations) evolving separately from others and with its own unitary evolutionary role and tendencies" (Simpson, 1961:153; see also Simpson, 1951). "A species is a single lineage of ancestral descendant populations of organisms which maintains its identity from other such lineages and which has its own evolutionary tendencies and historical fate" (Wiley, 1978:18; see also Wiley, 1981).

Ecological Species Concept (Van Valen, 1976). This term was proposed by Van Valen (1976) for his modification of Simpson's (1961) definition; it emphasizes the importance of ecologically based natural selection in maintaining species: "A species is a lineage (or a closely related set of lineages) which occupies an adaptive zone minimally different from that of any other lineage in its range and which evolves separately from all lineages outside its range" (Van Valen, 1976:233; see also Andersson, 1990).

Cohesion Species Concept (Templeton, 1989). This term was proposed by Templeton (1989) for his own definition, which synthesizes components of the evolutionary, ecological, isolation, and recognition definitions. It emphasizes the mechanisms that maintain evolutionary lineages by promoting genetic relatedness and determining the boundaries of populations with respect to micorevolutionary processes such as gene flow, genetic drift, and natural selection. "The cohesion concept [of] species is the most inclusive population of individuals having the potential for phenotypic cohesion through intrinsic cohesion mechanisms" (Templeton, 1989:12). See Templeton (1989, table 2) for a summary of proposed cohesion mechanisms.

Phylogenetic Species Concept (e.g., Cracraft, 1983). This term has been used for at least three distinct classes of species definitions associated with the taxonomic ideology known as Phylogenetic Systematics or Cladistics (e.g., Hennig, 1966; Eldredge and Cracraft, 1980; Nelson and Platnick, 1981).

Phylogenetic species concept I (Panchen, 1992). Also referred to as the "cladistic species concept" (Ridley, 1989) and the "Hennigian species concept" (Nixon and Wheeler, 1990), definitions in the first group stem from Hennig's (1966) discussion of species considered in the time dimension, which emphasizes cladogenesis (the splitting of lineage) and its implications concerning the limits of species. "The limits of [a] species in a longitudinal section through time [are] determined by two processes of speciation: the one through which it arose as an independent reproductive community, and the other through which the descendants of this initial population ceased to exist as a homogeneous reproductive community" (Hennig, 1966:58). "A species is . . . that set of organisms between two speciation events, or between one speciation event and one extinction event, or that are descended from a speciation event" (Ridley, 1989:3).

Phylogenetic species concept II (Donoghue, 1985; Mishler, 1985). Also referred to as "species concepts based on monophyly" (de Queiroz and Donoghue, 1988), the "autapomorphic species concept" (Nixon and Wheeler, 1990), and the "monophyletic species concept" (Smith, 1994), definitions in the second group derive from Hennig's (e.g., 1966) distinction between monophyly and paraphyly and its application to species. The designation "(aut)apomorphic" describes the evidence commonly used to infer monophyly—derived or apomorphic characters. Some examples are as follows: "a population or group of populations defined by one or more apomorphous features" (Rosen, 1979:277); "monophyletic groups of organisms, recognized as lineages on the . . . basis of . . . shared, derived characters and ranked as species because of causal factors . . . that maintain the lineages as the smallest important monophyletic group recognized in a formal classification" (Mishler, 1985:213).

Phylogenetic species concept III (Cracraft, 1983). Also referred to as the "diagnostic approach" (Baum and Donoghue, 1995), definitions in the third group emphasize diagnosability, regardless of whether the diagnostic characters are apomorphic: "A species is the smallest diagnosable cluster of individual organisms within which there is a parental pattern of ancestry and descent" (Cracraft, 1983:170); "the smallest aggregation of populations (sexual) or lineages (asexual) diagnosable by a unique combination of character states in comparable individuals (semaphoronts)" (Nixon and Wheeler, 1990: 218). Some advocates of definitions in this group do not consider the concepts of monophyly and paraphyly applicable to species (e.g., Nixon and Wheeler, 1990); others apply the concepts to (diagnosable) species and conclude that species are not necessarily monophyletic (e.g., Eldredge and Cracraft, 1980).

Genealogical Species Concept (Baum and Shaw, 1995). This term was proposed by Baum and Shaw (1995) for a their own definition, which draws from the perspectives of systematics and population biology: "species [are] basal, exclusive groups of organisms" where "[a] group of organisms is exclusive if their genes coalesce [unite in a common ancestral gene] more recently within the group than between any member of the group and any organisms outside the group" (Baum and Shaw, 1995:291, 296). Baum and Shaw (1995; see also Baum and Donoghue, 1995) classify their definition as a "phylogenetic species concept," and Luckow (1995) classifies it as a "monophyletic species concept," which is in keeping with the interpretation of monophyly as exclusivity of common ancestry relationships (see de Queiroz and Donoghue, 1990; Baum and Shaw, 1995; Baum and Donoghue, 1995).

Phenetic Species Concept (Sokal and Crovello, 1970). This term has been applied to species definitions that emphasize the evidence and operations used to recognize species in taxonomic practice, particularly those formulated within the context of the taxonomic ideology known as Phenetics or Numerical Taxonomy (e.g., Sokal and Sneath, 1963; Sneath and Sokal, 1973). "A species is a group of organisms not itself divisible by phenetic gaps resulting from concordant differences in character states (except for morphs such as those resulting from sex, caste, or age differences), but separated by such phenetic gaps from other such groups" (Michener, 1970:28). "We may regard as a species (a) the smallest (most homogeneous) cluster that can be recognized upon some given criterion as being distinct from other clusters, or (b) a phenetic group of a given diversity somewhat below the subgenus category" (Sneath and Sokal, 1973:365).

Genotypic Cluster [Species] Definition (Mallet, 1995). This term was proposed by Mallet (1995) for his own definition, which is intended to be independent of theories concerning the origin and maintenance of species. Like the phenetic definitions, it emphasizes the evidence used to recognize species, but it places greater emphasis on genetics: "species . . . are . . . identifiable genotypic clusters . . . recognized by a deficit of intermediates, both at single loci (heterozygote deficits) and at multiple loci (strong correlations or disequilibria between loci that are divergent between clusters)" (Mallet, 1995:296).

The General Lineage Concept of Species

Despite the diversity of perspectives represented by the definitions quoted above and their designation as species concepts, the differences among those definitions do not reflect fundamental differences with regard to the general concept of species. I do not mean to say that there are no conceptual differences among the diverse contemporary species definitions but rather that the differences in question do not reflect differences in the general concept of what kind of entity is designated by the term *species*. All modern species definitions either explicitly or implicitly equate species with segments of population level evolutionary lineages. I will hereafter refer to this widely accepted view as the *general lineage concept of species*. Before substantiating the claim that all modern species definitions are special cases of the general lineage concept, it is first useful to clarify some things about lineages and the segments of them that we call species.

Lineages and Species

I use the term *lineage* for a single line of direct ancestry and descent (see also Simpson, 1961; Hull, 1980). Biological entities at several different levels of organization form lineages; for example, genes, cells, and organisms all replicate or reproduce to form lineages. Lineages at one level of organization often make up, or are contained within, lineages at higher levels of organization; for example, numerous cell lineages often make up an organism lineage. Definitions that equate species with segments of lineages refer to lineages at a still higher level of organization—to groups of organism lineages that are united to form lineages at what is commonly known as the population level. At this level of organization, a lineage is a population extended through time, and conversely, a population is a short segment, a more or less instantaneous cross section, of a lineage (see Simpson, 1951, 1961; Meglitsch, 1954; George, 1956; Newell, 1956; Rhodes, 1956; Westoll, 1956). The population level is really a continuum of levels, from the deme to the species. Lineages at the lower levels in this continuum often separate and reunite over relatively brief time intervals and generally are not considered species. Species are more inclusive population level lineages, though the exact level of inclusiveness differs among authors.

Although the term *lineage* is often used interchangeably with both *clade* and *clone*, it is used here for a distinct concept (figure 5.1). A clade is a unit consisting of an ancestral species and its descendants; a clone is its (asexual) organism level counterpart. Either can be represented on a phylogenetic tree as a set of branches composed of any given branch and all of the branches distal to it (figure 5.1a). In contrast, a lineage—at the level of both species and organisms—is a single line of descent. It can be represented on a phylogenetic tree as a set of branches that forms a pathway from the root of the tree (or some other internal point) to a terminal tip (figure 5.1b). Thus, both clades and clones can be branched, but lineages, though they pass through branch points, are unbranched. And though clades and clones originate from lineages, they are themselves composed of multiple lineages. Furthermore, clades and clones are (by definition) monophyletic in terms of their component species and organisms, respectively, but lineages can be paraphyletic or even polyphyletic in terms of their lower level components. For example, the organisms making up the later

part of a population lineage may share a more recent common ancestor with organisms in a recently diverged but now separate lineage than with the earlier organisms of their own lineage. The fact that lineages can be paraphyletic or polyphyletic does not mean that the same is true for the segments of lineages called species; some species definitions permit paraphyly and polyphyly while others require monophyly.

Species do not correspond with entire population level lineages. If they did, species would be partially overlapping and *Homo sapiens* would be part of the same species as the common ancestor of all living things. Just as an organism lineage is composed of a series of ancestral and descendant organisms, a species lineage is composed of a series of ancestral and descendant species. Therefore, a species is not strictly equivalent to a lineage but rather to a lineage segment. Consider the three general models of speciation (figure 5.2) described by Foote (1996), some of which are implied by different species definitions. The models differ with respect to whether lineage splitting is equated with speciation and whether species are considered to persist through lineage splitting events, but all three equate species with lineage segments.

The formation of species level lineages (figure 5.3) is easiest to visualize in the case of sexual organisms. Here organism lineages continually anastomose as a result of sexual reproduction to create a higher level lineage whose component organism lineages are unified by that very process (figure 5.3a). If sexual reproduction is the only process that unifies collections of organism lineages to form higher level lineages, then lineages of asexual organisms do not form species (e.g., Dobzhansky, 1937, Hull, 1980). But perhaps there are other processes that unite collections of asexual organism lineages to form higher level lineages (figure 5.3b) that are comparable to those formed by sexual organism lineages in certain evolutionarily significant respects (e.g., Meglitsch, 1954; Templeton, 1989). Whether asexual organisms do in fact form such higher level lineages is not important to my analysis; what is important is that species definitions that are intended to be applicable to asexual organisms assume that they do. Thus, in describing species as segments of population-level evolutionary lineages, I use *population* in the general sense of an organizational level above that of the organism rather than the specific sense of a reproductive community of sexual organisms.

Alternative Species Definitions as Variants of the General Lineage Concept

With these clarifications in mind, let me reiterate that all contemporary species definitions describe variations of the general concept of species as evolutionary lineages. This concept was adopted by Darwin (1859) in the passage where he most explicitly described the origin of species (pp. 116–125), and it underlies virtually every species definition published during or after the period

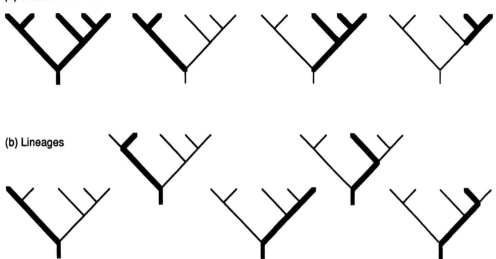

Figure 5.1. Clades and clones versus lineages. All nine branching diagrams represent the same (species or asexual organism) phylogeny, with the clades or clones highlighted in (a) and the lineages highlighted in (b). Additional lineages can be counted for the pathways from various internal nodes to the terminal tips.

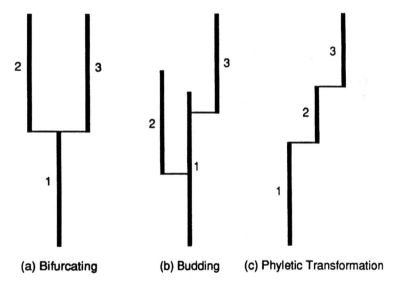

Figure 5.2. Species as lineage segments. In these diagrams illustrating three general models of speciation described by Foote (1996), species are represented as vertical lines (numbered) and speciation "events" as horizontal ones. (a) In the bifurcating model, species correspond precisely with the segments of lineages between speciation events. (b) In the budding model, species extend beyond speciation events and thus do not correspond with the segments of lineages between those events, though they still correspond with lineage segments. (c) In the phyletic transformation model, species once again correspond precisely with the segments of lineages between speciation events.

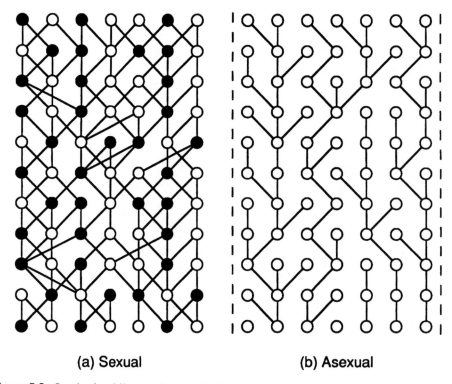

(a) Sexual   **(b) Asexual**

Figure 5.3. Species level lineages in sexual and asexual organisms. (a) In sexual (biparental) organisms, organism level lineages are united to form species-level lineages by the process of reproduction itself, which continually reconnects temporarily separated organism lineages. (b) In asexual (uniparental) organisms, reproduction does not bring together organism level lineages, which are thus constantly diverging. Therefore, species-level lineages in asexual organisms, if they exist at all, must result from processes other than reproduction. Circles represent organisms, some of which are filled and others unfilled to represent different sexes. Solid lines represent the reproductive connections between parents and offspring; dashed lines bounding the collection of organism lineages in (b) represent boundaries resulting from unifying processes other than sexual reproduction.

of the Evolutionary Synthesis. Indeed, any definition that is inconsistent with this general evolutionary concept of species is probably sufficiently removed from the mainstream of contemporary biology that it need be considered no further—but this statement is not intended to dismiss definitions that might contradict my thesis, for although such definitions can be identified in the history of biology, they are not advocated by any contemporary biologists.

Evidence of the general lineage concept of species can be found in, or associated with, every one of the species definitions quoted above, though it is easier to discern in some cases than in others. Several of the definitions make the equation of species with evolutionary lineages very explicit, beginning with the phrase "a species is a lineage" or some variant thereof (e.g., Simpson, 1951, 1961; Wiley, 1978, 1981; Van Valen, 1976). In other cases, such explicit statements are not actually contained in the species definitions themselves, but the authors of those definitions make statements that are equally explicit in their

discussions (e.g., Mishler, 1985; Ridley, 1989, 1990; Templeton, 1989, this volume; Nixon and Wheeler, 1990; Baum and Shaw, 1995; see also Eldredge and Cracraft, 1980; Donoghue, 1985; Mishler and Brandon, 1987; McKitrick and Zink, 1988; Kluge, 1990; Baum and Donoghue, 1995). In addition, several authors have published diagrams (similar to figure 5.3a) that clearly represent species as unified collections of organism lineages (e.g., Hennig, 1966; Nixon and Wheeler, 1990; Baum and Shaw, 1995; see also Eldredge and Cracraft, 1980; de Queiroz and Donoghue, 1988, 1990; Kluge, 1990; Davis and Nixon, 1992; O'Hara, 1993, 1994; Frost and Kluge, 1994; Graybeal, 1995).

Many modern species definitions do not explicitly equate species with lineages, but they nevertheless do so implicitly by equating species with populations—either in the definitions themselves (e.g., Wright, 1940; Mayr, 1942, 1963, 1982; Dobzhansky, 1950, 1970; Rosen, 1979; Paterson, 1985; see also Templeton, 1989; Nixon and Wheeler, 1990) or in associated discussions (e.g.,

Cracraft, 1983; Michener, 1970; Sneath and Sokal, 1973; Mallet, 1995). As noted above, a population is itself a lineage, or at least a segment of a lineage. That it to say, the concept of a population necessarily incorporates a temporal component in that the processes that determine the limits of populations are themselves temporally extended. For example, there is no population in which all organism lineages are simultaneously connected by interbreeding at any given instant (O'Hara, 1993). Thus, definitions that equate species with populations and those that equate them with lineages do not describe different species concepts; they describe time-limited and time-extended perspectives on the same species concept.

Even the seemingly most radical modern species definitions are at least consistent with the general lineage concept of species. Phenetic species definitions, for example, though developed within the context of a taxonomic ideology that attempted to formulate its concepts without an evolutionary basis (e.g., Sokal and Sneath, 1963; Sneath and Sokal, 1973), do not contradict the equation of species with populations or lineages. Instead, they explicitly or implicitly assume such an equation but emphasize the evidence and procedures that are used to recognize species in practice (e.g., Rogers and Appan, 1969; Michener, 1970; Sokal and Crovello, 1970; Sneath and Sokal, 1973; Doyen and Slobodchikoff, 1974; see also Mallet, 1995). Likewise, species definitions based on monophyly, which seem to deny that species differ in any important respect from higher taxa (e.g., Mishler and Donoghue, 1982; Nelson, 1989), also assume the equation of species with lineages or populations (e.g., Donoghue, 1985; Mishler, 1985). However, because the taxonomic ideology within whose context the definitions in question were formulated prohibits the recognition of paraphyletic taxa, and because ancestral taxa are necessarily paraphyletic, those definitions require either that only terminal branches be recognized as species (de Queiroz and Donoghue, 1988) or that considerations of species be restricted to single temporal planes (Baum and Shaw, 1995).

Thus, despite the diversity of alternative species definitions, there is really only one general species concept in modern systematic and evolutionary biology—species are segments of population level evolutionary lineages. But if all contemporary species definitions are merely variations on this general theme, to what can their manifest differences be attributed? The answer to this question becomes clear when one attempts to relate the various definitions to the general lineage concept of species. By performing this exercise, one finds that all of the definitions can be related to the general lineage concept using only three general categories. Some definitions describe the general lineage concept of species itself, others describe criteria for identifying or delimiting species taxa (while explicitly or implicitly adopting the general lineage concept), and still others do both.

Thus, the "evolutionary" species definitions of Simpson (1951, 1961) and Wiley (1978, 1981) and the "phylogenetic" definitions of Hennig (1966) and Ridley (1989) describe the general concept of species as lineages. The "biological" definition of Dobzhansky (1950) and the "cohesion" definition of Templeton (1989) are similar, though they describe the lineage over a shorter time interval (i.e., as a population). In contrast, the "biological" definition of Wright (1940), the "phylogenetic" definitions of Cracraft (1982) and Nixon and Wheeler (1990), the "phenetic" definitions of Michener (1970) and Sneath and Sokal (1973), and the "genotypic cluster" definition of Mallet (1995) emphasize criteria for identifying or delimiting species taxa. Finally, several of the definitions describe both the general lineage species concept and one or more species criteria. This is most evident in the case of Van Valen's (1976) "ecological" definition, but it also applies to the "biological" or "isolation" definitions of Mayr (1942, 1963, 1970) and Dobzhansky (1970), the "recognition" definition of Paterson (1985), the "phylogenetic" definitions of Rosen (1979) and Mishler (1985), and the "genealogical" definition of Baum and Shaw (1995). Several of the definitions also include explicit or implicit statements about the processes responsible for uniting organism lineages to form species (e.g., interbreeding, natural selection, common descent, developmental and other constraints), which effectively restrict or broaden application of those definitions within the context of the general lineage species concept.

To the extent that contemporary species definitions conform to a single general species concept, most of the fundamental differences among those definitions are related to species criteria. The great majority of the alternative species definitions attempt to identify such criteria, and they certainly differ in the criteria identified (table 5.1). Even authors who do not include explicit descriptions of species criteria in their species definitions nonetheless discuss such criteria in considerable detail. How is it that so many different species criteria can be identified within one general concept of species? The answer to this question becomes evident when one considers alternative species definitions in the context of the process or processes through which new species come into existence.

Species Definitions and the Process of Speciation

The process of speciation can be represented diagrammatically, under the general lineage concept of species, as a single line or trunk splitting into two (figure 5.4). In this diagram, the process is represented as if the ancestral population has divided equally into two descendants, but this is not meant to imply that the division could not have been highly unequal or polytomous. Numerous "events" or, more accurately, subprocesses, occur as a

Table 5.1. Species criteria.

Initial separation (regardless of cause)

Cohesion
  Interbreeding (reproductive isolation)
    Actual interbreeding (intrinsic or extrinsic isolation)
    Potential interbreeding (crossability/intrinsic isolation)
      Recognition (prezygotic isolation)
      Viability ⎫
             ⎬ (postzygotic isolation)
      Fertility ⎭
  Adaptive zone (niche)

Monophyly
  Apomorphy
  Exclusive coalescence of gene trees

Distinguishability
  Diagnosability (fixed difference)
  Phenetic cluster
  Genotypic cluster

This classification summarizes alternative species criteria, most of which are described by species definitions discussed in this chapter. The criteria are not necessarily mutually exclusive.

lineage divides, many of which are related. My purpose in this section is not to describe these events or subprocesses in detail, but only to summarize them for the purpose of relating them to species criteria.

One event or subprocess of obvious importance is the initial separation or split of the ancestral lineage. This may be caused by an extrinsic barrier or by an intrinsic one. If caused by an extrinsic barrier, the relationship of separation to another process, divergence, is likely indirect; if it is intrinsic, the relationship is presumably direct. Various phenomena are responsible for the divergence of the lineages, including the origins, changes in frequency, fixations, and extinctions of alleles, some of which underlie similar changes in the states of qualitative phenotypic characters and shifts in the frequency distributions of quantitative ones. Such changes can occur, for a given character, in one or the other or both descendant lineages, and differences between those lineages can accumulate both within and among characters. Alleles or character states that change in frequency or are lost or fixed can, of course, originate before as well as after the initial split, and divergence itself can precede the split, as in the cases of clinal and habitat divergence (see Templeton, 1981). The process of divergence affects other properties of the lineages, including their passage through polyphyletic, paraphyletic, and monophyletic stages in terms of their component organisms and genes (e.g., Neigel and Avise, 1986; Avise and Ball, 1990). Regarding their effects on the intrinsic separation of lineages, changes in different classes of characters presumably form a continuum, with some having virtually no effect and others—such as those influencing reproductive compatibility in sexual organisms—having profound

ones. At some point during divergence, the lineages cross a threshold beyond which their separation becomes irreversible: they can no longer fuse, which is not to say that there is an absolute barrier to gene exchange between them. Of course, divergence continues after the lineages cross this threshold.

The diversity of alternative species definitions—or more specifically, the diversity of alternative species criteria—is directly related to the diversity of events or subprocesses that occur during the process of speciation. Each criterion corresponds with one of the events that occurs during that process. Thus, "biological species concepts" (in the broad sense) use criteria based on the effects of divergence on potential interbreeding. For the subset of "biological" definitions corresponding with the "isolation concept," the criterion is a level of divergence beyond which organisms no longer mate under natural conditions to produce viable and fertile offspring. For the subset of "biological" definitions corresponding with the

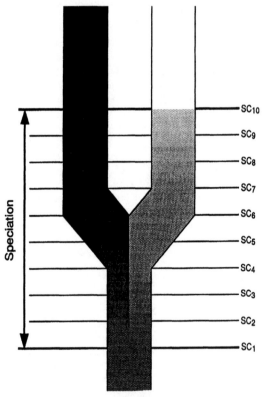

Figure 5.4. Speciation and species criteria. In this generalized diagram, speciation is equated with the entire set of events whose individual members serve as the basis for different species criteria; it is bounded by the first and last events in that set and is represented as a broad zone within which different species criteria, represented by horizontal lines (SC1–10), will result in different conclusions about the number of species.

"recognition concept," the criterion is a level of divergence beyond which organisms no longer recognize one another as potential mates—whether behaviorally or physiologically—so that fertilization does not occur. The "ecological species concept" uses a criterion based on the divergence of characters related to ecology—a level of divergence beyond which the lineages are considered to occupy different adaptive zones. In the case of the "cohesion species concept," the first primary category of cohesion mechanisms corresponds with a criterion based on the effects of character divergence on potential interbreeding, the second a criterion based on the divergence of characters related to ecology (see below). The "monophyletic" (including the "genealogical") version of the "phylogenetic species concept" uses a criterion based on the sorting of component lineages (itself a cause of character divergence)—a level beyond which all component gene or organism lineages share a more recent common ancestor with a member of their own population lineage than with one of another population lineage. The "diagnostic" version of the "phylogenetic species concept" uses a criterion based on the divergence of characters itself (rather than its cause or its effects on interbreeding or ecology)—a level of divergence beyond which the lineages exhibit a fixed character difference. The "phenetic species concept" and the "genotypic cluster definition" also use criteria based on the divergence of characters itself—in these cases, a level beyond which the lineages are distinguishable in terms of either qualitative or quantitative differences.

The timing of the events described above depends on various factors, such as demography, geography, genetics, gene flow, drift, and selection (for reviews, see Bush, 1975; Templeton, 1981), which differ from one situation to the next. For example, sorting of gene or organism lineages resulting in monophyly, and the loss or fixation of alleles resulting in diagnosability, might occur relatively early in a lineage segment originating from a founder event, but the same properties might evolve relatively late in one originating as a large subdivision of the ancestral lineage. The evolution of distinct specific mate recognition systems might occur relatively late if divergence in the relevant characters results from adaptation to the preferred habitat (e.g., Paterson, 1986) and the habitat does not change, but it might occur early if the habitats are altered or if divergence results from a coevolutionary "arms race" between males and females (e.g., Rice, 1996, this volume; Palumbi, this volume). In cases where fixation of certain alleles or karyotypes is itself responsible for setting up the initial reproductive barrier, diagnosability and reproductive incompatibility presumably arise simultaneously, but in cases where the initial reproductive barrier is set up by an extrinsic factor, a diagnostic character that had no effect on reproductive compatibility might arise long before one that had such an effect. And if isolation by distance is an important factor, divergence in all kinds of characters can precede the initial split

of the ancestral lineage. Consequently, there is no reason to expect that the various events that form the bases for alternative species criteria must always occur in the same order, and thus there is no reason to expect a fixed sequence of "types" of species (e.g., "phylogenetic" → "evolutionary" → "biological"; see Haffer, 1986), though there may be tendencies (Harrison, this volume).

Species Concepts and Species Criteria: A Revised Terminology

I argued above that alternative species definitions, although commonly viewed as descriptions of alternative species concepts, are more appropriately viewed either as alternative descriptions of the general lineage concept of species or as descriptions of alternative species criteria. This distinction between species concepts and species criteria is central to understanding both the unity and the diversity among species definitions. A *species concept* is an idea about the kind of entity represented by the species category, that is, about the kind of entity designated by the term *species*. A *species criterion* is a standard for judging whether a particular entity qualifies as a member of the species category, that is, for judging whether a particular entity is or is not a species. Species concepts and species criteria are clearly related in that a species concept underlies the formulation of species criteria; nevertheless, the distinction is an important one. It is analogous to the distinction between a disease and one of its symptoms.

Although the distinction between species concepts and species criteria is present to varying degrees in the writings of many authors, it has not been adopted consistently. Most importantly, it has not been incorporated into the names of the various ideas described by alternative species definitions. As a consequence, the current terminology regarding species definitions is ambiguous, if not downright misleading. On the one hand, reference to alternative species definitions as "species concepts" suggests disagreement at a more fundamental level than actually exists. On the other hand, the adjectives combined with that term often seem to have been chosen more for their persuasive than for their descriptive utility. General adjectives such as "biological," "evolutionary," and "phylogenetic," for example, can legitimately be applied to almost all modern species definitions, yet they are most often used to designate small subsets of them. Sometimes the terms also obscure important distinctions, as in the case of the so-called "phylogenetic" definitions. In general, the terms used to describe species definitions provide little insight concerning the distinctive characteristics of the definitions to which they refer.

By applying the distinction between species concepts and species criteria consistently and comprehensively to the diversity of ideas described by alternative species definitions, it is possible to develop a more informative terminology. In this section, I outline such a terminology, reclassifying the ideas described by alternative species definitions in accordance with that distinction and renaming them so as to describe their distinctive characteristics more accurately. This exercise will also demonstrate that the diversity of contemporary ideas about species can be unified under a single conceptual framework. Hereafter, I will use the term *species concept* only when referring to general ideas concerning the kind of entity designated by the term *species*, or when referring to the terminology of previous authors; otherwise, I will use the neutral term *species definition* for statements that describe either species concepts or species criteria.

Biology and Interbreeding

The term *biological species concept* is currently used to designate at least three distinct ideas, one of which is appropriately designated by this term, the other two of which should be given different names. The term was originally used to distinguish concepts of species that applied uniquely to biological entities from earlier essentialistic and nominalistic ones, according to which species were conceptualized as classes of similar objects and thus could be applied not only to organisms but also to chemicals, minerals, and other inanimate objects (e.g., Mayr, 1942, 1957, 1963, 1969, 1982). It is still appropriate to use *biological species concept* in this sense, in which case it applies to all modern species definitions— that is, to all definitions that equate species with populations or lineages (rocks don't form lineages). Consequently, the term will be useful for discussing the history of ideas about species but not—as it has most commonly been used—for distinguishing among contemporary species definitions.

Early and influential descriptions of a biological species concept emphasized interbreeding and the nature of species as reproductive communities (Mayr, 1969), gene pools (Dobzhansky, 1950), and fields for gene recombination (Carson, 1957). The equation of species with interbreeding groups or gene pools can legitimately be referred to as a species concept, which is useful for contrasting species definitions that apply only to sexual organisms with those that apply to both sexual and asexual organisms. It is thus a restricted version of the general lineage species concept and can be termed the *interbreeding [species] concept* (e.g., Mayr, 1963), the *gene pool [species] concept*, or the *sexual species concept*. The adjectives "biological," "genetical," and "reproductive" are ambiguous in that they apply equally to both sexual and asexual organisms and thus should not be used for this concept. The distinctive characteristic of several definitions based on the gene pool concept is a criterion of interbreeding, or more specifically, potential interbreeding (e.g., Mayr, 1942, 1963). This should be called the *potential interbreeding criterion*. The breeding system of sexual organisms is composed of several classes of components in which organisms must be compatible to satisfy the potential interbreeding criterion (table 5.2).

Isolation and Recognition

In the current terminology, both "isolation concept" and "recognition concept" refer to definitions based on the gene pool concept and the potential interbreeding criterion; their distinctive characteristics are secondary criteria for assessing potential interbreeding. Definitions currently referred to as examples of the "isolation concept" use a secondary criterion of reproductive incompatibility (i.e., intrinsic reproductive isolation) to infer the limits of potential interbreeding. This should be called the *isolation criterion*. It is satisfied by an incompatibility in one or more components of the breeding system (table 5.2). Viewed from the perspective of species unity (rather than separation), this criterion can be termed the *crossability criterion* (see Mayr, 1942:119), which is satisfied by compatibility in all components of the breeding system. In contrast, species definitions currently referred to as examples of the "recognition concept" use a secondary criterion of reproductive compatibility (i.e., common specific mate recognition or fertilization systems) to infer the limits of potential interbreeding. This should be called the *recognition criterion*. It is satisfied by compatibility in all components of the fertilization system, that is, all

Table 5.2. Components of the sexual breeding system (factors influencing potential interbreeding).

Prezygotic components = fertilization system[1]
  Premating components
    Habitat components: habitat in which organisms mate
    Temporal components: times when organisms mate[2]
    Ethological components: courtship behavior
  Mating components = mechanisms of gamete transfer[3]

  Postmating components = mechanisms of gamete union[4]
Postzygotic components = developmental system
  Somatic components: hybrid[5] viability
  Germ line components: hybrid[5] fertility

This classification is an attempt to restate the classifications of "isolating mechanisms" (e.g., Dobzhansky, 1937, 1970; Mayr, 1942, 1963) in a way that is neutral with respect to the isolation and recognition perspectives. It is most applicable to multicellular organisms, where pre- and postzygotic components are relatively clearly distinguishable.

[1]In the broad sense (e.g., Paterson, 1985).

[2]In relation to annual, lunar, and daily cycles and other environmental cues, such as rainfall.

[3]Including genitalia, flower parts, pollen, and pollinators.

[4]These components constitute the fertilization system in the narrow sense of syngamy.

[5]Including F1, F2, and backcross hybrids.

prezygotic components of the breeding system (table 5.2). Viewed from the perspective of species separation (rather than unity), this criterion can be termed the *prezygotic isolation criterion*, which is satisfied by incompatibility in one or more prezygotic components of the breeding system.

The isolation and recognition criteria should not be confused with ideas about the adaptive versus nonadaptive nature of the differences that prevent interbreeding between organisms of different species and the evolutionary processes that may have produced those differences. The differences in question have commonly been called *isolating mechanisms* (e.g., Dobzhansky, 1937, 1970; Mayr, 1942, 1963). In the context of Williams's (1966) subsequently proposed distinction between adaptations and fortuitous effects, this term implies (since mechanisms are adaptations) that selection has produced the differences in question for the very reason that they protect the integrity of separate gene pools (e.g., Paterson, 1981, 1986, 1988, 1993b). The term *isolating effects* (Paterson, 1986) implies that the differences are incidental by-products of selection for some other evolved function. The neutral term *isolating barriers* (Chandler and Gromko, 1989) avoids these connotations, as do *intrinsic reproductive barriers* and *intrinsic barriers to gene flow*. The view that intrinsic reproductive barriers between species are true isolating mechanisms implies a particular model of speciation (see Paterson, 1978, 1986; Lambert and Paterson, 1982), in which initial divergence of postmating breeding system components results in selection against hybrids, thus favoring the evolution of premating reproductive barriers and ultimately complete reproductive isolation (e.g., Dobzhansky, 1940). The isolation criterion should not be confused with this model of speciation (Chandler and Gromko, 1989), which has already been termed [*speciation by*] *reinforcement* (Blair, 1955; Howard, 1993). Nor should the recognition criterion be confused with the model of speciation favored by advocates of that criterion, in which adaptation of the specific mate recognition and fertilization systems of allopatric populations to new or modified habitats leads to the evolution of differences in those systems, with isolation occurring as an incidental effect (Paterson, 1978, 1985, 1986). This model can be termed *speciation by primary fertilization system divergence*. Both it and the reinforcement model are special cases of [*speciation by*] *adaptive divergence* (Templeton, 1981).

Terms are also useful for the different perspectives in the controversy about isolation versus recognition (e.g., Coyne et al., 1988; White et al., 1990). The term *isolation perspective* can be used for the interrelated and historically associated set of ideas including an emphasis on reproductive isolation in species definitions, the view that barriers to gene flow between species are true isolating mechanisms, the theory of speciation by reinforcement, and the idea that the species category is a relational con-

cept. Similarly, the term *recognition perspective* can be used for the interrelated and historically associated set of ideas proposed as an alternative to the isolation perspective, including an emphasis on specific mate recognition and fertilization systems in species definitions, the view that barriers to gene flow between species are incidental isolating effects, the theory of speciation by adaptive divergence in allopatry, and the idea that the species category is not a relational concept (e.g., Paterson, 1985, 1986, 1988).

Evolution and Ecology

Simpson (1951, 1961) used the term *evolutionary species* to emphasize the explicitly evolutionary formulation of his species definition, which was one of the first attempts to describe the general lineage concept of species (rather than a species criterion) in the form of an explicit definition. Wiley (1978) added the term *concept* to Simpson's "evolutionary species," but he still used the term in a general sense. Later, however, he contrasted the "evolutionary species concept" with the "biological species concept" (Wiley, 1981; see also Mayr and Ashlock, 1991; King, 1993), and the former term has been used subsequently to designate the species definitions of Simpson and Wiley (e.g., Haffer, 1986; Templeton, 1989; Frost and Hillis, 1990; Panchen, 1992; Ridley, 1993). This usage is misleading in that all modern species definitions are evolutionary. The term *evolutionary species concept* should be used for the general concept of species as evolutionary lineages, that is, in contrast with truly nonevolutionary species concepts, such as those based on the metaphysics of essentialism (see Mayr, 1957, 1963, 1969, 1982). Specific formulations of the evolutionary (general lineage) species concept can simply be referred to as "Simpson's species definition" or "Wiley's species definition." Viewed retrospectively, the origin of separate lineages will trace back to their initial separation (e.g., Sober, 1984), whether caused by intrinsic or extrinsic factors; if used as a species criterion, this can be termed the *initial split criterion*.

The definition currently referred to as the "ecological species concept" describes, first and foremost, the general lineage concept of species; therefore, it can simply be called "Van Valen's species definition." Van Valen's definition incorporates the occupation of a distinct adaptive zone as a species criterion, and this should be termed the *adaptive zone* or *niche criterion*. Van Valen (1976) used the adaptive zone criterion in his species definition because he believed that ecologically based natural selection was more important than reproductive isolation for maintaining separate evolutionary lineages. This theory about species maintenance should not be confused with the adaptive zone criterion for recognizing entities as species, which can be adopted even if species are maintained by other processes.

Cohesion

The term *cohesion* can be used for the general phenomenon or class of phenomena responsible for the unification of organism lineages to form species level lineages. If so, then the concept of cohesion, emphasized in Templeton's species definition, is implicit in all variants of the general lineage concept of species, and there is probably no need for another term to describe that concept. The various "cohesion mechanisms" described by Templeton (1989) correspond with species criteria, including several used in the species definitions of other authors. Thus, *genetic exchangeability*, the first of Templeton's two primary categories of cohesion mechanisms, "refers to the ability to exchange genes via sexual reproduction" (Templeton, 1989:14), that is, the potential interbreeding criterion. It consists of two primary subcategories, one of which corresponds with the isolation criterion, the other with the recognition criterion plus what may be termed the *viability* and *fertility criteria* (the capability of producing viable and fertile offspring). Together, the latter three criteria (recognition, viability, and fertility) correspond with the crossability criterion (see Mayr, 1942:119).

Demographic exchangeability, the other primary category of cohesion mechanisms, refers to the fact that every organism in the population is a "potential common ancestor to the entire population at some point in the future" (Templeton, 1989:15), which is simply a statement of the general lineage concept of species. It depends on conspecific organisms sharing "the same fundamental niche" (Templeton, 1989:14), which corresponds with the niche or adaptive zone criterion. One of the central features of Templeton's (1989) species definition is its explicit applicability to the entire reproductive continuum, from asexuals to syngameons. The term *reproductive continuum species concept* might therefore be used to contrast species definitions that apply explicitly to both sexual and asexual organisms with those that apply only to sexual ones.

Phylogenetic Systematics

The term *phylogenetic species concept* accurately describes all modern species definitions, which explicitly or implicitly equate species with branches, or branch segments, of phylogenetic trees. The term should not, therefore, be restricted to species definitions developed within the context of Phylogenetic Systematics (Cladistics). For historical purposes, those definitions can be called *phylogenetic systematic* or *cladistic species definitions* (as opposed to *concepts*), but no one of them should be singled out as *the* phylogenetic systematic or cladistic species definition. Different species definitions in this historically defined category describe at least two different species criteria, and they are associated with at least two different general models of speciation. Conse-

quently, for the purposes of biological theory and practice, it will be more useful to use entirely different terms for the ideas in question.

In most respects, definitions in the first group of phylogenetic systematic species definitions are simply statements of the general lineage concept of species. As such, they can simply be referred to as "Hennig's (1966) species definition" and "Ridley's (1989) species definition." These definitions have two distinctive characteristics. First, speciation is equated with lineage cleavage or cladogenesis; anagenetic change within an unbranched lineage is not considered speciation. Thus, *successive* or *successional species* (Imbrie, 1957; Simpson, 1961)—often incorrectly referred to as "chronospecies" and "paleospecies" (see Sylvester-Bradley, 1956)—are not considered true species. This characteristic implies a *cladogenetic model of speciation*, which subsumes both the *bifurcating model of speciation* (Wagner and Erwin, 1995) and the *budding model of speciation* (Foote, 1996) (figure 5.2a,b; table 5.3). The second distinctive characteristic of these definitions is that ancestral species are not considered to persist after giving rise to descendants, which implies the bifurcating model (figure 5.2b).

The distinctive characteristic of the second group of species definitions developed within the context of phylogenetic systematics is a criterion of monophyly (e.g., Bremer and Wanntorp, 1979; Donoghue, 1985; Mishler, 1985; de Queiroz and Donoghue, 1990). This should be termed the *monophyly criterion* (e.g., Baum, 1992). The presence of a derived character state (e.g., Rosen, 1979) is often used as a secondary criterion for inferring monophyly, which can be termed the *autapomorphy* or the *apomorphy criterion*. A different but related secondary criterion is exclusivity of common ancestry relationships in multiple gene trees, that is, concordant coalescence of gene genealogies (Baum and Shaw, 1995; see also Avise and Ball, 1990); this should be termed the *concordant* or *exclusive coalescence criterion*. In any case, the general property of monophyly should not be confused with the specific kind of evidence by which it is inferred. Because ancestors are (by definition) nonmonophyletic, the monophyly criterion implies that ancestral lineages cannot be species, which limits the application of the monophyly criterion to terminal (though not necessarily recent) lineages; alternatively, the criterion can be applied to subterminal lineages in a relative sense (i.e., ignoring their descendants).

The distinctive characteristic of the third group of phylogenetic systematic species definitions is the idea of diagnosability in the sense of "unique combinations of primitive and derived characters" (Cracraft, 1983:170), where characters are attributes that do not vary among organisms of comparable age, sex, and so on (Nixon and Wheeler, 1990). If this idea is interpreted as a species criterion, it can be termed the *diagnosability criterion*. On the other hand, if it is interpreted as a procedure for identifying taxa for use in phylogenetic analysis (i.e., with

Table 5.3. General models of speciation.

Cladogenetic: speciation corresponds with lineage splitting (figure 5.2a, b)

    Bifurcation: ancestral species does not persist through cladogenetic event (figure 5.2a)

    Budding: ancestral species persists through cladogenetic event (figure 5.2b)

Anagenetic: speciation corresponds with lineage modification (figure 5.2c)

    Phyletic transformation: speciation within an unbranched lineage (figure 5.2c)

Modified from Foote (1996).

no claims that they are unitary lineages), then there is no need to call those taxa species; they can be called *terminal taxa* (Farris, 1977) or *operational taxonomic units* (see below). Some authors, rather than interpreting diagnosability as only a necessary property of species, interpret it as a necessary and sufficient property and therefore recognize every diagnosable lineage segment as a separate species (e.g., Nixon and Wheeler, 1992). This interpretation effectively equates speciation with the fixation of traits. It creates the potential for recognizing a succession of species in an unbranched lineage thus implying an *anagenetic* or *phyletic transformation* (Foote, 1996) *model of speciation* (figure 5.2c; table 5.3), which is in direct opposition to the views of Hennig (1966) and Ridley (1989).

Phenotypic and Genotypic Clusters

The term *phenetic species concept* is currently used for a set of species definitions whose distinctive characteristic is a criterion of detectable phenetic clusters; it can therefore be termed the *phenetic cluster criterion* or simply the *phenetic criterion*. Of course, not all phenetic clusters correspond with species; some correspond with groups of species (whether clades or para- or polyphyletic groups), and others correspond with parts of species (whether morphs or differentiated subpopulations). Phenetic clusters are often treated as *operational taxonomic units* or *OTUs* (Sokal and Sneath, 1963), a term that is best used to designate any units that are defined by a set of taxonomic procedures, not only phenetic clusters.

The distinctive characteristic of Mallet's (1995) "genotypic cluster" definition is a criterion of identifiable genotypic clusters, in particular, those that can coexist with other such clusters without fusing. This should be termed the *genotypic cluster criterion*. Mallet (1995) proposed his definition as an alternative to species definitions based on interbreeding, which emphasize one form of species cohesion and might therefore bias hypotheses of speciation to favor models that involve extrinsic barriers to gene flow. The genotypic cluster definition was supposed to provide a definition that is useful "however species are maintained and however they have come to be" (Mallet, 1995:295–296). This concern is satisfied by the general lineage concept of species, which is sufficiently general to be consistent with a diversity of hypothesized mechanisms both for generating and for maintaining species as separate lineages.

Kinds of Species

Several additional terms have been coined for classes of species that satisfy (or fail to satisfy) a particular species criterion or set of criteria—for example, "cladospecies" and "paraspecies" (Ackery and Vane-Wright, 1984), "metaspecies" (Donoghue, 1985; see also Archibald, 1994), and "ferespecies" (Graybeal, 1995). Although these terms serve the useful purpose of abbreviating more complete descriptions of the categories in question, in many cases the abbreviation is slight and does not seem to offset the cost of learning an unfamiliar term and its corresponding definition. That is to say, it will often be more straightforward to use descriptive adjectives in conjunction with the term *species*. Thus, alternatives to the terms listed above are (in the same order) *monophyletic species, paraspecies, questionably monophyletic species,* and *nonmonophyletic interbreeding species*. Given that it may often be useful to describe a species in terms of multiple criteria, this approach seems preferable to coining a name for every class of species that could be recognized for a different combination of species criteria. Similarly, vague adjectives can be replaced with ones that describe the criteria satisfied by particular lineages more explicitly; for example, the term "phylogenetic species" would be replaced with either *monophyletic species* or *diagnosable species*.

Classes of Species Definitions: Their Significance and Limitations

In addition to providing the basis for a more useful terminology, the distinction between species concepts and species criteria provides insight into the significance and limitations of different classes of species definitions. Although many contemporary species definitions combine descriptions of the general lineage concept of species with descriptions of particular species criteria, most emphasize one or the other. These different emphases reflect different goals, and consequently, definitions of one kind should not be criticized for failing to fulfill the goals of the other. For example, the species definitions of Simpson (1951, 1961) and Wiley (1978, 1981) have often been criticized for being vague because they fail to specify causal mechanisms or explicit criteria for delimiting species (e.g., Sokal and Crovello, 1970; Mayr, 1982; Haffer, 1986; Templeton, 1989; Ridley, 1993). These criticisms are inappropriate. The definitions in question do not attempt to describe operational criteria for delimiting species taxa but only the general concept of species as evolutionary lineages.

In contrast with definitions that describe the general lineage species concept, definitions that describe species criteria must be operational to some degree. Species criteria provide the bridge between the general theoretical concept of species and the practical operations and empirical data used to recognize and delimit the entities conforming to that concept. This does not mean, however, that the criteria themselves have to be easy to use, universally applicable, or definitive; instead, they only have to be useful for investigating the separation of lineages. For example, we should not be troubled by the fact that certain criteria, such as interbreeding and monophyly, often have to be inferred using secondary criteria. Nor should we be troubled by the fact that the potential interbreeding criterion cannot be used in the case of asexual organisms; it is only one of several possible lines of evidence. Nor should we consider it problematical that organisms making up separate sexual lineages are sometimes able to interbreed; the separation may be maintained by other factors, such as natural selection. Nor should we view as a difficulty the fact that the organisms making up separate lineages do not always form mutually exclusive monophyletic groups; the separation may be too recent for monophyly to have evolved. Because every species criterion will probably fail to identify separate lineages under certain conditions, the best inferences about lineage separation will be based on lines of evidence described by several different species criteria.

The Interpretation of Species Criteria and the Resolution of Their Conflicts

In the preceding section, I have interpreted species criteria as lines of evidence relevant to analyzing the separation of lineages—that is, for inferring whether particular organisms or local populations are parts of the same or different species (as lineages). An alternative and commonly adopted interpretation is that species criteria are defining (necessary) properties of the species category—that is, properties that populations or lineages *must have* to merit recognition as species. It is worthwhile to consider the second interpretation and its implications further, because this interpretation turns out to be largely responsible for a perceived conflict among alternative species definitions.

As I argued above, the various species criteria correspond with different "events" that occur during lineage separation and divergence. Picking one those events as the defining property of the species category amounts to imposing a certain level of divergence as an arbitrary line of demarcation on a continuous process. There seems to be general agreement that speciation is a temporally ex-

tended process rather than an instantaneous event, but judging from the different species criteria, there is less agreement about when that process begins and ends, or about which of its component events or subprocess is the most significant. Viewed impartially, the entire set of events that serve as the bases for alternative species criteria defines a broad gray zone that can be equated with the process of speciation (figure 5.4). Selecting one of those events as the property that a lineage must have to be considered a species narrows the gray zone, and thus also the meaning of "speciation," by replacing a relatively protracted process with a relatively brief one. This narrowing may be convenient for taxonomic purposes, but it does not make the process of lineage separation and divergence any narrower or less continuous. Moreover, it sets up a potential for conflict because different events in that process can be selected as representing the critical level of divergence.

Of course, particular criteria are usually chosen because of their theoretical significance. Despite my emphasizing the arbitrariness of picking a particular event in the process of lineage divergence as a necessary property of species, I do not mean to imply that any of the criteria are theoretically or biologically meaningless. On the contrary, all of them are significant. Nevertheless, no one criterion has primacy over the others in the context of general evolutionary theory; instead, the significance of the various criteria depends on the question being addressed. Thus, for reconstructing phylogeny and analyzing historical biogeography, diagnosable lineages, particularly monophyletic ones, will be most significant. For studying hybrid zones, differences in breeding systems are more relevant. And for examining host races, niche differences are obviously important. Components of the fertilization system may be critical for studying the evolution of intrinsic reproductive barriers in some cases, but those of the developmental system may be critical in others. Because different criteria are useful for addressing different questions, it is not surprising that the criteria advocated by different authors tend to reflect their research interests. In any case, no single criterion is optimal for all questions.

Criteria are sometimes chosen because of their relevance to the evolutionary fates of lineages; however, because of the historically contingent nature of evolutionary fate, no criterion is definitive (O'Hara, 1994). For example, intrinsic reproductive isolation might be considered to indicate that lineages have become irreversibly separated, a seemingly important event in determining their fates. But premating barriers based on habitat differences can disappear when environmental conditions change, and postmating barriers can disappear through elimination of the genetic elements responsible for the reduced fitness of hybrids. Separation probably does become irreversible eventually, but precisely when that threshold is crossed is not only difficult to determine but depends on the unique circumstances of each situation.

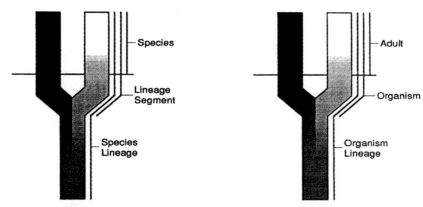

Figure 5.5. Terminology for species-level entities under a common interpretation of species criteria (left) compared with the terminology for organism-level entities (right). The equation of species with a stage in the existence of lineage segments, which makes the concept of the species analogous with that of the adult, is implied by the interpretation of species criteria as defining (necessary) properties of the species category.

Moreover, irreversible separation is not the only factor that has important effects on the fate of lineages. For example, whatever factor initiates the separation of lineages, whether intrinsic or extrinsic, plays an important role in determining their fates.

Finally, no matter which event is chosen as the critical level of divergence, the defining property interpretation of species criteria implies that a species is a stage in the existence of a lineage segment. To use an organism level analogy, a species is like an adult. Just as an organism is considered an adult after it reaches a certain stage in its existence, a segment of a population level lineage is considered a species when it reaches a certain stage in its existence (figure 5.5). This situation explains the common interpretation of species definitions as descriptions of alternative species concepts. Just as different events in the process of organismal maturation (e.g., production of functional gametes, development of a certain secondary sexual characteristic, cessation of growth) can be treated as necessary properties for defining alternative concepts of the adult, different events in the process of lineage separation and divergence (e.g., initial separation, monophyly, reproductive isolation) can be treated as necessary properties for defining alternative concepts of the species. But perhaps this interpretation of species criteria should be reconsidered.

If the species category is to have the general theoretical significance that we so often claim for it, then it probably should not be treated as analogous to the category *adult*; instead, it should be treated as analogous to the category *organism* (figure 5.6). The concept of the organism is, after all, more general than that of a particular stage in the existence of organisms. If the concept of the species is to have comparable theoretical importance, it must refer not to a stage in the separation and divergence of lineages but to entire lineage segments, from initial sepa-

ration to extinction. An important consequence of this minor yet fundamental conceptual and terminological shift is that the various criteria discussed above would no longer be species criteria—at least not in the sense of standards for granting lineages taxonomic status *as species*. Instead, they would be criteria for different stages in the existence of species—the diagnosable stage, the monophyletic stage, the reproductively isolated stage, and so on. Under this view of species, the various criteria would no longer be in competition with one another, and

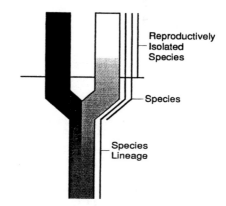

Figure 5.6. Proposed terminology for species-level entities. The equation of species with entire lineage segments, from origin to extinction, would bring the general terminology for species-level entities into line with that for organism-level entities (see figure 5.5, right). It would also remove the conflict between alternative "species" criteria by making them criteria for different stages in the existence of species (the reproductively isolated stage in this example) rather than criteria for species status.

their conflicts would vanish. Of course, there would still be problems related to determining the limits of species in practice, but there would no longer be any greater controversy about the concept of species than currently exists for the concept of organism.

Conclusion

The answer to the question, "What is a species?" is considered one of the central issues of biology as well as one of its most vexing problems. Numerous definitions have been proposed as attempts to answer this question, of which the ones summarized in this chapter constitute only a fraction. The problem is that no single definition of the species category has proved optimal for all of its different uses. Consequently, although one definition or class of definitions has often come to be favored for a certain period of time or by a certain group of biologists, none of them has enjoyed universal endorsement within biology as a whole. This situation has come to be known as "the species problem."

The realization that all modern species definitions are variations on the same general species concept helps to clarify the nature of the differences between them. It reveals that the perception of a major unresolved problem concerning the fundamental nature of species stems, in a large part, from a failure to distinguish clearly between species concepts and species criteria. When viewed as descriptions of species concepts, the fundamental agreement among alternative species definitions is obscured, and they tend to be seen as incompatible. In contrast, when viewed as descriptions of species criteria, the fundamental agreement among alternative species definitions is more evident, and they tend to be seen as complementary. Each criterion provides a different kind of information about the separation (or lack thereof) between lineages, or it describes a different stage in the divergence of lineages. In either case, there is nearly universal agreement about the general nature of the entities called species.

In arguing that almost all contemporary biologists adopt the same general species concept, I do not mean to imply that there are no conceptual differences in their views on species. Differences of opinion are numerous and include such important issues as whether species can persist though lineage-splitting events, whether more than one successive species can exist in an unbranched lineage, and whether asexual organisms form species. Numerous differences also exist concerning mechanistic hypotheses about the origin and maintenance of species in terms of geography, demography, genetics, gene flow, drift, and natural selection (see Bush, 1975; Templeton, 1981). But all these manifest differences do not concern the concept of the kind of entity designated by the term *species*—there is virtually universal agreement that species are segments of population-level evolutionary lineages. In this chapter, I have tried to develop a conceptually unified terminology that clarifies both the general agreement and the specific differences among alternative species definitions. My hope is that this terminology will facilitate communication among biologists with different research emphases and thus promote the study of species and speciation, a field that, by its very nature, lies at the intersection of several biological disciplines.

Afterword

I would like to end my chapter with a statement about its relevance to the work of Guy Bush, in whose honor it is being published. Despite devoting his professional career to the study of speciation, Guy Bush has carefully avoided invoking one of the contemporary species definitions, which he considers "putting the cart before the horse" (Bush, 1994:286, see also Bush, 1993, 1995). To his critics (e.g., Claridge, 1995), this position seems incongruous, and under the view that different species definitions represent alternative species concepts, it would be. That is, it would be very difficult to formulate appropriate questions about how species are formed without having a clear concept of what species are. But the perspective implicit in this criticism is not, as I have argued above, the most appropriate way to view alternative species definitions. Those definitions do not represent alternative concepts of the general kind of entity designated by the species category but merely alternative criteria for granting entities conforming to that concept taxonomic recognition as species. Moreover, as taxonomic standards, those criteria are arbitrary lines of demarcation imposed on the continuous process of speciation. Interpreted in this context, Bush's position is not incongruous at all.

For one thing, Guy Bush does have a clear idea about what kind of entity species are; they are, not surprisingly, "evolutionary lineages" (e.g., Bush, 1993:242). Furthermore, he has incorporated this concept of species into his concept of speciation, which he defines "as the differentiation of taxa into lineages irrevocably committed to distinct evolutionary fates" (Bush 1995:38). Guy Bush is also keenly aware of the continuous nature of that process (e.g., Bush, 1993), and he recognizes that it is "impossible to pinpoint the precise time, place, or circumstance when two or more sister populations . . . become irrevocably committed to different evolutionary paths" (Bush, 1993:243). For these reasons, abstaining from advocating a particular species criterion in no way compromises Guy Bush's research on speciation, and it may even be the better approach. As Bush (1995) recognizes, adopting a particular criterion can interfere with the study of speciation by focusing undue attention on one of the many significant events in that process. In short, his position makes perfect sense.

Although I did not meet Guy Bush until I attended the symposium in his honor for which these chapters form the proceedings, I was touched by the warmth of his personality and the obvious affection held for him by his close colleagues. I was also impressed by his perseverance despite the difficulty of reconciling his views with once prevailing doctrines about speciation, as well as by the work itself and other research efforts that it inspired. I feel honored to have been asked to contribute a chapter to a volume recognizing his contributions to the biology of speciation, and I am pleased, but not surprised, that my conclusions happen to support his views on species definitions.

Acknowledgments   I thank Rosemary Grant, Jim Mallet, and Molly Morris for references, clarifications, and discussions. I am also grateful to Stewart Berlocher and Dan Howard for the invitation to contribute this chapter, and to Stewart Berlocher, Phil Cantino, Rick Harrison, Jim Mallet, Kerry Shaw, and David Wake for thoughtful comments on an earlier draft.

References

Ackery, P. R., and Vane-Wright, R. I. 1984. Milkweed butterflies. Their cladistics and biology. Ithaca, N.Y.: Cornell University Press.

Andersson, L. 1990. The driving force: Species concepts and ecology. Taxon 39:375–382.

Archibald, J. D. 1994. Metataxon concepts and assessing possible ancestry using phylogenetic systematics. Syst. Biol. 43:27–40.

Avise, J. C., and Ball, R. M., Jr. 1990. Principles of genealogical concordance in species concepts and biological taxonomy. Oxford Surv. Evol. Biol. 7:45–67.

Baum, D. 1992. Phylogenetic species concepts. Trends Ecol. Evol. 7:1–2.

Baum, D. A., and Donoghue, M. J. 1995. Choosing among alternative "phylogenetic" species concepts. Syst. Bot. 20:560–573.

Baum, D. A., and Shaw, K. L. 1995. Genealogical perspectives on the species problem. In P. C. Hoch and A. G. Stephenson, eds., Experimental and Molecular Approaches to Plant Biosystematics. St. Louis: Missouri Botanical Garden, pp. 289–303.

Blair, W. F. 1955. Mating call and stage of speciation in the *Microhyla olicacea–M. carolinensis* complex. Evolution 9:469–480.

Bremer, K., and Wanntorp, H.-E. 1979. Geographic populations or biological species in phylogeny reconstruction. Syst. Zool. 28:220–224.

Bush, G. L. 1975. Modes of animal speciation. Annu. Rev. Ecol. Syst. 6:339–364.

Bush, G. L. 1993. A reaffirmation of Santa Rosalia, or why are there so many kinds of *small* animals? In D. R. Lees and D. Edwards, eds., Evolutionary Patterns and Processes. London: Academic Press, pp. 229–249.

Bush, G. L. 1994. Sympatric speciation in animals: New wine in old bottles. Trends Ecol. Evol. 9:285–288.

Bush, G. L. 1995. Reply from G. L. Bush. Trends Ecol. Evol. 10:38.

Carson, H. L. 1957. The species as a field for gene recombination. In E. Mayr, ed., The Species Problem. Washington, D.C.: American Association for the Advancement of Science, pp. 23–38.

Chandler, C. R., and Gromko, M. H. 1989. On the relationship between species concepts and speciation processes. Syst. Zool. 38:116–125.

Claridge, M. 1995. Species and speciation. Trends Ecol. Evol. 10:38.

Coyne, J. A., Orr, H. A., and Futuyma, D. J. 1988. Do we need a new species concept? Syst. Zool. 37:190–200.

Cracraft, J. 1983. Species concepts and speciation analysis. Curr. Ornithol. 1:159–187.

Darwin, C. 1859. On the Origin of Species by Means of Natural Selection. London: John Murray.

Davis, J. I., and Nixon, K. C. 1992. Populations, genetic variation, and the delimitation of phylogenetic species. Syst. Biol. 41:421–435.

de Queiroz, K., and Donoghue, M. J. 1988. Phylogenetic systematics and the species problem. Cladistics 4:317–338.

de Queiroz, K., and Donoghue, M. J. 1990. Phylogenetic systematics or Nelson's version of cladistics. Cladistics 6:61–75.

Dobzhansky, T. 1937. Genetics and the Origin of Species. New York: Columbia University Press.

Dobzhansky, T. 1940. Speciation as a stage in evolutionary divergence. Am. Nat. 74:312–321.

Dobzhansky, T. 1950. Mendelian populations and their evolution. Am. Nat. 84:401–418.

Dobzhansky, T. 1970. Genetics of the Evolutionary Process. New York: Columbia University Press.

Donoghue, M. J. 1985. A critique of the biological species concept and recommendations for a phylogenetic alternative. Bryologist 88:172–181.

Doyen, J. T., and Slobodchikoff, C. N. 1974. An operational approach to species classification. Syst. Zool. 23:239–247.

Eldredge, N., and Cracraft, J. 1980. Phylogenetic Patterns and the Evolutionary Process. New York: Columbia University Press.

Farris, J. S. 1977. Phylogenetic analysis under Dollo's Law. Syst. Zool. 26:77–88.

Foote, M. 1996. On the probability of ancestors in the fossil record. Paleobiology 22:141–151.

Frost, D. R., and Hillis, D. M. 1990. Species in concept and practice: Herpetological applications. Herpetologica 46:87–104.

Frost, D. R., and Kluge, A. G. 1994. A consideration of epistemology in systematic biology, with special reference to species. Cladistics 10:259–294.

George, T. N. 1956. Biospecies, chronospecies and morphospecies. In P. C. Sylvester-Bradley, ed., The Species Concept in Palaeontology. London: The Systematics Association, pp. 123–137.

Graybeal, A. 1995. Naming species. Syst. Biol. 44:237–250.

Haffer, J. 1986. Superspecies and species limits in vertebrates. Z. Zool. Syst. Evol. Forsch. 24:169–190.

Häuser, C. L. 1987. The debate about the biological species concept—a review. Z. Zool. Syst. Evol. Forsch. 25:241–257.

Hennig, W. 1966. Phylogenetic Systematics. Urbana: University of Illinois Press.

Howard, D. J. 1993. Reinforcement: Origin, dynamics, and fate of an evolutionary hypothesis. In R. G. Harrison, ed., Hybrid Zones and the Evolutionary Process. New York: Oxford University Press, pp. 46–69.

Hull, D. L. 1980. Individuality and selection. Annu. Rev. Ecol. Syst. 11:311–332.

Hull, D. L. 1996. The ideal species concept and why we can't get it. In M. F. Claridge, H. A. Dawah, and M. R. Wilson, eds., Species: The Units of Diversity. London: Chapman and Hall, pp. 357–380.

Huxley, J., ed. 1940. The New Systematics. London: Oxford University Press.

Imbrie, J. 1957. The species problem with fossil animals. In E. Mayr, ed., The Species Problem. Washington, D.C.: American Association for the Advancement of Science, pp. 125–153.

Jordan, K. 1905. Der Gegensatz zwischen geographischer und nichtgeogeographischer Variation. Z. wiss. Zool. 83:151–210.

King, M. 1993. Species Evolution. The Role of Chromosomal Change. Cambridge: Cambridge University Press.

Kluge, A. G. 1990. Species as historical individuals. Biol. Philos. 5:417–431.

Lambert, D. M., and Paterson, H. E. 1982. Morphological resemblance and its relationship to genetic distance measures. Evol. Theory 5:291–300.

Lambert, D. M., and Paterson, H. E. H. 1984. On "Bridging the gap between race and species": The isolation concept and an alternative. Proc. Linn. Soc. New South Wales 107:501–514.

Lambert, D. M., and Spencer, H. G. 1995. Speciation and the Recognition Concept. Theory and Application. Baltimore: Johns Hopkins University Press.

Luckow, M. 1995. Species concepts: Assumptions, methods, and applications. Syst. Bot. 20:589–605.

Mallet, J. 1995. A species definition for the Modern Synthesis. Trends Ecol. Evol. 10:294–299.

Masters, J. C., Rayner, R. J., McKay, I. J., Potts, A. D., Nails, D., Ferguson, J. W., Weissenbacher, B. K., Allsopp, M., and Anderson, M. L. 1987. The concept of species: Recognition versus isolation. S. Afr. J. Sci. 83:534–537.

Mayr, E. 1942. Systematics and the Origin of Species. New York: Columbia University Press.

Mayr, E. 1955. Karl Jordan's contribution to current concepts in systematics and evolution. Trans. Roy. Entomol. Soc. London 107:45–66.

Mayr, E. 1957. Species concepts and definitions. In E. Mayr, ed., The Species Problem. Washington, D.C.: American Association for the Advancement of Science, pp. 1–22.

Mayr, E. 1963. Animal Species and Evolution. Cambridge, Mass.: Harvard University Press.

Mayr, E. 1969. The biological meaning of species. Biol. J. Linn. Soc. 1:311–320.

Mayr, E. 1970. Populations, Species, and Evolution. Cambridge, Mass.: Harvard University Press.

Mayr, E. 1982. The Growth of Biological Thought. Diversity, Evolution, and Inheritance. Cambridge, Mass.: Harvard University Press.

Mayr, E. 1988. The why and how of species. Biol. Philos. 3:431–441.

Mayr, E., and Ashlock, P. D. 1991. Principles of Systematic Zoology. New York: McGraw-Hill.

Mayr, E., and Provine, W. B. 1980. The Evolutionary Synthesis: Perspectives on the Unification of Biology. Cambridge, Mass.: Harvard University Press.

McKitrick, M. C., and Zink, R. M. 1988. Species concepts in ornithology. Condor 90:1–14.

Meglitsch, P. A. 1954. On the nature of species. Syst. Zool. 3:49–68.

Michener, C. D. 1970. Diverse approaches to systematics. Evol. Biol. 4:1–38.

Mishler, B. D. 1985. The morphological, developmental, and phylogenetic basis of species concepts in bryophytes. Bryologist 88:207–214.

Mishler, B. D., and Brandon, R. N. 1987. Individuality, pluralism, and the phylogenetic species concept. Biol. Philos. 2:397–414.

Mishler, B. D., and M. J. Donoghue. 1982. Species concepts: A case for pluralism. Syst. Zool. 31:491–503.

Neigel, J. E., and Avise, J. C. 1986. Phylogenetic relationships of mitochondrial DNA under various demographic models of speciation. In E. Nevo and S. Karlin, eds., Evolutionary Processes and Theory. London: Academic Press, pp. 515–534.

Nelson, G. 1989. Species and taxa. Systematics and evolution. In D. Otte and J. A. Endler, eds., Speciation and Its Consequences. Sunderland, Mass.: Sinauer, pp. 60–81.

Nelson, G., and Platnick, N. 1981. Systematics and Biogeography. Cladistics and Vicariance. New York: Columbia University Press.

Newell, N. D. 1956. Fossil populations. In P. C. Sylvester-Bradley, ed., The Species Concept in Palaeontology. London: The Systematics Association, pp. 63–82.

Nixon, K. C., and Wheeler, Q. D. 1990. An amplification of the phylogenetic species concept. Cladistics 6:211–223.

Nixon, K. C., and Wheeler, Q. D. 1992. Extinction and the origin of species. In M. J. Novacek and Q. D. Wheeler, eds., Extinction and Phylogeny. New York: Columbia University Press, pp. 119–143.

O'Hara, R. J. 1993. Systematic generalization, historical fate, and the species problem. Syst. Biol. 42:231–246.

O'Hara, R. J. 1994. Evolutionary history and the species problem. Am. Zool. 34:12–22.

Panchen, A. L. 1992. Classification, Evolution, and the Nature of Biology. Cambridge: Cambridge University Press.

Paterson, H. E. H. 1978. More evidence against speciation by reinforcement. S. Afr. J. Sci. 74:369–371.

Paterson, H. E. H. 1980. A comment on "mate recognition systems." Evolution 34:330–331.

Paterson, H. E. H. 1981. The continuing search for the unknown and the unknowable: A critique of contemporary ideas on speciation. S. Afr. J. Sci. 77:113–119.

Paterson, H. E. H. 1985. The recognition concept of species. In E. S. Vrba, ed., Species and Speciation. Pretoria: Transvaal Museum, pp. 21–29.

Paterson, H. E. H. 1986. Environment and species. S. Afr. J. Sci. 82:62–65.

Paterson, H. E. H. 1988. On defining species in terms of sterility: Problems and alternatives. Pacific Sci. 42:65–71.

Paterson, H. E. H. 1993a. Evolution and the Recognition Concept of Species. Collected Writings. Baltimore: Johns Hopkins University Press.

Paterson, H. E. H. 1993b. The term "isolating mechanism" as a canalizer of evolutionary thought. In S. F. McEvey, ed., Evolution and the Recognition Concept of Species. Baltimore: Johns Hopkins University Press, pp. 1–10.

Poulton, E. B. 1903. What is a species? Proc. Entomol. Soc. London 1903:77–116.

Rhodes, F. H. T. 1956. The time factor in taxonomy. In P. C. Sylvester-Bradley, ed., The Species Concept in Palaeontology. London: The Systematics Association, pp. 33–52.

Rice, W. R. 1996. Sexually antagonistic male adaptation triggered by experimental arrest of female evolution. Nature 381:232–234.

Ridley, M. 1989. The cladistic solution to the species problem. Biol. Philos. 4:1–16.

Ridley, M. 1990. Comments on Wilkinson's commentary. Biol. Philos. 5:447–450.

Ridley, M. 1993. Evolution. Cambridge, Mass.: Blackwell Science.

Rogers, D. J., and Appan, S. G. 1969. Taximetric methods for delimiting biological species. Taxon 18:609–752.

Rosen, D. E. 1979. Fishes from the uplands and intermontane basins of Guatemala: Revisionary studies and comparative geography. Bull. Am. Mus. Nat. Hist. 162:267–376.

Simpson, G. G. 1951. The species concept. Evolution 5:285–298.

Simpson, G. G. 1961. Principles of Animal Taxonomy. New York: Columbia University Press.

Smith, A. B. 1994. Systematics and the Fossil Record. Documenting Evolutionary Patterns. Oxford: Blackwell Scientific Publications.

Sneath, P. H. A., and Sokal, R. R. 1973. Numerical Taxonomy. The Principles and Practice of Numerical Classification. San Francisco: Freeman.

Sober, E. 1984. Sets, species, and evolution: Comments on Philip Kitcher's "Species." Philos. Sci. 51:334–341.

Sokal, R. R., and Crovello, T. J. 1970. The biological species concept: A critical evaluation. Am. Nat. 104:127–153.

Sokal, R. R., and Sneath, P. H. A. 1963. Principles of Numerical Taxonomy. San Francisco: Freeman.

Sylvester-Bradley, P. C. 1956. The Species Concept in Palaeontology. London: The Systematics Association.

Templeton, A. R. 1981. Mechanisms of speciation—a population genetic approach. Annu. Rev. Ecol. Syst. 12:23–48.

Templeton, A. R. 1989. The meaning of species and speciation: A genetic perspective. In D. Otte and J. A. Endler, eds., Speciation and Its Consequences. Sunderland, Mass.: Sinauer, pp. 3–27.

Van Valen, L. 1976. Ecological species, multispecies, and oaks. Taxon 25:233–239.

Vrba, E. S. 1995. Species as habitat-specific, complex systems. In D. M. Lambert and H. G. Spencer, eds., Speciation and the Recognition Concept. Theory and Application. Baltimore: Johns Hopkins University Press, pp. 3–44.

Wagner, P. J., and Erwin, D. H. 1995. Phylogenetic patterns as tests of speciation models. In D. H. Erwin and R. L. Anstey, eds., New Approaches to Studying Speciation in the Fossil Record. New York: Columbia University Press, pp. 87–122.

Westoll, T. S. 1956. The nature of fossil species. In P. C. Sylvester-Bradley, ed., The Species Concept in Palaeontology. London: The Systematics Association, pp. 53–62.

White, C. S., Michaux, B., and Lambert, D. M. 1990. Species and neo-Darwinism. Syst. Zool. 39:399–413.

Wiley, E. O. 1978. The evolutionary species concept reconsidered. Syst. Zool. 27:17–26.

Wiley, E. O. 1981. Phylogenetics: The Theory and Practice of Phylogenetic Systematics. New York: Wiley.

Williams, G. C. 1966. Adaptation and Natural Selection. A Critique of Some Current Evolutionary Thought. Princeton, N.J.: Princeton University Press.

Wright, S. 1940. The statistical consequences of Mendelian heredity in relation to speciation. In J. Huxley, ed., The New Systematics. London: Oxford University Press, pp. 161–183.

Part III

Geography, Ecology, and Population Structure

6

Theory and Models of Sympatric Speciation

Paul A. Johnson
Urban Gullberg

In his 1963 book, *Animal, Species and Evolution*, Ernst Mayr threw down the gauntlet to proponents of sympatric speciation, challenging them to theoretically and empirically support their conjectures of nonallopatric divergence and speciation. Mayr's basic challenge to theory was to demonstrate that ecological divergence within a population can lead to speciation in the face of gene flow that is uninhibited except by genetic isolating mechanisms. He presented careful arguments showing that proponents would have to do better than past defenders of the "faith," or else they should cede to the universality of allopatric divergence.

John Maynard Smith (1962, 1965, 1966), A. D. Bazykin (1965), Guy Bush (1969), and many other biologists have accepted Mayr's challenge. Our intent in this chapter is not to exhaustively review the research on the subject of sympatric speciation over the ensuing years, except to say that progress has been made in understanding the population genetics of divergence under sympatric conditions. Rather, we focus on models of sympatric speciation with the objective of highlighting and integrating some of our theoretical understanding of the process. For empirical understanding and work, we refer the reader to other chapters in this book.

The Process of Sympatric Speciation

Sympatric speciation has been defined by Kondrashov and Mina (1986) as "a formation of species out of a population whose spatial structure is not important genetically" (p. 201), which can occur when individuals of diverging morphs within a population share the same "cruising range" (sympatry; Mayr, 1963). Population morphs diverge sympatrically in an ideal sense when the probability of mating between two individuals depends only on their genotypes. The outcome of a sympatric divergence process can be speciation in the strict sense, which occurs if the diverging habitat morphs become isolated reproductively such that gene flow between them ceases and becomes essentially impossible due to genotype-dependent homogamous mating.

The notion of sympatric speciation is based on the idea that a speciation event might somehow occur even when individuals of a population have ready access to alternative habitats within the same region of space. Here we use the word "habitat" loosely, such that two distinct resources in a homogeneous environment represent two "habitats" in same region of space. Models of sympatric speciation generally assume that each individual in the population selects one alternative habitat and then lives and utilizes resources in that habitat exclusively, which is the case for many species of parasites and herbivorous insects whose individuals have access to alternative host species. Models used by Seger (1985) and Doebeli (1996) are examples of exceptions that include no distinct habitats, but rather a resource that varies over a continuum.

Selection of habitat is assumed to be random under conditions of sympatry, unless a mechanism for nonrandom selection of habitat exists, such as habitat preference. Also, mate pairing of individuals in the population is assumed to be random, unless individuals mate within their selected habitat or some form of assortative mating exists.

Requirements and Agents for Sympatric Speciation

Sympatric speciation requires constraints that allow natural selection to drive divergence within the population. These constraints include the existence of habitat-specific factors involved in the regulation of population size independently within alternative habitats, which allows selection to be divergent or disruptive: divergent if mating partitions individuals into separate mating pools that correspond to different habitats (see Pimentel et al., 1967; Soans et al., 1974; Bush, 1982; Shaw and Platenkamp, 1993), and disruptive if mating takes place in a single mating pool (e.g., Felsenstein, 1981; Kondrashov, 1983a,b). (Another less traditional constraint is considered in the discussion section below.) The existence of alternative habitats and the nature of the effects of independent regulatory factors for each habitat depend considerably on the genetic and phenotypic variation within the population, which is ultimately largely determined by the population's genotype-phenotype map. Such maps are a product of (a) developmental constraints caused by the structure, character, composition, or dynamics of the developmental system (Maynard Smith et al., 1985), (b) architectures of epigenetic systems (Wagner, 1996) that make a finely integrated adult organism possible, and (c) "modularity" that imparts structure to genotype-phenotype maps through the evolution of reduced pleiotropic effects among characters serving different functions (Wagner and Altenberg, 1996).

Perhaps the next most fundamental requirement for the strict form of sympatric speciation is that of genotype-dependent structuring of mating to produce reproductively isolated population morphs even if individuals of different morphs happen to find themselves within a common habitat. This requirement can be met by means of genotype-dependent "homogamous mating," where individuals tend to pair and mate with individuals of the same or similar genotype or phenotype, due to ethologically determined preferences. Homogamous mating is often referred to as simply "assortative mating" in models of sympatric speciation that include only that form of assortative mating.

Other means of structuring mating to promote the development of reproductive isolation between population morphs are less direct, but can play a vital role. For example, habitat assortative mating can be a means of structuring mating, where each individual tends to mate within its selected habitat. If selection of habitat is genotype dependent, then habitat assortative mating can facilitate natural selection in driving sympatric divergence, since there can be variation between genotypes in a population with respect to "habitat fidelity," which is the probability of an individual of a genotype being found in the habitat in which it is most fit. The genetics of a population can cause selection of habitat to deviate from randomness under sympatry, not only by means of genotype-dependent habitat selection, hereafter called "habitat preference," where an individual prefers one or the other habitat independent of the habitat in which it is reared, but also by means of genotype-dependent habitat conditioning, where an individual's genotype influences the probability that an individual selects the habitat within which it was reared.

First Phase of Sympatric Speciation (Diversification)

Sympatric speciation involves two genetically different phases: first, diversification of portions of an ancestral population, followed by the evolution of reproductive isolation between those portions (Maynard Smith, 1966). Barton and Charlesworth (1984) describe the first phase as the splitting of an ancestral population, under selection, into populations located at different equilibria.

Grant (1963, pp. 448–451) outlines the fundamental nature of the genetics for sympatric formation of a new species via disruptive (or divergent) selection within a population whose individuals have access to alternative habitats. The population contains "habitat-divergent loci," or hd-loci for convenience (our terminology). Each hd-locus includes two alleles for facilitating adaptation, one allele for each habitat. Consequently, an hd-locus can contribute directly to divergence of morphs within the population. Hd-loci define "m-genotypes" (marginal genotypes), one for each alternative habitat (see figure 6.1).

Each m-genotype in the population is geared more than any other genotype for successful utilization of a given alternative habitat, which normally means that each

M-GENOTYPES IMPLEMENTED BY DISRUPTIVE OR DIVERGENT SELECTION

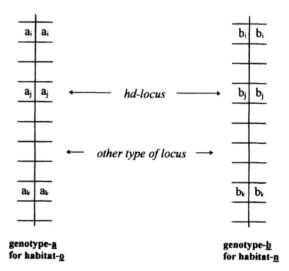

genotype-<u>a</u>
for habitat-<u>o</u>

genotype-<u>b</u>
for habitat-<u>n</u>

Figure 6.1. For a population experiencing disruptive or divergent selection because its individuals have access to two alternative habitats, two m-genotypes (marginal genotypes) exist, genotype-a and genotype-b, with optimal combinations of alleles at hd-loci (habitat-divergent loci) for habitat-specific utilization of habitat-o and habitat-n, respectively. At each hd-locus, of which there are three in the figure, an m-genotype is expected to be homozygous for an allele that most benefits utilization of its "designated habitat."

of its hd-loci is homozygous for an allele that, more than any other allele of the locus, contributes to the genotype's success in utilizing resources in that habitat. All genotypes other than the two m-genotypes are phenotypically intermediate (i-genotypes), with at least one of the hd-loci being heterozygous.

Natural selection is able to drive the process of sympatric divergence via disruptive or divergent selection because across-habitat fitness of each m-genotype is greater than that of any i-genotype. Hence, the dynamics of sympatric speciation essentially center around the capacity for the two m-genotypes to coexist and, with the assistance of reproductive isolating mechanisms, to allow disruptive or divergent selection to overcome recombination's opposition to progress in eliminating the i-genotypes.

Colonization

Utilization of both alternative habitats often requires that one of the habitats be colonized. Ludwig (1950) and Rosenzweig (1978) describe a process of colonization when mating is panmictic. That process is driven by disruptive or divergent selection. The idea has been elaborated by Pimm (1978), Udovic (1980), and more recently by Wilson and Turelli (1986) and Wilson (1989).

The colonization process can be outlined as follows (see table 6.1 for terms). Because selection is disruptive or divergent, two m-genotypes exist within the population: genotype-a, the genotype most capable of utilizing habitat-o, the original habitat, and genotype-b, the genotype most capable of utilizing habitat-n, the new alternative habitat. Genotype-a and genotype-b have across-habitat fitness values W_a and W_b, respectively. The across-habitat fitness value W_c for genotype-c, an i-genotype, is lower by definition.

Initially, genotype-b can be crudely adapted to habitat-n and maladaptive in habitat-o, due to fitness trade-offs. Its initial frequency can be very low compared to the frequency of genotype-a. However, if habitat-o is "full," and therefore highly competitive, and habitat-n is relatively "empty," and therefore less competitive, as in documented cases (Feder, 1995; Feder et al., 1995), then genotype-b's advantage in utilizing habitat-n where competition is low can compensate for its being crudely adapted to habitat-n. Genotype-b's across-habitat fitness value W_b may then be large enough that genotype-b becomes established and coexists with genotype-a. But polymorphism supported by disruptive or divergent selection is not inherently stable, unlike a balanced polymorphism, which is supported by stabilizing selection due to superior across-habitat fitness of intermediate genotypes (heterozygous superiority). The following model addresses means by which disruptive or divergent selection under sympatry can overcome this first major obstacle to driving sympatric divergence, namely, the lack of intrinsic stability of polymorphism.

Table 6.1. List of terms and names.

Term	Definition
hd-locus	Habitat-divergent locus—a locus tailored for coadaptation of a population to two alternative habitats under disruptive or divergent selection by including for each habitat a major alllele that facilitates utilization of that habitat through habitat-specific fitness, habitat preference, or mate preference.
m-genotype	Marginal genotype—a genotype, defined by a combination of alleles at hd-loci, that under disruptive or divergent selection is, more than any other genotype, geared to utilizing one of the alternative habitats and coexisting with a complementary m-genotype for the other alternative habitat in such a manner that together they have a disruptive or divergent selection advantage over all other genotypes (called i-genotypes). An m-genotype is expected to be homozygous at each hd-locus.
i-genotype	Intermediate genotype—any genotype in the population that is not an m-genotype and that is expected to be phenotypically intermediate to the m-genotypes and to include at least one hd-locus that is heterozygous. An i-genotype has a disruptive or divergent selection disadvantage to an m-genotype.
habitat-o	The alternative habitat that was the original habitat
habitat-n	New alternative habitat
genotype-k	An arbitrary genotype
genotype-a	The m-genotype for utilizing of habitat-o
genotype-b	The m-genotype for utilizing of habitat-n
genotype-c	A particular or arbitrary i-genotype for consideration in a theoretical analysis

Competition between Genotypes (m-Genotypes in Particular)

Wilson (1989) draws on earlier work, such as MacArthur (1972) and Rosenzweig (1978) to model the process of colonization described above and to show how stable polymorphism can be established when selection is not only disruptive but also density and frequency dependent. Here we elaborate further to illustrate other mechanisms, especially habitat preference, for reducing competition between m-genotypes so that polymorphism becomes and remains stable.

The across-habitat absolute fitness value W_k for genotype-k, an arbitrary genotype in the population, is defined by the equation

$$W_k = P_{k,n}(1 + s_{k,n}) + P_{k,o}(1 + s_{k,o}), \qquad (6.1)$$

where $s_{k,n}$ and $s_{k,o}$ are the within-habitat selection factors for genotype-k in habitat-n and habitat-o, respectively. $P_{k,n}$ and $P_{k,o}$ are the probabilities of an individual of genotype-k being found in habitat-n and habitat-o, respectively. Those probabilities are determined by the nature of selection of habitat for genotype-k, defined as:

$$u_{k,n,n} = A_k G_{k,n,n} + (1 - A_k)H_{k,n,n}, \qquad (6.2)$$

$$u_{k,n,o} = 1 - u_{k,n,n}, \qquad (6.3)$$

where $u_{k,n,n}$ is the probability that if genotype-k is in habitat-n when it is born, it will remain in that habitat, and $u_{k,n,o}$ is the probability that instead it will migrate to habitat-o.

$H_{k,n,n}$ is the probability that individuals of genotype-k if born in habitat-n stay in that habitat given that they do not take into account resource abundances in the two habitats. The value of $H_{k,n,n}$ depends upon which habitat genotype-k prefers (see table 6.2). In the expressions defining $H_{k,n,n}$ in table 6.2, the parameter $g_{k,n,n}$ is the probability that the phenotype of an individual of genotype-k will express its genotype regarding preference for a habitat. If $g_{k,n,n} = 0$, then even if individuals of genotype-k have a preferred habitat, that preference is never expressed, so in effect, the genotype has no habitat preference, and therefore individuals of the genotype born in habitat-n are as likely to leave as to remain in that habitat. At the other extreme, if $g_{k,n,n} = 1$, an individual of genotype-k not only has a preference for habitat-n, but is also certain to express that preference, meaning that an individual born in habitat-n is virtually certain to remain in that habitat. Genotype-dependent habitat preference is commonly defined in this manner (Felsenstein, 1981; Diehl and Bush, 1989; Johnson et al., 1996).

Let R_n and R_o denote resource abundance (or resource density) within habitat-n and habitat-o, respectively, and

let r_n and r_o denote the proportion of total resources, per unit region of space, provided by habitat-n and habitat-o, respectively [i.e., $r_n = R_n/(R_n + R_o)$, and $r_o = 1 - r_n$]. With probability A_k, an individual of genotype-k selects a habitat based upon the relative abundance of resources in the two alternative habitats. This manner of selecting a habitat represents a form of genotype-dependent habitat selection that is neither habitat preference nor habitat conditioning, but rather "resource-abundance habitat selection." Ecologists include various forms of resource-abundance habitat selection in theories of interspecific density-dependent optimal habitat selection (e.g., Rosenzweig, 1979, 1981, 1985), intraspecific habitat selection (e.g., Fretwell and Lucas, 1970; Fretwell, 1972), and optimal patch use (e.g., Charnov, 1976; Brown, 1988).

If an individual of genotype-k takes only into account resource abundance in selecting a habitat ($A_k = 1$), then the probability $u_{k,n,n}$ of staying in habitat-n (see equation 6.2) is given by

$$G_{k,n,n} = \frac{|0.5 - r_n|}{0.5}[r_n(1 - r_n A_k)H_{k,n,n} + r_n A_k] + \left(1 - \frac{|0.5 - r_n|}{0.5}\right)H_{k,n,n} \qquad (6.4)$$

which depends on the distribution of resources in the two habitats in terms of resource abundance, r_n and r_o, in habitat-n and habitat-o, respectively. Given that $A_k = 1$, if $r_n = 0$ so that all the resources are in habitat-o, then $G_{k,n,n} = 0$. Inversely, if $r_n = 1$, so that all the resources are in habitat-n, then $G_{k,n,n} = 1$. However, if $r_n = r_o = 0.5$, so that resource abundance is exactly the same in both habitats, then individuals of genotype-k select a habitat in the same manner as they would if resource abundance in the two habitats was not taken into account (i.e., if $A_k = 0$), in which case $G_{k,n,n} = H_{k,n,n}$.

We define $u_{k,o,o}$ to be the probability that an individual of genotype-k born in habitat-o stays in that habitat, and $u_{k,o,n}$ to be the probability that it leaves instead of stays. Analogous to equations 6.2 and 6.3, we have

$$u_{k,o,o} = A_k G_{k,o,o} + (1 - A_k)H_{k,o,o}, \qquad (6.5)$$

$$u_{k,o,n} = 1 - u_{k,o,o}. \qquad (6.6)$$

Here $H_{k,o,o}$, which is analogous to $H_{k,n,n}$, is the probability that an individual of genotype-k if born in habitat-o stays in that habitat given that it disregards the distribution of amounts of resources between the two habitats (see table 6.2 for computation of $H_{k,o,o}$).

Following MacArthur's (1972) definitions of across-habitat fitness for consumer population dynamics, the selection factors $s_{k,n}$ and $s_{k,o}$ in equation 6.1 can be defined as

$$s_{k,n} = ec_{k,n}R_n - T_{k,n}, \qquad (6.7)$$

$$s_{k,o} = ec_{k,o}R_o - T_{k,o}, \qquad (6.8)$$

Table 6.2. How $H_{k,n,n}$ is calculated, depending on which habitat individuals of genotype-k prefer.

	Prefers Habitat-n	Prefers Habitat-o
$H_{k,n,n} =$	$0.5(1 - g_{k,n,n}) + g_{k,n,n}$	$0.5(1 - g_{k,n,n})$
$H_{k,o,o} =$	$0.5(1 - g_{k,o,o})$	$0.5(1 - g_{k,o,o}) + g_{k,o,o}$

For definitions of $H_{k,n,n}$, $H_{k,o,o}$, $g_{k,n,n}$, and $g_{k,o,o}$, see text.

where, in habitat-o for example, the genotype-dependent death rate $T_{k,o}$ of genotype-k is constant. The expected birth rate $ec_{k,o}R_o$ for individuals of genotype-k within habitat-o is a product of the resource abundance R_o in that habitat, the genotype-dependent consumption rate $c_{k,o}$ for genotype-k in that habitat, and a factor e that converts units of resource into units of consumer.

Consider now competition between the two m-genotypes. If individuals of the genotype-b can consume resources in habitat-n faster than individuals of the genotype-a, and analogously for individuals of genotype-a in habitat-o (i.e., $c_{b,n} > c_{a,n}$, and $c_{a,o} > c_{b,o}$), then frequency-dependent selection is an important factor in promoting stable polymorphism for fitness hd-loci: when genotype-b has a low frequency across habitats relative to genotype-a, it then has a competitive advantage over genotype-a because resources in habitat-n are relatively plentiful due to the low frequency of genotype-b relative to that of genotype-a. Pimm (1978), Udovic (1980), Wilson and Turelli (1986), and Wilson (1989) analyze the nature of such complementation of frequency-dependent selection and disruptive selection in promoting stable polymorphisms.

Following Wilson (1989), the equilibrium density $\hat{R}_n$ of resources in habitat-n is defined by

$$\hat{R}_n = \frac{K_n}{N \sum_k x_k c_{k,n}} \qquad (6.9)$$

so that the equilibrium density is a function of (a) genotype frequencies (x_k for genotype-k), (b) resource consumption rates ($c_{k,n}$ for genotype-k), and (c) the density N of the consumer population. K_n is the density of resource n in the absence of consumption. Implicit in equation 6.9 is an assumption, namely, change in densities of resources is fast relative to change in densities of consumer genotypes, so resource densities equilibrate to genotype densities, even when the later is not in equilibrium (see MacArthur, 1972). The equilibrium density $\hat{R}_o$ of resources in habitat-o is defined analogously.

Mechanisms for Increasing the Stability of Polymorphism

The model illustrates several mechanisms for increasing the ability of the m-genotypes to coexist, which as stated earlier, increases the stability of polymorphism for the population when selection is disruptive or divergent.

Habitat Preference

Levene (1953) suggested that when the individuals of genotypes move to habitats where their fitness is greatest, then conditions are more favorable for stable polymorphism under disruptive selection. Later, Maynard Smith (1966) confirmed Levene's conclusion.

Habitat preference is a fundamental means for individuals of an m-genotype to move to the alternative habitat most favorable for their fitness. Since individuals of the two m-genotypes prefer different alternative habitats, and since population size is regulated independently in each habitat, habitat preference reduces competition between the m-genotypes. When habitat preference is strong, each m-genotype succeeds independently by utilizing "its own" habitat.

More formal and detailed treatments of the role of genotype-dependent habitat selection in establishing stable polymorphisms include Levene (1953), Dempster (1955), Maynard Smith (1962, 1966), Maynard Smith and Hoekstra (1980), Jones and Probert (1980), Templeton and Rothman (1981), Hoekstra et al. (1985), Garcia-Dorado (1986, 1987), Hedrick (1986, 1990a,b, 1993), Diehl and Bush (1989), Liberman and Feldman (1989), and De Meeus et al. (1993).

Habitat-Specific Survival Rates

Large survival differences within each habitat between m-genotypes ($T_{b,n} >> T_{a,n}; T_{a,o} >> T_{b,o}$) also increases the stability of polymorphism (see Maynard Smith, 1962). Consider the extreme case: if individuals of each m-genotype can survive, but only within "their own" habitat, competition between the two genotypes is in name only, for there is no significant contest, because neither m-genotype can replace the other.

Habitat-Specific Consumption Rates

Habitat-specific differences between m-genotypes in rates of consumption of resources ($c_{b,n} >> c_{a,n}; c_{a,o} >> c_{b,o}$) also increases the stability of polymorphism for essentially the same reason as habitat-specific survival rates: if individuals of each m-genotype can consume resources, but only within "their own" habitat, competition between the m-genotypes disappears.

Also, if habitat-specific consumption rates influence abundance of resources within each habitat separately, as in the model above, then frequency-dependent resource abundance $\hat{R}_n$ results in frequency-dependent selection (see Wilson, 1989) that contributes further to the capacity of the two m-genotypes to coexist. Specifically, if one m-genotype becomes more frequent than the other, the resources for which it enjoys a consumption advantage can be depleted, leaving resources for which it has a consumption disadvantage. Hence, increase in frequency of either m-genotype beyond a point is then impossible, and therefore replacement of either by the other is impossible. This mechanism for increasing stability of polymorphism is enhanced by resource-abundance habitat selection.

Mutual Enhancement of Factors Contributing to Stable Polymorphism

As the strength of habitat preference increases for an m-genotype, selection pressure for increase in habitat-specific survival rates and consumption rates increases, and vice versa. Again, consider the extremes. If individuals of each m-genotype are virtually certain to select their preferred habitat, there is a high selection pressure for fitness and consumption that is specific to that habitat. Similarly, if individuals of each m-genotype can only survive within "their own" habitat, or can only consume resources in that habitat, there is a high selection pressure for habitat preference.

In other words, an increase in any one of the three habitat-specific adaptation traits for the m-genotypes, habitat preference, habitat-specific survival rates, or habitat-specific consumption rates, increases the selection advantage of the other two traits. Any accommodation of those selection advantages by successful mutations improves the capacity of coexistence for the m-genotypes and therefore the stability of polymorphism.

Second Phase of Sympatric Speciation (Reproductive Isolation)

As stable polymorphisms develop, the shape of the adaptive landscape evolves and becomes two peaks separated by a valley of low fitness (see Rosenzweig, 1978; Wilson, 1989). The valley of low fitness represents disruptive (or divergent) selection against the i-genotypes. The slopes to the peaks signify directional selection.

Wilson (1989) points out that this form of adaptive landscape does not by itself produce a structuring of the population in the form of a bimodal distribution of the phenotypes. Although natural selection acts as a force to eliminate phenotypically intermediate individuals (i-genotypes and therefore heterozygotes), sexual reproduction, through the randomness of segregation and recombination, repeatedly recreates them (see Felsenstein, 1979; Rice, 1984, 1987; Kondrashov, 1986; Diehl and Bush, 1989). Hence, individuals of i-genotypes can continue to outnumber those of the m-genotypes. An "isolating mechanism" is needed for sympatric divergence to overcome its second major obstacle: the opposition of recombination. Isolating mechanisms assist natural selection in eliminating i-genotypes such that two population morphs located at different equilibria (peaks) emerge, each consisting of a single m-genotype.

Mechanisms for Reproductive Isolation

Sympatric speciation in the strict sense is possible only if (a) the two m-genotypes are reproductively isolated from one another (Kondrashov, 1986; Johnson et al., 1996), and (b) the selective advantage of those two genotypes over i-genotypes, together with isolating mechanisms, are sufficient to eliminate the i-genotypes despite the opposition of recombination.

Same Loci for Selection and Reproductive Isolation

In theory, alleles at fitness hd-loci can also drive reproductive isolation, if they also influence mate pairing or habitat preference. For example, in models studied by Kondrashov (1983a,b), the alleles at fitness hd-loci determine not only a genotype's habitat-specific fitness, but also its genotype-dependent bias in mate pairing. More specifically, a population of birds may have access to two habitats, one habitat with large seeds that might require a large broad beak for optimal foraging efficiency (rates of consumption), and the other habitat with small seeds that might require a small narrow beak for optimal foraging efficiency. Certain fitness hd-loci responsible for variation in beak size and shape may also be responsible for variation in female display characters, as a correlated trait, if those characters include beak size and shape. Other hd-loci, which may or may not be fitness hd-loci, are then needed for variation in beak preference by males with regard to beaks' of females (see Grula and Taylor, 1979; Roelofs et al., 1987).

Also, the alleles at habitat preference hd-loci can be partially responsible for both fitness and habitat preference, if m-genotypes, which have strong preference for an alternative habitat, experience superior fitness to i-genotypes, with weaker preference for either habitat (see Rice, 1984). However, many biologists (see Maynard Smith, 1966; Udovic, 1980; Felsenstein, 1981; Diehl and Bush, 1989) believe that, in general, loci controlling habitat-specific fitness are likely to be separate from loci controlling isolation agents.

Different Loci for Selection and Reproductive Isolation

If reproductive isolation is controlled by hd-loci other than fitness hd-loci, two possibilities exist (see Felsenstein, 1981). In the "one-allele" case for isolating mechanisms, both m-genotypes carry the same isolating allele, since isolation proceeds by substituting the same allele in both population morphs. A one-allele isolating locus for reproductive isolation can be regarded a modifier locus if it has no influence on genotype viabilities (Liberman and Feldman, 1989).

In the "two-allele" case, the m-genotypes carry different isolating alleles, since isolation proceeds by substituting different alleles for the two population morphs. As an isolating mechanism, habitat preference is expected to be two-allele, inasmuch as it is hard to imagine how the same allele will produce effective preference for different habitats. On the other hand, isolating mechanisms such as habitat conditioning, homogamous mating, and resource-abundance habitat selection can be either one- or two-allele.

For one-allele isolating systems or models (e.g., Maynard Smith, 1966; Balkau and Feldman, 1973; Udovic,

1980; Kondrashov, 1986; De Meeus et al., 1993), recombination does not oppose increase in frequency of isolating alleles (at "modifier loci") if those alleles have the same frequencies in both population morphs. Hence, recombination need not oppose reproductive isolation in one-allele models. Consequently, selection can be effective even when weak (see Balkau and Feldman, 1973).

In contrast, for two-allele models (e.g., Maynard Smith, 1966; Felsenstein, 1981; Rice, 1984; Diehl and Bush, 1989; Johnson et al., 1996), recombination opposes reproductive isolation, since it opposes substitution of different alleles for the two morphs of a population when there is gene flow between them. Nevertheless, many biologists find the occurrence in nature of two-allele isolating mechanisms easier to imagine than one-allele mechanisms.

Reproductive Isolation without Habitat Fidelity

Wilson (1989) suggests three means of eliminating i-genotypes when there is no habitat fidelity, namely, (a) closer linkage, (b) evolution of dominance (see De Meeus et al., 1993) and other epistasis modifiers that cause heterozygotes to resemble homozygotes phenotypically, and (c) genotype-dependent homogamous mating. Wilson goes on to explain that, regarding the first means, although linkage can reduce the number of intermediate phenotypic forms, due to segregation it can not alone eliminate them if mating is random. Regarding the second means, although evolution of dominance and epistatic modifiers can lead to just two phenotypic forms, it reduces selection pressure for homogamous mating, which is ultimately needed for complete elimination of i-genotypes. On the other hand, closer linkage will normally increase that selection pressure.

Genotype-dependent homogamous mating, the third suggested means of eliminating i-genotypes, is a fundamental mechanism of assisting natural selection in overcoming recombination's opposition to increase in frequency of the m-genotypes at the expense of frequency of i-genotypes. However, models that include homogamous mating as the only isolating mechanism reveal that a relatively high threshold amount of across-habitat fitness advantage of the two m-genotypes over i-genotypes can be necessary to overcome the swamping effect of recombination (Felsenstein, 1981; Rice 1984, 1987; Kondrashov, 1986), if homogamous mating is incomplete for the m-genotypes such that individuals of those genotypes sometimes mate with individuals of i-genotypes or individuals of the other m-genotype.

One-Allele Systems. As discussed earlier, for one-allele systems or models (e.g., Maynard Smith, 1966; Dickinson and Antonovics, 1973; Udovic, 1980; Kondrashov, 1986; Failkowski, 1988, 1992), establishment of homogamous mating alleles is unopposed by recombination. Nevertheless, without habitat fidelity, then even with reasonably high levels of homogamous mating, Kondrashov (1986) found with a Monte Carlo model that the amount of disruptive selection required to overcome recombination's opposition to elimination of i-genotypes was no less than 10%, and grew with the number of fitness hd-loci.

Two-Allele Systems. For two-allele systems or models (see Felsenstein, 1981; Diehl and Bush, 1989) that only include homogamous mating as an isolating mechanism, the across-habitat fitness of the m-genotypes must be sufficiently greater than that of i-genotypes to overcome recombination acting to break up the allele combinations of m-genotypes not only across the fitness hd-loci, but also across homogamous mating hd-loci. Indeed, for two-allele systems, each m-genotype is vulnerable to reduction in "true breeding" due to recombination events that occur freely across the entire set of hd-loci.

Linkage disequilibrium between homogamous mating hd-loci and fitness hd-loci is a measure of progress achieved by disruptive or divergent selection in reducing the proportion of i-genotypes to m-genotypes. However, Felsenstein (1981) and Diehl and Bush (1989) found that no linkage disequilibrium develops between homogamous mating hd-loci and fitness hd-loci, irrespective of the strength of selection against i-genotypes, if the fitness hd-loci interact additively. Nevertheless, if there are at least two fitness hd-loci that interact multiplicatively, then very strong disruptive or divergent selection and nonhabitat assortative mating can generate some linkage disequilibrium. But as Felsenstein points out, such cases are unrealistic.

On the other hand, if reproductive isolation is already pronounced, as in cases of secondary contact after allopatric divergence, then the remaining very small proportion of i-genotypes can be eliminated under sympatric conditions. The conditions and requirements for such completion of speciation under sympatry has been analyzed by Kondrashov (1983a,b, 1986).

Reproductive Isolation with Habitat Fidelity

To assess the capacity of habitat preference to facilitate elimination of i-genotypes, consider again the model defined earlier. The across-habitat fitness values W_c for genotype-c, an arbitrary i-genotype, and W_b for genotype-b, an m-genotype, are given by

$$W_c = P_{c,n}(1 + s_{c,n}) + P_{c,o}(1 + s_{c,o}), \qquad (6.10)$$

$$W_b = P_{b,n}(1 + s_{b,n}) + P_{b,o}(1 + s_{b,o}), \qquad (6.11)$$

(see equation 6.1), with $W_b \geq W_c$, by definition. The across-habitat fitness advantage of genotype-b over genotype-c is then

$$W_b - W_c = P_{b,n}s_{b,n} + P_{b,o}s_{b,o} - P_{c,n}s_{c,n} - P_{c,o}s_{c,o}. \quad (6.12)$$

To avoid unnecessary complication, assume that the carrying capacities, which are represented by amounts of resource, are the same in both habitats, and that $r_b = r_c = 0.5$, so that there is no resource-abundance habitat selection. Consider first the case of both genotypes having no habitat preference (i.e., $g_{b,n,n} = g_{b,o,o} = g_{c,n,n} = g_{c,o,o} = 0$), so that habitat selection is random, and therefore $P_{b,n} \cong P_{c,n} \cong P_{b,o} \cong P_{c,o} \cong 0.5$. Then the across-habitat fitness advantage for genotype-b over genotype-c is given by

$$q = \hat{W}_b - \hat{W}_c \cong 0.5(s_{b,n} + s_{b,o}) - 0.5(s_{c,n} + s_{c,o}), \quad (6.13)$$

which represents the difference between the across-habitat fitness values for genotype-b and genotype-c. Compare this case to the case of fullest possible benefit of habitat preference for increasing the across-habitat selective advantage of genotype-b over genotype-c: that is, habitat preference is complete for genotype-b (i.e., $g_{b,n,n} = g_{b,o,o} = 1$ so that $P_{b,n} \cong 1$, $P_{b,o} \cong 0$) and absent for genotype-c (i.e., $g_{c,n,n} = g_{c,o,o} = 0$ so that $P_{c,n} = P_{c,o} = 0.5$). The across-habitat fitness advantage of genotype-b over genotype-c is then

$$Q = \hat{W}_b - \hat{W}_c \cong s_{b,n} - 0.5(s_{c,n} + s_{c,o}). \quad (6.14)$$

This means that habitat preference can increase the selective advantage of an m-genotype over an i-genotype by as much as

$$Q - q \cong 0.5(s_{b,n} - s_{b,o}), \quad (6.15)$$

which, in our example, is half the difference between the fitness values of genotype-b in the two alternative habitats.

Several models confirm the capacity of habitat fidelity to act effectively as the only isolating mechanism in assisting natural selection in driving sympatric divergence (e.g., Maynard Smith, 1966; Rice, 1984; Diehl and Bush, 1989), especially during early phases of the process. Seger (1985) takes the interesting next step of examining the synergism between resource fidelity and homogamous mating acting together as isolating mechanisms for promoting sympatric divergence of a haploid population when selection is disruptive. In a somewhat similar study, Johnson et al. (1996) examine the synergism between habitat fidelity and homogamous mating acting together as isolating mechanisms for a diploid population undergoing divergent selection.

Large Numbers of Habitat-Divergent Loci (Trade-Offs)

Models (Kondrashov, 1986; Johnson et al., 1996) show that for any habitat-divergent trait, an increase in number of hd-loci controlling the trait allows a lower contri-

bution per locus to achieve sympatric speciation within a given amount of time. However, as the number of hd-loci increases, the across-habitat fitness values of more and more i-genotypes become closer and closer to the across-habitat fitness values of the m-genotypes, so that fitness differences between the two m-genotypes and i-genotypes becomes increasingly blurred. Hence, the effects of disruptive or divergent selection and homogamous mating are weakened, as confirmed by Dickinson and Antonovics (1973), Kondrashov (1986) and Johnson et al. (1996). This weakening results in a longer time for eliminating the i-genotypes and a higher required threshold difference between the across-habitat fitness values of the two m-genotypes and the mean across-habitat fitness value of the i-genotypes. Also, the capacity of recombination to oppose selection increases as numbers of hd-loci increases, because the number of recombination sites increases. Therefore, beyond a point (but see Johnson et al., 1996), increase in number of hd-loci contributing to sympatric divergence provides diminishing returns.

Origin of Marginal Genotypes

To eliminate intermediates and the possibility of gene flow between habitat morphs, speciation must already have partially occurred inasmuch as two m-genotypes must exist within the population, which are unable to mate with one another due to homogamous mating. Hence, critical questions are how, when, and why m-genotypes, unable to mate with one another, come into existence or already exist within a population.

These questions have yet to be explicitly addressed, as far as we know, although the work of Doebeli (1996) does indirectly pertain to the questions by showing how a population utilizing a single habitat or a resource distributed unimodally can include a bimodal phenotype distribution represented by population morphs that are partially reproductively isolated through homogamous mating. Thus, should an alternative habitat become available, the parent population might already contain m-genotypes that are reproductively isolated.

The sympatric divergence mechanism described by Doebeli (1996) includes the aesthetic of clarifying the role of ecological character displacement (Slatkin, 1980) in sympatric divergence. At the same time, the existence of that mechanism rests on the assumption that assortative mating is based on the character that determines ecological interactions (outcomes of competition). Likely satisfaction of that assumption in nature is questioned by many evolutionary biologists, as mentioned earlier regarding the feasibility of an allele or locus controlling both fitness and isolating mechanisms.

In the course of a process of sympatric speciation, naturally occurring evolution of selection pressures is a direct and perhaps fundamental means for a population to acquire m-genotypes that are unable to mate successfully with one another. Selection pressures and consequences of those pressures for habitat-specific fitness, habitat fidelity, and homogamous mating are synergistic: increase in the level of each, due to a successful mutation, increases the selection pressure for success of mutant genotypes that increase the level of the other two. Thus, a natural spiraling of increasing levels of all three traits is expected for the m-genotypes of a population experiencing conditions that initiate sympatric divergence. Due to the spiraling increase in levels of the three primary traits, as long as the conditions for sympatric divergence continue, the sympatric speciation process will include the evolution of m-genotypes that are unable to mate successfully with one another, and sympatric divergence will be driven to completion by elimination of all i-genotypes.

For simplicity and mathematical tractability especially, models for sympatric speciation, like most evolutionary models, assume that selection pressures are constant rather than coevolving. Thus, models have yet to explicitly incorporate the fundamental dynamic of spiraling of trait levels for sympatric divergence processes, although the models of Seger (1985) and Johnson et al., (1996) point distinctly in that direction.

Discussion

As stated early in this chapter, sympatric speciation requires constraints that allow natural selection to drive divergence within the population. A fundamental constraint is assumed for most theoretical considerations of sympatric divergence, namely, the existence of independent factors involved in the regulation of population size within alternative habitats, which allows selection to be truly disruptive or divergent.

Doebeli (1996) has identified another alternative basic configuration of constraints: the joint effect of (a) genotype-dependent homogamous mating and (b) frequency-dependent competition between genotypes. This joint effect extends the theory of competitive speciation (Rosenzweig, 1978, 1995; Wilson, 1989). Even when a single resource is unimodally distributed, this configuration of constraints can drive a bimodal character distribution, and also then sympatric divergence of population morphs whose "m-genotypes" are at least partially reproductively isolated through homogamous mating. For this configuration of constraints, it may well be that selection pressure for increase in strength of homogamous mating for each of the two m-genotypes can be sufficient to isolate those genotypes reproductively, in which case sympatric speciation in the strictest sense is in effect the result, even though i-genotypes are able to continue to exist and allow gene flow between the population morphs due to underuse of resources.

We return to the more traditional conception of a fundamental constraint underlying the dynamics of sympatric speciation, namely, the existence of habitat-specific factors independently regulating population size within alternative habitats, with disruptive or divergent selection driving divergence. With respect to this constraint, Maynard Smith (1965, p. 23) argues for the plausibility of sympatric speciation based on three observations. (1) If mating is random and individuals do not select habitats, then stable polymorphism for fitness hd-loci, driven by disruptive selection, is unlikely, since it normally can only be caused by very intense selection pressures. (2) However, if there is some degree of preference in selecting alternative habitats, then the chances of a stable polymorphism are greatly improved. (3) Once a stable polymorphism of this kind is established, there will be strong selection in favor of reproductive isolation by mating preferences, and plenty of time for such selection to be effective in bringing about speciation. He then concluded that there is no theoretical reason why sympatric speciation should not take place.

Maynard Smith's (1965) rationale for the dynamics of sympatric divergence has basically been confirmed. The likely presence of habitat preference to assist disruptive or divergent selection in sympatric divergence is supported by Kawecki (1996), who shows that even transient polymorphism caused by beneficial mutations spreading to fixation at fitness loci, other than hd-loci, results in selection pressure favoring habitat preference over random dispersal, provided the mutations affect fitness in a habitat-specific manner. Also, Liberman and Feldman (1989), who studied a model involving a multiple-allele modifier of migration rate between two habitats, found that although the modifier locus and a fitness hd-locus may achieve linkage equilibrium in each of the habitats (Hardy-Weinberg equilibrium), the equilibrium is externally unstable to alleles that reduce migration rates. Models for haploid (Diehl and Bush, 1989) and diploid populations (Johnson et al., 1996) clarify the role of habitat preference in facilitating sympatric speciation, and demonstrate its importance in implementing conditions required for homogamous mating to drive the process to completion.

Research on natural host races and sympatric sister species, comparative phylogenetic analyses, and laboratory experiments (see Rice and Hostert, 1993) have strengthened the case for sympatric speciation playing a significant role in the origin of species (Bush, 1993) and have generally supported Maynard Smith's conception of the dynamics. Bush (1994) points out that sympatric speciation can occur because traits evolving in response to divergent selection experienced by subpopulations adapting to different habitats provide sufficient intrinsic premating isolation for initiating sympatric divergence through habitat shift in animals that mate within a preferred habitat.

Maynard Smith's last point, that once a stable polymorphism is established, and supported by habitat preference, there will be strong selection in favor of reproductive isolation by mating preferences (homogamous mating), is confirmed indirectly with models by Seger (1985) and Johnson et al. (1996). Moreover, models show that if homogamous mating by the marginal genotypes is strong enough to make sympatric speciation in the strict sense possible, either due to genotype-dependent habitat preference or genotype-dependent homogamous mating, then the process can, with realistic levels of selection, take fewer than 1,000 generations, and often considerably fewer (Rice, 1984; Kondrashov, 1986; Johnson et al., 1996).

Acknowledgments   Drs. Jeffrey Feder, Stewart Berlocher, Guy Bush, and Alexey Kondrashov provided helpful criticism and comments. This work was supported by the Swedish National Board for Industrial and Technical Development.

References

Balkau, B., and Feldman, M. W. 1973. Selection from migration modification. Genetics 74:171–174.

Barton, N. H., and Charlesworth, B. 1984. Genetic revolutions, founder effects, and speciation. Annu. Rev. Ecol. Sys. 15:133–164.

Bazykin, A. D. 1965. On the possibility of sympatric species formation (Russian). Byulleten Moskovskogo Obshchestva Ispytateley Prirody Otdel Biologicheskly 70(1):161–165.

Brown, J. S. 1988. Patch use as an indicator of habitat preference, predation risk, and competition. Behav. Ecol. Sociobiol. 22:37–47.

Bush, G. L. 1969. Sympatric host race formation and speciation in frugivorous flies of the genus *Rhagoletis*. Evolution 23:237– 251.

Bush, G. L. 1982. What do we really know about speciation? *In* Perspectives on evolution (R. Milkman, ed.), pp. 119–128. Sinauer, Sunderland, Mass.

Bush, G. L. 1993. A reaffirmation of Santa Rosalia, or why are there so many kinds of small animals? *In* Evolutionary patterns and processes (D. Edwards and D. R. Lees, eds.), pp. 229–249. Academic Press, New York.

Bush, G. L. 1994. Sympatric speciation in animals: new wine in old bottles. TREE 9:285–288.

Charnov, E. L. 1976. Optimal foraging: the marginal value theorem. Theor. Pop. Biol. 9:129–136.

De Meeus, T., Michalakis, Y., Renaud, F., and Olivieri, I. 1993. Polymorphism in heterogeneous environments, evolution of habitat selection and sympatric speciation: soft and hard selection models. Evol. Ecol. 7:175–198.

Dempster, E. R. 1955. Maintenance of genetic heterogeneity. Cold Spring Harbor Symp. Quant. Biol. 20:25–32.

Dickinson, H., and Antonovics, J. 1973. Theoretical considerations of sympatric divergence. Am. Nat. 107:256–274.

Diehl, S. R., and Bush, G. L. 1989. The role of habitat preference in adaptation and speciation. In Speciation and its consequences (D. Otte and J. Endler, eds.), pp. 345–365. Sinauer, Sunderland, Mass.

Doebeli, M. 1996. A quantitative genetic competition model for sympatric speciation. J. Evol. Biol. 9:893–909.

Feder, J. L. 1995. The effects of parasitoids on sympatric host races of *Rhagoletis pomonella* (Diptera: Tephritidae). Ecology 76:801–813.

Feder, J. L., Reynolds, K., Go, W., and Wang, E. C. 1995. Intra- and interspecific competition and host race formation in the apple maggot fly, *Rhagoletis pomonella* (Diptera: Tephritiedae). Oecologia (Berlin) 101:416–425.

Felsenstein, J. 1979. Excursions along the interface between disruptive and stabilizing selection. Genetics 93:773–795.

Felsenstein, J. 1981. Skepticism towards Santa Rosalia, or why are there so few kinds of animals. Evolution 35:124–138.

Fialkowski, K. R. 1988. Lottery of sympatric speciation-a computer model. J. Theor. Biol. 130:379–390.

Fialkowski, K. R. 1992. Sympatric speciation: a simulation model of imperfect assortative mating. J. Theor. Biol. 157:9–30.

Fretwell, S. D. 1972. Populations in a seasonal environment. Princeton University Press, Princeton, N.J.

Fretwell, S. D., and Lucas, H. L. 1970. On territorial behavior and other factors influencing habitat distribution in birds. I. Theoretical development. Acta Biotheor. 14:16–36.

Garcia-Dorado, A. 1986. The effect of niche preference on polymorphism protection in a heterogeneous environment. Evolution 40:936–945.

Garcia-Dorado, A. 1987. Polymorphism from environmental heterogeneity: some features of genetically induced niche preference. Theor. Pop. Biol. 32:66–75.

Grant, V. 1963. The origin of adaptations. Columbia University Press, New York.

Grula, J. W., and Taylor, O. R., Jr. 1979. The inheritance of pheromone production in the sulphur butterflies *Colias eurytheme* and *C. Philodice*. Heredity 42:359–371.

Hedrick, P. W. 1986. Genetic polymorphism in heterogeneous environments: a decade later. Annu. Rev. Ecol. Syst. 17:535–566.

Hedrick, P. W. 1990a. Theoretical analysis of habitat selection and maintenance of genetic variation. *In* Ecological and evolutionary genetics of *Drosophila* (J. S. F. Barker, W. T. Starmer, and R. J. MacIntyre, eds.), pp. 209–227). Plenum, New York.

Hedrick, P. W. 1990b. Genotype-specific habitat selection: a new model. Heredity 65:145–149.

Hedrick, P. W. 1993. Sex-dependent habitat selection and genetic polymorphism. Am. Nat. 141:491–500.

Hoekstra, R. F., Bijlsma, R., and Dolman, A. J. 1985. Polymorphism from environmental heterogeneity: models are only robust if the heterozygote is close in fitness to the favored homozygote in each environment. Genet. Res. 45:299–314.

Johnson, P. A., Hoppensteadt, F. C., Smith, J. J., and Bush, G. L. 1996. Conditions for sympatric speciation: a diploid model incorporating habitat fidelity and nonhabitat assortative mating. Evol. Ecol. 10:187–205.

Jones, J. S., and Probert, R. F. 1980. Habitat selection maintains a deleterious allele in a heterogeneous environment. Nature (London) 287:632–633.

Kawecki, T. J. 1996. Sympatric speciation driven by beneficial mutations. Proc. R. Soc. Lond. B 263:1515–1520.

Kondrashov, A. S. 1983a. Multilocus model of sympatric speciation. I. One character. Theor. Pop. Biol. 24:121–135.

Kondrashov, A. S. 1983b. Multilocus model of sympatric speciation. II. Two characters. Theor. Pop. Biol. 24:121–135.

Kondrashov, A. S. 1986. Multilocus model of sympatric speciation. III. Computer simulations. Theor. Pop. Biol. 29: 1–15.

Kondrashov, A. S., and Mina, M. V. 1986. Sympatric speciation: when is it possible? Biol. J. Linn. Soc. 27:201–223.

Levene, H. 1953. Genetic equilibrium when more than one ecological niche is available. Am. Nat. 87:331–333.

Liberman, U., and Feldman, M. W. 1989. The reduction principle for genetic modifiers of the migration rate. *In* Mathematical evolutionary theory (M. W. Feldman, ed.), pp. 111–144. Princeton University Press, Princeton, N.J.

Ludwig, W. 1950. Zur Theorie der Konkurrenz. Die Annidation (Einnischung) als funfter Evolutionsfaktor. Neue Erbeg. Probleme Zool, Klatt-Restschrift 1950:516–537.

MacArthur, R. H. 1972. Geographical Ecology. Harper and Row, New York.

Maynard Smith, J. 1962. Disruptive selection, polymorphism and sympatric speciation. Nature 195:60–62.

Maynard Smith, J. 1965. Mr. J. Maynard Smith (comments). Proc. R. Entomol. Soc. Lond. 30:22–23.

Maynard Smith, J. 1966. Sympatric speciation. Am. Nat. 100: 637–650.

Maynard Smith, J., and Hoekstra, R. 1980. Polymorphism in a varied environment: how robust are the models? Genet. Res. Camb. 35:45–57.

Maynard Smith, J., Burian, R., Kauffman, S., Alberch, P., Campbell, J., Goodwin, B., Lande, R., Raup, D., and Wolpert, L. 1985. Developmental constraints and evolution. Q. Rev. Biol. 60:265–287.

Mayr, E. 1963. Animal Species and Evolution. Harvard University Press, Cambridge, Mass.

Pimentel, D., Smith, G. J. C., and Soans, J. S. 1967. A population model of sympatric speciation. Am. Nat. 101:493–504.

Pimm, S. L. 1978. Sympatric speciation: a simulation model. Biol. J. Linn. Soc. 11:131–139.

Rice, W. R. 1984. Disruptive selection on habitat preference and the evolution of reproductive isolation: a simulation study. Evolution 38:1251–1260.

Rice, W. R. 1987. Selection via habitat specialization: the evolution of reproductive isolation as a correlated character. Evol. Ecol. 1:301–314.

Rice, W. R., and Hostert, E. E. 1993. Laboratory experiments on speciation: what have we learned in 40 years? Evolution 47:1637–1653.

Roelofs, W. R., Glover, T., Tang, X., Sreng, L., Robbins, D., Eckenrode, C., Lofstedt, G., Hannon, B. S., and Bengtsson, B. O. 1987. Sex pheromone production and perception in European corn borer moths is determined by both autosomal and sex-linked genes. Proc. Natl. Acad. Sci. USA 84:7585–7589.

Rosenzweig, M. L. 1978. Competitive speciation. Biol. J. Linn. Soc. 10:275–289.

Rosenzweig, M. L. 1979. Optimal habitat selection in two-species competitive systems. Fortschr. Zool. 25:283–293.

Rosenzweig, M. L. 1981. A theory of habitat selection. Ecology 62:327–335.

Rosenzweig, M. L. 1985. Some theoretical aspects of habitat selection. *In* Habitat Selection in Birds (M. L. Cody, ed.), pp. 517–540. Academic Press, New York.

Rosenzweig, M. L. 1995. Species diversity in space and time. Cambridge University Press, Cambridge.

Seger, J. 1985. Intraspecific resource competition as a cause of sympatric speciation. *In* Evolution (P. J. Greenwood, P. H. Harvey, and M. Slatkin, eds.), pp. 43–53. Cambridge University Press, Cambridge.

Shaw, R. G., and Platenkamp, G. A. J. 1993. Quantitative genetics of response to competitors in *Nemophila menziesii*: a greenhouse study. Evolution 47:801–812.

Slatkin, M. 1980. Ecological character displacement. Ecology 61:163–177.

Soans, A. B., Pimentel, D., and Soans, J. S. 1974. Evolution of reproductive isolation in allopatric and sympatric populations. Am. Nat. 108:117–124.

Templeton, A. R., and Rothman, E. D. 1981. Evolution in fine-grained environments. II. Habitat selection is a homeostatic mechanism. Theor. Pop. Biol. 19:326–340.

Udovic, D. 1980. Frequency-dependent selection, disruptive selection, and the evolution of reproductive isolation. Am. Nat. 116:621–641.

Wagner, A. 1996. Does evolutionary plasticity evolve? Evolution 50:1008–1023.

Wagner, G. P., and Altenberg, L. 1996. Perspective: complex adaptations and the evolution of evolvability. Evolution 50:967–976.

Wilson, D. S. 1989. The diversification of single gene pools by density- and frequency-dependent selection. *In* Speciation and Its consequences (D. Otte and J. Endler, eds.), pp. 345–365. Sinauer, Sunderland, Mass.

Wilson, D. S., and Turelli, M. 1986. Stable underdominance and the evolutionary invasion of empty niches. Am. Nat. 127:835–850.

Wright, S. L. 1921. Systems of mating I–IV. Genetics 6:111–178.

7

On the Sympatric Origin of Species by Means of Natural Selection

Alexey S. Kondrashov
Lev Yu. Yampolsky
Svetlana A. Shabalina

[T]he more diversified the descendants from any one species become in structure, constitution, and habits, by so much will they be better enabled to seize on many and widely diversified places in the polity of nature, and so be enabled to increase in numbers. . . . The competition will generally be most severe between those forms which are most nearly related to each other in habits, constitution, and structure. Hence all the intermediate forms . . . , as well as the original parent-species itself, will generally tend to become extinct.

(Darwin, 1859, ch. 4)

Gradually accumulating evidence (Bush, 1994; Kreslavskiy, 1994; Kreslavskiy and Mikheyev, 1994; Lazarus et al., 1995; Mina et al., 1996; Schliewen et al., 1994; Taylor and Bentzen, 1993; Theron and Combes, 1995; see Kondrashov and Mina, 1986, for a review; see also several chapters in this volume) leaves no doubt that sympatric speciation does occur in nature. However, we still do not know what conditions can lead to it and how common these conditions are.

Theoretical analysis is necessary to answer the first of these questions. Theory of sympatric speciation should have blossomed by now because (1) sympatric speciation by natural selection was proposed by Darwin (1859), (2) speciation is supposed to be the central subject of evolutionary biology, and (3) only sympatric speciation certainly requires specific theory at the population level. While allopatric speciation may involve nothing more than fixations of different favorable alleles in geographically isolated and independently evolving large populations, sympatric speciation depends on interactions, with uncertain outcome, among selection, recombination, and nonrandom mating, all acting within the same population.

In reality, however, the theory of sympatric speciation is still rudimentary. Most models (e.g., Johnson et al.,

1996; Kawecki, 1996) assume that reproductive isolation and adaptation to different ecological niches are caused by genetic differences at very few loci, which is hardly the common situation. Because the number of loci involved may be critical for the outcome, such models are of only limited applicability. More general and useful treatments must deal with polygenic traits, with arbitrary numbers of loci controlling their variability. Currently, the corresponding models were studied analytically only at the final stage of sympatric speciation, when almost all individuals belong to one or the other incipient species, which profoundly simplifies the dynamics of the population (Kondrashov, 1983a,b).

In most cases sympatric speciation apparently follows the same two-stage scenario (Kondrashov and Mina, 1986; but see Turner and Burrows, 1995). First, at least a pair of genotypes reproductively isolated from each other must appear within the range of the intrapopulational genetic variability. Second, the genotypes intermediate between those that are reproductively isolated must be eliminated. Reproductive isolation may be selected for as the consequence of selection against the intermediates (see Kondrashov, 1984), or may simply be a by-product of some profound differences between the isolated genotypes (e.g., they may reproduce in different habitats and/or at different times, or may be too morphologically different to interbreed). Some overlap is possible between these two stages, but complete reproductive isolation is an obvious prerequisite for total elimination of the intermediates and, thus, for the completion of sympatric speciation.

If reproductive isolation is caused by differences at just one locus (homozygotes AA and aa do not mate), elimination of the intermediates (heterozygotes Aa) proceeds due to Mendelian segregation alone. The same is true under some very strict modes of polygenic reproductive isolation (see Kondrashov and Mina, 1986). However, in more realistic cases elimination of the intermediate genotypes requires some selection against them. This selection can be either disruptive selection acting uniformly on every individual (e.g., due to the presence of two substantially different resources within the same habitat) or divergent selection (see Johnson et al., 1996) in two or more habitats, accompanied by habitat choice, which allows the extreme genotypes to chose the habitats where they are most fit. Divergent selection without habitat choice (Felsenstein, 1981) causes the intermediates to have, on average, more or less the same fitness as the extremes and thus does not lead to speciation (Kondrashov and Mina, 1986; Bush, 1994).

The Problem

Consider the second stage of sympatric speciation, assuming that a pair of reproductively isolated genotypes already exists but this isolation alone does not lead to elimination of the intermediates. What intensities of selection against the intermediates will lead to their eventual elimination? According to Mayr (1963), "The facts that zygotes in sexually reproducing organisms are diploid and that sexual reproduction maintains the genetic cohesion of every local population present serious obstacles to all hypotheses of sympatric speciation" (p. 478). Thus, we need to know when selection can destroy this cohesion.

We report here some simple analytical results on the intensities of selection required to start the population split, and compare them with what is known regarding elimination of the last intermediates. An individual-based computer model is used to bridge the gap between these two extremes. One polygenic trait x, influenced by n loci, will be considered, with habitatless disruptive selection.

The intensity of selection may be characterized by the ratio r of the lowest fitness over the highest one. With $r = 1$ there is no selection, and the smaller r is, the stronger the selection may be. Best genotypes will be that of the two incipient species, which must have the same fitness (assumed to be 1, without loss of generality), because otherwise one would oust the other. In the course of speciation, genetic composition of the population undergoes drastic changes, and r is a suitable parameter because it depends only on genotype fitnesses, and not on their fre-

quencies. Obviously, $1 - r$ provides the upper limit for the genetic load L, reached if all the genotypes present have the lowest fitness (see Shnol and Kondrashov, 1994). Of course, different modes of selection (fitness functions), all characterized by the same r, can act differently. We find the highest value of r under which selection still can split the population. Some results on the minimal genetic load required in the course of speciation are also reported.

Selection Required for Elimination of the Last Intermediates

At the very end of sympatric speciation, the most efficient mode of selection, in terms of splitting the population under a given r, is such that both incipient species have fitness 1, while all intermediate genotypes have fitness r. In this case, the already rare intermediates will certainly be eliminated if $r \leq 0.5$. To prove this, it is sufficient to observe that when every intermediate individual mates with a partner from either one or the other incipient species (the last intermediates are too rare to mate each other), without selection the total number of intermediates would less than double each generation, because in this case each mating contributes exactly two offspring to the next generation, and some of the offspring belong to the incipient species, due to segregation and recombination. Of course, with less efficient modes of disruptive selection, lower values of r may be required (Kondrashov, 1983a). Obviously, selection against rare intermediates induces only a small genetic load, such that at the very end of sympatric speciation L approaches 0.

Selection Required for Beginning of the Population Splitting

Suppose that initially the alleles at all n loci that affect our trait are distributed independently of each other (disruptive selection has not yet started, and reproductively isolated extreme genotypes are so rare that their inability to interbreed does not matter). If n is large, such that x can be treated as a continuous variable, this implies that the distribution of the trait p(x) is close to Gaussian.

Assume now that disruptive selection begins to operate. Will it split the population? With random mating and free recombination, sexual reproduction halves linkage disequilibria every generation. Thus, if our organisms are haploid, deviation of the variance of p(x) from its value when all alleles are distributed independently is also halved (e.g., Bulmer, 1985). With diploid organisms the effect of sex is larger, because Hardy-Weinberg equilibrium is reached instantly, but the difference is small with large n [there are n statistical interactions between alleles at the same loci, compared to $n(n-1)/2$ interactions between alleles at different loci].

Thus, in order to initiate the split of a haploid population, each action of disruptive selection must at least double the variance of the trait. Nonrandom mating of intermediates and/or linkage will make a weaker selection sufficient, but we concentrate on the situation that requires the most intensive selection.

What is the maximal possible r of a fitness function w(x) that doubles the variance of a trait with Gaussian distribution? Without loss of generality we can assume that p(x) is standardized. It can be shown (Shnol and Kondrashov, unpublished) that with a given r the maximal increase of the variance occurs when w(x) has a rectangular "fitness valley" of the width 2a in the middle:

$$w(x) = \begin{cases} 1; & x < -a \\ r; & -a \le x < a \\ 1; & x \ge a \end{cases} \quad (7.1)$$

We also consider two less efficient forms of disruptive selection: inverted Gaussian fitness valley,

$$w(x) = 1 - (1 - r)\exp[-x^2/(2a^2)], \quad (7.2)$$

and triangular fitness valley,

$$w(x) = \begin{cases} 1; & x < -a \\ r + |x|(1 - r)/a; & -a \le x < a \\ 1; & x \ge a. \end{cases} \quad (7.3)$$

All these fitness functions depend on two parameters: depth 1 − r and width a (figure 7.1).

Selection replaces p(x) with $\bar{p}(x)$:

$$\bar{p}(x) = p(x)w(x) \div \int_{-\infty}^{\infty} p(x)w(x)dx \quad (7.4)$$

For each fitness function we can find $a_{max}(r)$, the value of a for which the variance of $\bar{p}(x)$, $\hat{V}$, is maximal under a given r. Obviously, a_{max} must satisfy the condition $\frac{dV}{da_{max}} = 0$. Because $\hat{V}$ is expressed through integrals that involve p(x) and w(x), a_{max} cannot be always found explicitly. With Gaussian p(x) this is possible only for the fitness function (equation 7.2), where the formula giving a_{max} as a root of a cubic equation is not very useful. However, a_{max} can be always found numerically (THINK C program is available on request). When a_{max} is known, it is easy to calculate the corresponding $\hat{V}$ (figure 7.2).

For an arbitrary Gaussian p(x) the results on a_{max} must be interpreted as the deviations from the mean of p(x) in units of its standard deviation. We can see that, unless r is very small, selection has maximal efficiency when the width of the fitness valley slightly exceeds the standard deviation of p(x). With fitness functions (7.1) and (7.2) doubling of the trait variance with a_{max} requires r no more than 0.20 ($a_{max} = 1.42$) and 0.08 ($a_{max} = 1.42$), respectively. The corresponding values of the genetic load are 0.68 and 0.75. With fitness function (7.3) the variance can be doubled only when r = 0 while a → ∞, leading to L → 1. With fitness function (7.1) and a = $0.5a_{max}$, the variance is doubled with r = 0.07, and L = 0.57, such that the required genetic load is actually lower than that under a_{max}. In contrast, too wide a fitness valley (a = $2a_{max}$) is inefficient even with the fitness function (7.1): with realistic genetic loads the variance is never doubled (data not reported).

Under a given L, the maximal $\hat{V}$ is achieved under the fitness function (7.1) with r = 0. In this case, in order to double the variance of p(x), a must be 0.75 of its standard deviation, leading to L = 0.55.

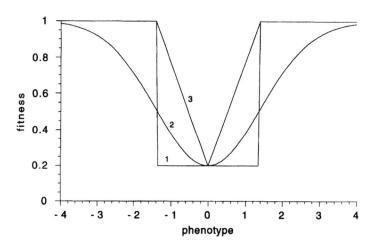

Figure 7.1. Fitness functions with rectangular, inverted Gaussian, and triangular fitness valleys, defined by (7.1), (7.2), and (7.3) with a = 1.4 and r = 0.2 (phenotype is measured in the units of the standard deviation of p(x) from its mean).

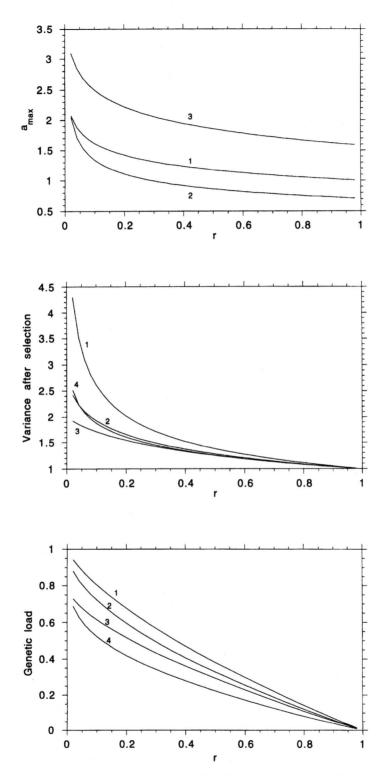

Figure 7.2. (a) The values of a_{max} as the function of r for the fitness functions. (b) The values of $\tilde{V}$ that correspond to a_{max} under fitness functions (7.1), (7.2), and (7.3), as well as the values of $\tilde{V}$ that correspond to $0.5a_{max}$ under the fitness function (7.1) (1–4). (c) The values of L under the same conditions as in (b) (1–4).

The Whole Process of Elimination of the Intermediates

It is reasonable to assume that p(x) will remain close to Gaussian even after selection has begun to work, as long as the range of intrapopulation variability remains narrow compared to the difference between the extreme genotypes (each generation sex will level out the deviations of p(x) from Gaussian at the scale below √ n). However, we do not know how to estimate the accuracy of the Gaussian approximation. After the split of the population advances enough, p(x) will no longer be close to Gaussian, making the above theory inapplicable, while the linear theory of the elimination of the last intermediates also cannot be used until they become so rare that matings between them can be ignored. Thus, we resorted to individual-based computer simulations to trace the whole process.

Earlier, such simulations led one of us to conclude that when n is not large, the elimination of the last intermediates required the strongest selection, while with many loci (n > 10) sympatric speciation is impossible because the split cannot be initiated (Kondrashov, 1986). Now we realize that this conclusion depends on the assumption that w(x) remains the same throughout the whole process of population splitting.

Figure 7.3 illustrates why this is the case. An efficient selection must discriminate between some frequent genotypes, and at the beginning of the split this range is much narrower. Thus, if applied at the end of splitting, $w_1(x)$ would cause high fitness of most of the intermediates present. With large n, $w_1(x)$ and $w_2(x)$ are so different that neither of them can carry out the whole process of

splitting. In particular, applying selection similar to $w_2(x)$ at the beginning of the splitting (which was done in Kondrashov, 1986) can lead to speciation only when n is small.

Thus, with large n sympatric speciation may be possible only if the fitness valley gets wider in the course of the population splitting. A plausible scenario which leads to such selection was proposed by Darwin (1859, pp. 111–116) and is employed here. We assume that there are two resources, and that individuals with smaller x outcompete others for the first resource, while those with larger x outcompete others for the second resource. Each resource can support fraction q of the population. Thus, the fractions q of the individuals with the smallest and largest x have fitness 1, while the rest have fitness r. Actually, the fitness of one of the two extreme fractions, the one toward which the population mean was shifted, was slightly reduced, in order to keep the population polymorphic.

Apart from this, our model is essentially the same as that in Kondrashov (1986). Generations were discrete, with the life cycle consisting of mating, reproduction, and selection. There were just two alleles, 0 and 1, at each locus, and x was determined as the number of alleles 1 at all n loci. Two mating algorithms were considered. With both, the prospective partners were chosen randomly. Then, in the first case (no assortativity, NA) the mating always occurred, except when the prospective partners had the opposite extreme genotypes. In the second case (linear assortativity, LA) the probability of mating of the prospective partners decreased linearly with the increase of the difference between their trait values, reaching zero under the maximal difference (2n). Each individual made

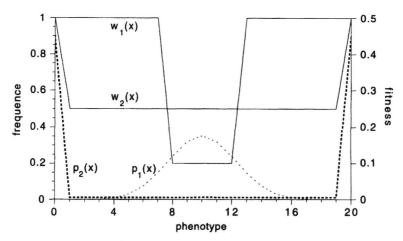

Figure 7.3. Fitness functions $w_1(x)$ (r = 0.2) and $w_2(x)$ (r = 0.5) are the most efficient ones (with a given r) at the beginning and at the end of the process of population splitting, when the distributions of the trait are $p_1(x)$ and $p_2(x)$, respectively. Selection acts at the diploid phase with n = 10. Distribution $p_1(x) = \binom{20}{x} 0.5^x$, while $p_2(x)$ was obtained using linear theory of Kondrashov (1983a) under assumption that the incipient species represent 90% of the population.

no more than 100 attempts to mate, and mated no more than once. Recombination was free. Organisms were either haploid (with reproduction consisting of meiosis followed by syngamy) or diploid (with reproduction consisting of syngamy followed by meiosis). (THINK C program available on request.)

Figure 7.4 shows the equilibrium state of the population as the function of r with q = 0.1 [under r = 0.20 and a = a_{max} = 1.42 with fitness function (7.1) q = 0.08]. In the NA case r = 0.15 is required for speciation, while higher values of r only lead to some increase of the equilibrium variance but are not enough to split the popula-

tion. This is probably caused by the fact that in the course of the population splitting, p(x) becomes platikurtic, in which case disruptive selection generally causes a smaller increase in variance than when p(x) is Gaussian. Thus, the middle of the splitting requires the strongest selection, but the increase over what is required at the beginning of the process, when p(x) is Gaussian, is small.

The LA case is different in the two respects. First, a much higher r = 0.29 is sufficient, because assortativity of mating among the intermediates weakens the resistance of a sexual population to splitting. Second, while r declines, gradual increase of the equilibrium variance is

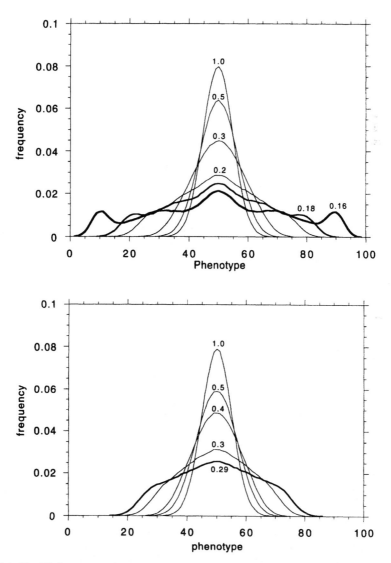

Figure 7.4. Equilibrium states (data averaged from generations 450 to 500) of the population of haploid organisms with n = 100 as functions of r (values of r are shown in the figures) with q = 0.10 and NA (a) or LA (b) mating. With r = 0.15 (NA) and r = 0.28 (LA) speciation occurs, such that the equilibrium population contains only the equal proportions of 00 . . . 0 and 11 . . . 1 genotypes (not shown).

suddenly followed by speciation, because the resistance to splitting always declines when the variance of p(x) grows, leading to more nonrandom mating among the intermediates. In contrast, with NA the resistance begins to decline only when reproductively isolated incipient species with the extreme genotypes (00 . . . 0 and 11 . . . 1) become abundant.

Figure 7.5 shows the dynamics of splitting in a typical run (NA, r = 0.12). Although the selection is only slightly more intense than what is necessary for speciation, the process is very fast.

The results were almost identical with n = 100 and n = 50. Thus, such numbers of loci appear to already lead to the asymptotical behavior characteristical of large n. With diploid organisms the necessary value of r was 0.14 in the NA case, while in the LA case it remained un-

changed. With small n ≤ 10 the required intensities of selection were lower than with higher n, both in haploid and diploid cases (data not reported).

Discussion

Surprisingly, our chapter appears to be the first attempt to quantitatively address a very simple question: when can disruptive selection overcome the unifying force of sex and lead to sympatric speciation in the general multilocus case? We have found that a suitable fitness function must meet the following conditions:

1. Selection must be intense: at the beginning of the process a significant fraction of the intermediate

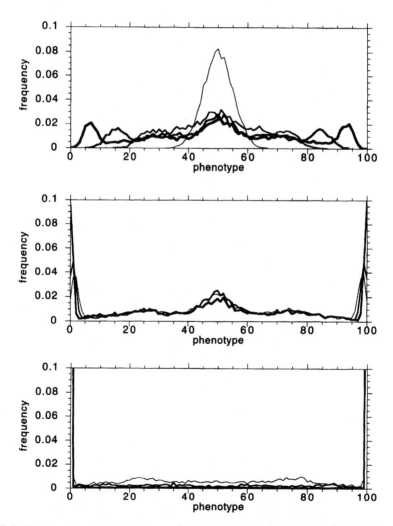

Figure 7.5. Distribution p(x) in the course of speciation in a haploid population with n = 100. (a) Generations 1, 10, 20, and 30. (b) Generations 40, 42, and 44. (c) Generations 45, 46, and 47. In a–c the increase of the width of the curve corresponds to the increased generation number.

phenotypes must have fitnesses below 0.15 of the highest one, while the genetic load must be above 0.5. However, no catastrophic selection (with, say, L > 0.90) is necessary. Weaker selection is sufficient to eliminate the last intermediates.

2. The fitness valley in the middle of the fitness function must have steep walls: the fitness function with the triangular valley cannot double the phenotypic variance, while the functions with rectangular or inverted Gaussian valleys (figure 7.1) can do so (figure 7.2).

3. Selection must be frequency dependent: the width of the fitness valley must remain of the order of the standard deviation of p(x) in the course of the population splitting (figure 7.3).

How often such selection occurs in nature is the question that has to be addressed experimentally. However, it seems certain that competition is the only factor that can lead to such selection. Thus, the mechanism of sympatric speciation proposed by Darwin (1859; see epigraph) is the most plausible one (Doebeli, 1996).

Quantitatively, our conclusions apply to the worst case for sympatric speciation: a large number n of unlinked loci controlling the trait and no assortative mating between the intermediate phenotypes. Otherwise, the required conditions may be much less stringent (figure 7.4), but qualitatively they remain the same, unless n is very small. Still, we can reject the claim of Kondrashov and Mina (1986) that sympatric speciation can only produce incipient species that differ at very few loci.

Elimination of the last intermediates requires weaker selection because sex is much less efficient in resisting the population splitting when almost all individuals have extreme genotypes and mate mostly within themselves (either $00 \ldots 0 \times 00 \ldots 0$ or $11 \ldots 1 \times 11 \ldots 1$). Therefore, when n is large the theory of Kondrashov (1983a,b) provides only necessary, but not sufficient conditions for sympatric speciation. However, it may provide the sufficient conditions for speciation by reinforcement, which begins with the extreme genotypes already abundant and, thus, under large n requires less stringent conditions than sympatric speciation in the strict sense.

When at least the opposite extreme genotypes are completely reproductively isolated, sympatric speciation, if possible, usually proceeds very fast (figure 7.5). Therefore, the pattern known as punctuated equilibrium may be caused by sympatric speciation, among other mechanisms. The only possibility for slow splitting of a population occurs when it proceeds simultaneously with perfecting of still incomplete reproductive isolation between the incipient species.

Historically, the theory of speciation is full of confusion. Contrary to Mayr's (1963) claim that Darwin was "rather vague on the subject" (p. 449), Darwin (1859, chapter 4) definitely accepted sympatric speciation; in particular, his hypothetical examples involving a carnivorous quadruped and a grass assume explicitly that speciation proceeds under strict sympatry. Responding to Wagner (1868), who argued that geographical isolation is necessary for speciation, Darwin (1869) inserted the following statement into the fifth edition of *The Origin*: "from reasons already assigned I can by no means agree with this naturalist, that migration and isolation are necessary for the formation of new species" (p. 120). During the time of the Modern Synthesis the attitude toward sympatric speciation was unstable; for example, three editions of *Genetics and the Origin of Species* contain contradictory statements on the subject (Dobzhansky, 1937, p. 257; 1941, p. 284; 1952, p. 206). Later, it somehow became accepted by many that the neodarwinian view on speciation "requires geographic barriers between conspecific populations" (Coyne, 1992, p. 511).

Even more surprisingly, a lot of effort was spent on models of speciation which, contrary to the main premise of Darwin, depend on episodes of counteradaptive changes, pushed through by random genetic drift. Speciation due to peak shifting, genetic transilience, genetic revolutions, founder effect, founder flush, and so on, although not impossible theoretically (Gavrilets and Hastings, 1996; Slatkin, 1996), requires rather restrictive conditions and so far is not supported by any data (Moya et al., 1995). Remarkably, Darwin (1859) held that "fewness of individuals will greatly retard the production of new species through natural selection, by decreasing the chance of the appearance of favorable variations" (p. 105), an assertion corroborated by a recent analysis (Orr and Orr, 1996). Sympatric speciation obviously must be adaptive throughout, because fairly strong selection is required at every stage.

The sympatric origin of species by means of natural selection alone should no longer be neglected by theorists. It is time to address a lot of very basic questions on it. What would happen if fitness depends on more than one trait, and/or if fitness and mate preferences depend on different traits? How would habitat choice and fitness coevolve in a general polygenic case if selection is divergent, instead of disruptive? How does the pattern of dispersal among the habitats, under strict sympatry, affect the process of speciation? We currently do not know.

Acknowledgment This chapter was supported by NSF Grant DEB-9417753.

References

Bulmer, M. G. 1985. The Mathematical Theory of Quantitative Genetics (2nd ed.). Oxford University Press, Oxford.

Bush, G. L. 1994. Sympatric speciation in animals: new wine in old bottles. Trends in Ecology and Evolution 9, 285–288.

Coyne, J. A. 1992. Genetics and speciation. Nature 355, 511–515.

Darwin, C. 1859. The Origin of Species by Means of Natural Selection (1st ed.). John Murray, London. (1869, 5th ed.).

Dobzhansky, T. G. 1937. Genetics and the Origin of Species (1st ed.). Columbia University Press. (1941, 2nd ed.; 1951, 3rd ed.).

Doebeli, M. 1996. A quantitative genetic competition model for sympatric speciation. Journal of Evolutionary Biology 9, 893–909.

Felsenstein, J. 1981. Skepticism towards Santa Rosalia, or why are there so few kinds of animals? Evolution 35, 124–138.

Gavrilets, S., and Hastings, A. 1996. Founder effect speciation. American Naturalist 147, 466–491.

Johnson, P. A., Hoppensteadt, F. C., Smith, J. J., and Bush, G. L. 1996. Conditions for sympatric speciation: a diploid model incorporating habitat fidelity and non-habitat assortative mating. Evolutionary Ecology 10, 187–205.

Kawecki, T. J. 1996. Sympatric speciation driven by beneficial mutations. Proceedings of the Royal Society of London Series B 263, 1515–1520.

Kondrashov, A. S. 1983a. Multilocus model of sympatric speciation. I. One character. Theoretical Population Biology 24, 121–135.

Kondrashov, A. S. 1983b. Multilocus model of sympatric speciation. II. Two characters. Theoretical Population Biology 24, 136–144.

Kondrashov, A. S. 1984. On the intensity of selection for reproductive isolation at the beginning of sympatric speciation. Genetika 20, 408–415.

Kondrashov, A. S. 1986. Multilocus model of sympatric speciation. III. Computer simulations. Theoretical Population Biology 29, 1–15.

Kondrashov, A. S., and Mina, M. V. 1986. Sympatric speciation: when is it possible? Biological Journal of Linnean Society 27, 201–223.

Kreslavskiy, A. G. 1994. Sympatric speciation in animals: disruptive selection or ecological segregation? Zhurnal Obshchei Biologii 55, 404–419.

Kreslavskiy, A. G., and Mikheyev, A. V. 1994. Gene geography of racial differences in Lochmaea capreae L. (Co-leoptera, Chrysomelidae), and the problem of sympatric speciation. Entomological Review (English Translation of Entomologicheskoye Obozreniye) 73, 85–92.

Lazarus, D., Hilbrecht, H., Spencer-Cervato, C., and Thierstein, H. 1995. Sympatric speciation and phyletic change in Globorotalia truncatulinoides. Paleobiology 21, 28–51.

Mayr, E. 1963. Animal Species and Evolution. Belknap Press, Cambridge: Mass.

Mina, M. V., Mironovsky, A. N., and Dgebuadze, Yu. Yu. 1996. Lake Tana large barbs: phenetics, growth and diversification. Journal of Fish Biology 48, 383–404.

Moya, A., Galiana, A., and Ayala, F. J. 1995. Founder-effect speciation theory: failure of experimental corroboration. Proceedings of the National Academy of Sciences USA 92, 3983–3986.

Orr, H. A., and Orr, L. H. 1996. Waiting for speciation: the effect of population subdivision on the time to speciation. Evolution 50, 1742–1749.

Schliewen, U. K., Tautz, D., and Paabo, S. 1994. Sympatric speciation suggested by monophyly of crater lake cichlids. Nature 368, 629–632.

Shnol, E. E., and Kondrashov, A. S. 1994. On some relations between different characteristics of selection. Journal of Mathematical Biology 32, 835–840.

Slatkin, M. 1996. In defense of founder-flush speciation. American Naturalist 147, 493–505.

Taylor, E. B., and Bentzen, P. 1993. Evidence for multiple origins and sympatric divergence of trophic ecotypes of smelt Osmerus in northeastern North America. Evolution 47, 813–832.

Theron, A., and Combes, C. 1995. Asynchrony of infection timing, habitat preference, and sympatric speciation of schistosome parasites. Evolution 49, 372–375.

Turner, G. F., and Burrows, M. T. 1995. A model of sympatric speciation by sexual selection. Proceedings of the Royal Society of London Series B 260, 287–292.

Wagner, M. 1868. Die Darwinische Theorie und das Migrationsgesetz der Organismen. Verlag von Duncker und Humblot, Leipzig.

8

Can Sympatric Speciation via Host or Habitat Shift Be Proven from Phylogenetic and Biogeographic Evidence?

Stewart H. Berlocher

For the last five decades the theory of geographic or allopatric speciation has enjoyed overwhelming acceptance; the burden of proof, with the exception of polyploidy and similar mechanisms in plants, has been on sympatric speciation (speciation that does not require geographic isolation at any stage). This period of relative agreement has its genesis in Ernst Mayr's 1942 *Systematics and the Origin of Species* (also Mayr, 1947, 1963). Despite a recent renaissance of interest in, and an increase in support for, sympatric speciation (Bush, 1975, 1992, 1994; Bush and Smith, in press; Diehl and Bush, 1984; Tauber and Tauber, 1989; Asquith, 1993; Schliewen et al., 1994; Emelianov et al., 1995; see also chapters by Schluter, Menken and Roessingh, and Feder, this volume), many workers remain unconvinced.

Here I investigate how the hypothesis of sympatric speciation via host or habitat shift might be tested with biogeographic and phylogenetic evidence (see also Bush, 1969; Smith and Todd, 1984; Bush and Howard; 1986; Lynch, 1989; Chesser and Zink, 1994; Asquith, 1993). (Evidence from experiments and observations in recent, historical time are discussed in Feder, and Menken and Roessingh, this volume.) My predictions are completely testable. However, they require complete or almost complete phylogenetic, ecological (host or habitat), and biogeographic data, and relatively few taxa are known at this level. Thus, I do not test all predictions here, and results of those tests I do carry out are, to varying degrees, preliminary.

Background

Terminology

My choice of particular definitions for terms has been guided by the basic model of host- or habitat-shift spe-

ciation (Bush, 1969; see also Johnson and Gullberg, and Feder, this volume). Consider a parasite species in which mating occurs on the host. Imagine that a few individuals of parasite species A on host α have genotypes that cause them to seek new host β instead of α. If *mating occurs on the host*, then these few individuals on host β start a population that is at least partially reproductively isolated from the original population on host α. As natural selection increases the ability of the new population to find host β, it also concomitantly reduces gene flow from the original population on host α. The population on host β is more and more recognizable as species B as gene flow decreases (Bush, 1969).

For sympatric species, the biological or isolation species concept (see definition in Harrison, this volume) is thus the most appropriate, as restriction of gene flow in sympatry is the key "product" of the sympatric speciation model. The various phylogenetic concepts (Harrison, this volume) are inappropriate, as sympatric speciation may initially produce polyphyletic species (Berlocher, 1989). For allopatric species included in my tests of predictions, I have simply accepted the judgment of the taxonomists involved.

Two populations are considered to have speciated sympatrically if they evolved reproductive isolation while at all times being located within the "normal cruising range" of an individual organism. (For elaboration of this definition, see Futuyma and Mayer, 1980, pp. 254–255.)

The three kinds of sympatric speciation I consider are (1) speciation of a parasite by a shift to a new host (host-shift speciation; Bush, 1969; Feder, this volume); (2) speciation by a shift in habitat or an important niche parameter (habitat-shift speciation; Schluter, this volume); and (3) collectively, autopolyploidy, allopolyploidy, and hybrid or recombinational (Reiseberg et al., 1996) speciation. For brevity, the term *sympatric speciation* is henceforth restricted to host- or habitat-shift sympatric speciation; allopolyploidy and similar modes will be specifically identified when necessary. As a further step in reducing verbal complexity, I will frequently use the terminology of host-shift speciation for both habitat-shift and host-shift speciation; the two modes are theoretically almost identical (Johnson and Gullberg, this volume). All modes of speciation requiring an initial episode of geographic isolation will be referred to collectively as allopatric speciation.

Given the importance of mating on the host for sympatric speciation, it is surprising that a specific term for this life history feature was not coined until recently. For phytophagous insects, the characteristic of mating on the host plant has been termed *host fidelity* (see Feder, this volume). I define the term *host faithful* as the corresponding adjective. Host fidelity is an absolute requirement for sympatric speciation. In the absence of host fidelity, even the most extreme host specialization is very unlikely to produce a new species, as adaptation to a new host will be greatly retarded by gene flow from the ancestral population (Diehl and Bush, 1989; Johnson and Gullberg, this volume).

However, the hypothesis that host-faithful species *can* speciate sympatrically is not the same as hypothesizing that all such species *must always* do so. Even the most host-faithful species imaginable can still have its range divided geographically, and speciate allopatrically. For brevity I here term host-faithful species that live in *different* hosts as *host shifters*. Species that either lack host fidelity, or display host fidelity but use the same host, I term *nonshifters*. Generally, I will be applying the terms host shifters and nonshifters to pairs of sister species.

Sympatric speciation can be tested by contrasting biogeographic and phylogenetic attributes of host shifters with those of nonshifters. It is critical to stress that the logic of such tests is not circular; sympatric speciation is *not* assumed simply by classifying species as host shifters. Classifying species pairs as host shifters assumes only that host fidelity and use of different hosts are requirements of the basic theoretical model of sympatric speciation.

Stages of Sympatric Speciation

One approach that Mayr has taken in explaining and testing allopatric speciation has been to break down the continuous process of speciation into a set of stages (Mayr, 1940; 1942, pp. 159–162, figure 16; 1963, pp. 489–491). These stages have varied somewhat over the years, but a representative set is (1) uniform single species with a wide, continuous range; (2) geographically variable (broken into subspecies) single species with a wide, continuous range; (3) geographically isolated (allopatric) subspecies; (4) allopatric subspecies with partial reproductive isolation; (5) allopatric species with complete isolation; and (6) sympatric species. Overall, this approach has been successful. Demonstrating the existence of intermediate stages, particularly stage 4 taxa with their partial repro-

ductive isolation, was very important in creating confidence in the allopatric speciation hypothesis during the Modern Synthesis of the 1930s and 1940s.

The stage approach has not previously been applied to the hypothesis of sympatric speciation. Bush (1969) developed a model for the evolution of host races in sympatry and stated that continued adaptation to hosts could eventually produce strong reproductive isolation, but did not propose hypothetical stages. To aid in developing my predictions, I here describe a set of hypothetical stages for the true fruit flies of the *Rhagoletis pomonella* species group (Bush, 1966, 1969, 1992, 1994). These flies have historically been the most discussed case of possible sympatric speciation (Mayr, 1963; Futuyma and Mayer, 1980).

The *pomonella* species group is a complex of described species, undescribed species, host races, and some incompletely understood geographic isolates (Bush, 1966, 1992; Berlocher et al., 1993). All species are very similar morphologically and, with the exception of *R. zephyria* Snow, are broadly sympatric in eastern North America. All have a 1-year life cycle and show great host fidelity. An allozyme phylogeny (Berlocher et al., 1993) and two mitochondrial DNA sequence phylogenies (McPheron and Han, 1997; Smith and Bush, 1997) are similar; the allozyme phylogeny is shown in figure 8.1. Some additional taxa that are not completely characterized (e.g., the "sparkleberry fly," Payne and Berlocher, 1995) are not shown in figure 8.1, but their inclusion would not change the interpretation of the stages. The taxa that I use as examples for each hypothetical stage are listed below in terms of increasing divergence from the ancestral hawthorn (*Crataegus*, Rosaceae) race of *R. pomonella* (Walsh).

Stage 1: Host Race. Represented by the apple race of *R. pomonella*, derived ~160 years ago from the ancestral hawthorn race (Feder, this volume). The apple race is characterized by significant gene flow from the hawthorn race (and vice versa), which is counterbalanced by strong selection. No prezygotic or postzygotic isolation independent of host fidelity occurs. Isolation due to host fidelity is caused by differences in postdiapause emergence time and in host choice behavior. No race-specific allozymes exist, and allozyme frequency differences, while consistent, are small (~0.15). Mean Nei distance between a representative set of paired apple and hawthorn race geographic populations is (mean ± S.D.) 0.005 ± 0.004 (data from McPheron, 1990).

Stage 2: Species Isolated Only by Host Fidelity. Represented well, with one possible exception, by the "flowering dogwood fly", an undescribed species infesting only *Cornus florida* (Cornaceae) (Berlocher et al., 1993). Large (up to ~0.80) allozyme frequency differences exist, but do not exclude the possibility of gene flow with *R. pomonella* (Berlocher et al., 1993; Berlocher, un-

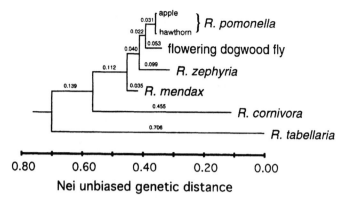

Figure 8.1. Neighbor-joining tree of 29 locus (allozyme) unbiased Nei distances in the *R. pomonella* species group (Berlocher et al., 1993). *R. tabellaria* is an outgroup in a different species complex. The apple race of *R. pomonella* was simply added to the completed tree.

published). A very small amount of postzygotic isolation observed in one experiment (a ~10% reduction in egg hatch in dogwood fly × *R. pomonella* backcrosses; Smith, 1986) keeps the dogwood fly from being a perfect example of an idealized stage 2; when caged together without fruit, dogwood flies and *R. pomonella* mate randomly (Smith, 1986). Differences in host selection behavior are greater than those between host races; the flowering dogwood fly prefers smaller fruit than the hawthorn race of *R. pomonella* (Berlocher et al., 1993; Smith, 1986), while no fruit size preferences distinguish the apple and ancestral hawthorn *R. pomonella* host races. Because of latitudinal allozyme clines in both *R. pomonella* (Berlocher and McPheron, 1996) and the dogwood fly, the smallest Nei distance between these taxa is not the 0.084 indicated from figure 8.1 (Illinois populations), but 0.029 between a pair of Texas populations (Berlocher, unpublished). No species-specific allozymes exist; totally fixed loci could conceivably be lacking throughout the genomes of stage two species.

Stage 3: Species with Prezygotic and/or Postzygotic Isolation Unrelated to Host Fidelity.   This stage can be represented by *R. mendax* Curran, the blueberry (*Vaccinium* spp., Ericaceae) maggot. *R. mendax* and *R. pomonella* do not interbreed in nature even in "microsympatry" (Feder and Bush, 1989). Large differences in fruit size preference exist (Prokopy and Bush, 1973). Some assortative mating is seen in cages even in the absence of the host plant (Smith, 1986). Egg hatch is reduced in both F_1 crosses (Smith, 1986), and viability reduction in the F_1 is ~50% (depending on host; Bierbaum and Bush, 1990), but F_1s and backcrosses are fertile. Genetic distances are greater than those in stage 2; mean Nei distance between *R. mendax* and the clade containing *R. pomonella*, the dogwood fly, and *R. zephyria* (figure 8.1) is 0.151 ± 0.053 (Berlocher et al., 1993). Species-specific allozymes exist but are not fixed, so some very low level

of gene flow with relatives cannot be excluded (Feder and Bush, 1989; Berlocher, 1995). Nonfixed morphological autapomorphies exist (Bush, 1966; Jenkins, 1996).

Stage 4: Totally Isolated Species.   This stage is represented by *R. cornivora* (infesting shrubby *Cornus* spp., Cornaceae), characterized by great genetic divergence (mean Nei distance between it and the rest of the group is 0.663 ± 0.097, with fixed autapomorphic differences at five loci; Berlocher et al., 1993), no gene flow with relatives (Smith et al., 1993), very strong postzygotic isolation unrelated to host adaptation (Smith et al., 1993), and morphological autapomorphies (Jenkins, 1996).

From the standpoint of substantiating the sympatric speciation hypothesis, the key stages are 1 and 2, and intermediates between them. These stages correspond in importance to stage 4 of the allopatric hypothesis, in that a complete set of intermediates could reveal how the transition from host race to species occurs.

Does Sympatric Speciation Make Testable Predictions?

Before making formal predictions, a review of several issues that bear on sympatric speciation is needed. I discuss these as a series of questions, with suggestions as to answers.

Does Geographic Range Tell Us Anything about Mode of Speciation?   It has long been observed that sister species in many taxa occur in adjacent geographic areas (Wagner, 1868, cited in Mayr, 1963; Jordan, 1908; Mayr, 1942, 1963). However, most of the taxa studied by these workers are nonshifters—there is little evidence that birds, for example, have the life history characteristics required of habitat shift models (Chesser and

Zink, 1994). On the other hand, clusters of congeneric species of host-faithful phytophagous insects and other parasites have often been observed to utilize different hosts, and to show broad sympatry (Bush, 1969, 1975; Diehl and Bush, 1984; Bush and Howard, 1986).

To evaluate the validity of the observations above, a formal, testable prediction for the sympatric speciation hypothesis can be framed: early stages of taxa that meet the requirements of sympatric speciation models will be sympatric. To maximize the probability of studying early stages of sympatric speciation, sister species (or sister host races) must be studied. To test the prediction, one needs to first identify many sister species pairs of host shifters, *independent of any information on geographic range overlap*. One could then test statistically whether mean range overlap for host shifter sister pairs is greater than mean overlap for a similar set of nonshifter sister pairs; if so, the hypothesis of sympatric speciation would be supported. However, to be completely convincing additional factors need to be considered.

A great problem in inferring anything about a speciation event, whether it concerns geography, genetics, or morphology, is distinguishing between the initial species differences, and those added by postspeciational evolution. Postspeciation change in geographic range, an undeniable reality (Chesser and Zink, 1994), has allowed critics of sympatric speciation (e.g., Mayr, 1947, 1963) to deny that sympatry of sister species supports sympatric speciation. Mayr's basic argument is that allopatric speciation followed by dispersal (figure 8.2a) can explain all sympatric sister species.

In its simplest form, Mayr's secondary contact explanation for sympatric sister species is not totally convincing. The problem with using dispersal in an ad hoc manner is that it can be used to explain almost any distribution, to support almost any process. Dispersal could even be used to support sympatric speciation, as a sympatric origin followed by postspeciation dispersal could

"explain" allopatric sister species (figure 8.2b). However, Mayr has also, as discussed in following sections, made more complex, less easily ignored antisympatric arguments involving dispersal.

Before discussing these extensions to Mayr's argument, though, I must point out that a very direct approach is available for ascertaining whether the existence of sympatric sister species is evidence for sympatric speciation. This is to study modes of speciation that absolutely must happen in sympatry (figure 8.2c): allopolyploidy, autopolyploidy, or hybrid (recombinational; Reiseberg et al., 1996) speciation. If, for example, allopolyploid species and their immediate ancestors do not have greater range overlap than nonallopolyploid, nonshifter sister species pairs, then sympatry of sister species is clearly a poor indicator of any mode of sympatric speciation, and more complex arguments become moot.

Does Interspecific Competition Confound Interpretations of Range Overlap? Mayr early on recognized the challenge that sympatric sister species pairs of phytophagous or parasitic insects posed to the universality of allopatric speciation. To meet this challenge, he proposed that sister species that occupied different hosts would compete less, and thus overlap more in range, than species that did not change hosts (Mayr, 1947, p. 282). The current consensus is that interspecific competition is in fact important in phytophagous insects (Denno et al., 1995), and that it may affect range overlap (Conner and Bowers, 1987). However, Mayr's argument is in principle refutable. Not all host-specific insects show host fidelity; site of mating may be unrelated to larval host plant location. Insects lacking host fidelity, no matter how host-specific as larvae, are incompatible with models of host-shifting sympatric speciation, but sister species specialized on different larval hosts, whether host faithful or not, should have ranges similarly unaffected by interspecific competition. Thus, a prediction of sympatric

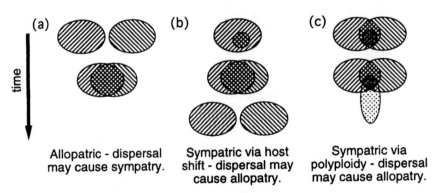

Figure 8.2. Change of geographic range and mode of speciation. (a) Allopatric speciation. (b) Sympatric speciation. (c) Polyploidy or hybrid speciation.

speciation is that range overlap between *host faithful*, host specific, sister species will be greater than range overlap between sister species of similarly host specific, but non-hostfaithful insects.

Does Allopatry Always Precede Sympatry?   Mayr's stage approach to proving allopatric speciation involves more than just finding hypothesized intermediate stages. Mayr has argued that reconstructing all stages into "the correct chronological sequence" (Mayr, 1963, p. 488) "proves" allopatric speciation. Under the allopatric speciation hypothesis, pairs of sister taxa in the earliest, allopatric stages should on average have diverged more recently than pairs in the last, sympatric stage. Mayr's argument is valid—as long as a reliable way to determine chronological sequence is available.

If the stages I have proposed for sympatric speciation do occur in sequence from stage 1 to stage 4, then sympatry will be maximal in the earliest stages, and may be less extensive in later stages. As time continues, post-speciation dispersal (as might be caused by range change of the host plants) of the sister species should on average *reduce* the initial complete range overlap. Thus, the allopatric model predicts minimal sympatry of the youngest sister pairs, while the sympatric speciation hypothesis predicts maximal sympatry of the youngest sister species pairs.

For both the allopatric and sympatric hypotheses, determining chronological order of hypothesized stages is thus critical. Several approaches to inferring the chronological sequence exist. The most general and most accurate is to use the "molecular clock" (Hillis et al., 1996) to estimate divergence dates between pairs of species representing each stage. For allopatric speciation, then, the prediction is that both genetic distance and probability of range overlap should increase with time between pairs of taxa, such that genetic divergence and range overlap should be positively correlated. This approach is similar to that of Coyne and Orr (1997), who used genetic distance to analyze rates of evolution of reproductive isolation. Several studies of the early electrophoretic era (see Avise, 1994) analyzed genetic divergence in the context of Mayr's stages of speciation (generally agreeing with the allopatric model), but no compilation of genetic distance and range overlap across many taxa has been made. The only available tests of the allopatric prediction that sympatry should increase with age of taxa use "speciation level" (Lynch, 1989), or phylogenetic branching order. One study found some evidence for greater sympatry as time elapses (Chesser and Zink, 1994), one found no correlation (Lynch, 1989), and one (Anderson and Evensen, 1978) found a negative correlation (more allopatry of distantly related taxa; see my discussion of measurement of overlap below). This approach is of course impossible without including older, nonsister species in the comparisons (as

discussed later, it may be useful to include nonsister species when using the molecular clock approach as well).

Sympatric speciation makes the opposite prediction that, in host shifters, genetic distance and probability of range *nonoverlap* should tend to increase with time between pairs of taxa, such that genetic divergence and range overlap should be *negatively* correlated. This prediction has not to my knowledge been tested in any form.

Does Phylogeny Tell Us Anything About Mode of Speciation?   In some cases taxa exist in a constrained geographic area and cannot easily disperse beyond that area. For example, a set of related fish species endemic to a small lake may all have essentially the same geographic range. In such cases, predictions based on range overlap cannot be tested. However, other predictions are possible. Allopatric speciation predicts that species "flocks" can be explained by either (1) geographic fragmentation and unification by fluctuating water levels, (2) allopatric speciation occurring within the lake (as on different reefs), or (3) multiple colonization from a non-lake source area or areas (Mayr, 1984, p. 5). Smith and Todd (1984, pp. 46–48) discuss these predictions and how they could be tested using geological, phylogenetic, and ecological data; here I focus only on the multiple colonization prediction and how it could be tested.

Sympatric speciation predicts that if the species (of a given taxonomic group) in a lake are habitat specialists, and show habitat fidelity, then the species in a lake will be monophyletic. This is just the basic biogeographic prediction of sympatric speciation, that sister species should be sympatric, but inverted: sympatric species should be sister species. This is an eminently testable prediction using phylogenetic evidence.

A perfect example of how phylogeny can contribute to a test of sympatric speciation is the work of Schliewen et al. (1994) on flocks of ecologically specialized cichlid fish restricted to very small West African volcanic crater lakes. The topography of the lakes is essentially an inverted cone, so changes in water level do not subdivide the lakes. Of particular interest, the mitochondrial DNA phylogeny of species in two different lakes rejected the multiple colonization explanation, as all species in a lake were monophyletic, with species from nearby rivers (connected by streams draining from a lake) being basal. As Schliewen et al. (1994) argued, the only way in which allopatric speciation is consistent with the phylogeny is if a precisely repeated pattern of colonizations, allopatric speciations, and extinctions has occurred. A simplified, hypothetical version of their argument is shown in figure 8.3. Sympatric speciation emerges as the most parsimonious explanation in this case. For a counterexample in which sympatric speciation is rejected, see Mayden et al. (1992).

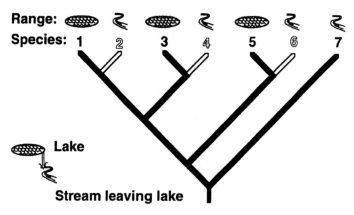

Figure 8.3. A simplified allopatric explanation for a species swarm of lake fish (Schliewen et al., 1994). A crater lake drains into a stream system containing related cichlids. Multiple colonization from the stream, subsequent allopatric speciation, and extinction (species in outline) of lake species could generate an apparently monophyletic group of lake species (see text).

Five Predictions of Sympatric Speciation, How They Could Be Tested, and Preliminary Tests

Four specific predictions can be based on the considerations discussed above. My fifth "prediction," involving phylogeographic analysis of modes of speciation, is really a brief discussion of the promise of this new approach for testing mode of speciation.

Prediction 1:   Mean range overlap of ancestor/descendant taxa in a set of species arising via allopolyploidy and other unambiguously sympatric modes will be greater than mean range overlap of sister species pairs in a set of nonallopolyploid, etc., nonshifter taxa.

This prediction presents several technical problems. One is making comparable range overlap measurements in two-species (allopatric) and three-species (allopolyploid, hybrid speciation) cases. Calculation of range overlap for a pair of sister species is straightforward (Anderson and Evensen, 1978, p. 424): overlap is O/R_1, where O is the area of overlap and R_1 is the smaller of the two species ranges. For three species cases, my initial proposal is to reduce the problem to a two-species case by defining the area of overlap of the two ancestor species as the range of one "species" of a sister pair, and the descendant species as the other member of the pair. The logic is that the descendant must arise in the area of overlap of the ancestor species; other parts of the range of the ancestor species are irrelevant to the calculation. However, this approach is not perfect. Cases in which the ancestor species are allopatric due to a small gap, but the descendant overlaps broadly with both ancestors, are scored as zero overlap—which seems intuitively wrong. Further work on methods for measuring overlap is needed.

Another concern is parapatry. One reason that Anderson and Evensen (1978) reached different conclusions from Lynch (1989) and Chesser and Zink (1994) is that Anderson and Evensen categorized abutting ranges as parapatric, separate from allopatry or sympatry, while in the latter two studies (and my analysis) abutting distributions are considered to represent zero overlap, that is, allopatry. Eventually it will be necessary to separate parapatry from allopatry, but at present relatively few taxonomic studies of, for example, phytophagous insects have sampled on a geographic scale fine enough to resolve the locations of species borders precisely. A related problem for future study is treatment of species separated by hybrid zones; I have simply omitted such cases in my analyses.

A critical requirement is that systematic information be complete, with sister pairs and ancestors of polyploids and hybrid species identified correctly. This implies that all extant species have been discovered, that a modern, explicit phylogeny estimation including all extant species is available, and that genetic data for determining ancestors of hybrids are available. Reliable geographic range data are also an absolute requirement.

Ideally, comparisons should involve paired representatives from particular taxa, so that dispersal abilities are similar within each test set. In other words, in an ideal test one would choose, say, a genus, take one sister pair of nonshifters and one ancestor/descendant pair (or triad, for allopolyploids and hybrid speciation) to include in the test, and then repeat this process with different genera, until a sufficient sample size is attained. Also, to the extent knowable, neither the polyploid/hybrid speciation nor nonshifter cases should be biased toward very young or old taxa (my preliminary data are in fact biased, including a number of polyploids that have evolved in the last 100 years). Finally, all tests should be based upon very large numbers of comparisons, as dispersal over evolutionary time clearly does weaken the original geographic "signature" of a given mode of speciation.

I carried out a preliminary test of prediction one using a data set compiled from the literature (see the appendix). Prediction one is not rejected (figure 8.4), as mean overlap for allopolyploids, autopolyploids, and recombinational species at 56.3% is greater than for the nonshifter, nonallopolyploid, etc., pairs at 14.5% (p = 0.006, Mann-Whitney U test). Postspeciational range change does not appear to completely eradicate the initial sympatry of sympatric speciation. However, postspeciational change in geographic range is very striking in some cases. For several allopolyploids the ancestor species no longer occurred sympatrically, in some cases being hundreds of kilometers apart (e.g., *Gymnocarpium dryopteris*, Pryer and Haufler, 1993). For nonshifters, the pattern is as expected from earlier analyses (Jordan, 1908; Anderson and Evensen, 1978; Wiley and Mayden, 1985; Lynch, 1989; Chesser and Zink, 1994); across a broad sampling of taxa including birds, frogs, fish, both predatory and non-host-shifting phytophagous insects, angiosperms, a fern, a moss, and a fungus, fully 62% of sister species pairs are allopatric (mean overlap = 14.5%).

Prediction 2: Mean range overlap of sister species of host shifters will be greater than mean range overlap between sister species of nonshifters.

The issues involved here are similar to those for prediction one. Range overlap calculation is simpler in that only two-species calculations are involved. But phytophagous insects and habitat specialists pose a new problem, that of knowing the host or habitat for all taxa. For many phytophagous insect groups, described species with no host information are common.

To represent host-shifting insects in a preliminary test, I chose the membracid treehopper genus *Enchenopa* (Guttman and Weigt, 1989), the tephritid fruit fly genus *Rhagoletis* (Bush, 1966, 1969; Berlocher et al., 1993), and the chrysomelid beetle genus *Ophaella* (Futuyma et al., 1995; Funk et al., 1995), as these genera all contain host-shifter sister species pairs, and all three have received much attention. (I became aware of Asquith [1993, 1995] too late for inclusion). I included only described species or apparent species-level taxa (*Enchenopa*) in the test. Sister host races could also be included in a test, but care would have to be taken to include only verified host races.

Using the same allopatric taxa as in prediction 1, prediction 2 is not rejected (mean overlap for the host shifters

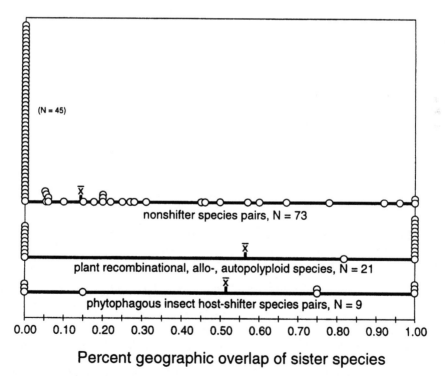

Figure 8.4. Geographic range overlap of sister species pairs (each pair represented by a dot). Data are shown for sister species pairs of nonshifter, host-shifter, and a combination of allopolyploid, autopolyploid, and hybrid speciation species. See text for discussion of measurement of range overlap; see the appendix for data.

= 51.6%, p = 0.023, Mann-Whitney U test). I stress that species pairs were initially classified as host shifters entirely on the ecological data. Thus, *R. pomonella* and *R. zephyria*, which were considered by Bush (1969) to be an example of allopatric speciation because of largely nonoverlapping (14.8%) ranges, were categorized as host shifters here because these host-faithful species use different hosts. (*R. pomonella* and *R. zephyria* are not sister species in figure 8.1, but trees based on more extensive population sampling [Berlocher, unpublished] suggest that they are.)

Too few data were available to control for the potential effect of larval interspecific competition on range overlap. However, specialist phytophagous insects that do not use the larval host plant as a mate rendezvous site do exist. For example, in saturniid moths females release sex-attractant pheromones to attract males. The large North American genus *Hemileuca* (Tuskes et al., 1996) would be ideal for study of range overlap in non-host-faithful host specialists. Another group that could be very informative is cerambycid beetles, which contains many species that are host specific as larvae but that in some cases mate at sites such as inflorescenses of non-larval-host species (Linsley, 1959). Unfortunately, almost no phylogenetic analysis of saturniid and cerambycid genera has been done. Comparisons similar to those described above should be possible for habitat specialists in general.

Prediction 3:   For a set of sister species pairs of host shifters, mean genetic divergences and range overlap should be negatively correlated.

The basic prediction is very robust—the slope of the relationship of range overlap and genetic distance should be negative for sympatric speciation, while positive for allopatric speciation (figure 8.5a). The prediction would be worthless if it were *not* robust, as the distribution of points around the predicted relationships can be expected to be very large (Lynch, 1989). The lines in figure 8.5 indicate only trends. Many factors could affect the exact form of the slopes, but none should affect the *sign*, at least early in divergence. At some point long after speciation, however, ranges should become totally randomized (lose all information on mode of speciation; figure 8.5b). The effect of interspecific competition on the relationship for nonshifters should be to change the slope, but not the sign (figure 8.5c; in host-shifting speciation new species will not normally be competitors). The most significant factor is gene flow between host races (very significant in *R. pomonella*; Feder, this volume) and closely related species (possibly in stage 2 species), which could retard genetic divergence in early stages (figure 8.5d). However, bizarre conditions would be necessary for gene flow to change the sign of the slope. Basing divergence estimates solely on genes directly involved in speciation (Clarke et al., 1996) though impossible for most groups at present, would minimize problems due to gene flow. Alterna-

tively, for a conservative test one could include only species with very strong reproductive isolation (my stages three and four), although this would limit sample sizes, and test only part of the predicted pattern.

Unlike predictions 1 and 2, which focus entirely on the most closely related taxa, pairs of taxa at later stages of divergence can be used in the analyses. Inclusion of older taxa (i.e., nonsister species) may be necessary if range change is so slow that substantial change from initial ranges does not occur until late stages, and may also be useful in increasing the number of data points (as long as suitable procedures for ensuring statistical independence are used). Insular species and species on different continents should not be analyzed together with species pairs from the same continent, because of the great disparities in dispersal rates involved. Because post-speciation dispersal will almost certainly be episodic, and not simply stochastic, very large sample sizes will be needed. As with the previous predictions, an ideal test would involve paired comparisons of host-shifter and nonshifter subsets of each taxon represented, which would ensure comparisons with not only similar average dispersal abilities, but also similar average rates of molecular evolution. While the confidence limits of even a perfect molecular clock are large enough to make absolute dating difficult, Hillis et al. (1996) point out that molecular divergence can be used to estimate "relative times of divergence" with some confidence—and this is all that is needed for a test of prediction 4. Too few data for host-shifting taxa are available for a test at present.

Prediction 4:   Given three host shifter species A, B, and C, if A and B occur in one geographic area and C occurs in another, then the phylogeny will be (AB)C, and not (CB)A or (CA)B.

While I initially discussed this prediction in terms of species flocks in lakes and other restricted areas, in some instances it may be practical to test this prediction for larger geographic areas. Many published phylogenies applicable to testing prediction 3 concern taxa on archipelagoes, which have both advantages and disadvantages for testing sympatric speciation. On the one hand, a cluster of lakes or islands constitutes a set of replicate areas, so a replicated pattern of within-lake or within-island sister species distributions strongly supports sympatric speciation (Taylor et al., 1996; see chapters by Schluter, and McCune and Lovejoy, this volume). On the other hand, archipelagos offer perfect opportunities for allopatric speciation (e.g., the Hawaiian Islands; Wagner and Funk, 1995). Thus, phylogenies of a series of taxa on an archipelago, all with a life history conducive to sympatric speciation, will likely reveal both within-island *and* between-island sister species pairs (Asquith, 1995). Endemic species clusters restricted to *single* isolated islands or other small isolated areas are not uncommon (White, 1978, pp. 244–249), but such potentially informative taxa have received remarkably little study. The work of

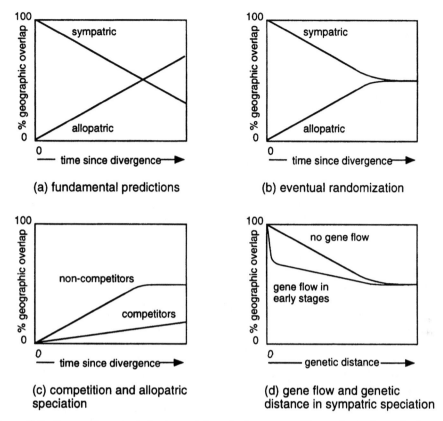

Figure 8.5. Proposed approach for determining whether geographic overlap and genetic distances between pairs of taxa can be used to determine mode of speciation. (a) Fundamental prediction for divergence time between pairs of species assuming allopatric or sympatric speciation. (b) Eventual randomization (loss of informativeness on mode of speciation) of ranges. (c) Effect of interspecific competition on range overlap under allopatric speciation (less overlap if species compete). (d) Gene flow in early stages of sympatric speciation could retard genetic divergence.

Schliewen et al. (1994), already discussed, illustrates the potential informativeness of such taxa.

Ultimately, arguments of the kind shown in figure 8.3 will need to be recast as weighted parsimony (or eventually, likelihood) analyses, to allow quantitative comparison of the probabilities of sympatric speciation with probabilities of allopatric speciation. A start on this problem has been made by vicariance biogeographers (Wiley, 1988), but much additional work is needed.

Prediction 5:   Additional tests of the sympatric speciation hypothesis can be designed using the intraspecific phylogeography approach.

Intraspecific phylogeography (Avise et al., 1987; Avise, 1994) is the biogeographic study of allele phylogenies (genealogies). Phylogeography can in theory tell us about past population subdivision, past changes in geographic ranges, ancestor–descendant relationships among closely related species, and ordering and polarity of ecological niches, behavior, and other characteristics among closely related species (Avise, 1994; Templeton et al., 1995)—things we need to know to infer mode of speciation. The power of phylogeography for testing the sympatric speciation hypothesis in phytophagous insects is just being realized; the pioneering works are the review of Harrison (1991) and empirical studies by two groups (Funk et al., 1995; Brown et al., 1996, 1997).

Mayr (1942, p. 199; see also Mayr 1963, pp. 461–464) proposed, in a modification of his multiple colonization argument, that clusters of related, sympatric sister species of parasites using different hosts, historically used to support sympatric speciation (Mayr, 1947), could come into existence via host shifts during allopatric isolation. The phylogeographic approach, however, can under the right circumstances refute such an argument. One such circumstance is the occurrence of a pair of sister species, on different hosts, that are neither completely allopatric nor broadly sympatric. To show how such a situation could be informative, I present in figure 8.6 a hypothetical phylogeographic analysis of the evolution of the

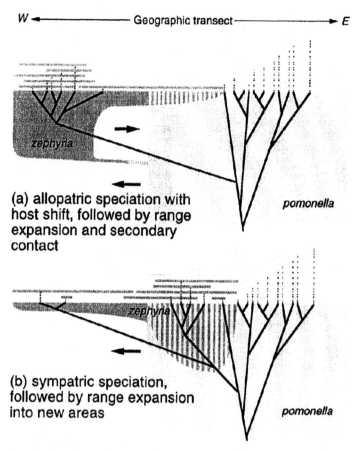

Figure 8.6. Two possible speciational origins and phylogeographic consequences for *R. zephyria*, largely allopatric to its sister species *R. pomonella*. The trees illustrate the phylogeny (genealogy) of a gene; the light and dark gray horizontal bars demark the geographic range of particular alleles of *R. pomonella* and *R. zephyria*, respectively. Time runs from bottom to top. (a) Pattern expected from allopatric speciation. (b) Pattern expected from host-shifting sympatric speciation. For further details, see text.

largely allopatric *R. zephyria*, which feeds on snowberry, from *R. pomonella*, infesting hawthorn. In both cases, the ancestral host is hawthorn, as inferred from the paraphyly of *R. pomonella* (see Brown et al., 1997). In figure 8.6a, *R. zephyria* has speciated by a geographic founder event, with a concomitant shift to snowberry. *R. pomonella*, having more time to reach genetic equilibrium than *R. zephyria*, has accumulated a larger number of more divergent sequences than *R. zephyria*. The area of overlap has resulted from a rapid episode of range expansion by *R. zephyria*, which has also left it genetically depauperate in the newly colonized area. In figure 8.6b, on the other hand, *R. zephyria* has arisen sympatrically in just part of the range of *R. pomonella*, and has then spread rapidly west out of the area of sympatry. The east-to-west decline in genetic variation resulted from the rapid expansion. If phylogeographic theory becomes robust enough to be reliably predictive, patterns like those in figure 8.6 could be used to infer mode of speciation with

some confidence (although some phylogeographic patterns may not be capable of rejecting any hypothesis).

Conclusions

I have frequently heard the statement, "Yes, sympatric speciation may be possible, but you can explain all cases equally well by allopatric mechanisms." I have attempted to demonstrate here that this is not so: some cases may in principle be best explained by a sympatric mode. Both the sympatric and the allopatric speciation hypotheses make testable predictions.

What conclusions can be drawn at present? Within the very narrow limits of the available data, I argue that some cases are in fact more parsimoniously explained by host/habitat-shifting sympatric speciation than by allopatric speciation. Oddly enough, considering that the primary focus of the sympatric speciation debate has been phytophagous and other parasitic insects, the strongest case can be made for lake fish, the premier example being the work of Schliewen et al. (1994) on West African crater lake cichlids. Other cases of lake speciation (Wood and Foote, 1996; see also Schluter, and McCune and Lovejoy, this volume) are also compelling.

The problem with phytophagous and parasitic insects is incomplete data. Despite substantial effort, effort that has taught us much about host races (see Feder, and Menken and Roessingh, this volume) and the evolutionary ecology of adaptation to new hosts (Futuyma et al., 1995), no genus or even species group is known completely enough to test the predictions I have discussed here. Consider the frequently discussed genera *Rhagoletis*, *Enchenopa*, and *Ophraella*. Among the problems are (1) uncertainty about geographic variation and species limits (*Ophraella*, *conferta-sexvittata*, Futuyma, 1990); (2) insufficient collection of all three genera in Mexico, which alone could easily contain as many species as are now known from the rest of North America; (3) incomplete information to apply to the (admittedly sometimes difficult) distinction between host race and sympatric species (*O. americana*, Futuyma, 1990; several *Enchenopa* populations, Guttman and Weigt, 1989); (4) species with ranges based on so few localities that ranges cannot be reliably estimated (*O. arctica*, *O. artemesiae*, Funk et al., 1995; several *Rhagoletis* species, Bush, 1966); (5) lack of compilation of range maps (*Enchenopa*, Guttman and Weigt, 1989); (6) phylogenies lacking known species that could have significant impact on biogeographic interpretations (*O. californica*, Futuyma et al., 1995; *R. terpinae*, Hernández-Ortiz, 1993); (7) poor phylogenetic resolution with available molecular data (*R. pomonella* group, Berlocher et al., 1993; Smith and Bush, 1997; McPheron and Han, 1997); and (8) no information on possible host plants (several *Rhagoletis* species, Bush, 1966). While the evidence for host-shifting sympatric speciation provided by host races is strong (see Feder, and Menken and Roessingh, this volume), the case from biogeography and phylogeny cannot yet be judged complete.

The limitations above are especially vexing when phytophagous insects are compared with the West African crater lake cichlids, where all the species in a lake have been discovered, all geographic ranges are known, all species are included in the phylogeny, and at least basic ecologies of most species are known. Of course, Lake Bermin has a total area of 0.6 km^2 (Schliewen et al., 1994), while the *R. pomonella* group ranges across millions of square kilometers from Canada to central Mexico, with dozens of unstudied potential host plants. Nonetheless, if the sympatric speciation debate is ever to be resolved, one or more genera of host-shifting insects must become known as well as, say, a typical bird genus. This may not seem feasible given the current funding climate for evolutionary biology, but the information potentially obtainable from arthropods is simply enormous. The number of insect species alone is vast, presenting a huge array of combinations of different levels

and types of interspecific competition and varying degrees and kinds of host fidelity, an ecological diversity that could be used to test many alternative hypotheses with great statistical power. Without better data the sympatric speciation debate is doomed to continue ad infinitum. If we can have a complete genome sequence for a bacterium or a yeast or a nematode, why not a complete systematics for a leaf beetle, leaf hopper, or fruit fly genus?

Appendix

Ranges were estimated by weighing areas cut out from photocopies of published maps, with overlap calculated as described in the text. Ranges that were estimated from verbal descriptions, or incomplete, contradictory, or otherwise confusing data are indicated by "~" before the overlap value. Data from the compilations of Lynch (1989, all data), Chesser and Zink (1994, table 2), and Wiley and Mayden (1985) formed the core of the non-shifter cases (percentage overlap from the former two reports were used without recalculating). Data used from Wiley and Mayden (1985) are as follows (values in % overlap):

Fundulus pulvereous and *confluentus*, 0
F. catenatus and *stellifera*, 0
Ammocrypta beani and *bifasciata*, 0
A. vivax and *pellucida*, 0
Hybopsis winchelli and cf. *winchelli*, 0
H. x-punctata and *cahni*, 0
Notropis euryzonus and *hypselopterus*, 50.0
N. atrapicilus and *rosepinnis*, 0
N. umbratilis and *ardens*, 5.3
N. pillsbryi and *zonatus*, 0
N. coccogenis and *zonistius*, 0
N. scepticus and *semperasper*, 0
N. nubilus and *leucoides*, 0
Percina sciera and *nigrofasciata*, 19.6
P. nasutata and sp. nova, 0
P. oxyrhyncha and *squamata*, 0
Etheostoma chlorosomum and *davisoni*, 0
E. tetrazonum and *euzonum*, 0
E. osburni and *kanawhae*, 5.5.

Additional species pairs were added from computer and text literature searches:

Nonshifter

Pleurotis (fungi) groups III and VII, 0
R. Vilgalys and B. L. Sun. 1994. Proc. Natl. Acad. Sci. USA. 91:4599–4603.

Limnoporus (predatory water strider bugs) *rufoscutellatus* and *genitalis*, 45.2
N. M. Andersen and J. R. Spence. 1992. Can. J. Zool. 70:753–785.

Elliptoleus (predatory carabid beetles) *crepericornis* and *whiteheadi, balli* and *tequilae*, and *acutesculpus* and *olisthopoides*, all 0

Sericoda (predatory carabid beetles) *obsoleta* and *bogemannii*, 57.6, and *quadripunctata* and *lissoptera*, 100

Anchomenus (predatory carabid beetles)*funebris* and *capensis*, 0
  J. K. Liebherr. 1991. Bull. Am. Mus. Nat. Hist. 202;
  J. K. Liebherr. 1994. Can. Entomol. 126:841–860.

Dalbulus (plant hoppers), *guevari* and *elimnatus*, 95.0, and *charlesi* and *tripsacoides*, 0
  B. W. Triplehorn and L. R. Nault. 1985. Ann. Entom. Soc. Am. 78:291–315.

Orellia (true fruit flies)*palposa* and*occidentalis*, 45.7
  G. J. Steck. 1981. North American Terrellinae (Diptera: Tephritidae): Biochemical systematics and evolution of larval feeding niches and adult life histories. Ph.D. thesis, University of Texas at Austin;
  R. H. Foote, F. L. Blanc, and A. L. Norrbom. 1993. Handbook of the Fruit Flies (Diptera: Tephritidae) of America North of Mexico. Comstock Publishing, Ithaca, N.Y.

Rhagoletis (true fruit flies) *completa* and *boycei*, 0
  Bush, 1966; Smith and Bush, 1997.

Ophraella (leaf beetles), *conferta* and *sexvitta*, ~10
  L. A. LeSage. 1986. Mem. Entomol. Soc. Can. 133: 1–75; Funk et al., 1995.

Vaccinium (diploid blueberry) *boreale* and *myrtilloides*, 91.7, and *tenellum* and *darrowi*, 24.9
  L. P. Bruederle and N. Vorsa. 1994. Syst. Bot. 19: 337–349.

Pinus (pine) *discolor* and *johannis*, 0
  J. Malusa. 1992. Syst. Bot. 17:42–64.

Asclepias (milkweed) *texana* and *perennis*, 0
  A. L. Edwards and R. Wyatt. 1994. Syst. Bot. 19: 291–307.

Salix (willow) *silicola* and *alaxensis*, 0
  B. G. Purdy and R. J. Bayer. 1995. Syst. Bot. 20: 179–190.

Coreopsis (daisy) *nuensensoides* and *nuensis*, 17.9
  D. J. Crawford and E. D. Smith. 1982. Evolution 36:379–386.

Hybrid speciation

Helianthus (sunflower) *anomalus* and (*annuus/petiolaris*), ~100
  Rieseberg et al., 1996.

Stephanomeria (composite plant)*diagensis* and (*exigua/virgata*), 82.4

L. D. Gottlieb. 1971. Evolution 25:312–329; G. P. Gallez and L. D. Gottlieb. 1982. Evolution 36:1158–1167.

Iris nelsonii and (*fulva/hexagona/brevicaulis*), 100
  M. L. Arnold. 1993. Am. J. Bot. 80:577–583.

Allopolyploid

Plagiomnium (moss) *medium* and (*ellipticum/insigne*), 100
  R. I. Wyatt, J. Odrzykoski, and A. Stoneburner. 1992. Syst. Bot. 17:532–550.

Gymnocarpium (fern) *dryopteris* and (*appalachianum/distjunctum*), 0
  Pryer and Haufler, 1993.

Asplenium (fern) *ebenoides* and (*rhizophyllum/platyneuron*), *bradleyi* and (*platyneuron/montanum*), and *pinnatifidum* and (*rhizophyllum/montanum*), all 0
  C. R. Werth, S. I. Guttman, and W. H. Eshbaugh. 1985. Science 228:731–733.

Tragopogon (salsify) *mirus* and (*dubius/porrifolius*), 100, and *miscellus* and (*dubius/pratensis*), 100
  S. J. Novak, D. E. Soltis, and P. S. Soltis. 1991. Am. J. Bot. 78:1586–1600.

Talium (rock-pink) *teretifolium* and (*mengesii/parviflorum*), 0
  W. H. Murdy and M. E. B. Carter. 1985. Am. J. Bot. 72:1590–1597.

Clarkia (annual flowering plant)*delicata* and (*unguiculata/epilobiodes*), *similis* and (*modesta/epilobiodes*), and *rhomboidea* and (*virgata/mildrediae*), all 0
  N. L. Huerta-Smith. 1986. J. Hered. 77:349–354;
  H. Lewis and M. E. Lewis. 1955. Univ. Cal. Pub. Bot. 20:241–363.

Lathyrus (vetch), *venosus* and (*palustris/ochroleucus*)
  J. F. Gutierrez, F. Vaquero, and F. J. Vences. 1994. Heredity 73:29–40; C. L. Hitchcock. 1952. Univ. Wash. Pub. Biol. 15:1–104, 100.

Gilia (annual flowering plant) *transmontana* and (*minor/clokeyi*), 0, and*malinor* and (*minor/aliquant*), 100
  A. Day. 1965. El Aliso 6:25–75.

Galeopsis (annual flowering plant) *tetrahit* and (*speciosa/pubescens*), 100
  Müntzing, 1932, cited in V. Grant. 1981. Plant Speciation (2nd ed.) Columbia University Press.

Spartina (marsh grass) *anglica* and (*townsendii/alternifolia*), 0
  A. F. Raybould, A. J. Gray, M. J. Lawrence, and D. F. Marshall. 1991. Biol. J. Linn. Soc. 43:111–126, 44:369–380.

Autotetraploid

Aster (annual flowering plant) *kantoensis* and *asagrayi*, 0
>M. Maki, M. Masuda, and K. Inoue. 1996. Am. J. Bot. 83:296–303.

Coreopsis (annual flowering plant) *longipes* and *grandiflora*
>D. J. Crawford and E. B. Smith. 1984. Syst. Bot. 9:219–225, 0.

Host shifter

Enchenopa (tree hoppers) "celastrus" and "ptelea," and "viburnum" and "cercis," both ~75
>Guttman and Weigt, 1989.

Rhagoletis (true fruit flies) *pomonella* and *zephyria*, 14.8
>McPheron and Han, 1997

"Sparkleberry" and "dogwood" flies, 100 (while the smallest Nei distance involving the dogwood fly involves *R. pomonella*, the bulk of the samples form a sister group to the undescribed "sparkleberry fly"
>Berlocher, unpublished

R. chionanthus and *osmanthus*, 100
>McPheron and Han, 1997, Smith and Bush, 1997

R. indifferens and *cingulata*, 0
>McPheron and Han, 1997

Ophraella (leaf beetle), *notulata* and *slobodkini*, 100, *arctica* and *bilineata*, 0, and *artemisiae* and *nuda*, 0
>Funk et al., 1995.

Acknowledgments   I thank Ron Prokopy, Gary Steck, Bruce McPheron, Steve Sheppard, D. Courtney Smith, Jeff Feder, Dorothy Prowell, Jerry Payne, and of course, the inimitable Guy Bush, for many wonderful hours in the field, in the lab, and at some great parties. Jeff Feder, Dan Howard, Robert Zink, and Douglas Futuyma suggested many improvements. This chapter is dedicated to Jasper Loftus-Hills, who should have been there.

References

Anderson, S., and M. K. Evensen. 1978. Randomness in allopatric speciation. Systematic Zoology 27:421–430.

Asquith, A. 1993. Patterns of speciation in the genus *Lopidea* (Heteroptera: Miridae: Orthotylinae). Systematic Entomology 18:169–180.

Asquith, A. 1995. Evolution of *Sarona* (Heteroptera, Miridae). In W. L. Wagner and V. A. Funk (eds.), Hawaiian biogeography: evolution over a hot spot. Washington, D.C.: Smithsonian Institution Press, pp. 90–120.

Avise, J. C. 1994. Molecular markers, natural history, and evolution. New York: Chapman and Hall.

Avise, J. C., J. Arnold, R. M. Ball, E. Bermingham, T. Lamb, J. E. Neigel, C. A. Reeb, and N. C. Saunders. 1987. Intra-specific phylogeography: the mitochondrial DNA bridge between population genetics and systematics. Annual Review of Ecology and Systematics 18:489–522.

Berlocher, S. H. 1989. The complexities of host races and some suggestions for their identification by enzyme electrophoresis. In Hugh D. Loxdale and J. den Hollander (eds.), Electrophoretic studies on agricultural pests. Systematics Association special volume No. 39. Oxford: Clarendon Press, pp. 51–68.

Berlocher, S. H. 1995. Population structure of the blueberry maggot, *Rhagoletis mendax*. Heredity 74:542–555.

Berlocher, S. H., and B. A. McPheron. 1996. Population structure of *Rhagoletis pomonella*, the apple maggot fly. Heredity 77:83–99.

Berlocher, S. H., B. A. McPheron, J. L. Feder, and G. L. Bush, 1993. Genetic differentiation at allozyme loci in the *Rhagoletis pomonella* (Diptera: Tephritidae) species complex. Annals of the Entomological Society of America 86:716–727.

Bierbaum, T. J., and G. L. Bush. 1990. Genetic differentiation in the viability of sibling species of *Rhagoletis* fruit flies on host plants, and the influence of reduced hybrid viability on reproductive isolation. Entomologia experimentalis et applicata 55:105–108.

Brown, J. M., W. G. Abrahamson, and P. A. Way. 1996. Mitochondrial DNA phylogeography of host races of the goldenrod ball gallmaker *Eurosta solidaginis* (Diptera: Tephritidae). Evolution 50:777–786.

Brown, J. M., J. H. Leebens-Mack, J. N. Thompson, O. Pellmyr, and R. G. Harrison. 1997. Phylogeography and host association in a pollinating seed parasite, *Greya politella* (Lepidoptera: Prodoxidae). Molecular Ecology 6:215–224.

Bush, G. L. 1966. The taxonomy, cytology, and evolution of the genus *Rhagoletis* in North America (Diptera, Tephritidae). Bulletin of the Museum of Comparative Zoology 134:431–562.

Bush, G. L. 1969. Sympatric host race formation and speciation in frugivorous flies of the genus *Rhagoletis* (Diptera, Tephritidae). Evolution 23:237–251.

Bush, G. L. 1975. Modes of animal speciation. Annual Review of Ecology and Systematics 6:339–364.

Bush, G. L. 1992. Host race formation and sympatric speciation in *Rhagoletis* fruit flies (Diptera: Tephritidae). Psyche 99:335–358.

Bush, G. L. 1994. Sympatric speciation in animals: new wine in old bottles. Trends in Ecology and Evolution 9:285–288.

Bush, G. L., and D. J. Howard. 1986. Allopatric and non-allopatric speciation: assumptions and evidence. In S. Karlin and E. Nevo (eds.), Evolutionary process and theory. New York: Academic Press, pp. 411–438.

Bush, G. L., and J. J. Smith. 1997. The sympatric origin of phytophagous insects. In K. Dettner, G. Bauer, and W. Völkl (eds.), Vertical food web interactions: evolutionary patterns and driving forces. Ecological Studies 130:3–19. Heidelberg: Springer-Verlag.

Chesser, R. T., and R. M. Zink. 1994. Modes of speciation in birds: a test of Lynch's method. Evolution 48:490–497.

Clarke, B., M. S. Johnson, and J. Murray. 1996. Clines in the genetic distance between two species of island land snails: how molecular leakage can mislead us about speciation.

Philosophical Transactions of the Royal Society of London Series B 351:773–784.

Conner, E. F., and M. A. Bowers. 1987. The spatial consequences of interspecific competition. Annals Zoologica Fennicia 24: 213–226.

Coyne, J. A., and H. A. Orr. 1997. "Patterns of speciation in *Drosophila*" revisited. Evolution 51:295–303.

Denno, R. F., M. S. McClure, and J. R. Roth. 1995. Interspecific interactions in phytophagous insects: competition reexamined and resurrected. Annual Review of Entomology 40:297–332.

Diehl, S. R., and G. L. Bush. 1984. An evolutionary and applied perspective of insect biotypes. Annual Review of Entomology 29:471–504.

Diehl, S. R., and G. L. Bush. 1989. The role of habitat preference in adaptation and speciation. In D. Otte and J. A. Endler (eds.), Speciation and its consequences. Sunderland, Mass.: Sinauer, pp. 527–553.

Emelianov, I., J. Mallet, and W. Baltensweiler. 1995. Genetic differentiation in *Zeiraphera diniana* (Lepidoptera: Tortricidae, the larch budmoth): polymorphism, host races, of sibling species? Heredity 75:416–424.

Feder, J. L., and G. L. Bush. 1989. A field test of differential host-plant usage between two sibling species of *Rhagoletis pomonella* fruit flies (Diptera: Tephritidae), and its consequences for sympatric models of speciation. Evolution 43:1813–1819.

Funk, D. J., D. J. Futuyma, G. Orti, and A. Meyer. 1995. A history of host associations and evolutionary diversification for *Ophraella* (Coleoptera: Chrysomelidae): new evidence from mitochondrial DNA. Evolution 49:1008–1017.

Futuyma, D. J. 1990. Observations on the taxonomy and natural history of *Ophraella* Wilcox (Coleoptera: Chrysomelidae), with a description of a new species. Journal of the New York Entomological Society 98:163–186.

Futuyma, D. J., and G. C. Mayer. 1980. Non-allopatric speciation in animals. Systematic Zoology 29:254–271.

Futuyma, D. J., M. C. Keese, and D. J. Funk. 1995. Genetic constraints on macroevolution: the evolution of host affiliation in the leaf beetle genus *Ophraella*. Evolution 49:797–809.

Guttman, S. I., and L. A. Weigt. 1989. Macrogeographic genetic variation in the *Enchenopa binotata* complex (Homoptera: Membracidae). Annals of the Entomological Society of America 82:156–165.

Harrison, R. G. 1991. Molecular changes at speciation. Annual Review of Ecology and Systematics 22:281–308.

Hernández-Ortiz, V. 1993. Description of a new *Rhagoletis* species from tropical Mexico (Diptera: Tephritidae). Proceedings of the Entomological Society of Washington 95:418–424.

Hillis, D. M., B. K. Mabel, and C. Moritz. 1996. Applications of molecular systematics. In D. M. Hillis, C. Moritz, and B. K. Mable (eds.), Molecular systematics (2nd ed.). Sunderland, Mass.: Sinauer, pp. 407–514.

Jenkins, J. 1996. Systematic studies of *Rhagoletis* and related genera. Ph.D. thesis, Michigan State University.

Jordan, D. S. 1908. The law of geminate species. American Naturalist 42:73–80.

Linsley, E. G. 1959. Ecology of Cerambycidae. Annual Review of Entomology 4:99–138.

Lynch, J. D. 1989. The gauge of speciation: on the frequencies of modes of speciation. In D. Otte and J. A. Endler (eds.), Speciation and its consequences. Sunderland, Mass.: Sinauer, pp. 527–553.

Mayden, R. L., R. H. Matson, and D. M. Hillis. 1992. Speciation in the North American genus *Dionda* (Teleosti: Cypriniformes). In R. L. Mayden (ed.), Systematics, historical ecology, and North American fishes. Stanford, Calif.: Stanford University Press, pp. 710–746.

Mayr, E. 1940. Speciation phenomena in birds. American Naturalist 74:249–278.

Mayr, E. 1942. Systematics and the origin of species. From the viewpoint of a zoologist. New York: Columbia University Press.

Mayr, E. 1947. Ecological factors in evolution. Evolution 1:263–288.

Mayr, E. 1963. Animal Species and Evolution. Cambridge, Mass.: Belknap Press.

Mayr, E. 1984. Evolution of fish species flocks: A commentary. In A. A. Echelle and I. Kornfield (eds.), Evolution of fish species flocks. Orono: University of Maine, pp. 3–12.

McPheron, B. A. 1990. Genetic structure of apple maggot fly (Diptera: Tephritidae) populations. Annals of the Entomological Society of America 83:568–577.

McPheron, B. A., and H. Y. Han. 1997. Phylogenetic analysis of North American *Rhagoletis* (Diptera: Tephritidae) and related genera using mitochondrial DNA sequence data. Molecular Phylogenetics and Evolution 7:1–16.

Payne, J. A., and S. H. Berlocher. 1995. Phenological and electrophoretic evidence for a new blueberry-infesting species in the *Rhagoletis pomonella* (Diptera: Tephritidae) sibling species complex. Entomologia experimentalis et applicata 75:183–187.

Prokopy, R. J., and G. L. Bush. 1973. Ovipositional responses to different sizes of artificial fruit by flies of the *Rhagoletis pomonella* species group. Annals of the Entomological Society of America 66:927–929.

Pryer, K. M., and C. H. Haufler. 1993. Isozymic and chromosomal evidence for the allotetraploid origin of *Gymnocarpium dryopteris* (Dryopteridaceae). Systematic Botany 18:150–172.

Rieseberg, L. H., B. Sinervo, C. R. Linder, M. C. Ungerer, and D. M. Arias. 1996. Role of gene interactions in hybrid speciation: evidence from ancient and experimental hybrids. Science 272:741–745.

Schliewen, U. K., D. Tautz, and S. Pääbo. 1994. Sympatric speciation suggested by monophyly of crater lake cichlids. Nature 368:629–632.

Smith, D. C. 1986. Genetics and reproductive isolation of *Rhagoletis* flies. Ph.D. thesis, University of Illinois at Urbana-Champaign.

Smith, D. C., S. A. Lyons, and S. H. Berlocher. 1993. Production and electrophoretic verification of F_1 hybrids between the sibling species *Rhagoletis pomonella* and *Rhagoletis cornivora*. Entomologia experimentalis et applicata 69: 209–213.

Smith, G. R., and T. N. Todd. 1984. Evolution of species flocks of fishes in North Temperate Lakes. In A. A. Echelle and

I. Kornfield (eds.), Evolution of fish species flocks. Orono: University of Maine, pp. 45–68.

Smith, J. J., and G. L. Bush. 1997. Phylogeny of the genus *Rhagoletis* (Diptera: Tephritidae) inferred from DNA sequences of mitochondrial cytochrome oxidase II. Molecular Phylogenetics and Evolution 7:33–43.

Tauber, C. A., and M. J. Tauber. 1989. Sympatric speciation in insects. In D. Otte and J. A. Endler (eds.), Speciation and its consequences. Sunderland, Mass.: Sinauer, pp. 307–344.

Taylor, E. B., C. J. Foote, and C. C. Wood. 1996. Molecular genetic evidence for parallel life-history evolution within a pacific salmon (Sockeye and Kokanee, *Oncorhynchus nerka*). Evolution 50:401–416.

Templeton, A. R., E. Routman, and C. A. Philips. 1995. Separating population structure from population history: a cladistic analysis of the geographic distribution of mitochondrial DNA haplotypes in the tiger salamander, *Ambystoma tigrinium*. Genetics 140:767–782.

Tuskes, P. M., J. P. Tuttle, and M. M. Collings. 1996. The wild silk moths of North America. A natural history of the Saturniidae of the United States and Canada. Ithaca, N.Y.: Comstock.

Wagner, W. L., and V. A. Funk, eds. 1995. Hawaiian biogeography: evolution over a hot spot. Washington, D.C.: Smithsonian Institution Press.

White, M. J. D. 1978. Modes of speciation. San Francisco: Freeman.

Wiley, E. O. 1988. Vicariance biogeography. Annual Review of Ecology and Systematics 19:513–542.

Wiley, E. O., and R. L. Mayden. 1985. Species and speciation in phylogenetic systematics, with examples from the North American fish fauna. Annals of the Missouri Botanical Garden 72:596–635.

Wood, C. C., and C. J. Foote. 1996. Evidence for sympatric divergence of anadromous and nonanadromous morphs of Sockeye salmon (*Oncorhynchus nerka*). Evolution 50:1265–1279.

9

Ecological Causes of Speciation

Dolph Schluter

The genotype of a species is an integrated system adapted to the ecological niche in which the species lives. Gene recombination in the offspring of species hybrids may lead to formation of discordant gene patterns.

T. Dobzhansky (1951)

How environments cause new species to form remains a fundamental problem in the search for diversity's origins. Major ideas on the subject were elaborated and discussed earlier this century (Fisher 1930; Wright 1940; Muller 1940; Mayr 1942, 1963; Simpson 1944, 1953; Dobzhansky 1951) but were not resolved. Few tests of ecological mechanisms in speciation have been carried out since then, and consequently our understanding of them has advanced little. Interest in the topic waned somewhat through recent decades as researchers turned their attention to purely genetic causes of speciation such as drift and founder events (Carson and Templeton 1984; Barton and Charlesworth 1984; Barton 1989). During this time, discussion of ecological factors was kept alive chiefly in the section of literature devoted to the possibility of speciation in sympatry (Maynard Smith 1966; Wilson and Turelli 1986; Diehl and Bush 1989; Tauber and Tauber 1989; Rice and Hostert 1993; Bush 1994). However, the question of environmental causes is a general one, crucial to understanding speciation whatever the geographical arrangement of diverging populations.

Here I review the two principal ideas about how ecological processes drive speciation. The first is the hypothesis of "ecological speciation": divergent natural selection pressures drive the accumulation of differences causing reproductive isolation between populations in distinct environments (Mayr 1942, 1963; Dobzhansky 1951). This is a general hypothesis that applies to both sympatric and allopatric speciation, regardless of whether prezygotic isolation evolves entirely as a by-product of selection or additionally involves reinforcement. The second view is that of "ecological persistence": ecological processes principally affect speciation via their influ-

ence on the viability of populations undergoing speciation rather than by an influence on the rate of evolution of reproductive isolation within these populations (Mayr 1963). Under this view, reproductive isolation evolves by mechanisms less closely tied to the evolution of ecological differences between populations.

I address these two views with recent evidence from very young species in nature, particularly from vertebrate lineages undergoing adaptive radiation. I draw frequently on my own work and that of others on fishes in postglacial lakes, where systematic investigations of the hypothesis of ecological speciation are underway. I summarize accumulating evidence that resource-based divergent natural selection indeed plays a major role in the evolution of reproductive isolation. The evidence is nevertheless incomplete, and I outline some types of data still needed. Less evidence is available to test the hypothesis of ecological persistence. Environmental effects on population viability are likely to be present, but they remain poorly documented.

By speciation I mean the evolution of reproductive isolation. Reproductive isolation is defined as the complete absence of interbreeding between individuals from different populations (should they encounter one another), or the strong restriction of gene flow sufficient to prevent collapse of genetically distinct populations that interbreed occasionally. This is the biological species concept (Mayr 1942) relaxed to accommodate the fact that a great many sexual species hybridize (e.g., Grant 1981; Gill 1989; Grant and Grant 1992; Rieseberg and Wendel 1993; Mallet 1995), yet existing levels of assortative mating do not decay. "Ecological processes" are the interactions between individual organisms and their environment that determine individual fitness and thereby generate population dynamics and natural selection.

Two Views of How Ecology Drives Speciation

The idea that ecology must have *some* relevance to speciation is prompted by the frequent observation that the rate of speciation (or speciation rate minus extinction rate) varies greatly with ecological circumstances. The most conspicuous example is adaptive radiation, whereby a lineage experiences a burst of speciation and rapid phenotypic evolution under conditions of high ecological opportunity, such as colonization of resource-rich environments free of competitors and/or predators, including remote archipelagoes and in newly formed lakes (Huxley 1942; Simpson 1944, 1953; Lack 1947; Amadon 1950; Mayr 1963; Fryer and Iles 1972); after mass extinctions (Stanley 1979; Jablonski 1986); following the acquisition of novel adaptations for consuming previously untapped resources (Simpson 1944, 1953; Heard and Hauser 1994); and after massive nutrient inputs to the biosphere (Vermeij 1995).

Figure 9.1 gives an example of the correlation. It shows that speciation rates of recent lineages in novel environments (remote archipelagos and newly formed lakes) are elevated by a factor of about 2 over rates in control taxa inhabiting environments more saturated with species (continents) or having a lower diversity of resources (rivers) (Wilcoxon paired-sample test, $n = 7$, $P = 0.015$). A "control" taxon is the hypothesized sister taxon or, when this is unknown, a group closely related and ecologically similar to the taxon in the novel environment.

This estimate of a twofold difference is probably biased by the disproportionate representation in the literature of highly spectacular cases of adaptive radiation. A more systematic coverage of taxa might reduce the effect in figure 9.1, but probably not eliminate it. Note also that speciation rates in control taxa are often high— for example, speciation rate in the *Carduelis-Acanthis-Serinus* clade is nearly as high as in the related Hawaiian honeycreepers. However, this rate is achieved by accumulation of species across several continents. The extraordinary aspect of the Hawaiian and other bursts is not merely the high rate of speciation but the extremely confined geographical areas in which they occurred. Large differences between taxa in speciation rate are also evident (figure 9.1), but this may be a spurious consequence of different rates of molecular evolution.

Efforts to explain the correlation between ecological opportunity and speciation rate highlight the two major views of environments and speciation. The first hypothesis, *ecological speciation*, is that reproductive isolation evolves from the same forces that cause phenotypic differentiation, namely, divergent selection stemming from use of alternative environments and from resource competition (Huxley 1942; Simpson 1944, 1953; Dobzhansky 1951; Mayr 1942, 1963). Reproductive isolation evolves as a by-product of phenotypic differentiation and may involve reinforcement of prezygotic isolation in the later stages of divergence. Speciation by sexual selection is a variant of ecological speciation if mating preferences leading to reproductive isolation evolve because of divergent ecological selection pressures (Schluter and Price 1993). Under this view, speciation rates are high in adaptive radiation because reproductive isolation evolves most quickly when divergent selection is strongest.

The second view, *ecological persistence*, is that ecological processes affect speciation mainly through their influence on the viability of populations undergoing speciation (Mayr 1963; Farrell et al. 1991; Allmon 1992; Heard and Hauser 1994). Under this view, species accumulate rapidly in novel environments because more populations there avoid extinction for long enough to evolve reproductive isolation. This is in contrast to ecological speciation, where reproductive isolation evolves more quickly within populations in novel environments. For example, the absence of competitors and predators in novel environments may lead to large population densities, reducing the chances of extinction, with the result that more populations evolve reproductive isolation. Reproductive isolation itself may evolve via a number of nonecological mechanisms including drift (Wright 1940) (including nonadaptive divergence in mating preferences; Fisher 1930; Lande 1981), founder events (Mayr 1954, 1963), and the fixation of alternative advantageous alleles in populations experiencing identical selection pressures (Muller 1940).

This second hypothesis may be attributed to Mayr (1963), who felt that the principal significance of niche shift was that it enhanced population persistence, and thereby allowed the lengthy speciation process to go to completion: "We see again and again that an incipient species can complete the process of speciation only if it can find a previously unoccupied niche" (p. 574). Mayr (1963) considered both ecological and nonecological mechanisms as causes of speciation, although he emphasized, and is best remembered for, the founder event model. Mayr (1963) also recognized that variation in speciation rate could be caused by variation in both selection and persistence times, although he emphasized the latter. The persistence view of ecology and speciation rate is probably the predominant one among contemporary evolutionists (Farrell et al. 1991; Allmon 1992; Heard and Hauser 1994).

Not all aspects of these two hypotheses are mutually exclusive, since ecological speciation admits an influence of environment on population viability; and selection may contribute to the evolution of reproductive isolation even under the persistence view. The major clash between them is over their emphasis on different mechanisms respon-

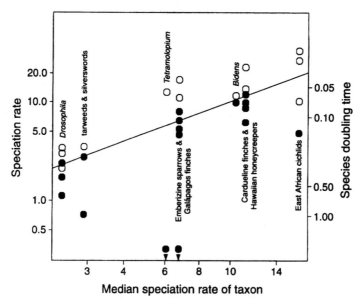

Figure 9.1. Per capita speciation rates of clades in novel environments (○) and in closely related "control" lineages inhabiting other environments (●). Rate estimates (y; left axis) are plotted against a dummy variable, the median of y-values. The solid line indicates $y = x$; points above the line therefore exhibit high rates. Rates were calculated from phylogenies based on allozyme frequencies. Time is measured in units of Nei's (1972) or Rogers's (1972) genetic distance D. The calculation of rate y assumes exponential growth of species number: $y = \ln(N)/t$, where N is the number of extant species in a clade and t is its estimated time of origin. If the time at which a clade originated was not known, I used the time of the first splitting event within the clade instead and set N equal to half the number of species in the clade. Corresponding times required for species number to double are indicated on the right; the number of species in a clade doubles after $\ln(2)/y$ time units.

1. *Drosophila.* Hawaiian (○) versus mainland taxa. Individual points are (ordered from highest rate to lowest) *planitibia* subgroup (Johnson et al. 1975); Hawaiian picture wing group, assumed monophyletic (Ayala 1975); species in the *willstoni* group (Ayala et al. 1974); Hawaiian white-tipped group, assumed monophyletic (Ayala 1975); *obscura* subgroup (Lakovaara et al. 1972; Cabrera et al. 1983); *melanogaster* subgroup (Eisses et al. 1979).
2. Hawaiian silverswords (○) versus two possible sister taxa from North America, the tarweeds *Madia* and *Raillardiopsis* (Baldwin et al. 1991). I assume that the latter are monophyletic, but they may be paraphyletic (Baldwin et al. 1991), in which case speciation rate in the true sister group to the Hawaiian clade is overestimated.
3. Hawaiian asters in the genus *Tetramolopium* (○) versus its probable sister clade, *T. alinae* from New Guinea (Lowrey 1995; Lowrey and Crawford 1985; T. Lowrey, personal communication).
4. The Galápagos finches (○) versus mainland emberizine finches. Individual points are (ordered from highest to lowest) species in the genus *Geospiza* only; all Galápagos finch species (Yang and Patton 1981); species in the "greater *Zonotrichia*" clade; *Zonotrichia* species only (Zink 1982); *Ammodramus* only (Zink and Avise 1990); *Melospiza* only; *Passerella* only (Zink 1982). Rates of speciation in mainland emberizines and carduelines (see 6 below) combined showed a declining relationship with increasing age of the taxon ($r = -0.26$). All rates (including the island groups) were therefore adjusted for age using a linear regression of speciation rate on time.
5. Hawaiian *Bidens* (beggar's ticks) (○) versus probable mainland sister *Bidens* clade (Helenurm and Ganders 1985; F. Ganders, personal communication). The count of Hawaiian species does not include derived Polynesian forms from the same clade. The mainland count excludes two Jamaican species.
6. The Hawaiian honeycreepers (○) versus mainland cardueline finches. Individual points are (ordered from highest to lowest) species in the "greater *Psittirostra*" clade, including subfossil species (Olson and James 1982; Johnson et al. 1991); all Hawaiian honeycreepers (Olson and James 1982; Johnson et al. 1991); species in the clade including *Carduelis*, *Acanthis*, and *Serinus* (Marten and Johnson 1986); species in the *Loxia* clade, where *L. curvirostra* is assumed to represent 12 species (Marten and Johnson 1986; Groth 1990; C. W. Benkman,

sible for the evolution of reproductive isolation, and I therefore concentrate on this issue. Genetic drift and other nonecological mechanisms eventually lead to reproductive isolation between populations isolated geographically for a long time, but divergent selection can cause this to happen very rapidly. The question is: Does it commonly do so?

A third explanation for the patterns in figure 9.1 is that they result from differences between environments in opportunities for geographical isolation. This is unlikely. Geographic isolation might indeed aid speciation in these environments (Lack [1947] presents evidence of this for the Galápagos finches), but opportunities for geographic isolation are surely no greater in Hawaii and Galápagos than in whole continents and their satellite islands, where the control taxa have diversified (figure 9.1). The East African lakes are not more subdivided spatially than adjacent rivers.

Evidence of Ecological Speciation

I evaluate five kinds of data. First, I review evidence that reproductive isolation evolves rapidly during adaptive radiation, when niche diversification also takes place at a high rate. Second, I summarize evidence that young, ecologically different species persist despite gene flow, in some cases having originated in sympatry. Third, I show that selection against hybrids of very young species may result from ecological mechanisms rather than from genetic incompatibilities between parental genomes. Fourth, I summarize comparative evidence that premating isolation has diverged along with the phenotype during a niche shift. Finally, I ask whether genetic mechanisms of postzygotic isolation, when they occur, evolved by divergent selection instead of by nonecological mechanisms.

Niche Shift and the Rate of Evolution of Reproductive Isolation

A straightforward test of ecological speciation compares the rate at which reproductive isolation evolves between regions differing in the frequency and strength of divergent natural selection. For example, does reproductive isolation evolve more quickly in novel environments such as on remote archipelagos (figure 9.1) than on mainlands? Limited data on this are available from laboratory measures of prezygotic isolation in *Drosophila* (table 9.1).

Average strength of isolation is greater between species or subspecies of Hawaiian picture winged flies (*planitibia* subgroup) than between continental *Drosophila* of similar age, but the pattern is weak. These observations are few, they are not independent, and the pattern is certainly not statistically significant. Nevertheless, they show the kind of data needed to compare rates of evolution of reproductive isolation. Additional caution is warranted in this case given that so little information exists on the frequency and strength of divergent selection on ecologically relevant traits in Hawaiian and continental *Drosophila*.

A second test compares the strength of pre- and postzygotic isolation among equal-age pairs of sister populations varying in the degree of similarity of their environments. If ecological speciation is the norm, then we would predict that reproductive isolation should evolve most quickly between sister populations undergoing extensive ecological differentiation, assuming that the latter reflects strong divergent natural selection. Unfortunately, forces other than divergent selection can also produce a correlation between niche shift and reproductive isolation. Chance dispersal can create a habitat shift in one of two sister populations and instantly produce prezygotic isolation through an alteration of breeding time or location. Tests of the second prediction should therefore consider only cases in which selection is the probable cause of niche shift. Alternatively, if the niche shift itself was caused by other agents, then the test should compare only the portion of reproductive isolation that accumulated *after* the shift.

This second prediction has not been systematically tested with species in nature. The best evidence in support is the observation that many of the youngest species on Earth are strongly differentiated ecologically (e.g., Grant 1986; Bush 1994). This pattern lacks a control and hence is preliminary, but it nevertheless provides an indication. Below I summarize recent examples of this phenomenon in fishes of postglacial lakes, which are among the youngest species in nature. The cases are interesting because the same pattern of ecological separation is so often repeated, and also because divergence is mainly in foraging habitat rather than in breeding habitat. I review evidence that selection is indeed the cause of ecological differentiation. I give more details of some of these cases in later sections where further evidence for ecological speciation is reviewed.

Many examples of very young fish species in lakes of previously glaciated areas of the northern hemisphere

---

personal communication); *Carpodacus* (Marten and Johnson 1986); *Coccothraustes* (Marten and Johnson 1986). See note in item 4 concerning adjustments for taxon age.
7. Cichlid fishes of the East African Rift Valley lakes (○; Victoria, Malawi and Tanganyika) versus the riverine sister genus, *Astatotilapia*, of one of the lacustrine clades (Meyer 1993; Sage et al. 1984). Speciation rate in the true sister clade is overestimated because *Astatotilapia* is probably paraphyletic, not monophyletic (Meyer 1993).

Table 9.1. Levels of premating isolation (0 = no, 1 = full) between *Drosophila* species in Hawaii and on continents.

Hawaiian Species Pair	Isolation Index	Continental Species Pair	Isolation Index
Sympatric species			
heteroneura–sylvestris	.83	*americana–texana*	.24
		ananassae–palidosa	.90
		athabasca EA–*athabasca* EB	.17
		paulistorum AB–*paulistorum* T	.71
Mean isolation	.83		.50
Allopatric species			
differens–sylvestris	.77	*pseudoobscura* Bog–*pseudoobscura* USA	.22
differens–heteroneura	.43	*equinoxialis caribbensis–e. equinoxialis*	.21
differens–planitibia	.20	*willistoni quechua–w. willistoni*	.30
heteroneura–planitibia	.55		
planitibia–sylvestris	.52		
Mean isolation	.50		.24

All comparisons are of species or subspecies of similar age, as judged by degree of differentiation at allozyme loci. Sympatric and allopatric species are compared separately. Sympatric species used are 0.0–0.1 units of Nei's (1972) genetic distance apart; allopatric species are 0.1–0.3 units apart. Data were obtained from lab measurements of assortative mating compiled by Coyne and Orr (1989). Laboratory measures of postmating isolation were similar in Hawaii and on mainlands.

have recently been confirmed. Table 9.2 lists sympatric pairs between which genetic distances (Nei's *D* or equivalent) are small but significant (e.g., $0 < D \leq 0.06$), independent evidence of assortative mating exists, and morphological differences are likely inherited rather than environmentally induced (Schluter 1996). The lakes in which they occur are usually less than 15,000 years old, generally putting an upper limit on the duration of sympatry. Still other cases are known that were not included in table 9.2 (Behnke 1972; Svärdson 1979; Schluter and McPhail 1993; Robinson and Wilson 1994; Schluter 1996) because genetic confirmation of species status is still lacking.

As is true in many adaptive radiations, the sympatric species are highly differentiated ecologically (table 9.2). Most remarkably, pairs usually divide resources in the same way: typically, one species is a pelagic zooplanktivore whereas the other consumes benthic prey from the littoral zone or deeper sediments. A consistent set of morphological differences is associated with this habitat split: planktivores are smaller and more slender than benthivores and tend to have narrower mouths and longer, more numerous gill rakers.

There is strong evidence that divergent selection is responsible for foraging habitat differences. First, morphological divergence has rendered sympatric species differently capable of exploiting pelagic and littoral or benthic habitats (Schluter 1993; Malmquist 1992; see also Robinson et al. 1996). In the open water habitat (transplanted to large aquaria) the "limnetic" species of stickleback captured plankton at three times the rate achieved by the "benthic" species; this advantage was reversed in

the littoral zone (Schluter 1993). Second, growth rates measured in a transplant experiment in the wild mirrored results on foraging efficiency: limnetic sticklebacks grew at double the rate of benthics in open water, whereas benthics held an equivalent advantage in the littoral zone (Schluter 1995). Third, observations and experiments suggest that ecological and morphological divergence in sympatry is greater now than it was in the past, and that it was driven by competition for food. Stickleback species occurring alone in small lakes ("solitary") are morphologically intermediate between limnetics and benthics and exploit both habitats (Schluter and McPhail 1992). In a pond experiment, natural selection favored the more benthiclike individuals within a solitary species following introduction of a planktivore (Schluter 1994).

Species may also exhibit differences in breeding habitats, but these are often less than dissimilarities in foraging habitat. For example, sympatric stickleback species breed side by side in the littoral zone, although they place their nests in somewhat different microhabitats (Hatfield 1995). Anadromous sockeye salmon breed in many of the rivers in which kokanee also spawn (Ricker 1940). This dissociation between feeding and breeding contrasts with many specialized insects that mate where they fed as larvae, such that premating isolation evolves in step with foraging habitat specialization (Bush 1994).

Ecological and Genetic Mechanisms of Hybrid Fitness

Contrasting natural selection on phenotypes implies that hybrids, if they are intermediate in phenotype, will be

Table 9.2. Examples of very young sympatric species pairs in lakes and rivers of recently glaciated areas.

Nominal Species	Region	Trophic Characteristics	Genetic Difference
Threespine stickleback *Gasterosteus aculeatus*	British Columbia	Limnetic (planktivore) Benthic (benthivore)	0.02/—/0.02–0.10
Lake whitefish *Coregonus clupeaformis*	E. Canada, Maine	Dwarf (planktivore) Normal (benthivore)	0.01/0.5/—
Lake whitefish *C. clupeaformis*	Yukon, Alaska	High gill rakers (planktivore) Low gill rakers (benthivore)	0.01–0.02/—/0.02–0.30
Sockeye salmon *Oncorhynchus nerka*	W. Canada, Alaska	Sockeye (anadromous) Kokanee (freshwater resident)	0.02/—/—
Atlantic salmon *Salmo salar*	Newfoundland	Anadromous Freshwater resident	0.06/—/—
Brown trout *Salmo trutta*	Ireland	Sonaghen (planktivore) Gillaroo (benthivore)	0.04/—/0.08
Brown trout *S. trutta*	Sweden	Planktivore Bbenthivore	0.03/—/—
Arctic charr *Salvelinus alpinus*	Scotland	Planktivore Benthivore	0.02/—/—
Arctic charr *S. alpinus*	Iceland	Planktivore and piscivore Small and large benthivore	0.001/—/0.01
Rainbow smelt *Osmerus mordax*	E. Canada, Maine	Dwarf (planktivore) Normal (benthivore, piscivore)	—/—/0.01—0.10

Genetic differences are given as $x/y/z$, where x is Nei's distance based on electrophoresis; y is % mtDNA sequence divergence estimated using restriction enzymes; z is % mtDNA nucleotide divergence (Nei and Miller 1990), which combines differences in both nucleotide sequence and haplotype frequency. In the Icelandic char example, four forms are present in Lake Thingvallavatn, and their relationships and species status are still unclear. Preliminary indications are that two genetically distinct lineages are present, each of which has two developmental morphs (Skúlason et al. 1992). This table is updated from table 1 in Schluter (1996, 1998), and includes additional information on mtDNA divergence in Irish sonaghan and gillaroo (Hynes et al. 1996) and lake whitefish in eastern Canada (Pigeon et al. 1997).

less fit than the parent species. This reduction in fitness amounts to direct "ecological" postzygotic isolation stemming from the presence in the hybrid of traits that are disadvantageous in the parental environments (Price and Waser 1979; Shields 1982; Waser 1993; equivalent to Rice and Hostert's [1993] "environment-dependent" postzygotic isolation). Ecological postzygotic isolation may arise for example because an intermediate gape size or body form renders the hybrid inefficient at capturing the most profitable prey in the two main environments, or because intermediate defenses leave the hybrid susceptible to predation and parasitism. In contrast, "genetic" mechanisms of postzygotic isolation (Dobzhansky's "discordant gene patterns"; Rice and Hostert's [1993] "unconditional" postzygotic isolation) result from the breakup of favorable gene combinations that have positive epistatic interactions in the parent species, and from interactions between parental alleles leading to underdominance (Lynch 1991; Waser 1993). The practical difference between the two kinds of mechanisms is that ecologically based postzygotic isolation occurs only in a special natural setting and may vanish in the laboratory environment (e.g., one with prey of all sizes and no

predators), whereas genetic mechanisms are largely independent of environment and should be detectable in the laboratory as well as in the wild. Tests of ecological speciation will vary depending on whether the mechanism of postzygotic isolation is ecological or genetic.

Genetic mechanisms of postzygotic isolation between the youngest species of an adaptive radiation are frequently weak or lacking, as indicated by high viability and fertility of hybrids in the laboratory environment. Examples include Hawaiian and many other *Drosophila* (Templeton 1989; Coyne and Orr 1989), Galápagos finches (Grant and Grant 1992), Hawaiian silverswords (Carr and Kyhos 1981) (and indeed many perennial flowering plants; Grant 1981; Gill 1989; Rieseberg and Wendel 1993; Mcnair and Gardner, this volume), some East African cichlid fishes (Fryer and Iles 1972) as well as sympatric fish taxa in postglacial lakes (McPhail 1984; Wood and Foote 1990; Hatfield 1995). Unfortunately, most of these studies have examined only the F_1 hybrids, whereas genetic mechanisms of postmating isolation are typically most pronounced in the F_2 hybrids and backcrosses (e.g., Lynch 1991). However, in some of these cases fitness is also high in backcrosses and/or the

F_2 generation, such as the Galápagos finches and stickleback fishes (Grant and Grant 1992; Hatfield 1995). Thus, speciation can occur well before significant genetic mechanisms of postmating isolation evolve.

Genetic mechanisms of postmating isolation may nevertheless accumulate rapidly in some cases, such as between lines of *Drosophila* in some laboratory settings (Rice and Hostert 1993; see below). Complete sterility or inviability of at least one sex from at least one combination of parental crosses occurs in many pairs of wild *Drosophila* species, but such drastic postzygotic isolation tends to occur late in the speciation process, by which time most pairs of lineages have already evolved premating isolation (Coyne and Orr 1989). Genetic mechanisms of postzygotic isolation between sister species, combined with incomplete prezygotic isolation, frequently leads to the formation of hybrid zones where geographic ranges of the species abut (Hewitt 1989). Such zones are also indication that genetic mechanisms of postzygotic isolation play a role in speciation in nature.

Origin and Persistence of Species Despite Gene Flow

I summarize evidence for ecological mechanisms of selection against hybrids in two parts. In the present section I give examples of fish species that persist despite occasional interbreeding, and despite high viability and fertility of hybrids. The persistence of such species implies that some form of ecological selection against hybrids must be present. In extreme cases these sympatric species may have originated entirely in sympatry in the face of gene flow. In the subsequent section I review direct measurements of ecological selection against hybrids.

Gene flow is commonly detected by genetic and observational studies of young species in adaptive radiation (Grant 1998). For example, molecular studies indicate a history of gene flow between sympatric fish species in postglacial lakes. At the low end of the scale are sympatric "dwarf" and "normal" whitefishes in lakes of eastern Canada and northern Maine (Bernatchez and Dodson 1990). In one lake the two species are fixed for alternative mtDNA haplotype groups whose main geographic distributions correspond to late-Pleistocene refugia (Bernatchez and Dodson 1990). Pairs of species in other lakes within the same river basin, however, contain both haplotype groups (Bernatchez and Dodson 1990; Pigeon et al. 1997), indicating that limited gene flow has occurred following secondary contact between them.

At the other extreme are instances of apparent sympatric or parapatric speciation. Sockeye salmon represent the clearest case (Taylor et al. 1996). Anadromous sockeye spend the first two years of their life in large lakes before migrating to sea where they attain large size. In several drainages they have given rise to a form ("kokanee") that resides permanently in the lakes and matures at a smaller size. The two forms overlap broadly in the time and location of breeding, yet are genetically distinct. Phylogenies based on allozymes, minisatellite DNA, and mitochondrial DNA (mtDNA) all indicate that the kokanee form is polyphyletic and arose independently in separate drainages. This conclusion is strengthened by the fact that the geographic distribution of kokanee is completely nested within the range of anadromous sockeye, and that spontaneous appearence of kokanee within new drainages has followed the artificial transplants of only anadromous sockeye (Taylor et al. 1996).

But in the majority of cases it is not certain whether the species evolved entirely in sympatry or result instead from secondary contact between previously allopatric forms that subsequently hybridized extensively. For example, a phylogeny of stickleback species pairs based on mtDNA restriction fragments (figure 9.2) suggests very recent sympatric speciation in at least two of the lakes (Taylor et al. 1997). This evidence is in conflict, however, with studies of allozyme frequencies and larval salinity tolerance, which showed that the limnetic species in two of the lakes are similar to the contemporary marine threespine stickleback, whereas the benthic species are more distant (McPhail 1984, 1992; Kassen et al. 1995). The discrepancy might be explained by mtDNA gene flow; morphological data indicate that hybridization still occurs at a very low rate in the wild (McPhail 1984, 1992).

Regardless of which scenario is correct in these uncertain cases—full sympatric speciation or secondary contact followed by gene flow—the implications are similar: drift alone cannot be responsible for the evolution and/or maintenance of reproductive isolation. Natural selection is required for stable coexistence in the face of gene flow. A demonstration of such selection and the elucidation of its mechanisms would be strong evidence for ecological speciation.

Ecological Mechanisms of Postzygotic Isolation

Additional evidence for ecological speciation is gained by the demonstration that ecological mechanisms directly reduce the fitness of hybrids between species that occasionally crossbreed yet lack genetic mechanisms of postzygotic isolation. The significance of such mechanisms to speciation is twofold. First, any environmental agent that preferentially removes hybrids helps forestall the collapse of sympatric species by hybridization. Second, such mechanisms favor further divergence between species.

The presence of ecological mechanisms of postzygotic isolation in adaptive radiation is suggested by comparative and experimental evidence for ecological character displacement in Galápagos finches (Schluter and Grant 1984; Schluter et al. 1985) and postglacial fishes (Schluter and McPhail 1992, 1993; Robinson and Wilson 1994;

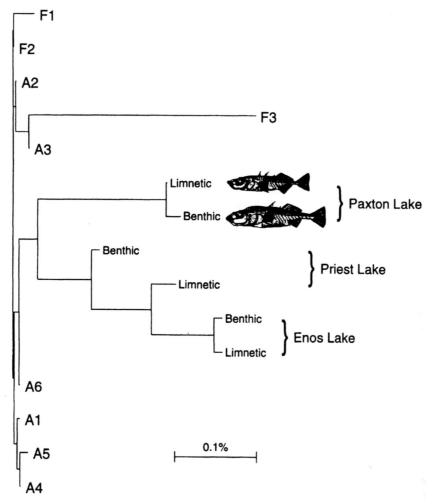

Figure 9.2. Unrooted neighbor-joining tree of threespine stickleback populations (*Gasterosteus sp.*) of southwestern British Columbia, Canada, based on estimates of mtDNA nucleotide divergences. A1–A6 are marine (anadromous) populations, F1 and F2 are populations resident in freshwater streams, and F3 is a solitary lake population. The benthic and limnetic species are from Paxton, Priest, and Enos lakes. Paxton and Priest lakes are in different drainages on Texada Island, a large island in the Strait of Georgia. Enos Lake is on Vancouver Island not far from Texada. Modified from Taylor et al. (1997).

Schluter 1994). In these groups competition for food has favored divergence between species in habitat use and in phenotypic traits used to harvest resources, such as beak size and gill raker number. Character divergence is the outcome of selection against intermediate phenotypes (and therefore hybrids) because of their competitive disadvantage.

This idea was taken a step further in a field study of F_1 hybrid fitness in sticklebacks (Hatfield 1995; Hatfield 1996). Growth rate of F_1 hybrids in the laboratory slightly exceeded that of their midparent (average of limnetic and benthic parent species), a pattern that was reversed when parents and hybrids were transplanted to enclosures in the

wild. Hybrids grew more slowly than the benthic parent in the littoral zone of a two-species lake, and more slowly than the limnetic parent in the open water habitat (Hatfield 1995). This growth deficiency in the wild matched consumption rates of intermediate phenotypes (in this case F_{10} hybrids, which are morphologically intermediate like the F_1 hybrids) foraging in open water and littoral zone environments transplanted from the same lake to large aquaria (Schluter 1993). Benthics were more successful at acquiring food from littoral sediments than hybrids, mainly because they could ingest larger prey; limnetics were superior to hybrids at seizing and retaining small, evasive plankton in open water. Equivalent experiments are lacking for other fish species pairs, although measurements of saltwater physiology and development time of F_1 hybrids between anadromous sockeye and kokanee suggest similar ecological mechanisms are at work (Wood and Foote 1990).

An extension of this method would be to mitigate the putative agent of selection in the environment and observe whether hybrid fitness is increased—and indeed whether the ultimate consequence is the collapse of the species pair. Such an experiment has never been done, but something like it began to happen in a "natural" experiment in the Galápagos finches. Grant and Grant (1992, 1993) recorded the fate of offspring of crosses between *Geospiza fuliginosa* (small ground finch) and *G. fortis* (medium ground finch) over 20 years on Daphne Major Island (figure 9.3). *G. fuliginosa* is an uncommon but regular immigrant to the island, and the majority of adults hybridize with *G. fortis*. The offspring are intermediate in beak size between the parent species and consume mainly small, soft seeds also eaten by *G. fuliginosa* (Grant and Grant 1996). Small seeds are typically much less abundant than the large, hard seeds eaten by *G. fortis*, and hybrid survival is correspondingly poor (figure 9.3). However, food conditions were dramatically changed in the years after record rains associated with an El Niño event, and this elevated hybrid survival to a level not less than pure *G. fortis*. Hybrids also suffered no reduc-

tion in fertility through this period (they mated mainly to *G. fortis*). Hybrid fitness seems here to be governed entirely by ecological conditions rather than by genetic breakdown.

These examples suggest that hybrids in nature may often be selected against because they fall between the niches of their parents—they are best adapted to intermediate environments that do not exist. Additional examples can be found. The beak of each species of red crossbill (*Loxia curvirostra* complex) is adapted to consuming seeds of a single type of conifer, and intermediate phenotypes are less efficient at handling these seeds (Benkman 1993). Life histories of host races of the apple maggot *Rhagoletis pomonella*, particularly the timing of diapause, are adapted to the divergent phenologies of their different host plants (Feder, this volume). As a result, the offspring of individuals that switch hosts are heavily disadvantaged. Hybrid irises (*Iris fulva* × *I. hexagona*) are intermediate between the parent species in salinity and shade tolerance and do not persist in either of the parental habitats, although they do occupy a narrow intermediate habitat (Arnold and Bennett 1993). Genetic incompatibilities also result in postzygotic isolation between the parent species (hybrid *Iris* have low pollen fertility and poor germination); nevertheless, the case illustrates that ecological mechanisms also contribute to reproductive isolation. Conceivably, many young

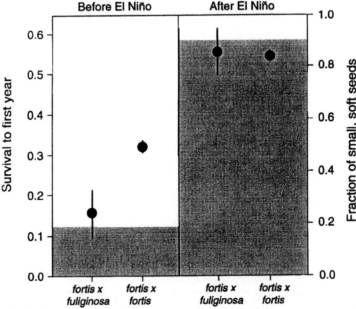

Figure 9.3. Survival of F_1 hybrid Galápagos ground finches (*Geospiza fortis* × *G. fuliginosa*) before and after the El Niño event of 1982–1983. Survival is the fraction of fledglings still alive after one year. Hybrid survival is compared with that of pure *G. fortis* (pure offspring of *G. fuliginosa* were rare). Vertical bars denote standard errors. Shaded area (right scale) indicate the abundance (mg/m²) of small, soft seeds as a proportion of all available seeds before and after the El Niño. Data are from Grant and Grant (1992, 1993).

species in nature only exist as separate entities because of the continuous action of natural selection against intermediate phenotypes.

The Origin of Genetic Mechanisms of Postzygotic Isolation

Divergent selection between environments may also lead to genetic mechanisms of postzygotic isolation. For example, reduced F_1 hybrid fitness evolved between *Drosophila* lines subjected to divergent ecological conditions in three of four laboratory experiments reviewed by Rice and Hostert (1993). The fitness reduction was evident even under laboratory conditions, implying that the mechanism is not ecological as in my earlier examples. The question is then whether past divergent selection is responsible for genetic mechanisms of hybrid breakdown between species in nature.

One way to test this is to identify in the parent species the loci responsible for hybrid breakdown, and measure their adaptive significance in the parent environments. This is not a simple task and has rarely been carried out. One example is in the monkey flower (*Mimulus guttatus*), in which alleles conferring tolerance to soils contaminated with copper are lethal when combined in the offspring of crosses with plants from uncontaminated soils (Macnair and Christie 1983).

A second test examines geographical gradients in adaptive characters differing between species that form narrow hybrid zones where their ranges abut (Hewitt 1989). If postzygotic isolation is a by-product of interactions between alleles at the loci responsible for adaptive trait differences between species, then the location at which trait means shift from predominantly one extreme to predominantly the other extreme should coincide with the location of the zone in which hybrid breakdown occurs (e.g., Orr 1996). Alternatively, if the genes responsible for hybrid breakdown are not those responsible for adaptive trait differences between the species, then the geographical gradients in ecologically important traits should wander away from the locations of hybrid zones. The majority of hybrid zones are broadly associated with environmental transitions (Hewitt 1989), and the locations of transitions in the means of ecologically important traits often coincide with the location of these zones. This supports the hypothesis that hybrid breakdown is the result of fixation of genes responsible for differences between the species in adaptive characters. The problem with this evidence, however, is that nonecological hypotheses predict the same pattern if there has not been sufficient time for geographical gradients in ecologically significant traits to become uncoupled from zones of hybrid breakdown.

A third test examines whether natural selection is able to rebuild genetic mechanisms of postzygotic isolation between two or more species after their genomes have been blended by hybridization. This experiment, which has never been carried out, has two steps. The first crosses two or more ecologically distinct species, creating a hybrid population. This population is then taken through a series of generations intended to eradicate positive linkage disequilibrium between genes inherited from the same parent species. These hybrids should show some postzygotic isolation from both parent species, but its severity should be less than that between the parent species themselves. The second step places lines established from this hybrid blend into environments similar to one or other of those of their wild parents. The hypothesis of ecological speciation predicts that divergent selection will cause postzygotic isolation to build between hybrid lines placed in different environments; and that reproductive isolation should decay simultaneously between the parent species and those hybrids raised in the same environment. This experiment requires that the genetic mechanisms of postzygotic isolation are not too strong so that hybrids can be formed and bred with little loss of parental alleles. The design could also be used to test whether other modes of reproductive isolation between species (e.g., premating isolation) have evolved by means of divergent natural selection.

A fourth test examines whether genetic mechanisms of postzygotic isolation evolve in parallel between independent lineages experiencing similar environmental selection pressures ("parallel speciation"; Schluter and Nagel 1995). For example, consider an ancestral species that independently gives rise to a series of populations in two different types of environment at the periphery of its range. Parallel speciation by natural selection occurs when reproductive isolation evolves between descendant forms adapting to different environments, but not between descendants adapting to similar environments. The phenomenon is strong evidence for divergent natural selection for the same reason that parallel evolution generally is evidence for adaptation: drift is unlikely to cause reproductive isolation to evolve in the same way multiple times in a specific environmental setting. The test has never been carried out on levels of postzygotic isolation between species in nature. Kilias et al. (1980) carried out the test with replicate laboratory lines of *Drosophila* subjected to contrasting selection pressures for a large number of generations, but little postzygotic isolation evolved. However, prezygotic isolation evolved in parallel in this experiment; this topic is considered in the following section.

Theoretical models of the evolution of postzygotic isolation by the accumulation of "complementary genes" in different populations—genes that are advantageous by themselves but deleterious when combined in a hybrid—successfully account for several known properties of genetic mechanisms of postzygotic isolation (e.g., Haldane's Rule and the large effect of the X-chromosome in *Drosophila*; Turelli and Orr 1995). This supports the idea that genetic mechanisms of postzygotic isolation have evolved by selection, but the result does not distinguish whether genetic differences were favored by divergent selection or whether they accumulated in spite of uniform selection.

Divergent Selection and Mate Discrimination

At least three methods test whether prezygotic isolation between species is the outcome of divergent natural selection rather than drift, founder events, or divergent responses to similar selection pressures. The first examines whether assortative mating between species is based on the same phenotypic traits that are responsible for efficient exploitation of alternative ecological environments. If so, then divergent selection is implicated: mate recognition most likely evolved as a pleiotropic effect of divergent selection on those phenotypic traits; or mate recognition on the basis of these phenotypic traits was directly favored by natural selection during divergence (e.g., by reinforcement or because of selection for efficient mate finding).

For example, body size is strongly divergent between sympatric stickleback species, and several lines of evidence suggest that this difference is the result of contrasting natural selection pressures between the habitats they mainly exploit (Schluter and McPhail 1992; Schluter 1993; Nagel and Schluter 1998). In no-choice laboratory mating trials body size was found to strongly affect the probability of interspecific hybridization (figure 9.4). Interspecific spawnings occurred only between the largest individuals of the smallest species (limnetics) and the smallest individuals of the largest species (benthics) (Nagel and Schluter 1997). Similar results have been found in other instances of size differentiation between closely related species. Body size is also an important component of assortative mating between sockeye and kokanee salmon (Foote and Larkin 1988). Similarly, size and shape of the beak and body of Galápagos finch species, which are strongly selected for efficient food exploitation, are also used as cues in interspecific mate discrimination (Ratcliffe and Grant 1983).

The second test measures whether prezygotic isolation has evolved in parallel between independent populations in similar environments ("parallel speciation"; see previous section). Sympatric species of threespine sticklebacks represent a possible example, assuming that limnetics and benthics have evolved independently multiple times in separate lakes (Schluter and McPhail 1992; Taylor et al. 1997; figure 9.2). Nagel (1994) showed that assortative mating between limnetics from different lakes was weak, and also that assortative mating was weak among benthics from these same two lakes and from Enos Lake (figure 9.2). In other words, closely related fish species that evolved similar body sizes and shapes

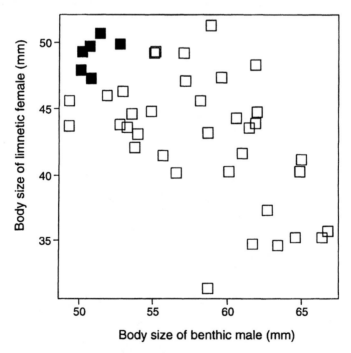

Figure 9.4. Spawning probability between female limnetic and male benthic sticklebacks compared with their body sizes. A solid symbol indicates that the pair spawned within 30 minutes; open symbols indicate no spawning. A trial involved introducing a single female to an aquarium containing a single nesting male. Data are from two lakes combined. After Nagel and Schluter (1998).

under similar environmental conditions have also evolved similar patterns of mate recognition. These conclusions are strongly contingent upon the hypothesis that benthics and limnetics from different lakes evolved independently multiple times, which is supported by the mtDNA phylogeny but requires further confirmation (Taylor et al. 1997). Two other possible cases of parallel evolution of prezygotic isolation in similar environments are sockeye/kokanee salmon and cave amphipods (Schluter and Nagel 1995).

A third test determines whether natural selection rebuilds prezygotic isolation between two or more species whose genomes have been blended by hybridization. This experiment is outlined in the previous section. No tests of this kind have yet been attempted.

High Rates of Speciation via Enhanced Population Persistence

The second major view of ecology's role in speciation is that ecological environments mainly influence the persistence of geographically isolated populations. Reproductive isolation does not necessarily evolve more rapidly in novel environments; rather, speciation rates rise there because more populations evolve reproductive isolation. Under this view the evolution of reproductive isolation is driven by nonecological mechanisms, or possibly by adaptive mechanisms less closely tied to niche shift. This view is widespread and probably represents the opinion of a majority of evolutionary biologists. However, its popularity stems mainly from its plausibility and simplicity rather than from direct evidence, since data on population persistence and speciation in nature are completely lacking.

Indirect evidence in favor of this hypothesis is the high population density of many species belonging to lineages undergoing adaptive radiation. Abundance of seed-eating Galápagos finches is substantially greater than the summed densities of ecologically similar finches in continental regions when seed supplies are similar (Schluter and Repasky 1991). The difference may be attributable to higher predation risk on mainlands and more competitors for seeds. A second example is by Farrell et al. (1991), who demonstrated that speciation rates of plant clades possessing latex or resin canals deterring herbivory by insects are significantly higher than speciation rates in their sister taxa that lack secretory canals. Field studies revealed that plants possessing canals are more abundant in tropical forests than are nonsecreting plants (Farrell et al. 1991; B. D. Farrell, personal communication). Further data of this kind are badly needed.

Actual estimates of rates of population persistence may be necessary to evaluate the hypothesis more thoroughly. One way this might be carried out is by comparison of complete phylogenies of extant populations. Harvey et al. (1994) have derived methods to estimate rates of speciation and species extinction from phylogenies, under the assumption that gains and losses of species follows a simple birth and death process. Working backward from the extant species to the past, the phylogenies allow one to generate a curve of species accumulation through time (Nee et al. 1992). Changes in inclination of the curve at the present and further back in time depend directly on speciation and extinction rates, such that the entire curve yields estimates of both parameters (Harvey et al. 1994). The same method could, in principle, be applied to populations in different types of environments.

Conclusion

The idea was prominent in the first half of this century that divergent natural selection on populations in distinct ecological environments was a principal cause of speciation (Mayr 1942; Dobzhansky 1951; Simpson 1953). Mayr (1963) later emphasized genetic causes of the evolution of reproductive isolation instead, but maintained a strong belief that ecological processes affected speciation rate by influencing a key prerequisite of speciation: formation and persistence of geographically isolated populations. Although one or both of these views still find adherents decades later, evidence in support of either one is only now beginning to emerge. The absence of an accepted explanation for high speciation rates in adaptive radiation is only a glaring reminder of the distance still to go.

Exactly why progress has been slow in identifying the ecological causes of speciation is unclear. Perhaps the subject is inherently difficult. Measuring ecological opportunity's effects on population establishment and persistence is not a trivial task. However, two other factors probably contributed (Schluter 1996). The first is a historical preoccupation with other issues in speciation, such as the contributions of founder events. Second, broad acceptance of the principle that speciation was chiefly allopatric probably discouraged study of the role of divergent natural selection. Given enough time, random drift and divergence under a common selection regime will eventually yield complete inviability or sterility in offspring between geographically isolated populations. Ecological selection pressures are therefore not *necessary* to explain the fact of species. However, divergent selection can cause reproductive isolation much sooner, and may allow stable sympatry after a period of time too brief for drift to effect much change.

This chapter focused on the most important distinction between the two views, namely, the causes of the evolution of reproductive isolation. The evidence for ecology's role here is as follows. Reproductive isolation evolves at a high rate when accompanied by rapid ecological differentiation among populations. An increasing number of examples are known in which stable sympatry between ecologically differentiated species is achieved despite gene flow between them; in extreme cases reproductive isolation appears to have evolved without allopatry. The context of these examples is often the novel environment, implying that ecological opportunity causes rates of evolution of reproductive isolation to rise in both sympatry and allopatry. In several examples, ecological mechanisms appear to be a major determinant of postzygotic reproductive isolation (see also Feder, this volume). Genetic mechanisms of postzygotic isolation between young sister species are often weak, at least in the examples drawn from studies of young species in adaptive radiations. The causes of the origin of genetic mechanisms of postzygotic isolation in nature, when present, are not clear. Comparative study of prezygotic isolation suggests that it has often evolved in response to divergent selection. For example, assortative mating between species is often based on the very traits that influence exploitation of resources in alternative environments. Premating isolation may evolve in parallel between independent, closely related lineages inhabiting similar environments (parallel speciation). The link between divergent selection and origin of genetic mechanisms of postmating isolation, when present, is less clear.

These results do not support the idea that niche shifts affect speciation solely or even mainly by increasing population longevity. Rather, ecological processes appear to greatly affect the rate at which reproductive isolation evolves. Ecological processes may also influence speciation via population establishment and persistence, but evidence from nature for or against this hypothesis is scarce.

My emphasis here was on speciation in adaptive radiation, in part to address the question raised by the correlation between speciation rate and ecological opportunity (figure 9.1), and because ecological speciation may be most common when divergent natural selection pressures are strongest. Such a selection of cases would inflate our perception of the importance of ecology in speciation only if adaptive radiation is rare—which is unlikely. They bias our view of speciation in general if the causes of speciation in adaptive radiation are qualitatively different from the causes of speciation unaccompanied by ecological and phenotypic divergence—which is possible. Ideally, we should eventually be able to identify the circumstances under which alternative causes of speciation are most prevalent.

I also focused on what I believe are the two main issues of ecology and speciation—the role of divergent selection in driving the evolution of reproductive isolation, and the influence of population persistence on speciation rate. Other important issues have not been dealt with here, especially the influence of population size on the rate of evolution of reproductive isolation. Small population size accompanying a bottleneck or a founder event is a crucial ingredient in "founder-effect" and related speciation models in which divergence is initially resisted by selection, but the importance of this mechanism to speciation in nature remains uncertain (Barton and Charlesworth 1984; Carson and Templeton 1984; Barton 1989; Rice and Hostert 1993; Templeton 1996). The rate at which alternative favorable mutations causing hybrid breakdown are fixed by selection is also thought to be influenced by population size and subdivision (Orr and Orr 1996). Unfortunately, there is as yet no evidence from nature on the influence of population size in any scenario.

Finally, it is worthwhile to ask how a greater understanding of ecological mechanisms in speciation might be achieved. A tendency of recent years has been to infer mechanisms of speciation in nature entirely from genetic measurements—particularly the genetic basis of reproductive isolation and the deduced strength of selection acting on hybrids. Such evidence is indeed crucial, but will of itself lead to a poor understanding of the causes of speciation. A complementary natural history of the evolution of reproductive isolation is needed in addition. Field studies of selection on body size, beak shape, and body coloration have told us a great deal about the causes of morphological diversity. It is therefore likely that field studies of natural selection on traits conferring degrees of reproductive isolation will contribute greatly to our knowledge of the ecological causes of speciation.

Acknowledgments   I'm grateful to many colleagues for help and insights: S. Berlocher, M. Blows, J. Feder, T. Hatfield, S. Heard, D. Kapan, J. Losos, J. McKinnon, A. Mooers, L. Nagel, J. Pritchard, S. Vamosi, and M. Whitlock. My research is supported by the Natural Sciences and Engineering Research Council (NSERC) of Canada.

References

Allmon, W. D. 1992. A causal analysis of stages in allopatric speciation. *Oxford Surveys in Evolutionary Biology* 8, 219–57.

Amadon, D. 1950. The Hawaiian honeycreepers (Aves, Drepaniidae). *Bulletin of the American Museum of Natural History* 95, 157–268.

Arnold, M. L., and Bennett, B. D. 1993. Natural hybridization in Louisiana irises: genetic variation and ecological determinants. In *Hybrid zones and the evolutionary process* (ed. R. G. Harrison), pp. 115–39. Oxford University Press, Oxford.

Ayala, F. 1975. Genetic differentiation during the speciation process. *Evolutionary Biology* 8, 1–75.

Ayala, F., Tracey, M., Hedgecock, D., and Richmond, R. C. 1974. Genetic differentiation during the speciation process in *Drosophila*. *Evolution* 28, 576–92.

Baldwin, B. G., Kyhos, D. W., Dvorak, J., and Carr, G. D. 1991. Chloroplast DNA evidence for a North American origin of the Hawaiian silversword alliance (Asteraceae). *Proceedings of the National Academy of Sciences, USA* 88, 1840–43.

Barton, N. H. 1989. Founder effect speciation. In *Speciation and its consequences* (ed. D. Otte and J. A. Endler), pp. 229–56. Sinauer, Sunderland, Mass.

Barton, N. H., and Charlesworth, B. 1984. Genetic revolutions, founder effects, and speciation. *Annual Review of Ecology and Systematics* 15, 131–64.

Behnke, R. J. 1972. The systematics of salmonid fishes of recently glaciated lakes. *Journal of the Fisheries Research Board of Canada* 29, 639–71.

Benkman, C. W. 1993. Adaptation to single resources and the evolution of crossbill (*Loxia*) diversity. *Evolution* 63, 305–25.

Bernatchez, L., and Dodson, J. J. 1990. Allopatric origin of sympatric populations of lake whitefish (*Coregonus clupeaformis*) as revealed by mitochondrial-DNA restriction analysis. *Evolution* 44, 1263–71.

Bush, G. 1994. Sympatric speciation in animals—new wine in old bottles. *Trends in Ecology and Evolution* 9, 285–88.

Cabrera, V., Gonzales, A. M., Larruga, J. M., and Gullon, A. 1983. Genetic distance and evolutionary relationships in the *Drosophila obscura* group. *Evolution* 37, 675–89.

Carr, G. D., and Kyhos, D. W. 1981. Adaptive radiation in the Hawaiian silversword alliance (Compositae: Madiinae). I. Cytogenetics of spontaneous hybrids. *Evolution* 35, 543–56.

Carson, H. L., and Templeton, A. R. 1984. Genetic revolutions in relation to speciation phenomena: the founding of new populations. *Annual Review of Ecology and Systematics* 15, 97–131.

Coyne, J. A., and Orr, H. A. 1989. Patterns of speciation in *Drosophila. Evolution* 43, 362–81.

Diehl, S. R., and Bush, G. L. 1989. The role of habitat preference in adaptation and speciation. In *Speciation and its consequences* (ed. D. Otte and J. A. Endler), pp. 345–65. Sinauer, Sunderland, Mass.

Dobzhansky, T. 1951. *Genetics and the origin of species* (3rd ed.). Columbia University Press, New York.

Eisses, K, VanDijk, H., and VanDelden, W. 1979. Genetic differentiation within the *melanogaster* species subgroup of the genus *Drosophila. Evolution* 33, 1063–68.

Farrell, B. D., Dussourd, D. E., and Mitter, C. 1991. Escalation of plant defense: do latex and resin canals spur plant diversification? *American Naturalist* 138, 881–900.

Fisher, R. A. 1930. *The genetical theory of natural selection*. Oxford University Press, Oxford.

Foote, C. J., and Larkin, P. A. 1988. The role of male choice in the assortative mating of anadromous and non-anadromous sockeye salmon (*Oncorhynchus nerka*). *Behaviour* 106, 43–62.

Fryer, G., and Iles, T. D. 1972. *The cichlid fishes of the Great Lakes of Africa*. Oliver and Boyd, Edinburgh.

Gill, D. E. 1989. Fruiting failure, pollinator inefficiency, and speciation in orchids. In *Speciation and its consequences* (ed. D. Otte and J. A. Endler), pp. 458–81. Sinauer, Sunderland, Mass.

Grant, B. R., and Grant, P. R. 1993. Evolution of Darwin's finches caused by a rare climatic event. *Proceedings of the Royal Society of London B, Biological Sciences* 251, 111–17.

Grant, B. R., and Grant, P. R. 1996. High survival of Darwin's finch hybrids: effects of beak morphology and diets. *Ecology* 77, 500–9.

Grant, P. R. 1986. *Ecology and evolution of Darwin's finches*. Princeton University Press, Princeton, N.J.

Grant, P. R. 1998. Speciation. In *Evolution on islands* (ed. P. R. Grant and B. C. Clarke), pp. 83–101. Oxford University Press, Oxford.

Grant, P. R., and Grant, B. R. 1992. Hybridization of bird species. *Science (Washington, D.C.)* 256, 193–97.

Grant, V. 1981. *Plant speciation*. Columbia University Press, New York.

Groth, J. G. 1990. *Cryptic species of nomadic birds in the red crossbill Loxia curvirostra complex of North America*. Ph.D. thesis, University of California, Berkeley.

Harvey, P. H., May, R. M., and Nee, S. 1994. Phylogenies without fossils. *Evolution* 48, 523–29.

Hatfield, T. 1995. *Speciation in sympatric sticklebacks: hybridization, reproductive isolation and the maintenance of diversity*. Ph.D. thesis, University of British Columbia.

Hatfield, T. 1996. Genetic divergence in adaptive characters between sympatric species of sticklebacks. *American Naturalist* 149, 1009–29.

Heard, S. B., and Hauser, D. L. 1994. Key evolutionary innovations and their ecological mechanisms. *Historical Biology* 10, 151–73.

Helenurm, K., and Ganders, F. R. 1985. Adaptive radiation and genetic differentiation in Hawaiian *Bidens. Evolution* 39, 753–65.

Hewitt, G. M. 1989. The subdivision of species by hybrid zones. In *Speciation and its consequences* (ed. D. Otte and J. A. Endler), pp. 85–110. Sinauer, Sunderland, Mass.

Huxley, J. 1942. *Evolution, the modern synthesis*. Allen and Unwin, London.

Hynes, R. A., Ferguson A., and McCann, M. A. 1996. Variation in mitochndrial DNA and post-glacial colonization of north western Europe by brown trout. *Journal of Fish Biology* 48, 54–67.

Jablonski, D. 1986. Evolutionary consequences of mass extinctions. In *Patterns and processes in the history of life* (ed. D. M. Raup and D. Jablonski), pp. 313–29. Springer, Berlin.

Johnson, N. K., Marten, J. A., and Ralph, C. J. 1991. Genetic evidence for the origin and relationships of the Hawaiian honeycreepers (Aves: Fringillidae). *Condor* 91, 379–96.

Johnson, W., Carson, H. L., Kaneshiro, K., Steiner, W., and Cooper, M. 1975. Genetic variation in Hawaiian *Drosophila*. II. Allozyme differentiation in the *D. planitibia* subgroup. In *Isozymes IV. Genetics and evolution* (ed. C. L. Markert), pp. 563–84. Academic Press, New York.

Kassen, R., Schluter, D., and McPhail, J. D. 1995. Evolutionary history of threespine sticklebacks (*Gasterosteus spp.*) in British Columbia: insights from a physiological clock. *Canadian Journal of Zoology* 73, 2154–58.

Kilias, G., Alahiotis, S. N., and Pelecanos, M. 1980. A multifactorial genetic investigation of speciation theory using *Drosophila melanogaster. Evolution* 34, 730–37.

Lack, D. 1947. *Darwin's finches*. Cambridge University Press, Cambridge.

Lakovaara, S., Saura, A., and Falk, C. 1972. Genetic distance and evolutionary relationships in the *Drosophila obscura* group. *Evolution* 26, 177–84.

Lande, R. 1981. Models of speciation by sexual selection on polygenic traits. *Proceedings of the National Academy of Sciences, USA* 78, 3721–25.

Lowrey, T. K. 1995. Phylogeny, adaptive radiation, and biogeography of Hawaiian *Tetramolopium* (Compositae: Astereae). In *Hawaiian biogeography: evolution on a hot spot archipelago* (ed. W. L. Wagner and V. A. Funk), pp. 195–220. Smithsonian Institution Press, Washington, D.C.

Lowrey, T. K., and Crawford, D. J. 1985. Allozyme divergence and evolution in *Tetramolopium* (Compositae: Astereae) on the Hawaiian Islands. *Systematic Biology* 10, 64–72.

Lynch, M. 1991. The genetic interpretation of inbreeding depression and outbreeding depression. *Evolution* 45, 622–29.

Macnair, M. R., and Christie, P. 1983. Reproductive isolation as a pleiotropic effect of copper tolerance in *Mimulus guttatus? Heredity* 50, 295–302.

Mallet, J. 1995. A species definition for the modern synthesis. *Trends in Ecology and Evolution* 10, 294–99.

Malmquist, H. J. 1992. Phenotype-specific feeding behaviour of two arctic charr *Salvelinus alpinus* morphs. *Oecologia* 92, 354–61.

Marten, J. A., and Johnson, N. K. 1986. Genetic relationships of North American cardueline finches. *Condor* 88, 409–20.

Maynard Smith, J. 1966. Sympatric speciation. *American Naturalist* 100, 637–50.

Mayr, E. 1942. *Systematics and the origin of species*. Columbia University Press, New York.

Mayr, E. 1954. Change of genetic environment and evolution. In *Evolution as a process* (ed. J. Huxley, A. C. Hardy, and E. B. Ford), pp. 157–80. Allen and Unwin, London.

Mayr, E. 1963. *Animal species and evolution*. Harvard University Press, Cambridge, Mass.

McPhail, J. D. 1984. Ecology and evolution of sympatric sticklebacks (*Gasterosteus*): morphological and genetic evidence for a species pair in Enos Lake, British Columbia. *Canadian Journal of Zoology* 62, 1402–8.

McPhail, J. D. 1992. Ecology and evolution of sympatric sticklebacks (*Gasterosteus*): evidence for a species pair in Paxton Lake, Texada Island, British Columbia. *Canadian Journal of Zoology* 70, 361–69.

Meyer, A. 1993. Phylogenetic relationships and evolutionary processes in East African cichlid fishes. *Trends in Ecology and Evolution* 8, 279–84.

Muller, H. J. 1940. Bearings of the *Drosophila* work on systematics. In *The new systematics* (ed. J. S. Huxley), pp. 185–268. Clarendon Press, Oxford.

Nagel, L. M. 1994. *The parallel evolution of reproductive isolation in threespine sticklebacks*. M.S. thesis, University of British Columbia.

Nagel, L. M., and Schluter, D. 1998. Body size, natural selection, and speciation in sticklebacks. *Evolution* 52, 209–18.

Nee, S., Mooers, A. O., and Harvey, P. H. 1992. Tempo and mode of evolution revealed from molecular phylogenies. *Proceedings of the National Academy of Sciences, USA* 89, 8322–26.

Nei, M. 1972. Genetic distance between populations. *American Naturalist* 106, 283–92.

Nei, M., and Miller, J. C. 1990. A simple method for estimating average number of nucleotide substitutions within and between populations from restriction data. *Genetics* 125, 873–9.

Olson, S. L., and James, H. F. 1982. Prodromus of the fossil avifauna of the Hawaiian Islands. *Smithsonian Contributions to Zoology* 365, 1–59.

Orr, H. A., and Orr, L. H. 1996. Waiting for speciation: the effect of population subdivision on the time to speciation. *Evolution* 50, 1742–49.

Orr, M. R. 1996. Life-history adaptation and reproductive isolation in a grasshopper hybrid zone. *Evolution* 50, 704–16.

Pigeon, D., Chouinard, A., and Bernatchez, L. 1997. Multiple modes of speciation involved in the parallel evolution of sympatric morphotypes of lake whitefish (*Coregonus clupeaformis*, Salmonidae). *Evolution* 51, 196–205.

Price, M. V., and Waser, N. M. 1979. Pollen dispersal and optimal outcrossing in *Delphinium nelsoni*. *Nature (London)* 277, 294–97.

Ratcliffe, L. M., and Grant, P. R. 1983. Species recognition in Darwin's finches (*Geospiza*, Gould). I. Discrimination by morphological cues. *Animal Behaviour* 31, 1139–53.

Rice, W. R., and Hostert, E. E. 1993. Laboratory experiments on speciation: what have we learned in 40 years? *Evolution* 47, 1637–53.

Ricker, W. E. 1940. On the origin of kokanee, a fresh-water type of sockeye salmon. *Transactions of the Royal Society of Canada* 34, 121–35.

Rieseberg, L. H., and Wendel, J. F. 1993. Introgression and its consequences in plants. In *Hybrid zones and the evolutionary process* (ed. R. G. Harrison), pp. 70–109. Oxford University Press, Oxford.

Robinson, B. W., and Wilson, D. S. 1994. Character release and displacement in fishes: a neglected literature. *American Naturalist* 144, 596–627.

Robinson, B. W., Wilson, D. S., and Shea, G. O. 1996. Trade-offs of ecological specialization: an intraspecific comparison of pumpkinseed sunfish phenotypes. *Ecology* 77, 170–78.

Rogers, J. S. 1972. Measures of genetic similarity and genetic distance. *University of Texas Publications* 7213, 145–53.

Sage, R. D., Loiselle, P. V., Basasibwaki, P., and Wilson, A. C. 1984. Molecular versus morphological change amnong cichlid fishes of Lake Victoria. In *Evolution of fish species flocks* (ed. A. A. Echelle and I. Kornfield), pp. 185–97. University of Maine Press, Orono, Maine.

Schluter, D. 1993. Adaptive radiation in sticklebacks: size, shape, and habitat use efficiency. *Ecology* 74, 699–709.

Schluter, D. 1994. Experimental evidence that competition promotes divergence in adaptive radiation. *Science (Washington, D.C.)* 266, 798–801.

Schluter, D. 1995. Adaptive radiation in sticklebacks: trade-offs in feeding performance and growth. *Ecology* 76, 82–90.

Schluter, D. 1996. Ecological speciation in postglacial fishes. *Philosophical Transactions of the Royal Society of London Series B* 351, 807–14.

Schluter, D. 1998. Ecological speciation in postglacial fishes. In *Evolution on islands* (ed. P. R. Grant and B. C. Clarke), pp. 163–80. Oxford University Press, Oxford.

Schluter, D., and Grant, P. R. 1984. Determinants of morphological patterns in communities of Darwin's finches. *American Naturalist* 123, 175–96.

Schluter, D., and McPhail, J. D. 1992. Ecological character displacement and speciation in sticklebacks. *American Naturalist* 140, 85–108.

Schluter, D., and McPhail, J. D. 1993. Character displacement and replicate adaptive radiation. *Trends in Ecology and Evolution* 8, 197–200.

Schluter, D., and Nagel, L. 1995. Parallel speciation by natural selection. *American Naturalist* 146, 292–301.

Schluter, D., and Price, T. 1993. Honesty, perception and population divergence in sexually selected traits. *Proceedings of the Royal Society of London Series B* 253, 117–22.

Schluter, D., and Repasky, R. R. 1991. Worldwide limitation of finch densities by food and other factors. *Ecology* 72, 1763–74.

Schluter, D., Price, T. D., and Grant, P. R. 1985. Ecological character displacement in Darwin's finches. *Science (Washington, D.C.)* 227, 1056–59.

Shields, W. M. 1982. *Philopatry, inbreeding, and the evolution of sex*. State University of New York Press, Albany, N.Y.

Simpson, G. G. 1944. *Tempo and mode in evolution*. Columbia University Press, New York.

Simpson, G. G. 1953. *The major features of evolution*. Columbia University Press, New York.

Skulason, S., Antonsson, T., Gudbergsson, G., Malmquist, H. J., and Snorrason, S. 1992. Variability in Icelandic charr. *Icelandic Agricultural Science* 6, 142–53.

Stanley, S. M. 1979. *Macroevolution*. Freeman, San Francisco.

Svärdson, G. 1979. Speciation of Scandinavian *Coregonus*. *Report from the Institute of the Freshwater Research, Drottningholm* 57, 1–95.

Tauber, C. A., and Tauber, M. J. 1989. Sympatric speciation in insects: perception and perspective. In *Speciation and its consequences* (ed. D. Otte and J. A. Endler), pp. 307–44. Sinauer, Sunderland, Mass.

Taylor, E. B., Foote, C. J., and Wood, C. C. 1996. Molecular genetic evidence for parallel life-history evolution within a Pacific salmon (sockeye salmon and kokanee, *Oncorhynchus nerka*). *Evolution* 50, 401–16.

Taylor, E. B., McPhail, J. D., and Schluter, D. 1997. History of ecological selection in sticklebacks: uniting experimental and phylogenetic approaches. In *Molecular evolution and adaptive radiation* (ed. T. J. Givnish and K. J. Sytsma), pp. 511–34. Cambridge University Press, Cambridge.

Templeton, A. R. 1989. The meaning of species and speciation: a genetic perspective. In *Speciation and its consequences* (ed. D. Otte and J. A. Endler), pp. 3–27. Sinauer, Sunderland, Mass.

Templeton, A. R. 1996. Experimental evidence for the genetic-transilience model of speciation. *Evolution* 50, 909–15.

Turelli, M., and Orr, H. A. 1995. The dominance theory of Haldane's rule. *Genetics* 140, 1805–13.

Vermeij, G. J. 1995. Economics, volcanoes, and Phanerozoic revolutions. *Paleobiology* 21, 125–52.

Waser, N. M. 1993. Population structure, optimal outbreeding, and assortative mating in angiosperms. In *The natural history of inbreeding and outbreeding* (ed. N. H. Thornhill), pp. 173–99. Chicago University Press, Chicago.

Wilson, D. S., and Turelli, M. J. 1986. Stable underdominance and the evolutionary invasion of empty niches. *American Naturalist* 127, 835–50.

Wood, C. C., and Foote, C. J. 1990. Genetic differences in the early development and growth of sympatric sockeye salmon and kokanee (*Oncorhynchus nerka*) and their hybrids. *Canadian Journal of Fisheries and Aquatic Sciences* 47, 2250–60.

Wright, S. 1940. The statistical consequences of Mendelian heredity in relation to speciation. In *The new systematics* (ed. J. S. Huxley), pp. 161–83. Clarendon Press, Oxford.

Yang, S. Y., and Patton, J. L. 1981. Genic variability and differentiation in the Galapagos finches. *Auk* 98, 230–42.

Zink, R. M. 1982. Patterns of genic and morphologic variation among sparrows in the genera *Zonotrichia, Melospiza, Junco,* and *Passerella. Auk* 99, 632–49.

Zink, R. M., and Avise, J. C. 1990. Patterns of mitochondrial DNA and allozyme evolution in the avian genus *Ammodramus. Systematic Zoology* 39, 148–61.

10

The Apple Maggot Fly, *Rhagoletis pomonella*

Flies in the Face of Conventional Wisdom about Speciation?

Jeffrey L. Feder

It is often said that Darwin never really tackled the problem posed by the title of his thesis *On the Origin of Species*. Instead, he offered a proof of organismal evolution and one long argument that natural selection was a prime engine for evolution. But the full title of Darwin's work was *On the Origin of Species by Means of Natural Selection or the Preservation of Favoured Races in the Struggle for Life*. Darwin therefore pondered how new species formed, and his answer was by natural selection, the same process responsible for change within populations.

Darwin could be rightly accused of being somewhat ambivalent as to the geographic context of speciation. But Benjamin Walsh was not. As early as 1864, Walsh proposed that certain host-specific phytophagous insects could speciate sympatrically (i.e., in the absence of complete geographic isolation) by shifting and adapting to new host plants. In particular, Walsh (1867) cited the shift of the apple maggot fly, *Rhagoletis pomonella* (Diptera: Tephritidae, Walsh), from its native host hawthorn (*Crataegus* L. spp.) to introduced, domestic apple (*Malus pumila* L.), an event that occurred in the Hudson Valley Region of New York in the mid-1800s, as an example of an incipient sympatric speciation event (for further discussion of the natural history of this host shift, see Bush, 1966, 1992; Bush et al., 1989). Subsequently, the term *host race* has been used to describe this purported initial stage in sympatric divergence, host races being defined as partially reproductively isolated, conspecific populations that owe their isolation to host-associated adaptations (Diehl and Bush, 1984). Host races therefore represent a special class of ecological polymorphism, one in which the polymorphism pleiotropically results in reproductive isolation. This is in contrast to the situation where an ecological polymorphism does not affect the pattern of mating or gene flow within a species, instead being maintained by some form of balancing, frequency-

or density-dependent selection (Wilson, 1989). It is interesting to note that like *R. pomonella*, many reported cases of host races appear to involve introduced plants (Diehl and Bush, 1984), a situation where a new potential host interaction suddenly becomes available.

In this chapter, I examine the evidence for sympatric host race formation in *R. pomonella*. It seems most appropriate to concentrate on *R. pomonella* because it is the fly that prompted Walsh (1864, 1867) to propose, and later Guy Bush (1966, 1969a,b, 1975a,b) to refine, the concept of sympatric speciation via host race formation. Before delving into the data, however, I first outline Bush's (1966, 1969a,b, 1975a,b) general model for sympatric host race formation, placing his model in context with the known biology of *R. pomonella* and in juxtaposition with previous thinking about postzygotic reproductive isolation.

Bush's Model for Sympatric Host Race Formation

Bush's model for sympatric race formation and speciation rests on two main pillars:

Host (habitat) specific mating. Host-specific mating for *Rhagoletis* translates into adult flies mating on and ovipositing into the same species of host fruit that they fed within as larvae (figure 10.1). I shall refer to any such tendency as *host fidelity*. Host fidelity is important because it establishes a system of positive assortative mating that acts as a premating barrier to gene flow between demes specialized on alternative plants.

Host-associated fitness trade-offs. Fitness trade-offs are necessary to offset any "leakiness" that may exist in host fidelity (figure 10.1). As such, they act as post-

The Life Cycle of *Rhagoletis pomonella*

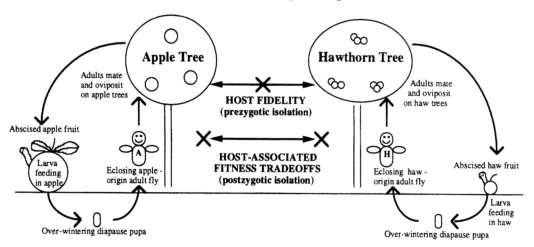

Figure 10.1. Summary of the life cycle of *R. pomonella* emphasizing the roles that host fidelity and fitness trade-offs play in isolating apple- and hawthorn-infesting races of the fly.

zygotic barriers to gene flow. Postzygotic isolation in animals has become synonymous with sterility or inviability problems in hybrids or backcrosses due to the detrimental mixing of incompatible genomes (Mayr, 1963). But this is not what Bush is arguing for. Rather, he is saying that in the earliest stages of divergence, postzygotic isolation is primarily the by-product of divergent selection pressures imposed by different habitats. Hybrids between very recently formed races or incipient species will therefore usually not die strictly because of inherently deleterious gene–gene interactions. Such hybrids will often thrive under laboratory conditions (see Schluter, this volume), as is the case for offspring between apple and hawthorn flies (Reissig and Smith, 1978; Smith, 1988; S. H. Berlocher and G. L. Bush, unpublished). Rather, hybrids will survive poorly because the genes they possess produce phenotypes that make them ill-suited for the niches occupied by their parents. Hybrid sterility and inviability problems that are intrinsic to the genetic environment will therefore tend to be secondary phenomena in speciation, completing a process already well under way. It is adaptation to variation in the external physical and biotic environment that provides the critical push that gets the process rolling. One need only look at the explosive nature of species genesis associated with adaptive radiations to find supporting evidence for this view (see chapters by Schluter and by McCune and Lovejoy, this volume).

To have a fitness trade-off in *Rhagoletis*, then, the same trait(s) or gene(s) that confer an advantage to a fly on one host plant must simultaneously incur a cost on alternative hosts (Dethier, 1954). Fitness trade-offs are equivalent to antagonistic pleiotropy. However, observing negative performance correlations in insects across different plants does not guarantee the existence of antagonistic pleiotropy (Futuyma and Philippi, 1987; Via, 1991; Mackenzie, 1996). This is because linkage disequilibrium can also produce negative correlations. (I define linkage disequilibrium in this instance as repulsion phase gametic disequilibrium between derived genes [i.e., nonancestral alleles] fixed at different loci in host-associated populations.) Such linkage disequilibrium may often be the cause for negative performance correlations when comparisons involve different insect species or geographically separated demes specialized on different plants. But negative correlations due to linkage do not constitute fitness trade-offs in the strict sense. This is because recombination can theoretically generate a "jack of all trades" genotype that has high fitness on all host plants. The genetic architecture of host specialization is important because it is difficult for negative performance correlations to evolve de novo in sympatry via linkage. Very simply, an allele giving an insect an advantage on one plant, but having no detrimental consequence on other hosts, has the capacity to spread and rapidly fix through the metapopulation as a whole. This is true unless sympatric demes exchange few migrants and new mutations increasing performance arise almost simultaneously in different host demes at tightly linked loci. Consequently, to initiate sympatric speciation in *Rhagoletis*, a new mutation, preexisting allele and/or combination of genes, must in all likelihood result in antagonistic pleiotropy.

Few would argue that when host-specific mating and fitness trade-offs exist, substantial reproductive isolation can evolve in sympatry as a pleiotropic by-product of host-associated adaptation (Rice and Hostert, 1993). Consequently, the critical question is not whether sympatric speciation is theoretically possible, because it is (see Kondrashov et al., and Johnson and Gullberg, this volume). Rather, it is whether sympatrically formed hosts races exist in nature. That is, do host-specific mating and fitness trade-offs ever evolve in tandem within geographically contiguous insect populations?

Genetic Evidence for *R. pomonella* Host Races

The first question to tackle is whether partially reproductively isolated and genetically differentiated host races of *R. pomonella* actually exist on apples and hawthorns. The answer to this question is yes. Sympatric pairs of apple- and hawthorn-infesting fly populations collected from field sites across eastern North America consistently displayed significant allele frequency differences at six allozyme loci: *Malic enzyme (Me), Aconitase-2 (Acon-2), Mannose phosphate isomerase (Mpi), NADH-Diaphorase-2 (Dia-2), Aspartate amino transferase-2 (Aat-2),* and *Hydroxyacid dehydrogenase (Had)* (Feder et al., 1988, 1990; Feder and Bush, 1989; McPheron et al., 1988; Berlocher and McPheron, 1996). These six allozyme loci map to only three different regions of the genome (*Aat-2* and *Dia-2* map to linkage group I; *Me, Acon-2,* and *Mpi* are tightly linked on group II; and *Had* is on group III; Berlocher and Smith, 1983; Feder et al. 1989). Significant levels of linkage disequilibrium have been found between nonallelic genes within, but not between, each of these three genomic regions (Feder et al., 1988, 1990). Seven other polymorphic allozyme loci displayed little differentiation between the host races (Feder et al., 1990). Walsh and Bush therefore appear vindicated in their claim of host races in *R. pomonella*. Furthermore, the documented historical time frame of the shift from hawthorns to apples argues for a sympatric origin for the apple race (Walsh, 1867; Bush et al., 1989).

Host Fidelity in *R. pomonella*

The next issue is whether Bush's model for sympatric race formation is true. Regarding host fidelity, the first pillar of his model, the answer again is yes. Ron Prokopy and co-workers (Prokopy et al., 1971, 1972) showed in a series of field experiments that *R. pomonella* mate exclusively on or near the fruit of their host plants. Studies on the oviposition acceptance behaviors of naive apple- and hawthorn-origin flies have also suggested that genetically based differences in host preference exist between the races (Prokopy et al., 1972, 1988). Although females of both host races prefer to oviposit into hawthorns, hawthorn-origin females are much more averse to ovipositing into apples than are apple-origin flies (Prokopy et al., 1972, 1988). Mark-release-recapture studies conducted at a field site near the town of Grant, Michigan,

indicated that host fidelity limits interhost movement of adults between apple and hawthorn trees to ~6% per generation (Feder et al., 1994). A number of factors contributed to this host fidelity, including inherent differences in host preference, apple flies being averse to emigration when they eclose from beneath apple trees, and eclosion time differences between the races causing allochronic isolation (Feder et al., 1994; Smith, 1988). Data from the mark-release-recapture study also showed that the mating success of interhost migrants was not statistically different from that of resident flies, suggesting an absence of ethological isolation (Feder et al., 1994, 1998). In addition, the oviposition behavior of immigrant females was not significantly different from that of non-migrants (Feder et al., 1994, 1998). Finally, experimental crosses have given no indication of any sterility or inviability barriers between apple and hawthorn flies (Reissig and Smith 1978; Smith, 1988; Berlocher and Bush, unpublished). Consequently, estimates of interhost migration derived from the mark-release-recapture study accurately reflect levels of genetic exchange between the host races.

The Search for Host-Associated Fitness Trade-offs in *R. pomonella*

The mark-release-recapture study cannot be the complete story, however. Although the results confirm the existence of fairly strong host fidelity in *R. pomonella*, they also indicate that host fidelity is not absolute. If the effective level of gene flow between apple and hawthorn populations is actually 6% per generation (the level suggested by the mark-release-recapture study), then allozyme frequency differences between the races would quickly disappear. Some form of host-associated selection is therefore required to counteract this gene flow. But there is little evidence for host-related fitness trade-offs in phytophagous insects (Futuyma and Moreno, 1988; Jaenike, 1990; Via, 1990; Futuyma and Keese, 1992; but for possible exceptions, see Gould, 1979; Mitter et al., 1979; Fry, 1990; Karowe, 1990; Via, 1991; Mackenzie, 1996). This would seem to cut the foundation from beneath Bush's second pillar for sympatric divergence. Nevertheless, the allozyme data imply that divergent selection is acting on apple and haw races, as allele frequencies for *Me, Acon-2, Mpi, Aat-2, Dia-2,* and *Had* have remained consistently different between apple and hawthorn populations at the Grant, Michigan, site, despite gene flow, over at least an 11-year period since they were first monitored in 1984 (Feder et al., 1990, 1993, 1997a). What then is the source of the selection maintaining the genetic integrity of the host races for the three regions of the genome displaying host-associated differentiation?

One logical choice would be some sort of feeding adaptation affecting larval survivorship within host fruits (Bush, 1969a,b, 1975a,b). But reciprocal egg transplant experiments performed by Prokopy et al. (1988) gave no evidence for any feeding specialization in larvae related to chemical or nutritional differences between apple and hawthorn fruits. Apple and hawthorn larvae survived equally well in hawthorn fruits. Both races fared equally poorly in apples. While these results accord with hawthorns being the ancestral host of *R. pomonella*, they are distressing regarding the issue of fitness trade-offs.

So, why do *R. pomonella* females continue to oviposit into apples when apples are nutritionally inferior to hawthorns for larval survival? Part of the reason is enemy-free space (Hairston et al., 1960; Bernays and Graham, 1988; Jaenike, 1990). Levels of braconid parasitism, interspecific competition (from a number of different moth species and plum curculio weevils), and intraspecific competition are much lower for flies infesting apples than for those infesting hawthorns (Feder, 1995; Feder et al., 1995). These factors were excluded from Prokopy et al.'s (1988) larval survivorship estimates and greatly compensate for the nutritional handicap of feeding within apples (Feder, 1995; Feder et al., 1995). But an escape from parasitoids and competitors does not constitute a fitness tradeoff in the strict sense for *R. pomonella*, for if a hawthorn-origin female were to oviposit into apples, then her offspring would receive the same beneficial escape from parasitoids that apple-origin larvae enjoy. In other words, aside from traits contributing to host fidelity, there is no genetic basis for the apple race's escape from natural enemies. Without a genetic basis for such a trait, there can be no genetic cost and hence no antagonistic pleiotropy or fitness trade-off. Some other factor besides enemy-free space must be responsible for maintaining genetic differentiation between the host races at the allozyme loci.

Life-History Traits, Host Plant Phenology, and Fitness Trade-offs in *R. pomonella*

Detoxification of plant secondary compounds and enemy-free space are but two potential avenues for host-associated adaptation. There are others. A number of observations suggest that *Me, Acon-2, Mpi, Dia-2, Aat-2,* and *Had* either encode or are linked to genes affecting development rates in *R. pomonella*. Flies possessing the alleles *Me 100, Acon-2 95, Mpi 37, Dia-2 100, Aat-2 +75,* and *Had 100* leave host fruits, pupate, and eclose earlier than other flies (Feder et al., 1993; Berlocher, unpublished). The allozymes present in higher frequencies in the hawthorn race at the Grant site therefore appear to be associated with faster rates of development in *R. pomonella*.

Patterns of geographic and temporal variation observed for the allozymes also implicate ambient temperature as a prime factor affecting the genetics of the host races. First, *Me, Acon-2, Mpi, Dia-2, Aat-2,* and *Had* all display latitudinal allele frequency clines in both apple and hawthorn races (Feder and Bush, 1989; Feder et al., 1990; Berlocher and McPheron, 1996). Second, allozyme frequencies in the hawthorn race at the Grant site correlate with spring temperatures over an 11-year period beginning in 1984 (Feder et al., 1993, 1997a).

The action of ambient temperature on fly development is a prime suspect for selection acting on the allozymes or linked loci. But what mediates this selection such that it differentially affects the host races? A likely candidate is host plant phenology. An insect such as *R. pomonella* that is univoltine, overwinters in the soil in a facultative pupal diapause, and has a limited adult longevity (Boller and Prokopy, 1976) must be developmentally synchronous with host acceptability to maximize fitness. Asynchrony in any stage of the life cycle is disastrous for a fly, leading to either its immediate death or lower viability / fecundity because it is in the wrong developmental state for the season. This could be due, for example, to the larva or pupa not entering diapause before the onset of winter or to an adult being active before or after host fruits are present.

Apples and hawthorns represent different temporal resources. Fruits on apple varieties favored by *R. pomonella* peak ~3 weeks earlier in the season than hawthorns (Feder et al., 1993). This difference has many important consequences to the developmental profiles of apple and hawthorn flies. One consequence that I will initially focus on is that larvae leave abscised apple fruits and pupate in the soil an average of over 16 days earlier in the season than they do from hawthorns (Feder et al., 1994). Because *R. pomonella* are facultative diapausers, apple-origin pupae that develop too rapidly in the summer run the risk of breaking diapause prematurely and directly developing into adults. Almost all *R. pomonella* larvae do, in fact, fail to diapause when held at temperatures above 28° C (Prokopy, 1968), and small second generations of apple race flies have been reported in the field (Caesar and Ross, 1919; Porter, 1928; Phipps and Dirks, 1933; Dean and Chapman, 1973). Nondiapausing flies are inevitably doomed; either they eclose at times when suitable host fruits are no longer available or they commit to, but do not complete, adult development before the onset of winter and subsequently freeze to death. The greater exposure of apple-pupae to warm weather preceding winter may therefore favor flies with deeper diapauses (or lower basal metabolic/development rates) in the apple than the hawthorn race. Selective pressures are likely to be different for hawthorn flies. The relatively late phenology of hawthorns means that slow-developing hawthorn flies may not enter pupal diapause quickly enough before the onset of winter and freeze to death. I shall henceforth refer to this idea as the diapause trade-off hypothesis.

The Prewinter Experiment—An Empirical Test of the Diapause Trade-off Hypothesis

In order to test the diapause trade-off hypothesis, co-workers and I (Feder et al., 1997a) experimentally manipulated environmental rearing conditions for hawthorn flies to determine whether we could induce a genetic response at the six allozymes displaying host-associated differentiation. In particular, we systematically altered the time period preceding winter for different subsamples of hawthorn-origin pupae collected from the Grant site (figure 10.2). Our rationale was that lengthening the prewintering period would expose hawthorn pupae to extended periods of warm weather, conditions they would face if they were infesting a host plant with an earlier fruiting phenology like apples. Our expectation was that such treatments would selectively eliminate pupae in shallow diapauses or with high metabolic/development rates from the surviving hawthorn-fly population. Conversely, brief prewinter treatments (e.g., 2 days), would favor rapidly developing pupae that quickly enter diapause.

What pattern of genetic response does the diapause trade-off hypothesis predict for the prewinter selection experiment? Actually, there are two predictions. First, that populations of diapausing hawthorn flies that survive in-

creasing periods of prewinter heating and the ensuing chilling period should become increasing more similar to populations of apple flies in their genetic constitution. After all, we are exposing hawthorn pupae to environmental conditions that we believe mimic those faced by apple flies. If these conditions are affecting the allozymes then we should see a response in the direction of the apple race. Second, nondiapausing flies that eclose prior to winter should display high frequencies of electromorphs more common to the hawthorn race at the Grant site (i.e., *Me 100, Acon-2 95, Mpi 37, Dia-2 100, Aat-2 +75*, and *Had 100*), as these alleles have been previously correlated with fast development rates (Feder et al., 1993, unpublished).

Analysis of the prewinter experiment supported the diapause trade-off hypothesis (figure 10.3; Feder et al., 1997a). The predicted genetic response was observed for *Me, Acon-2, Mpi, Dia-2,* and *Aat-2*, as allele frequencies in surviving flies became more "applelike" with longer prewinter heat treatments (see figure 10.3 for results for *Me 100*). In fact, we came very close to genetically trans-

Experimental Design for Pre-Winter Experiment

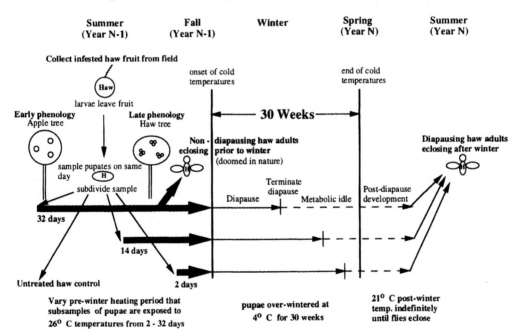

Figure 10.2. Overview of the experimental design for the prewinter study (for a detailed description of methods, see Feder et al., 1997a). Infested fruits were collected from beneath a hawthorn tree at the Grant, Michigan, study site on September 15, 1989, and placed on wire screens above plastic trays in the laboratory. Puparia were collected from the plastic trays on a daily basis and divided into subsamples. One daily subsample was immediately frozen to serve as an untreated genetic control. The remaining subsamples were held at 26°C in a constant temperature room for 2, 7, 14, 21, 28, or 32 days, after which time they were placed in cold storage for 30 weeks at 4°C to simulate winter. Adult flies were collected as they eclosed during the periods preceding (nondiapausing flies) and following chilling (diapausing flies). Flies were also sampled from an apple tree at the Grant site on August 15, 1989, to provide a baseline control for the apple race. Flies were genetically scored for *Me, Acon-2, Mpi, Dia-2, Aat-2,* and *Had* using standard starch gel electrophoretic techniques.

Me 100 results for Hawthorn Race

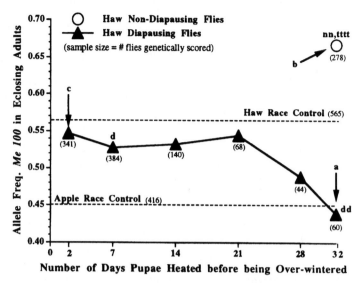

Figure 10.3. Allele frequencies for *Me 100* in diapausing (triangles) and nondiapausing (circle) hawthorn fly adults for the 2–32-day treatments in the prewinter experiment. The *Me 100* frequencies in the untreated apple and hawthorn fly control samples are designated by the dashed lines. d = significant allele frequency difference between diapause flies and hawthorn control sample as determined by one-tailed, randomized Fisher exact tests. n = significant difference between nondiapause flies and hawthorn control sample. t = significant difference between diapause and nondiapause flies for the same heat treatment. Number of letters designates significance level (one letter = P < 0.05, two = P < 0.01, four = P < 0.0001). Arrow "a" indicates how similar the allele frequency for *Me 100* for diapausing hawthorn flies in the 32-day treatment is to the untreated apple fly control. Arrow "b" points to *Me 100* frequency for nondiapausing hawthorn flies in the 32-day treatment being much greater than that for either diapausing flies or the untreated hawthorn fly control. Arrow "c" shows that the frequency of *Me 100* in the 2-day treatment was slightly lower than that of the untreated hawthorn fly control.

forming the hawthorn race into the apple race after only a single generation of mass selection by the 32-day treatment (see arrow "a" in figure 10.3). A significant proportion of pupae also eclosed as nondiapausing adults in the 32-day treatment. As expected, these nondiapausing flies had very high frequencies of *Me 100, Acon-2 95, Mpi 37, Dia-2 100, Aat-2 +75,* and *Had 100* (see arrow "b" in figure 10.3).

The prewinter experiment answers many questions. Most importantly, it establishes a direct, empirical link among ambient temperature, host phenology, fly development, and the six allozyme loci. The experiment also implies that the apple race was derived from standing genetic variation in the hawthorn race and, in particular, from pupae comprising the slowest developing, tail end of the hawthorn fly distribution.

Issues Raised by the Prewinter Experiment

The results from the prewinter experiment also raise several issues, however. I will attempt to answer what I feel are a number of the most pertinent and thought-provoking questions below. The reader is referred to Feder et al. (1997a) for further discussion of the data.

Does the selection that occurs in the hawthorn race also occur in the apple race? If not, then extrapolating the results for the hawthorn race to the apple race would be like comparing oranges (or haws) to apples. Could it be that the remainder of the genome of apple flies is sufficiently different from that of hawthorn flies to cause relative fitnesses at the allozyme loci to differ between the races? If true, then hybrids between the races might have similar fitnesses to apple flies if they were to infest apples.

The preponderance of evidence argues against the above scenario, at least as far as the allozymes are concerned. Recall that only the three genomic regions containing *Me, Acon-2, Mpi, Dia-2, Aat-2,* and *Had* show host-related differentiation. Other polymorphic allozyme loci display little host-associated or geographic variation (Feder et al., 1990). This suggests that gene flow between the races is sufficient to homogenize allele frequencies at loci not directly under selection or linked to such genes.

Alternatively, balancing selection could also account for the similarity of allozyme frequencies at these other loci in the races, but this seems a less likely explanation given the number of loci involved and the known 6% level of genetic exchange between apple and hawthorn populations each generation. Either way, genetic differences between the races do not appear to be evenly dispersed throughout the genome but clumped in specific regions. Consequently, polymorphism in the remainder of the genome does not appear to be acted upon in a substantively different way between, as opposed to within, the races. Clinal patterns of variation for *Me, Acon-2, Mpi, Dia-2, Aat-2,* and *Had* in the apple race underscore this point by indicating that the apple race is responding to the same selection pressures in a similar manner as the hawthorn race (see next question). Finally, the apple race at the Grant site showed a qualitatively similar genetic response to selection as the hawthorn race in a just completed, parallel prewinter experiment (Feder et al., unpublished).

Does the prewinter experiment accurately reflect the situation in nature? After all, the experiment was conducted under artificial laboratory conditions. There is no guarantee that selection detected under such a situation also operates in nature. In this regard, an important implication of the diapause trade-off hypothesis (DTH) is that the same selective factors differentiating the host races should also operate within the races. The results from the prewinter experiment predict that allozyme frequencies should track local ambient temperature conditions, since temperature affects both the diapause status of flies and the fruiting times of host trees (Phipps and Dirks, 1933; Lathrop and Dirks, 1945; Glass, 1960; Oatman, 1964; Reissig et al., 1979).

Earlier, I presented two lines of evidence supporting a relationship between ambient temperature and the allozymes. The first was that *Me, Acon-2, Mpi, Dia-2,* and *Had* display latitudinal allele frequency clines among both apple- and hawthorn-fly populations in the United States (Feder et al., 1990; Feder and Bush, 1989; Berlocher and McPheron, 1996). It so happens that populations from northern (i.e., colder) latitudes possess higher frequencies of *Me 100, Acon-2 95, Mpi 37, Dia-2 100, Aat-2 +75,* and *Had 100,* the alleles associated with faster development rates. In addition, the clines show several perturbations in allele frequencies that coincide with differences in ambient temperature conditions among local collecting sites (Feder and Bush, 1989, 1991). Second, allozyme frequencies in the haw race at the Grant site have tracked spring temperature conditions since 1984 (Feder et al., 1993, 1997a). What I did not mention earlier, however, is that they track temperatures in the spring of the preceding year. Because most pupae break diapause after their first winter, flies sampled in year N are representative of pupae that survived the previous season (year N − 1). High spring temperatures in year N − 1 meant an early field season and correlated with increased frequencies in year N of alleles common to the apple race (or southern populations). Conversely, cold springs selected for alleles more common in northern populations.

Is the selection generated in the prewinter experiment strong enough to account for the continued differentiation of the host races in the face of gene flow? The answer is yes, at least with respect to gene flow from the hawthorn into the apple race. The apple race pupates about 16 days on average earlier than the hawthorn race. We could therefore get a relative gauge of the force of selection by comparing treatments in the prewinter experiment that differ by approximately 2 weeks; the 26°C, 15:9-hour light:dark conditions used to rear pupae in the experiment being close to those experienced by apple flies at the height of the field season in late July to early August. Based on this criterion, selection coefficients (s values) against *Me 100, Acon-2 95, Aat-2 +75,* and *Had 100* homozygotes (or homozygotes at closely linked genes) calculated for hawthorn flies from a comparison of the untreated control and the 14-day treatment under an additive fitness model are 0.264, 0.254, 0.083, and 0.552, respectively. These coefficients are greater than those needed to counteract the estimated 6% gene flow from the hawthorn into the apple race each generation for the three genomic regions displaying differentiation (*Me* = 0.063, *Acon* = 0.136, *Aat* = 0.074, *Had* = 0.041; these values were calculated assuming that apple and hawthorn control samples represent allele frequencies in the races after interbreeding but prior to selection). The apple race therefore continues to persist as a diverged genetic entity from the hawthorn race due, in part, to developmental adaptations in pupae stemming from differences in the fruiting phenologies of apples and hawthorns.

What is maintaining the genetic integrity of the hawthorn race? The prewinter experiment really only explains the source of selection impeding introgression of certain regions of the genome from the hawthorn into the apple race. For the sympatric host race formation model to work, selection must also be operating in the opposite direction or the metapopulation as a whole would simply become more "applelike." And, the metapopulation is not becoming more "applelike," as allele frequencies for the six allozyme loci have remained consistently different between the host races at the Grant, Michigan, site since 1984 (Feder et al., 1993, 1997a). Earlier, I stated that the late phenology of hawthorns should select for faster rates of development in hawthorn pupae in order for them to achieve diapause before the first frost. This should favor individuals possessing the alleles for faster development rates (i.e., *Me 100, Acon-2 95, Mpi 37, Dia-2 100, Aat-2 +75,* and *Had 100*). However, the shortest heat treatment in the prewinter experiment (2 days) did not result in any marked increase in the frequencies of "hawthorn race" alleles in surviving adults. In fact, the frequency of *Me 100* in the 2-day treatment was actually slightly lower than that of the untreated hawthorn race control (see arrow "c" in figure 10.3). This appears to contradict an important component of the diapause trade-off hypothesis. If it is true that short prewintering periods do not select for fast fly development, then what is limiting effective gene flow from the apple into the hawthorn race?

First, the overwintering conditions that were used in the prewinter experiment (30 weeks of chilling at 4°C) turn out to constitute what would be a very long winter for hawthorn pupae at the Grant site (Feder et al., 1997b). Experiments in which we have varied the length of the overwintering period for pupae indicate that 30 weeks of chilling at 4°C select against *Me 100, Acon-2 95, Mpi 37, Dia-2 100, Aat-2 +75*, and *Had 100* (Feder et al., 1997b). These data show that we actually did select for fast fly development in the 2-day prewinter treatment, but the long overwintering period used in the prewinter experiment counteracted the effects of this selection.

Second, the need to synchronize adult eclosion with host fruit availability also limits gene flow. Apple flies eclose an average of ~10 days earlier than hawthorn flies at the Grant site (Feder, 1995), a difference coinciding with the earlier phenology of apples. So, the races are being pulled apart allochronically as a consequence of selection for different mean eclosion times. Due to a quirk in the ecology and genetics of eclosion, this selection is likely to act against "hybrids" (offspring from apple × hawthorn fly matings) at the Grant site. The allozymes *Me 100, Acon-2 95, Mpi 37, Dia-2 100, Aat-2 +75*, and *Had 100* are all found in higher frequencies in the first individuals to eclose in both host races (Feder et al., 1993), and the apple race possesses lower frequencies of these "fast development" alleles than the hawthorn race at the Grant site. Taken together, this suggests that apple flies are actually genetically predisposed to eclose later, not earlier, than hawthorn flies if the two races were to be reared under identical environmental conditions. And, in fact, the period between pupation and eclosion is ~1 week longer for apple flies at the Grant site; they have a head start in that they pupate an average of 16 days earlier than hawthorn flies in the fall, but eclose only 10 days sooner the following summer. The reason that apple flies eclose earlier in nature is therefore environmental. The inherently slower development rate of apple flies is offset by the higher temperatures and longer photoperiods that these flies experience as larvae and pupae. Hybrids between the races should therefore have genotypes that translate into intermediate eclosion times (development rates) when reared under controlled conditions. But in nature, of course, this would not be the case, as apples and hawthorns are seasonally asynchronous. Consequently, hybrids should eclose earlier than apple flies if they were to infest apples and later than hawthorn flies if they were to infest hawthorns. The repercussions of this are straightforward. Hybrids would eclose too early in the

season to optimally utilize apples and too late to attack hawthorns. They would not eclose at the happy median that would let them use both hosts. This is another example of how gene × environmental interactions, rather than deleterious gene × gene effects, are responsible for isolating the races.

A third piece of the puzzle is that selection also appears to be acting on larval development rates. The alleles *Me 100, Acon-2 95* and *Mpi 37* not only correlate with adult eclosion time, but also with when larvae complete feeding, exit host fruits, and pupate (Filchak and Feder, unpublished data). We have found that the quality of apple and hawthorn fruits kept outdoors under natural, field conditions deteriorates faster than that for fruits kept in a protective, open-air garage (Filchak and Feder, unpublished data). Consequently, larval mortality was significantly greater in the field compared to garage, suggesting that field conditions favor rapidly developing larvae that quickly leave host fruits (Filchak and Feder, unpublished data). Consistent with this hypothesis, allele frequencies for *Me 100, Acon-2 95*, and *Mpi 37* were also significantly higher in surviving field- than garage-reared larvae for both host races (Filchak and Feder, unpublished data). These results imply that genotype specific mortality is, after all, occurring during the larval life-history stage of *R. pomonella*. The caveat is that the selection is not happening in the manner originally envisioned by Bush (1969a,b; 1975a,b), who thought feeding adaptations of *R. pomonella* to chemical and/or nutritional differences between apple and hawthorn fruits would be the key to larval survivorship. Rather, the prime factor appears to be how fast fruits necrose. Also, the selection is directional, favoring the alleles *Me 100, Acon-2 95*, and *Mpi 37* (and/or linked genes) in both host races. As we discussed, these same alleles (or linked blocks of genes) can be discovered in pupae, especially in the apple-fly race, as they correlate with premature diapause termination. Host-dependent fitness trade-offs in *Rhagoletis* may therefore be due as much to differences in the relative strengths of directional selection pressures acting on different life-history stages, as disruptive selection affecting any one particular stage (i.e., divergent selection results from the summation of directional selection vectors across the life-histories of the races being of different sign). The necessity to consider details of the entire life-cycle highlights one of the difficult challenges posed to documenting fitness trade-offs for phytophagous insects.

What is becoming increasingly clear is that the host races represent semiautonomous populations residing in a balanced state on alternative adaptive peaks. Gene flow between the populations is not sufficient to counteract selection and perturb the races from their respective peaks. The interaction of ambient temperature conditions, host-plant phenology, fruit decay, and fly development is a major factor shaping the adaptive landscape. The valley between the peaks is not the result of any inherently negative genetic interaction between the genomes of the host races but stems from divergent selection pressures exerted externally by the environment. Finally, spatial and temporal variation in ambient temperature conditions provide an explanation for the existence of standing developmental variation within the hawthorn race, sufficient genetic variation to allow a portion of the hawthorn fly population to shift and adapt to the earlier phenology of apples.

Why can't flies evolve a compromise combination of developmental characters that permit them to effectively use both apples and hawthorns? Why isn't the range of developmental responses plastic and fine-tuned to prevailing environmental conditions? Why can't modifier genes evolve and expunge any negative pleiotropic consequences of the allozymes? For instance, why can't a modifier gauge and counteract any detrimental consequences that fast larval development has on individuals that subsequently experience warm prewinter temperatures as pupae? Why do the six allozymes have such similar effects across so many different life-history stages of *R. pomonella*?

As for the last question, the answer is probably that the allozymes encode or are linked to genes whose products have fundamental consequences to basal metabolic levels in *R. pomonella*. (I use the word *linked* here because there is no proof that selection is working directly on the allozymes. Also recall that the six allozymes map to three different regions of the genome and that linkage disequilibrium has been observed in each of these regions [Feder et al., 1988, 1990], thereby increasing the chances that selection may actually be acting on a gene[s] linked to the allozymes.) As to why development is not more plastic, one must always remember that not everything is possible. Sometimes developmental constraints really do exist in nature. After all, individuals can have only a single ontogeny. Nevertheless, sensory inputs from the environment could still be used to modify the unfolding of this ontogeny. But such information must be both reliable (an accurate predictor of things to come) and consistent (provide the same message about what to do for the future) for insects infesting different host plants if plasticity rather then specialization is to evolve. For example, think about a modifier gene that slows down pupal development during warm weather periods at the Grant site. Such a modifier locus could be beneficial for apple flies, as they develop relatively early in the season when one hot day is likely to be followed by others. But such a gene could be detrimental to hawthorn flies that develop later in the season when weather is more erratic and when maximal utilization of every warm day is more critical to ensure the completion of prediapause development.

The above scenario emphasizes that it may be much easier for nonspecific modifiers to differentially evolve in each of the races than for a specific modifier conferring developmental plasticity to arise and spread through both races. (By a nonspecific modifier I mean an allele whose phenotypic effect is expressed regardless of the genetic background. A specific modifier refers to a gene that affects a trait only in the presence of certain other nonallelic genes.) After all, the races are already semi-autonomous entities. The reduction in gene flow could only facilitate this process and further hasten the continued divergence of the races at auxiliary loci.

If nonspecific modifiers can fix independently in the host races, then why hasn't the apple race evolved an obligate diapause to counteract its longer prewinter exposure to warm weather and elevated day lengths? The reason is probably one of balance and timing, specifically, a balance between the depth of pupal diapause and the timing of adult eclosion. Although pupae in deep diapauses avoid the risk of premature eclosion in the fall, they also emerge significantly later than other flies during the following summer (Feder et al., unpublished). There seems to be no mechanism to entirely decouple these two aspects of development in *R. pomonella*. They are correlated characters. Consequently, I suspect that obligately diapausing apple flies would eclose too late in the field season to effectively utilize apples as a host resource.

Does the general paucity of trade-offs found for phytophagous insects argue against a high incidence for sympatric divergence? Could R. pomonella *be the exception rather than the rule?* Yes, if the paucity were real. But Rausher (1992) has commented that most tests for trade-offs have concentrated on metabolic costs associated with detoxification of host secondary compounds, while neglecting costs "associated with coordinating life-history events with host plant phenology, with tolerance of microclimatic conditions and with escape from predators via crypsis" (p. 70). In addition, experimental designs have often been restricted so that only one or a few factors influencing fitness were actually measured. The results from *R. pomonella* certainly point to the relationship between host phenology and insect development as an overlooked source of trade-offs. They also point to the need to take a "holistic" approach to the study of fitness trade-offs examining an insect's entire life cycle, not just feeding stages. A plethora of fitness trade-offs may therefore await discovery in phytophagous insects, sending a cautionary note about overinterpreting the current shortfall of examples as an indicator of the unlikelihood of nonallopatric speciation (Butlin, 1987).

But why should developmental synchrony be a rich source for fitness trade-offs when plant secondary compounds appear not? I have argued that traits involved in sympatric race formation must be "nonlabile," that is, inflexible traits that at some point become incapable of compensatory adjustments to changes in the environment (Futuyma and Moreno, 1988). If they were not, then phenotypic plasticity (a phytophagous generalist) would be a common evolutionary solution to host plant heterogeneity. For reasons discussed above, developmental specialization may often be rigid and host specific. For instance, once a *Rhagoletis* pupa breaks diapause, there is no turning back—the fly is committed to adult development. This being the case, life-history traits adapting flies to the phenology of one host plant will limit their opportunity to utilize alternative hosts with different phenologies. In contrast, traits associated with metabolic detoxification may be more labile. For instance, it may be possible to induce detoxification pathways only when they are needed or to ameliorate their costs when they are not needed. Also, a wider repertoire of behaviors may be available in the arsenal of insects to combat plant defenses than there are to modify development. The end result of this being that life-history traits are better candidates for fitness tradeoffs than are traits involved in the detoxification of plant compounds.

How much of the genome must be impervious to gene flow for two populations to be considered species? The host races differ in at least three regions of the genome that impart partial postmating reproductive isolation by developmentally adapting the races to their respective host plants. We have yet to study the genetics of host preference, so there may well be more than three regions. Whatever the final number, however, a substantial part of the *R. pomonella* genome would seem to be "open" to interrace gene flow. We also do not know how many regions differ between recognized sibling species in the *R. pomonella* group. But again, the answer is not the entire genome. Taxa in the *R. pomonella* complex are not fixed for alternative alleles at allozyme loci, the only exception being the distantly related *R. cornivora* (Berlocher and Bush, 1982; Berlocher et al., 1993). Instead, they share most allozymes in common, differing in allele frequencies at certain loci (Berlocher and Bush, 1982; Berlocher et al., 1993; Berlocher and Feder, unpublished). This pattern could be explained by shared ancestral polymorphism. But it could also be due to a low level of gene flow among taxa. If the latter, then most *R. pomonella* species are not quantitatively different from host races, especially the more recently derived taxa. In both cases selection counteracts gene flow to varying degrees across the genome. The main difference is that at the species level gene flow is lower and/or the intensity of selection is greater. *R. pomonella* group species do not appear to be unique in this regard. Numerous examples of the introgression of genomic regions between species can be found in the literature (Harrison, 1990), suggesting that "the field of genetic recombination is broader than the taxonomic species and groups that are behaving as evolutionarily independent entities" (Templeton, 1989, p. 10). The possession of a completely autonomous, self-cohesive gene pool is therefore not a required calling card for a "good species." This is not to say that genetic incompatibilities and negative epistatic interactions will not be found extending across large portions of the genome between certain species pairs (see chapters by Naviera and Maside, and Wu and Hollocher, this volume), just that such extensive closure of two genomes may often accumulate after speciation.

A New or Old Species Definition?

The realization that the entire genome need not crystallize or be completely impervious to gene flow when new species form and that disruptive selection can be a prime impetus for divergence necessitates a rethinking of species concepts. One recent suggestion is that species are genotypic clusters that can overlap spatially without fusing (Mallet, 1995). In essence, this represents a genetic updating of Darwin's original species definition. While this definition may not be an end-all, it does free us from the self-referential problem of species as reproductively isolated entities (Wallace, 1865; Mallet, 1995), a concept that confuses cause and effect (Paterson, 1985; Mallet, 1995; see Templeton, this volume).

Application of the cluster definition is not without its difficulties, however. Exactly how different do two genetic clusters have to be to be considered different species? How many genes and how much differentiation at these loci? For successful speciation, a minimum of two complementary genes are needed according to Dobzhansky (1937) and Mayr (1963). Could our criterion therefore be differentiation for at least two unlinked loci (J. Mallet, personal communication)? But if just a couple of unlinked genes are required for species designation, then shouldn't the *R. pomonella* host races be considered different species? This seems a bit premature to me. But where to draw the line on a seeming continuum in levels of divergence is not a problem unique to the genotypic cluster definition. Rather, the problem is an inherent result of the speciation process itself.

One possible solution would be to calculate likelihood scores that individuals with a particular genotype belong to one taxon as opposed to another based on a representative sample of loci spaced throughout the organism's genome (a variant of this idea was first suggested to me by J. Mallet). If (1) a sample of individuals can be sorted into two distinguishable genotypic clusters such that there is less than a 5% chance in misassigning a randomly chosen individual from one cluster into the opposing cluster and (2) the population of individuals comprising each of the clusters is in Hardy-Weinberg equilibrium for a majority of loci, then the clusters could be considered species. A corollary of finding distinguishable sympatric clusters will frequently be that geographically distant populations of the same species exchange more genes, and hence are more genetically similar to each other, than sympatric populations of different species. In other words, populations of the same species will eventually form what appears to be a monophyletic clade, even if the populations had multiple origins. This corollary provides one basis for distinguishing host races from species (see Berlocher, ch. 8 this volume); enough gene flow occurs between local host race populations that although the races maintain their genetic differences in sympatry, local host race populations are genetically more similar to one another than each is to other populations infesting the same host plant in other areas of the species range. (Note: genetic similarity refers to a representative sample of loci evenly spaced throughout the genome. It is possible that if only those loci under selection where considered, then populations infesting the same host plant would be most similar across the entire species range.)

An example of a likelihood analysis is shown in figure 10.4 for the *R. pomonella* complex using a 17-locus allozyme data set. Graphs of the distributions of lod scores are given for pairwise comparisons among the apple- and hawthorn-infesting populations of *R. pomonella*, the undescribed *Cornus florida* fly (collected from flowering dogwood), the undescribed sparkleberry fly (collected from *Vaccinium aboreum*), and *R. mendax* (collected from high bush blueberries, *V. corymbosum*, and deerberries, *V. stamineum*; see Berlocher, ch. 8 this volume, for further discussion of the biology of these flies). (Lod score = log10, of the likelihood ratio that a fly comes from one taxa versus another given the fly's genotype, i.e., the likelihood that a fly has a given genotype given that the fly belongs to taxon 1 divided by the likelihood that the fly has this genotype and came from taxon 2). Such an analysis is well suited for *R. pomonella* group flies because their host plant affiliations provide an a priori basis for assigning individuals to different taxa (e.g., the genotypes of flies collected from apples are tested against flies collected from hawthorns at sympatric sites). The results indicate that apple and hawthorn populations of *R. pomonella* constitute host races and not sibling species. Although apple and hawthorn flies form discernible genetic clusters in pairwise tests of sympatric populations collected from across the Northeast (figure 10.4a), the clusters overlap to such an extent that ~20% of the flies sampled from one host plant would be misclassified as belonging to the other race based on their genotypes, far above the 5% threshold set for species status. In addition, local populations of apple and hawthorn flies are sometimes more genetically similar to one another than to other geographically distant populations of flies infesting either apples or hawthorns, respectively (Berlocher, unpublished; Feder, unpublished; McPheron, unpublished). The likelihood analysis does suggest that several closely related taxa to apple and hawthorn flies in the *R. pomonella* group should be classified as species (figure 10.4b–g), although the *Cornus florida* fly and sparkleberry fly appear to be just at the 5% cutoff point for species status. Finally, there were few instances of significant single locus deviations from Hardy-Weinberg equilibrium for any of the five taxa included in the analysis (total of 19 significant tests out of 393 conducted, none significant on a tablewise basis using the sequential Bonferroni method; Rice, 1989).

Figure 10.4a–d (first column of graphs in the figure) highlights the continuum in the levels of genetic differentiation existing between *R. pomonella* host races and sibling species. This observation lends credence to Bush's (1966, 1969a,b, 1975a,b, 1992) argument for the fluid nature of sympatric race and species formation in the *R. pomonella* group, that host races represent the initial, formative stage in a gradual, sequential process triggered by ecological adaptation that results in sympatric speciation. Further work is still needed, however, on the transition period from host races to sibling species to confirm the rate and chronology of sympatric speciation in the *R. pomonella* group (see Berlocher, ch. 8 this volume).

The genotypic cluster method is not foolproof, however. Further work is needed to establish guidelines for sorting individuals into clusters when populations do not differ by some a priori criterion, such as their host affiliation. An exhaustive genetic data set is also required to perform the analysis that may be prohibitive for certain

---

Figure 10.4. Genotypic cluster analysis for *R. pomonella* group flies based on a 17-locus allozyme data set (see h for a list of the 17 loci). Flies included in the study were the apple- and hawthorn-infesting populations of *R. pomonella*, the undescribed *Cornus florida* fly (collected from flowering dogwood), the undescribed sparkleberry fly (collected from *Vaccinium aboreum*) and *R. mendax* (collected from high bush blueberries [*V. corymbosum*] and deerberries [*V. stamineum*]). Lod scores (= log10 of the likelihood ratio that a fly comes from one taxon versus another given the fly's genotype) were calculated for individual flies collected from pairs of different host plants at sympatric sites or sites in close geographic proximity (see h for a list of study sites and collecting years). The results for a given pair of taxa were then combined across sites to generate the composite distributions of lod scores shown in a–g. To calculate lod scores, genotype frequencies were first determined at each of the 17 allozyme loci in the two populations (p1, p2) infesting alternative host plants (h1, h2, respectively) at a site. For a given fly collected from either h1 or h2, the frequency of the genotype that the fly possessed at locus 1 in p1 was divided by the frequency of that genotype in p2. Next, the product of these individual locus ratios was taken across all 17 loci. The logarithm base 10 of this product represents the log10 likelihood ratio (or lod score) that a fly of a given genotype infested h1 versus h2. Plots of the distributions of these ratio scores over 0.5 unit intervals appear in a–g. Also given are the mean lod scores averaged across sites for each taxa in the various pairwise comparisons (designated by X), the number of individuals collected from a given host plant that would be mistyped as belonging to the alternative taxa over the total number of individuals sampled, the percentage of misclassifications, and the sites included in the analysis.

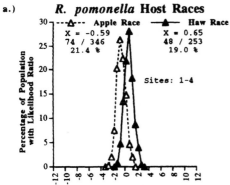

a.) R. pomonella Host Races

b.) C. florida fly / R. pomonella

c.) Sparkleberry fly / R. pomonella

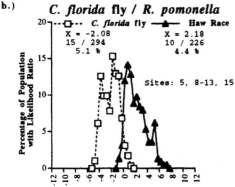

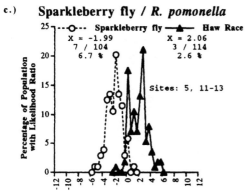

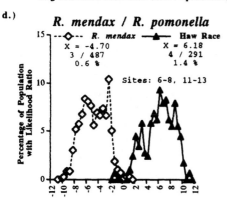

d.) R. mendax / R. pomonella

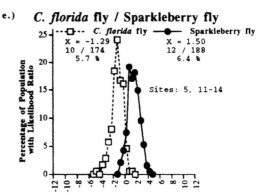

e.) C. florida fly / Sparkleberry fly

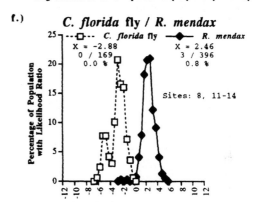

f.) C. florida fly / R. mendax

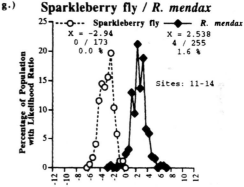

g.) Sparkleberry fly / R. mendax

h.)

Collecting Sites:	Allozymes (17 Loci):
1. Grant, MI. 1984	Aconitase-1&2
2. Ephriam, WI. 1984	Adenylate kinase-2
3. Gas City, IN. 1987	Alcohol dehydrogenase-1
4. Urbana, IL. 1986	Aldolase
5. Fairfield, IL. 1981	Aspartate aminotransferase-2
6. Sawyer, MI. 1985	NADH-Diaphorase-1&2
7. Amherst, MA. 1986	Fumarase
8. Princeton, N.J. 1988	Glucosephosphate isomerase
9. Beltsville, MD. 1989	Hydroxy acid dehdrogenase
10. Bowling Gr., VA. 1989	Isocitrate dehydrogenase
11. Clark Hill, S.C. 1989	Malate dehydrogenase-1&2
12. Fort Valley, GA. 1989	Mannose phosphate isomerase
13. Tuskegee, AL. 1989	Phosphoglucomutase
14. Gainesville, FL. 1989	Triosephosphate isomerase
15. Nacogdoches, TX. 1989	

taxa. Along these lines, additional thought is required to control for the number and type of genes included in an analysis. Finally, the method has its own unique peculiarities. For example, it is possible for two subpopulations that recently became completely reproductively isolated to still be considered the same species because they have yet to accumulate sufficient genetic divergence to sort themselves into discernible clusters. But the genetic cluster definition does provide a criterion for species demarcation that at least in principle could be uniformly applied across taxa. The definition does not constrain species to monophyletic or bifurcate origins, nor does it bias our thinking as to how clusters (species) form (Mallet, 1995). And it has the added bonus of potentially working for both sexual and asexual taxa.

An Open Mind toward Speciation

In conclusion, *Rhagoletis pomonella* may not be that unconventional after all, at least not if you were to have asked Charles Darwin. Darwin said from the beginning that speciation was the natural outcome of populations adapting to different environments. This certainly appears to be the case in *Rhagoletis*. Also, the apparently fluid nature of races and species in the *R. pomonella* group (see Berlocher, ch. 8 this volume) is what he said would be expected from an active process like speciation. Benjamin Walsh certainly would not disagree and would add the caveat that complete geographic isolation is not always required to initiate speciation.

So why in these current times do so many consider *R. pomonella* so unconventional and sympatric speciation so controversial? The following passage from C. I. Wu (1996) denotes a commonly held belief that

> among all differences between species, the traits contributing to reproductive isolation are the most intriguing. Although in itself the phenomenon seems to make no sense (what good does it do to produce sterile progeny?), reproductive isolation is crucial for nascent species to continue their divergence without their unique innovations being lost by blending through gene migration. (p. 105)

Instead of dwelling on the history of this view, I think it more insightful to say that it is "the production of nascent species, rather than the acquisition of hybrid inviability or sterility, that seems the most intriguing event in speciation. To achieve this, presumably adaptive innovations must be protected from gene flow by selection" (J. Mallet, personal communication). *Rhagoletis* flies are controversial because they argue that speciation can be triggered by natural selection. This is not to say that reproductive isolation itself is directly selected for, but that isolation evolves as an inadvertent by-product of adaptation to the environment (Rice and Hostert, 1993). We therefore should not think of reproductive isolation in purely genetic terms divorced from external ecological pressures. The fitness of a phenotype is, after all, a product of the interaction of genes with the environment as mediated through development. Consequently, the ruckus over sympatry is somewhat tangential to the central issue of disruptive selection. For given that new innovations can be sheltered from gene flow by selection, the complete allopatric separation of populations no longer becomes an absolute prerequisite for divergence. This does not mean that geographic considerations do not and have not often facilitated the adaptive divergence of populations. Allopatry certainly relaxes the initial restriction of antagonistic pleiotropy associated with sympatry, making it easier for divergence to occur via linkage (i.e., the sequential fixation of alternative sets of alleles at linked genes that each confer a selective advantage in one habitat but that have no or even detrimental effects on fitness in alternative environments relative to the ancestral allelic state). The possibility of reproductive isolation caused by negative epistasis and drift is also enhanced. But sometimes divergent selection pressures may be strong enough to pull populations apart in sympatry. Is it any wonder that Darwin was ambivalent when it came to geography?

The reascension of Darwin's view for a direct role of the environment and natural selection in speciation may be Guy Bush's and *Rhagoletis*'s most long-lasting legacy. As Guy is wont to say, "to understand the critical early stages of speciation, one must fully understand the biology of one's study organism(s). This is how one can identify the key traits under selection that initiate population divergence. Each case is likely to be different and have its own particular set of circumstances and rules" (Bush, personal communication).

If nothing else, this volume testifies to Guy's view that a great diversity of patterns and processes underlies speciation, a view that was fostered and nurtured by Bush through his constant reminders to us to keep an open mind concerning speciation.

Acknowledgments   I thank Stewart Berlocher, Jim Mallet, Joe Roethele, Kristin Lewis, Uwe Stolz, and an anonymous reviewer for thoughtful discussions and comments on an earlier draft of this chapter. Also, I thank the commodore himself, my mentor, Guy Bush, for his useful and concise research advice and steadfast support through the years. Finally, I thank Dorie Bush, who always had a kind word, sandwich, and drink (usually a strong one) ready for any situation. My research has primarily been supported from grants by the United States Department of Agriculture and by the National Science Foundation.

References

Berlocher, S. H., and G. L., Bush. 1982. An electrophoretic analysis of *Rhagoletis* (Diptera: Tephritidae) phylogeny. Syst. Zool. 31:136–155.

Berlocher, S. H., and D. C. Smith. 1983. Segregation and mapping of allozymes of the apple maggot fly. J. Hered. 74: 337–340.

Berlocher, S. H., and B. A. McPheron. 1996. Population structure of *Rhagoletis pomonella*, the apple maggot fly. Heredity 77:83–99.

Berlocher, S. H., B. A. McPheron, J. L. Feder, and G. L. Bush. 1993. Genetic differentiation at allozyme loci in the *Rhagoletis pomonella* (Diptera: Tephritidae) species complex. Ann. Entomol. Soc. Am. 86:716–727.

Bernays, E., and M. Graham. 1988. On the evolution of host specificity in phytophagous arthropods. Ecology 69:886–892.

Boller, E. F., and R. J. Prokopy. 1976. Bionomics and management of *Rhagoletis*. Ann. Rev. Entomol. 21:223–246.

Bush, G. L. 1966. The taxonomy, cytology and evolution of the genus *Rhagoletis* in North America (Diptera: Tephritidae). Museum of Comparative Zoology, Cambridge, Mass.

Bush, G. L. 1969a. Sympatric host race formation and speciation in frugivorous flies of the genus *Rhagoletis* (Diptera: Tephritidae). Evolution 23:237–251.

Bush, G. L. 1969b. Mating behavior, host specificity, and the ecological significance of sibling species in frugivorous flies of the genus *Rhagoletis* (Diptera: Tephritidae). Am. Nat. 103:669–672.

Bush, G. L. 1975a. Sympatric speciation in phytophagous parasitic insects. pp. 187–206 *in* Evolutionary strategies of parasitic insects and mites (P. W. Price, ed.). Plenum, New York.

Bush, G. L. 1975b. Modes of animal speciation. Ann. Rev. Ecol. Syst. 6:339–364.

Bush, G. 1992. Host race formation and sympatric speciation in Rhagoletis fruit flies (Diptera: Tephritidae). Psyche 99:335–358.

Bush, G. L., J. L. Feder, S. H. Berlocher, B. A. McPheron, D. C. Smith, and C. A. Chilcote. 1989. Sympatric origins of *R. pomonella*. Nature 339:346.

Butlin, R. K. 1987. A new approach to sympatric speciation. Trends in Ecol. Evol. 2:310–311.

Caesar, L., and W. A. Ross. 1919. The apple maggot. Ontario Dept. Agr. Bull. 271:1–32.

Dean, R. W., and P. J. Chapman. 1973. Bionomics of the apple maggot in eastern New York. Search Agric. Entomol. Geneva No. 3., Geneva, N.Y.

Dethier, V. G. 1954. Evolution of feeding preference in phytophagous insects. Evolution 8:33–54.

Diehl, S. R., and G. L. Bush. 1984. An evolutionary and applied perspective of insect biotypes. Ann. Rev. Entomol. 29: 471–504.

Dobzhansky, Th. 1937. Genetics and the origin of species (1st ed.). Columbia Univ. Press, New York.

Feder, J. L. 1995. The effects of parasitoids on sympatric host races of the apple maggot fly, *Rhagoletis pomonella* (Diptera: Tephritidae). Ecology 76:801–813.

Feder, J. L., and G.L. Bush. 1989. Gene frequency clines for host races of *Rhagoletis pomonella* (Diptera: Tephritidae) in the midwestern Unites States. Heredity 63:245–266.

Feder, J. L., and G. L. Bush. 1991. Genetic variation among apple and hawthorn host races of *Rhagoletis pomonella* (Diptera: Tephritidae) across an ecological transition zone in the mid-western United States. Entomol. Exp. Appl. 59:249–265.

Feder, J. L., C. A. Chilcote, and G. L. Bush. 1988. Genetic differentiation between sympatric host races of *Rhagoletis pomonella*. Nature 336:61–64.

Feder, J. L., C. A. Chilcote, and G. L. Bush. 1989. Inheritance and linkage relationships of allozymes in the apple maggot fly. J. of Hered. 80:277–283.

Feder, J. L., C. A. Chilcote, and G. L. Bush. 1990. The geographic pattern of genetic differentiation between host associated populations of *Rhagoletis pomonella* (Diptera: Tephritidae) in the eastern United States and Canada. Evolution 44:570–594.

Feder, J. L., T. A. Hunt, and G. L. Bush. 1993. The effects of climate, host plant phenology and host fidelity on the genetics of apple and hawthorn infesting races of *Rhagoletis pomonella*. Entomol. Exp. Appl. 69:117–135.

Feder, J. L., S. Opp, B. Wlazlo, K. Reynolds, W. Go, and S. Spisak. 1994. Host fidelity is an effective pre-mating barrier between sympatric races of the apple maggot fly. Proc. Natl. Acad. Sci. 91:7990–7994.

Feder, J. L., K. Reynolds, W. Go, and E. C. Wang. 1995. Intra- and interspecific competition and host race formation in the apple maggot fly, *Rhagoletis pomonella* (Diptera: Tephritidae). Oecologia 101:416–425.

Feder, J. L., J. B. Roethele, B. Wlazlo, and S. H. Berlocher. 1997a. The selective maintenance of allozyme differences between sympatric host races of the apple maggot fly. Proc. Natl. Acad. Sci. U.S.A. 94:11417–11421.

Feder, J. L., Stolz, U., Lewis, K. M., Perry, W., Roethele, J. B., and Rogers, A. 1997b. The effects of winter length on the genetics of apple and hawthorn races of *Rhagoletis pomonella* (Diptera: Tephritidae). Evolution 51:1862–1876.

Feder, J. L., S. H. Berlocher, and S. B. Opp. 1998. Sympatric host race formation and speciation in *Rhagoletis* (Diptera: Tephritidae): a tale of two species for Charles D. Pg. 408–411. *In* Genetic structure and local adaptation in natural insect populations: effects of ecology, life history, and behavior (S. Mopper and S. Strauss, eds.). Chapman and Hall, London.

Fry, J. D. 1990. Trade-offs in fitness on different hosts: Evidence from a selection experiment with a phytophagous mite. Am. Nat. 136:569–580.

Futuyma, D. J., and M. C. Keese. 1992. Evolution and coevolution of plants and phytophagous arthropods. pp. 439–475 *in* Herbivores: their interaction with secondary plant metabolites, Vol. 2 (G. A. Rosenthal and M. R. Berenbaum, eds.). Academic Press, San Diego.

Futuyma, D. J., and G. Moreno. 1988. The evolution of ecological specialization. Annu. Rev. Ecol. Sys. 19:207–233.

Futuyma, D. J., and T. E. Philippi. 1987. Genetic variation and covariation in responses to host plants by *Alsophila pometaria* (Lepidoptera: Geometridae). Evolution 41:269–279.

Glass, E. H. 1960. Apple maggot fly emergence in western New York. New York State Exp. Sta. (Cornell Univ.), Bull. 789.

Gould, F. 1979. Rapid host range evolution in a population of the phytophagous mite *Tetranycchus urticae* Koch. Evolution 33:791–802.

Hairston, N. G., F. E. Smith, and L. B. Slobodkin. 1960. Community structure, population control and competition. Am. Nat. 94:421–425.

Harrison, R. G. 1990. Hybrid zones: Windows on evolutionary process. Oxford Surveys in Evolutionary Biology 7:69–128.

Jaenike, J. 1990. Host specialization in phytophagous insects. Annu. Rev. Ecol. Sys. 21:243–273.

Karowe, D. N. 1990. Predicting host range evolution: colonization of *Coronnilla varia* by *Colios philodice* (Lepidoptera: Pieridae). Evolution 44:1637–1647.

Lathrop, F. H., and C. O. Dirks. 1945. Timing the seasonal cycles of insects: the emergence of *Rhagoletis pomonella*. J. Econ. Entomol. 38:330–334.

Mackenzie, A. 1996. A trade-off for host plant utilization in the black bean aphid, *Aphis fabae*. Evolution 50:155–162.

Mallet, J. 1995. A species definition for the Modern Synthesis. Trends Ecol. Evol. 10:294–299.

Mayr, E. 1963. Animal species and evolution. Harvard Univ. Press, Cambridge, Mass.

McPheron, B. A., D. C. Smith, and S. H. Berlocher. 1988. Genetic differences between *Rhagoletis pomonella* host races. Nature 336:64–66.

Mitter, C., D. J. Futuyma, J. C. Schneider, and J. D. Hare. 1979. Genetic variation and host plant relations in a parthenogenic moth. Evolution 33:770–790.

Oatman, E. R. 1964. Apple maggot emergence and seasonal activity in Wisconsin. J. Econ. Entomol. 57:676–679.

Paterson, H. E. H. 1985. The recognition concept of species. Transvaal Mus. Monogr. 4:21–29.

Phipps, C. R., and C. O. Dirks. 1933. Notes on the biology of the apple maggot fly. J. Econ. Entomol. 26:349–358.

Porter, B. A. 1928. The apple maggot. U.S. Dept. Agr. Tech. Bull. 66.

Prokopy, R. J. 1968. The influence of photoperiod, temperature and food on the initiation of diapause in the apple maggot. Can. Entomol. 100:318–329.

Prokopy, R. J., E. W. Bennett, and G. L. Bush. 1971. Mating behavior in *Rhagoletis pomonella* (Diptera: Tephritidae). I. Site of assembly. Can. Entomol. 103:1405–1409.

Prokopy, R. J., E. W. Bennett, and G. L. Bush. 1972. Mating behavior in *Rhagoletis pomonella* (Diptera: Tephritidae). II. Temporal organization. Can. Entomol. 104:97–104.

Prokopy, R. J., S. R. Diehl, and S. S. Cooley. 1988. Behavioral evidence for host races in *Rhagoletis pomonella* flies. Oecologia 76:138–147.

Rausher, M. D. 1992. Natural selection and the evolution of plant-insect interactions. pp. 20–88 *in*: Insect chemical ecology: an evolutionary approach. (B. D. Roitberg and M. B. Isman, eds.). Chapman and Hall, New York.

Reissig, W. H., and D. C. Smith. 1978. Bionomics of *Rhagoletis pomonella* in *Crataegus*. Ann. Entomol. Soc. Am. 71:155–159.

Reissig, W. H., J. Barnard, R. W. Weires, E. H. Glass, and R. W. Dean. 1979. Prediction of apple fly emergence from thermal unit accumulation. Environ. Entomol. 8:51–54.

Rice, W. R. 1989. Analyzing tables of statistical tests. Evolution 43:223–225.

Rice, W. R., and E. E. Hostert. 1993. Laboratory experiments on speciation: what have we learned in 40 years? Evolution 47:1637–1653.

Smith, D. C. 1988. Heritable divergence of *Rhagoletis pomonella* host races by seasonal asynchrony. Nature 336:66–67.

Templeton, A. R. 1989. The meaning of species and speciation: A genetic perspective. pp. 3–27 *in* Speciation and its consequences (D. Otte and J. A. Endler, eds.). Sinauer, Sunderland, Mass.

Via, S. 1990. Ecological genetics and host adaptation in herbivorous insects: the experimental study of evolution in natural and agricultural systems. Annu. Rev. Entomol. 35:421–426.

Via, S. 1991. The genetic structure of host plant adaptation in a spatial patchwork: demographic variability among reciprocally transplanted pea aphid clones. Evolution 45:827–852.

Wallace, A. R. 1865. On the phenomena of variation and geographical distribution as illustrated by the papilionidae of the Makyan region. Trans. Linn. Soc. Lond. 25:1–71.

Walsh, B. J. 1864. On phytophagous varieties and phytophagous species. Proc. Entomol. Soc. Philadelphia 3:403–430.

Walsh, B. J. 1867. The apple-worm and the apple maggot. J. Hort. 2:338–343.

Wilson, D. S. 1989. The diversification of single gene pools by density- and frequency-dependent selection. pp. 366–385 *in* Speciation and its consequences (D. Otte and J. A. Endler, eds.). Sinauer, Sunderland, Mass.

Wu, C.-I. 1996. Now blows the east wind. Nature 380:105–106.

11

Evolution of Insect–Plant Associations

Sensory Perception and Receptor Modifications
Direct Food Specialization and Host Shifts
in Phytophagous Insects

Steph B. J. Menken
Peter Roessingh

Speciation, the process of becoming a new species, is the most important phenomenon in biology as it has been the source of the staggering variety of fossil and extant life forms. It is central to biology since it connects microevolution with macroevolution. The debate on the conditions that promote speciation has long been dominated by postulating allopatric speciation as null hypothesis: geographic isolation is the prerequisite to allow lineages to diverge. However, some 130 years ago Walsh (1867) proposed the concept of sympatric speciation, that is, divergence without an extrinsic geographic barrier to gene flow. Over the past 30 years it has become evident that sympatric speciation is theoretically possible (Bush, 1969; see chapters by Johnson and Gullberg, and Kondrashov et al., this volume). The question remains of whether we can find evidence that host races (i.e., conspecific populations that are partially reproductively isolated due to adaptation to different food plants) actually evolve sympatrically in nature (Feder, this volume).

Sympatric speciation has been frequently examined in host-associated herbivorous insects. Basic to the survival of a phytophage is host-plant choice, especially since most insect herbivores are specialists who, under penalty of extinction, need to choose correctly their own food plant from among the vast number of hostile plants. Basic to the evolution of host races is a polymorphism for host-plant choice that pleiotropically leads to assortative mating if mating takes place on the food plant. Through disruptive selection, different segments of one interbreeding population can thus diverge on different food plants and eventually become new species.

Here we review aspects of the evolution of insect–plant relationships, including mechanics and genetics of host choice, in a phylogenetic context with the small ermine moth genus *Yponomeuta* as a model. The predominance of specialist insect herbivores is discussed in the context of a lack of genetic variation to modify the nervous system and/or physiology necessary for shifts and adaptation to phytochemically different plants. Examples of all genetic mechanisms proposed in this host shift scenario can be found in nature. Although discussed in relation to sympatric speciation, similar scenarios can be envisioned in allopatry if the plant of choice of a group of colonizing insects appears to be rare or absent and whenever the host plant of a phytophage is going extinct.

Value of Phylogenetic Reconstruction for Interpreting Evolutionary Hypotheses about Insect–Host Associations

Understanding the evolution of associations between phytophagous insects and their host plants can only be achieved by an integration of macroevolutionary (phylogenetic and biogeographic) and microevolutionary (population genetical, ecological, and behavioral) studies. Progress in phylogenetic methodology (Hennig 1966; Forey et al. 1992) and molecular biology (Hillis et al. 1996), allowing estimation of phylogenetic relationships from molecular data, has greatly facilitated the growing recognition of the significance of an historical perspective in ecology (Brooks and McLennan 1991).

Normally, morphological, allozyme, and/or molecular data are used in establishing the phylogeny of a group. Whether ecological characters, such as host-plant associations, contain phylogenetic information is controversial. Just like behavioral characters (Paterson et al. 1995), it is not easy to prove that ecological similarities among species are homologous. Furthermore, phylogenies can be constructed on the basis of extant species only because, with the exception of leaf-miners and gall formers, host associations do not fossilize (some morphological features of the insect might indicate mode of feeding or oviposition though). Finally, although on average ecological characters are no more homoplasious than morphological traits (de Queiroz and Wimberger 1993), some may be intrinsically labile (Miller 1987; Löfstedt et al. 1991; Ward 1991), phytophage–host-plant interactions being an example (e.g., Radtkey and Singer 1995). Rather than using ecological data as primary input for phylogenetic reconstruction, one can evaluate these associations against existing, independently derived cladograms (Mitter and Brooks 1983; Mitter et al. 1991; Menken et al. 1992).

Major Evolutionary Hypotheses about Insect–Host Associations

Ehrlich and Raven (1964) assumed that reciprocal selection between plants and insects had induced both chemical diversification and resistance in plants and diet specialization in insects. A plant species that has evolved novel chemical defences due to selection by a range of herbivores alleviates herbivory, and this leads to species diversification. Phytophagous insects that adapt to these new defences can subsequently diversify. This so-called escape and radiation coevolution is often referred to as cospeciation (Miller and Wenzel 1995), which would result in a strict concordance between the cladograms of plants and insects. However, tandem speciation is only one of the possible outcomes of evolving species interactions (Thompson 1994). Moreover, cospeciation is not necessarily the result of reciprocal evolution but might simply be the fortuitous outcome of an interaction. Interpretation is also complicated by the frequent lack of phylogenetic information about the plants and by asynchronous speciation or extinction, which result in a poor agreement of the cladograms, despite the fact that coevolution has occurred. Finally, a continuous association between plants and insects has only occurred if the two clades are of similar age, since host shifts might be "mediated by plant characters strictly concordant with plant phylogeny" (Mitter et al. 1991, p. 290) such that a recent diversification of an insect lineage results in a phylogeny strictly congruent with plant phylogeny.

To date, only two well-documented cases of cospeciation in phytophages exist, namely, the beetle genera *Phyllobrotica* (Farrell and Mitter 1990) and *Tetraopes* (Farrell and Mitter 1993). These genera share major features of their biology (Farrell and Mitter 1993), in particular, the fact that larvae feed on the roots and adults on the leaves and flowers of the same single plant species. In the monophagous and oligophagous species of the beetle genus *Ophraella*, however, where both larvae

and adults feed on leaves, cospeciation has apparently not occurred (Futuyma et al. 1995). These beetle life cycles contrast markedly with nectar feeding and mobile butterflies and moths and with many other phytophagous insects, and thus generalizations for phytophagous insects cannot be made easily.

As an alternative to coevolution, Jermy (1984) proposed that via host shifts phytophagous insects have evolved against a preexisting background of plants and chemical diversity. Instead of reciprocal evolutionary change, the evolution of phytophages followed that of their food plants without much affect on plant evolution. This so-called sequential evolution results in cladograms of insects and hosts that are mostly incongruent (e.g., type A in Jermy 1984).

Heritable changes in the insects' plant recognition mechanism are proposed as the primary event in the evolution of insect–plant associations, because plant chemistry is presumably the most important source of information used by females to decide where to oviposit (Städler 1992; Renwick and Chew 1994). This is particularly true of herbivorous insects with relatively immobile larvae (like most Lepidoptera) where host choice largely operates through female oviposition behavior (Courtney and Kibota 1990). Consequently, host-plant shifts are usually constrained in chemical channels, with some notable exceptions (e.g., Chew and Robbins 1984), and the predicted course of evolution is colonization of preexisting plant species that are presumably phytochemically similar, but not necessarily taxonomically related, to their extant hosts.

Accumulated evidence from cladistic analyses across orders of insects indicates that colonization is the predominant mode for the evolution of host affiliations of insect herbivores (reviewed in Miller and Wenzel 1995). Understanding such evolutionary host transfers requires insight into relevant plant aspects (with plant chemistry as most likely candidate) as well as the sensory capabilities of the insect. In this review, we concentrate on the chemosensory side of the interaction.

Predominance of Specialists

Diet specialists predominate among phytophagous insects (e.g., Farrell and Mitter 1993; Thompson 1994). Gravid females often restrict oviposition to, and their larvae only can feed on, plants from one genus (monophagy) or one family (oligophagy); a minority of species feed on plants from different families (polyphagy). Many species are even associated with only one plant species over all of their distribution area (so-called strict monophagy) such as most *Yponomeuta* species (Menken et al. 1992).

However, our knowledge about host-plant associations is still rather patchy, thus hampering generalizations about patterns of diet breadth. Most data come from the temperate zones, and our current understanding of host-plant relationships in the tropics is very fragmentary. In the only detailed, long-term study on feeding specialization of butterflies and moths in Costa Rica, Janzen (1988) estimated that more than half of the species feed on only one plant species and that oligophagy predominates among the remainder, similar to what is found in the temperate areas.

Moreover, unreliable data due to misidentifications of both plants and insects and incomplete taxonomic knowledge are a further complication. Polyphagous species, for instance, might comprise a sibling species complex with individual species being mono- or oligophagous (Hagen et al. 1991; Menken et al. 1992). On the other hand, quite a number of phytophagous insects have been described solely on the basis of their host-plant association, leading to oversplitting of a generalist species into specialized taxa (e.g., the *Stigmella aurella* complex; Menken and Raijmann 1996).

Whereas field data on host utilization can therefore be unreliable, larval feeding experiments in the laboratory might also be inaccurate. Such experiments can yield overestimates of diet breadth because of unnatural conditions and because in Lepidoptera ovipositing females usually have a narrower host spectrum then their larvae (Wiklund 1974; Karowe 1990; Futuyma et al. 1995). Conversely, females sometimes oviposit on unsuitable hosts (Wiklund 1975; see also below). Thus, oviposition preference and larval performance are not necessarily correlated, and the extent of the correlation might even differ between closely related species (Roininen and Tahvanainen 1989).

In spite of the need for more and better data, it is generally agreed that host specialization is common and widespread. A number of mainly ecological hypotheses have been put forward to explain this ubiquity of specialist herbivores, such as escape from interspecific competition, enemy-free space, and trade-offs in feeding efficiency (Jermy 1993; Denno et al. 1995; Dyer 1995; Menken 1996). There is no hypothesis, however, that stands out as providing a general explanation for specialization. The trade-off hypothesis (i.e., an evolutionary increase in offspring performance on a new host concomitantly results in a reduction in adaptation to its former host), which has been considered to be a powerful explanation for reduced diet breadth (Jaenike 1990), has, with few exceptions (Mackenzie 1996 and references therein), not been supported by positive evidence (Jaenike 1990; Futuyma and Keese 1992). Thompson (1996) has argued that it is not surprising that evidence for trade-offs in single components of larval performance has been so scarce because "trade-offs . . . are more likely to involve coordination among the various components of performance together with ecological factors that allow higher fitness on one host than on others" (p. 133). Another approach has been taken by Bernays and Wcislo (1994), who developed the idea that host specialization is a result of the need for fast decision making. Processing complex information ultimately entails costs associated with long decision times (e.g., increased predation risk) or larger, less economic brains. As specialization appears to reduce decision time, it will be favored by natural selection (Bernays 1996).

All the above-mentioned hypotheses assume unconstrained evolution. However, host specialization as well as frequent shifts to related plants and long associations with a particular group of plants (so-called phylogenetic conservatism; Holloway and Hebert 1979; see also table 11.1) suggest that insects often lack adequate selectable variation to adapt to a greater range of host species (Jermy 1993). Constraints on the evolution of the insects' nervous system or physiology probably make insects particularly preadapted to phytochemically similar plants, resulting in specialization as well as phylogenetic conservatism, as closely related plant species are on average phytochemically more similar than distantly related ones. The importance of genetic constraints for adaptive evolution in outbred populations is, however, still a matter of debate (Rausher 1992; Futuyma et al. 1995). Besides genetic or developmental constraints, the widespread occurrence of diet specialization and phylogenetic conservatism can be explained by stabilizing selection. However, this is unlikely considering the vast spatial and temporal variation in food plant availability (Futuyma et al. 1995).

If, besides ecological circumstances, genetic constraints have guided the evolution of host shifts, it is of interest to analyze whether an extant species contains heritable variation for adaptation to a plant on which it does not occur but which is the host of conspecific populations or closely related species. Several studies found genetic variation for host-plant use (references in Futuyma et al. 1993). In a very detailed study, however, Futuyma and colleagues showed that four *Ophraella* species generally lack variation to adapt to a wider range of food plants (overview in Futuyma et al. 1995) and concluded that the trajectory of evolution of this complex character is severely constrained by the absence of genetic variation or by high canalization.

Both monophagy and phylogenetic conservatism can be nicely demonstrated within the moth family Nepticulidae (table 11.1). With more than 700 described species worldwide, it is the largest family of non-Dytrisian Lepidoptera. Their larvae usually mine in the leaves of dicotyledon trees and shrubs. Over 90% of the Nepticulidae (85% of the hosts are known) are monophages, and those that are oligophages feed on taxa in related plant genera (van Nieukerken 1986b). There is predominantly phylogenetic conservatism at the species group level, but not so much at the generic and higher levels (Scoble 1983). Species of a species group (which are usually monophyletic; e.g., van Nieukerken 1986a,b; Menken 1990 and references therein, and unpublished) often feed on the same or related hosts, and most species groups are confined to one host family or even one genus, with very few outliers (e.g., *S. myrtilella* on *Vaccinium* in *salicis* group).

Table 11.1. Food plant relations in some Nepticulidae species groups.

Species Group	N	Distribution	Food Plant Family	Outliers
E. angulifasciella	±20	Holarctic	Rosaceae (Maloideae + Prunoideae + Rosoideae)	
E. subbimaculella	±20	Palaearctic	Fagaceae	
S. aurella	±15	Palaearctic	Rosaceae (Rosoideae)	
S. oxyacanthella	±25	Holarctic	Rosaceae (Maloideae + Prunoideae)	
S. ruficapitella	>30	Palaearctic	Fagaceae	Betulaceae (1), Caprifoliaceae (2)
S. salicis	±17	Holarctic	Salicaceae	Ericaceae (1)
S. ulmivora	8	Palaearctic	Ulmaceae	
T. immundella	±30	West Palaearctic	Fabaceae (tribe Genistae)	

The diet breadth per group of species, about 90% of which are monophages, indicates strong phylogenetic conservatism. E., Ectoedemia; S., Stigmella; T., Trifurcula; N, number of described species.

Yponomeuta

General Biology and Host Associations

For more than 20 years we have addressed questions concerning the evolution of insect–plant associations and genetical differentiation in phytophages in the small ermine moth genus Yponomeuta (Menken et al. 1992; Menken 1996). The genus has a worldwide distribution and comprises some 70 species. With few exceptions, species are univoltine; mating takes place on the food plant. Host affiliations are known for 37 species, 20 of which feed on only one plant species. There is a long-standing association with Celastraceae: 22 species feed on plants of the genus Euonymus (Gershenson and Ulenberg, 1998), and within the subfamily Yponomeutinae, the genus Yponomeuta belongs to a monophyletic clade, of which the common ancestor fed and the great majority of extant species still feed on Celastraceae (Ulenberg and Gershenson, unpublished manuscript). The pattern of food plant affiliations has led to the following working hypothesis (Menken et al. 1992): the present-day associations in Yponomeuta evolved from an ancestral relation with Celastraceae through speciation in allopatry, mostly on Euonymus species, and through host shifts in sympatry or allopatry to new relationships with mainly Rosaceae.

Mapping host-plant associations onto an independently derived estimate of the phylogeny of 14 Yponomeuta species, based on allozyme and molecular information, suggests that a unique shift from Celastraceae to Rosaceae has occurred in the evolution of the genus (Menken 1996). The alternative hypothesis of an ancestral association with Rosaceae requires at least six shifts to Celastraceae and is thus less likely. Minor shifts have occurred to Crassulaceae (Y. vigintipunctatus) and Salicaceae (Y. rorellus and Y. gigas). Host-plant relations in Yponomeuta clearly support the model of sequential evolution: the cladograms of insects and hosts are incongruent, and the radiation of Yponomeuta postdated the divergence of their host plants. The hypothesis that Yponomeuta diversified in concert with its hosts would, after all, require that this genus be virtually as old as the angiosperms. The age of Yponomeuta, however, is estimated at between 10 and 20 million years (Menken 1982; Menken and Ulenberg 1987).

Like most phytophagous insects, the genus Yponomeuta is made up of food specialists, but there is no phylogenetic conservatism at the genus level. Yponomeuta species seem to be committed to specialization per se rather than to a particular plant group: whatever shift they have made in the evolutionary past, the descendant species remained monophagous (Menken et al. 1992). However, it is also conceivable that (some of the) specialists have originated from now extinct transient generalists. Yponomeuta padellus, which currently feeds on a number of rosaceous plants, could be an example of such a transient oligophagous taxon (Menken et al. 1992; Menken and Raijmann 1996). During its evolution it has probably broadened its host range, but selection might favor future populations with a restricted host range, leading to renewed monophagy. A similar "oligophagous bridge" scenario was recently proposed by Feeny (1995) for swallowtail butterflies. The tiger swallowtail Papilio glaucus uses a large number of phylogenetically diverse plant species as hosts and appears to have escaped from some of the chemical contraints that limited the host range of its ancestors. However, P. eurymedon, P. rutulus, and P. multicaudatus, species believed to derive quite recently from (an immediate ancestor of) P. glaucus, are more specialized than P. glaucus, which can thus be viewed as an unspecialized bridge species.

Possible Chemical Trigger of Sequential Evolution in Yponomeuta

If genetic constraints indeed generally restrict the potential for host-plant shifts and subsequent host-race formation and speciation (Bush 1994), we can ask what circumstances allowed the historical shift in *Yponomeuta* from Celastraceae to the quite dissimilar Rosaceae. It should be noted that what "phytochemically similar" means to an insect is not necessarily the same as what it means to a phytochemist. For an insect with a limited set of chemoreceptors, plants that share certain key compounds will be perceived as similar notwithstanding many differences. Then, even host shifts that are clearly not phylogenetically (taxonomically) conservative might be viewed as *sensorily* conservative. Phytochemical similarity should therefore be defined from the perspective of the insect as those plant taxa that elicit a similar sensory response in a particular insect species.

In discussing possible mechanisms that might lead to host-race formation and premating reproductive isolation, we concentrate on chemosensory stimuli from the plant. We thus ignore stimuli such as vision (predominant in the precontact phase and alightment on a plant) and presence of conspecifics, competitors, parasitoids, and predators, all of which may affect the final behavioral reaction. The chemosensory input affecting oviposition and feeding preferences consists of positive and negative chemical stimuli, and changes in the balance of these stimuli might change preference (Dethier 1982). Consequently, host-plant shifts and diet extension in phytophagous insects can be facilitated by gain of stimulant sensitivity or loss of deterrent sensitivity. Both possibilities have been suggested for *Yponomeuta* species (Menken et al. 1992).

Gain of Sensitivity for Stimulants

New plants might be incorporated in the diet because the stimulating compounds important for accepting the old host are also present in the new one. It has been hypothesized that the presence in *Prunus* spp. of low levels of dulcitol (a characteristic compound of Celastraceae and a strong phagostimulant for Celastraceae-feeding *Yponomeuta*) facilitated the ancestral host shift from Celastraceae to Rosaceae (Peterson et al. 1990; Menken et al. 1992).

Although the dulcitol concentration in *Prunus* species is about 10 times lower than it is in Celastraceae, this level is—given the logarithmic nature of sensory responses—certainly within the sensitivity range of present-day *Euonymus* feeders (figure 11.1). In addition, variation for dulcitol sensitivity is apparently present (figure 11.2). To what extent this variation has a genetic basis in *Yponomeuta* is still unknown. Unfortunately, this lack of knowledge is almost universal, since chemosensory sensitivity has commonly been viewed as a fixed property of the species. Although individual variation due to, for instance, age or physiological state has been addressed (Blaney et al. 1986), the possibility of polymorphism in a population has rarely been considered and analysis of true genetic variability of sensory responses has been neglected (but see Wieczorek 1976 for an exception). Due

to time constraints originating from the recording technique, researchers usually investigate a limited number of (frequently related) individuals and nonresponding insects, which could be indicative of a polymorphism, are often discarded. Sensory polymorphism is further partly masked by the common expression of sensitivity as population means, which tends to swamp the few extreme-responding insects, the very ones that might be preadapted to perform a host shift.

We suggest a route from dulcitol-containing Celastraceae to dulcitol-containing Rosaceae (i.e., *Prunus* spp.) taken by preadapted Celastraceae-feeding individuals with above-average sensitivity for dulcitol. In Rosaceae, the predominant sugar alcohol and phagostimulant for *Yponomeuta* species associated with Rosaceae is sorbitol, a stereoisomer of dulcitol. Once on the new host, selection for acquiring sensitivity to a novel compound (viz., sorbitol) will operate, a process that is probably more complicated than losing sensitivity. Interestingly, one of the present-day *Prunus* feeders (*Y. evonymellus*) possesses the dual sensitivity that is predicted as the outcome of this scenario. However, in other present-day Rosaceae feeders, dulcitol sensitivity is absent or severely reduced as it is in *Y. mahalebellus* and *Y. padellus* (figure 11.2).

Loss of Sensitivity for Deterrents

Along with increased sensitivity to phagostimulants, loss of sensitivity to deterrents has been proposed as a driving force for host-plant shifts and a determinant of diet breadth (Jermy 1993; Bernays and Chapman 1994). Several Rosaceae (e.g., *Prunus padus*, the host plant of *Y. evonymellus*) contain prunasin, a feeding deterrent for most *Yponomeuta* larvae (van Drongelen 1980), but prunasin sensitivity is relative low in Rosaceae feeders (e.g., *Y. evonymellus*; van Drongelen 1979), suggesting that reduced sensitivity to this compound facilitated the shift in an ancestral *Yponomeuta* population to a *Prunus* host. A similar loss of sensitivity for deterrents has been suggested as the basis for some of the minor host shifts in *Yponomeuta*. For instance, insensitivity to phloridzin in *Y. malinellus* or to salicin in *Y. rorellus* might have opened the possibility to accept apple or willow, respectively (van Drongelen 1979; Menken et al. 1992).

In conclusion, these observations are consistent with a host shift driven by sensory modifications and resulting in an oligophagous "bridge" species, accepting both Celastraceae and Rosaceae. Due to a variety of mechanisms (for mechanisms, see Jermy 1993 and above), the present-day Rosaceae specialists might have evolved from this hypothetical taxon.

The Molecular Basis of Chemoreception

Progress in understanding the molecular basis of chemosensory systems (for reviews, see Stengl et al. 1992; Lancet and Ben-Arie 1993; Shephard 1994) now offers

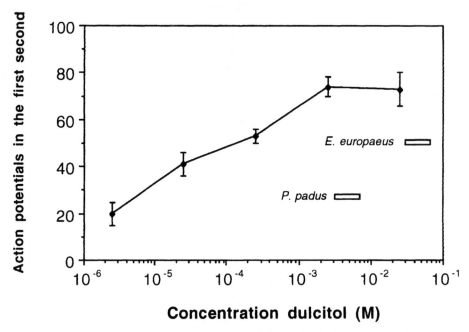

Figure 11.1. Dose-response curve for dulcitol, as perceived by the lateral sensilla styloconica on the mouthparts of larvae of *Yponomeuta cagnagellus*. Dulcitol is the primary transport sugar in *Euonymus europaeus*, the host of *Y. cagnagellus*. The range of the dulcitol concentration in the host (assuming all sugar is available to the insect) is indicated by the horizontal bar. The dulcitol concentration in *Prunus padus*, the host of *Y. evonymellus*, is about an order of magnitude lower.

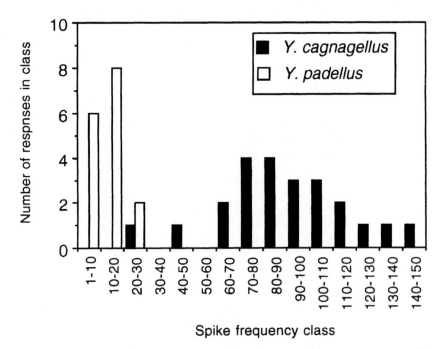

Figure 11.2. Histogram of responses to 10^{-2} M dulcitol in the lateral sensilla styloconica on the galea of individual larvae of *Yponomeuta cagnagellus* and *Y. padellus*. Dulcitol is present in the host of *Y. cagnagellus* and a phagostimulant for this species. The hosts of *Y. padellus* contain only traces of dulcitol or dulcitol is completely lacking, and the species is not very sensitive to this compound. Note the large variation in response intensity in both species.

good opportunities for explaining the above-postulated changes in *Yponomeuta* in terms of the underlying genetical mechanisms. The sequence of events during the perception of chemical stimuli can be summarized as follows (table 11.2). Stimulus molecules enter the sensory hair via pores in the cuticula and bind relatively unspecifically to proteins present in the receptor lymphe of olfactory and gustatory receptors (Steinbrecht et al. 1995; Ozaki et al. 1995). These receptor lymphe-borne proteins are thought to function as carriers for the stimulus molecules and may also play a role in the deactivation of the stimulus. While these binding proteins might confer some specificity on the receptor by acting as a filter, the sensitivity spectrum of sensory cells is commonly ascribed to specific g-protein-coupled receptor proteins embedded in the dendritic membrane of the receptor neuron. These receptor proteins belong to the seven transdomain receptor family. Binding of a stimulus molecule causes, via a second messenger cascade, the flow of a receptor current that generates a series of action potentials traveling over the axon of the receptor neuron into the central nervous system.

Connections in the nervous system are highly conserved among species (Kutsch and Breidbach 1994), probably indicating strong constraints on central modifications. However, there is now ample evidence that relative simple changes in the periphery of the sensory system can have dramatic behavioral effects. For instance, a candidate receptor gene for the odorant diacetyl was recently identified in the nematode *Caenorhabditis elegans* (Sengupta et al. 1996). One G-to-A transition in this gene, causing the replacement of histidine by tyrosine in the third transmembrane section of the protein, abolishes the normal response to diacetyl. Such a loss of sensory capability (a so-called knockout mutation) might in case of a deterrent cell lead to diet extension (see previous section). Furthermore, Arora et al. (1987) described a *Drosophila* mutant that apparently expresses a receptor protein for salt in a sugar-sensitive neuron (i.e., in a cell with phagostimulatory central connections). As a result, these mutants are strongly stimulated by NaCl at concentrations that are repellent to the wild-type fly. Finally, Joshi et al. (1989) studied several *Drosophila* mutants with modified salt sensitivity that, depending on the central connections of the affected cells, display either decreased or increased feeding responses.

The expression in neurons of previously untranscribed receptor protein genes that might change the behavioral effect of a stimulus is also illustrated by the observation of Strotmann et al. (1995). In a comparative analysis of several rodent species, these authors found that specific subtypes of an olfactory receptor family shifted from their normal topographic region in the mucosa to a new location, suggesting that the receptor genes came under the control of different regulatory elements during evolution. Similar mechanisms might explain the (low) sensitivity to oviposition stimulants present in deterrent cells in *Pieris* butterflies (van Loon 1996) and the occurrence of sugar receptor sites on salt and water neurons in flies (see Schoonhoven et al. 1992).

It should be noted that, with the exception of the receptor gene in *C. elegans*, none of the above-mentioned peripheral changes involve the actual structure of the receptor proteins. More likely, changes in the genes that control expression of receptor proteins (cf. Chess et al. 1994; Sengupta et al. 1994) or genes that code and control second messenger cascades are causing the observed effects. In the silkworm *Bombyx mori*, a polyphagous mutant has been found with reduced sensitivity to glycosides, while other deterrents are still perceived (Asaoka 1994 and personal communication). Interestingly, this mutation affects the styloconic sensilla on the galea as well as the epipharyngeal sensilla. These sensilla belong to separate sensory systems, and are involved in slightly different phases of the host acceptance process.

Although the idea of dulcitol (and other as yet unidentified compounds) acting as a phytochemical stepping stone from Celastrace to Rosaceae seems attractive, all information on electrophysiological sensitivity and behavior in *Yponomeuta* is confined to experiments with larvae. The critical phase in evolution, however, is the oviposition choice of a female. In the simplest case, larval food preference should be reflected in adult oviposition preference. This is an old assumption, dating back to Hopkins (1917), and used in a modified way by Corbet (1985). However attractive, it must be concluded that very little evidence exists to support this idea (e.g., Veltman and Corbet 1991; but see Anderson et al. 1995). One could speculate about a more direct mechanism, though. It seems feasible that genetic changes that alter the expression of receptor proteins, and hence chemosensory response profiles in larvae, might also be active in adult neurons (cf. Riesgo-Escovar et al. 1992, who provided evidence for genes expressed in the olfactory systems of

Table 11.2. Events during perception of chemical cues.

Step	Event
1	Chemical stimulus (olfactory or gustatory) present in environment enters sensory hair
2	Binding of stimulant to relatively unspecific binding proteins in the receptor lymphe
3	Binding of stimulant to specific g-protein coupled seven transdomain receptors embedded in the dendritic membrane, causing generation of a receptor current in the sensory cell
4	Receptor current causes action potential generation, coding the intensity of the stimulus via the frequency of the impulses
5	CNS interpretation and integration with other peripheral and central information finally results in a behavioral response

both larval and adult *Drosophila*). When oviposition and feeding are at least partly controlled by the same phytochemicals, both life stages might be affected simultaneously. In many phytophagous insects, larvae and adults indeed use the same compounds as host-plant cues. In *Pieris,* for instance, glucosinolates are stimulating both larval feeding and adult oviposition (Schoonhoven 1969; van Loon et al. 1992). More examples can be found in Städler (1992). Unfortunately, in *Yponomeuta* very little information on adult host choice and its chemosensory basis is yet available and investigating adult host evaluation (currently in progress in our laboratory) will be a critical test for the role of ducitol in the proposed shift from Celastraceae to Rosaceae.

The Significance of Receptor Gene Numbers

The above-mentioned new insights in the genetic basis of sensory responses show that examples of the mechanisms proposed in the host shift scenario are indeed found in nature. However, it is as yet unclear how easy it is to evolve modified receptor proteins for stimuli previously not perceived (e.g., generate a sorbitol receptor from a dulcitol receptor). In rodents the receptor protein family appears to contain thousands of genes coding large arrays of slightly modified proteins. Invertebrates probably have a much more limited set of receptor genes (Lancet and Ben-Arie 1993; Troemel et al. 1995). While this will likely constrain the evolution of new receptor types, it has also been suggested that in species with small receptor repertoires, polymorphisms will have a more defined phenotypic manifestation (Lancet et al. 1993). This might not only explain the observed variation in dulcitol sensitivity in *Yponomeuta*, but could also be one of the ultimate causes of the evolution of the large number of phytophagous insect species. This argument is closely related to that of Chapman (1982), who suggested that the reduction in receptor cell numbers might have been one of the key events in the evolution of feeding specialization in insects. The reduction in numbers of sensory cells, and the implicit reduction in the number of compounds that can be perceived, might facilitate specialization and constrain host-plant shifts to phytochemically similar plants.

The probably relatively small number of receptor genes in phytophagous insects also underpins the remark of Fox and Lalonde (1993) that sensory information is a limited basis for decision making. If an insect has a restricted view of phytochemistry, a female might not discriminate against an unsuitable plant species that has a "signature" similar to its preferred host (Fox and Lalonde 1993; Chew and Renwick 1995). Such confusion particularly holds for recently introduced or manipulated, cultivated plants for which the insect has not had enough time to evolve the capacity for discrimination (Morill 1982). As an alternative explanation, Larsson and Ekbom (1995) stated that it does not pay to discriminate between good and poor hosts with similar signatures if the proportion of poor hosts is large and ovipositing females are constrained in time. This effect is reinforced when oviposition is on the plant of eclosion, a situation not uncommon for insects with an intimate association with their plants like gall-formers and leaf-miners, which are also often poor dispersers (Menken and Wiebosch-Steeman 1988).

Conversely, plant species exist on which females do not oviposit but which are highly suitable as hosts. This could result from the inability (due to sensory constraints) to recognize this plant or from the fact that confusion of this potential host with a highly unsuitable plant led to selection against its use (Fox and Lalonde 1993).

Thus, the state of peripheral sensitivity and/or metabolic pathways might interfere with the correlation between preference and performance. Although these two complex traits are not necessarily encoded for by the same genes, a taxon can only be evolutionary successful if some genetic relationship between the two traits exists (Thompson 1994). When, for conceptual simplicity, binary adaptive states are considered, four combinations are possible for a particular plant (table 11.3). The insect will (1) be fully adapted and correctly use the plant, (2) ignore the plant, (3) be confused in the sense of Larsson and Ekbom (1995) and make oviposition "mistakes," or (4) be sensorily constrained, underutilizing its potential host-plant range. In the latter two cases, selection for better adapted individuals will not necessarily work because the required genetic variation might not be available or opposing selective forces on the new host might counteract the food plant change. A successful host-plant shift therefore requires, besides ecological opportunity, that genetic variation for preference and performance co-occur.

Host Races in *Yponomeuta padellus*

Geographic populations that freely exchange genes, partially reproductively isolated host races, and completely reproductively isolated sister species are stages in a continuous process from panmixia to speciation completed. Studying the microevolutionary aspects of host race formation supplies the insights necessary to understand macroevolutionary patterns. It is often possible to distinguish in (population) genetical terms between host races and species (Berlocher, ch. 8 this volume; see also Feder, this volume), but this does not automatically provide conclusive evidence. For instance, complete reproductive isolation could have rapidly been established without much genetic differentiation across the genome, and therefore, additional information about, among other things, mating behavior is required for a final judgment on the evolutionary status of the studied taxa.

Yponomeuta padellus feeds on a variety of rosaceous plants (e.g., *Crataegus* spp., *P. spinosa*, *P. domestica*, and *P. cerasifera*) and might be in an active state of differentiation (Menken et al. 1992; Raijmann 1996). Two questions are briefly considered here: Do host races exist

Table 11.3. The state of adaptation of a phytophagous insect on a particular plant.

Sensory Information from Host	Metabolic Suitability of Host	Needs/Selects (Pre)Adaptation	Status
1	1	None	Adapted (plant correctly used)
1	0	Metabolic	Confused (oviposition mistakes)
0	1	Sensory	Constrained (underutilizing plants)
0	0	None	Adapted (plant correctly ignored)

This table shows the state of adaptation as the result of the interaction between the sensory information obtained from the plant (1, positive cue; 0, negative cue) and the metabolic and detoxifying capabilities of the insect (1, present; 0, absent).

in this moth species, and if so, did they evolve once or were they initiated independently in various places of the phytophage distribution? An analysis of the genetical population structure of *Y. padellus* with allozymes (based on a maximum of 23 genetic loci belonging to no more than 7 linkage groups; Raijmann et al. 1997) reveals a rather complex pattern. The general picture that emerges after a study of approximately 15 years reveals that in most localities populations on different food plants comprise a panmictic unit. One consistent exception are populations on *P. cerasifera* (Menken 1981; Raijmann and Menken 1992), the very ones with an oviposition preference for their own food plant and best larval performance on that host (Kooi 1990 and unpublished). Usually, *P. spinosa* is the preferred host irrespective of the host of origin of the insects (Gerrits-Heybroek et al. 1978; Kooi 1990). In a phylogeographic analysis, however, populations on *P. cerasifera* do not cluster (different loci are involved in the differentiation in different localities) and there is only a weak phylogenetic component to geography (Raijmann 1996). It appears as if different preadapted populations of *Y. padellus* have shifted onto *P. cerasifera* in different localities.

In the few other cases studied, phylogeographic analyses mostly showed a good correspondence between host use and patterns of genetic variation, indicating that a single rather than multiple shifts have occurred (*Rhagoletis*, Feder et al. 1990; *Eurosta*, Waring et al. 1990; *Euura*, Roininen et al. 1993), after which the shifted insects migrated to other populations of the newly colonized host. Alternatively, repeated reversals in host associations were found in *Euphydryas editha* (Radtkey and Singer 1995), and host association is a labile character in this species. As a consequence, populations of this species cluster more according to geographic isolation than to host-plant association, although, again, a tendency for evolutionary conservatism of preference was encountered. An important difference with the previously mentioned insects is the fact that mating in *E. editha* is independent of food plant. The correlated response between host (habitat) choice and mate choice seems to be a prerequisite for nonallopatric host race formation and speciation to occur (Bush 1994).

Population structures such as those found in *Y. padellus* could also develop each generation from a panmictic adult population, with differentiation being due to disruptive selection in the larval stages (due to host-plant-specific selection [Menken 1981] or to some other aspect of the host-plant environment such as host-related parasitoids or predators) and then being swept away by random mating. Sampling of mating and ovipositing females is required to be able to choose between the two alternatives (McPheron et al. 1988). In *Y. padellus* there is no evidence that direct selection from the food plant can explain the observed differentiation on *P. cerasifera*, and reduced gene flow due to different host fidelities seems to best account for the observed population subdivision (Menken 1981).

The macro- and microevolutionary patterns of host use and patterns of chemoreceptor responses in *Yponomeuta* suggest that host-plant shifts are both constrained and facilitated by changes in peripheral sensitivity to phytochemicals. Examples of the genetic mechanisms underlying the proposed sensory changes can be found in nature. However, even with large constraints operating, in evolutionary time "bizarre" shifts can sometimes be realized (e.g., shifts from angiosperms to mosses and liverworts). We would like to conclude that an integration of phylogenetic and comparative approaches with genetic, ecological, behavioral, sensory phytochemical, and evolutionary biological research lines is essential for a complete understanding of the processes of population differentiation and speciation.

Acknowledgments   We thank Stewart Berlocher, Katja Hora, Tibor Jermy, Stig Larsson, Erik van Nieukerken,

Leon Raijmann, Erich Stadler, Louis Schoonhoven, and Sandrine Ulenberg for their constructive criticisms. SBJM thanks Guy and Dorie Bush for permanently creating a home with a hospitable climate for scientific and societal exchange and optimal feeding physiology.

References

Anderson, P., Hilker, M., and Lofquist, J. 1995. Larval diet influence on oviposition behaviour in *Spodoptera littoralis*. Entomologia experimentalis et applicata 74:71–82.

Arora, K., Rodrigues, V., Swati, J., Shanbhag, S., and Siddiqi, O. 1987. A gene affecting the specificity of the chemosensory neurons of *Drosophila*. Nature 330:62–63.

Asaoka, K. 1994. Different spectrum in responses of deterrent receptor cells in Sawa-J, a strain of the silkworm, *Bombyx mori*, with abnormal feeding habit. Zoological Science 11(Supplement): 102.

Bernays, E. A. 1996. Selective attention and host-plant specialization. Entomologia experimentalis et applicata 80:125–131.

Bernays, E. A., and Chapman, R. F. 1994. Host Plant Selection of Phytopagous Insects. Chapman and Hall, New York.

Bernays, E. A., and Wcislo, W. T. 1994. Sensory capabilities, information processing and resource specialization. Quarterly Review of Biology 69:187–204.

Blaney, W. M., Schoonhoven, L. M., and Simmonds, M. S. J. 1986. Sensitivity variations in insect chemoreceptors; a review. Experientia 42:13–19.

Brooks, D. R., and McLennan, D. 1991. Phylogeny, Ecology and Behavior: A Research Program in Comparative Biology. University of Chicago Press, Chicago.

Bush, G. L. 1969. Sympatric host race formation and speciation in frugivorous flies of the genus *Rhagoletis* (Diptera: Tephritidae). Evolution 23:237–251.

Bush, G. L. 1994. Sympatric speciation in animals: new wine in old bottles. Trends in Ecology and Evolution 9:285–288.

Chapman, R. F. 1982. Chemoreception: the significance of receptor numbers. Advances in Insect Physiology 16:247–356.

Chess, A., Simon, I., Cedar, H., and Axel, R. 1994. Allelic inactivation regulates olfactory receptor gene expression. Cell 78:823–843.

Chew, F. S., and Renwick, J. A. A. 1995. Host plant choice in *Pieris* butterflies. In R. T. Carde and W. J. Bell (eds). Chemical Ecology of Insects, Volume 2. New York: Chapman and Hall, pp. 214–238.

Chew, F. S., and Robbins, R. K. 1984. Egg laying in butterflies. In R. I. Vane-Wright and P. Ackery (eds). The Biology of Butterflies. London: Academic Press, pp. 65–79.

Corbet, S. A. 1985. Insect chemosensory responses: a chemical legacy hypothesis. Ecological Entomology 10:143–153.

Courtney, S. P., and Kibota, T. T. 1990. Mother doesn't know best: selection of hosts by ovipositing insects. In E. A. Bernays (ed.). Insect–Plant Interactions, Volume 2. Boca Raton: CRC Press, pp. 161–188.

Denno, R. F., McClure, M. S., and Ott, J. R. 1995. Interspecific interaction in phytophageous insects: competition reexamined and resurrected. Annual Review of Entomology 40:297–331.

De Queiroz, A., and Wimberger, P. H. 1993. The usefulness of behavior for phylogeny estimation: levels of homoplasy in behavioral and morphological characters. Evolution 47:46–60.

Dethier, V. G. 1982. Mechanism of host-plant recognition. Entomologia experimentalis et applicata 31:49–56.

Dyer, L. A. 1995. Tasty generalists and nasty specialists? Antipredator mechanisms in tropical lepidopteran larvae. Ecology 76:1483–1496.

Ehrlich, P. R., and Raven, P. 1964. Butterflies and plants: a study in coevolution. Evolution 18:586–608.

Farrell, B., and Mitter, C. 1990. Phylogenesis of insect–plant interactions: have *Phyllobrotica* (Coleoptera: Chrysomelidae) and the Lamiales diversified in parallel? Evolution 44:1389–1403.

Farrell, B., and Mitter, C. 1993. Phylogenetic determinants of insect/plant community diversity. In E. Ricklefs and D. Schluter (eds). Historical and Geographical Determinants of Community Diversity. Chicago: Chicago University Press, pp. 253–266.

Feder, J. L., Chilcote, C. A., and Bush, G. L. 1990. Regional, local and microgeographic allele frequency variation between apple and hawthorn populations of *Rhagoletis pomonella* in western Michigan. Evolution 44:595–608.

Feeny, P. 1995. Ecological opportunism and chemical constraints on the associations of swallowtail butterflies. In J. M. Scriber, Y. Tsubaki, and R. C. Lederhouse (eds). Swallowtail Butterflies: Their Ecology and Evolutionary Biology. Gainesville, Fla.: Scientific Publishers, pp. 9–15.

Forey, P. L., Humphries, C. J., Kitching, I. L., Scotland, R. W., Siebert, D. J., and Williams, D. M. 1992. Cladistics. A Practical Course in Systematics. Oxford: Clarendon Press.

Fox, C. W., and Lalonde, R. G. 1993. Host confusion and the evolution of insect diet breadths. Oikos 67:577–581.

Futuyma, D. J., and Keese, M. C. 1992. Evolution and coevolution of plants and phytophagous arthropods. In G. A. Rosenthal and M. R. Berenbaum (eds). Herbivores: Their Interactions with Secondary Plant Metabolites. San Diego: Academic Press, pp. 439–475.

Futuyma, D. J., Keese, M. C., and Scheffer, S. J. 1993. Genetic constraints and the phylogeny of insect plant associations: responses of *Ophraella communa* (Coleoptera: Chrysomelidae) to host plants of its congeners. Evolution 47: 488–905.

Futuyma, D. J., Keese, M. C., and Funk, D. J. 1995. Genetic constraints on macroevolution: the evaluation of host affiliation in the leaf beetle genus *Ophraella*. Evolution 49:797–809.

Gerrits-Heybroek, E. M., Herrebout, W. M., Ulenberg, S. A., and Wiebes, J. T. 1978. Host plant preference of five species of small ermine moths (Lepidioptera:Yponomeutidae). Entomologia experimentalis et applicata 24:360–368.

Gershenson, Z. S., and Ulenberg, S. A. (1998). The Yponomeutinae (Lepidoptera) of the world exclusive of the Americas. Verhandelingen van de Koninklijkhe Nederlandse

Academie van Wetenschappen, afdeling Natuurkunde, 2ᵉ reeks, deel 99, 200 pp.

Hagen, R. H., Lederhouse, R. C., Bossart, J. L., and Scriber, J. M. 1991. *Papilio canadensis* and *P. glaucus* (Papilionidae) are distinct species. Journal of the Lepidopterists' Society 45:245–258.

Hennig, W. 1966. Phylogenetic Systematics. Urbana: University of Illinois Press.

Hillis, D. M., Moritz, C., and Mable, B. K. 1996. Molecular Systematics. Sunderland: Sinauer Associates.

Holloway, J. D., and Hebert, P. D. N. 1979. Ecological and taxonomic trends in macrolepidopteran host plant selection. Biological Journal of the Linnean Society 112:229–251.

Hopkins, A. D. 1917. Entomologists' discussions. Journal of Economic Entomology 10:92–93.

Jaenike, J. 1990. Host specialisation in phytophagous insects. Annual Review of Ecology and Systematics 21:243–273.

Janzen, D. 1988. Ecological characterization of a Costa Rican dry forest caterpillar fauna. Biotropica 20:120–135.

Jermy, T. 1984. Evolution of insect–host plant relationships. American Naturalist 124:609–630.

Jermy, T. 1993. Evolution of insect-plant relationships—a devil's advocate approach. Entomologia experimentalis et applicata 66:3–12.

Joshi, S., Arora, K., and Siddiqi, O. 1989. Cationic acceptor sites on the labellar chemosensory neurons of *Drosophila melanogaster*. In R. N. Sing and N. J. Strausfeld (eds). Neurobiology of Sensory Systems. New York: Plenum Press, pp. 439–448.

Karowe, D. N. 1990. Predicting host range evolution: colonization of *Coronilla varia* by *Colias philodice* (Lepidoptera: Pieridae). Evolution 44:1637–1647.

Kooi, R. E. 1990. Host-plant selection and larval food-acceptance by small ermine moths. Thesis, University of Leiden.

Kutsch, W., and Breidbach, O. 1994. Homologous structures in the nervous systems of arthropoda. Advances in Insect Physiology 24:1–113.

Lancet, D., and Ben-Arie, N. 1993. Olfactory receptors. Current Biology 3:668–674.

Lancet, D., Sadovsky, E., and Seidemann, E. 1993. Probability model for the molecular recognition in biological receptor repertoires: significance to the olfactory system. Proceedings of the National Academy of Sciences USA 90:3715–3719.

Larsson, S., and Ekbom, B. 1995. Oviposition mistakes in herbivorous insects: confusion or a step towards a new host plant? Oikos 72:155–160.

Löfstedt, C., Herrebout, W. M., and Menken, S. B. J. 1991. Sex pheromones and their potential role in the evolution of reproductive isolation in small ermine moths (Yponomeutidae). Chemoecology 2:20–28.

Mackenzie, A. 1996. A trade-off for host plant utilization in the black bean aphid *Aphis fabae*. Evolution 50:155–162.

McPheron, B. A., Smith, D. C. and Berlocher, S. H. 1988. Microgeographic genetic variation in the apple maggot *Rhagoletis pomonella*. Genetics 119:445–451.

Menken, S. B. J. 1981. Host races and sympatric speciation in small ermine moths, Yponomeutidae. Entomologia experimentalis et applicata 30:280–292.

Menken, S. B. J. 1982. Biochemical genetics and systematics of small ermine moths. Zeitschrift für zoologische Systematik und Evolutionsforschung 20:131–143.

Menken, S. B. J. 1990. Biochemical systematics of the leaf-mining moth family Nepticulidae (Lepidoptera). III. Allozyme variation patterns in the *Ectoedemia subbimaculella* group. Bijdragen tot de Dierkunde 60:189–197.

Menken, S. B. J. 1996. Pattern and process in the evolution of insect-plant associations: *Yponomeuta* as an example. Entomologia experimentalis et applicata 80:297–305.

Menken, S. B. J., and Raijmann, L. E. L. 1996. Biochemical systematics: principles and perspectives for pest management. In W.O.C. Symondson and J. E. Liddell (eds). The Ecology of Agricultural Pests: Biochemical Approaches. London: Chapmann and Hall, pp. 7–29.

Menken, S. B. J., and Ulenberg, S. A. 1987. Biochemical characters in agricultural entomology. Agricultural Zoology Reviews 2:305–360.

Menken, S. B. J., and Wiebosch-Steeman, M. 1988. Clonal diversity, population structure, and dispersal in the parthenogenetic moth *Ectoedemia argyropeza*. Entomologia experimentalis et applicata 49:141–152.

Menken, S. B. J., Herrebout, W. M., and Wiebes, J. T. 1992. Small ermine moths, *Yponomeuta*: their host relations and evolution. Annual Review of Entomology 37:41–66.

Miller, J. S. 1987. Host-plant relationships in the Papilionidae (Lepidoptera): parallel cladogenesis or colonization? Cladistics 3:105–120.

Miller, J. S., and Wenzel, J. W. 1995. Ecological characters and phylogeny. Annual Review of Entomology 40:389–415.

Mitter, C., and Brooks, D. R. 1983. Phylogenetic aspects of coevolution. In D. J. Futuyma and M. Slatkin (eds). Coevolution. Sunderland: Sinauer Associates, pp. 65–98.

Mitter, C., Farrell, B., and Futuyma, D. J. 1991. Phylogenetic studies of insect–plant interactions: insights into the genesis of diversity. Trends in Ecology and Evolution 6:290–293.

Morill, W. L. 1982. Hessian fly: host selection and behavior during oviposition, winter biology, and parasitoids. Journal of the Georgia Entomological Society 17:156–167.

Ozaki, M., Morisaki, K., Idei, W., Ozaki, K., and Tokunaga, F. 1995. A putative lipophilic stimulant carrier protein commonly found in the taste and olfactory systems. European Journal of Biochemistry 230:298–308.

Paterson, A. M., Wallis, G. P., and Gray, R. D. 1995. Penguins, petrels, and parsimony: does cladistic analysis of behavior reflect seabird phylogeny? Evolution 49:974–989.

Peterson, S. C., Herrebout, W. M., and Kooi, R. E. 1990. Chemosensory basis of host-colonization by small ermine moth larvae. Proceedings Koninklijke Nederlandse Academie van Wetenschappen 93:287–294.

Radtkey, R. R., and Singer, M. C. 1995. Repeated reversals of host-preference evolution in a specialist insect herbivore. Evolution 49:351–359.

Raijmann, L. E. L. 1996. In search for speciation: genetical differentiation and host race formation in *Yponomeuta padellus* (Lepidoptera, Yponomeutidae). Thesis, University of Amsterdam.

Raijmann, L. E. L., and Menken, S. B. J. 1992. Population genetical evidence for host-race formation in *Yponomeuta padellus.* In S. B. J. Menken, J. H. Visser, and P. Harrewijn (eds). Proceedings of the 8th International Symposium on Insect–Plant Relationships. Dordrecht: Kluwer, pp. 209–211.

Raijmann, L. E. L., van Ginkel, W., Heckel, D. G., and Menken, S. B. J. 1997. Inheritance and linkage of isozymes in *Yponomeuta padellus* (Lepidoptera, Yponomeutidae). Heredity 78:645–654.

Rausher, M. D. 1992. Natural selection and the evolution of plant-insect interactions. In B. D. Roitberg and M. B. Isman (eds). Insect Chemical Ecology: An Evolutionary Approach. New York: Chapman and Hall, pp. 20–88.

Renwick, J. A. A., and Chew, F. S. 1994. Oviposition behaviour in Lepidoptera. Annual Review of Entomology 39:377–400.

Riesgo-Escovar, J., Woodard, C., Gaines, P., and Carlson, J. 1992. Development and organization of the *Drosophila* olfactory system: an analysis using enhancer traps. Journal of Neurobiology 23:947–964.

Roininen, H., and Tahvanainen, J. 1989. Host selection and larval performance of two willow-feeding sawflies. Ecology 70:129–136.

Roininen, H., Vuorinen, J., Tahvanainen, J., and Ulkunen-Tiitto, R. 1993. Host preference and allozyme diffentiation in shoot galling sawfly, *Euura atra.* Evolution 47:300–307.

Schoonhoven, L. M. 1969. Gustation and foodplant selection in some lepidopterous larvae. Entomologia experimentalis et applicata 12:555–564.

Schoonhoven, L. M., Blaney, W. M., and Simmonds, M. S. J. 1992. Sensory coding of feeding deterrents in phytophageous insects. In E. A. Bernays (ed.). Insect–Plant Interactions, Volume 4. Boca Raton: CRC Press, pp. 59–79.

Scoble, M. J. 1983. A revised cladistic classification of the Nepticulidae (Lepidoptera) with descriptions of new taxa mainly from South Africa. Transvaal Museum Monograph 2:1–105.

Sengupta, P., Colbert, H. A., and Bargmann, C. I. 1994. The *C. elegans* gene odr-7 encodes an olfactory-specific member of the nuclear receptor superfamily. Cell 79:971–980.

Sengupta, P., Chou, J. H., and Bargmann, C. I. 1996. Odr-10 encodes a seven transmembrane domain olfactory receptor required for responses to the odorant diacetyl. Cell 84:899–909.

Shephard, G. M. 1994. Discrimination of molecular signals by the olfactory receptor neuron. Neuron 13:771–790.

Stadler, E. 1992. Behavioural responses of insects to plant secondary compounds. In G. A. Rosenthal and M. R. Berenbaum (eds). Herbivores: Their Interaction with Secondary Plant Metabolites, Volume 2. New York: Academic Press, pp. 45–88.

Steinbrecht, R. A., Laue, M., and Ziegelberger, G. 1995. Immunolocalization of pheromone-binding protein and general odorant-binding protein in olfactory sensilla of the silk moths *Antheraea* and *Bombyx.* Cell Tissue Research 282:203–217.

Stengl, M., Hatt, H., and Breer, H. 1992. Peripheral processes in insect olfaction. Annual Review of Physiology 54:665–681.

Strotmann, J., Beck, A., Kubick, S., and Breer, H. 1995. Topographic patterns of odorant receptor expression in mammals: a comparative study. Journal of Comparative Physiology A 177:659–666.

Thompson, J. N. 1994. The Coevolutionary Process. Chicago: University of Chicago Press.

Thompson, J. N. 1996. Trade-offs in larval performance on normal and novel hosts. Entomologia experimentalis et applicata 80:133–139.

Troemel, E. R., Chou, J. H., Dwyer, N. D., Colbert, H. A., and Bargmann, C. I. 1995. Divergent seven transmembrane receptors are candidate chemosensory receptors in *C. elegans.* Cell 83:207–218.

Van Drongelen, W. 1979. Contact chemoreception of host plant specific chemicals in larvae of various *Yponomeuta* species (Lepidoptera). Journal of Comparative Physiology 134:265–279.

Van Drongelen, W. 1980. Behavioural responses of two small ermine moth species (Lepidoptera: Yponomeutidae) to plant constituents. Entomologia experimentalis et applicata 28:54–58.

Van Loon, J. J. A. 1996. Chemosensory basis of feeding and oviposition behaviour in herbivorous insects: a glance at the periphery. Entomologia experimentalis et applicata 80:7–13.

Van Loon, J. J. A. Blaakmeer, A., Griepink, F. C., van Beek, T. A. Schoonhoven, L. M., and de Groot, A. 1992. Leaf surface compounds from *Brassica oleracea* (Cruciferae) induces oviposition by *Pieris brassicae* (Lepidoptera: Pieridae). Chemoecology 3:39–44.

Van Nieukerken, E. J. 1986a. A provisional phylogenetic check-list of the western palaearctic Nepticulidae, with data on hostplants (Lepidoptera). Entomologica scandinavica 17:1–27.

Van Nieukerken, E. J. 1986b. Systematics and phylogeny of Holarctic genera of Nepticulidae (Lepidoptera, Heteroneura: Monotrysia). Zoologische Verhandelingen 236:1–93.

Veltman, C. J., and Corbet, S. A. 1991. In search of a model system for exploring the chemical legacy hypothesis: *Drosophila melanogaster* and geraniol. Journal of Chemical Ecology 17:2459–2468.

Walsh, B. J. 1867. The apple-worm and the apple maggot. Journal of Horticulture 2:338–343.

Ward, P. S. 1991. Phylogenetic analysis of pseudomyrmecine ants associated with domatia-bearing plants. In C. R. Huxley and D. F. Cutler (eds). Ant-Plant Interactions. New York: Oxford University Press, pp. 335–352.

Waring, G. L., Abrahamson, W. G., and Howard, D. J. 1990. Genetic differentiation among host-associated populations of the gall-maker *Eurosta solidaginis* (Diptera: Tephritidae). Evolution 44:1648–1655.

Wieczorek, H. 1976. The glycoside receptor of the larvae of *Mamestra brassicae* L. (Lepidoptera, Noctuidae). Journal of Comparative Physiology 106:153–176.

Wiklund, C. 1974. Oviposition preference in *Papilio machaon* in relation to the host plant of the larvae. Entomologia experimentalis et applicata 17:198–198.

Wiklund, C. 1975. The evolutionary relationship between adult oviposition preferences and larval host plant range in *Papilio machaon* L. Oecologia 18:185–197.

12

The Evolution of Edaphic Endemics

Mark R. Macnair
Mike Gardner

The distribution of most plant species is profoundly affected by edaphic factors, features of the soil such as nutrient status, chemical composition, water retention ability, and so on. Species differ both in the mean of their preferred environment and in their range of tolerance to differences in edaphic factors. The rapid change in species composition of communities that can be achieved in the short term by changes in the soil chemistry has been most dramatically demonstrated at the Park Grass Experiment, Rothamsted. In 1856 Lawes divided a uniform park grassland into a number of plots that have received various fertilizer treatments ever since (AFRC, 1991). The different plots support very different plant communities, and individual species show a range of abilities to grow on the range of environments that have been created during the 150 years of the experiment (Silvertown, personal communication). However, while most plant species can be found in a range of habitats, albeit with obvious differences in optima, some species can only be found growing on soils with a particular edaphic feature. Such species are known as *edaphic endemics*. It is often clear that these species are specifically adapted to this particular environment; if we could understand the processes leading to the evolution of such species, it could assist in understanding the processes of plant speciation, and the relationship between adaptation and the formation of new species.

One of the most dramatic environments in which edaphic endemics are found is serpentine. Serpentinization is a common process associated with the hydration of ultramaphic (ultrabasic) rocks. They are found throughout the world, but they are particularly associated with orogenesis (Malpas, 1991). They are fast weathering, producing basic soils that are free draining, have a high heavy metal content (particularly nickel, chromium, and cobalt), and also tend to have an abnormal ratio of calcium to magnesium. Normal soils tend to have more Ca than Mg, while serpentine soils typically have a Mg:Ca ratio that is substantially greater than 1. Serpentine soils are typically nutrient poor. These characteristics mean that serpentine soils tend to have low productivity and tend to be toxic to many plant species. Particularly in ancient serpentine soils, the level of endemism is high; for instance, in New Caledonia, two monotypic families, over 30 genera, and more than 60% of the island's flora are restricted to serpentine soils (Jaffré, 1981). Serpentine soils share many characteristics with the phytotoxic wastes produced as a consequence of mining for various heavy metals: a toxic soil that is free draining because of poor soil structure, with low nutrient status. These sites range in age from less than 100 years to several thousand years old, and it is tempting to use the colonization of these soils by plants tolerant to the metals as a model for the evolution of plants on serpentine soils (Kruckeberg, 1984).

Narrow endemic species have classically been classified as palaeoendemics or neoendemics, though in practice it may not be possible to distinguish between the two in many cases. Palaeoendemics are relict species, in which the current limited distribution of a taxon is a reflection of the contraction of a once much more widespread distribution. In the context of edaphic endemics, one can envisage a situation in which a widespread species can be found both on and off the substrate (say, serpentine); the environment, climatic or biological, changes and causes the populations of the species on normal soils to go extinct. The populations on serpentine, however, are not faced with the same level of competition and persist. If these populations change subsequently, the process is essentially an example of classic allopatric speciation. Neoendemics are recent species that have evolved from existing species in their present environment and either are expanding or have reached their ecological limits. Where they coexist with sister or progenitor species, they could have evolved parapatrically or sympatrically, though allopatric speciation cannot be ruled out a priori. Allopatric speciation would involve a model in which the nonserpentine populations became locally extinct, following which the serpentine population evolved into the novel neoendemic species. Secondary contact was reestablished when the progenitor reinvaded the area. Such models will therefore require an ad hoc hypothesis as to why the progenitor population became both locally extinct for sufficient time for the neoendemic to evolve, and was then able to reinvade the habitat from which it had been excluded. Postulating a sympatric or parapatric speciation event may be more parsimonious, if a satisfactory model of the process can be established.

This chapter considers the problem of the evolution of edaphic neoendemics in the context of potential sympatric/parapatric speciation, concentrating on ongoing research in this laboratory on neoendemics in the *Mimulus guttatus* complex.

Endemic or Ecotype?

Kruckeberg (1984) reviewed the flora of Californian serpentines. He found that 215 taxa are endemic to serpentine soils; though serpentine makes up less than 1% of Californian soils, more than 10% of the plant species endemic to California are found on this substrate. Another 221 species are regional indicators. These species are only found on the serpentine in any particular area, but can be found on normal soils in a different region. This phenomenon can be seen very clearly in some European species that are commonly found on old mine workings. Thus, *Armeria maritima* is generally restricted to the maritime zone but is found inland in northern Europe growing on metal-contaminated sites; *Thlaspi caerulescens* is an alpine species found at sea level on lead and zinc mine workings. These species may illustrate the principle underlying the palaeoendemic model of speciation, that these extreme sites provide a protected environment that can support a species that would otherwise be unable to grow in the prevailing climatic and competitive conditions. In the context of this chapter, such species will not be considered further. Another class of species found on serpentines are ones that appear able to grow on both serpentine and normal soils in a particular area. Kruckeberg called these *bodenvag* species, and found 1,108 Californian species in this category. But these three classes of species able to colonize serpentine soils are still a minority of all taxa found in California; the remainder are excluded from this edaphic environment.

So, what is the difference between endemics, *bodenvag*, and excluded species? If we consider all the edaphic and biotic features of normal and serpentine soils and were able to express them in a single dimension, we can visualize the difference in figure 12.1. The range of phenotypes present in populations of excluded species have high fitness in normal soils but have low or zero fitness on serpentine; the converse is true of serpentine endemics. For *bodenvag* species, there are two possibilities. First, such species may show considerable phenotypic plasticity, such that individual genotypes can grow in either

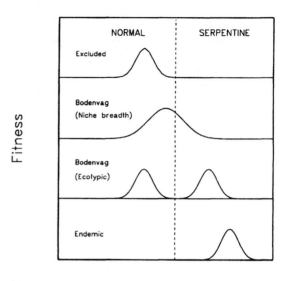

Figure 12.1. Possible relationships between fitness and phenotypic distribution within species either excluded from serpentine, endemic to it, or indifferent (*bodenvag*). For further details, see text.

environment. Second, the species may colonize serpentine by forming ecotypes, such that populations growing on serpentine are genetically different from those growing on normal soils. There is considerable evidence that the second model is very commonly true. Thus, Kruckeberg (1951,1967) has shown ecotypic differentiation in many Californian species, and Marrs and Proctor (1976) did likewise in *Agrostis stolonifera* from Scottish serpentines. In *Mimulus guttatus*, we have shown that the survivorship of a nonserpentine population was very low on serpentine soils from Lake County, California (Gardner, 1995). Failure to find ecotypes has been reported more rarely (but see Griffin, 1965).

So, the difference between an excluded species and a *bodenvag* may be in the ability of the species to form ecotypes. Research into the evolution of plants onto metal-contaminated sites has thrown considerable light on the factors affecting species' ability to evolve ecotypes (for reviews, see Macnair and Baker, 1994; Macnair, 1987). The most important step appears to be the ability to tolerate the toxic metal present. Macnair (1993) has reviewed the evidence on the genetics of metal tolerance, and it appears likely that this adaptation is normally controlled by one or a small number of genes. Possession of this adaptation enables the colonization of the new environment, following which there will be further selection for genes that increase metal tolerance and provide tolerance to other adverse conditions of the mine (e.g., drought, low nutrient status, reduced biotic competition). Early work (Walley, Khan, and Bradshaw, 1974; Wu, Bradshaw, and Thurman, 1975) showed that if seed from normal populations of species commonly found on mines, *Agrostis capillaris* and *A. stolonifera*, was sown on mine soil, most seedlings died (because of the toxicity), but a small number of survivors was normally found. These were tolerant, so clearly normal populations of these species have the gene(s) for tolerance present at low frequency. Similar experiments with a range of common grassland species (Ingram, 1988) found that while all species that grow on mines showed a pattern similar to that of *A. capillaris*, there were a number of species that appeared to have the genes for metal tolerance in normal populations but are not typically found on mines. It is likely that these species are excluded because of their inability to tolerate some other feature of the mine environment (e.g., low pH or nutrient levels). Thus, the presence of appropriate genetic variance is a necessary but not a sufficient condition for evolving a tolerant ecotype. Bradshaw (1991) has argued that the lack of appropriate genetic variance is what defines the limits to a species' ability to adapt and has called this phenomenon *genostasis*.

So, the question may resolve to what the difference is between an ecotype and an endemic. What are the factors leading to the evolution of an endemic, and why don't all ecotypes speciate to form endemics? Kruckeberg (1986) proposed a model for the evolution of an endemic in which he visualized a number of steps:

0. Some preadaptation for serpentine tolerance exists in normal populations of a species. This provides the conditions that will permit the evolution of a serpentine tolerant ecotype.
1. Disruptive selection causes the separation of the species into serpentine-tolerant and -intolerant gene pools. This is the formation of an ecotype.
2. Further genetic divergence in structural and functional traits occurs within the ecotype.
3. Isolation between the two races becomes genetically fixed, such that the two gene pools are unable to exchange genes.
4. Further divergence of the incipient species occurs, "put in motion by the initial genetic discontinuity" (p. 459).

But, while this may represent a more or less accurate description of the steps through which evolution proceeds from ecotype to endemic, it does not indicate why some ecotypes evolve through all the steps and others do not. The crucial step differentiating an ecotype from an endemic is likely to be the acquisition of a partial or complete reduction in gene flow between normal, ancestral population and ecotype, allowing an independent gene pool to develop (stage 3). How this occurs, in the absence of an extrinsic barrier to gene flow between the two contiguous races, is, of course, the standard problem of all nonallopatric speciation models.

Edaphic Endemics in the *Mimulus Guttatus* Complex

Description of the Species and Their Taxonomy

We have been investigating this question in the *M. guttatus* complex. *M. guttatus* in the strict sense is a hydrophilic species that grows ubiquitously in streams and damp places in Western North America from Alaska to Mexico, from the Rockies to the Pacific. It is generally perennial, except in seasonal streams, where it grows as a facultative annual. Rarely, obligate annual populations have been described (Vickery, 1959). It is normally primarily outcrossing (Ritland and Ritland, 1989; Fenster and Ritland, 1992), though because it is self-compatible, it will suffer geitonogamous selfing (geitonogamy is selfing caused by the transfer of pollen from one flower to another on the same plant by a pollinator, as opposed to autogamy, which is the transfer of pollen from anther to stigma within a single flower), and, if it has not been previously crossed, it will self-fertilize with variable efficiency on corolla abscission (Dudash and Ritland, 1991). In Calaveras County, California, it has colonized a number of old copper mines, forming a typical metal-tolerant ecotype (Allen and Sheppard, 1971; Macnair et al., 1993). On the mines, it is morphologically indistinguishable from *M. guttatus* elsewhere, and is apparently as dependent as elsewhere on pollinators for good seed set. The size of these mine plants depends on the wetness of the local microenvironment: in the streams and wet places the plants grow over a meter tall, while on the dry tailings they are small, unbranched facultative annuals. Most of our work with this species has concerned the Copperopolis population, where it is one of the dominant species of the mine community.

M. guttatus is the probable progenitor of a number of species with more limited geographic distribution and more specialized habitat. Three species are relevant to this chapter (see figure 12.2 and table 12.1; pictures of all the species discussed here can also be found at www.ex.ac.uk/~MRMacnai). *M. nudatus* Curran is a serpentine endemic restricted to the serpentines of Lake and Napa counties, California (in the coastal mountains). *M. pardalis* Pennell is another serpentine endemic restricted to the serpentines of the Sierra Nevada from Amador County to Mariposa County (Pennell, 1950). *M. cupriphilus* Macnair is a species discovered on a small copper mine located some 10 km from Copperopolis (Macnair, 1989). The exact age of this mine is uncertain, but most of the mines in this area were opened around 1861 and were worked until the beginning of the twentieth century. Some mines were reopened during the 1920s, and some were worked during World War II (Heyl, 1954). These species, and various other segregants, are closely related to *M. guttatus* (Ritland and Ritland, 1989; Ritland et al., 1993; Fenster and Ritland, 1994; Macnair, unpub-

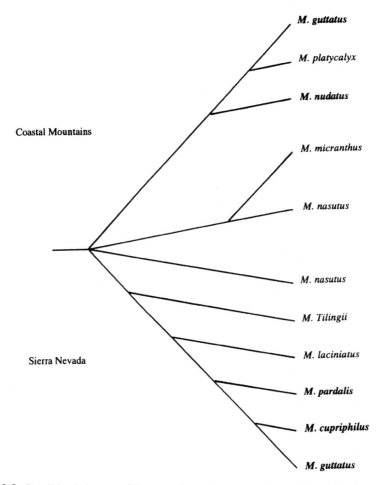

Figure 12.2. Possible phylogeny of the *Mimulus guttatus* complex, based on morphology, bioge-ography, and molecular evidence. Species studied in this chapter are in boldface.

lished) on the basis of allozyme, chloroplast (cp) DNA, and ribosomal DNA sequence similarity, and thus the suggestion that *M. guttatus* is ancestral is supported by the geographical distribution of these species (see figure 12.2).

All three endemics are obligate annuals that are smaller vegetatively, are more floriferous, and have smaller flowers than *M. guttatus* (table 12.1). *M. nudatus* is an outcrosser (Ritland and Ritland, 1989) but the other two appear to be inbreeders (both set seed readily without pollinators and have very little allozyme polymorphism).

It should be noted that the formal taxonomy of this complex is still unresolved. Pennell (1950) recognized many species within the complex, but Campbell (1950) took a different view, lumping all the early-flowering variants of *M. guttatus* into *M. guttatus* var. *gracilis,* but retaining specific status for *M. nudatus* and *M. laciniatus* (another local annual species). Thompson (1992), in the latest edition of the Jepson Manual, has lumped many of the segregants previously recognized into *M. guttatus*

with the comment, "Exceedingly complex: local populations may be unique but their forms intergrade over geog or elevation: variants not distinguished here" (p. 1043). He has followed Campbell (1950), however, in retaining specific status for *M. nudatus* and *M. laciniatus*. However, there is no objective reason for recognizing *M. laciniatus* and not *M. pardalis* or *M. cupriphilus*, except perhaps that *M. laciniatus* is easier to score in herbarium sheets because of distinctive leaf shape. In the field, the taxa discussed here are easily distinguished from *M. guttatus* and have distinctive ecologies. Thompson (1992) is obviously concerned that when the full geographical range of *M. guttatus* is taken into account, intermediates between, say, *M. pardalis* and *M. guttatus* can be discovered, particularly when only herbarium specimens are studied.

However, evolutionary biologists have long studied this group because they recognize that there are various morphologically and ecologically distinct forms, the evo-

Table 12.1. Summary of the principal morphological and physiological differences between the taxa referred to in this chapter.

Character	*M. guttatus*		*M. nudatus* Serpentine endemic	*M. pardalis* Serpentine endemic	*M. cupriphilus* Copper mine endemic
	Normal population	Copper mine ecotype			
Metal tolerance	None	Copper tolerant	Nickel tolerant	Nickel tolerant	Copper tolerant
Life history	Perennial, facultatively annual	Perennial, facultatively annual[1]	Obligate annual	Obligate annual	Obligate annual
Size and shape of plant	Tall, apically dominant	Tall, apically dominant	Short and branched	Short and branched	Short and branched
Flowering time	Relatively late	Earlier than normal population	Early	Early	Early
Flower size	Large	Large	Small	Small	Small
Dominant pollinator	Bumble bee	Bumble bee	Sweat bee	(None)[2]	None
Breeding system	Outcrossing	Outcrossing	Outcrossing	Selfing	Selfing
Leaf shape	±length = width	±length = width	Length > width	Length > width	±length = width

[1]Some obligate annuals are also found in the Copperopolis population.

[2]This is assumed; no field study has been performed.

lution of which may prove very fruitful to investigate (e.g., Vickery, 1978; Ritland and Ritland, 1989; Fenster and Ritland, 1992; Dole, 1992). To lump all these forms into *M. guttatus* may make a tidy taxonomy, but it does not describe the situation in the field. Where two sympatric forms differ in ecology, breeding system, and morphology, and these differences are governed by more than a single gene difference, then selection or some other process must be maintaining the disparate gene pools (Mallet, 1995), and it seems more sensible to recognize this taxonomically.

Adaptation of the Complex to Edaphic Features

We can compare the morphological and physiological features of typical normal populations of *M. guttatus* with copper- and serpentine-tolerant ecotypes and the three endemics. In particular, we can ask whether the endemics show greater adaptation to their typical environment than the ecotypes, and whether these differences in adaptation explain the evolution of the endemics. All ecotypes and endemics appear to be metal tolerant either to copper (Copperopolis and *M. cupriphilus*) or nickel (serpentine ecotypes, *M. nudatus* and *M. pardalis*) compared to populations of *M. guttatus* from normal environments (Macnair, 1989, 1992, and unpublished). *M. nudatus* is more tolerant of an increased Mg:Ca ratio in artificial feeds in the glasshouse than either a serpentine-tolerant or -intolerant population of *M. guttatus* (Macnair, 1992; Gardner, 1995). However, when the growth of a normal and a serpentine-tolerant population of *M. guttatus* were compared with *M. nudatus* in normal soil and a native serpentine soil from Lake County, the responses to the change in soil by the serpentine-tolerant *M. guttatus* ecotype and *M. nudatus* were similar, while the nontolerant population died in the serpentine soil. In addition, an attempt to relate the distribution of *M. guttatus* and *M. nudatus* on serpentine soils to the detailed chemical composition (Ni, Ca, Mg, Cr) of the soils on which they were growing failed to find any difference between soils supporting populations of the two species (Gardner, 1995). Thus, there appears to be little evidence that eco-

types or endemics differ significantly in their degree of adaptation to the chemical environment of these unusual soils.

The distribution of these species in the field strongly suggests that relative availability of soil water is the most important factor controlling plant fitness. *M. cupriphilus* was originally found on the top of a dry copper mine, while *M. guttatus* was only found lower down the slope and in the stream bed leading away from the mine. *M. pardalis* can be found growing alongside *M. guttatus* along small streams traversing the Sierran serpentines, but is also typically found a long way from streams growing in cracks in rock outcrops (Pennell, 1950; Macnair, unpublished). In Lake County, *M. nudatus* is typically found farther up slopes than *M. guttatus*. Figure 12.3 shows the result of a belt transect taken down a slope into a colluvial fan in Lake County. At the top of the transect, almost all the plants are *M. nudatus*; at the bottom *M. guttatus* predominates. As the environment improves, both species increase their size (figure 12.3b), but *M. guttatus* gets much bigger than *M. nudatus*, which appears unable to compete in the wetter environment.

We are currently studying the serpentine endemics to ascertain whether they show any physiological adaptations to drought, but have so far failed to find any difference between the species in the complex in their inability to tolerate relatively mild water deficits. *M. nudatus* and *M. pardalis* both have elongated, narrow leaves compared to *M. guttatus* (table 12.1). Reduced leaf area is a typical adaptation found in xerophytic species. However, the data strongly suggests that the primary adaptation of these species to their relatively dry environment is to escape the drought by early flowering and rapid seed set before drought-induced death. In common garden experiments, all three endemic species are early flowering compared to local populations of *M. guttatus* (see table 12.2). In *M. cupriphilus* and *M. nudatus*, early-flowering appears to be partially dominant and may be partially under major gene control (Macnair and Cumbes, 1989; Macnair, 1992). Note that the Copperopolis population of *M. guttatus*, a copper-tolerant ecotype, also flowers earlier than a local nontolerant population (table 12.2) but not as early as *M. cupriphilus* or *M. pardalis*.

Table 12.2. Mean flowering time (in days from an arbitrary date) of the four taxa considered in this chapter.

Mimulus guttatu					
Copperopolis	Normal population	*M. nudatus*	*M. pardalis*	*M. cupriphilus*	Reference
50.6 ± 0.4				45.2 ± 0.95	Macnair and Cumbes (1989)
69.4 ± 3.4		45.7 ± 1.3		58.4 ± 2.7	Macnair (1992)
36 ± 2	51 ± 4			29 ± 2	McCombie (1995)
			27.7 ± 0.4	25.6 ± 0.2	Macnair (unpublished data)

In each experiment, plants were grown under similar conditions in the greenhouse.

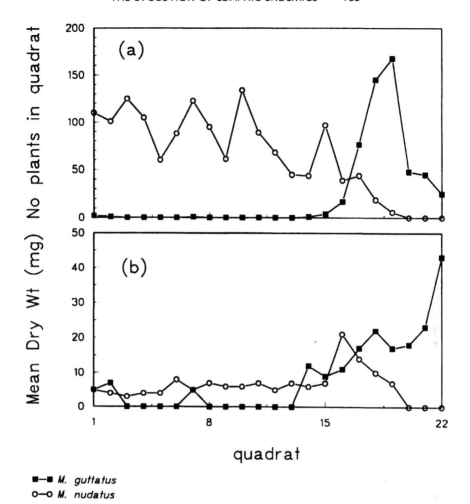

Figure 12.3. The number of plants per quadrat of *M. nudatus* and *M. guttatus* (a), and the mean dry weight of each species (b), in a belt transect down a shallow slope from a predominantly *M. nudatus* population into a damp colluvial soil supporting grassland and *M. guttatus*.

Macnair, Macnair, and Martin (1989) compared the fitness (as measured by seed set) of *M. guttatus* and *M. cupriphilus*. It is obvious that when the plants are growing in optimal conditions the fitness of *M. guttatus* is much greater: it can grow much larger and produce a vastly greater number of seeds. However, these species only coexist in an environment that is not optimal for *M. guttatus*. All plants are very small and produce relatively small numbers of seeds. Figure 12.4 shows the relationship between the size of the plant and the average number of seeds it can produce for *M. cupriphilus* and *M. guttatus*. The size of a plant is determined by the microenvironment in which it is growing: figure 12.4 shows that where the microenvironment is sufficiently poor to produce only plants of less than about 9 cm in height, an *M. cupriphilus* individual will be fitter than a

similarly sized *M. guttatus* plant. Conversely, as the environment improves and the plants can get bigger, *M. guttatus* will be fitter than *M. cupriphilus*. Similar relationships can be found for *M. nudatus* and *M. pardalis*. Figure 12.5 shows a plot of seed production versus plant dry weight for field-collected plants of both *M. pardalis* and *M. nudatus* and sympatric *M. guttatus*. For both endemic species the slope of the relationship between seed production and plant size is greater than for *M. guttatus*, but neither species can attain the size of *M. guttatus* in favored environments (see figure 12.3b).

Thus, all three endemic species appear to have evolved an early-flowering strategy in which seed production as a small plant has been optimized at the expense of the vegetative growth that would enable greater seed production at a larger size.

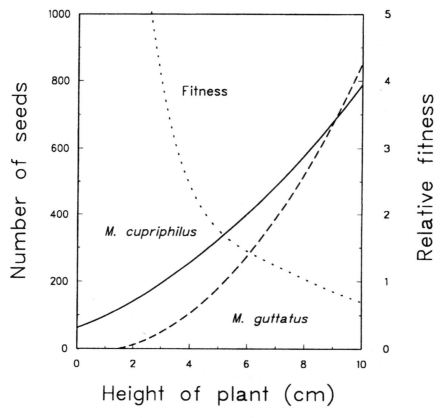

Figure 12.4. The estimated mean number of seeds produced by small plants of *M. guttatus* (dashed curve) and *M. cupriphilus* (solid curve) in the wild on copper mines in Calaveras County, California. Dotted line: the relative fitness of *M. cupriphilus* compared to *M. guttatus*. Redrawn from data in Macnair et al. (1989).

Reproductive Isolation

Reproductive isolation between the endemics and local populations of *M. guttatus* has been achieved differently in the three taxa. *M. cupriphilus* and *M. pardalis* are self-fertilizing taxa with much smaller flowers than *M. guttatus*. In the Sierras, *M. guttatus* is fertilized primarily by bumble bees (particularly *Bombus californicus*) and honey bees. When we visited Copperopolis in May, 1985, pollinators were abundant and visiting *M. guttatus* in great numbers. Seed set did not appear to be pollinator limited. In contrast, in April 1987, when *M. cupriphilus* was in flower and *M. guttatus* was only just coming into flower, few worker bees were seen and the rate of flower visitation (as measured by closure of the sensitive stigma of these species) was relatively low. Observation of *M. cupriphilus* flowers indicated that they were even less visited than sympatric *M. guttatus* (Macnair et al., 1989). The evolution of self-fertilization from outcrossing ancestors is one of the prevailing themes in plant evolution (Stebbins, 1970) and one of the selective forces postulated to account for this in many cases is reproductive assurance (Jain 1976). We therefore suggested

(Macnair et al., 1989) that early flowering had been the primary adaptation of *M. cupriphilus*, and that self-fertilization had evolved subsequently to improve seed set in the absence of abundant pollinators. This was achieved primarily by a reduction in flower size, which is also commonly found in self-fertilising species (Wyatt, 1988). A reduction in flower size has the effect of reducing the distance between the anthers and the stigma, making pollen transfer easier in the absence of a pollinator. Floral size is a recessive character, so the F1 between *M. cupriphilus* and *M. guttatus* has flowers indistinguishable from *M. guttatus*. This will mean that, when occasional crosses between *M. cupriphilus* and *M. guttatus* do occur, the F1 will tend to cross with *M. guttatus* and genes will flow primarily from *M. cupriphilus* to *M. guttatus*. This characteristic of the genetics of flower size may also be important in protecting the *M. cupriphilus* gene pool. We do not yet have any information on the genetic basis of the differences between *M. pardalis* and *M. guttatus*, or on the behavior of pollinators to this species in the field.

In contrast to *M. pardalis* and *M. cupriphilus*, *M. nudatus* is an outcrossing species (Ritland and Ritland,

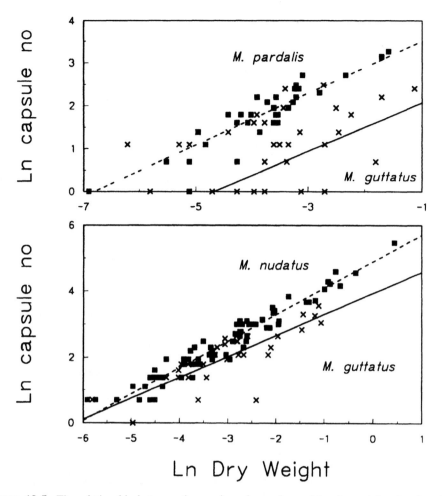

Figure 12.5. The relationship between the number of capsules and the dry weight of a plant for *M. pardalis* and *M. guttatus* (top) or *M. nudatus* and *M. guttatus* (bottom). All plants were collected in mixed populations in the field. In both graphs the plotted lines differ significantly: (top) $F_{1,65}$ = 15.84, P < 0.001; (bottom) $F_{1,103}$ = 10.08, P = 0.002.

1989). Gene flow between *M. nudatus* and *M. guttatus* is reduced by at least two factors. First, crosses between the species are highly inviable, with only a small percentage of hybrids being produced, and no hybrids have been observed in the field. Second, pollinator behavior induces a partial premating barrier. *M. nudatus* grows sympatrically with *M. guttatus*, and the bumble and honey bee pollinators of *M. guttatus* have been observed to fly from *M. guttatus* to *M. nudatus*. However, *M. nudatus* appears to be also, and possibly primarily, pollinated by a sweat bee (*Dialictus* sp.). These are generalist foragers and often scavenge residual pollen left by larger insects. Studies of pollinator behaviour in a mixed population (Gardner, 1995) showed that both honey bees and sweat bees showed nonrandom transitions between the species, with most transitions being between flowers of the same species. However, movements of both between the spe-

cies did occur, which will allow pollen transfer between them. In practice, this will tend to be greater from *M. guttatus* to *M. nudatus* than vice versa for two reasons: first, *M. guttatus* has a larger flower with larger anthers and therefore produces more pollen, and second, the sweat bees are too small to touch the stigma of an *M. guttatus* flower but can still pick up pollen from one. This asymmetry in gene flow can be detected in the field by a greater reduction in viable seed produced by *M. nudatus* plants when surrounded by *M. guttatus* plants than vice versa. Figure 12.6b shows the result of an experiment in which individual "target" plants of *M. nudatus* were identified in a mixed patch, and the number of individual *M. nudatus* and *M. guttatus* flowers in a circle of 24 cm diameter enumerated. The proportion of viable seeds for each plant was determined, and it is clear that this proportion is adversely affected by a high proportion of *M. guttatus* in the immediate vicinity. The same experiment performed with *M. guttatus* as the "target" plants found no reduction in fertility (figure 12.6a).

The change in pollinator syndrome is associated with changes in floral size and subtle changes in flower shape. It is possible to speculate that this could also have been a response to selection for reproductive assurance following the change in flowering date if sweat bees are present in greater numbers early in the season than bumble or honey bees. Only detailed fieldwork can test this hypothesis. It is likely that the number of gene changes required to effect the change is relatively small: Macnair and Cumbes (1989), using quantitative genetic techniques (Lande, 1981), showed that the number of genes causing the flower size difference between *M. guttatus* and *M. cupriphilus* was probably in the order of five, while recent work by Bradshaw et al. (1995), has shown by

QTL (quantitative trait locus) mapping that a similar number of genes may be involved in various aspects of flower size and shape in a cross between *M. cardinalis* and *M. lewisii*.

Postmating barriers normally involve complementary gene interactions between two or more loci. As noted by early theorists (e.g., Dobzhansky, 1936; Muller, 1942), if an ancestral species has the genotype *aabb*, the evolution in one taxon of *AAbb*, and in another of *aaBB*, can cause postmating isolation if *AaBb* is wholly or partially inviable. No "adaptive valley" needs to be crossed in the evolution of either taxon. As pointed out by Orr (1995), interactions leading to isolation will tend to arise between derived and ancestral alleles: the derived alleles can either have all evolved in one taxon (i.e., the ancestral species remaining unchanged), or both taxa can have evolved. Important questions about the evolution of this

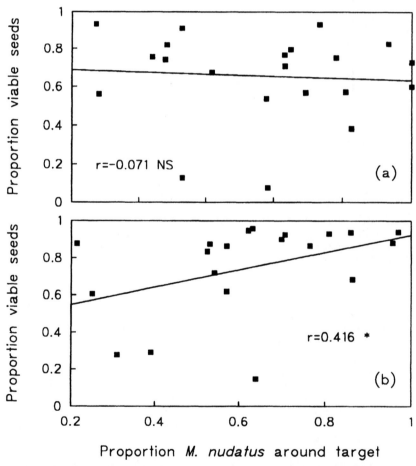

Figure 12.6. Graphs showing the proportion of viable seeds produced by 21 *M. guttatus* (a) or 21 *M. nudatus* (b) plants in mixed populations of *M. nudatus* and *M. guttatus* as a function of the proportion of *M. nudatus* flowers in a circle of 24 cm radius around the plant (data from Gardner, 1995). The regression lines are plotted, and the significance of the correlation coefficients (calculated on arcsine transformed data) are given.

sort of barrier to gene exchange include the magnitude of individual gene effects (i.e., how many gene changes are required to produce isolation), and what causes the spread of the genes. In *Drosophila*, there is considerable evidence that there is a large number of interacting genes, each of which can have a significant effect on male sterility (Cabot et al., 1994; Wu and Hollocher, this volume). Note, though, that there are single genes of large effect that can rescue hybrids (Hutter and Ashburner, 1987). Other systems in which many genes have been demonstrated include *Bombina* (Szymura and Barton, 1981) and *Podisma* (Barton and Hewitt, 1981). In plants, however, though complex systems involving many genes are known (e.g., rice, Sano et al., 1979), major genes causing hybrid inviability are well documented both within and between species (see Christie and Macnair, 1987, for references). Macnair and Christie (1983) showed that the major gene giving copper tolerance in the Copperopolis ecotype of *M. guttatus* also acted as one of the genes involved in complementary inviability of this type. It is not known whether the gene gives this effect as a pleiotropic effect of copper tolerance, or whether the original mutation for copper tolerance fortuitously arose in close proximity to a gene with this effect. Christie and Macnair (1984, 1987) and Vickery (1978) have shown that such genes are commonly segregating in normal populations of *M. guttatus*, and so this latter possibility is not improbable. However, whether due to pleiotropy or hitchhiking, it is clear that natural selection for a clearly adaptive feature has caused a gene for postmating isolation to spread through a population.

We have been investigating whether the postmating barrier between *M. nudatus* and *M. guttatus* has as simple a basis as those we have found within *M. guttatus*. It seems probable that most of the gene changes involved in evolving the barrier have occurred in *M. nudatus*. Table 12.3 shows the proportion of viable seeds found in crosses between *M. nudatus* from Lake County (coastal mountains) with local populations of *M. guttatus* and with populations from the Sierras. The proportion of inviable seeds is less in the crosses to the Sierran populations but is still substantial. If the Sierran populations could be considered ancestral to both *M. nudatus* and *M. guttatus* from Lake County, this would indicate that some gene changes have occurred in both lineages but, assuming genes of equal effect, more in the lineage leading to *M. nudatus*. In fact, of course, it is unlikely that the Sierran populations are ancestral and that the differences between Sierran and Coastal *M. guttatus* represent divergence in this species that is completely unconnected with the evolution of *M. nudatus*. Thus, all the gene changes causing the postmating barrier could have occurred in the *M. nudatus* lineage. Using crosses to *M. cupriphilus* (with which both species are interfertile) as an intermediate, Gardner (1995) has found some evidence that genes of large effect may be involved in the barrier. It remains to be determined whether the genes are associated with any of the adaptive features already identified in the evolution of *M. nudatus*, so we do not know if the barrier has spread by pleiotropy or hitchhiking in the way that the isolating barrier associated with copper tolerance has done in the Copperopolis population (Macnair and Christie, 1983).

A Simple Model

The process of adaptation to this novel environment can be formalized by the following model. Consider a single gene, A/a, that gives adaptation to some aspect of the serpentine, with aa being ancestral. aa has a fitness of 1 on normal soil, and AA a fitness of 1 on serpentine (figure 12.7). aa and AA have reduced fitnesses on serpentine and normal soil, respectively (with selection coeffi-

Table 12.3. The proportion of viable seeds produced in crosses.

Allopatric	% Viability	Sympatric	% Viability
Copperopolis	4.07	Colusa Co. Line 1	2.56
Napoleon 1	3.74	Colusa Co. Line 2	1.73
Napoleon 2	3.15	Colusa Co. Line 3	2.48
McNulty	5.20	Colusa Co. Line 4	1.99
Star & Excelsior	2.44	Colusa Co. Line 5	2.39
Quail	2.43	Butts Canyon	3.40
Buckham Gulch	5.75	Knoxville, Public Access	3.47
Sutter Creek	10.96		
Chinese Camp	12.06		

Seeds were produced in crosses between *M. nudatus* (from Lake Co.) and *M. guttatus* from seven local Lake County populations (sympatric with *M. nudatus*) and from nine allopatric populations from the Sierra Nevada (data from Gardner, 1995). Difference between allopatric and sympatric is significant ($F_{1,14} = 5.55$, $p = 0.034$).

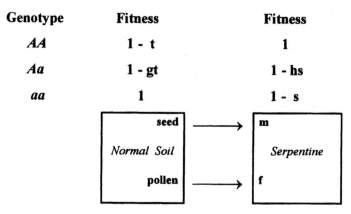

Genotype	Fitness	Fitness
AA	1 - t	1
Aa	1 - gt	1 - hs
aa	1	1 - s

Figure 12.7. The parameters of a simple model for the spread of a single gene giving adaptation to serpentine. *AA* is the fittest genotype on serpentine soil; *aa*, the fittest on normal soil. Dominance of the gene in each habitat is given by g and h. Since normal soils are more prevalent than serpentine, and in the first stages of evolution population sizes of the ancestral population will be larger than the derived, gene flow from normal to serpentine population will be much larger than vice versa. Seed flow can be either via seeds (m) or pollen (f).

cients of s and t, respectively). g and h define the degree of dominance of A. If this trade-off did not occur, the favored gene would obviously spread throughout the area. Assuming that, at least initially, the ancestral population is much larger than the ecotype (and that certainly the geographical area of the serpentine is smaller than that of normal soil), gene flow is going to be predominantly from the normal population onto the ecotype. Gene flow will occur in two forms. Seed from the ancestral population (primarily or exclusively *aa*) will be carried onto the serpentine with a rate m; pollen (genotype a), with rate f. Whether A can spread in the novel environment will be a function of (m,s,f,h). If s is large enough, it will spread whatever m and f. Thus, for instance, the gene for metal tolerance is likely to have a value for s of close to 1, and so will spread easily.

If gene flow by pollen is eliminated between the two populations, then whether genes will spread within the serpentine population depends solely on s and h. But if the gene does spread, such that the serpentine population is *AA* and the ancestral population is *aa*, whether the two forms can persist together in the face of seed flow (m) will depend on a function of (m,s). It is obvious that $f(m,s)$ must be easier to satisfy than $f(m,s,f,h)$. Thus, genes that cannot spread in the face of gene flow can spread if gene flow through pollen is reduced.

A likely example of this principle is given by the genes that alter the life history characteristics of the endemic *Mimulus* species. Their obligate annual life history, together with the greater branching and rate of production of flowers of the endemics, means that they produce more seeds (i.e., are fitter) when growing in microhabitats that produce small plants but are less fit when growing in higher quality microhabitats (see figure 12.4). Thus, s will depend on the balance of the number of good and bad microsites but is likely to be much less than the selection

on physiological characters (such as metal tolerance), which will be more uniformly distributed.

In the Copperopolis population of *M. guttatus* we have found evidence of within-population variation in life history characteristics. Plants range from obligate annuals to perennials, with the annuals flowering faster and having lower overall seed production over a long season than the perennials (McCombie, 1995). At Copperopolis, two distinct habitats for *M. guttatus* can be seen: along the stream that runs through the mine site and in other damp places the plant grows as typical large plants producing hundreds of thousands of seeds, while on the drier tailings it grows as small plants that die early in the season. The seed production of these plants is low (figure 12.4). We have studied the degree of spatial differentiation within the mine for the life history variation, and find effectively none. It appears that the small plants on the dry areas are genetically the same as the large plants in the wet areas. Presumably the gene flow, by both seed and pollen, from the central stream will prevent any differentiation of the population that would allow the plants on the dry areas to perform better on these areas at the expense of growing well in the optimal environment.

It is likely that this gene flow between the productive perennial forms and the less productive annuals will prevent the evolution of the annual. If, however, genes spread that prevent pollen flow from occurring, then there is a greater opportunity for the spread of genes that allow the annuals to colonize the drier areas effectively.

The Model Applied to the *Mimulus Guttatus* Complex

The analysis of the ecological and genetical features of these endemics in the *M. guttatus* complex therefore suggests the following generalization. The difference between an ecotype and an endemic is that in an ecotype none of the adaptations involved in the colonization of the novel habitat have resulted in the reduction of gene flow between the ancestral form and the evolving ecotype. The population therefore remains at stage 2 of Kruckeberg's scenario (see above). Time alone is not a sufficient reason for a population to go from stage 2 to stage 3: if gene flow between the populations remains high, genes unassociated with the adaptations to the environment will spread throughout the two populations, and genes adaptive in only one of the populations cannot spread. In an endemic, however, one or more of the specific adaptations to the novel environment have the pleiotropic effect of reducing, though not necessarily eliminating, gene flow between ecotype and ancestral population. Once this has happened, genes that can further adapt the ecotype, which could not have spread if the populations had remained in contact, will also spread. The population has moved into stage 3 of Kruckeberg's scenario and can in time evolve further. The crucial point is that the adaptations that reduce gene flow also enable other adaptations to the local environment to evolve, such that a population develops that has a number of correlated adaptations, which can be maintained by natural selection in the face of some gene flow from the ancestral population.

In the case of *M. pardalis* and *M. cupriphilus* the suggested scenario is thus as follows. Metal/serpentine tolerance permitted the colonization of the contaminated habitat, which is on average drier and more nutrient poor than the normal habitat, and permits a shorter growing period producing smaller plants that die early. Genes that permit early flowering to maximize seed production would be at an advantage, but pollination efficiency would drop because maximal flowering would precede maximal bumble bee activity. Self-fertilization to provide reproductive assurance would then be at an advantage, producing pleiotropically the reduction in pollen flow that would have prevented greater changes in life history previously. The change in life history characteristics to maximize seed production from small plants followed. Note that we are proposing that self-fertilization evolves to promote reproductive assurance following a change in flowering time, and has the effect of reducing gene flow as a by-product. This is not the same as the scenario modeled by Antonovics (1968), where selfing evolves as an adaptation to reduce gene flow, and thus as a premating barrier (i.e., reinforcement).

In the case of *M. nudatus*, the difference is that reproductive assurance was achieved by a change in pollinator, and by the fact that a postmating barrier had fortuitously arisen during the fixation of other genes giving either physiological adaptation, or the change in pollinator. We are not suggesting that this postmating barrier arose in direct response to selection for reduced hybridization, as was suggested by Grant (1966) for *Gilia*, but that genes giving adaptation to the serpentine or the pollinator were synthetic lethals, either pleiotropically or because a synthetic lethal was carried to fixation by hitchhiking.

These scenarios are obviously tentative and speculative at present, but make predictions that can be tested by genetic analysis and by the production of recombinant inbred lines that can be contrasted in the field for the relative fitness of different characteristics.

Conclusion

Edaphic endemics provide a fascinating example of plant speciation. It is almost certain that they evolve by a number of different modes—speciation is not a uniform process that always happens the same way. Thus, the classic allopatric model must be true in many cases, particularly of palaeoendemics where a population has speciated over time when its nonserpentine progenitor has become locally or globally extinct. Regional indicator species may offer a currently observable intermediate in this process. Large areas of uniform habitat separated by a sharp ecotone would also provide the opportunity for parapatric speciation (Endler, 1977): the short dispersal distance of many plant species offers excellent opportunities for this process. However, the examples discussed in this chapter indicate that it may be possible for speciation to occur while the populations remain in more intimate contact, and that whether a population colonizing a novel habitat does so as an ecotype or proceeds to full speciation depends on the particular sequence of adaptations that occur to the local environment. Levin (1993) has recently argued that local speciation in plants is the rule rather the exception; his argument is based on the likely patterns of gene flow and the lack of a mechanism for maintaining the integrity of widespread populations evolving in tandem. Thus, the phenomenon of edaphic endemics may well be a particularly clear example of more general principles pertaining to plant speciation, and the patterns and processes that lead to these species will not be confined to this relatively unusual group of plants.

References

AFRC. 1991. Rothamsted Experimental Station: Guide to the classical field experiments. Harpenden: AFRC.

Allen, W. R., and Sheppard, P. M. 1971. Copper tolerance in some Californian populations of the monkey flower *Mimulus guttatus*. Proc. R. Soc. Lond. B 177:177–196.

Antonovics, J. 1968. Evolution in closely adjacent plant populations. V. Evolution of self-fertility. Heredity 23:219–238.

Barton, N. H., and Hewitt, G. M. 1981. The genetic basis of hybrid inviability in the grasshopper *Podisma pedestris*. Heredity 47:367–383.

Bradshaw, A. D. 1991. Genostasis and the limits to evolution. Phil. Trans. Roy. Soc. Lond. B. 333:289–305.

Bradshaw, H. D., Wilbert, S. M., Otto, K. G., and Schemske, D. W. 1995. Genetic mapping of floral traits associated with reproductive isolation in monkey flowers (*Mimulus*). Nature 376:762–765.

Cabot, E. L., Davis, A. W., Johnson, N. A., and Wu, C.-I. 1994. Genetics of reproductive isolation in the *Drosophila simulans* clade: complex epistasis underlying male sterility. Genetics 137:175–189.

Campbell, G. R. (1950) *Mimulus guttatus* and related species. El Aliso 2:319–335.

Christie, P., and Macnair, M. R. 1984. Complementary lethal factors in two North American populations of the yellow monkey flower. J. Hered. 75:510–511.

Christie, P., and Macnair, M. R. 1987. The distribution of postmating reproductive isolating genes in populations of the yellow monkey flower, *Mimulus guttatus*. Evolution 41:571–578.

Dobzhansky, T. 1936. Studies on hybrid sterility. II. Localization of sterility factors in *Drosophila pseudoobscura* hybrids. Genetics 21:113–135.

Dole, J. A. 1992. Reproductive assurance mechanisms in 3 taxa of *the Mimulus guttatus* complex (Scrophulariaceae) Am. J. Bot. 79:650–659.

Dudash, J. R., and Ritland, K. 1991. Multiple paternity and self-fertilisation in relation to floral age in *Mimulus guttatus* (Scrophulariaceae). Am. J. Bot. 78:1746–1753.

Endler, J. A. 1977. Geographic variation, speciation and clines. Princeton: Princeton UP.

Fenster, C. B., and Ritland, K. 1992. Chloroplast DNA and isozyme diversity in 2 *Mimulus* species (Scrophulariaceae) with contrasting mating systems. Am. J. Bot. 9:1440–1447.

Fenster, C. B., and Ritland, K. 1994. Evidence for natural selection on mating system in *Mimulus* (Scrophulariaceae). Int. J. Plant Sci. 155:588–596.

Gardner, M. 1995. The evolution of *Mimulus nudatus* from *M. guttatus*. Ph.D. thesis, University of Exeter.

Grant, V. 1966. The selective origin of incompatibility barriers in the plant genus *Gilia*. Am. Nat. 100:99–118.

Griffin, J. R. 1965. Digger pine seedling response to serpentinite and non-serpentinite soil. Ecology 46:801–807.

Heyl, R. 1954. The zinc-copper mines of the Quail Hill area, Calaveras County, California. Calif. J. Mines Geol. 50:111–125.

Hutter, P., and Ashburner, M. 1987. Genetic rescue of inviable hybrids between sibling species of *Drosophila*. Nature 327:331–333.

Ingram, C. 1988. The evolutionary basis of ecological amplitude of plant species. Ph.D. thesis, Liverpool University.

Jaffré, T. 1981. Etude ecologique du peuplement vegetal des sols derivés des roches ultrabasiques en Nouvelle Caledonie. Office Rech. Sci. Technol. Outre Mere: Paris.

Jain, S. K. 1976. The evolution of inbreeding in plants. Annu. Rev. Ecol. Syst. 7:469–95.

Kruckeberg, A. R. 1951. Intraspecific variability in the response of certain native plant species to serpentine soil. Am. J. Bot. 38:408–419.

Kruckeberg, A. R. 1967. Ecotypic response to ultramafic soils by some plant species of northwestern United States. Brittonia 19:133–151.

Kruckeberg, A. R. 1984. Californian serpentines: Flora, vegetation, geology, soils and management problems. Univ. Calif. Publ. Bot. 78:1–180.

Kruckeberg, A. R. 1986. An essay: the stimulus of unusual geologies for plant speciation. Syst. Bot. 11:455–463.

Lande, R. 1981. The minimum number of genes contributing to quantitative variation between and within populations. Genetics 99:541–553.

Levin, D. A. 1993. Local speciation in plants: the rule not the exception. Syst. Bot. 18:197–208.

Macnair, M. R. 1987. Heavy metal tolerance in plants: a model evolutionary system. Trends Ecol. Evol. 2:354–359.

Macnair, M. R. 1989. A new species of *Mimulus* endemic to copper mines in California. Bot. J. Linn. Soc. 100:1–14.

Macnair, M. R. 1992. Preliminary studies on the genetics and evolution of the serpentine endemic *Mimulus nudatus* Curran. In A. J. M. Baker, J. Proctor, and R. D. Reeves (eds.). The vegetation of Ultramaphic (serpentine) soils. Andover Intercept: pp. 409–420.

Macnair, M. R. 1993. The genetics of metal tolerance in vascular plants. New Phytol. 124:541–559.

Macnair, M. R., and Baker, A. J. M. 1994. Metal tolerance in plants: evolutionary aspects. In M. E. Farago (ed.). Plants and the chemical elements. New York: VCH Publishers. pp. 67–86.

Macnair, M. R., and Christie, P. 1983. Reproductive isolation as a pleiotropic effect of copper tolerance in *Mimulus guttatus*? Heredity 50:295–302.

Macnair, M. R., and Cumbes, Q. J. 1989. The genetic architecture of interspecific variation in *Mimulus*. Genetics 122:211–222.

Macnair, M. R., Cumbes, Q. J., and Smith, S. 1993. The heritability and distribution of variation in degree of copper tolerance in *Mimulus guttatus* on a copper mine at Copperopolis, California. Heredity 71:445–455.

Macnair, M. R., Macnair, V. E., and Martin, B. E. 1989. Adaptive speciation in *Mimulus*: an ecological comparison of *M. cupriphilus* with its presumed progenitor, *M. guttatus*. New Phytol. 112:269–279.

Mallet, J. 1995. A species definition for the modern synthesis. Trends Ecol. Evol. 10:294–299.

Malpas, J. 1991. Serpentine and the geology of serpentinized rocks. In B. A. Roberts and J. Proctor (eds.). The ecology of areas with serpentinized rocks: a world view. Dortrecht: Kluwer, pp. 7–30.

Marrs, R. H., and Proctor, J. 1976. The response of serpentine and non-serpentine *Agrostis stolonifera* to magnesium and calcium. J. Ecol. 64:953–964.

McCombie, H. 1995. Life history variation in *Mimulus guttatus* (Scrophulariaceae), the importance of ecological pressures in space and time. Ph.D. thesis, University of Exeter.

Muller, H. J. 1942. Isolating mechanisms, evolution and temperature. Biol. Symp. 6:71–125.

Orr, H. A. 1995. The population genetics of speciation: the evolution of hybrid incompatibilities. Genetics 139:1805–1813.

Pennell, F. W. 1950. *Mimulus*. In L. Abrams (ed.). Illustrated flora of the Pacific states. Stanford: Stanford UP, pp. 688–731.

Ritland, C. E., and Ritland, K. 1989. Variation in sex allocation among eight taxa of the *Mimulus guttatus* species complex (Scrophulariaceae). Am. J. Bot. 76:1731–1739.

Ritland, C. E., Ritland, K., and Straus, N. A. 1993. Variation in the ribosomal internal transcribed spacers (ITS1 and ITS2) among 8 taxa of the *Mimulus guttatus* species complex. Mol. Biol. Evol. 10:1273–1288.

Sano, Y., Chu, Y. E., and Oka, H. I. 1979. Genetic studies of speciation in cultivated rice. 1. Genic analysis for the F1 sterility between *O. sativa* L. and *O. glaberrima* Steud. Jap. J. Genet. 54:121–132.

Stebbins, G. L. 1970. Adaptive radiation in Angiosperms. I. Pollination mechanisms. Annu. Rev. Ecol. Syst. 1:307–326.

Szymura, J. M., and Barton, N. H. 1991. The genetic structure of the hybrid zone between the fire-bellied toads *Bombina bombina* and *Bombina variegata*: comparisons between transects and between loci. Evolution 45:237–261.

Thompson, D. M. 1992. *Mimulus*. In J. C. Hickman (ed.). The Jepson Manual. Berkeley: University of California Press, pp. 1037–1046.

Vickery, R. K., Jr. 1959. Barriers to gene exchange within *Mimulus guttatus* (Scrophulariaceae). Evolution 13:300–310.

Vickery, R. K., Jr. 1978. Case studies in the evolution of species complexes in *Mimulus*. Evol. Biol. 11:405–507.

Walley, K. A., Khan, M. S. I., and Bradshaw, A. D. 1974. The potential for the evolution of heavy metal tolerance in plants. Heredity 32:309–319.

Wu, L., Bradshaw, A. D., and Thurman, D. A. 1975. The potential for the evolution of heavy metal tolerance in plants. III. The rapid evolution of copper tolerance in *Agrostis stolonifera*. Heredity 34:165–187.

Wyatt, R. 1988. Phylogenetic aspects of the evolution of self-pollination. In: L. D. Gottlieb and S. K. Jain (eds.). Plant evolutionary biology. London: Chapman and Hall, pp. 109–131.

13

The Relative Rate of Sympatric and Allopatric Speciation in Fishes

Tests Using DNA Sequence Divergence between Sister Species and among Clades

Amy R. McCune
Nathan R. Lovejoy

The diversification of endemic fishes within lakes is one of the most peculiar and interesting aspects of speciation in vertebrates. Cichlid fishes in the African great lakes and the cottoid (sculpins) fishes of Lake Baikal, Russia, are perhaps the best-known examples of such "species flocks," that is, complexes of closely related species endemic to a geographically confined area (Brooks 1950; Fryer and Iles 1972; Greenwood 1974; Echelle and Kornfield 1984). However, there are several other less familiar lacustrine species flocks, such as the cyprinids of Lake Lanao in the Philippines (Kornfield and Carpenter 1984); killifishes in Lake Titicaca in the Andes (Parenti 1984; Parker and Kornfield 1995); pupfishes in Lake Chichancanab, Mexico (Humphries 1984); cyprinids in Lake Tana, Ethiopia (Nagelkerke et al. 1994, 1995); and even extinct semionotid fishes from Mesozoic lakes in North America (McCune et al. 1984; McCune 1987, 1996). The endemism of monophyletic clades within lakes has led many to believe that diversification occurred *in situ,* leading in turn to controversy over whether this intralacustrine speciation was allopatric, microallopatric (microgeographic of Mayr 1947), or sympatric (e.g., Worthington 1954; Kosswig 1963; Ribbink 1994).

It has long been recognized that speciation by African cichlids must have been very rapid, even "explosive" (e.g., Stanley 1979; Futuyma 1986), but it has been documented only recently that rapid speciation is characteristic of most endemic radiations of lacustrine fishes (McCune 1997). Time for speciation (henceforth TFS) in endemic lacustrine fishes ranges from 1500 to 300,000 years, in contrast to a TFS of 600,000 to 1 million years (MY) for bird and arthropod (including Hawaiian drosophilids) island endemics (McCune

1997). Do fishes really speciate faster than other animals? There are several possible explanations for this difference. One possibility might simply be that lakes are more conducive to speciation than islands (or less conducive to extinction). Alternatively, the difference might be an artifact of the relative recency of fish radiations or of more precise estimates of clade origin for fishes (McCune 1997). These hypotheses are not mutually exclusive and each might be tested. However, in this chapter, we consider another possibility, that the short TFS for endemic lacustrine fishes is a result of mode of speciation.

It is little appreciated that sympatric speciation, when and if it occurs, should be faster than allopatric speciation (Bush 1975, 1993; Kondroshov et al., this volume). Allopatric speciation depends on the separation of a population by an extrinsic barrier for long enough that genetic differences accumulate (due to local adaptation and/or genetic drift), leading to reproductive isolation. As such, allopatric speciation is a by-product of local adaptation or drift that may eventually and indirectly lead to reproductive isolation. For sympatric speciation to occur, there must be disruptive selection on a polymorphic trait, as well as some kind of nonrandom mating (Kondrashov and Mina 1986) such as assortative mating, sexual selection, or the probability of mating being determined by habitat choice. For sympatric speciation, a decreased probability of matings between diverging populations depends primarily on individual genotypes, whereas with allopatric speciation, a decreased probability of matings between populations depends on the existence of an extrinsic barrier. If reproductive isolation begins to develop between sympatric populations, the potentially direct role of selection in modifying the probability of mating may result in faster evolution of reproductive isolation than generally occurs between allopatric populations. Thus, the more direct effects of selection on the probability of mating should mean that when it occurs, sympatric speciation proceeds more rapidly than allopatric speciation and we expect that $TFS_{sympatric} < TFS_{allopatric}$.

Given this theoretical expectation, we consider the possibility that the rapid speciation by fishes in species flocks reflects a high incidence of sympatric speciation. We evaluate this hypothesis in two ways. First, we compare the degree of divergence between sister species pairs of fishes with presumed allopatric origin to pairs of fishes with possible sympatric origin. We also compare TFS, estimated from maximum sequence divergence for a clade and assuming a symmetrical model of branching, for clades of fishes with presumed allopatric origins to clades with possible sympatric origins.

Methods

Through extensive computerized searches (using BIOSIS Previews of Biological Abstracts, Inc.) we attempted to find all potentially relevant publications on molecular systematics of fishes. These publications were then reviewed for applicability according to criteria explained below.

Classifying Cases as Allopatric or Sympatric

To compare TFS for allopatric versus sympatric speciation in fishes, we identified probable cases of allopatric speciation and possible cases of sympatric speciation. Given the considerable empirical and conceptual support for the preeminence of allopatric speciation among vertebrates (Mayr 1963; Wiley and Mayden 1985; Lynch 1989; Slatkin 1996), we assumed speciation was allopatric unless an hypothesis of sympatric speciation had been previously implicated. Among allopatric cases, most identifiable pairs of sister species showed a predominately nonoverlapping geographic range, except in four cases, where an allopatric species living in a distinct habitat relative to its sister species (e.g., upland streams versus lowland streams) occurs within a very restricted geographic range included within the limits of the range of its common, widespread sister species. In the absence of any suggestion in the literature that speciation was sympatric in these instances, we still assumed allopatric speciation, recognizing that the populations are separated by an extrinsic barrier and/or that the current distribution could be attributable to a range expansion of the common species (cf. Lynch 1989; Berlocher, ch. 8, this volume). Note that we lump various allopatric models together (dumbbell and others) on the grounds that they all rely at least initially on an extrinsic interruption of gene flow. The efficacy of allopatric models such as peripatric speciation (Mayr 1954), founder-flush (Carson 1975), or genetic transilience (Templeton 1980) is controversial

(Moya et al. 1995; Barton 1996; Slatkin 1996), but if they occur, all imply more rapid speciation than the classical dumbbell model. If allopatric speciation by these other modes occurs frequently in fishes, it will be more difficult to detect a difference in rate between allopatric and sympatric speciation.

The possible cases of sympatric speciation used in our comparisons are all "species flocks," or complexes of closely related species endemic to a single lake (Brooks, 1950; Echelle and Kornfield 1984), the best-known examples being the cichlid fishes of the African lakes. In these cases, the occurrence of an endemic monophyletic clade within a lake (e.g., Meyer et al. 1990; Schliewen et al. 1994) has led to general agreement that speciation has been intralacustrine, though it has long been controversial whether cichlid speciation within each of the great lakes was sympatric or allopatric (Kosswig 1947, 1963; Mayr 1947, 1963; Dominey 1984; Martens et al. 1994). Whether speciation should be considered sympatric depends in part on what is meant by sympatric speciation. For example, one widely cited hypothesis to explain the evolution of the rock-dwelling mbuna in Lake Malawi is that populations of mbuna species have been isolated in patches of rocky shore separated by patches of sandy shore (Fryer and Iles 1972). This might be regarded as sympatric speciation because so-called isolated populations are within "the cruising range of each other" (Endler 1977:14) and isolation might be a result of habitat selection. Others might consider the same situation micro-allopatric (Mayr 1947, 1963). We have not attempted to resolve the conceptual issue or identify the mechanism of speciation in any particular case. Instead, for this chapter we recognize all controversial cases of intralacustrine speciation as *possibly* sympatric. Importantly, in none of these possible sympatric cases has the rate of speciation or the amount of sequence divergence been used as evidence for sympatric speciation.

Estimating TFS

We gathered divergence data for both cytochrome b and the control region of the mitochrondrial genome because these were the most commonly available sequences. Data from the control region are included in the tables (see below) but are not used in analyses because rates of divergence in the control region appear highly variable relative to cytochrome b (see tables 13.1, 13.2, 13.4, 13.5) and because control region sequence data are not abundant enough to analyze separately. We estimated TFS from sequence divergence between sister species, and for entire clades.

Estimating TFS for Species Pairs

To estimate TFS for species pairs we used sequence divergence in cytochrome b of the mitochondrial genome. Although the concept of a molecular clock is controversial (Avise 1994), sequence divergence is accepted here as providing at least an approximate measure of separation time. It is highly unlikely that the genes used here would be affected by selection on traits conferring reproductive isolation. In order to translate divergence directly to TFS, a taxon-specific and gene- or region-specific calibration factor is required (see McCune 1997 for discussion). However, to the extent that similar genes and similar taxa are used for divergence calculations, divergence can be directly compared. We used uncorrected estimates to facilitate comparison between different studies; at the low levels of divergence involved, multiple hits are unlikely to affect our results. For all cases, maximum sequence divergence between individuals (either from different species, or the same species; see below) was calculated.

To estimate the amount of sequence divergence accumulated during allopatric speciation, we use both sequence divergence between sister species and among populations within species. The range of sequence divergence associated with speciation lies approximately at the intersection of the minimum divergence between species and the maximum divergence between geographic populations within species (figure 13.1). Minimum interspecific divergence provides a lower bound. Fully separated species have most likely continued to diverge since spe-

Estimating TFS $_{allopatric}$

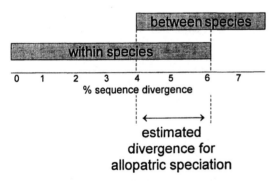

Figure 13.1. Estimating TFS$_{allopatric}$. The sequence divergence associated with allopatric speciation is estimated as the intersection of (1) the minimum sequence divergence between sister species and (2) the maximum sequence divergence among geographic populations within a species. Thus, in the figure, if allopatrically derived sister species differ by 4–8% sequence divergence, and the maximum divergence among geographic populations within a species is 0–6%, then we estimate that the divergence associated with allopatric speciation is about 4–6%. Such an estimate of sequence divergence can be used as a proxy for TFS$_{allopatric}$ or converted to TFS$_{allopatric}$ by some calibration factor, assuming constant average rates of sequence evolution.

ciating; thus, interspecific divergence alone may overestimate TFS. On the other hand, maximum intraspecific divergence provides an upper bound for the range of divergence associated with speciation; for any greater degree of divergence, taxa have been described as separate species. For sympatric cases, we estimated TFS directly from interspecific sequence divergence because intraspecific divergence was not generally available. Presumably, the effect of using only interspecific data for sympatric cases was that our estimated TFS$_{sym}$ is higher than the actual TFS$_{sym}$ and thus conservative with respect to our test.

Several criteria were established for the selection of studies from the literature for comparison of species pairs:

1. The availability of sequence data from cytochrome b and/or control region. Our emphasis on these two most commonly studied mitochondrial genes data facilitated broad comparisons between taxa.
2. Availability of a phylogeny including all relevant members of the clade in question to allow selection of sister species pairs. Of particular concern was that omission of one or more species from molecular systematic analyses may result in erroneously identified sister species relationships because the actual sister species had been omitted from taxa surveyed (see figure 13.2). Where possible, morphological phylogenies were used to evaluate this possibility. For example, Rauchenberger's (1989) morphological study of *Gambusia* was used in conjunction with a molecular study (Lydeard et al. 1995).
3. Availability of geographic distributional data for the taxa in question.
4. For estimates of allopatric intraspecific divergence, we use studies that presented sequence data from a number of widely distributed individuals.

Estimating TFS at the Clade Level

For monophyletic groups in which species-level relationships have not been resolved, we estimated an average TFS for species within the clade by the formula TFS = t (ln 2)/ln n, where t = time of clade origin, and n = number of species in the clade (McCune 1997). This estimate assumes a fully symmetrical or balanced tree, correcting for speciation events occurring simultaneously in separate lineages within the same clade. We made no attempt to estimate extinction. Thus, if extinction rates were unusually high in one group, the amount of speciation would be underestimated and TFS would be overestimated.

Sequence divergence was calibrated for cytochrome b sequences by the commonly used rate of 2.5%/MY (actually based on cytochrome b divergence in ungulates from Irwin et al. [1991] cited in Meyer et al. [1990]), although we recognize that this is not ideal, given the likely differences in rate between different taxonomic

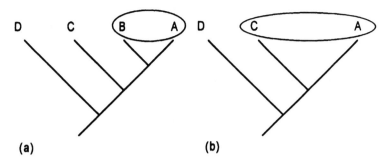

Figure 13.2. Effects of taxon sampling on determination of sister species. In (a), all species have been sampled and A and B are correctly inferred to be sister species. In (b), species B has not been included in the analysis, causing A and C to appear erroneously to be sister species.

groups and, to some extent, the differences in rate within a particular gene (Avise 1994; Li and Graur 1991).

For clade-level computations of TFS, we again restricted our scope to studies reporting sequence data from cytochrome b and/or control region for various groups of fishes. In contrast to species pair comparisons, we included studies of more sparsely sampled clades that provided maximum sequence divergence between distantly related species in the clade.

A Note on Species Concepts

The dominant conceptual basis for defining species of fishes has involved interbreeding or reproductive isolation in some form (Hubbs 1941, 1943; Rosen 1960), although its implementation has surely varied. For this study, we generally accepted species designations as previously defined by the workers on various taxa, recognizing that morphological distinctiveness resulting from reproductive isolation was the operational criterion used to define species in most cases. More recently, use by some ichthyologists of Rosen's (1979) autapomorphic species concept is a radical departure in the meaning of species that can dramatically affect the results of this or other comparative studies at the species level (see discussion below). Rosen (1979) and Rauchenberger et al. (1990) split several species of *Xiphophorus* (swordtails) previously defined using the biological species concept (Rosen 1960), basing their decisions solely on autapomorphies characterizing each small population. Because we are interested specifically in the amount of divergence associated with reproductive incompatibility, we have reverted here to using Rosen's (1960) earlier species designations for *Xiphophorus*, which were based on the biological species concept. For species pair comparisons, this means that divergence between autapomorphically defined species is categorized as intraspecific divergence. For clade-level comparisons, reverting to reproductively defined species almost halved the number of species

(from 22–24 to 13–14) in *Xiphophorus* (Rosen 1960, 1979; Rauchenberger et al. 1990; Nelson 1984).

Taxa Sampled

Taxa used for sister species comparisons were not necessarily the same as those used for clade-level comparisons. However, for both sets of comparisons, data from a fairly broad range of teleosts fishes are included. For sister species comparisons, data are included for Acipenseridae: *Acipenser* (sturgeon), and the teleosts, Cyprinidae: *Notropis* (shiners); Salmonidae: *Salmo* (trout); Osmeridae: *Osmerus* (smelt), *Mallotus* (capelin); Fundulidae; *Fundulus* (killifishes); Poecilidae: *Gambusia* (mosquitofishes), *Poecilia* (livebearers), *Xiphophorus* (swordtails); Melanotaeniidae: *Melanotaenia* (rainbowfishes); Aplocheilidae: *Rivulus* (rivulines); Gasterosteidae: *Gasterosteus* (sticklebacks); Cottoidei: *Cottocomephorus* (Baikal sculpins), *Batrachocottus* (Baikal sculpins), *Comephorus* (Baikal oilfishes); and Cichlidae: *Konia, Myaka, Sarotherodon, Stomatepia*. Clade-level comparisons involve Osmeridae: *Osmerus* (smelt); Poecilidae: *Gambusia* (mosquitofishes), *Xiphophorus* (swordtails); Melanotaeniidae: *Melanotaenia* (rainbowfishes); Aplocheilidae: *Rivulus* (rivulines); Cyprinodontidae: *Cyprinodon* (pupfishes); and Cichlidae: Ectodini, *Tropheus*, Lamprologini, tilapiines, and haplochromines.

Results

Comparisons of Species Pairs

For allopatric cases, DNA sequence divergence between pairs of sister species and among geographic populations within species are summarized in tables 13.1 and 13.2. Interspecific divergences for possible sympatric pairs are listed in table 13.3. The ranges of divergences associated with allopatric speciation, using the method described above, is about 2.0–5.7% sequence divergence of cytochrome b (figure 13.3, tables 13.1, 13.2). In contrast, for sympatric cases, interspecific divergence is much lower, having a range of 0% to 1.25% (figure 13.3; table 13.3). Unfortunately, it is impossible to evaluate this difference statistically because the range of divergence estimated for allopatric speciation is the intersection of within and between species divergence and does not constitute any sort of population sample. However, biologically, these data are very highly suggestive that divergence associated with allopatric speciation is considerably higher than divergence associated with possible sympatric speciation.

Comparisons of Clade-Level TFS

Tables 13.4 and 13.5 summarize clade-level data on species number, age of the clade, and TFS for probable allopatric and possible sympatric cases, respectively. TFS for

allopatric clades is estimated to be approximately 0.88–2.27 MY/species, whereas TFS for possible sympatric clades is considerably less, ranging from virtually 0 to 0.83 MY/species, and generally falling below 0.5 MY/species. A two-sample t-test comparing means with separate variances (Systat version 6.0, SPSS Inc.) yields a highly significant difference ($p = 0.004$) between mean TFS for allopatric and sympatric cases (figure 13.4).

Discussion

The Importance of Species Concepts and the Interpretation of Taxonomic Literature

In the course of compiling data for this study, the practical consequences of using varied species concepts became painfully clear. Most of the fish species used in this study were originally described (explicitly or implicitly) using

Table 13.1. Maximum interspecific divergence between sister species with presumed allopatric origin.

Taxa	Maximum Divergence	Sequence	Notes/References
Salmo salar/S. trutta	6.17–7.50%	610 bp cyt b + ATPase	Average difference between haplotypes; Giuffra et al. 1994
Salmo salar/S. trutta	4.6%	302 bp cyt b	Pálsson and Árnason 1994
Salmo salar/S. trutta	3.8%	378 bp cyt b	From Patarnello et al. 1994, which includes data from McVeigh et al. 1991; only this value used for *Salmo* in figure 13.3
Osmerus dentax/O. mordax	6.3%	300 bp cyt b	Taylor and Dodson 1994
Notropis topeka/ N. stramineus	7.8%	512 bp cyt b	Distributions overlap but species occupy different habitats; probably sister taxa but some species missing from analysis; Schmidt and Gold 1995; Lee et al. 1980.
Fundulus heteroclitus/ F. grandis	7.2%	263 bp cyt b	Probably sister taxa but some species missing from analysis; Bernardi and Powers 1995; Cashner et al. 1992
Gambusia affinis/G. holbrooki	4.0%	402 bp cyt b	Lydeard et al. 1995; Rauchenberger 1989
Gambusia melapleura/ G. wrayi	3.2%	402 bp cyt b	Distributions overlap but species occupy different habitats; Lydeard et al. 1995; Rauchenberger 1989; Fink 1971
Gambusia eurystoma/ G. sexradiata	2.0%	402 bp cyt b	Distributions overlap but species occupy different habitats; Lydeard et al. 1995; Rauchenberger 1989; Miller 1975
Rivulus fuscolineatus/ R. isthmensis	3.9 %	360 bp cyt b	Distributions overlap but species occupy different habitats; good sister pair (Murphy, pers. comm.); Murphy and Collier 1996 (data set provided by Murphy)
Salmo salar/S. trutta	5.44–6.44%	640 bp control region	Average difference; Giuffra et al. 1994

bp, base pairs; cyt b, cytochrome b.

Table 13.2. Maximum intraspecific divergence of geographic populations.

Taxon	Distribution	Maximum Intraspecific Divergence	Sequence	Notes/Reference
Melanotaenia eachamensis	2 populations in Australia	0.01%	351 bp cyt b	Zhu et al. 1994
Melanotaenia splendida	Many populations in Australia, New Guinea	10%	351 bp cyt b	Zhu et al. 1994
Gambusia affinis	3 populations in Texas	0.25%	402 bp cyt b	Lydeard et al. 1995
Gambusia holbrooki	2 populations in Florida	1.0%	402 bp cyt b	Lydeard et al. 1995
Gambusia rhizophorae	3 populations in Florida and Cuba	0.75%	402 bp cyt b	Lydeard et al. 1995
Gambusia yucatana	2 populations in Mexico	1.0%	402 bp cyt b	Lydeard et al. 1995
Salmo trutta	18 populations in N. Italy	1.36%	259 bp cyt b	Giuffra et al. 1994
Mallotus villosus	Norwegian & Newfoundland populations	5.68%	253 bp cyt b	Birt et al. 1995
Gasterosteus aculeatus	25 localities in Europe, Japan, North America	3.08%	747 bp cyt b	Orti et al. 1994
Rivulus cylindraceus	2 populations, Isle of Pines and Cuba	2.2%	360 bp cyt b	Murphy and Collier 1996
Rivulus tenuis	2 populations in Mexico and Guatemala	5.8%	360 bp cyt b	Murphy and Collier 1996
Xiphophorus clemenciae	?	1.1%	330 bp cyt b	Meyer et al. 1994
Xiphophorus maculatus	?	1.5%	330 bp cyt b	Meyer et al. 1994
Xiphophorus birchmanni	?	0.6%	330 bp cyt b	Meyer et al. 1994
Xiphophorus montezumae	?	0.8%	330 bp cyt b	Meyer et al. 1994
Xiphophorus multilineatus/ X. nigrensis	2 populations (not species) under BSC	0.3%	330 bp cyt b	Meyer et al. 1994; Rosen 1960, 1979; Rauchenberger et al. 1990
Xiphophorus couchianus/ X. gordoni	2 populations (not species) under BSC	0.8%	330 bp cyt b	Meyer et al. 1994; Rosen 1960, 1979; Rauchenberger et al. 1990
Salmo trutta	18 populations in N. Italy	1.92%	640 bp control region	Bernatchez et al. 1992
Acipenser transmontanus	Columbia and Frazier Rivers	4.29%	462 bp control region	Brown et al. 1993
Poecilia reticulata	9 populations from 7 rivers in Trinidad and Tobago	5.6%	465 bp control region	Fajen and Breden 1992
Xiphophorus couchianus/ X. gordoni	2 populations (not species) under BSC	0.5%	402 bp control region	Meyer et al. 1994; Rosen 1960, 1979; Rauchenberger et al. 1990
Xiphophorus multilineatus/ X. nigrensis	2 populations (not species) under BSC	0.6%	402 bp control region	Meyer et al. 1994; Rosen 1960, 1979; Rauchenberger et al. 1990

BSC, Biological Species Concept.

Table 13.3. Maximum interspecific divergence between sister species with possible sympatric origin.

Taxa	Distribution	Divergence	Sequence	Reference
Cottocomephorus grewinki and *C. inermis*	Lake Baikal	1.25%	420 bp cyt b	Slobodyanyuk et al. 1995
Comephorus baicalensis/ C. dybowsii	Lake Baikal	0.5%	804 bp cyt b and ATPase 8 +6	Slobodyanyuk et al. 1995
Batrachocottus multiradiatus and *B. nikolskii*	Lake Baikal	0.5%	804 bp cyt b and ATPase 8 +6	Slobodyanyuk et al. 1995
Konia dikume and *K. eisentrauti*	Barombi Mbo, Cameroon	0.3%	340 bp cyt b	Schliewen et al. 1994
Sarotherodon caroli and *S. linnellii*	Barombi Mbo, Cameroon	0%	340 bp cyt b	Schliewen et al. 1994
Myaka myaka and *S. linnellii/S. caroli*	Barombi Mbo, Cameroon	0.6%	340 bp cyt b	Schliewen et al. 1994
Stomatepia mariae/ S. pindu	Barombi Mbo, Cameroon	0.9%	340 bp cyt b	Schliewen et al. 1994
Sarotherodon steinbachi/ S. lohbergeri	Barombi Mbo, Cameroon	0.6%	340 bp cyt b	Schliewen et al. 1994

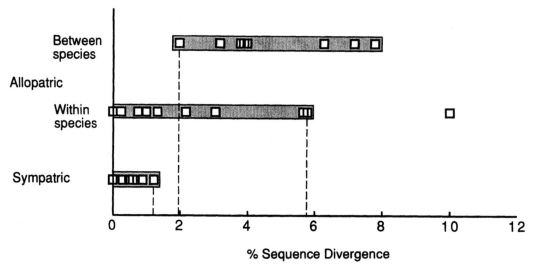

Figure 13.3. Percentage sequence divergence associated with presumed allopatric speciation versus possible sympatric speciation fishes. Boxes show individual data points. Shaded bars show ranges of divergences. Actual maximum sequence divergence between sister species of fishes ranges from about 2% to 8%, whereas divergence among geographic populations within species ranges from 0% to 5.6%. An outlier of 10% sequence divergence within *Melanotaenia splendida* (table 13.2) was excluded from the shaded bar for intraspecific comparisons because the "species" is apparently polyphyletic and there is possible introgression (Zhu et al. 1994). Thus, estimated divergence associated with allopatric speciation is 2.0–5.6%, in contrast to the much lower divergence of 0–1.25% between sympatrically derived sister species. Data are given in tables 13.1–13.3.

Table 13.4. Maximum divergence and average TFS for clades with possible sympatric origin.

Taxa	# Species	Maximum Divergence	Sequence	TFS (MY)	Notes/Reference
"Haplochromine" cichlids, Lake Victoria	300	0%	363 bp cyt b	0	7 of 300 species sampled; Meyer et al. 1990
Cichlids: "mbuna" (group B) + "haplochromines" (group A), Lake Malawi	400	2.2%	363 bp cyt b	0.10	4 of 400 species sampled; divergence between consensus sequences of groups A, B; Meyer et al. 1990
Tropheus, Lake Tanganyika	6	5.1%	402 bp cyt b	0.79	All 6 species sampled; Sturmbauer and Meyer 1992, 1993
Ectodini, Lake Tanganyika	30	9.5%	402 bp cyt b	0.77	12 of 30 species sampled; Sturmbauer and Meyer 1993
Lamprologini, Lake Tanganyika	65	8.0%	402 bp cyt b	0.53	16 of 65 species sampled, Sturmbauer et al. 1994
Tilapiine cichlids, Barombi Mbo, Cameroon	11	2.8%	340 bp cyt b	0.32	All 11 species sampled; Schliewen et al. 1994
Tilapiine cichlids, Lake Bermin, Cameroon	9	0.6%	340 bp cyt b	0.07	All 9 species sampled; only 2 haplotypes found; Schliewen et al. 1994
Cyprinodon (pupfish), Lake Chicancauab, Mexico	5	0.24%	420 bp control region and parts of 2 tRNA genes	not cyt b	All 5 endemics sampled; complete reproduction isolation certain for only 1 species; Strecker et al. 1996
"Haplochromine" cichlids, Lake Victoria	300	1.1%	440 bp control region and parts of 2 tRNA genes	not cyt b	14 of 300 species sampled; Meyer et al. 1990
Tropheus, Lake Tanganyika	6	13.0%	442 bp control region and parts of 2 tRNA genes	not cyt b	Sturmbauer and Meyer 1992, 1993
Ectodini, Lake Tanganyika	30	11.0%	450 bp control region and parts of 2 RNA genes	not cyt b	Sturmbauer and Meyer 1993; Sturmbauer et al. 1994
Lamprologini, Lake Tanganyika	65	20.0%	452 bp control region and parts of 2 tRNA genes	not cyt b	25 species sampled, Sturmbauer et al. 1994
Cichlids: "mbuna" (group B) + "haplochromines" (group A), Lake Malawi	400	4.6%	440 bp control region and parts of 2 tRNA genes	not cyt b	19 of 400 species sampled; divergence between consensus sequences of groups A, B; Meyer et al. 1990

Age of clade for calculation of TFS was estimated using maximum sequence divergence in cytochrome b calibrated by 2.5%/MY.

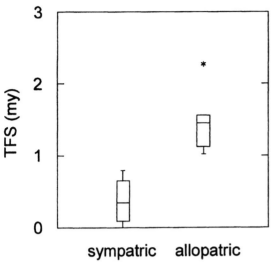

Figure 13.4. Boxplot of average TFS estimated at the clade level for probable allopatrically derived species versus possible sympatrically derived clades. The box shows the interquartile range containing the central 50% of values, with the center horizontal line marking the median. Vertical bars show the range of values falling within 1.5 times the interquartile range, and the asterisk marks a value outside the range of the bars. TFS is a logarithmic estimate, assuming a fully symmetrical pattern of branching (McCune 1997), using maximum percentage sequence divergence of cytochrome b within the clade, calibrated by 2.5%/MY (after Meyer et al. 1990). Divergence associated with possible sympatric speciation is significantly lower (two-sample t-test with separate variances; $p = 0.004$) than divergence associated with presumed allopatric speciation (Systat version 6.0, SPSS Inc.). Data for presumed allopatric clades and possible sympatric clades are given in tables 13.4 and 13.5.

a species concept based on assumed reproductive isolation (i.e., the biological species concept or an approximation thereof). However, some of the swordtail species that we considered had been elevated from subspecific to specific status based on an autapomorphic species concept. Rosen (1979:277) stated in the introduction to his revision of *Xiphophorus*, "I am, thus, compelled to reject both the biological species as a conceptual tool and the subspecies as a methodological one, and this argument constitutes my reason for now recognizing as species forms that were hitherto recognized as subspecies." Our goal here is not to defend the empirical or theoretical validity of a particular species concept. However, for our comparative purposes, it is evident that a relatively uniform standard for species must be used. Excluding Rosen's autapomorphically defined swordtail species, the divergences between sister species pairs range from 2.0% to 7.5% (table 13.1). In contrast, divergences between

autapomorphically defined swordtail species range from only 0.3% to 0.8% (table 13.2), thus falling within the range of intraspecific divergence of 0% to 5.6% for other taxa (table 13.2). The inclusion of these *Xiphophorus* divergences in our interspecific category would have erroneously made the ranges of interspecific and intraspecific divergences identical. This problem highlights the need for vigilance in interpreting the taxonomic literature and the adoption of a consistent species concept, especially for comparative and synthetic studies.

Allopatric Speciation Involves Greater Divergence and More Time Than Sympatric Speciation

The sequence divergence of cytochrome b associated with allopatric speciation appears to be greater than the divergence associated with possible sympatric speciation. This is evident from both sister-species and clade-level comparisons. If divergence can be assumed to provide even a very approximate estimate of age, our results strongly suggest that sympatric speciation is faster than allopatric speciation. This empirical result supports theoretical predictions that sympatric speciation, directly driven by selection, will necessarily be more rapid than indirect effects conferring reproductive isolation evolved in allopatry.

Sister Species Comparisons

To our knowledge, the method of estimating divergence associated with allopatric speciation as the range bracketed by maximum intraspecific and minimum interspecific divergence between species pairs is novel. The combined use of interspecific and intraspecific divergence provides an important advantage over using only interspecific divergence to estimate the time speciation takes. With only interspecific divergence, differences accumulated since speciation cannot be factored out. The incorporation of intraspecific levels of divergence, however, provides a more accurate estimate of the window of divergence inside which reproductive incompatibility (pre- or postmating) is likely to have evolved. In some ways, the species pair method has advantages over the clade-level comparisons: a particular clade topology (i.e., balanced or imbalanced) does not have to be assumed for estimates of TFS, and because divergence is not averaged over an entire clade, sister species comparisons may be more accurate.

However, the species pair method also has difficulties. One problem is that the approach itself prevents statistical assessment of the difference in divergence between sympatrically and allopatrically derived species. The allopatric estimate is based on the intersection of two sets of samples, while the sympatric estimate is based on a single set only (interspecific divergence) and is therefore not strictly comparable. Ideally, intraspecific divergence data for possible sympatrically speciating taxa

Table 13.5. Maximum divergence and average TFS for clades with presumed allopatric origins.

Taxa	# Species	Divergence	Sequence	TFS (MY)	Notes/Reference
Melanotaenia	32	17.9%	351 bp cyt b	1.43	9 of 32 species sampled; Zhu et al. 1994
Osmerus	3	9.0%	300 bp cyt b	2.27	All 3 species sampled; Taylor and Dodson 1994
Gambusia	45	15.2%	402 bp cyt b	1.10	24 of 45 species sampled; Lydeard et al. 1995
Xiphophorus	13	9.3%	360 bp cyt b	1.00	Species number based on reproductive criteria; autapomorphically defined species number 22–24 (Rosen 1979; Rauchenberger et al. 1990); sequence data from Meyer et al. 1994
Rivulus	~70	23.6%	360 bp cyt b	1.54	14 of 70 species sampled; divergence may be low since geographic sampling restricted to north; Murphy and Collier 1996
Xiphophorus	13	9.4%	402 bp control region	not cyt b	Meyer et al. 1994

Age of clade for calculation of TFS estimated using maximum sequence divergence in cytochrome b calibrated by 2.5% /MY.

would be included. However, the few studies on within-species variation in African lake cichlids (Bowers et al. 1994; Sturmbauer and Meyer 1992; Meyer et al. 1996) have indicated that levels of intraspecific divergence may be as high as, or higher than, interspecific divergence, and thus do not provide a lower bound for divergence associated with speciation as they do for the allopatric category.

A related consideration is the potential effect of incomplete lineage sorting. In this analysis, most interspecific comparisons were based on only one or two individuals per species. However, if more individuals were sampled, and if speciation was recent, we might sometimes see that species are not monophyletic according to allele phylogenies (Tajima 1983; Neigel and Avise 1986). For example, McMillan and Palumbi (1995) observed that cytochrome b sequences from several Indo-West Pacific butterflyfishes did not form monophyletic groups that corresponded to species boundaries. In some haplochromine cichlid clades, incomplete lineage sorting appears to be epidemic (Moran and Kornfield 1993; Parker and Kornfield 1997). In these situations, the interpretation of interspecific and intraspecific divergence is difficult. We would like to be able to interpret intraspecific divergence as having arisen within a species, but if the variation predates the origin of the species, our estimates will be incorrect. It is more feasible to consider these cases using the clade level method.

Also problematic is the potential effect of hybridization, which could allow the transfer of mitochondrial DNA lineages between different species. Sister species that share mitochondrial haplotypes due to hybridization would falsely appear to have diverged extremely recently. This issue is of particular concern because hybridization may be most likely among sympatric species; if hybridization is common in lake endemics, our sympatric species pairs may be systematically biased in favor of their having low levels of divergence. However, field studies on the cichlid species flocks suggest that hybridization is very rare in nature (e.g., Greenwood 1974; Ribbink et al. 1983; but see Clarke et al. 1996 on the low levels of hybridization required to cause genetic homogenization between species). Also, as discussed below, hybridization is less likely to be a problem for the clade-level comparisons.

Finally, of particular concern in an era of molecular systematics is the need for completely sampled species-level phylogenies; poor sampling of taxa was a serious limitation on data available for this study. Generally, sequence data were collected for only a handful of taxa in large clades, giving no reason to believe that what appeared to be sister species in a published phylogeny actually were, because so many species were omitted from the analysis (figure 13.2). Such studies could only be used for clade-level analyses, and even for these, maximum divergence may be underestimated due to inadequate sampling. This situation is common among molecular systematic studies and acute in the species flock literature, in which sometimes as few as 1% of the known species in a clade were sampled (table 13.4).

Clade-Level Comparisons

Comparisons of TFS calculated from clades show a strong statistical difference between the allopatric and sympatric categories, thus reinforcing our finding from sister-species comparisons. Furthermore, the clade-level calculations are relatively immune to some of the problems affecting the sister-species comparisons. Incomplete lineage sorting, for example, is unlikely to have effects at this level—small amounts of ancestral polymorphism initially present within the ancestor of a clade are likely to be swamped by the larger amounts of divergence that subsequently evolve within the clade itself. This is particularly true for the slow allopatrically speciating clades. In the case of some cichlid clades from the sympatric category, retained ancestral polymorphisms may represent most of the observed divergence (Parker and Kornfield 1997). However, this would mean that actual divergence and the corresponding TFS is lower for these taxa, making the difference between allopatric and sympatric categories even more significant. Also, clade-level comparisons should avoid the potential difficulties caused by hybridizing taxa. Unless hybridization is rampant among most members of species flocks, sampling multiple species should provide a good estimate of the divergence present in the clade.

Unfortunately, the taxonomic sample for clade comparisons was not as broad as one would like. All putative sympatrically speciating taxa in our clade-level comparisons are African cichlids, and cichlids are absent from our allopatric speciation group. We are therefore unable to rule out the possibility that cichlid cytochrome b genes evolve extremely slowly. This seems to be an unlikely alternative, however, as divergences for African tilapiine cichlids (not members of species flocks) appear to show levels of divergence comparable to other allopatrically speciating clades (Schliewen et al. 1994). The future addition of cichlids to our allopatric group and of noncichlid species flocks, such as cyprinodontids from Lake Titicaca and Baikal sculpins, to the sympatric group will help to determine whether low average levels of divergence are unique to African lacustrine cichlids, or whether other species flocks show the same pattern. For the moment, the low level of divergence of cytochrome b between sister species of Baikal sculpins (table 13.3), and of control region in Chichancanab cyprinodontids (table 13.4) together with paleolimnological evidence for rapid speciation (six species in <8,000 years) in semionotid fishes (McCune 1996, 1997) suggest that rapid speciation is not merely a cichlid phenomenon.

Allopatric Speciation Takes Too Long for It to Account for Species Flocks of African Cichlids

One unanticipated result of this study is that it suggests that allopatric speciation alone can not account for speciation by cichlids in Lake Victoria and Lake Malawi. Divergences between species pairs with probable allopat-

ric origins range from 2.0% to 5.7% or, if calibrated by 2.5% sequence divergence/MY, an estimated 0.8–2.3 MY per speciation event. Using clade-level data, again calibrated by 2.5%/MY, the estimated time course of allopatric speciation increases slightly, to 0.88–2.3 MY per speciation event. However, Lake Malawi is, at most, an estimated 1–2 MY old (Fryer and Iles 1972; McCune 1997), which is enough time for allopatric speciation to occur only one to three times, producing a maximum of eight species. Assuming a fully branched, symmetrical phylogeny, allopatric evolution of the Malawi cichlid clade of 400 species would have required enough time for eight or nine speciation events, or a minimum of 7.2 MY. Allopatric evolution of the monophyletic clade of 300 cichlids endemic to Lake Victoria should have taken about the same amount of time, and yet, Lake Victoria has recently been shown to be only 12,400 years old (Johnson et al. 1996), which allows about 1,500 years per speciation event. Various models, including fragmenting populations due to fluctuating lake levels, or patchy habitats coupled with strong habitat selection, have been proposed to explain the evolution of cichlids in a manner consistent with allopatric speciation (Fryer and Iles 1972; Echelle and Kornfield 1984; Martens et al. 1994), but the fact is that even with such barriers, there has not been enough time for allopatric speciation, unassisted by other factors (e.g., secondary reinforcement in sympatry [see Coyne and Orr 1989], selection on mate recognition, etc.), to have produced the present diversity of cichlid fishes.

Summary

An earlier study (McCune, 1997) suggested that species flocks of fishes have speciated more rapidly than many other groups of vertebrates and invertebrates. To test whether rapid speciation is a property of fish in general or only of species flocks, we attempted to determine whether the time necessary for speciation (TFS) in fishes depended on the mode of species origination. Fishes were classed as having a (1) probable allopatric origin or (2) possible sympatric origin (lacustrine species flocks). We estimated the approximate amount of cytochrome b sequence divergence associated with allopatric speciation as the intersection of divergence between sister species pairs and divergence within species. For the sympatric category, only divergence between sister species was used. Using this method, the putative sympatrically speciating fishes show lower levels of divergence and, we therefore assume, faster speciation than the allopatric set. A similar result obtained from clade-level comparisons: clades of putative sympatrically speciating fishes show a lower average sequence divergence and correspondingly lower TFS values than clades with probable allopatric origins. These findings agree with theoretical predictions that sympatric speciation, when it occurs, should be faster than allopatric speciation (Bush 1975).

As more thoroughly sampled and taxonomically diverse molecular studies become available, we should be able to strengthen our tests of the effect of mode on rate of speciation.

Acknowledgments   We thank R. Harrison, A. Kondrashov, and D. Winkler for interesting discussions. S. Berlocher, J. Feder, and D. Winkler provided many helpful and provocative comments on the manuscript. E. Gournis prepared one of the figures. This study was funded in part by the National Science Foundation (BSR 8707500) and Hatch Project 421 to ARM.

References

Avise, J. C. 1994. Molecular Markers, Natural History and Evolution. New York: Chapman and Hall.

Barton, N. H. 1996. Natural selection and random genetic drift as causes of evolution on islands. Phil. Trans. R. Soc. Lond. B 351:785–795.

Bernardi, G., and Powers, D. A. 1995. Phylogenetic relationships among nine species from the genus *Fundulus* (Cyprinodontiformes, Fundulidae) inferred from sequences of the cytochrome *b* gold. Copeia 1995:469–473.

Bernatchez, L., Guyomard, R., and Bonhomme, F. 1992. DNA sequence variation of the mitochondrial control region among geographically and morphologically remote European brown trout *Salmo trutta* populations. Mol. Ecol. 1:161–173.

Birt, T. P., Friesen, V. L., Birt, R. D., Green, J. M., and Davidson, W. S. 1995. Mitochondrial DNA variation in Atlantic capelin, *Mallotus villosus:* a comparison of restriction and sequence analyses. Mol. Ecol. 4:771–776.

Bowers, N., Stauffer, J. R., and Kocher, T. D. 1994. Intra- and interspecific mitochondrial DNA sequence variation within two species of rock-dwelling cichlids (Teleostei: Cichlidae) from Lake Malawi, Africa. Mol. Phylogen. Evol. 3:75–82.

Brooks, J. L. 1950. Speciation in ancient lakes. Q. Rev. Biol. 25:30–176.

Brown, J. R., Beckenbach, A. T., and Smith, M. J. 1993. Intraspecific DNA sequence variation of the mitochondrial control region of white sturgeon (*Acipenser transmontanus*). Mol. Biol. and Evol. 10:326–341.

Bush, G. L. 1975. Modes of animal speciation. Annu. Rev. Ecol. Syst. 6:339–365.

Bush, G. L. 1993. A reaffirmation of Santa Rosalia, or why are there so many kinds of *small* animals? In D. R. Lees and D. Edwards (eds.). Evolutionary Patterns and Processes. New York: Academic Press, pp. 228–249.

Carson, H. L. 1975. The genetics of speciation at the diploid level. Am. Nat. 109:83–92.

Cashner, R. C., Rogers, J. S., and Grady, J. M. 1992. Phylogenetic studies of the genus *Fundulus*. In R. L. Mayden (ed.). Systematics, Historical Ecology and North American Freshwater Fishes. Stanford: Stanford University Press, pp. 421–437.

Clarke, B., Johnson, M. S., and Murray, J. 1996. Clines in the genetic distance between two species of island land snails: how 'molecular leakage' can mislead us about speciation. Phil. Trans. R. Soc. Lond. B 351:773–784.

Coyne, J. A., and Orr, H. A. 1989. Pattern of speciation in *Drosophila*. Evolution 43:362–381.

Dominey, W. 1984. Effects of sexual selection and life history on speciation: species flocks in African cichlids and Hawaiian drosophila. In A. A. Echelle and I. Kornfield (eds.). Evolution of Fish Species Flocks. Orono: University of Maine Press, pp. 231–249.

Echelle, A. A., and Kornfield, I., eds. 1984. Evolution of Fish Species Flocks. Orono: University of Maine Press.

Endler, J. 1977. Geographic Variation, Speciation and Clines. Princeton, N.J.: Princeton University Press.

Fajen, A., and Breden, F. 1992. Mitochondrial DNA sequence variation among natural populations of the Trinidad guppy *Poecilia reticulata*. Evolution 46:1457–1465.

Fink, W. L. 1971. A revision of the *Gambusia nicaraguensis* complex (Pisces: Poeciliidae). Publ. Gulf Coast Res. Lab. Mus. 2:47–77.

Fryer, G., and Iles, T. D. 1972. The Cichlid Fishes of the Great Lakes of Africa. Neptune City, N.J.: TFH.

Futuyma, D. 1986. Evolutionary Biology (2nd ed.). Sunderland, Mass.: Sinauer.

Giuffra, E., Bernatchez, L., and Guyomard, R. 1994. Mitochondrial control region and protein coding genes sequence variation among phenotypic forms of brown trout *Salmo trutta* from northern Italy. Mol. Ecol. 3:161–171.

Greenwood, P. H. 1974. Cichlid fishes of Lake Victoria, East Africa: the biology and evolution of a species flock. Bull. B. Mus. (Nat. Hist.) Zool. Suppl. 6.

Hubbs, C. L. 1941. The relation of hydrological conditions to speciation in fishes. In A symposium on Hydrobiology. Madison: University of Wisconsin Press, pp. 182–195.

Hubbs, C. L. 1943. Criteria for subspecies, species and genera, as determined by researches on fishes. Ann. N.Y. Acad. Sci. 44:109–121.

Humphries, J. M. 1984. Genetics of speciation in pupfishes from Laguna Chichancanab, Mexico. In A. A. Echelle and I. Kornfield (eds.). Evolution of Fish Species Flocks. Orono: University of Maine Press, pp. 129–140.

Irwin, D. M., Kocher, T. D., and Wilson, A. C. 1991. Evolution of the cytochrome *b* gene of mammals. J. Mol. Evol. 32:128–144.

Johnson, T. C., Scholz, C. A., Talbot, M. R., Kelts, K., Ricketts, R. D., Ngobi, G., Beuning, K., Ssemmanda, I., and McGill, J. W. 1996. Late Pleistocene dessiccation of Lake Victoria and rapid evolution of cichlid fishes. Science 273:1091–1093.

Kondrashov, A., and Mina, M. V. 1986. Sympatric speciation: when is it possible? Biol. J. Linn. Soc. 7:201–233.

Kornfield, I., and Carpenter, K. E. 1984. Cyprinids of Lake Lanao, Philippines: taxonomic validity, evolutionary rates and speciation scenarios. In A. A. Echelle and I. Kornfield (eds.). Evolution of Fish Species Flocks. Orono: University of Maine Press, pp. 69–84.

Kosswig, C. 1947. Selective mating as a factor for speciation in cichlid fish of East African lakes. Nature 159:604–605.

Kosswig, C. 1963. Ways of speciation in fishes. Copeia 1963: 238–244.

Lee, D. S., Gilbert, C. R., Hocutt, C. H., Jenkins, R. E., McAllister, D. E., and Stauffer, J. R., Jr. 1980. Atlas of North American Freshwater Fishes. Raleigh: North Carolina State Museum of Natural History.

Li, W. H., and D. Graur. 1991. Fundamentals of molecular evolution. Sunderland, Mass.: Sinauer.

Lydeard, C., Wooten, M. C., and Meyer, A. 1995. Cytochrome *b* sequence variation and a molecular phylogeny of the livebearing fish genus *Gambusia* (Cyprinodontiformes: Poeciliidae). Can. J. Zool. 73:213–227.

Lynch, J. D. 1989. The gauge of speciation: on the frequencies of modes of speciation. In D. Otte and J. A. Ender (eds.). Speciation and Its Consequences. Sunderland, Mass.: Sinauer, pp. 527–553.

Martens, K., Goddeeris, B., and Coulter, G., eds. 1994. Advances in Limnology, Vol. 44, Speciation in Ancient Lakes.

Mayr, E. 1947. Ecological factors in speciation. Evolution 1:263–288.

Mayr, E. 1954. Change in genetic environment and evolution. In J. Huxley, A. C. Hardy, and E. B. Ford (eds.). Evolution as a Process. New York: Macmillan, pp. 157–180.

Mayr, E. 1963. Animal Species and Evolution. Cambridge, Mass.: Harvard University Press.

McCune, A. R. 1987. Lakes as laboratories of evolution: Endemic fishes and environmental cyclicity. Palaois 2:446–454.

McCune, A. R. 1996. Biogeographic and stratigraphic evidence for rapid speciation in semionotid fishes. Paleobiology 22:34–48.

McCune, A. R. 1997. How fast do fishes speciate? Molecular, geological, and phylogenetic evidence from adaptive radiations of fishes. In T. J. Givnish and K. J. Sytsma (eds.). Molecular Evolution and Adaptive Radiation. Cambridge: Cambridge University Press, pp. 585–610.

McCune, A. R., Thomson, K. S., and Olsen, P. E. 1984. Semionotid fishes from the Mesozoic great lakes of North America. In A. Echelle and I. Kornfield (eds.). Evolution of Fish Species and Flocks. Orono: University Maine Press, pp. 27–44.

McMillan, W. O., and Palumbi, S. R. 1995. Concordant evolutionary patterns among Indo-West Pacific butterflyfishes. Proc. R. Soc. Lond. B Biol. Sci. 260:229–236.

McVeigh, H. P., Bartlett, S. E., and Davidson, W. D. 1991. Polymerase chain reaction—direct sequence analysis of the cytochrome *b* gene in *Salmo-Salar*. Aquaculture 95: 225–234.

Meyer, A., Kocher, T. D., Basasibwaki, P., and Wilson, A. C. 1990. Monophyletic origin of Lake Victoria cichlid fishes suggested by mitochondrial DNA sequences. Nature 347: 550–553.

Meyer, A., Morrissey, J. M., and Schartl, M. 1994. Recurrent origin of a sexually selected trait in *Xiphophorus* fishes inferred from a molecular phylogeny. Nature 368:539–542.

Meyer, A., Knowles, L. L., and Verheyen, E. 1996. Widespread geographical distribution of mitochondrial haplotypes in rock-dwelling cichlid fishes from Lake Tanganyika. Mol. Ecol. 5:341–350.

Miller, R. R. 1975. Five new species of Mexican poeciliid fishes of the genera *Poecilia*, *Gambusia*, and *Poeciliopsis*. Occas. Pap. Mus. Zool. Univ. Mich. 672:1–44.

Moran, P., and Kornfield, I. 1993. Retention of an ancestral polymorphism in the mbuna species flock (Pisces: Cichlidae) of Lake Malawi. Mol. Biol. Evol. 12:1085–1093.

Moya, A., Galiana, A., and Ayala, F. J. 1995. Founder-effect speciation theory: failure of experimental corroboration. Proc. Nat. Acad. Sci. USA 92:3893–3986.

Murphy, W. J., and Collier, G. E. 1996. Phylogenetic relationships within the aplocheiloid fish genus *Rivulus* (Cyprinodontiformes, Rivulidae): implications for Caribbean and Central American biogeography. Mol. Biol. Evol. 13:642–649.

Nagelkerke, L. A. J., Sibbing, F. A., Van-Den-Boogaart, J. G. M., Lammens, E. H., and Osse, J. W. M. 1994. The barbs (*Barbus* spp.) of Lake Tana: a forgotten species flock? Environ. Biol. Fishes 39:1–22.

Nagelkerke, L. A. J., Sibbing, F. A., and Osse, J. W. M. 1995. Morphological divergence during growth in the large barbs (*Barbus* spp.) of Lake Tana, Ethiopia. Neth. J. Zool. 45:431–454.

Neigel, J. E., and Avise, J. C. 1986. Phylogenetic relationsips of mitochondrial DNA under various demographic models of speciation. In E. Nevo and S. Karlin (eds.). Evolutionary Processes and Theory. New York: Academic Press, pp. 515–534.

Nelson, J. T. 1984. Fishes of the World (2nd ed.). New York: Wiley.

Orti, G., Bell, M. A., Reimchen, T. E., and Meyer, A. 1994. Global survey of mitochondrial DNA sequences in the threespine stickleback: evidence for recent migrations. Evolution 48:608–622.

Pálsson, S., and Árnason, E. 1994. Sequence variation for cytochrome *b* genes of three salmonid species from Iceland. Aquaculture 128:29–39.

Parenti, L. R. 1984. Biogeography of the Andean killifish genus *Orestias* with comments on the species flock concept. In A. A. Echelle and I. Kornfield (eds.). Evolution of Fish Species and Flocks. Orono: University of Maine Press, pp. 85–92.

Parker, A., and Kornfield, I. 1995. A molecular perspective on evaluation and zoogeography of cyprinodontid killifishes. Copeia 1995:8–21.

Parker, A., and Kornfield, I. 1997. Evolution of the mitochondrial DNA control region of the *mbuna* (Cichlidae) species flock of Lake Malawi, East Africa. J. Mol. Evol. 45:70–83.

Patarnello, T., Bargelloni, L., Caldara, F., and Colombo, L. 1994. Cytochrome *b* and 16S rRNA sequence variation in the *Salmo trutta* (Salmonidae, Teleostei) species complex. Mol. Phylogen. Evol. 3:69–74.

Rauchenberger, M. 1989. Systematics and biogeography of the genus *Gambusia* (Cyprinodontiformes: Poeciliidae) Am. Mus. Novit. 2951:1–74.

Rauchenberger, M., Kallvian, K. D., and Morizot, D. C. 1990. Morphology and geography of the R'o Pánuco basin swordtails (genus *Xiphophorus*) with descriptions of four new species. Am. Mus. Novit. 2975:1–41.

Ribbink, A. J. 1994. Alternative perspectives on some controversial aspects of cichlid fish speciation. Adv. Limnol. 44:101–125.

Ribbink, A. J., Marsh, B. A., Marsh, A. C., Ribbink, A. C., and Sharp, B. J. 1983. A preliminary survey of the cichlid fishes of rocky habitats in Lake Malawi. S. Afr. J. Zool. 18:149–310.

Rosen, D. E. 1960. Middle American poeciliid fishes of the genus *Xiphophorus*. Bull. Fl. State Mus. Biol. Sci. 5:57–242.

Rosen, D. E. 1979. Fishes of the uplands and intermontane basins of Guatemala: revisionary studies and comparative geography. Bull. Am. Mus. Nat. Hist. 162:267–376.

Schliewen, U. K., Tautz, D., and Paabo, S. 1994. Sympatric speciation suggested by monophyly of crater lake cichlids. Nature 368:629–632.

Schmidt, T. R., and Gold, J. R. 1995. Systematic affinities of *Notropis topeka* (Topeka Shiner) inferred from sequences of the cytochrome *b* gene. Copeia 1995:199–204.

Slatkin, M. 1996. In defense of founder-flush theories of speciation. Am. Nat. 147:493–505.

Slobodyanyuk, S. Ja., Kirilchik, S. V., Pavlova, M. E., Belikov, S. I., and Novitsky, A. L. 1995. The evolutionary relationships of two families of cottoid fishes of Lake Baikal (East Siberia) as suggested by analysis of mitochondrial DNA. J. Mol. Evol. 40:392–399

Stanley, S. M. 1979. Macroevolution: pattern and process. San Francisco: Freeman.

Strecker, U., Meyer, C. G., Sturmbauer, C., and Wilkens, H. 1996. Genetic divergence and speciation in an extremely young species flock in Mexico formed by the genus *Cyprinodon* (Cyprinodontidae, Teleostei). Molecular Phylogenetics and Evolution 6:143–149.

Sturmbauer, C., and Meyer, A. 1992. Genetic divergence, speciation and morphological stasis in a lineage of African cichlid fishes. Nature 358:578–581.

Sturmbauer, C., and Meyer, A. 1993. Mitochondrial phylogeny of the endemic mouthbrooding lineages of cichlid fishes from Lake Tanganyika in Eastern Africa. Mol. Biol. Evol. 10:751–768.

Sturmbauer, C., Verheyen, E., and Meyer, A. 1994. Mitochondrial phylogeny of the Lamprologini, the major substrate spawning lineage of cichlid fishes from Lake Tanganyika in Eastern Africa. Mol. Biol. Evol. 11:691–703.

Tajima, F. 1983. Evolutionary relationships of DNA sequences in finite populations. Genetics 105:437–460.

Taylor, E. B., and Dodson, J. J. 1994. A molecular analysis of relationships and biogeography within a species complex of Holarctic fish (genus *Osmerus*). Mol. Ecol. 3:235–248.

Templeton, A. 1980. The theory of speciation via the founder principle. Genetics 94:1011–1038.

Wiley, E. O., and Mayden, R. L. 1985. Species and speciation in phylogenetic systematics, with examples from the North American fish fauna. Ann. Mo. Bot. Gard. 72:596–635.

Worthington, E. B. 1954. Speciation of fishes in African lakes. Nature 173:1064–1067.

Zhu, D., Jamieson, B. G. M., Hugall, A., and Moritz, C. 1994. Sequence evolution and phylogenetic signal in control-region and cytochrome *b* sequences of rainbow fishes (Melanotaeniidae). Mol. Biol. Evol. 11:672–683.

14

The First Stage of Speciation as Seen in Organisms Separated by the Isthmus of Panama

H. A. Lessios

The study of speciation, like the study of many other dynamic historical processes, consists of assessing the status of fixed stages and placing them in a chronological series, so that a trajectory can be reconstructed. Mayr (1954a) has pointed out that in this respect the methodology of evolutionary biology is not unlike that of cytology or any other process-oriented branch of science. In this idealized view, populations at various stages of speciation are analogous to results of natural experiments allowed to run for different lengths of time. In practice, however, the study of speciation is more similar to an experiment in which most traces of materials and methods have been lost and need to be reconstructed from the results themselves. Thus, the nature, efficacy, and timing of an extrinsic barrier thought to have resulted in geographic speciation are rarely self-evident and must be deduced from present-day distributions and patterns of divergence. Such reconstructions by necessity have to rest on many assumptions. It is therefore not surprising that critics of Mayr's insistence on the primacy of geographic isolation (e.g., Bush 1975, 1994; Bush and Howard 1986; White 1978; Tauber and Tauber 1989; King 1993) often complain that many speciation studies shoehorn their interpretation of the mode of speciation into a preconceived geographic model.

One setting in which the barrier that split populations can be identified with confidence consists of the tropical marine species on the two sides of Central America. Geological (Woodring 1966; Emiliani et al. 1972; Saito 1976; Keigwin 1978, 1982; Jones and Hasson 1985; Keller et al. 1989; Duque-Caro 1990; Coates et al. 1992; Coates and Obando 1996) as well as biogeographic (Jordan 1908; Ekman 1953; Briggs 1974; Vermeij 1978) evidence makes it clear that the tropical Atlantic and Pacific oceans were connected until the Pliocene. As the Central American land barrier emerged, it removed any potential for gene exchange between populations of tropical marine organisms on either side and allowed for the independent evolution in allopatry considered necessary

for geographic speciation. Successive sedimentological (Duque-Caro 1990; Coates et al. 1992) and paleontological studies of both marine (Woodring 1966; Saito 1976; Keigwin 1978, 1982; Jones and Hasson 1985; Coates et al. 1992) and terrestrial (Marshall et al. 1979, 1982; Marshall 1985) organisms have narrowed the estimates of the timing of the final isthmus completion to 3.0–3.5 million years (but see Crouch and Poag [1979] and Keller et al. [1989] for estimates of more recent closure). Thus, this particular natural experiment is one in which materials and methods have not been completely lost, or rather one in which they can be reconstructed from independent evidence. It should therefore have a great deal to tell us about the allopatric stage of geographic speciation. The isolates on the two sides of the isthmus, known as "geminates"—a term coined by Jordan (1908) and equally useful whether or not Atlantic and Pacific populations are recognized as separate species—should be helpful in shedding light on the factors that affect rates of divergence and the emergence of reproductive isolation, information that is difficult to gather in other marine or terrestrial organisms.

In this chapter I first give a general overview of the advantages of studying speciation on the two sides of the Isthmus of Panama, a brief historical account of the studies that have used the isthmus as a backdrop for such studies, and a list of the possible pitfalls that can cause misinterpretation of the results. I then discuss the existing data on divergence and reproductive isolation between marine populations from the Caribbean and the eastern Pacific and their relevance to understanding allopatric speciation.

Why Study Speciation on the Two Coasts of Central America?

As pointed out above, the very fact that an identified and fairly well-dated barrier to gene flow exists to the present day is sufficient motive for conducting studies of geographic speciation on the two coasts of tropical America. As Vermeij (1993) has put it, "tropical America will continue to be perhaps the finest laboratory in which to answer the big questions about what controls biological diversity" (p. 1604). However, there are additional reasons that bring this particular setting closer to the ideal for providing information on the first stage of geographic speciation.

(1) The time scale is right. The elapsed time since the final completion of the isthmus is not only well defined geologically but also sufficient for the accumulation of measurable interpopulational differentiation. Yet the vicariant event is not so ancient as to lead to the emergence of higher taxa and major uncertainties about the relationships of the species in question.

(2) There is no question about the efficacy of the barrier. Unlike most obstacles to migration that restrict, but do not completely sever, gene flow, the presence of a strip of land, interrupted only by the recently completed (1914) freshwater Panama Canal, ensures that nearly all contemporary populations do not exchange propagules.

(3) Populations of organisms belonging to many phyla were separated by the same barrier. The presumably simultaneous separation of populations from many taxa, with different life histories, population densities, generation times, fertilization systems, and dispersal abilities, provides ample opportunities for comparisons that could help tease apart the contribution of each of these factors to the probability of differentiation and speciation.

(4) There are well-defined differences in the environments on the two sides of the isthmus. Neotropical Atlantic and Pacific environments differ both in physical (Glynn 1972) and biotic (Glynn 1982) parameters. Most of these dissimilarities owe their existence to the presence of the isthmus itself, and have thus been in place for the last 3 million years (Cronin 1985; Jackson et al. 1993; Cronin and Dowsett 1996). Thus, comparisons between sister species can provide information on how adaptation to specific environmental differences can affect genetic divergence and speciation.

A Brief History of Studies Conducted with Geminate Species

For more than a century, evolutionary biologists have recognized the advantages of the isthmus for evolutionary studies and have employed the geminate species of Panama as examples of vicariant separation and geographic speciation (table 14.1). Günther (1868) was the first to note the faunal similarities of fishes on the two sides of the Isthmus of Panama. Based on this evidence, he proposed a prior connection between the Caribbean and the eastern Pacific. Following this study, the initial work in each taxonomic group consisted of recognizing and enumerating sister species in the two oceans (Jordan

1885; Meek and Hildebrand 1923–1928; Mortensen 1928–1950; de Laubenfels 1936; Hedgpeth 1948; Bayer et al. 1970; Chesher 1972; Vermeij 1978; Thomson et al. 1979; Voight 1988). Then came generalizations regarding factors responsible for divergence and speciation through the examination of distributions of geminate species among taxonomic lines (Rosenblatt 1963, 1967) or ecological habitats (Ekman 1953; Rosenblatt 1963; Vermeij 1978). Quantitative estimates of morphological divergence started early (Rubinoff 1963) but—considering the ease with which they can be conducted—are still limited in number (Lessios 1981a; Weinberg and Starczak 1988, 1989; Lessios and Weinberg 1994). Estimates of genetic divergence through protein (Gorman et al. 1976; Gorman and Kim 1977; Lessios 1979a,b, 1981a; Vawter et al. 1980; West 1980; Laguna 1987; Bermingham and Lessios 1993; Knowlton et al. 1993; Lessios and Weinberg 1994) or mitochondrial DNA (mtDNA) comparisons (Collins 1989; Bermingham and Lessios 1993; Knowlton et al. 1993) appeared as these techniques became available. With the exception of the pioneering work of Rubinoff and Rubinoff (1971), studies of the emergence of reproductive isolation between the geminates (Lessios 1984; Lessios and Cunningham 1990; Knowlton et al. 1993; Lessios and Weinberg 1993) were late in coming—despite their importance for understanding speciation—probably because they required the permanent establishment of evolutionary biologists in the tropics. Studies of adaptation of each member of a geminate pair to its respective environment (Graham 1971; Lessios 1979a, 1981b; Graves et al. 1983; Lessios 1990), though not intended to address questions of speciation per se, provided information on some of the factors responsible for divergence in particular traits. The possibility of a saltwater sea-level canal, taken seriously in the late 1960s and early 1970s, was the catalyst of many speculative articles regarding the biological effects of renewed faunal exchange between the oceans (Menzies 1968; Briggs 1969; Topp 1969; Rubinoff 1968, 1970; Glynn 1982), but of little original research. Though not directly addressing speciation issues, several works (e.g., Bowen et al. 1991; Martin et al. 1992; Cunningham and Collins 1994; Lessios et al. 1995; Shulman and Bermingham 1995; Collins 1996) have taken advantage of the well-defined times of splitting between geminate species to calibrate rates of molecular divergence.

Potential Problems

Although the geminate species hold many advantages as model organisms for the study of the first stage of speciation, the interpretation of observed patterns still encounters problems. As with any other reconstruction of evolutionary processes, one has to consider various possible pathways through which the present-day patterns of distribution and divergence have come to be. The in-

Table 14.1. Studies of genetic divergence and speciation using the setting of the Central American isthmus.

Reference	Emphasis of Study	Study Organisms
Günther (1868)	Biogeography	Fishes
Jordan (1908)	Mode of speciation	Fishes
Mayr (1954a)	Mode of speciation	Sea urchins
Rosenblatt (1963)	Rates of divergence	Fishes
Rubinoff (1963)	Morphological divergence	Fishes
Rubinoff and Rubinoff (1971)	Reproductive isolation	Fishes
Gorman et al. (1976)	Molecular (protein) divergence	Fishes
Gorman and Kim (1977)	Molecular (protein) divergence	Fishes
Vermeij (1978)	Morphological adaptation	Gastropods
Lessios (1979a)	Molecular, morphological, and ecological divergence	Sea urchins
Lessios (1979b)	Molecular (protein) divergence	Sea urchins
Vawter et al. (1980)	Molecular (protein) divergence	Fishes
West (1980)	Molecular (protein) divergence	Crabs
Lessios (1981a)	Molecular (protein) and morphological divergence	Sea urchins
Lessios (1984)	Temporal reproductive isolation	Sea urchins
Cronin (1985)	Rates of speciation	Ostracodes
Collins (1989)	Molecular (mtDNA) divergence	Gastropods
Weinberg and Starczak (1989)	Morphological divergence	Isopods
Lessios and Cunningham (1990)	Gametic reproductive isolation	Sea urchins
Bermingham and Lessios (1993)	Molecular (protein, mtDNA) divergence	Sea urchins
Knowlton et al. (1993)	Molecular (protein, mtDNA) and behavioral divergence	Alpheid shrimp
Lessios and Weinberg (1993)	Reproductive isolation	Isopods
Lessios and Weinberg (1994)	Genetic (protein) and morphological divergence	Isopods

formation we have is pairs of species, apparently closely related, with each member of the pair on one side of the isthmus. The most parsimonious explanation is that they were created by the splitting of an ancestral range by the isthmus (I will call this the "true geminate" model), but other possibilities exist. There is no doubt that the final closure of the Isthmus of Panama separated many species at roughly the same time, but any given pair of alleged geminate species may not represent a "typical" case. The potential problems of incorrectly assuming that present-day patterns have resulted from the true geminate model are obvious and have already been considered on an ad hoc basis in the literature (e.g., Gorman et al. 1976; Lessios 1979b, 1981a; Vawter et al. 1980; Selander 1982; Weinberg and Starczak 1989; Lessios and Cunningham 1990; Knowlton et al. 1993; Lessios and Weinberg 1994; Collins 1996). However, it may be advantageous to mention them all in one place and to examine the degree to which they can affect conclusions.

The first problem is one of identifying which of many potential candidates in each ocean are geminate species. This requires evidence supporting not just their status as sister species, but also their splitting through the erection of the Central American isthmus. Traditionally both the phylogenetic relationships and the assumption about the timing of the split have rested on morphology (e.g., Meek and Hildebrand 1923–1928; Mortensen 1928–1950; Hedgpeth 1948; Rubinoff 1963; Chesher 1972; Vermeij 1978; Thomson et al. 1979). However, lacking a "mor-

phological clock," we have no means of estimating the amount of morphological change expected to occur in 3 million years. Thus, lists of geminate pairs prepared by various authors (e.g. Jordan 1908; Mayr 1954a; Bayer et al. 1970; Chesher 1972; Thomson et al. 1979) do not always include the same species. In groups such as sea urchins, in which most of the genera have only one species on each side of the isthmus (Chesher 1972), the potential for confusion of true sister species relationships (figure 14.1A) is minimized. The danger of inclusion of a species into the wrong pair increases as the gaps between lineages leading to the supposed geminates become smaller. Thus, in genera such as the fish *Abudefduf* or *Kyphosus*, with at least two species on each coast, and in the much more speciose ones such as the shrimp genus *Alpheus*, with a minimum of 23 species in the eastern Pacific (Kim and Abele 1988) and 22 in the Caribbean (Chace 1972), or the gastropod genus *Conus*, with 22 proposed geminate pairs (Vermeij 1978), the probability of error in presumed sister pair affiliation is considerably higher.

The second possible source of error involves the assumption of simultaneous splitting in all pairs. Species on either side of the isthmus may have been separated by the same barrier, but the final interruption of gene flow may not have occurred at the same time (figure 14.1B). As Ekman (1953) concluded on biogeographic considerations alone, deep water species were probably split much earlier than shallow water ones. The shoaling of

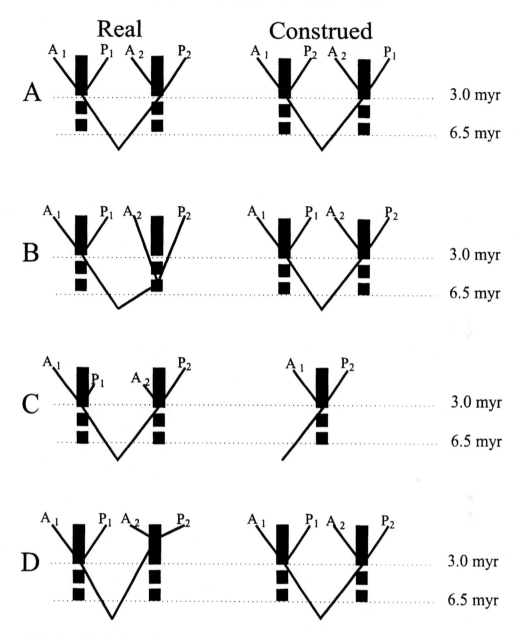

Figure 14.1. Possible mistakes in interpretation arising from making wrong assumptions regarding phylogeny and times of splitting of organisms that appear, on the basis of present-day distribution and divergence, to be geminate species. Cladograms on the left depict real phylogenies; cladograms on the right, phylogenies as they may be mis-construed. Letters with the same subscript indicate true members of geminate pairs. Vertical bar indicates the Central American isthmus. A, species in the Atlantic; P, species in the Pacific.

the isthmus is thought to have proceeded for a period longer than the length of time between the final closure and the present (for reviews, see Jones and Hasson 1985; Coates and Obando 1996). In the Late Miocene (6–7 mya) the channels connecting the Caribbean to the Pacific were 150 meters deep. The restrictions to circulation presum-ably increased with time, finally becoming complete in the late Pliocene, approximately 3 mya (Holcombe and Moore 1977; Coates and Obando 1996). Thus, the pos-sibility of different times of final interruption of gene flow, because of differences in mode of dispersal, habi-tat preferences, physiological tolerances, and vagility of adults and larvae, is one that needs to be considered, even for shallow water species.

Erroneous times of separation may also arise from replacement of a species in one of the two oceans by another species invading the region from somewhere else. This is a more likely possibility for the eastern Pacific, where many species of clear Indo-Pacific origin are found (Rosenblatt et al. 1972; Glynn and Wellington 1983; Vermeij 1991; Lessios et al. 1996). Indeed, extinctions and incomplete sampling of species can cause many errors in presumed phylogenies. The most extreme case would occur if members of two different geminate pairs in opposite oceans became extinct (or were not sampled), and if the clades leading to each pair were not so dissimilar as to preclude assumptions of geminate relationships between the remaining species (figure 14.1C). Then both present-day distributions and divergence would lead one to think that species that had speciated before the rise of the isthmus for other reasons and coexisted in the same ocean during part of their history are members of a geminate pair.

The final alternative to the true geminate model is that there was gene flow between members of the geminate species after the completion (or near-completion) of the Central American Isthmus (figure 14.1D). There are various ways this could have happened. For example, it is possible that with sea level fluctuations occurring over the last 3 million years (Golik 1968; Haq et al. 1987), the lowest parts of the isthmus were breached by salt water incursions (Coates and Obando 1996; Cronin and Dowsett 1996). Even though such breaches must have been limited in extent and duration (otherwise they would have obliterated the biogeographic signature of the geminate species), they may have resulted in reticulate evolution in some species pairs. Limited foraminiferan evidence suggests that "incipient littoral-neritic leakage" may have occurred across the Panama Isthmus as recently as 1.8 mya (Crouch and Poag 1979; Keller et al. 1989). More recent introductions of lineages of euryhaline species into the wrong ocean may have occurred through the fresh waters of the Panama Canal (Hildebrand 1939; Rubinoff and Rubinoff 1969; McCosker and Dawson 1975). Ballast water taken by ships transiting the canal may transport planktonic organisms and larvae (Chesher 1968; Carlton and Geller 1993). A similar possibility of anthropogenic conveyance exists for fouling organisms attached to ship hulls (Spivey 1976). Finally, cosmopolitan species may maintain gene flow between the two coasts of America via circumglobal dispersal.

Do these potential confounding factors negate the advantages of geminate species as objects of speciation studies? I think not. Although it is dangerous to assume, without any corroborating evidence, that any two species on either side of the isthmus are ipso facto geminates separated for 3 million years, there are solutions to most of these problems. The greatest advantage of the historical setting of the Isthmus of Panama for evolutionary studies is that it separated marine organisms of a vast array of taxa. Many of the possible errors in inference can be avoided through the judicious choice of organisms with characteristics tailored to the question of a particular study. Thus, it would be foolhardy to use holoplanktonic or fouling organisms to calibrate rates of genetic divergence under the assumption that they have not exchanged genes for 3 million years. However, it is easy to find species for which the assumption of lack of passage through the Panama Canal is likely to hold true. There are many marine invertebrates with larvae unlikely to survive a trip through pump impellers and the effects of antifouling paint in sufficient numbers to have a measurable probability of reaching sexual maturity and thus of transporting genes from one ocean to the other (Rubinoff 1970). Similarly, choice of species with similar modes of dispersal can increase the possibility of simultaneous times of splitting, when this assumption is crucial (Lessios 1979b). What is more, most of the erroneous reconstructions of phylogenetic topology depicted in figure 14.1 can be avoided by adequate sampling of species and characters. The misidentification of geminate pairs depicted in figure 14.1A will disappear through simple pairwise comparisons of divergence when any phylogenetically informative character is measured, as long as all closely related species on each side of the isthmus are sampled. Differences in separation times (figure 14.1A–14.1D) can also be detected through the use of multiple character sets to estimate divergence (e.g., Lessios 1981a; Bermingham and Lessios 1993; Knowlton et al. 1993). It is likely that species pairs that were split earlier than the rest will show more differentiation between their members in several unrelated sets of characters even if these characters do not evolve in a clocklike manner. Such pairs can be excluded from comparisons requiring the assumption of simultaneous separation. Conversely, each additional character set showing equivalent degrees of divergence between members of several pairs increases confidence that they conform to the true geminate model. For example, general agreement in mtDNA, protein, and behavioral compatibility divergence values in alpheid shrimp on the two sides of the Panama isthmus (Knowlton et al. 1993) has produced compelling evidence that three presumed species pairs were split at various times (and possibly due to different causes) before the closure of the isthmus, but it has also produced evidence that four additional species pairs, which cluster around a central value for each measure of intrapair divergence, were split at roughly the same time (Cunningham and Collins 1994). Future studies of geminate alpheid shrimp that need to compare evolution between species split simultaneously will have the benefit of knowing which species to use and which to avoid. Problems will undoubtedly remain. For example, when various lines of evidence suggest different times of splitting between species on either side of Central America (e.g., Knowlton et al. 1993), it is still not possible to distinguish whether this is due to the gradual shoaling of the Isthmus (figure 14.1B), or whether it might be due to incorrect assumptions regarding sister species relationships (i.e., figure 14.1A,C). Only extensive fossil evidence (such as that presented in Jackson et al. [1993] for strombinid gastropods) can help with the complications arising from extinctions. However, the important point is that the geminate species provide the means with which to verify the assumptions on which conclusions will rest.

What Questions about Speciation Have the Geminate Species Helped Answer?

The first (and possibly most important) conceptual advance to which geminate species contributed was support for geographic speciation. Using the sister species of Panama along with other examples, D. S. Jordan (1908) proposed the "law of geminate species":

> Given any species in any region, the nearest related species is not to be found in the same region nor in a remote region, but in a neighboring district separated from the first by a barrier of some sort or at least by a belt of country, the breadth of which gives the effect of a barrier. (p. 73)

In today's intellectual climate in which the debate has shifted toward the question of whether sympatric speciation is even possible (Bush 1975; Bush and Howard 1986), Jordan's (1905, 1908) advocacy of Wagner's (1868) hypothesis of geographic speciation may not seem a great insight, but coming at a time in which DeVries's (1901) ideas about saltational emergence of new species through macromutations reigned supreme among experimental biologists, this was an important step in the history of speciation research (Mayr 1963, p. 487).

Of course, if support for the geographic speciation model was all that geminate species had to offer to speciation research, there would be little need to write about them today, except as a note in the history of science. However, there are few questions in speciation that have been closed, and thus many hypotheses for which the geminate species can provide pertinent evidence. The rest of this paper is devoted to exploring the relevance of data from geminate species to the two major components of speciation: the accumulation of overall genomic divergence and the emergence and perfection of reproductive isolation. How the two are related to each other and what factors affect the evolution of each are questions central to the study of speciation.

Divergence and Reproductive Isolation in Geminate Species

Because both the magnitude of genetic divergence and the probability of emergence of reproductive isolation between allopatric populations are related to the time two

isolates have remained separate, they are often correlated with each other (Coyne and Orr 1989, 1997). For this reason it is difficult to determine whether divergence has caused reproductive isolation, as most models of geographic isolation tacitly assume (Wright 1940, 1982a,b; Muller 1942; Mayr 1954b, 1963, 1982; Carson 1968, 1975, 1982, 1985; Templeton 1980, 1982; Carson and Templeton 1984; Barton and Charlesworth 1984) or whether reproductive isolation can appear independently of major genetic restructuring, as a minority has postulated (Lewontin 1974; Bush 1975; Nei et al. 1983; Wu 1985). The geminate species pairs that were split simultaneously can provide information on this relation between reproductive isolation and genetic distance by eliminating the confounding effect of time from the comparisons. Because sister species are allopatric but some of them are in the same ocean as other congeners, they can also help evaluate the importance of reinforcement (Dobzhansky 1940, 1970; Butlin 1989) in perfecting prezygotic reproductive isolation. Finally, because the species pairs that have been compared across the Isthmus of Panama differ in dispersal ability, and because informed guesses can be made about the past ranges of some of them, they can help evaluate the importance of vagility and of restrictions in effective population size in determining whether geographically separated populations will speciate.

Table 14.2 presents data on divergence between members of geminate pairs of sea urchins, fish, shrimp, and isopods in which reproductive isolation has been assessed or can be reasonably inferred. Three species pairs of shrimp from the Knowlton et al. (1993) comparisons have been omitted, because both isozymes and mtDNA indicate that they may have been split earlier than the rest. To avoid circularity, it is necessary to use at least one character set as an external clock that would indicate whether the assumption of simultaneous splitting holds. Mitochondrial DNA appears to fit this requirement for the three species of sea urchins studied by Bermingham and Lessios (1993), and for four species of shrimp studied by Knowlton et al. (1993) (though not for all presumed geminate species; see Lessios et al. [1995] for a much smaller value of mtDNA percentage dissimilarity in a species pair of the fish genus *Abudefduf*). The uniformity of mtDNA divergence values even holds fairly well across phyla and assaying techniques. Though the estimates of percentage dissimilarity in this molecule come from restriction fragment length polymorphisms (RFLPs) in sea urchins and from sequencing of 681 base pairs in the Cytochrome Oxidase I (COI) region in shrimp, their range is rather narrow (5.3–8.5% dissimilarity), thus suggesting that it is reasonable to assume that the timing of bifurcation in all of these independent lineages is (on a geologic time scale) comparable. Preliminary data from approximately 620 base pairs in the COI region in sea urchins (Gonzalez and Lessios, unpublished) suggest that *Diadema* (4.7% sequence dissimilarity between *D. antillarum* and *D. mexicanum*) may have diverged more re-

Table 14.2. Morphological, isozyme, and mtDNA divergence between geminate species of Panama in which reproductive isolation was assessed or can be inferred.

Ocean		Morphology		Isozymes		mtDNA		Reproductive Isolation	
		Within Species	Between Species	Within Species	Between Species	Within Species	Between Species	Prezygotic	Postzygotic
Atlantic	Pacific								
Sea urchins									
Diadema antillarum	*D. mexicanum*	8.041[1]	7.827[1]	0.013[3]	0.051[3]	0.0044[4]	0.053[4]	Temporal, bidirectional	—
Echinometra lucunter	*E. vanbrunti*	5.325[1]	6.412[1]	0.023[3]	0.341[3]	0.003[4]	0.061[4]	Gametic, unidirectional	—
Echinometra viridis	*E. vanbrunti*	4.788[1]	12.913[1]	0.015[1]	0.524[3]	0.006[4]	0.081[4]	None known	Complete
E. lucunter–E. viridis		4.803[1]	11.746[1]	0.021[3]	0.201[3]	0.004[4]	0.065[4]	Gametic, unidirectional	
Fish									
Bathygobius soporator	*B. ramosus*	—	6.515[2]	—	0.420[3]	—	—	Behavioral, incomplete	—
Bathygobius soporator	***B. andrei***	—	**8.402**[2]	—	**0.146**[3]	—	—	**Behavioral, incomplete**	—
	B.andrei–B. ramosus	—	7.980[2]	—	0.418[3]	—	—	Behavioral, complete	Complete
Shrimp									
Alpheus paracrinitus sp. b	*A. rostratus*	—	—	—	0.028[3]	—	0.077[5]	Behavioral, incomplete	—
Alpheus paracrinitus sp. a	*A. paracrinitus*	—	—	—	0.114[3]	—	0.066[5]	Behavioral, incomplete	—
Alpheus formosus sp. a	*A. panamensis*	—	—	—	0.109[3]	—	0.077[5]	Behavioral, incomplete	—

	Alpheus cylindricus	A. cylindricus	Excirolana braziliensis morph C	E. braziliensis morph C'				Behavioral, incomplete	Complete?
	—					0.121[3]	0.085[5]		
Isopods									
Excirolana braziliensis morph C	—	0.778[1]	2.263[1]	0.059[3]	0.215[3]	0.121[3]	0.085[5]	?	**Complete?**
Excirolana braziliensis morph P		0.649[1]	7.035[1]	0.122[3]	0.763[3]	—	—	?	Complete?
E. braziliensis morph C–morph P		0.650[1]	8.134[1]	0.125[3]	0.665[3]	—	—	?	Almost complete

Where more than one value of divergence is available because of multiple intraspecific sampling, the mean is shown. Where there is ambiguity of geminate relationship, rows in boldface indicate most likely geminate pair. The column labeled "postzygotic" includes information on whether genetic data indicate the absence of introgression, even though it is not known whether this is due to truly postzygotic isolation or an unstudied mechanism of prezygotic isolation. Data from Rubinoff and Rubinoff (1971), Gorman et al. (1976), Lessios (1981a, 1984), Lessios and Cunningham (1990), Bermingham and Lessios (1993), Knowlton et al. (1993), Lessios and Weinberg (1994), Bermingham and Lessios (unpublished).

[1]Mahalanobis's (1936) √D²

[2]Coefficient of difference (Mayr et al. 1953).

[3]Nei's (1987) D.

[4]Nei and Miller's (1990) D_{xy}.

[5]Kimura's (1980) corrected percentage sequence divergence.

cently than *Echinometra* (9.8% dissimilarity between *E. lucunter* and *E. vanbrunti*), but if true, this would only serve to strengthen the inferences from geminate comparisons regarding the role of genetic divergence in the emergence of reproductive isolation (see below). Because no mtDNA data exist for the isopod *Excirolana braziliensis*, the question of whether the Caribbean morph C and the eastern Pacific morph C', which are actually different species (see Lessios and Weinberg 1993, 1994), split at the same time as the sea urchin and shrimp geminates must be regarded as open. However, the extensive sampling of *Excirolana* on both coasts of Central America by Weinberg and Starczak (1988, 1989), Lessios et al. (1994), and Lessios and Weinberg (1993, 1994) suggests that the only other extant possible geminate is a morph called P' on the Atlantic coast of Brazil, which is likely to be a sister species of the Pacific P morph (Weinberg and Starczak 1989). Thus, the possibility that an unsampled species might result in confusion between members of different geminate pairs in *Excirolana* is remote. Weinberg and Starczak (1988) and Lessios and Weinberg (1994) considered the possibility that the C–C' pair may be the result of a recent introduction through the Panama Canal, but rejected it in part due to the distributions of the two forms.

Information on reproductive isolation and genetic divergence from additional species, congeneric and sympatric with one of the geminates, is also included in table 14.2 when available. For sympatric species, genetic data can indicate whether or not they exchange genes. Thus, two allozyme loci fixed for different alleles (Bermingham and Lessios, unpublished) indicate that the sympatric sea urchin species *Echinometra lucunter* and *E. viridis* are not hybridizing in nature, even though they show only unidirectional gametic isolation in the laboratory (Lessios and Cunningham 1990). Lack of introgression is likely to be the result of postzygotic isolation, because the annual reproductive cycles of these species overlap (Lessios 1981b, 1985a) and because they show no lunar cycles in their spawning (Lessios 1991). Similarly, the sympatric species of the goby *Bathygobius* will not hybridize in captivity (Rubinoff and Rubinoff 1971), and genetic data indicate that they do not exchange genes in nature, because they show no shared alleles at five loci (Gorman et al. 1976). Evidence for reproductive isolation in *Excirolana* comes only from the genetics of natural populations (Lessios and Weinberg 1993, 1994), and can thus not be conclusive in the case of allopatric morphs. However, reproductive isolation in this species complex, as measured by the excess of migration over gene flow, is present even between adjacent populations of the same morph (Lessios and Weinberg 1993). It would therefore be very surprising if the allopatric and divergent C and C' morphs hybridized freely if they were to find themselves in the same ocean.

What can these eight species pairs tell us about factors that affect genetic divergence and reproductive isolation in allopatry? The first question they can help answer

is whether overall genetic divergence is a requirement for the emergence of reproductive isolation. It should be remembered that even though speciation is the emergence of reproductive isolation, most models of speciation only deal with genomic divergence. This is because it is implicitly assumed that reproductive isolation is the product of small effects of many loci (Mayr 1963; Barton and Charlesworth 1984; Paterson 1985), and can thus come about only as the result of a general overhaul of the genome. In part, this general belief seems to have its origins on the overreliance of evolutionary biology on evidence from terrestrial vertebrates and insects. In these groups complicated courtship rituals and the exchange of multiple behavioral cues, the necessary prerequisites to successful mating, are likely to have a polygenic basis. Change in such traits usually (but not always) may require genetic substitutions in many loci (see Zouros 1991; Coyne 1992), or else may be controlled by relatively few loci but with major developmental effects (Lewontin 1974; Templeton 1981; Bush and Howard 1986). The evidence from the geminate species, though still very limited, indicates that species that depend on courtship show a correlation between divergence and premating reproductive isolation, whereas species lacking courtship may show prezygotic reproductive isolation that is independent of divergence. In shrimp, for which behavioral interactions are important, the pair consisting of *Alpheus paracrinitus* sp. b and *A. rostratus*, which has the aberrantly smallest value of Nei's D (table 14.2), is also the pair in which laboratory experiments show the highest compatibility between members (Knowlton et al. 1993). However, among free-spawning sea urchins, complete reproductive isolation through gamete release at different lunar phases has evolved among the geminates of *Diadema*, the genus in which both isozyme and morphological transisthmian differentiation is not substantially larger than intraspecific variability (Lessios 1984). The geminate species of *Echinometra*, on the other hand, *E. lucunter* and *E. vanbrunti*, even though they are one order of magnitude more divergent in allozymes than the species of *Diadema* (and somewhat more divergent in morphology and mtDNA as well), show only unidirectional gametic isolation. Even this partial mechanism is absent between the allopatric *E. viridis* and *E. vanbrunti*, the most differentiated pair in the trio. Although other, as yet undiscovered, mechanisms of reproductive isolation may exist in *Echinometra* (Lessios and Cunningham 1990, 1993), the evidence at hand suggests that in these organisms with external fertilization the emergence of reproductive isolation may be unrelated to the amount of accumulated genetic divergence. Such decoupling is easy to understand if gametic recognition depends on mutations in only two loci, those controlling the gamete recognition molecules bindin (Palumbi and Metz 1991) and bindin receptor (Foltz et al. 1993), or if spawning time is under simple genetic control. Lack of correlation between genetic divergence and reproductive isolation in organisms with simple genetic control of reproduction does not necessarily mean that speciation will be either frequent or rapid. Because there will be strong selection against mutant genotypes that produce incompatible gametes or spawn asynchronously with the rest of the population, the rise of new species through such mechanisms requires the convergence of many events of low probability to occur (Lessios and Cunningham 1990).

A similar dichotomy between species with and without courtship is suggested when one considers the role of reinforcement of reproductive isolation through selection to avoid hybridization in sympatry. In the fish *Bathygobius*, single-pair no-choice experiments in aquaria suggested that mating discrimination between the two geminates, and between each of them and the more distantly related *B. ramosus*, was complete. However, when males of a second species were introduced in a tank containing both sexes of another, some heterospecific spawnings occurred between allopatric species, but not between sympatric ones. This led Rubinoff and Rubinoff (1971) to conclude that reinforcement had perfected reproductive isolation in sympatry. This conclusion remains robust if one considers the possible effects of time, because the outgroup species, *B. ramosus*, is equidistant in both morphology and isozymes from the two members of the geminate pair (table 14.2). These results from fish contrast with results from sea urchins. In *Echinometra*, eggs of *E. lucunter* are less likely to be fertilized by sperm of the allopatric *E. vanbrunti* than by sperm of the sympatric *E. viridis* (Lessios and Cunningham 1990), the opposite of what the reinforcement hypothesis would expect. Time cannot be factored out in this comparison. Even though the RFLP data shown in table 14.2 cannot resolve the phylogeny of the three species of *Echinometra*, sequencing of the COI region of mtDNA (Gonzalez and Lessios, unpublished) indicates that the two sympatric species split from a common stock that had been already separated from the Pacific *E. vanbrunti*. There was, therefore, less time for *E. lucunter* to evolve reproductive isolation toward the sympatric *E. viridis* than toward *E. vanbrunti*. However, if gametic isolation depended on time alone, one would expect not just *E. lucunter* but also *E. viridis* to show gametic incompatibility with *E. vanbrunti*, and this is not so (Lessios and Cunningham 1990). Thus, reinforcement seems to have been important in the evolution of prezygotic isolation of fish, but not of sea urchins, suggesting that in the latter the emergence of reproductive isolation is more a matter of chance appearance in a lineage than of selection to avoid wasting gametes in inferior hybrids.

The Importance of Population Size in Divergence and Reproductive Isolation

The second question about which geminate species can provide relevant data concerns the importance of population size restrictions in causing rapid speciation. Most speciation models which assume that reproductive isolation results from overall genetic divergence also postulate passage through small population size as a mechanism of rapid genetic change. Authors of such models envision different ways in which population bottlenecks will destabilize the genome, but they all believe that the consequences will be drastic; they have accordingly coined appropriately radical terms for these events. Thus, there is "genetic revolution" (Mayr 1954b, 1963), "founder-flush cycles" (Carson 1975), and "genetic trasilience" (Templeton 1980). Although none of these authors deny that, given enough time, separation of large populations can also cause speciation, the assumption is that divergence in the absence of bottlenecks will proceed at much slower rates and that large isolates have a much higher chance of fusing if the geographic barrier is removed (e.g., Grant 1963; Carson 1982). A necessary consequence is the belief that most species are the products of peripatric speciation, that is, speciation that involves small isolates at the periphery of the species range (see Bush 1975, 1994; Mayr 1982; Lynch 1989). However, Barton and Charlesworth (1984) and Barton (1989), even though they also believe that reproductive isolation requires genomic divergence, have argued that restrictions in population size are unimportant in speciation, because passage from one adaptive peak to another can be achieved even in large populations.

Coyne and Orr (1989) have concluded from a review of extensive data on electrophoretic differentiation and reproductive isolation that speciation in *Drosophila* requires 1.5–3.5 million years to be completed, and Coyne and Orr (1997) have estimated that allopatric species in this genus require 2.7 million years to speciate. If small population size is important in the completion of reproductive isolation, most speciation in *Drosophila* presumably has involved restrictions in population size. Geographic isolation in the geminate species of Panama, however, is more likely to conform to what has come to be known as the dumbbell (Mayr 1982), vicariance (Lynch 1989), or dichopatric (Bush 1994) model, that is, one that involves populations sufficiently large to exclude inbreeding as a factor, and would thus be expected to show few, if any, isolating mechanisms. Mayr (1967) has speculated the bisection of the Panamic biogeographic province must have left "two colossal gene pools" on either side of the isthmus, and that, for this reason, "differences are still either nonexistent or they are so slight that one doesn't really like to rank these as species" (p. 49).

Mayr's estimate of the size of gene pools was probably correct. Fractionation of large populations is suggested by the present-day ranges of geminate species. If the completion of the isthmus separated only a peripheral deme of each genus in one of the oceans, is 3 million years enough time for all such demes to have expanded their populations, such that they range in the entire tropical portion of each ocean as most of them do? Even if this is a possibility for other groups, fossil evidence of echinoids—despite this group's propensity to fossilize poorly (Kier 1977; Gordon 1991)—suggests that the ranges of both isolates of various genera that include presumed geminates (Chesher 1972) have remained wide from the Miocene to the Recent (figure 14.2).

Genetic variability in extant populations also provides evidence for a long history of large population size. Both isozyme heterozygosity (18.1–27.8%, Lessios 1979b) and intraspecific mtDNA diversity (0.237–0.614%, Bermingham and Lessios, unpublished) in *Diadema* and *Echinometra* are high and approximately equal in all species. Heterozygosities in shrimp (1.6– 8.0%, Knowlton et al. 1993) and in *Bathygobius* (2.7–6.9%, Gorman et al. 1976) are lower than in sea urchins, but still not suggestive of bottlenecks in either ocean. Despite the apparent lack of drastic reductions in history of population size, the allopatric species of *Diadema*, *Alpheus*, and *Bathygobius* show premating reproductive isolation, thus giving credence to Barton and Charlesworth's (1984) claim that speciation does not require small demes.

Another piece of evidence suggesting that the importance of population bottlenecks as a cause of divergence may have been overemphasized comes from a catastrophe that befell one of the geminate species. *Diadema antillarum* suffered mass mortality in 1983, which reduced populations throughout the western Atlantic by more than 97% (Lessios et al. 1984a,b; Lessios 1988a,b). This drastic reduction had no effects on the average heterozygosity, number of alleles, or gene frequencies of *D. antillarum* populations (Lessios 1985b). In 1993 the populations in Panama were still at less than 3.5% of their premortality levels (Lessios 1995a), but genetic variability and gene frequencies remained unaltered despite the prolonged reduction in population size (Lessios 1995b).

In addition to historical fluctuations in population size, genetic divergence and the emergence of reproductive isolation should theoretically be affected by the mode of dispersal of the organisms in question. Though he gave no quantitative estimates, Rosenblatt (1963) stated that there are more recognized geminate species among families of large fish with a long-lived planktonic larval stage than among families of small fish that lead a sedentary

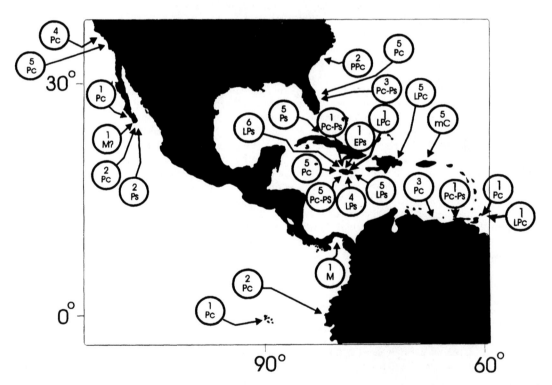

Figure 14.2. Localities and epochs in which sea urchin fossils of species considered to be members of geminate pairs have been found. Compiled from Mortensen (1928–1950), Grant and Hertlein (1938), Kier (1963), Cooke (1961), Gordon (1991), Gordon and Donovan (1992), Donovan and Gordon (1993), Donovan and Embden (1996). Genera: 1, *Eucidaris*; 2, *Arbacia*; 3, *Lytechinus*; 4, *Tripneustes*; 5, *Echinometra*; 6, *Diadema* or *Astropyga*. Epochs: M, Miocene; Pc, Pliocene; Lpc, late Pliocene; Ps, Pleistocene; EPs, early Pleistocene; LPs, late Pleistocene; mC, mid-Cenozoic; PPc, post-Pliocene.

existence as adults and have a short-lived larval stage. He suggested that the pattern is due to accelerated rates of divergence among the latter, produced by lower levels of intraoceanic gene flow and more pronounced genetic structuring. However, more recent genetic data from other geminate species, with much more pronounced differences in vagility than large and small fish, have failed to support this view. Sea urchins, shrimp, and fish have planktonic larvae that can disperse over wide ranges. Isopods, on the other hand, are brooders with no dispersal phase other than possible rafting of adults. This sedentary habit of all life stages, along with infrequent extinctions and colonizations of intertidal environments from a single source (Lessios et al. 1994), has resulted in populations of *Excirolana braziliensis* that are highly structured genetically. Populations of the same morph less than 1 km apart are fixed for alternate alleles (Lessios and Weinberg 1994) and show incipient reproductive isolation (Lessios and Weinberg 1993). Despite these high levels of intraoceanic genetic divergence, however, the allozyme transisthmian divergence between geminate morphs of isopods is no larger than that of most sea urchins, shrimp, or fish (table 14.2), possibly because ancestral alleles are preserved in low frequencies in each ocean by occasional hybridization between populations and morphs (Lessios and Weinberg 1994). Thus, the expectation that organisms with low dispersal would tend to have accelerated rates of divergence and higher probability of speciation is not fulfilled. Whether this is a general result, or whether it is the consequence of the peculiarities of the population genetics of *Excirolana*, remains to be seen.

Conclusions

The geminate species on the two sides of Central America have had a long history of contributing evidence relevant to speciation, though they have yet to be used to their full potential. From eight species pairs for which there are multiple measures of genetic divergence and assessments of reproductive isolation, I have concluded that in organisms lacking copulation and courtship, reproductive isolation in allopatry may arise even in the absence of major reorganization of the genome, and that reinforcement is not important in perfecting reproductive isolation. It is obvious that comparisons between eight species pairs, no matter how well their history of splitting is known, cannot provide definitive answers to questions that have not been answered in 100 years of speciation research. It is entirely possible, indeed it is likely, that the next study using geminate species may generate evidence that would lead to exactly the opposite conclusions from the ones reached here. A result that may remain unaltered, on the other hand, is that Atlantic and Pacific populations in each pair have acquired varying degrees of reproductive isolation despite the absence of bottlenecks, which suggests that restrictions in population size may not be of primary importance in speciation. No matter how future studies turn out, if this chapter has provided a flavor of the kinds of questions geminate species can help answer and convinced some evolutionary biologists that they are a very underutilized tool for understanding vicariant speciation, it will have served its purpose. As Jones and Hasson (1985) have put it, "The evolutionary experiment begun in the late Pliocene by the uplift of Central America has been performed; more investigators are needed to interpret the data" (p. 351).

Acknowledgments I thank M. McCartney, W. G. Eberhard, B. R. Grant, D. J. Howard, B. Kessing, I. Rubinoff, M.-J. West-Eberhard, and an anonymous reviewer for comments on the manuscript, and D. Howard and S. Berlocher for the invitation to participate in the "Endless Forms" conference. The Smithsonian Tropical Research Institute has supported 18 out of the 20 post-1954 studies listed in table 14.1, so it is accurate to state that this chapter would not have been written without it.

References

Barton, N. H. 1989. Founder effect speciation. In D. Otte and J. A. Endler (eds.). Speciation and Its Consequences. Sunderland, Mass.: Sinauer, pp. 229–256.

Barton, N. H., and B. Charlesworth. 1984. Genetic revolutions, founder effects, and speciation. Annu. Rev. Ecol. Syst. 15:133–164.

Bayer, F. M., G. L. Voss, and C. R. Robins. 1970. Bioenvironmental and Radiological Safety Feasibility Studies Atlantic-pacific Interoceanic Canal. Miami: University of Miami Rosenstiel School of Marine and Atmospheric Sciences.

Bermingham, E. B., and H. A. Lessios. 1993. Rate variation of protein and mtDNA evolution as revealed by sea urchins separated by the Isthmus of Panama. Proc. Natl. Acad. Sci. USA 90:2734–2738.

Bowen, B. W., A. B. Meylan, and J. C. Avise. 1991. Evolutionary distinctiveness of the endangered Kemp's Ridley sea turtle. Nature 352:709–711.

Briggs, J. C. 1969. The sea-level Panama Canal: potential biological catastrophe. Bioscience 19:44–47.

Briggs, J. C. 1974. Marine Zoogeography. New York: McGraw-Hill.

Bush, G. L. 1975. Modes of animal speciation. Annu. Rev. Ecol. Syst. 6:339–364.

Bush, G. L. 1994. Sympatric speciation in animals: new wine in old bottles. Trends Ecol. Evol. 9:285–288.

Bush, G. L., and D. J. Howard. 1986. Allopatric and nonallopatric speciation: assumptions and evidence. In S. Karlin and E. Nevo (eds.). Evolutionary Process and Theory. Orlando: Academic Press, pp. 411–438.

Butlin, R. 1989. Reinforcement of premating isolation. In D. Otte and J. A. Endler (eds.). Speciation and Its Consequences. Sunderland, Mass.: Sinauer, pp. 158–179.

Carlton, J. T., and J. B. Geller. 1993. Ecological roulette: the global transport of nonindigenous marine organisms. Science 261:78–82.

Carson, H. L. 1968. The population flush and its genetic consequences. In R. C. Lewontin (ed.). Population Biology and Evolution. New York: Syracuse University Press, pp. 123–137.

Carson, H. L. 1975. The genetics of speciation at the diploid level. Am. Nat. 109:83–92.

Carson, H. L. 1982. Speciation as a major reorganization of polygenic balances. In C. Barigozzi (Ed.). Mechanisms of Speciation. New York: Liss, pp. 411–433.

Carson, H. L. 1985. Unification of speciation theory in plants and animals. Syst. Bot. 10:380–390.

Carson, H. L., and A. R. Templeton. 1984. Genetic revolutions in relation to speciation phenomena: the founding of new populations. Annu. Rev. Ecol. Syst. 15:97–131.

Chace, F. A., Jr. 1972. The shrimps of the Smithsonian-Bredin Caribbean expeditions with a summary of the West Indian shallow-water species (Crustacea: Decapoda: Natantia). Smithsonian Contrib. Zool. 98:1–179.

Chesher, R. H. 1968. Transport of marine plankton through the Panama Canal. Limnol. Oceanogr. 13:387–388.

Chesher, R. H. 1972. The status of knowledge of Panamanian echinoids, 1971, with comments on other echinoderms. Bull. Biol. Soc. Wash. 2:139–158.

Coates, A. G., and J. A. Obando. 1996. The geologic evolution of the Central American Isthmus. In J. B. C. Jackson, A. F. Budd, and A. G. Coates (eds.). Evolution and Environment in Tropical America. Chicago: University of Chicago Press, pp. 21–56.

Coates, A. G., J. B. C. Jackson, L. S. Collins, T. M. Cronin, H. J. Dowset, L. M. Bybell, P. Jung, and J. A. Obando. 1992. Closure of the Isthmus of Panama: the near-shore marine record of Costa Rica and western Panama. Bull. Geol. Soc. Am. 104:814–828.

Collins, T. M. 1989. Rates of Mitochondrial DNA Divergence in Transisthmian Geminate Species. Ph.D. Dissertation, Yale University.

Collins, T. M. 1996. Molecular comparisons of transisthmian species pairs: rates and patterns of evolution. In J. B. C. Jackson, A. F. Budd, and A. G. Coates (eds.). Evolution and Environment in Tropical America. Chicago: University of Chicago Press, pp. 303–334.

Cooke, C. W. 1961. Cenozoic and Cretaceous echinoids from Trinidad and Venezuela. Smithsonian Misc. Coll. 142:1–35.

Coyne, J. A. 1992. Genetics and speciation. Nature 355:511–515.

Coyne, J. A., and H. A. Orr. 1989. Patterns of speciation in Drosophila. Evolution 43:362–381.

Coyne, J. A., and H. A. Orr. 1997. "Patterns of speciation in Drosophila" revisited. Evolution 51:295–303.

Cronin, T. M. 1985. Speciation and stasis in marine ostracoda: climatic modulation of evolution. Science 227:60–63.

Cronin, T. M., and H. J. Dowsett. 1996. Biotic and oceanographic response to the Pliocene closing of the Central American Isthmus. In J. B. C. Jackson, A. F. Budd, and A. G. Coates (eds.). Evolution and Environment in Tropical America. Chicago: University of Chicago Press, pp. 76–104.

Crouch, R. W., and W. C. Poag. 1979. Amphistegina gibbosa D'Orbigny from the California borderlands: the Caribbean connection. J. Forminif. Res. 9:85–105.

Cunningham, C. W., and T. M. Collins. 1994. Developing model systems for molecular biogeography: vicariance and interchange in marine invertebrates. In B. Schierwater, B. Streit, G. P. Wagner, and R. Desalle (eds.). Molecular Ecology and Evolution: Approaches and Applications. Switzerland: Birkhauser Verlag, pp. 405–433.

de Laubenfels, M. W. 1936. A comparison of the shallow-water sponges near the Pacific end of the Panama Canal with those at the Caribbean end. Proc. U.S. Nat. Mus. 83:441–466.

DeVries, H. 1901. Die Mutation und die Mutationsperioden bei der Entstehung der Arten. Verh. Ges. Deut. Naturf. Artzte 73:202–212.

Dobzhansky, T. 1940. Speciation as a stage in evolutionary divergence. Am. Nat. 74:312–321.

Dobzhansky, T. 1970. Genetics of the Evolutionary Process. New York: Columbia University Press.

Donovan, S. K., and B. J. Embden. 1996. Early Pleistocene echinoids of the Manchioneal formation, Jamaica. J. Paleont. 70:485–493.

Donovan, S. K., and C. M. Gordon. 1993. Echinoid taphonomy and the fossil record: supporting evidence from the Plio-Pleistocene of the Caribbean. Palaios 8:304–306.

Duque-Caro, H. 1990. Neogene stratigraphy, paleoceanography and paleobiogeography in northwest South America and the evolution of the Panama seaway. Palaeogeog. Palaeocl. Palaeoec. 77:203–234.

Ekman, S. 1953. Zoogeography of the Sea. London: Sidgwick and Jackson.

Emiliani, C., S. Gartner, and B. Lidz. 1972. Neogene sedimentation on the Blake Plateau and the emergence of the Central American Isthmus. Palaeogeog. Palaeocl. Palaeoec. 11:1–10.

Foltz, K. R., J. S. Partin, and W. J. Lennarz. 1993. Sea urchin egg receptor for sperm: sequence similarity of binding domain and HSP70. Science 259:1421–1425.

Glynn, P. W. 1972. Observations on the ecology of the Caribbean and Pacific coasts of Panama. Bull. Biol. Soc. Wash. 2:13–30.

Glynn, P. W. 1982. Coral communities and their modifications relative to past and prospective Central American seaways. Adv. Mar. Biol. 19:91–132.

Glynn, P. W., and G. M. Wellington. 1983. Corals and Coral Reefs of the Galápagos Islands. Berkeley: University of California Press.

Golik, A. 1968. History of Holocene transgression in the Gulf of Panama. Journal of Geology 76:497–507.

Gordon, C. M. 1991. The poor fossil record of Echinometra (Echinodermata: Echinoidea) in the Caribbean region. J. Geol. Soc. Jamaica 28:37–41.

Gordon, C. M., and S. K. Donovan. 1992. Disarticulated echinoid ossicles in paleoecology and taphonomy: the last interglacial Falmouth Formation of Jamaica. Palaios 7:157–166.

Gorman, G. C., and Y. J. Kim. 1977. Genotypic evolution in the face of phenotypic conservativeness: *Abudefduf* (Pomacentridae) from the Atlantic and Pacific sides of Panama. Copeia 1977:694–697.

Gorman, G. C., Y. J. Kim, and R. W. Rubinoff. 1976. Genetic relationships of three species of *Bathygobius* from the Atlantic and Pacific sides of Panama. Copeia 1976:361–364.

Graham, J. B. 1971. Temperature tolerances of some closely related tropical Atlantic and Pacific fish species. Science 172:861–863.

Grant, U. S., IV, and L. G. Hertlein. 1938. The west American Cenozoic Echinoids. University of California Publications in Mathematical and Physical Sciences 2:1–225.

Grant, V. 1963. The Origin of Adaptations. New York: Columbia University Press.

Graves, J. E., R. H. Rosenblatt, and G. N. Somero. 1983. Kinetic and electrophoretic differentiation of lactate dehydrogenase of teleost species-pairs from the Atlantic and Pacific coasts of Panama. Evolution 37:30–37.

Günther, A. 1868. An account of the fishes of the states of Central America, based on the collections made by Capt. J. M. Dow, F. Godman, Esq. and O. Salvin, Esq. Trans. Zool. Soc. Lond. 6:377–402.

Haq, B. U., J. Hardenbol, and P. R. Vail. 1987. Chronology of fluctuating sea level since the Triassic (250 million years ago to present). Science 235:1158–1167.

Hedgpeth, J. W. 1948. The Pycnogonida of the western north Atlantic and the Caribbean. Proc. U.S. Nat. Mus. 97:1–342.

Hildebrand, S. F. 1939. The Panama Canal as a passageway for fishes, with lists and remarks on the fishes and invertebrates observed. Zool.: N.Y. Zool. Soc. 24:15–45.

Holcombe, T., and W. S. Moore. 1977. Paleocurrents in the eastern Caribbean: geologic evidence and implications. Mar. Geol. 23:35–56.

Jackson, J. B. C., P. Jung, A. G. Coates, and L. S. Collins. 1993. Diversity and extinction of tropical American mollusks and emergence of the Isthmus of Panama. Science 260:1624–1626.

Jones, D. S., and P. F. Hasson. 1985. History and development of the marine invertebrate faunas separated by the Central American Isthmus. In F.G. Stehli and S.D. Webb (eds.). The Great American Interchange. New York: Plenum Press, pp. 325–355.

Jordan, D. S. 1885. A list of the fishes known from the Pacific coast of tropical America, from the Tropic of Cancer to Panama. Proc. U.S. Nat. Mus. 8:361–394.

Jordan, D. S. 1905. The origin of species through isolation. Science 22:545–562.

Jordan, D. S. 1908. The law of the geminate species. Am. Nat. 42:73–80.

Keigwin, L. D. 1978. Pliocene closing of the Isthmus of Panama, based on biostratigraphic evidence from nearby Pacific Ocean and Caribbean Sea cores. Geology 6:630–634.

Keigwin, L. D. 1982. Isotopic paleoceanography of the Caribbean and east Pacific: role of Panama uplift in Late Neogene time. Science 217:350–353.

Keller, G., C. E. Zenker, and S. M. Stone. 1989. Late Noegene history of the Pacific-Caribbean gateway. J. So. Am. Earth Sci. 2:73–108.

Kier, P. M. 1963. Tertiary echinoids from the Caloosahatcee and Tamiami formations of Florida. Smithsonian Misc. Coll. 149:1–68.

Kier, P. M. 1977. The poor fossil record of the regular echinoid. Paleobiology 3:168–174.

Kim, W., and L. G. Abele. 1988. The snapping shrimp genus *Alpheus* from the eastern Pacific (Decapoda: Caridea: Alpheidae). Smithsonian Contrib. Zool. 454:1–119.

Kimura, M. 1980. A simple method for estimating evolutionary rates of base substitutions through comparative studies of nucleotide sequences. J. Mol. Evol. 16:111–120.

King, M. 1993. Species evolution: the role of chromosome change. Cambridge: Cambridge University Press.

Knowlton, N., L. A. Weight, L. A. Solorzano, D. K. Mills, and E. Bermingham. 1993. Divergence in proteins, mitochondrial DNA and reproductive compatibility across the Isthmus of Panama. Science 260:1629–1632.

Laguna, J. E. 1987. *Euraphia eastropacensis* (Cirripedia, Chthamaloidea), a new species of barnacle from the tropical eastern Pacific: morphological and electrophoretic comparisons with *Euraphia rhizophorae* (de Olveira) from the tropical western Atlantic and molecular evolutionary implications. Pac. Sci. 41:132–140.

Lessios, H. A. 1979a. Molecular, Morphological and Ecological Divergence of Shallow-water Sea Urchins Separated by the Isthmus of Panama. Ph.D Dissertation, Yale University.

Lessios, H. A. 1979b. Use of Panamanian sea urchins to test the molecular clock. Nature 280:599–601.

Lessios, H. A. 1981a. Divergence in allopatry: molecular and morphological differentiation between sea urchins separated by the Isthmus of Panama. Evolution 35:618–634.

Lessios, H. A. 1981b. Reproductive periodicity of the echinoids *Diadema* and *Echinometra* on the two coasts of Panama. J. Exp. Mar. Biol. Ecol. 50:47–61.

Lessios, H. A. 1984. Possible prezygotic reproductive isolation in sea urchins separated by the Isthmus of Panama. Evolution 38:1144–1148.

Lessios, H. A. 1985a. Annual reproductive periodicity in eight echinoid species on the Caribbean coast of Panama. In B.F. Keegan and B.D. O'Connor (eds.). Echinodermata. Proceedings of the 5th International Echinoderm Conference, Galway. Rotterdam: Balkema, pp. 303–311.

Lessios, H. A. 1985b. Genetic consequences of mass mortality in the Caribbean sea urchin *Diadema antillarum*. Proc. 5th Coral Reef Congr. 4:119–126.

Lessios, H. A. 1988a. Mass mortality of *Diadema antillarum* in the Caribbean: what have we learned? Annu. Rev. Ecol. Syst. 19:371–393.

Lessios, H. A. 1988b. Population dynamics of *Diadema antillarum* (Echinodermata: Echinoidea) following mass mortality in Panamá. Mar. Biol. 99:515–526.

Lessios, H. A. 1990. Adaptation and phylogeny as determinants of egg size in echinoderms from the two sides of the isthmus of Panama. Am. Nat. 135:1–13.

Lessios, H. A. 1991. Presence and absence of monthly reproductive rhythms among eight Caribbean echinoids off the coast of Panama. J. Exp. Mar. Biol. Ecol. 153:27–47.

Lessios, H. A. 1995a. *Diadema antillarum* 10 years after mass mortality: still rare, despite help from a competitor. Proc. R. Soc. Lond. Ser. B. 259:331–337.

Lessios, H. A. 1995b. Direct evidence about bottlenecks in marine organisms: the 1983 *Diadema* pandemic. J. Cell. Biochem. (Suppl.) 19B:333.

Lessios, H. A., and C. W. Cunningham. 1990. Gametic incompatibility between species of the sea urchin *Echinometra* on the two sides of the Isthmus of Panama. Evolution 44:933–941.

Lessios, H. A., and C. W. Cunningham. 1993. The evolution of gametic incompatibility in neotropical *Echinometra*: a reply to McClary. Evolution 47:1883–1885.

Lessios, H. A., and J. R. Weinberg. 1993. Migration, gene flow and reproductive isolation between and within morphotypes of the isopod *Excirolana* in two oceans. Heredity 71:561–573.

Lessios, H. A., and J. R. Weinberg. 1994. Genetic and morphological divergence among morphotypes of the isopod *Excirolana* on the two sides of the Isthmus of Panama. Evolution 48:530–548.

Lessios, H. A., J. D. Cubit, D. R. Robertson, M. J. Shulman, M. R. Parker, S. D. Garrity, and S. C. Levings. 1984a. Mass mortality of *Diadema antillarum* on the Caribbean coast of Panama. Coral Reefs 3:173–182.

Lessios, H. A., D. R. Robertson, and J. D. Cubit. 1984b. Spread of *Diadema* mass mortality through the Caribbean. Science 226:335–337.

Lessios, H. A., J. R. Weinberg, and V. R. Starczak. 1994. Temporal variation in populations of the marine isopod *Excirolana*: how stable are gene frequencies and morphology? Evolution 48:549–563.

Lessios, H. A., G. R. Allen, G. M. Wellington, and E. Bermingham. 1995. Genetic and morphological evidence that the eastern Pacific damselfish *Abudefduf declivifrons* is distinct from *A. concolor* (Pomacentridae). Copeia 1995:277–288.

Lessios, H. A., B. D. Kessing, G. M. Wellington, and A. Graybeal. 1996. Indo-Pacific echinoids in the tropical eastern Pacific. Coral Reefs 15:133–142.

Lewontin, R. C. 1974. The Genetic Basis of Evolutionary Change. New York: Columbia University Press.

Lynch, J. D. 1989. The gauge of speciation: on the frequencies of modes of speciation. In D. Otte and J. A. Endler (Eds.). Speciation and Its Consequences. Sunderland, Mass.: Sinauer, pp. 527–553.

Mahalanobis, P. C. 1936. On the generalized distance in statistics. Proc. Nat. Inst. Sci. India 2:49–55.

Marshall, L. G. 1985. Geochronology and land-mammal biochronology of the transamerican fauna interchange. In F. G. Stehli and S. D. Webb (eds.). The Great American Biotic Interchange. New York: Plenum Press, pp. 49–85.

Marshall, L. G., R. F. Butler, R. E. Drake, G. H. Curtis, and R. H. Telford. 1979. Calibration of the great American interchange. Science 204:272–279.

Marshall, L. G., S. D. Webb, J. J. Sepkoski Jr., and D. M. Raup. 1982. Mammalian evolution and the great American interchange. Science 215:1351–1357.

Martin, A. P., G. J. P. Naylor, and S. R. Palumbi. 1992. Rates of mitochondrial DNA evolution in sharks are slow compared with mammals. Nature 357:153–155.

Mayr, E. 1954a. Geographic speciation in tropical echinoids. Evolution 8:1–18.

Mayr, E. 1954b. Change of genetic environment and evolution. In J. Huxley (ed.). Evolution as a Process. London: Allen and Unwin, pp. 157–180.

Mayr, E. 1963. Animal Species and Evolution. Cambridge, Mass.: Harvard University Press.

Mayr, E. 1967. Evolutionary challenges to the mathematical interpretation of evolution. In P. S. Moorhead and M. M. Kaplan (Eds.). Mathematical Challenges to the Neo-Darwinian Interpretation of Evolution. Philadelphia: Wistar Institute Press, pp. 47–58.

Mayr, E. 1982. Process of speciation in animals. In C. Barigozzi (Ed.). Mechanisms of Speciation. New York: Liss, pp. 1–20.

Mayr, E., E. G. Linsley, and R. L. Usinger. 1953. Methods and Principles of Systematic Zoology. New York: McGraw Hill.

McCosker, J. E., and C. E. Dawson. 1975. Biotic passage through the Panama Canal, with particular reference to fishes. Mar. Biol. 30:343–351.

Meek, S. E., and S. F. Hildebrand. 1923–1928. The Marine Fishes of Panama. 2 parts. Chicago: Field Museum of Natural History.

Menzies, R. J. 1968. Transport of marine life between oceans through the Panama Canal. Nature 220:802–803.

Mortensen, T. 1928–1950. A Monograph of the Echinoidea. 5 vols. Copenhagen: Reitzel.

Muller, H. J. 1942. Isolating mechanisms, evolution and temperature. Biol. Symp. 6:71–125.

Nei, M. 1987. Molecular Evolutionary Genetics. New York: Columbia University Press.

Nei, M., and J. C. Miller. 1990. A simple method for estimating average number of nucleotide substitutions within and between populations from restriction data. Genetics 125:873–879.

Nei, M., T. Maruyama, and C.-I. Wu. 1983. Models of evolution of reproductive isolation. Genetics 103:557–579.

Palumbi, S. R., and E. C. Metz. 1991. Strong reproductive isolation between closely related tropical sea urchins (genus *Echinometra*). Mol. Biol. Evol. 8:227–239.

Paterson, H. E. H. 1985. The recognition concept of species. In E.S. Vrba (Ed.). Species and Speciation. Pretoria: Transvaal Museum, pp. 21–29.

Rosenblatt, R. H. 1963. Some aspects of speciation in marine shore fishes. Syst. Assoc. Publ. No. 5:171–180.

Rosenblatt, R. H. 1967. The zoogeographic relationships of the marine shore fishes of tropical America. Stud. Trop. Oceanogr. 5:579–592.

Rosenblatt, R. H., J. E. McCosker, and I. Rubinoff. 1972. Indo-West Pacific fishes from the Gulf of Chiriqui, Panama. Contrib. Sci. Nat. Hist. Mus. L.A. Co. 234:1–18.

Rubinoff, I. 1963. Morphological Comparisons of Shore Fishes Separated by the Isthmus of Panama. Ph.D. Thesis, Harvard University.

Rubinoff, I. 1968. Central American sea-level canal: possible biological effects. Science 161:857–861.

Rubinoff, I. 1970. The sea-level canal controversy. Biol. Conserv. 3:33–36.

Rubinoff, R. W. and, I. Rubinoff. 1969. Observation on the migration of a marine goby through the Panama Canal. Copeia 1969:395–397.

Rubinoff, R. W., and I. Rubinoff. 1971. Geographic and reproductive isolation in Atlantic and Pacific populations of Panamanian *Bathygobius*. Evolution 25:88–97.

Saito, T. 1976. Geologic significance of coiling direction in the planktonic foraminifera *Pulleniatina*. Geology 4:305–312.

Selander, R. K. 1982. Phylogeny. In R. Milkman (Ed.). Perspectives on Evolution. Sunderland, Mass.: Sinauer, pp. 32–59.

Shulman, M. J., and E. Bermingham. 1995. Early life histories, ocean currents, and the population genetics of Caribbean reef fishes. Evolution 49:897–910.

Spivey, H. R. 1976. The cirripeds of the Panama Canal. Corrosion Marine-Fouling 1:43–49.

Tauber, C. A., and M. J. Tauber. 1989. Sympatric speciation in insects: perception and perspective. In D. Otte and J. A. Endler (Eds.). Speciation and Its Consequences. Sunderland, Mass.: Sinauer pp. 307–344.

Templeton, A. R. 1980. The theory of speciation by the founder principle. Genetics 92:1011–1038.

Templeton, A. R. 1981. Mechanisms of speciation—a population genetic approach. Annu. Rev. Ecol. Syst. 12:23–48.

Templeton, A. R. 1982. Genetic architectures of speciation. In C. Barigozzi (Ed.). Mechanisms of Speciation. New York: Liss, pp. 105–122.

Thomson, D. A., L. T. Findley, and A. N. Kerstitch. 1979. Reef Fishes of the Sea of Cortez. Tucson: University of Arizona Press.

Topp, R. W. 1969. Interoceanic sea-level canal: effects on the fish faunas. Science 165:1324–1327.

Vawter, A. T., R. Rosenblatt, and G. C. Gorman. 1980. Genetic divergence among fishes of the eastern Pacific and the Caribbean: support for the molecular clock. Evolution 34:705–711.

Vermeij, G. J. 1974. Marine faunal dominance and molluscan shell form. Evolution 28:656–664.

Vermeij, G. J. 1978. Biogeography and Adaptation. Cambridge, Mass.: Harvard University Press.

Vermeij, G. J. 1991. When biotas meet: understanding biotic interchange. Science 253:1099–1104.

Vermeij, G. J. 1993. The biological history of a seaway. Science 260:1603–1604.

Voight, J. R. 1988. Trans-Panamanian geminate octopods (Mollusca: Octopoda). Malacologia 29:289–293.

Wagner, M. 1868. Die Darwin'she Theorie und das Migrationsgesetz der Organismen. Leipzig: Duncker und Humblot.

Weinberg, J. R., and V. R. Starczak. 1988. Morphological differences and low dispersal between local populations of the tropical beach isopod, *Excirolana braziliensis*. Bull. Mar. Sci. 42:296–309.

Weinberg, J. R., and V. R. Starczak. 1989. Morphological divergence of eastern Pacific and Caribbean isopods: effects of a land barrier and the Panama Canal. Mar. Biol. 103:143–152.

West, D. A. 1980. Genetic Variation in Transisthmian Geminate Species of Brachyuran Crabs from the Coasts of Panama. Ph.D. Dissertation, Yale University.

White, M. J. D. 1978. Modes of Speciation. San Francisco: Freeman.

Woodring, W. P. 1966. The Panama land bridge as a sea barrier. Proc. Am. Philos. Soc. 110:425–433.

Wright, S. 1940. Breeding structure of populations in relation to speciation. Am. Nat. 74:232–248.

Wright, S. 1982a. Character change, speciation and the higher taxa. Evolution 36:427–443.

Wright, S. 1982b. The shifting balance theory and macroevolution. Annu. Rev. Genet. 16:1–19.

Wu, C.-I. 1985. A stochastic simulation study on speciation by sexual selection. Evolution 39:66–82.

Zouros, E. 1991. Searching for speciation genes in the species pair *Drosophila mojavensis* and *D. arizonae*. In G. M. Hewitt, A. W. B. Johnston, and J. P. W. Young (Eds.). Molecular Techniques in Taxonomy. Berlin: Springer-Verlag, pp. 33–71.

15

Rivers, Refuges, and Ridges

The Geography of Speciation of Amazonian Mammals

James L. Patton
Maria Nazareth F. da Silva

This chapter examines the general issue of the geography of evolutionary divergence, at both the species and infraspecies levels, and seeks to determine the mode of species formation from information about the geographic and phylogenetic relationships of extant populations. The combination of geography and genealogy is the emergent field of "phylogeography," as defined by Avise and his co-workers a decade ago (Avise et al., 1987; Avise, 1989), and emphasized in Riddle's (1996) recent review. By examining the spatial distribution and genealogical propinquity of individual genes, and by the separation of population genetic structure from population history by explicit analytical methods (e.g., Templeton et al., 1995), it is possible to test alternative models of speciation and to search for generality for any given group of organisms, or for any specific geographic region, in ways not previously possible. Our area of focus in this chapter is the lowland tropical forested region of the Amazon basin; our taxa of interest are small-bodied terrestrial and arboreal mammals.

The means and timing of the diversification of the biota of Amazonia have received considerable attention in the past several decades, in part because the basin is perhaps home to the world's most diverse flora and fauna, whether diversity is measured at the level of local communities or across geography. Several models have been developed as explanations of the large diversity measures at both levels. We emphasize the utility of the phylogeographic perspective as a methodological approach for evaluating specific speciation modes. While available data yet remain too sparse for the vast majority of mammalian taxa, some patterns are beginning to emerge that call into question the current dogma regarding a recency of Amazonian diversification in general and specific models underlying divergence. Rather, these patterns are suggestive of a much deeper history, one that is concordant with episodes of Andean uplift.

The Phylogeographic Approach

Phylogeography involves the principles and processes governing the geographical distributions of genealogical lineages. Importantly, this includes lineages at the intraspecific level, for it is through the reconstruction of ancestral from current patterns that explicit models of diversification can be identified. An excellent example of this approach was presented by Harrison (1991) in his review of the molecular changes at speciation (see also Brooks and McLennan [1991] and Frey [1993], both of whom emphasize interspecific patterns only). Harrison showed that the combination of phyletic and geographic distribution of alleles (=haplotypes) between populations and closely related species can be used to infer the geographical relationships of ancestral populations prior to divergence. Put another way, each of the geographical models of species formation (allopatric, parapatric, or sympatric) has explicit expectations relative to any given gene tree in the descendent units (daughter populations or species).

Figure 15.1 illustrates the combination of gene tree and geographic position of populations under three different allopatric divergence models: vicariance (figure 15.1A,B), peripheral isolation (figure 15.1C), and colonization of "empty" habitat (figure 15.1D). In each, the derivative "daughter" lineage that represents a new species has a different set of phylogenetic and geographic relationships at the time of, and immediately following, speciation. This difference in expectation thus offers an opportunity to evaluate any particular allopatric scenario for a given set of closely related taxa. For example, disruption of a continuous range by a vicariant event may either be coincidental with the prior phylogeographic structure of the interacting populations (figure 15.1A) or not congruent with it (figure 15.1B). If the former is true, both genealogical and geographical components will be concordant, even at low levels of divergence, while for the latter, congruence will come only following the achievement of reciprocal monophyly at some future time (e.g., Tajima, 1983; Avise et al., 1983; Neigel and Avise, 1986). In a similar fashion, expectations for sympatric speciation include the combination of inclusive ranges with requisite sister relationships of the sympatric pair relative to others with allopatric or parapatric distributions (figure 15.2; see also Lynch, 1989, for further discussion and caveats). Finally, for parapatric divergence, daughter taxa are expected both to be phyletic sisters and to be distributed on opposite sides of a demonstrable gradient. While the phylogeographic pattern of a given diversification process, as illustrated in figures 15.1 and 15.2, is often simple, and while its signature might well be erased with time, these expectations nonetheless serve as guides to test geographic modes of speciation.

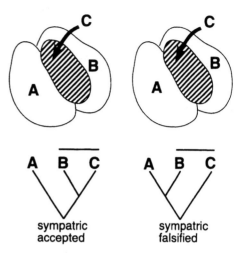

Figure 15.2. The linkage of phylogeny and geographic position can be used to distinguish between allopatric and sympatric speciation. Left, derivative species C is the sister taxon to species B, and its range is included within that of its progenitor; sympatric divergence is thus accepted. Right, while geographic position of derivative species C is contained within the range of species B, these two share a paraphyletic, not monophyletic, relationship; sympatric divergence is falsified. Taken from Lynch (1989).

Phylogeography, Reciprocal Monophyly, and Vicariance Biogeography

Avise and his co-workers (reviewed in Avise 1994) state that cases wherein populations are hierarchically organized into reciprocally monophyletic groups distinguished by large phylogeographic gaps probably result from long-term extrinsic barriers to gene flow. They go on to argue that the geographic placement of these phylogeographic gaps will likely be concordant among different species, such that the gaps both identify biogeographic boundaries and common historical events. In this context, the phylogeographic identification of reciprocally monophyletic assemblages with similar distributions among different taxa both identifies an allopatric origin and signals a vicariant history.

The theory of vicariant biogeography (Platnick and Nelson, 1978) argues that shared distribution patterns among diverse clades of organisms are more parsimoniously explained by the occurrence of extrinsic geological or climatic events, which affected all groups in the same way, than by intrinsic aspects of each group's history, which coincidentally produced parallel geographical patterns. Concordant distributions, or generalized tracks (Rosen, 1975), may thus be used as evidence to infer the biogeographical history of a region. As argued above, phylogeographic methodology can identify common patterns of both geographic placement

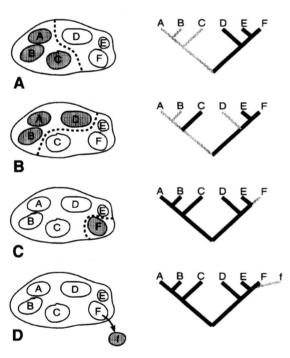

Figure 15.1. Four models of allopatric speciation and corresponding gene trees. A, vicariance subdivision that is congruent with phylogeographic structure at origin of barrier. B, vicariance subdivision that is initially incongruent with phylogeographic structure; congruence is expected in time with achievement of reciprocal monophyly. C, divergence by peripheral isolation. D, colonization of new habitat or region by a propagule from a single population of the ancestral species. Redrawn from Harrison (1991).

and genealogy, and thus may elucidate vicariant bio-geographic events. Elegant examples of this connection between phylogeographic pattern and vicariant history are the studies of Avise and his co-workers (e.g., Avise, 1992) of faunal assemblages along the coastal margins of the eastern and southern United States, and those of Moritz and his associates on rainforest vertebrates in northeastern Australia (e.g., Joseph et al., 1995).

Amazonian Diversification: Models and Problems

The high species diversity encountered in the lowland tropical forests of Amazonia has been hypothesized to result from riverine barriers (Wallace, 1852, 1853, 1876), Pleistocene habitat refuges (Haffer, 1969; Vanzolini and Williams, 1970), parapatric divergence due to sharp ecological gradients (Endler, 1977, 1982), long-term paleoclimatic shifts (Bush, 1994) or cycles (Haffer, 1993a), floodplain dynamics (Salo, 1988), and ecological heterogeneity (Tuomisto et al., 1995). There is no general consensus as to the validity of any of these processes, although the Refuge Hypothesis has received the widest attention (reviewed in Prance, 1982; Whitmore and Prance, 1987; evaluated in Bush, 1994; Salo, 1987). A lack of model consensus for any or all groups of organisms, however, should not be unexpected. For one, organisms have different inherent population structures and processes, and thus are likely to respond uniquely to the same historical event, be it the formation of a river or habitat island. Moreover, none of these hypotheses are mutually exclusive, and all are, therefore, of potential importance for any given set of taxa.

Both riverine barriers and habitat refuges will result in allopatric divergence, as each involves a vicariant process. Each, therefore, should produce phylogeographic patterns similar to those illustrated in figure 15.1. The gradient model is a parapatric divergence process, with its own explicit phylogeographic expectations (see below). The other models identified above either lack a specific geographic context in their formulation (e.g., long-term paleoclimatic shifts and cycles) or could potentially result in species formation under any geographic context (floodplain dynamics and ecological heterogeneity). A phylogeographic approach could be used to determine, for example, if the temporal shift of river channels and the ecological heterogeneity that results have generated allopatric, parapatric, or sympatric divergence for any taxon group. The phylogeographic approach, however, does not seem fruitful as an evaluator of long-term climate shifts or cycles.

One of the major problems with the achievement of any consensus for diversification processes in Amazonia, or elsewhere, has been the difficulty in falsifying a particular hypothesis, uniformly applied across a varied set of specific, or exemplar, organisms. However, some speciation models are amenable to falsification by phylogenetic methodologies. For example, the Gradient Hypothesis is falsified if the requisite phylogenetic propinquity of the taxa located on either side of an identifiable gradient is not verified by robust phylogenetic analysis (see Patton, 1987; Patton and Smith, 1992; but see Endler, 1983). Similarly, the underlying assumption of the Riverine Hypothesis is that rivers are primary barriers in an allopatric speciation process. If so, then the hypothesis expects genealogical similarity of populations along one side of a river as opposed to those on the opposite side and requires sister relationships between the opposite-bank populations relative to those elsewhere within the taxon range. Lack of the latter expectation falsifies the applicability of this model.

Unfortunately, although the Refuge Hypothesis has been the most widely applied to tropical systems, it is nearly impossible to test by phylogenetic methods. This is because the model does not specify any hierarchical temporal division of refuges that can be tracked by concordant taxon-splitting events. The hypothesis also does not specify a pattern of area relationships among the refuges that cannot be matched by expectations from other models (see discussion by Endler, 1982, 1983; Lynch, 1988). Moreover, because the glacial–interglacial cycles of the Pleistocene were repeated numerous times, each successive one could potentially override or erase the phylogeographic signal of those previous, leaving a record impossible to decipher and, hence, to test. This is also a major problem with Haffer's (1993a) more recent emphasis on long-term paleoclimatic cycles driven by Milankovitch oscillations. Even if true, this hypothesis simply cannot be tested, as it is not possible to discern the effects now of these types of cyclical changes without a priori expectations as to a resultant phylogeographic pattern.

The approach to this general problem of hypothesis testing that we and our colleagues have used has been to accumulate phylogeographical data for a diversity of unrelated small mammals throughout Amazonia. In so doing, we hope to determine the degree to which the independent patterns of geographic and genealogic distribution are concordant, and thereby indicate the degree to which there has been a common history underlying their divergences (Cracraft, 1988; Rosen, 1975). We can then address some of the models mentioned above by this phylogeographic approach, but, as stated, not all. This is a daunting task, as the potential numbers of mammalian taxa are great, our knowledge of their distribution and systematics is exceedingly poor, and the geographic span of Amazonia that should be sampled is very large (some 6 million km^2 of tropical rain forest). Nevertheless, we have been using comparative molecular sequence data, primarily from the mitochondrial genome (mtDNA), with some success to delineate the degree and pattern of phylogeographic structure in target taxa of mammals (e.g., da Silva and Patton, 1993; Patton et al., 1994, 1996, 1997; Peres et al., 1996). These types of data have added value in an analysis of the biogeographic history of any region since molecular distances scale approximately with time, and thus indicate the relative age of splitting events, in addition to their geographical location. Moreover, because the phylogeographic approach utilizes both intra- and interspecific gene trees, it is also possible, although certainly not automatic, that the geographic mode of species formation (allopatric, parapatric, or sympatric) can be ascertained.

Amazonian Mammals as Exemplars

The geographic mode of speciation for Amazonian mammals has not been explicitly examined for any group, with the exception of primates. Indeed, the oldest perspective on Amazonian diversification, that of Alfred Russel Wallace (1852, 1853, 1876), was based on distribution patterns of species of monkeys. While there is no reason to assume that any one geographic model might predominate (see, e.g., Bush, 1994), to our knowledge there has been no suggestion in the literature for either sympatric or parapatric divergence in relation to mammals. Indeed, we would argue that sympatric divergence would be very unlikely for any mammalian group, primarily because mammals in general are neither habitat nor food specialists, but rather exhibit relatively broad ranges in both. Consequently, diversifying selection due to differences in habitat patches within the "cruising range" of even small-bodied mammals is unlikely. The same can be concluded for parapatric divergence across sharp ecological gradients, although this is more problematic since many species of mammals have parapatric, or contiguously allopatric, ranges with boundaries at ecotones, which might suggest a parapatric process. Certainly, this geographical mode of speciation should be examined whenever possible, since it is amenable to falsification by phylogeographic methods.

Riverine Barriers

Wallace (1852, 1853, 1876) observed that geographic boundaries of primate species, or races, were coincidental with rivers, and he proposed that Amazonia was divided into four major biogeographic regions bounded by the Amazon/Solimões, Negro, and Madeira rivers. His model is especially appealing, since Amazonia is clearly dissected by some of the largest rivers in the world, in terms of width, water flow, and linear extent, with their origins in the highlands of the Andes and the Guianan and Brazilian plateaus. Ayres (1986) and Ayres and Clutton-Brock (1992) provided quantitative support for the model by documenting that the similarity of primate species communities on opposite banks of Amazonian rivers was strongly and negatively correlated with river

discharge, a measure of river size. These authors also showed that primate community similarity decreased along the length of the Amazon River, from upriver in Peru to near its mouth in eastern Brazil. Thus, rivers in general are potential barriers, and single rivers are expected to vary in their strength as barriers from headwaters to mouth as their width increases.

A general model of riverine diversification includes two predictions and one corollary. The predictions are (1) that populations on one side will be monophyletic relative to those on the opposite side, and (2) that the sets of opposite-bank populations/taxa will form a sister group relative to those from elsewhere in the species range. The corollary is that, if gene flow were to occur across the river, it would be more likely in the narrower headwater regions than at sites downriver. Consequently, successive downriver sites on opposite banks should exhibit increased degrees of genetic divergence (Hershkovitz, 1977; see also Haffer, 1993b). This model for Amazonian rivers also suggests that the strength of the potential barrier will be different for specialists of the floodplain forests (várzea) along river margins than for upland (terra firme) specialists, whose populations are separated by both river and floodplain width (discussed in Haffer, 1993b; Peres et al., 1996).

While the similarity in species community of primates is related to river width (Ayres, 1986; Ayres and Clutton-Brock, 1992), and the ranges of primate species often appear bounded by rivers (e.g., Hershkovitz, 1977), only two studies have examined the degree to which rivers serve as barriers to gene exchange between populations of Amazonian mammals. The explicit expectations of riverine patterns, with cross-river gene flow in the headwaters and progressive differentiation down both banks was observed for saddle-back tamarins (Saguinus fuscicollis) along the Rio Juruá in western Brazil (figure 15.3; Peres et al., 1996). This is a species with strongly marked races, the boundaries of which typically lie along opposite banks of major rivers within the western basin (Hershkovitz, 1977). However, while the pattern of haplotype sharing conforms to riverine expectations, the Rio Juruá was apparently not a primary barrier in the diversification of these tamarins, since phylogenetic analysis does not indicate that the two races found within the Rio Juruá basin (the left-bank fuscicollis and right-bank melanoleucus) share a sister relationship, as would be required (figure 15.3; Jacobs et al., 1995).

Spiny tree rats of the genus Mesomys, the second mammalian taxon explicitly examined for riverine effects, was also sampled along the Rio Juruá. Populations of this species, however, did not exhibit the pattern of cross-river divergence expected for the riverine model (Patton et al., 1994). Rather, opposite-bank samples were more similar in the mouth region than in the headwaters, contrary to expectations. Moreover, coalescence methodology applied to mitochondrial DNA haplotypes

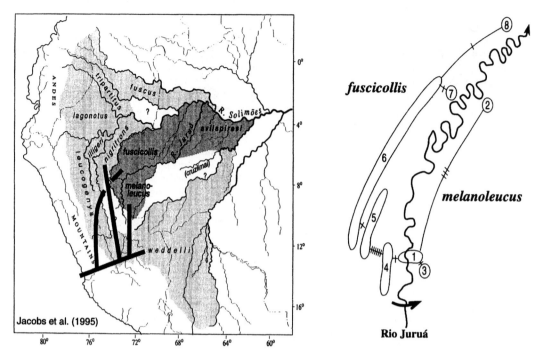

Figure 15.3. Right, haplotype network for upriver populations of the saddle-back tamarin (*Saguinus fuscicollis*) of the Rio Juruá, western Brazil. Individual cytochrome b haplotypes are numbered; the number of base substitution differences is indicated by tick marks on the interconnecting lines. The discordance of the haplotype network with river bank position in the headwaters region is interpreted to have resulted from the movement of right-bank *melanoleucus* individuals to the left bank, which preceded the genetic interaction of *melanoleucus* with left-bank populations of *fuscicollis*. Movement resulted from passive transfer due to shifting river channel, in the direction of the arrow. Historical records document the transfer of right-bank sections to the left-bank by meander shifts in the region (see Peres et al., 1996). Left, distribution of most races of the saddle-back tamarin, *Saguinus fuscicollis* (from Hershkovitz, 1977). The phylogenetic relationships among races bordering the Rio Juruá is indicated, based on mtDNA sequences (Jacobs et al., 1995).

indicated numerous historical movements across the river in both directions along its entire 1,000 km length.

Two studies, along the same river, and with somewhat contradictory results, are inadequate to assess the importance of rivers in the diversification process of Amazonian mammals. While patterns typical of saddle-back tamarins do suggest conformation to riverine expectations, only sister relationships among opposite-bank taxa would support rivers as primary barriers in the diversification process. Nonsister relationships would suggest that rivers are simply the convenient secondary meeting points for taxa that diverged elsewhere and subsequently spread to their present opposite-bank positions (Simpson and Haffer, 1978; Haffer, 1993b). Since primary and secondary divergence have explicit and generally opposite phylogenetic expectations, phylogeographic analysis can be used to distinguish between them.

Amazonian rivers are typically quite dynamic, with constantly shifting channels. Consequently, the passive transport of populations from one side to the other by lateral meanders, or ox-bow cutoffs, is expected. Individual organisms do not need to swim across rivers, just hitch a ride when chunks of habitat get shifted from one side to the other as a result of lateral channel migration (Hershkovitz, 1977, 1983). Consequently, riverine effects might be expected to be minimal for floodplain specialists and more important for taxa restricted to terra firme. However, even the nonflooded upland forested areas of Amazonia show well-preserved marks of earlier terraces and abandoned floodplains, indicating a complex and dynamic history (Räsänen et al., 1987; Salo et al., 1986). This apparent long-term dynamism of the positional placement of rivers renders it difficult to assess their importance in primary diversification, since constant shifting through time is likely to erase, or otherwise obscure phylogeographic pattern. Consequently, as appealing as the Riverine Hypothesis might be, there are many reasons why such an effect might not be immediately evident from a phylogeographic perspective.

Pleistocene Refuges

The refuge model was proposed independently by Haffer (1969) and Vanzolini (1970; see also Vanzolini and Williams, 1970) as an explanation for observed regional areas of high endemism in birds and lizards, respectively. Broadened by the inclusion of a number of angiosperm (e.g., Prance, 1973) and butterfly (e.g., Brown, 1982) taxa, the idea that the continuous lowland forest present today was fragmented into isolated patches at times coincidental with glacial cycles of the Pleistocene has received considerable interpretative support by numerous workers (see chapters in Prance, 1982). Refuge formation is an ecological vicariant process, and resulting diversification of taxa affected by refuges is thus necessarily allopatric. Geographic differentiation and speciation patterns of mammals that have been interpreted under this model have mostly included primates (e.g., Froehlich et al., 1991), although none of these studies have successfully falsified competing hypotheses explaining the observed patterns. As noted above, while the refuge hypothesis has had the greatest impact on recent discussions of diversification processes of Amazonian organisms, the model itself is generally untestable by the phylogeographic approach advocated here.

Bush (1994), in his excellent review of processes underlying Amazonian speciation, summarizes the available evidence in support of the refuge hypothesis and shows that it is flawed both on spatial and temporal scales (e.g., Cracraft and Prum, 1988; Lynch, 1988; Nelson et al., 1990) as well as by its climatological assumptions. He develops instead a paleoclimatic model that combines glacial age cooling, reduced atmospheric CO_2, and moderate reductions in precipitation. He also argues that Amazonian forests did not fragment during the Pleistocene but that the centers of endemism observed for modern taxa resulted from maximal disturbances within and between communities without modern analogues rather than to the stability of current communities as habitat refuges (see also Colinvaux et al., 1996). This hypothesis is somewhat testable by the approach we advocate here, for, if past communities of mammals were disharmonious assemblages of modern taxa, there should be little concordance in the phylogeographic structure exhibited by these taxa today. Consequently, strongly concordant patterns of geographic structure for unrelated organisms would represent partial falsification. However, other than this prediction, phylogeography cannot provide a means to assess any paleoclimatic model, this one or Haffer's original refuge idea, unless there is an explicit expectation of area relationships embedded within the model itself.

Andean Orogeny and Paleobasin Formation

Our initial phylogeographic analyses of arboreal echimyid rodents (da Silva and Patton, 1993) identified concordant patterns across a group of genera that divided the Amazon basin into at least three successive west-to-east phylogeographic units. While we recognized that such concordant patterns likely reflected a common history, we did not identify what that historical event (or events) might have been. Our sampling of both taxa and geography has now increased to the point where specific historical events can be implicated.

Figure 15.4 illustrates the patterns of clade divergence in the same mtDNA sequence (of the cytochrome b gene) for a group of western Amazonian small mammals, including one species of echimyid and three species of murid rodents. These, and 13 additional species complexes of small-bodied marsupials and other rodents, were sampled along the length of the Rio Juruá in western Brazil. The goal of the original fieldwork and subsequent phylogeographic analyses was to assess the validity of Wallace's Riverine Hypothesis (see above) for a set of commonly distributed small mammal taxa. The pattern that emerged, however, was an unexpected one (da Silva, 1995; da Silva and Patton, unpublished). None of these species of marsupials or rodents exhibited patterns of differentiation predicted from riverine divergence (see above). Rather, 11 of the 17 taxa exhibited a pattern of deeply divergent haplotype clades with concordant phylogeographic breaks in the central part of the Rio Juruá (figure 15.4). Haplotype clades within each taxon were strongly and regionally monophyletic, with divergences between clades averaging from more than 4% to nearly 14%, yet with divergences within them typically less than 1% (figure 15.4, table 15.1).

The singular phylogeographic break in haplotype assemblages across this broad spectrum of Amazonian small mammals supports a common vicariant history. While the geological documentation for the origin and history of the Amazon basin is poorly understood, at best, there are nonetheless features of the basin that provide tantalizing linkages between geological events and the phylogeographic record of these organisms. The Amazon basin is composed of several subbasins lying in different tectonic settings, each separated by structural arches. Those subbasins in the western Amazon were formed by subsidence apparently due to compressive foreland shortening resulting from Andean uplift during the mid to late Tertiary (figure 15.5; reviewed in Räsänen et al., 1987, 1992). One of these major structural arches, the Iquitos (or Jutai), cuts perpendicularly across the middle section of the Rio Juruá in a position coincidental with the phylogeographic breaks observed for the mammalian taxa we have examined, dividing the river into the paleo fluvial deposition systems of the upriver Acre Basin and the downriver Central Amazon Basin (figure 15.5). The exact timing of subbasin formation, and the ecological conditions that would have resulted across the area of the Iquitos Arch are unknown. However, the degree of haplotype clade structure is certainly deep, with estimated divergence times ranging between 1 and more

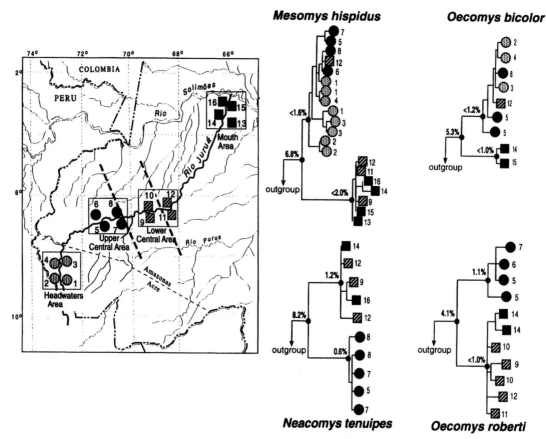

Figure 15.4. Phylogeographic concordance of mtDNA haplotype clades for four species of western Amazonian rodents along the Rio Juruá in western Brazil: the spiny tree rat, *Mesomys hispidus*; the semi-arboreal rice rats, *Oecomys bicolor* and *O. roberti*; and the terrestrial spiny mouse, *Neacomys* cf. *tenuipes*. Individual haplotypes identified in each of the four species trees are numbered according to the sample locality from which each was encountered, as indicated in the map of collecting localities (left). The average percentage sequence divergence (Kimura two-parameter distances) between each clade and relative to an outgroup, is indicated above the appropriate nodes. In each example, haplotype clades either break between the Upper Central and Lower Central geographic sample areas along the Rio Juruá, or overlap within the latter area. The approximate position of the Iquitos Arch underlying the Rio Juruá basin is indicated by heavy dashed lines.

than 3 million years (da Silva and Patton, 1993; table 15.1). Despite a substantial sampling error in these time estimates due to the relative short segment of DNA sequence available (approximately 800 base pairs), divergences largely predate the Quaternary and are within the presumptive time range for subbasin formation.

Other taxa, such as the terrestrial rice rat, *Oryzomys megacephalus*, which does not exhibit phylogeographic structure along the Rio Juruá (Patton et al., 1996), is still composed of strongly divergent haplotype clades across its distribution within Amazonia (figure 15.6). If our suggestion of a causal linkage between tectonic activity generating an earlier subdivision of the present Amazon basin with phylogenetic lineage splitting events is correct, the position of clades within taxa like *O. mega-*

cephalus may map to other structural arches, such as the Purus Arch, in central Amazonia (figure 15.5). In any event, a general role for subbasin formation and Amazonian speciation pattern and process is testable by focused phylogeographic sampling across other areas in the basin where subbasins and arches have been identified, as well as in other drainages that are explicitly intersected by the Iquitos Arch (such as the Rio Javari to the north of the Rio Juruá, and the Rio Purus to the south).

Conclusions and Prospectus

Bush (1994:15) concludes his review of speciation processes in Amazonian organisms with the statement that

Table 15.1. Pattern of haplotype divergence for 18 species of small-bodied mammals collected along the Rio Juruá in western Amazonian Brazil.

| Species | Regional Effect | | Riverine Effect | No Effect |
	% Divergence	Time[1]		
Proechimys sp. 3/4[2]	13.7%	2.81	Saguinus fuscicollis	Didelphis marsupialis
Isothrix bistriata[3]	11.2%	1.33		Micoureus demerarae
Proechimys sp. 2[2]	10.1%	2.52		Oryzomys megacephalus[4]
Dactylomys sp.[3]	9.5%	1.70		Oligoryzomys microtis[4]
Neacomys "tenuipes"	8.2%	3.17		Proechimys simonsi[2]
Mesomys hispidus	6.8%	1.05		Proechimys steerei[2]
Makalata didelphoides[3]	5.9%	0.22		
Oecomys bicolor	5.3%	1.11		
Oryzomys macconnelli	4.4%	1.76		
Oecomys roberti	4.1%	0.44		

Taxa listed under "Regional Effect" are those that exhibit strongly concordant monophyletic clade structure of the type illustrated in fig. 15.4. Those that exhibit "No Effect" show no discernable structure along the river (see, e.g., Patton et al., 1996). Kimura two-parameter distances between haplotypes of the mitochondrial cytochrome b gene (from 411 to 801 bases, depending upon the taxon; data from Patton et al., unpublished are given for those taxa exhibiting regional effects. Temporal divergence estimates for these taxa based on third position transversions are also given (see da Silva and Patton, 1993).

[1]Based on an estimated rate of 1.7% third position transversions per million years (see da Silva and Patton, 1993).

[2]Data from da Silva (1995).

[3]Data from da Silva and Patton (1993).

[4]Data from Patton et al. (1996).

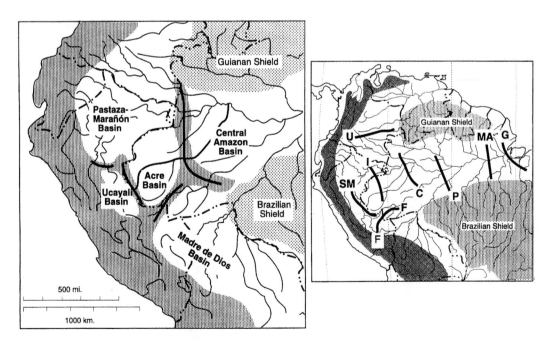

Figure 15.5. Left, major geological structures in western Amazonia, indicating position of molasse basins and intervening arches. The Iquitos (or Jutai) Arch cuts perpendicularly across the Rio Juruá (highlighted) and divided the present basin into the Acre and Central Amazonian paloebasins. Redrawn from Räsänen et al. (1987). Right, generalized positions of structural uplift arches across Amazonia. SM = Serra do Moa Arch; F = Fitzcarrald Arch; I = Iquitos Arch; C = Carauari Arch; U = Uaupes Arch; P = Purus Arch; MA = Monte Alegre Arch; G = Gurupa Arch. Redrawn from Räsänen et al. (1992).

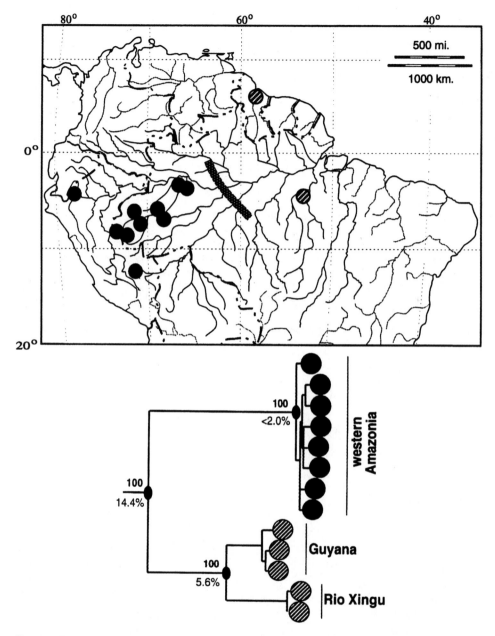

Figure 15.6. Phylogeography of sampled populations of the terrestrial rice rat, *Oryzomys mega-cephalus*, from the Amazon basin. There is a lack of structure for this species from throughout western Amazonia, including the Rio Juruá. However, two strongly divergent and reciprocally monophyletic haplotype clades are evident coincidental with western and eastern Amazonian positions. It remains to be determined if these clades border one another along one of the central Amazonian structural arches, such as the Purus Arch, the approximate position of which is indicated by the broad stippled band. The haplotype tree was rooted with *O. yunganus*, another member of the same species complex, as the outgroup. Average percentage sequence divergence (Kimura two-parameter distances) is indicated below each of the three major nodes (bottom); bootstrap values based on 500 replicates are indicated above each node.

"the need for a new biogeographic appraisal of the Amazon basin using paleoecological and modern phylogenetic [sic] techniques is clear." He goes on to suggest that "the history we reveal will almost certainly be complex and to some extent species-specific . . . with enough detail, the factors forcing speciation, and the timing of those events may help to resolve the riddle of tropical species richness."

The history of the Amazon basin is complex, with periods of Andean uplift and marine incursions (Hoorn, 1993; Räsänen et al., 1995), and with a redirection of the flow of the Amazon River from north into the Caribbean to its present eastward direction into the Atlantic (Hoorn, 1994). Nonetheless, common patterns of geographic positioning of phylogenetic units suggests the importance of single vicariant events in the evolution of organismal diversity. Thus, the coincidental position of phylogeographic breaks of extant mammals along the Rio Juruá in western Brazil with that of paleobasins separated by the Iquitos Arch, suggests a causal linkage between them. This linkage is strengthened by consideration of the timing of subbasin formation (somewhere between late Miocene through Pliocene) and clade formation estimated from molecular divergences. The dissection of the Rio Juruá basin by a structural arch, dividing the present river system into internal paleobasins, each with its own individual drainage patterns, also suggests that there may be an Amazonia-wide linkage between arch position, with interarch paleobasins forming the centers for diversification of Amazonian organisms. If the pattern observed for Rio Juruá mammals is repeated for other classes of organisms, and in other regions within Amazonia, then there may indeed be a relatively uniform mode of speciation for this large and diverse area. Continued phylogeographic analyses can thus provide a test for the generality of this model.

Acknowledgments   We thank our co-workers on the Projeto Juruá (Jay R. Malcolm, Carlos Peres, and Claude Gascon) through whose collective efforts the materials that form the basis for this chapter were collected and analyzed. Our work in Brazil has been under the auspices of the Conselho Nacional de Desenvolvimento Científico e Tecnológico (CNPq), the Instituto Brasileiro do Meio Ambiente e dos Recursos Naturais Renovavais (IBAMA), and the Instituto Nacional de Pesquisas da Amazônia (INPA); we are exceedingly grateful to each. Fieldwork was generously supported by grants from the National Geographic Society, Wildlife Conservation International, and the Museum of Vertebrate Zoology; laboratory analyses were underwritten by grants from the National Science Foundation. We thank Elizabeth Hadly, Márcia Lara, Meika Mustrangi, and Albert Ditchfield for helping to develop much of the content of this chapter. We also benefited greatly by discussions of our data and ideas with Jukka Salo and Hanna Tuomisto, and especially Jürgen Haffer, who, despite disagreements on interpretation, has been extremely generous with his time and knowledge. Finally, Rick Harrison and Amy McCune provided insightful comments on the manuscript, although neither should be held responsible for its contents.

References

Avise, J. C. 1989. Gene trees and organismal histories: a phylogenetic approach to population biology. Evolution 43:1192–1208.

Avise, J. C. 1992. Molecular population structure and the biogeographic history of a regional fauna: a case history with lessons for conservation biology. Oikos 63:62–76.

Avise, J. C. 1994. Molecular Markers, Natural History, and Evolution. Chapman and Hall, New York.

Avise, J. C., Shapira, J. F., Daniel, S. W., Aquadro, C. F., and Lansman, R. A. 1983. Mitochondrial DNA differentiation during the speciation process in *Peromyscus*. Mol. Biol. Evol. 1:38–56.

Avise, J. C., Arnold, J., Ball, R. M., Bermingham, E., Lamb, T., Neigel, J. E., Reeb, C. A., and Saunders, N. C. 1987. Interspecific phylogeography: the mitochondrial DNA bridge between population genetics and systematics. Annu. Rev. Ecol. Syst. 18:489–522.

Ayres, J. M. 1986. Uakaris and Amazonian flooded forests. Ph.D. dissertation, University of Cambridge.

Ayres, J. M., and Clutton-Brock, T. H. 1992. River boundaries and species range size in Amazonian primates. Am. Nat. 140:531–537.

Brooks, D. R., and McLennan, D. A. 1991. Phylogeny, Ecology, and Behavior. University of Chicago Press, Chicago.

Brown, K. S., Jr. 1982. Historical and ecological factors in the biogeography of aposematic neotropical butterflies. Am. Zool. 22:453–471.

Bush, M. B. 1994. Amazonian speciation: a necessarily complex model. J. Biogeogr. 21:5–17.

Colinvaux, P. A., De Oliveira, P. E., Moreno, J. E., Miller, M. C., and Bush, M. B. 1996. A long pollen record from lowland Amazonia: forest and cooling in glacial times. Science 274:85–88.

Cracraft, J. 1988. Deep-history biogeography: retrieving the historical pattern of evolving continental biotas. Syst. Zool. 37:221–236.

Cracraft, J., and Prumm, R. O. 1988. Patterns and processes of diversification: speciation and historical congruence in some Neotropical birds. Evolution 42:603–620.

da Silva, M. N. F. 1995. Systematics and phylogeography of Amazonian spiny rats of the genus *Proechimys* (Rodentia: Echimyidae). Ph.D. dissertation, University of California, Berkeley.

da Silva, M. N. F., and Patton, J. L. 1993. Amazonian phylogeography: mtDNA sequence variation in arboreal echimyid rodents (Caviomorpha). Mol. Phyl. Evol. 2:243–255.

Endler, J. A. 1977. Geographic Variation, Speciation, and Clines. Monogr. Pop. Biol. 10, Princeton University Press, Princeton, N.J.

Endler, J. A. 1982. Pleistocene forest refuges: Fact or fancy? In G. T. Prance, ed. Biological Diversification in the Tropics. New York: Columbia University Press, pp. 641–657.

Endler, J. A. 1983. Testing causal hypotheses in the study of geographic variation. In J. Felsenstein, ed. Numerical Taxonomy. NATO ASI Series, Vol. G1. Berlin: Springer-Verlag, pp. 424–443.

Frey, J. K. 1993. Modes of peripheral isolate formation and speciation. Syst. Biol. 42:373–381.

Froehlich, J. W., Supriatna, J., and Froehlich, P. H. 1991. Morphometric analysis of *Ateles*: systematic and biogeographic implications. Am. J. Primatol. 25:1–22.

Haffer, J. 1969. Speciation in Amazonian forest birds. Science 165:131–137.

Haffer, J. 1993a. Time's cycle and time's arrow in the history of Amazonia. Biogeographica 69:15–45.

Haffer, J. 1993b. On the "river effect" in some forest birds of southern Amazonia. Bol. Mus. Para. Emílio Goeldi, sér. Zool. 8:217–245.

Harrison, R. 1991. Molecular changes at speciation. Annu. Rev. Ecol. Syst. 22:281–308.

Hershkovitz, P. 1977. Living New World Monkeys (Platyrrhini) with an Introduction to Primates, Vol. 1. University of Chicago Press, Chicago.

Hershkovitz, P. 1983. Two new species of night monkeys, genus *Aotus* (Cebidae, Platyrrhini): a preliminary report on *Aotus* taxonomy. Am. J. Primatol. 4:209–243.

Hoorn, C. 1993. Marine incursions and the influence of Andean tectonics on the Miocene depositional history of northwestern Amazonia: results of a palynostratigraphic study. Palaeogeogr. Palaeoclim. Palaeoecol. 105:267–309.

Hoorn, C. 1994. An environmental reconstruction of the palaeo-Amazon River system (middle-late Miocene, NW Amazonia). Palaeogeogr. Palaeoclim. Palaeoecol. 112:187–238.

Jacobs, S. C., Larson, A., and Cheverud, J. M. 1995. Phylogenetic relationships and orthogenetic evolution of coat color among tamarins (genus *Saguinus*). Syst. Biol. 515–532.

Joseph, L., Moritz, C., and Hugall, A. 1995. Molecular support for vicariance as a source of diversity in rainforest. Proc. Roy. Soc., London B 260:177–182.

Lynch, J. D. 1988. Refugia. In A. A. Myers and P. S. Giller, eds. Analytical Biogeography. London: Chapman and Hall, pp. 311–342.

Lynch, J. D. 1989. The gauge of speciation: on the frequencies of modes of speciation. In D. Otte and J. A. Endler, eds. Speciation and Its Consequences. Sunderland, Mass.: Sinauer, pp. 527–553.

Neigel, J. E., and Avise, J. C. 1986. Phylogenetic relationships of mitochondrial DNA under various demographic models of speciation. In E. Nevo and S. Karlin, eds. Evolutionary Processes and Theory. New York: Academic Press, pp. 515–534.

Nelson, B. W., Ferreira, C. A. C., da Silva, M. F., and Kawasaki, M. L. 1990. Endemism centres, refugia and botanical collection density in Brazilian Amazonia. Nature 345:714–716.

Patton, J. L. 1987. Patrones de distribución y especiación de la fauna de mamíferos de los bosques nublados andinos del Perú. An. Mus. Hist. Nat., Valparaiso 17:87–94.

Patton, J. L., and Smith, M. F. 1992. mtDNA phylogeny of Andean mice: a test of diversification across ecological gradients. Evolution 46:174–183.

Patton, J. L., da Silva, M. N. F., and Malcolm, J. R. 1994. Gene genealogy and differentiation among arboreal spiny rats (Rodentia: Echimyidae) of the Amazon Basin: a test of the riverine barrier hypothesis. Evolution 48:1314–1323.

Patton, J. L., da Silva, M. N. F., and Malcolm, J. R. 1996. Hierarchical genetic structure and gene flow in three sympatric species of Amazonian rodents. Mol. Ecol. 5:229–238.

Patton, J. L., da Silva, M. N. F., Lara, M. C., and Mustrangi, M. A. 1997. Diversity, differentiation, and the historical biogeography of non-volant small mammals of the neotropical forests. In W. F. Laurance, R. O. Bierregaard, and C. Moritz, eds. Tropical Forest Remnants: Ecology, Management and Conservation of Fragmented Communities. Chicago: University of Chicago Press, pp. 455–465.

Peres, C. A., Patton, J. L., and da Silva, M. N. F. 1997. Riverine barriers and gene flow in Amazonian saddle-back tamarins. Folia Primatol. 67:113–124.

Platnick, N. I., and Nelson, G. 1978. A method for analysis of historical biogeography. Syst. Zool. 27:1–16.

Prance, G. T. 1973. Phytogeographic support for the theory of forest refuges in the Amazon Basin, based on evidence from distribution patterns in Caryocaraceae, Chrysobalanaceae, Dichapetalaceae and Lecythidaceae. Acta Amazonica 3:5–28.

Prance, G. T. (ed.). 1982. Biological Diversification in the Tropics. Columbia University Press, New York.

Räsänen, M., Salo, J. S., and Kalliola, R. J. 1987. Fluvial perturbance in the western Amazon basin: regulation by long-term sub-Andean tectonics. Science 238:1398–1401.

Räsänen, M., Neller, R., Salo, J., and Jungners, H. 1992. Recent and ancient fluvial deposition systems in the Amazonian foreland basin, Peru. Geol. Mag. 129:293–306.

Räsänen, M., Linna, A. M., Santos, J. C. R., and Negri, F. R. 1995. Late Miocene tidal deposits in the Amazonian foreland basin. Science 269:386–390.

Riddle, B. R. 1996. The molecular phylogeographic bridge between deep and shallow history in continental biotas. Trends Ecol. Evol. 11:207–211.

Rosen, D. E. 1975. A vicariance model for Caribbean biogeography. Syst. Zool. 24:431–464.

Salo, J. 1987. Pleistocene forest refuges in the Amazon: evaluation of the biostratigraphical, lithostratigraphical and geomorphological data. Ann. Zool. Fennica 24:203–211.

Salo, J. 1988. Rainforest diversification in the western Amazon basin: the role of river dynamics. Rep. No. 16, Dept. Biol., University of Turku, Finland.

Salo, J., Kalliola, R., Häkkinen, I., Mäkinen, Y., Niemelä, P., Puhakka, M., and Coley, P. D. 1986. River dynamics and the diversity of Amazonian lowland forest. Nature 322:254–258.

Simpson, B. B., and Haffer, J. 1978. Speciation patterns in the Amazonian forest biota. Annu. Rev. Ecol. Syst. 9:497–518.

Tajima, F. 1983. Evolutionary relationships of DNA sequences in finite populations. Genetics 105:437–460.

Templeton, A. R., Routman, E., and Phillips, C. A. 1995. Separating population structure from population history: a cladistic analysis of the geographical distribution of mitochondrial DNA haplotypes in the tiger salamander, *Ambystoma tigrinum*. Genetics 140:767–782.

Tuomisto, H., Ruokolainen, K., Kalliola, R., Linna, A., Danjoy, W., and Rodriguez, Z. 1995. Dissecting Amazonian biodiversity. Science 269:63–66.

Vanzolini, P. E. 1970. Zoologia sistématica, geografia e a origem das espécies. Inst. Geografico São Paulo. Serie Téses e Monografias 3:1–56.

Vanzolini, P. E., and Williams, E. E. 1970. South American anoles: geographic differentiation and evolution of *Anolis chrysolepis* species group (Sauria: Iguanidae). Arq. Zool., São Paulo, 19:1–298.

Wallace, A. R. 1852. On the monkeys of the Amazon. Proc. Zool. Soc. London 20:107–110.

Wallace, A. R. 1853. A Narrative of Travels on the Amazon and Rio Negro. London: Reeve.

Wallace, A. R. 1876. The Geographical Distribution of Animals. Vol. 1. London: Macmillan.

Whitmore, T. C., and Prance, G. T. (eds.). 1987. Biogeography and Quaternary History of Tropical America. Clarendon Press, Oxford.

Part IV

Reproductive Barriers

16

Songs, Reproductive Isolation, and Speciation in Cryptic Species of Insects

A Case Study Using Green Lacewings

Marta Martínez Wells
Charles S. Henry

While the concept of a species differs among evolutionary biologists (Templeton 1989), all definitions acknowledge the importance of reproductive isolation: the absence of gene flow between two populations. The existence of reproductive isolation may not be necessary to a particular species concept, but it is always a sufficient criterion to define species boundaries (Mayr 1963). In nature, there exists an array of possible obstacles to reproduction, subdivided into premating versus postmating barriers.

Animals behave, and therefore can possess premating barriers associated with courtship. Courtship is a test, administered by each member of the courting pair to the other partner. The questions asked by each sex are not always the same, because the sexes can have conflicting needs, dictated by fundamental sexual differences in the amount of present and future investment in each reproductive interaction (Rowe et al. 1994). Mating signals are the means to communicate such questions, through visual, tactile, chemical, or acoustical channels. At each stage of courtship, one individual signals, and the other receives the signal. Thus, it is more appropriate to refer to a mating signal system, consisting of both signal and receiver components, rather than to mating signals alone (Butlin and Ritchie 1994).

In this chapter, we discuss courtship songs and reproductive isolation in cryptic species of insects, which differ in their mating signals but are otherwise virtually indistinguishable. First, we point out several well-known characteristics of mating signal systems and outline the principal hypotheses explaining why signals diverge in different species. We then briefly summarize what is known about mating signal and song evolution in several complexes of cryptic insect species. Finally, we present a detailed case study of green lacewings in the genus *Chrysoperla*.

Characteristics of Mating Signals

Several characteristics of mating signals are familiar to evolutionary biologists. First, closely related, sympatric species typically have very different mating signals. This has been reported in a wide range of animal taxa, including birds (Baker 1991), fishes (McKaye et al. 1993), frogs (Gerhardt 1994a), salamanders (Uzendoski and Verrell 1993), and marine invertebrates (Knowlton 1993). Insects are particularly rich in such species, with examples from *Drosophila* (Tomaru and Oguma 1994), moths (Roelofs and Brown 1982), grasshoppers (Ingrisch 1995), crickets (Otte 1994), cicadas (Alexander 1960), hoppers (Claridge 1993; Den Hollander 1995), water bugs (Jansson 1979), and green lacewings (Henry et al. 1993).

Second, mating signals of sympatric species often are more distinct from one another than are other signals produced by the same species, such as aggressive signals. For example, in tree frogs and other anurans, aggressive calls are significantly less diverse acoustically than advertisement calls (Schwartz and Wells 1984; Gerhardt 1994b). Territorial crickets show a similar pattern (Boake 1983). This suggests greater differentiation in signals that function in species recognition.

Third, closely related allopatric species may show mating signals that are nearly identical. Illustrating this point are *Laupala* crickets of Hawaii, which have undergone sequential, rapid, intra-island species radiations on different islands of the archipelago (Shaw 1996b). Obviously, species confined to different regions have no possibility of confusing their signals.

Finally, experimental work has shown that, when given a choice, females typically respond more positively to conspecific than heterospecific signals (but see Ryan and Rand 1993; Shaw 1995). Data from frogs (Gerhardt

1994a,b) and salamanders (Verrell and Arnold 1989) support this finding, as do experiments on the vibrational signaling systems of spiders (Uetz and Stratton 1982). Insect examples include grasshoppers (Perdeck 1958), crickets (Hill et al. 1972), and green lacewings (Wells and Henry 1992). These data also generally support the idea that the receiving system is coordinated with the signaling system, such that responses are elicited reliably by conspecific signals.

Causes of Divergence in Mating Signals

What is not clear is the process by which mating signal systems diverge during or after speciation, to produce the patterns described above. The paradox is that species recognition will favor strong stabilizing selection on species-specific signal characteristics, but divergence of signals can only result from strong directional selection (Paterson 1978) or genetic drift. Changes in the receiver also must match changes in the signaler, or the mating signal system will fail; this requires genetic correlation or "genetic coupling" of changes between males and females (Lande 1981; Doherty and Hoy 1985). Several hypotheses for signal divergence have been proposed.

Widely accepted is the view of Mayr (1963), who treated signal divergence as a by-product of genetic divergence of populations in allopatry. Mating signals, like any other heritable feature, will gradually change in allopatric populations because of random mutation, chance (genetic drift), and environmental effects (natural selection). When these geographically isolated populations later become sympatric, their phenotypic characters, including mating signals, will differ to a degree that roughly corresponds to the duration of isolation. Selection will then stabilize the signal system phenotype in each new species, in response to the need for unambiguous species recognition.

Premating signals in pairs of closely related species often differ more in areas of sympatry than allopatry (Crane 1975; Waage 1979; Littlejohn and Watson 1985; Otte 1989; Gerhardt 1994a; Butlin 1995). Two hypotheses address the issue. The hypothesis of reproductive character displacement applies to cases in which postmating reproductive isolation between two such species is complete. Natural selection then favors enhancement of mating signal differences in sympatry, because this reduces the chances of unproductive mismatings (Butlin 1987). A more controversial hypothesis is that of reinforcement, which strictly applies only to incompletely isolated populations. Reinforcement actually completes the process of speciation by strengthening premating reproductive isolation through natural selection favoring assortative mating, for example, mating signal divergence (Dobzhansky 1951; Fisher 1958). However, to work effectively and completely, it requires very strong selection against hybrids, as well as linkage between the genes

for fitness and the genes for assortative mating (Maynard Smith 1966; Felsenstein 1981; Butlin 1989). Distinguishing the results of reinforcement from those of reproductive character displacement is problematic, because the predicted pattern of greater signal divergence in sympatry than in allopatry is the same for both models. Also, both models require close genetic coupling between signaler and receiver, but little evidence of such correlation or coupling has been found (Boake 1991).

A contrasting view is Paterson's recognition model, which asserts that mating signals within each species evolve to maintain an efficient Specific-Mate Recognition System (SMRS) (Paterson 1986). Here, emphasis is placed on the stabilization of intraspecific behavior (the "fertilization system") rather than on the development of interspecific barriers to mating. The model predicts that mating signal systems will diverge in allopatry as adaptive responses to local environmental conditions, for example, properties related to signal propagation (Bailey 1991). Thus, signal characteristics for each species will be tightly correlated with environmental parameters. Support for such correlations, however, is limited (Ryan et al. 1990; Endler 1992). The recognition model emphasizes physical rather than biotic factors and specifically excludes interactions among sympatric species (Paterson 1982). In contrast, other students of acoustical insects have argued that interactions among species, and the acoustic environment created by assemblages of sympatric species, are the principal forces leading to signal divergence (Alexander 1962; Otte 1994).

Much of the controversy about mating signal divergence and speciation concerns the speed with which premating barriers can arise between populations, relative to postmating barriers. For example, the process of reinforcement involves selection against hybrids, but hybrids will be produced only in the absence of premating barriers. On the other hand, if changes in mating signal systems in allopatry are rapid and precede the kinds of genetic changes responsible for postmating isolation, then speciation could result from premating barriers alone, or nearly complete speciation could be completed by stabilizing selection acting on the new signal within each population. West-Eberhard (1983) proposes sexual, or "social," selection as an hypothetical mechanism for such rapid divergence. In her model, divergence and speciation occur because of intraspecific competition for mates and other social interactions among individuals. Because a well-defined optimum does not exist for such interactions, mating or social signal systems are potentially free to change without limit and in any direction. Sexual or social selection provides a plausible mechanism for the rapid, nonadaptive acquisition of changes in mating signals, possibly leading to speciation without the need for reinforcement or other ad hoc assumptions (Kirkpatrick 1987). Like reinforcement, however, the process requires genetic correlation or coupling between signaler and receiver (Boake 1991). Putative examples of speciation via sexual selection are the complexes of sibling species of Hawaiian and neotropical *Drosophila*, the members of which possess highly distinctive courtship songs (Ritchie and Gleason 1995).

Cryptic Species and Their Mating Signals

The evolution of mating signals is particularly interesting in groups of cryptic species that exhibit few, if any, differences in morphology, but possess very distinct mating signals. Indeed, differences in mating signals have been used for many years to distinguish morphologically similar species (Mayr 1963). In some cases, genetic differentiation among these species is slight as well, suggesting either that mating signal evolution has been a rapid process, or that morphological evolution has been very conservative. There are many examples of cryptic species complexes in vertebrates, including mammals (Zimmermann 1990), birds (Johnson 1963), lizards (Webster and Burns 1973; Jenson and Gladson 1984), and frogs (Heyer et al. 1996). The most numerous and some of the most dramatic examples, however, are found in Arthropoda, and especially among the insects.

Courtship signals in cryptic species usually involve nonvisual modes of communication, such as sound (vibration) or chemicals (pheromones). Closely related organisms that rely on vision to tell each other apart tend to look different and are therefore not truly cryptic. Conspicuous exceptions to this rule include fireflies (beetles: Lampyridae), where cryptic species can be distinguished by differences in the pattern of light flashes (Barber 1951), and fiddler crabs (Knowlton 1986), which use the pattern of movement of the chelicerae for species discrimination.

Cryptic species of insects most commonly utilize sound in their mating signal systems, although airborne pheromones characterize many moths and other nocturnal insects (Cardé et al. 1977). Sound may be produced in the air, in a solid substrate, or under water, using a variety of stridulatory devices, tymbals, wing-flicking behaviors, and percussive and jerking motions. Species discrimination among close relatives can be achieved through differences in the frequency (pitch) of the signals, but it is more often associated with differences in temporal structure such as pulse duration, pulse rate, interpulse interval, or pulse-group organization (Bailey 1991).

As one might expect, complexes of cryptic species are particularly abundant in singing insects. Within Orthoptera, such complexes have been described from stridulating crickets (Fulton 1952; Walker 1963), katydids (Bailey 1976), and grasshoppers (Perdeck 1958). Several groups of Plecoptera (stoneflies) include morphologically cryptic species, defined by unique patterns of drumming behavior (Stewart et al. 1982, 1988). The order Hemiptera (bugs, hoppers, and cicadas) contains numerous examples of cryptic species that sing either using tymbals or stridulatory devices (Alexander and Moore 1958; Claridge and Reynolds 1973). Another rich source of cryptic species is *Drosophila* (order Diptera, true flies), where discrimination is based on differences in near-field songs produced by patterned wing pulsations (Ewing and Bennet-Clark 1968; Cowling and Burnet 1981).

In some of the insects mentioned above, experiments have shown that mating signals are used in species recognition and probably are responsible for maintaining reproductive isolation among sympatric populations. For example, Ulagaraj and Walker (1973) found that large numbers of two species of mole crickets, *Scapteriscus acletus* and *S. vicinus*, flew toward speakers broadcasting conspecific calling songs, but were not attracted to heterospecific songs. Walker (1957) tested the phonotactic responses of females of two species of tree crickets, *Oecanthus quadripunctatus* and *O. nigricornis*, that live in weedy fields, as well as the responses of *O. niveous* and *O. exclamationis*, two species that live in trees. He found that in all cases, females responded positively to male calling songs of their own species, but not to songs of other species singing in the same habitat.

Perdeck (1958) did behavioral experiments in which he placed females and males of two grasshoppers, *Chorthippus brunneus* and *C. bigutulus*, together in different combinations. He then recorded the responses of females to conspecific and heterospecific males and found that females were much more inclined to react to the songs of conspecifics. Hill et al. (1972) studied the Australian crickets *Teleogryllus commodus* and *T. oceanicus* and demonstrated through playback experiments that the absence of hybrids in a zone of sympatric overlap was due to a very selective phonotactic response by females to the distinctive calling songs of conspecifics.

Claridge et al. (1988) showed that differences in substrate-transmitted courtship signals seem to be the cause of low hybridization success between populations of the planthopper *Nilaparvata lugens* in Asia and Australia. When given a choice, females preferred to mate with males from their native region. Water boatmen (Corixidae) stridulate under water by rubbing their forelegs against the head. Jansson (1973) did playback experiments with eight cryptic species of *Cenocorixa* and found that females preferred signals from conspecific males. He concluded that stridulatory signals function as premating barriers among those species.

Mating Signal Systems of Green Lacewings

We use our work on lacewings to illustrate the evolution of mating signals and their possible role in speciation. We first describe differences in the courtship songs of cryptic species of the *carnea*-group in North America and Eurasia and then summarize the results of behavioral experiments and laboratory hybridization studies using the three song species of the *plorabunda* complex in North America to illustrate the role of courtship songs in species recognition and reproductive isolation. We then present preliminary data on the phylogenetic relationships of the cryptic species in North America and Eurasia, using mitochondrial DNA sequence data. Finally, we use the molecular phylogeny to trace the evolution of song characters, in order to understand patterns of song evolution and the relationship of those patterns to species formation in the *carnea* group.

Systematics of Green Lacewings (Chrysopidae)

Chrysopidae is a diverse insect family of over 1200 species. Of 75 genera and 11 subgenera currently recognized (Brooks and Barnard 1990), five are known to have substrate-borne mating signals, produced by abdominal vibration (Henry 1996). During this behavior, the abdomen does not strike the substrate, but instead shakes the stem or leaf upon which the individual is standing—a process known as tremulation (Michelsen et al. 1982). The vibrational signals are picked up by subgenual organs in the tibiae of the legs of potential mates (Devetak and Pabst 1994). Abdominal vibration is organized temporally into discrete volleys or syllables, to form species-specific mating songs (figures 16.1 and 16.2). In *Chrysoperla* Steinmann, courtship songs are especially well developed and have been studied in detail (Henry 1994). Within a given species of *Chrysoperla*, both males and females sing identically and must duet precisely with one another before copulation will occur.

Brooks (1994) recognizes 36 valid species of *Chrysoperla*, and places them in four species groups based on male genitalia and wing characters: the *carnea*-group, the *comans*-group, the *nyerina*-group, and the *pudica*-group. The *carnea*-group exhibits the best-developed singing behavior (Henry 1984). Songs have been shown to be especially important to premating reproductive isolation of closely related species in the *carnea*-group, and the development of premating barriers has apparently preceded that of postmating isolation in the clade (see below). The *carnea*-group itself consists of complexes of cryptic, sibling species of nearly identical morphology. Most of these "song species" have yet to receive formal recognition: nearly all have previously been considered part of a single Holarctic morphological species, *Chrysoperla carnea* (Stephens) (Tjeder 1960). Typically, each song species is acoustically invariant across a broad geographic range, and many are extensively sympatric with one another (see figures 16.1 and 16.2). The taxonomic affiliation of each lacewing species and song type is shown in table 16.1.

Within the *carnea*-group, at least two complexes of cryptic species exist in North America and an undetermined number of complexes in Eurasia (Henry et al. 1993, 1996). Based on their distinct song types, the six North American species can be placed either in the *plorabunda* complex (*C. plorabunda, C. adamsi,* and *C. johnsoni*) or in the *downesi* complex (*C. downesi* sp. 1, *C. downesi* sp. 2, and "*C. mohave*"). Eurasia presents a more complicated picture, with at least 10 song species grouped into complexes related to *C. carnea* s. str. or *C. mediterranea* or of uncertain affinities (Cianchi and Bullini 1992; Henry et al. 1996). Pending further study, we refer all Eurasian species to the *carnea* complex. The names temporarily assigned to undescribed song species (see table 16.1 and figures 16.1, 16.2, 16.7, 16.8) are simply descriptive and have no taxonomic validity.

Courtship Songs

The courtship songs of green lacewings are of low frequency, between 30 and 120 Hz. In the laboratory, lacewings will tremulate inside a cardboard cup covered with plastic wrap (the arena). Their vibrational signals are detected by a piezoelectric transducer touching the plastic wrap and recorded on cassette tapes (see Henry 1979, 1980b for details). The same arena is used for playback experiments. Recorded songs are played through a speaker placed just above the arena, causing the plastic wrap to vibrate in the speaker's near-field.

We analyzed five features of songs from several sympatric and allopatric populations of the three species in the *plorabunda* complex. We found significant differences in both temporal and frequency characteristics among all three species (Henry et al. 1993). In addition, the mode of dueting differs. Males and females of *C. plorabunda* and *C. adamsi* duet by exchanging one volley of vibration at a time. In contrast, *C. johnsoni* males produce three to five volleys before a potential mate responds with three to five volleys of her own (Wells and Henry 1992; Henry 1993).

Behavioral Isolation

To test the hypothesis that the courtship songs of green lacewings within the *plorabunda* complex are used for species recognition, we performed a series of playback experiments. We presented tape-recorded songs to females, adjusting the timing of song presentation to produce a duet. First, we compared responses of females to songs from their own population with responses to the same song type from other localities, to see whether the songs were functionally equivalent across all localities. Females gave similar responses to songs from their own population and those from other localities (Wells and Henry 1992). Second, we compared the responses that females gave to recordings of their own song type and to other song types from the same locality (figure 16.3). Females not only recognized their own song types, but preferred to duet with them when given a choice between conspecific and heterospecific songs (Wells and Henry 1992). These experiments clearly demonstrate that the

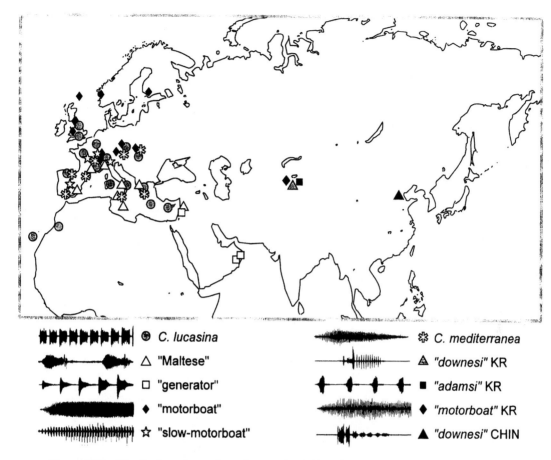

⊛ *C. lucasina*		❋ *C. mediterranea*	
△ "Maltese"		⊿ *"downesi"* KR	
□ "generator"		■ *"adamsi"* KR	
◆ "motorboat"		◆ *"motorboat"* KR	
☆ "slow-motorboat"		▲ *"downesi"* CHIN	

Figure 16.1. Distribution of green lacewing species in Eurasia. There are at least 10 distinct song types within the morphological species *Chrysoperla carnea*. Oscillographs of all songs are 12 seconds long. *C. lucasina* and *C. mediterranea* have been named as distinct species on the basis of song and color pattern differences.

songs of the lacewings, as in many other insects, are used for species recognition.

Hybridization Experiments and Postmating Isolation

We did hybridization experiments in the laboratory by crossing virgin males and females of *C. plorabunda* and *C. johnsoni* from the same locality in Idaho. The purpose of these studies was, first, to determine whether behavioral isolation prevents heterospecific matings when females are not given a choice of mates, and second, to determine the degree of postmating isolation between species due to reduced viability or fertility of hybrids. We compared the number of successful matings in conspecific and heterospecific crosses, the amount of time required to lay the first egg, the number of eggs produced, and the number that hatched. We raised the hybrid offspring and used them for F_2 crosses and backcrosses with each parental species. We were able to produce hybrids

in the laboratory in the no-choice crosses, but females in the heterospecific crosses took a week longer to lay their first eggs, indicating a reluctance to mate with the wrong species. The F_1 and F_2 hybrid progeny performed significantly less well than controls. Not only did they lay fewer eggs, but fewer of those hatched (figure 16.4). There was a reduction of 16% in hatching success in the F_1 generation compared to the control (conspecific) crosses, and another reduction of 15% in the F_2 generation (Wells 1993). From these studies we can see some reduction in genetic compatibility between these species, but postmating isolation is not complete.

Hybridization Experiments and Courtship Songs

The study of hybrid courtship songs and the behavioral responses of hybrids to their own songs and those of their parents is important for understanding not only the mode of inheritance of mating signals, but also the kinds of

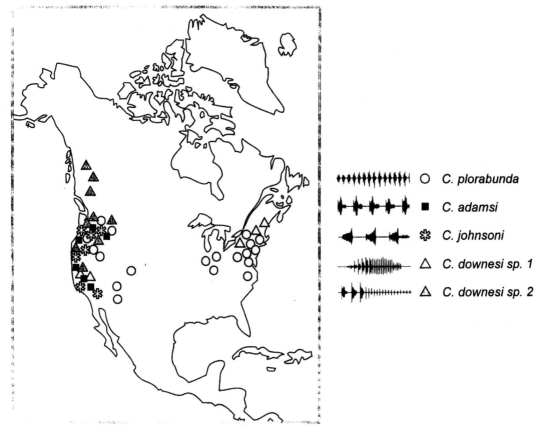

Figure 16.2. Distribution of green lacewing species in North America. There are two species complexes: the *plorabunda* complex, including *C. plorabunda*, *C. adamsi*, and *C. johnsoni*, and the *downesi* complex, including *C. downesi* sp. 1 and *C. downesi* sp. 2. Oscillographs of all songs are 12 seconds long.

changes required to produce reproductive isolation. Hybrid songs were intermediate between *C. plorabunda* and *C. johnsoni* songs in volley duration and intervolley interval (figure 16.5). The mode of dueting by hybrids was similar to that of *C. plorabunda*, in that hybrids exchanged one volley of vibration at a time.

Hybrid females responded preferentially to hybrid songs and gave few responses to the *C. plorabunda* song (figure 16.6). Hybrids responded to *C. johnsoni* song almost as much as they responded to their hybrid song, but the responses were not complete duets, because *C. johnsoni* males give a set of three to five volleys as a unit before the female responds with a similar number during a duet. The tapes had been made with enough space between the units of three to five volleys for individuals to insert their responses, but it seemed that hybrids were "interrupting" the *C. johnsoni* males, because they responded after each volley of the *C. johnsoni* song, not waiting for males in the recording to finish their three to five volley songs. Both males and females of parental species responded to their own songs, but discriminated

strongly against hybrid songs (Wells and Henry 1994). Therefore, any hybrid produced in nature would be at a disadvantage in acquiring mates.

In conclusion, the combined data from analysis of songs, behavioral experiments, and laboratory hybridization support the hypothesis that the different song-type populations within the *plorabunda* complex are distinct biological species that could have evolved through song divergence.

Phylogenetic Relationships of Green Lacewings

To understand modes of speciation and patterns of song evolution in the *carnea*-group, it is necessary to know the phylogenetic relationships among the North American species and their relationships to morphologically identical species in Europe and Asia. A pilot study using allozymes was undertaken to determine how much genetic divergence, if any, has occurred between two of the cryptic species within the *plorabunda* complex,

Table 16.1. Summary of classification of lacewing species and song types.

Taxon	Members
Chrysopa oculata Say	Outgroup of *Chrysoperla*
Chrysoperla	
pudica (*rufilabris*)-group	*C. rufilabris* (Burmeister)
	C. harrisii (Fitch)
carnea-group	
North America (≈6 cryptic species)	
plorabunda complex	*C. plorabunda* (Fitch)
	C. adamsi Henry et al.
	C. johnsoni Henry et al.
downesi complex	*C. downesi* sp. 1 (Smith)
	C. downesi sp. 2
	"*C. mohave* (Banks)"
Eurasia (≈10 cryptic species)	
carnea complex	*C. mediterranea* (Hölzel)
	C. lucasina (Lacroix)
	"Maltese"
	"Generator"
	"Motorboat"
	"Slow-motorboat"
	"Kyrghyzstan-*adamsi*"
	"Kyrghyzstan-*downesi*"
	"Kyrghyzstan-motorboat"
	"China-*downesi*"

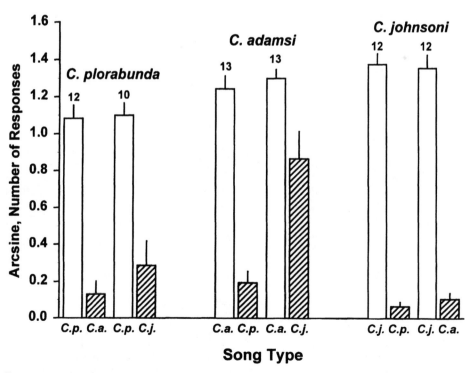

Figure 16.3. Sympatric comparisons: number of responses that females of *C. plorabunda, C. adamsi,* and *C. johnsoni* gave to recordings of their own songs (open bars) and songs from other species (hatched bars) in the same locality.

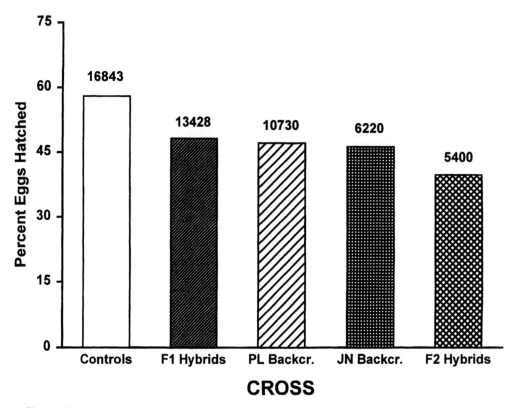

Figure 16.4. Hatching success for all crosses: percentage of eggs that hatched from hybridization experiments between *C. plorabunda* and *C. johnsoni*. The number on top of each bar is the total number of eggs laid in each type of cross.

and to try to reconstruct relationships among these taxa and other complexes in North America. We examined 17 enzyme loci in 350 individuals of several populations of *C. plorabunda*, *C. johnsoni*, and the outgroups *C. downesi* and *C. harrisii*. Analysis revealed small Nei's genetic distances (<0.005) between *C. plorabunda* and *C. johnsoni*, and between these species and *C. downesi* (<0.01). These data suggest very recent divergence of all three species. Unweighted pair-group method using arithmetic averages (UPGMA) clustering analysis showed only *C. harrisii* to be well separated from the other populations, a result consistent with *C. harrisii*'s known phylogenetic position outside of the *carnea*-group, in the *pudica*-group. We concluded that the species within the *plorabunda* complex are genetically very similar, and that a phylogenetic analysis using electrophoretic data is not possible (Wells 1994).

We then used mitochondrial DNA (mtDNA) to generate a phylogeny of green lacewings, using distinct song types from 18 populations in North America, Europe, and Asia. We used standard methods of DNA extraction (Dtab, Ctab; see Gustincich et al. 1991 for details) and amplification. We collected data from 516 base pairs of the cytochrome oxidase II (COII) gene. We chose COII

because there are many primers available, and, in some insects, enough information has been obtained to resolve relationships at the species level (Beckenbach et al. 1993; Brower 1994). At least two individuals of each species were sequenced. In all cases, the two individuals grouped together in the cladogram, so these were collapsed into a single taxon in our analyses.

Data were analyzed in several ways. We built a minimum evolution tree and obtained bootstrap values from 100 replicates (figure 16.7). The tree was calculated using log determinant corrected distances (Lockhart et al. 1994). We also built trees with a neighbor-joining Kimura two-parameter distance algorithm (Kimura 1980), a minimum evolution tree using HKY85 (Hasegawa et al. 1985) and a parsimony analysis using uniformly weighted characters in PAUP* (Swofford 1996). All analyses produced the same topology, indicating that there were no nucleotide biases among taxa affecting this topology. We also tested for among-site rate variation (ASRV) using the method of Sullivan et al. (1995) and found no significant ASRV in our data.

Chrysopa oculata was used as the outgroup for the genus *Chrysoperla*, and *Chrysoperla rufilabris* and *C. harrisii* (both of the *pudica*-group) were selected as outgroups

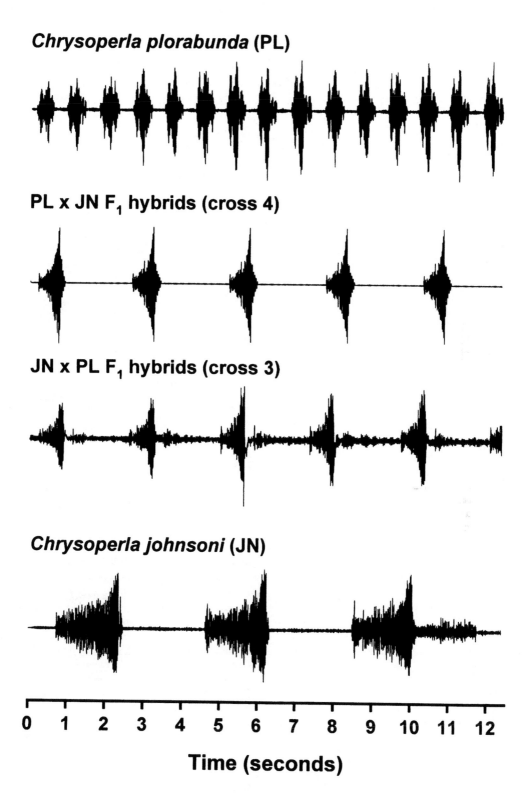

Figure 16.5. Digitized oscillograph of *C. plorabunda*, *C. johnsoni*, and their two types of F_1 hybrids.

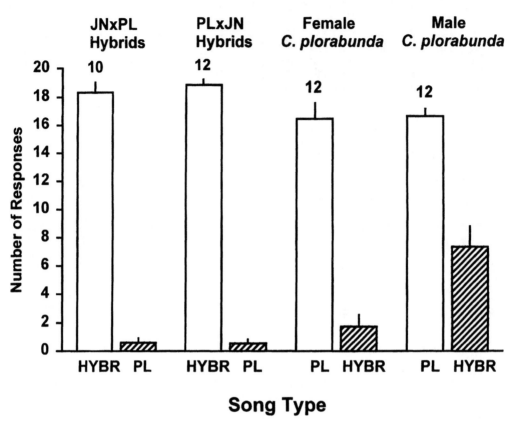

Figure 16.6. Playback experiments: number of responses that F₁ hybrids and *C. plorabunda* parents gave to recordings of their own songs (open bars) and each other's songs (hatched bars).

for the species complexes of the *carnea*-group in Eurasia and North America (table 16.1). Three consistent patterns emerged: (1) *C. rufilabris* and *C. harrisii* always appeared as sister taxa in the outgroup *pudica*-group; (2) the North American species formed a well-supported monophyletic clade, with the *plorabunda* complex forming a sister clade of the *downesi* complex; and (3) the Eurasian *carnea* complex formed a well-supported monophyletic clade, but with poorly resolved relationships within the clade (figure 16.7).

We also calculated percentage sequence divergence among Eurasian and North American species using the log determinant corrected distances. Percentage divergence between the *plorabunda* and *downesi* complexes in North America was 1.6%, whereas divergence between Eurasian and North American clades was 4.3%. Such low values indicate rapid divergence of species. When we compared the *carnea*-group to the *pudica*-group, divergence was 7.1%, while that between *Chrysoperla* and *Chrysopa* was 13.8%.

Green Lacewings and Speciation Models

As discussed above, a central issue in speciation research concerns the processes leading to reproductive isolation between sympatric populations. Green lacewings have been used to address that issue. In a widely cited study, Tauber and Tauber (1977a,b) hypothesized that *Chrysoperla downesi* in eastern North America evolved from *C. plorabunda* (then called *Chrysopa carnea*) through a process of sympatric speciation involving habitat differentiation and seasonal isolation. This model has not been universally accepted (Hendrickson 1978; Futuyma and Mayer 1980; Henry 1982; Tauber and Tauber 1982a,b). Later work by Henry (1980a, 1985, 1994) and Wells and Henry (1992) summarized above demonstrated the importance of courtship songs in the mating behavior of green lacewings and revealed a proliferation of valid song species among populations of "*C. carnea*" on three different continents, and among populations assigned to *C. plorabunda* and *C. downesi* in North America. This

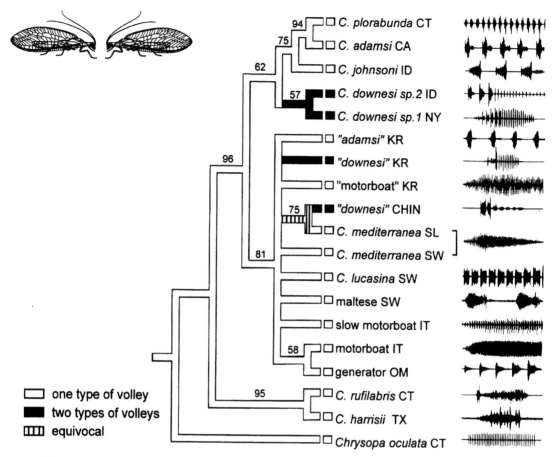

Figure 16.7. Songs character "type of volley," traced on a COII mtDNA phylogeny representing the bootstrap of 100 replicates of a minimum evolution tree calculated with log Determinant–corrected distances. The numbers on the nodes are the bootstrap values. The production of one type of volley is ancestral for the genus *Chrysoperla*. Two types of volleys is the derived condition (see text).

work demonstrated that habitat differentiation and seasonal isolation are not the sole differences among species of the *Chrysoperla carnea*-group, and that speciation is a more complicated process than originally envisioned by the Taubers.

Henry (1980a, 1985) proposed an alternative model in which song divergence, either in allopatry or in sympatry, could have given rise to new species. An indirect way to test the validity of either model of speciation is to compare the predicted phylogenetic relationships among species from each of the models to relationships obtained from our molecular phylogeny. The Tauber model predicts an ancestor–descendant relationship in which *C. downesi* is the sister species of *C. plorabunda*. The Henry model predicts that species in the *plorabunda* or *downesi* complex should be more closely related to other species within their own complex than to species outside their own complex, and that they should share uniquely derived song features. If our phylogeny based on the COII data is correct (figure 16.7), then *C. plorabunda* and *C. downesi* are not sister species, but instead are members of distinct but closely related clades. The *plorabunda* complex forms a monophyletic group, and *C. plorabunda* is the sister species of *C. adamsi*.

When we traced some of the features of the songs on the phylogeny to determine their evolutionary history, we found that one derived song feature, the production of two distinct types of volleys (see below and figure 16.7), apparently evolved once in North America and unites members of the *downesi* complex. In addition, *C. plorabunda* and *C. adamsi* share another derived trait, the production of single volleys between male and female during dueting. The basic structure of volleys in these species is similar to that of *C. johnsoni*, but the latter species differs in the way the volleys are delivered. With this evidence, we conclude that the proposed sympatric speciation model of the Taubers is unlikely, because it requires a less parsimonious sequence of song evolution.

Tracing Character Evolution

To understand the role of mating signals in the process of speciation requires an understanding of the patterns of signal evolution. One way to do this is to use the known phylogeny of a group to trace the evolution of its signal characters. Although a comparative systematic approach has long been used to elucidate patterns of character evolution (reviewed in Harvey and Pagel 1991), it is only very recently that larger studies have been undertaken, mapping changes in behavioral (or other) traits onto extensive phylogenetic trees. This approach can give unexpected results that provide insight into evolutionary processes.

We used the phylogeny based on COII data to trace the evolution of two song characters with MACCLADE (Maddison and Maddison 1992). The first character examined was the type of volleys produced. Most species of green lacewings produce songs consisting of a series of volleys, but all volleys are of the same type. For example, both *Chrysopa oculata* and "motorboat" have courtship songs made up of multiple short volleys, but all volleys are alike (figure 16.7). Similarly, *C. plorabunda*, *C. adamsi*, and *C. johnsoni* produce repeated series of longer volleys, but again all volleys are the same. A few species produce two distinct types of volleys. For example, in *C. downesi*, the first few volleys are longer and more widely separated, followed by a train of shorter volleys (figure 16.7). The single type of volley seems to be the ancestral condition in *Chrysoperla* and is characteristic of the songs of most species in North America and Eurasia. The apparent derived condition is to produce different kinds of volleys, which is seen in *C. downesi* in North America and "*downesi*-like" species in Kyrghyzstan and China (figure 16.7).

The second character we examined was pattern of volley production. Some species produce a long train of very short volleys given as a unit before the female responds. For example, in several Eurasian species such as *Chrysoperla mediterranea* or the outgroup species *Chrysopa oculata*, males and females exchange trains of volleys during dueting (figure 16.8). Others produce several long volleys as a repeated unit, followed by a similar response from the female, for example, *C. johnsoni* and *C. lucasina*. Finally, some species duet by answering after each volley, as in *C. plorabunda* and *C. adamsi* from North America and three Eurasian species (figure 16.8). The production of trains of very short volleys occurs in both outgroups and appears to be the ancestral condition in *Chrysoperla*. Multiple longer volleys and single volleys are probably derived character states. The production of single volleys during dueting apparently evolved just once in North America and unites *C. plorabunda* and *C. adamsi*, which are sister species according to the molecular phylogeny. This pattern also evolved at least once and possibly several times in Eurasia, being present in "*adamsi*-like" species from

Kyrghyzstan, "Maltese" from Europe, and "generator" from Oman. Multiple longer volleys seem to have evolved at least once in Eurasia (*C. lucasina*) and once in North America (*C. johnsoni*) (figure 16.8).

Convergent Evolution of Courtship Songs

At any single locality, the songs of sympatric lacewing species are very different. For example in Kyrghyzstan, there are three species with very distinct courtship songs: "*downesi*-like," "*adamsi*-like," and "motorboat." In Switzerland, "Maltese," "slow-motorboat," "motorboat," *C. lucasina*, and *C. mediterranea* all co-occur, but each sings very differently from the others (figure 16.1). North America shows the same pattern, with all three acoustically unique species of the *plorabunda* complex occurring at some of the same geographical sites and often on branches of the same plant (Henry 1993) (figure 16.2).

We did, however, find species with very similar courtship songs on different continents, but members of these pairs of species do not appear to be their own closest relatives. For example, there is a species with a courtship song similar to that of North American *C. adamsi*, but it is found in Kyrghyzstan, in central Asia. The DNA phylogeny clearly shows that these are not sister species, and in fact are members of two distinct monophyletic clades (figure 16.7). Another example is the *C. downesi*-like songs of two species in Kyrghyzstan and China. Both species produce two distinct types of volleys, behavior found only in *C. downesi* in North America. The song of the species in Kyrghyzstan is actually more similar to that of *C. downesi* than it is to the song of the species in China. These Asian species do not appear to be closely related to the North American *C. downesi* group, and it is not clear whether they are closely related to each other.

Three explanations can account for the existence of species with similar courtship songs that are not closely related. First, convergent evolution of courtship songs could have occurred across the continents. For instance, there might be some feature of the substrate upon which green lacewings are tremulating that determines what kind of songs are best transmitted through that substrate (Endler 1993), resulting in the evolution of similar songs in different regions. Convergence could also occur if the genetic basis of song structure is sufficiently simple that similar songs could evolve several times by chance (Henry 1985). Second, the phylogenetic trees obtained might be gene trees that reflect the history of the random assortment of mitochondrial genes, but not the evolution of species. Sequences obtained from nuclear genes could resolve this problem. Third, the song feature shared by two taxa might reflect retention of the plesiomorphic character state. In our case, both *C. adamsi* from North America and the "*adamsi*-like" species from Kyrghyzstan have one type of volley, an apparently ancestral condition for that song character.

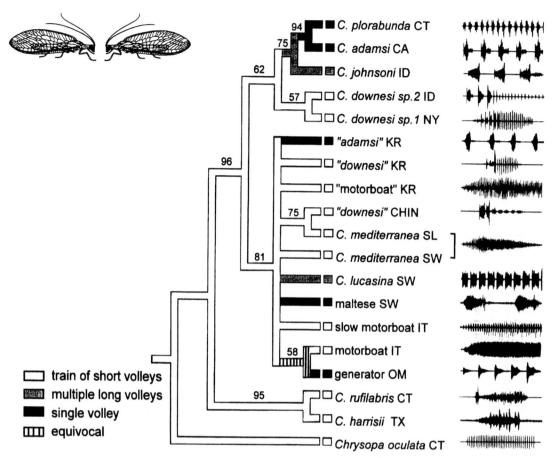

Figure 16.8. Song character "number of volleys before a response," traced on the same molecular phylogeny shown in figure 16.7. The production of a train of short volleys is ancestral for the genus *Chrysoperla*. The production of multiple long volleys and single volleys are derived conditions (see text).

There are other putative examples of convergent evolution in songs of insects. Alexander (1968) identified a group of cryptic species of crickets with almost identical songs, including *Gryllus veletis*, *G. pennsylvanicus*, and *G. campestris*. He proposed that *G. veletis* and *G. pennsylvanicus* were sister species that had evolved by allochronic speciation (i.e., divergence in mating seasons). However, more recent molecular work rejected the allochronic model of speciation by showing that *G. veletis* and *G. pennsylvanicus* are not sister species but instead could have similar songs due to convergence (Harrison 1979; Harrison and Bogdanowicz 1995). Another example is *Laupala* crickets of Hawaii. Shaw (1996b) produced a phylogeny using mtDNA and found that species that were grouped by Otte (1994) as sister species using morphological characters and similarities in songs were not each others' closest relatives. She suggested convergent evolution, in allopatry, as an explanation.

Concluding Remarks

The process of speciation is difficult to study, because we must infer a general sequence of events from the pooled observations of many different populations made at single points in time. Most such populations are reproductively isolated already, or are just in the initial stages of speciation.

Cryptic species that are distinguished mainly by their mating signals are good subjects for studies of speciation. Most other aspects of the biology of cryptic species do not differ, making the organisms into natural, controlled experiments. It is in such species that divergence of mating signals is likely to lead to the production of new species. Therefore, the key to understanding the processes by which mating signals diverge and their role in species formation is to understand the pattern of evolution of signal traits in cryptic species.

In our work with green lacewings, we gathered several kinds of information necessary for understanding the patterns and perhaps processes by which mating signals diverge. We first demonstrated that the courtship songs are distinct across populations in different regions of North America, Europe, and Asia. We then used behavioral experiments to demonstrate that these courtship songs function as reproductive barriers between sympatric populations. We also have evidence of partial postmating isolation through hybridization experiments that show reduction in hybrid fitness, as well as behavioral experiments demonstrating discrimination against hybrids by the parental species. Finally, we used an independently derived phylogeny to investigate the evolution of particular song traits.

Other recent studies have used the same approach. A pioneering work is that of Ewing and Miyan (1986), in which the evolution of near-field courtship songs in the *repleta* group of *Drosophila* is traced on a phylogeny reconstructed from cytological evidence. Subsequent studies have been directed at display behavior and social interactions in birds (Prum 1990), advertisement calls in anurans (Cocroft and Ryan 1995), and songs and life histories in crickets (Harrison and Bogdanowicz 1995). Harrison and Bognadowicz (1995) used an mtDNA-derived phylogeny of North American field crickets (*Gryllus*) to trace the evolution of song characters, life cycles, and habitat associations. As already described, their phylogenetic analysis forced them to reject an allochronic model of sympatric speciation proposed earlier. For similar reasons, we have rejected the Taubers' model for sympatric speciation of lacewings through habitat specialization and seasonal isolation. These types of studies emphasize the importance of understanding patterns of trait evolution before proposing processes of speciation.

We now need to combine phylogenetic studies with hybridization experiments, to reveal the genetic basis of song features. With this type of information we could determine how many genes are involved in changing a song to produce a new species. Shaw (1996a) has used this approach with crickets in Hawaii. She hybridized two closely related species, *Laupala kohalensis* and *L. paranigra*, and found polygenic inheritance of pulse rate, with approximately eight genetic factors accounting for the difference in pulse rate between the species.

For speciation to occur it is necessary to have coordinated evolutionary changes in the genes that control signal production and the genes that control signal reception. Studies of crickets and grasshoppers initially supported the existence of such genetic coupling between signaler and receiver (Hoy et al. 1977), but recent reviews of the evidence are less sanguine (Butlin and Ritchie 1989; Boake 1991). Genetic coupling (correlation) is not a problem in dueting insects such as lacewings, because both the production and the reception of signals are expressions of a common phenotype and genotype, present in all males and females that sing the same song. This special situation could make the divergence of mating signal systems more likely and rapid in lacewings, by allowing theoretically intractable processes such as reinforcement, reproductive character displacement, or sexual selection to operate more easily. Although much work remains to be done, our best working hypothesis is that song divergence is responsible for the origin of cryptic, sibling species in the *carnea*-group of the green lacewing genus *Chrysoperla*.

Acknowledgments   This study was supported in part by NSF Award DEB-9220579 to C. S. Henry and by grants from the Research Foundation of University of Connecticut to both authors. We thank Stephen J. Brooks (The Natural History Museum, London, U.K.), Peter Duelli (Swiss Federal Institute for Forest, Snow and Landscape Research, Birmensdorf, Switzerland), and James B. Johnson (University of Idaho, Moscow) for their collaboration in all Eurasian aspects of the research. We also thank Kentwood D. Wells (University of Connecticut, Storrs), who provided constructive criticism of the manuscript. We especially thank Chris Simon (University of Connecticut, Storrs) for use of her laboratory and help with analysis of DNA sequence data.

References

Alexander, R. D. 1960. Sound communication in Orthoptera and Cicadidae. In W. Lanyon and R. Tavolga, Eds., *Animal Sounds and Communication*, pp. 38–92. New York: AIBS Publications.

Alexander, R. D. 1962. Evolutionary change in cricket acoustical communication. Evolution 16:443–467.

Alexander, R. D. 1968. Life cycle origins, speciation, and related phenomena in crickets. Q. Rev. Biol. 43:1–41.

Alexander, R. D., and T. E. Moore. 1958. Studies on the acoustical behavior of seventeen-year cicadas. Ohio J. Sci. 58: 107–127.

Bailey, W. J. 1976. Species isolation and song types of the genus *Ruspolia* (Orthoptera: Tettigonioidea) in Uganda. J. Nat. Hist. 10:511–528.

Bailey, W. J. 1991. *Acoustic Behavior of Insects: An Evolutionary Perspective*. New York: Chapman and Hall.

Baker, M. C. 1991. Response of male indigo and Lazuli buntings and their hybrids to song playback in allopatric and sympatric populations. Behaviour 119:225–242.

Barber, H. S. 1951. North American fireflies of the genus *Photuris*. Smithsonian Misc. Coll. 117:1–58.

Beckenbach, A. T., Y. W. Wei, and H. Liu. 1993. Relationships in the *Drosophila obscura* species group, inferred from mitochondrial cytochrome oxidase II sequences. Mol. Biol. Evol. 10:619–634.

Boake, C. R. B. 1983. Mating systems and signals in crickets. In D. T. Gwynne and G. K. Morris, Eds., *Orthopteran Mating Systems: Sexual Competition in a Diverse Group of Insects*, pp. 28–44. Boulder, Colo.: Westview Press.

Boake, C. R. B. 1991. Coevolution of senders and receivers of sexual signals: genetic coupling and genetic correlations. Trends Ecol. Evol. 6:225–227.

Brooks, S. J. 1994. A taxonomic review of the common green lacewing genus *Chrysoperla* (Neuroptera: Chrysopidae). Bull. Br. Mus. Nat. Hist. (Entomol.) 63:137–210.

Brooks, S. J., and P. C. Barnard. 1990. The green lacewings of the world: a generic review (Neuroptera: Chrysopidae). Bull. Br. Mus. Nat. Hist. (Entomol.) 59:117–286.

Brower, A. V. Z. 1994. Phylogeny of *Heliconius* butterflies inferred from mitochondrial DNA sequences (Lepidoptera: Nymphalidae). Mol. Phylog. Evol. 3:159–174.

Butlin, R. K. 1987. Speciation by reinforcement. Trends Ecol. Evol. 2:8–13.

Butlin, R. K. 1989. Reinforcement of premating isolation. In D. Otte and J. A. Endler, Eds., *Speciation and Its Consequences*, pp. 158–179. Sunderland, Mass.: Sinauer.

Butlin, R. K. 1995. Genetic variation in mating signals and responses. In D. M. Lambert and H. G. Spencer, Eds., *Speciation and the Recognition Concept*, pp. 327–366. Baltimore: Johns Hopkins University Press.

Butlin, R. K., and M. G. Ritchie. 1989. Genetic coupling in mate recognition systems: what is the evidence? Biol. J. Linn. Soc. 37:237–246.

Butlin, R. K., and M. G. Ritchie. 1994. Mating behaviour and speciation. In P. J. Slater and T. R. Halliday, Eds., *Behaviour and Evolution*, pp. 43–79. Cambridge: Cambridge University Press.

Cardé, R. T., A. M. Cardé, A. S. Hill, and W. L. Roelofs. 1977. Sex pheromone specificity as a reproductive isolating mechanism among the sibling species *Archips argyrospilus* and *A. mortuanus* and other sympatric tortricine moths (Lepidoptera: Tortricidae). J. Chem. Ecol. 3:71–84.

Cianchi, R., and L. Bullini. 1992. New data on sibling species in chrysopid lacewings: the *Chrysoperla carnea* (Stephens) and *Mallada prasinus* (Burmeister) complexes (Insecta: Neuroptera: Chrysopidae). In M. Canard, H. Aspöck, and M. W. Mansell, Eds., *Current Research in Neuropterology. Proceedings of the Fourth International Symposium on Neuropterology, Bagnères-de-Luchon, France, 1991*, pp. 99–104. Toulouse: Sacco.

Claridge, M. F. 1993. Speciation in insect herbivores: the role of acoustic signals in leafhoppers and planthoppers. In D. R. Lees and D. Edwards, Eds., *Evolutionary Patterns and Processes*, pp. 285–297. London: Academic Press for the Linnaean Society of London.

Claridge, M. F., and W. J. Reynolds. 1973. Male courtship songs and sibling species in the *Oncopsis flavicollis* species group (Hemiptera: Cicadellidae). J. Entomol. Ser. B. Taxon 42:29–39.

Claridge, M. F., J. Den Hollander, and J. C. Morgan. 1988. Variation in host plant relations and courtship signals in weed-associated populations of the brown planthopper *Nilaparvata lugens* (Stål) from Australia and Asia: a test of the recognition species concept. Biol. J. Linn. Soc. 35: 79–93.

Cocroft, R. B., and M. J. Ryan. 1995. Patterns of advertisement call evolution in toads and chorus frogs. Anim. Behav. 49:283–303.

Cowling, D. E., and B. Burnet. 1981. Courtship song and genetic control of their acoustic characteristics in sibling

species of the *Drosophila melanogaster* subgroup. Anim. Behav. 29:924–935.

Crane, J. 1975. *Fiddler Crabs of the World (Ocypodidae: Genus Uca)*. Princeton, N.J.: Princeton Univ. Press.

Den Hollander, J. 1995. Acoustic signals as specific-mate recognition signals in leafhoppers (Cicadellidae) and planthoppers (Delphacidae) (Homoptera, Auchenorrhyncha). In D. M. Lambert and H. G. Spencer, Eds., *Speciation and the Recognition Concept*, pp. 440–463. Baltimore: Johns Hopkins University Press.

Devetak, D., and M. A. Pabst. 1994. Structure of the subgenual organ in the green lacewing, *Chrysoperla carnea*. Tissue and Cell 26:249–257.

Dobzhansky, T. 1951. *Genetics and the Origin of Species, 3rd edition*. New York, N.Y.: Columbia Univ. Press.

Doherty, J., and R. Hoy. 1985. Communication in insects. III. The auditory behavior of crickets: some views of genetic coupling, song recognition, and predator detection. Q. Rev. Biol. 60:457–472.

Endler, J. A. 1992. Signals, signal conditions, and the direction of evolution. Am. Nat. 139(S):125–153.

Endler, J. A. 1993. Some general comments on the evolution and design of animal communication systems. Philos. Trans. Roy. Soc. Lond. B 340:215–225.

Ewing, A. W., and H. C. Bennet-Clark. 1968. The courtship songs of *Drosophila*. Behaviour 31:288–301.

Ewing, A. W., and J. A. Miyan. 1986. Sexual selection, sexual isolation and the evolution of song in the *Drosophila repleta* group of species. Anim. Behav. 34:421–429.

Felsenstein, J. 1981. Skepticism towards Santa Rosalia, or why are there so few kinds of animals? Evolution 35:124–135.

Fisher, R. A. 1958. *The Genetical Theory of Natural Selection* (2nd ed.). New York: Dover Press.

Fulton, B. B. 1952. Speciation in the field cricket. Evolution 6:283–295.

Futuyma, D. J., and G. C. Mayer. 1980. Non-allopatric speciation in animals. Syst. Zool. 29:254–271.

Gerhardt, H. C. 1994a. The evolution of vocalization in frogs and toads. Annu. Rev. Ecol. Syst. 25:293–324.

Gerhardt, H. C. 1994b. Selective responsiveness to long-range acoustic signals in insects and anurans. Am. Zool. 34: 706–714.

Gustincich, S., G. Manfioletti, G. Del Sal, C. Schneider, and P. Carninci. 1991. A fast method for high-quality genomic DNA extraction from whole human blood. BioTechniques 11:298–301.

Harrison, R. G. 1979. Speciation in North American field crickets: evidence from electrophoretic comparisons. Evolution 33:1009–1023.

Harrison, R. G., and S. M. Bogdanowicz. 1995. Mitochondrial DNA phylogeny of North American field crickets: perspectives on the evolution of life cycles, songs, and habitat associations. J. Evol. Biol. 8:209–232.

Harvey, P. H., and M. D. Pagel. 1991. *The Comparative Method in Evolutionary Biology*. Oxford: Oxford University Press.

Hasegawa, M., Y. Iida, and T. Yano. 1985. Phylogenetic relationships among eukaryotic kingdoms inferred from ribosomal RNA sequences. J. Mol. Evol. 22:32–38.

Hendrickson, H. T. 1978. Sympatric speciation: evidence? Science 200:345–346.

Henry, C. S. 1979. Acoustical communication during courtship and mating in the green lacewing *Chrysopa carnea* (Neuroptera: Chrysopidae). Ann. Entomol. Soc. Am. 72: 68–79.

Henry, C. S. 1980a. The courtship call of *Chrysopa downesi* Banks (Neuroptera: Chrysopidae): its evolutionary significance. Psyche 86:291–297.

Henry, C. S. 1980b. The importance of low-frequency, substrate-borne sounds in lacewing communication (Neuroptera: Chrysopidae). Ann. Entomol. Soc. Am. 73:617–621.

Henry, C. S. 1982. Reply to Tauber and Tauber's "Sympatric speciation in *Chrysopa*: further discussion." Ann. Entomol. Soc. Am. 75:3–4.

Henry, C. S. 1984. The sexual behavior of green lacewings. In M. Canard, Y. Séméria, and T. R. New, Eds., *Biology of Chrysopidae*, pp. 101–110. The Hague: Dr W. Junk.

Henry, C. S. 1985. Sibling species, call differences, and speciation in green lacewings (Neuroptera: Chrysopidae: *Chrysoperla*). Evolution 39:965–984.

Henry, C. S. 1993. *Chrysoperla johnsoni* Henry (Neuroptera: Chrysopidae): acoustic evidence for full species status. Ann. Entomol. Soc. Am. 86:14–25.

Henry, C. S. 1994. Singing and cryptic speciation in insects. Trends Ecol. Evol. 9:388–392.

Henry, C. S. 1997. Modern mating systems in archaic Holometabola: sexuality in neuropterid insects. In J. C. Choe. and B. J. Crespi, Eds., *The Evolution of Mating Systems in Insects and Arachnids*, pp. 193–210. Cambridge, UK: Cambridge University Press.

Henry, C. S., M. M. Wells, and R. J. Pupedis. 1993. Hidden taxonomic diversity within *Chrysoperla plorabunda* (Neuroptera: Chrysopidae): two new species based on courtship songs. Ann. Entomol. Soc. Am. 86:1–13.

Henry, C. S., S. J. Brooks, J. B. Johnson, and P. Duelli. 1996. *Chrysoperla lucasina* (Lacroix): a distinct species of green lacewing, confirmed by acoustical analysis (Neuroptera: Chrysopidae). Syst. Entomol. 21:205–218.

Heyer, W. R., J. M. García-López, and A. J. Cardoso. 1996. Advertisement call variation in the *Leptodactylus mystaceus* species complex (Amphibia: Leptodactylidae) with description of a new sibling species. Amphibia-Reptilia 17:7–31.

Hill, K. G., J. J. Loftus-Hills, and D. F. Gartside. 1972. Premating isolation between the Australian field crickets *Teleogryllus commodus* and *T. oceanicus* (Orthoptera: Gryllidae). Austral. J. Zool. 20:153–163.

Hoy, R. R., J. Hahn, and R. C. Paul. 1977. Hybrid cricket auditory behavior: evidence for genetic coupling in animal communication. Science 195:82–84.

Ingrisch, S. 1995. Evolution of the *Chorthippus biguttulus* group (Orthoptera, Acrididae) in the Alps, based on morphology and stridulation. Rev. Suisse Zool. 102:475–535.

Jansson, A. 1973. Stridulation and its significance in the genus *Cenocorixa* (Hemiptera, Corixidae). Behaviour 46:1–36.

Jansson, A. 1979. Reproductive isolation and experimental hybridization between *Arctocorisa carinata* and *A. germari* (Heteroptera, Corixidae). Ann. Zool. Fennici 16:89–104.

Jenson, T. A., and N. L. Gladson. 1984. A comparative display analysis of the *Anolis brevirostris* complex in Haiti. J. Herpetol. 18:217–230.

Johnson, N. K. 1963. Biosystematics of sibling species of flycatchers in the *Empidonax hammondii-oberholserii-wrightii* complex. Univ. Cal. Publ. Zool. 66:79–238.

Kimura, M. 1980. A simple method for estimating evolutionary rates of base substitutions through comparative studies of nucleotide sequences. J. Mol. Biol. 16:111–120.

Kirkpatrick, M. 1987. Sexual selection by female choice in polygynous animals. Annu. Rev. Ecol. Syst. 18:43–70.

Knowlton, N. 1986. Cryptic and sibling species among the decapod Crustacea. J. Crustacean Biol. 6:356–363.

Knowlton, N. 1993. Sibling species in the sea. Annu. Rev. Ecol. Syst. 24:189–216.

Lande, R. 1981. Models of speciation by sexual selection on polygenic traits. Proc. Natl. Acad. Sci. USA 78:3721–3725.

Littlejohn, M. J., and G. F. Watson. 1985. Hybrid zones and homogamy in Australian frogs. Annu. Rev. Ecol. Syst. 16:85–112.

Lockhart, P. J., M. A. Steel, M. D. Hendy, and D. Penny. 1994. Recovering evolutionary trees under a more realistic model of sequence evolution. Mol. Biol. Evol. 11:605–612.

Maddison, W. P., and D. R. Maddison. 1992. *MacClade: Analysis of Phylogeny and Character Evolution* (ver. 3.01). Sunderland, Mass.: Sinauer.

Maynard Smith, J. 1966. Sympatric speciation. Am. Nat. 100: 637–650.

Mayr, E. 1963. *Animal Species and Evolution*. Cambridge, Mass.: Belknap Press.

McKaye, K. R., J. H. Howard, J. R. Stauffer, R. P. Morgan, and F. Shonhiwa. 1993. Sexual selection and genetic relationships of a sibling species complex of bower building cichlids in Lake Malawi, Africa. Jpn. J. Ichthyol. 40: 15–21.

Michelsen, A., F. Fink, M. Gogala, and D. Traue. 1982. Plants as transmission channels for insect vibrational songs. Behav. Ecol. Sociobiol. 11:269–281.

Otte, D. 1989. Speciation in Hawaiian crickets. In D. Otte and J. A. Endler, Eds., *Speciation and Its Consequences*, pp. 482–526. Sunderland, Mass.: Sinauer.

Otte, D. 1994. *The Crickets of Hawaii. Origin, Systematics and Evolution*. Philadelphia: The Orthopterist's Society.

Paterson, H. E. H. 1978. More evidence against speciation by reinforcement. S. Afr. J. Sci. 74:369–371.

Paterson, H. E. H. 1982. Perspectives on speciation by reinforcement. S. Afr. J. Sci. 78:53–57.

Paterson, H. E. H. 1986. The recognition concept of species. In E. S. Vrba, Ed., *Species and Speciation* (Transvaal Museum Monogr. No. 4), pp. 21–29. Pretoria: Transvaal Museum.

Perdeck, A. C. 1958. The isolating value of specific song patterns in two sibling species of grasshoppers (*Chorthippus brunneus* Thunb. and *C. biguttulus* L.). Behaviour 12: 1–75.

Prum, R. 1990. Phylogenetic analysis of the evolution of display behavior in the neotropical manakins (Aves: Pipridae). Ethology 84:202–231.

Ritchie, M. G., and J. M. Gleason. 1995. Rapid evolution of courtship song pattern in *Drosophila willistoni* sibling species. J. Evol. Biol. 8:463–479.

Roelofs, W. L., and R. L. Brown. 1982. Pheromones and evolutionary relationships of Tortricidae. Annu. Rev. Ecol. Syst. 13:395–422.

Rowe, L., G. Arnqvist, A. Sih, and J. Krupa. 1994. Sexual conflict and the evolutionary ecology of mating patterns: water striders as a model system. Trends Ecol. Evol. 9: 289–293.

Ryan, M. J., and A. S. Rand. 1993. Species recognition and sexual selection as a unitary problem in animal communication. Evolution 47:647–657.

Ryan, M. J., R. B. Cocroft, and W. Wilczynski. 1990. The role of environmental selection in intraspecific divergence of mate recognition signals in the cricket frog. Evolution 44:1869–1872.

Schwartz, J. J., and K. D. Wells. 1984. Interspecific acoustic interactions of the neotropical treefrog *Hyla ebraccata*. Behav. Ecol. Sociobiol. 14:211–224.

Shaw, K. 1995. Phylogenetic tests of the sensory exploitation model of sexual selection. Trends Ecol. Evol. 10:117–120.

Shaw, K. L. 1996a. Polygenic inheritance of a behavioral phenotype: Interspecific genetics of song in the Hawaiian cricket genus *Laupala*. Evolution 50:256–266.

Shaw, K. L. 1996b. Sequential radiations and patterns of speciation in the Hawaiian cricket genus *Laupala* inferred from DNA sequences. Evolution 50:237–255.

Stewart, K. W., S. W. Szczytko, B. P. Stark, and D. Zeigler. 1982. Drumming behavior of six North American Perlidae (Plecoptera) species. Ann. Entomol. Soc. Am. 75:549–554.

Stewart, K. W., S. W. Szczytko, and M. Maketon. 1988. Drumming as a behavioral line of evidence for delineating species in the genera *Isoperla*, *Pteronarcys*, and *Taeniopteryx* (Plecoptera). Ann. Entomol. Soc. Amer. 81:689–699.

Sullivan, J., K. Holsinger, and C. Simon. 1995. Among-site rate variation and phylogenetic analysis of the 12s rRNA in sigmodontine rodents. Mol. Biol. Evol. 12:988–1001.

Swofford, D. L. 1996. *PAUP*, Phylogenetic Analysis Using Parsimony (and Other Methods)* (Ver. 4.0, User's Manual). Sunderland, Mass: Sinauer.

Tauber, C. A., and M. J. Tauber. 1977a. A genetic model for sympatric speciation through habitat diversification and seasonal isolation. Nature 268:702–705.

Tauber, C. A., and M. J. Tauber. 1977b. Sympatric speciation based on allelic changes at three loci: evidence from natural populations in two habitats. Science 197:1298–1300.

Tauber, C. A., and M. J. Tauber. 1982a. Maynard Smith's model and corroborating evidence: no reason for misinterpretation. Ann. Entomol. Soc. Am. 75:5–6.

Tauber, C. A., and M. J. Tauber. 1982b. Sympatric speciation in *Chrysopa*: further discussion. Ann. Entomol. Soc. Am. 75:1–2.

Templeton, A. R. 1989. The meaning of species and speciation: a genetic perspective. In D. Otte and J. A. Endler, Eds., *Speciation and Its Consequences*, pp. 3–27. Sunderland, Mass.: Sinauer.

Tjeder, B. 1960. Neuroptera from Newfoundland, Miquelon, and Labrador. Opuscula Entomol. 25:146–149.

Tomaru, M., and Y. Oguma. 1994. Differences in courtship song in the species of the *Drosophila auraria* complex. Anim. Behav. 47:133–140.

Uetz, G. W., and G. E. Stratton. 1982. Acoustic communication and reproductive isolation in spiders. In P. N. Witt and J. S. Rovner, Eds., *Spider Communication*, pp. 123–159. Princeton, N.J.: Princeton University Press.

Ulagaraj, S. M., and T. J. Walker. 1973. Phonotaxis of crickets in flight: attraction of male and female crickets to male calling songs. Science 182:1278–1279.

Uzendoski, K., and P. Verrell. 1993. Sexual incompatibility and mate-recognition systems: a study of two species of sympatric salamanders (Plethodontidae). Anim. Behav. 46:267–278.

Verrell, P. A., and S. J. Arnold. 1989. Behavioral observations of sexual isolation among allopatric populations of the mountain dusky salamander, *Desmognathus ochrophaeus*. Evolution 43:745–755.

Waage, J. K. 1979. Reproductive character displacement in *Calopteryx* (Odonata: Calopterygidae). Evolution 33:104–116.

Walker, T. J. 1957. Specificity in the response of female tree crickets (Orthoptera, Gryllidae, Oecanthinae) to calling songs of the males. Ann. Entomol. Soc. Am. 50:626–636.

Walker, T. J. 1963. The taxonomy and calling songs of United States tree crickets (Orthoptera: Gryllidae: Oecanthinae). 2. The *nigricornis* group of the genus *Oecanthus*. Ann. Entomol. Soc. Am. 56:772–789.

Webster, T. P., and J. M. Burns. 1973. Dewlap color variation and electrophoretically detected sibling species in a Haitian lizard, *Anolis brevirostris*. Evolution 27:368–377.

Wells, M. M. 1993. Laboratory hybridization in green lacewings (Neuroptera, Chrysopidae, *Chrysoperla*): evidence for genetic incompatibility. Can. J. Zool. 71:233–237.

Wells, M. M. 1994. Small genetic distances among populations of green lacewings of the genus *Chrysoperla* (Neuroptera: Chrysopidae). Ann. Entomol. Soc. Am. 87:737–744.

Wells, M. M., and C. S. Henry. 1992. The role of courtship songs in reproductive isolation among populations of green lacewings of the genus *Chrysoperla* (Neuroptera: Chrysopidae). Evolution 46:31–42.

Wells, M. M., and C. S. Henry. 1994. Behavioral responses of hybrid lacewings (Neuroptera: Chrysopidae) to courtship songs. J. Insect Behav. 7:649–662.

West-Eberhard, M. J. 1983. Sexual selection, social competition, and speciation. Q. Rev. Biol. 58:155–182.

Zimmermann, E. 1990. Differentiation of vocalizations in bushbabies (Galaginae, Prosimiae, Primates) and the significance for assessing phylogenetic relationships. Z. Zool. Sist. Evol. Forsch. 28:217–239.

17

Reproductive Isolation in Sonoran Desert *Drosophila*

Testing the Limits of the Rules

Therese Ann Markow
Gregory D. Hocutt

Efforts to elucidate the mechanisms of speciation have focused on broadly observed patterns from which rules or hypotheses about these mechanisms are inferred. In one description of such patterns (Coyne and Orr 1989a), the vast data on hybrid sterility and inviability are distilled and the ubiquity of Haldane's Rule, in which the heterogametic sex is the one most seriously compromised when sex differences exist in hybrid fitness, is underscored. These authors argue convincingly for the importance of the X-chromosome in producing this pattern, utilizing the impressive data accumulated on species of *Drosophila* (Coyne and Orr 1989a; Charlesworth et al. 1987). In a second contribution and its recent update (Coyne and Orr 1989b, 1997), the *Drosophila* literature again is surveyed exhaustively, providing the basis for additional and robust patterns: an increase in pre- and postzygotic isolation with time, the evolution of sterility and inviability at similar rates in accordance with Haldane's Rule, the appearance of postzygotic isolation in hybrid males before females (Haldane 1922), and the apparent evolution of stronger prezygotic isolation compared to postzygotic isolation when in sympatry, but not allopatry, with a closely related species.

The identification of these patterns is extremely valuable because it provides a series of predictions that can be tested with observations on species not included in the original surveys. Below we describe the nature of reproductive isolation observed for *D. arizonae* and its sibling species *D. mojavensis*, as well as between geographic populations of *D. mojavensis*, some of which are sympatric with *D. arizonae*. We evaluate these observations in the context of the more general patterns for *Drosophila*, especially Haldane's Rule and the role of reinforcement in explaining patterns of sexual isolation in these two species.

Drosophila arizonae and *D. mojavensis*

Drosophila arizonae and *D. mojavensis* are cactophilic species found in North America. These two species provide the numerous experimental advantages of *Drosophila*, in addition to being among the most thoroughly studied flies of this genus with respect to their ecology. Qualities that make them attractive for studies of speciation include their recent divergence as well as the existence of geographically separated populations of *D. mojavensis* that utilize different host cacti and are considered to be in the early stages of speciation. *Drosophila arizonae* (Ruiz et al. 1990), formerly *D. arizonensis* (Patterson and Wheeler 1942) and *D. mojavensis* (Patterson and Crow 1940), are members of the mulleri complex of the repleta group (Wasserman 1982). The species were first described by Patterson and his associates at the University of Texas, during a highly productive era in *Drosophila* evolutionary biology.

Drosophila arizonae is a fairly widespread species (figure 17.1). It is sympatric with *D. mojavensis* in Sonora, southern Arizona, and the cape region of the Baja peninsula, although *D. arizonae* apparently occur only in low densities in the cape region. In Sonora it primarily uses cina cactus (*Stenocereus alamosensis*) as its host; however, *D. arizonae* is a generalist and its distribution is by no means limited to the occurrence of cina. It occasionally has been reared from saguaro (*Carnegiea gigantea*), organ pipe (*S. thurberi*), agria (*S. gummosus*), and various opuntias (Ruiz and Heed 1978) and from soil soaked with the juice of necrotic cardon (*Pachycereus pringlei*) (Breitmeyer and Markow, unpublished). In Arizona it has been collected from rotting citrus (Markow, unpublished).

Shifts in host utilization in different geographic regions are well characterized for its sibling species,

Figure 17.1. Distributions of sister species *Drosophila arizonae* and *D. mojavensis.*

D. mojavensis. These shifts underlie the ability of *D. mojavensis* to inhabit different regions of the desert (figures 17.1 and 17.2). Organ pipe is the principal host in Sonora and southern Arizona, but in Baja California, agria is preferred even though organ pipe is present. Agria is also found and utilized in a small area in coastal Sonora. Neither of these two hosts is present in southern California or northwestern Arizona, where *D. mojavensis* breed in barrel cactus (*Ferocactus cylindraceus*). *Drosophila mojavensis* has also been reared occasionally from cina, saguaro, and opuntia in Sonora (Ruiz and Heed 1978), creating the opportunity for host overlap with *D. arizonae.*

Genetic differentiation between *D. arizonae* and *D. mojavensis* has been assessed at the morphological, physiological, chromosome, protein, and DNA levels. Original clues to their being different species were the slight differences between them in their coloration patterns and more obvious differences in male genitalia (Patterson and Wheeler 1942; Patterson and Crow 1940). Chromosomal differences allow the direction of evolutionary relationships to be inferred (Wasserman 1982; Ruiz et al. 1990). Based on allozyme variation, Zouros (1973) estimated the genetic distance between them to be 0.212. The date of divergence is unclear. Mills et al. (1986) used ADH sequence data to estimate a divergence time of about 2–2.5 mya, but Pitnick et al. (1995) gave an earlier date, 6 mya, based on a combination of ADH and mitochondrial DNA (mtDNA) sequences. In contrast, based on mtDNA data, Park (1989) estimated a divergence time of between 0.15 and 1 mya depending on the genetic distance used to obtain the estimate. Inference is complicated by the fact that the ADH locus is duplicated and contains another coding region (Begun 1996), previously thought to be a pseudogene, and different strains of *D. arizonae* give different mtDNA results (Spicer, personal communication).

Chromosomal differences allow the direction of evolutionary relationships to be inferred (Wasserman 1982; Ruiz et al. 1990). Ruiz et al. (1990) describe a scenario for the evolution of these two species in southern Mexico and the subsequent colonization of Baja by *D. mojavensis* and mainland Mexico by *D. arizonae.* The evolutionary sequence is thought to have involved host shifts from an ancestral species using platyopuntia, or prickly pear cacti, to *D. arizonae* and *D. mojavensis* utilizing several different columnar cactus species. The ability to shift hosts enabled *D. mojavensis* to expand its range to southern California and the offshore islands and to Sonora, where it established contact with *D. arizonae.* A second zone of sympatry occurs in Baja California in

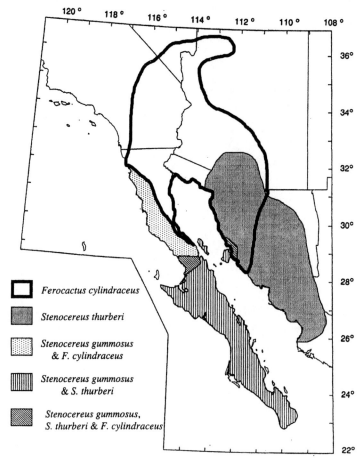

Figure 17.2. Distribution of major host cactus species of *Drosophila mojavensis* showing broad areas of overlap (compiled from Turner et al. 1995).

the region of La Paz. For purposes of this chapter, however, sympatric *D. mojavensis* will refer only to the population of the mainland Sonoran desert.

Interspecific Isolation: *D. arizonae* and *D. mojavensis*

Prezygotic Isolation

Sexual isolation between *D. arizonae* and *D. mojavensis* has been studied using a number of different experimental designs. Early studies of isolation between *D. arizonae* and *D. mojavensis* by Patterson (1947a) and by Baker (1947) were based on the numbers of females inseminated in reciprocal, no-choice pairings. Both studies utilized *D. mojavensis* from southern California and showed a significant degree of asymmetry in which fewer *D. arizonae* females mated with *D. mojavensis* males than the reverse,

although the magnitude of the difference was less in the Baker (1947) study.

Asymmetry is quite striking also in the results of "male choice" tests (Patterson 1947a). Using the same design, Wasserman and Koepfer (1977) not only confirmed the asymmetrical sexual isolation between *D. arizonae* and *D. mojavensis*, but reported character displacement for sexual isolation in regions where the two species are sympatric. The greatest influence of sympatry on sexual isolation is observed when the *D. mojavensis* female is from a region of sympatry. The situation is summarized in table 17.1.

This increase in preygotic isolation in sympatry conforms to the pattern described by Coyne and Orr (1989b, 1997). The hypothesis that increased sexual isolation observed with sympatric *D. mojavensis* has been created by reinforcement generates several predictions. These do not apply to *D. arizonae*, since it does not change its already strong sexual isolation in sympatry. First, sym-

Table 17.1. Influence of sympatry on sexual isolation.

Source of *mojavensis*	Isolation Direction	
Allopatric	AZ female × MO male > MO female × AZ male	
	(I = 0.85)	(I = 0.57)
Sympatric	AZ female × MO male = MO female × AZ male	
	(I = 0.94)	(I = 0.92)

Isolation indices (I) averaged from the male choice data of Wasserman and Koepfer (1977) are given in parentheses.

patric *D. mojavensis* should exhibit specific responsible behaviors that are measurably different from allopatric flies when paired with *D. arizonae* males. Furthermore, because the greatest increase in sexual isolation is observed with sympatric *D. mojavensis* females, these females should show the greatest behavioral difference when courted by heterospecific males.

Empirical results support these predictions. Males of both species preferentially court conspecific females, but this preference is more pronounced in *D. mojavensis* males from areas of sympatry with *D. arizonae* (Markow 1981). Furthermore, females of *D. arizonae* from all localities show extreme isolation from heterospecific males, while in *D. mojavensis*, only females from regions of sympatry behave differently toward *D. arizonae* males. Only half of the number of sympatric compared to allopatric *D. mojavensis* females exhibit the characteristic receptivity display to courting *D. arizonae* males, and of those ultimately indicating receptivity, effectively all attempted mountings fail when the female is from a sympatric strain (Markow 1981). The responsible male and female behaviors are under the control of different genetic factors (Markow 1981; Zouros 1981b).

If reinforcement is the explanation for the increased sexual isolation of sympatric *D. mojavensis* females, postmating isolation should be strongest in crosses in which the female parent is *D. mojavensis*, because this is where increased sexual isolation is strongest. Examining postmating isolation between these two species reveals, however, that testing this prediction is not simple, owing to the presence of postmating isolation at more than one level.

Postzygotic Isolation

Fertile females and sterile males are observed in the progeny of *D. arizonae* females mated to *D. mojavensis* males, consistent with Haldane's Rule. *Drosophila mojavensis* females crossed with *D. arizonae* males yield fertile female and male offspring, with the exception of *D. mojavensis* from one locality (Ruiz et al. 1990). Based solely upon hybrid sterility, the isolation can be summa-

rized as follows: AZ female × MO male > MO female × AZ male.

The genetics both of hybrid male sterility and of viability have been investigated. Hybrid male sterility in the progeny of *D. arizonae* females and *D. mojavensis* males has been extensively studied by Zouros and his colleagues (Zouros 1981a; Vigneault and Zouros 1986; Zouros et al. 1988; Goulielmos and Zouros 1995), because fertile hybrid progeny can be backcrossed to obtain different karyotypic combinations from the two species. They found that all major chromosomes carry genetic information that cause male sterility and that the expression of this sterility depends on the genetic background with respect to the other chromosomes. Four chromosome combinations produce male sterility: Y_a/IV_{mm}, X_a/III_{ma}, X_a/V_{ma}, and Y_m/III_{ma}, where roman numerals correspond to different autosome pairs, m = chromosome derived from *D. mojavensis*, and a = chromosome derived from *D. arizonae*. Furthermore, they have identified a speciation gene, SMF (sperm motility factor), in chromosome IV that causes sterility in conjunction with the Y (Pantazidis and Zouros 1988).

The reciprocal cross, between *D. mojavensis* females and *D. arizonae* males, yields significantly lower numbers of offspring, despite the fertility of both sexes in the F_1 (Ruiz et al. 1990; Baker 1947). Thus, based on productivity, the direction of isolation is opposite that found for the reciprocal cross: MO female × AZ male > AZ female × MO male. Possible explanations for the low progeny production from *D. mojavensis* females include asymmetrical sexual isolation (i.e., there are fewer progeny due to fewer interspecific matings involving *D. mojavensis* females), reproductive tract incompatibilities, postzygotic inviability, and selective sperm utilization by females. Behavioral isolation cannot account completely for the low productivity of these matings because it is observed even when the *D. mojavensis* females are from allopatric strains and the females are known to be inseminated. Remaining possible mechanisms involve either hybrid inviability or some sort of postmating but prezygotic phenomena in the female reproductive tract (Markow 1997).

Viability of hybrids (egg to adult mortality) has not been quantified, although larval to adult mortality does not appear to differ between the reciprocal crosses (Grant 1966). Postzygotic death, that is, dead larvae, from this cross (MO female × AZ male) has been observed in our laboratory and others (E. Zouros, personal communication) but the number of eggs laid, the proportion of eggs fertilized, and the level of embryonic lethality have never been quantified.

Zouros (1981b) found that progeny of backcrosses showed no breakdown of viability and that there was no sex difference in viability in any of the categories of progeny. This could explain why Nagle and Mettler (1969) were able to create interspecific hybrid populations in which certain chromosome combinations exhibited heterosis compared to the parental species. These observations question whether the cost to *D. mojavensis* females mating with *D. arizonae* males is significant enough to favor character displacement through reinforcement. In fact, these observations suggest that inviability alone may not constitute strong enough selection against hybridization in regions of sympatry (Zouros 1981b).

Postmating Prezygotic Isolation

What other mechanisms might explain the lower number of offspring from *D. mojavensis* mothers? And, are they strong enough to impose sufficient costs to heterospecifically mated *D. mojavensis* females to select for increased sexual isolation in the region of sympatry? Below we present evidence bearing upon potential interference with oviposition or normal remating by female *D. mojavensis* concerning this question. Females of both *D. arizonae* and *D. mojavensis* normally exhibit a large mass in the uterus following intraspecific mating (Patterson 1947b), called the insemination reaction, lasting for 7–9 hours. The onset of oviposition begins shortly after that time, but it is unclear if oviposition is responsible for the disappearance of the mass, or if the mass must degenerate before oviposition can begin. Remating in females of these species also coincides with reduction in the mass (Patterson 1947b; Krebs 1989). In heterospecific matings, the mass tends to be larger and of longer duration: the increase in size and persistence of the mass appear to be positively correlated with the level of divergence between the two species (Patterson 1947b).

Patterson (1947b) observed that in crosses between mulleri complex species more distantly related than *D. arizonae* and *D. mojavensis*, the mass often lasted indefinitely or caused severe damage to the female reproductive tract. In interspecific crosses between other species, Patterson (1947b) also observed disintegrated eggs in the mass as well as evidence of sperm incapacitation or death. These observations suggested to Patterson that the mass is a potential reproductive isolating mechanism. When *D. mojavensis* females are mated to *D. arizonae* males, the mass is still fairly large after 72 hours, while it disappears in the reciprocal cross after 9 hours. Because Baker (1947) observed that a large number of unproductive *D. mojavensis* females inseminated by *D. arizonae* males contained a large insemination reaction mass in their reproductive tracts, Ruiz et al. (1990) invoke the insemination reaction to explain the reduced productivity of *D. mojavensis* females mated to *D. arizonae* males, suggesting that the sperm of *D. arizonae* males are killed in the reproductive tracts of *D. mojavensis* females. Unfortunately, direct support for this hypothesis is lacking. Neither Baker (1947) nor Ruiz et al. (1990) reported motility of stored sperm, oviposition, or postzygotic mortality in these crosses.

We tested the hypothesis that the mass blocks oviposition in crosses between *D. mojavensis* females and *D. arizonae* males. Single females were confined with males until copulation was observed. Mated females were transferred daily for one week to fresh culture vials and the number of eggs was counted. Females then were dissected and the sperm storage organs were scored for the presence of sperm. Table 17.2 shows that oviposition is not prevented in *D. mojavensis* females inseminated by *D. arizonae* males, although in several combinations it is reduced. No reduction is observed in the reciprocal cross, and in fact, when the *D. mojavensis* males are from Sonora, there is an increase in oviposition by *D. arizonae* females. After one week, about three fourths of the females still had motile sperm and we observed no evidence of dead sperm in our dissections (table 17.3). While these motile sperm obviously are not killed in the female reproductive tract, their viability and recoverability are unknown. Obviously a large number are viable, since offspring are produced. About half of these females had a vestigial reaction mass in their uterine cavities, but it did not interfere with oviposition.

We also tested the hypotheses that sperm are inviable in the reproductive tract and that female remating is im-

Table 17.2. Mean number of eggs laid in interspecific crosses of *Drosophila arizonae* (Ariz) and *D. mojavensis* race A (MojA), race B$_I$ (MojB$_I$), and race B$_{II}$ (MojB$_{II}$).

Mating			Number of Eggs
Females	Males	n	(X ± SE)
MojA	MojA	16	157.6 ± 15.5
MojA	Ariz	14	108.4 ± 13.4
MojB$_I$	MojB$_I$	24	82.6 ± 7.6
MojB$_I$	Ariz	No matings	—
MojB$_{II}$	MojB$_{II}$	23	85.6 ± 9.1
MojB$_{II}$	Ariz	17	66.2 ± 7.7
Ariz	Ariz	33	82.8 ± 6.9
Ariz	MojA	24	81.7 ± 7.4
Ariz	MojB$_I$	8	116.0 ± 13.9
Ariz	MojB$_{II}$	10	76.2 ± 15.5

Table 17.3. Insemination reaction mass and storage of motile sperm.

Female	×	Male	n(♀♀)	Mass	Sperm	Reference
D. arizonae		D. mojavensis	39	16	19	Patterson
D. mojavensis		D. arizonae	154	97	80	Patterson
D. arizonae		D. mojavensis	76	20	?	Baker
D. mojavensis		D. arizonae	90	74	?	Baker
D. arizonae		D. mojavensis	17	0	15	Present study
D. mojavensis		D. arizonae	22	8[1]	17	Present study

Data are from present study and Baker (1947) assayed 7 days after mating, and from Patterson (1947b) assayed 96 hours after mating.

[1]Degenerate mass.

paired by the insemination reaction. Individual females from each species were placed with single males of the other species until a copulation was observed. Pairs were separated and the females were held for 24 hours before being tested for remating by placing them individually with males of their own species for one hour. Of 15 heterospecifically mated females from each species, all remated within the first 20 minutes, as did females mated initially with their own males 24 hours earlier. Although the reaction mass persists for 72 hours in one of the hybridizations, it does not appear to cause any delay in remating.

We conclude that the low productivity from *D. mojavensis* females is not due to a reduction in oviposition or to oocytes being trapped in the reaction mass. The differences in oviposition rates are insufficient by themselves to account for the productivity differences between reciprocal crosses. The presence of motile sperm in the ventral receptacles of all mated *D. mojavensis* females demonstrates that *D. arizonae* sperm are transferred, stored and still alive after one week. Finally, females suffer no consequences in their ability to remate with conspecific males following a heterospecific mating. The unanswered question is the fertility versus viability of the oviposited eggs from *D. mojavensis* females mated to *D. arizonae* males. If a proportion of the eggs laid are not fertilized, it would suggest that either the motile sperm are not viable, or that females can exercise some choice over the sperm they allow to fertilize their eggs. Experiments are needed to determine if the low productivity is due to embryonic inviability or to the deposition of unfertilized eggs.

In summary, crosses between *D. arizonae* and *D. mojavensis* exhibit both prezygotic and postzygotic isolation in the following directions: behavioral, AZ female × MO male > MO female × AZ male; productivity, AZ female × MO male < MO female × AZ male; sterility, AZ female × MO male > MO female × AZ male. All three types of isolation exhibit asymmetry, but in different directions. Only behavioral asymmetry changes in regions of sympatry. With respect to the character displacement hypothesis for the behavioral isolation observed in sympatric *D. mojavensis*, it is unlikely, given the fertility and via-

bility of the F_1 hybrids from *D. mojavensis* females, that selection against hybridization could provide the sole explanation for the observed patterns.

Reinforcement of premating isolation due to existing postzygotic isolation may cause character displacement like that observed in sympatric *D. mojavensis*. However, modeling indicates that the parameters under which reinforcement may occur are quite restrictive (Maynard Smith 1966; Moore 1979, 1981; Spencer et al. 1986; Butlin 1989; Liou and Price 1994). In fact, substantial niche differentiation and hybrid fitnesses at or close to zero are usually invoked. Maynard Smith (1966) determined that hybrid fitness as low as 0.25 would result in populations blending into a single hybrid swarm within eight generations, causing selection for more viable hybrids as opposed to increased premating isolation. If we assume that hybridizing *D. arizonae* females and *D. mojavensis* males produce roughly as many offspring as with homospecific mates, then male sterility in the progeny of this cross would reduce fitnesses of parents and average offspring fitness to 0.50 at a minimum. In the reciprocal hybrid cross, Baker (1947) and Ruiz et al. (1990) reported sevenfold and threefold reductions in numbers of offspring, respectively. At most, then, the fitness of hybridizing *D. mojavensis* females and *D. arizonae* males would be reduced by a factor of 7. This apparent fitness disadvantage, however, may be offset by heterosis (Nagle and Mettler 1969). Furthermore, lifetime fitness reductions could be far less substantial because of the high frequency of remating in both species. These, admittedly rough, estimates of the reduced fitness resulting from hybridization question whether reinforcement alone could have driven the putative sympatric character displacement in *D. mojavensis*.

Intraspecific Isolation and Differentiation: *D. mojavensis*

Mettler (1963) first proposed the existence of two races of *D. mojavensis*: race A in the deserts of southern California and race B from the Sonoran desert of Arizona,

Sonora, northern Sinaloa, and the Baja peninsula. Allozyme differences led Zouros (1973) to propose a further subdivision of race B into B_I including populations from Arizona and mainland Mexico, and race B_{II} from Baja. The populations in these separate geographic areas utilize different host plants (figures 17.1 and 17.2). While the role of host adaptation in the evolution of genetic differences between races is unknown, Starmer et al. (1977) suggest a functional relationship between allelic variation for ADH and host chemistry. In order to be considered true host races, populations should exhibit the greatest fitness on their respective hosts. In those populations of *D. mojavensis* where this has been examined, the expected relationship was not found. Agria is the preferred host of all races of *D. mojavensis* and the one on which fitness is the greatest (Ruiz and Heed 1978). Host shifts, therefore, are more likely to reflect host availability in those desert regions invaded by *D. mojavensis*. The genetic differentiation between races of *D. mojavensis* (Zouros 1973; Heed 1978; Cleland et al. 1996) may be a reflection of differences in host ecology.

Zouros and d'Entremont (1980) showed the existence of significant sexual isolation between the B_I and B_{II} subraces. Populations from southern California and from Santa Catalina Island are not behaviorally isolated from each other or from B_I or B_{II} flies (Markow 1991). Despite suggestions that the B_I–B_{II} isolation might be influenced by rearing media (Brazner 1983; Etges 1992), a number of independent investigations have found this sexual isolation to be repeatable and significant (Zouros and d'Entremont 1980; Wasserman and Koepfer 1977; Markow 1991; Krebs and Markow 1989; Krebs 1990) even when flies were reared on tissue of alternate host cacti (Etges 1992; Fogleman and Markow, unpublished). Furthermore, reciprocal hybridization (Krebs 1990) and artificial selection experiments (Koepfer 1987a,b) have demonstrated a clear genetic basis for the observed sexual isolation, with different genetic mechanisms underlying male and female behavior.

The sexual isolation within *D. mojavensis* is asymmetrical: females from race B_I show reduced receptivity to B_{II} males. Zouros and d'Entremont (1980) attribute this to the pressure on B_I *D. mojavensis* from sympatry with *D. arizonae*, in that the selection for increased mating discrimination by *D. mojavensis* females simultaneously increased their discrimination against males from Baja as well. This interpretation assumes that the same genetic mechanism underlies discrimination against conspecifics as well as against *D. arizonae* males. Zouros (1981b) has localized the genes for female isolation from *D. arizonae* males to specific *D. mojavensis* chromosomes. Because the chromosomal locations of the genes for discrimination against *D. mojavensis* males from B_{II} are unknown, we cannot tell if they are the same. It is unlikely that the mechanism is a general one, however, since Sonora females do not always discriminate against *D. mojavensis*

males from other isolated geographic areas (Markow 1991).

Another assumption of the character displacement explanation is that the behavioral "characters" or cues used for discrimination are also the same. There is no information bearing on this question either. In an earlier study (Markow and Toolson 1990), differences were reported in cuticular hydrocarbon composition along with evidence that these differences influence mating behavior of *D. mojavensis*. Studies of mating pairs, however, showed no association between hydrocarbon composition and mating success (Markow and Toolson, unpublished).

No decrement in F_1 viability has been detected in crosses between any of the races of *D. mojavensis* in which it has been studied, but fertility and backcross fertility have not been investigated (Etges and Heed 1987; Etges 1990, 1993). However a number of studies of divergence for quantitative traits such as development time (Etges and Heed 1987; Etges 1990, 1993) and locomotor activity (Krebs 1991) suggest that genetic differences between races may, in fact, be extensive. Whether enough time has elapsed for the accumulation of mutations with postzygotic intraspecific hybrid effects is a question that begs investigation.

Rules and Patterns: Conformity or Endless Forms?

It is clear from the foregoing that *D. arizonae* and *D. mojavensis* both conform to and challenge some of the proposed rules. Haldane's Rule, that the heterogametic sex is the one most seriously compromised when sex differences exist in hybrid fitness, certainly is true for male sterility in the cross between *D. arizonae* females and *D. mojavensis* males. Recent models of the evolution of sterility and inviability (Orr 1993; Turelli and Orr 1995) predict that the genetics of hybrid sterility will be largely sex specific, but that the genes causing hybrid inviability should tend to be expressed equally in both sexes. We observe this pattern for the *D. arizonae*/*D. mojavensis* species pair. While the causes of the low productivity of the *D. mojavensis* female × *D. arizonae* male cross are unknown, there is an apparent lack of sex differences in the hybrids produced (Zouros 1981c; Grant 1966). The chromosomal incompatibilities in these species previously discussed support the prediction of sex-specific sterility genes.

Efforts to explain Haldane's Rule have led to another rule, attributing the responsible genetic factors to the X-chromosome (Coyne and Orr 1989a; Charlesworth et al. 1987). This interpretation, however, may be a function of the more limited types of analyses typically performed, that is, hybridization and first generation backcrosses. When specific chromosomes or parts of chromosomes can be combined, through introgression, in either heterozygous or homozygous conditions, a broader array of chromosomal interactions can be examined for incompatibilities. The work of Zouros and his colleagues (Zouros 1981a; Vigneault and Zouros 1986, Zouros et al. 1988, Goulielmos and Zouros 1995) was the first to suggest that the importance of the X-chromosome in male sterility may not be as great as originally assumed. Although these investigators found that the X-chromosome plays a key role in some of the incompatibilities between *D. mojavensis* and *D. arizonae*, these same studies revealed that the Y-chromosome is involved in as many chromosomal incompatibilities causing sterility as the X. This result is unexpected given the proposed causes of the large X-effect: that the sex chromosomes experience more rapid evolutionary rates than the autosomes (Charlesworth et al. 1987; Coyne and Orr 1989a). Even assuming equal evolutionary rates between the X and Y chromosomes, if the role of each sex chromosome in determining fertility and viability is proportional to its size, we might expect the Y to be involved in far fewer incompatibilities than the X-chromosome. Instead, the Y-chromosome appears to play a disproportionately large role in determining male hybrid sterility in the *D. arizonae/D. mojavensis* system. Recent studies with other *Drosophila* species (Hollocher and Wu 1996; Lamnissou et al. 1996) also point to the importance of chromosomes other than the X-chromosome.

Wu and Davis (1993) have speculated that the highly specialized nature of spermatogenesis may cause this system to be less buffered against hybridization thus partially explaining Haldane's Rule in species with heterogametic males. Given the known importance of the Y-chromosome in determining *Drosophila* male fertility (Ashburner 1989) and the proposed heightened sensitivity of spermatogenesis to genetic perturbations, modeling and testing of the potential large Y effect of male hybrid sterility per Haldane's Rule may be a fruitful avenue for future investigation.

Prezygotic isolation also has been suggested to follow some rules. Coyne and Orr (1989a) have observed that species populations existing sympatrically with closely related species tend to express higher levels of premating isolation than do allopatric populations at comparable genetic distances. The *D. arizonae/D. mojavensis* species pair support this observation. Published genetic distances based on allozyme variability (Zouros 1973) indicate that each of the races of *D. mojavensis* are roughly genetically equidistant from *D. arizonae*, yet sympatric *D. mojavensis* Race B_I exhibits significantly stronger premating reproductive isolation from the sister species than either of the allopatric races. On the surface, the pattern with *D. arizonae/D. mojavensis* is consistent with the higher level of isolation in sympatry, but the interpretation that reinforcement is the most likely explanation is questioned on the basis of the available observations.

Given the theoretical (Maynard Smith 1966; Moore 1979; Butlin 1989) difficulty in invoking reinforcement as a significant cause of increased premating isolation (but see Howard 1993), additional explanations for the observed increased isolation in sympatry must be sought. For this species pair, the wealth of information on their ecology and mating systems may be informative. The observed patters of prezygotic isolation between *D. arizonae* and *D. mojavensis* and between populations of *D. mojavensis* could have originated as asymmetries having nothing to do with interspecific discrimination and mate recognition, instead reflecting coevolution between the sexes for different mating system components (West-Eberhard 1983). In the case of sexual behavior, male vigor and female receptivity coevolve in a population of the same species such that in strains where female receptivity is great, males are less vigorous. In strains where female receptivity thresholds are higher, males are more vigorous in their courtship. These differences may have nothing to do with discrimination, but are likely a function of the number of mating opportunities and the intensity of sexual selection in the recent evolutionary history of a population. For example, when the operational sex ratio is highly male biased, male vigor is favored. In less male-biased or in female-biased populations, female receptivity will be greater and males will not need to court as vigorously. Among *Drosophila* species the operational sex ratio varies widely due largely to enormous differences in female remating intervals and male maturation times (Markow 1996).

When two populations of the same species, having different male vigor and female receptivity, are placed together in a "choice test," asymmetries in the form of more matings between the vigorous males and receptive females are expected (Kence and Bryant 1978; Van den Berg et al. 1984). In the case of *D. mojavensis* from races B_I and B_{II}, males and females differ significantly in their mating speeds (Krebs and Markow 1989). Males from B_I court more quickly and vigorously than do B_{II} males, and females from the latter race are more receptive. Thus, the observed asymmetrical isolation between flies from these two races is expected on the basis of vigor/receptivity differences alone. It is clear that additional factors are responsible, however, since males do not initiate courtship of females from different strains or species at random, and it is possible to separate these components statistically (Zouros and d'Entremont, 1980).

How significant is this sort of observed sexual isolation to the development of biologically meaningful isolation? We do not mean to imply that deviations from random mating attributable to vigor/receptivity differences are unimportant. On the contrary, higher female receptivity thresholds may be a key prerequisite to the development of more specific, character-based behavioral divergence, such that females have more time to evaluate qualitative differences among suitors before mating. Thus, very early in the separation between two populations, asymmetrical sexual isolation may reflect vigor/receptivity differences, while later, additional, qualitative changes are expected. It is not clear, in the case of *D. arizonae* and *D. mojavensis*, exactly what the qualitative differences are.

Asymmetries in sperm utilization may also reflect differences in the coevolution of reproductive strategies. Male ejaculatory fluids contain proteins that stimulate oviposition, but females are known to exhibit differential responses to the proteins of different males (Markow 1982). It may not always be in the female's best interest to oviposit following mating, due either to lack of quality oviposition sites or to lack of quality sperm. Pairing males and females from strains that differ in this parameter should produce asymmetries in egg laying. Based on differences in resource availability for *D. mojavensis* from different host regions, we expect females of different populations to differ in the stimulation required to oviposit.

Generally, when multiple asymmetries exist at different levels or sites along the isolation continuum, certain combinations of them may facilitate the speciation process. These asymmetries can have their origins in the resource ecology of the populations in question without being a function of performance on particular hosts, per se. To what degree the observed isolation among *D. mojavensis* races is a reflection of nonspecific behavior versus behavior with the potential to significantly restrict gene flow is unclear.

These are just some suggested alternatives to reinforcement. The predictions of the reinforcement hypothesis, as it relates to both the inter- and intraspecific isolation, can be tested. One of these predictions, that the loci for sexual isolation within and between species are the same, ultimately can be tested by mapping the loci in both species. Second, if the responsible behaviors can be identified, they can be experimentally manipulated to see if they are the same in both species. Third, testing both species from the area of La Paz, Baja California, another region of sympatry, should give similar results to those observed in the sympatric flies from the mainland of Mexico. And finally, the degree to which both species share host cacti, and thus have the opportunity to hybridize, is something that can, contrary to many *Drosophila* species pairs whose hosts are less well known, actually be determined in this system.

The final pattern to which these two species appear to conform is the proposed earlier development of premating compared to postmating isolation in sympatry. Although we do not dispute the observed pattern in sympatry, we wonder how the general allopatric and sympatric patterns of pre- versus postmating isolation will hold up against future, more detailed scrutiny. We question whether these general patterns may be artifacts of the relative ease with which premating versus postmating isolation can be examined as opposed to a reflection of a true evolutionary pattern. Sexual isolation is far easier to test for and quantify than is postmating isolation. A large number of flies can easily be placed together in a mating chamber and the number of matings of different types scored in a short time. Large sample sizes permit even subtle behavioral isolation to be detected. Postmating isolation, on the other hand, is more labor intensive to measure, and if it is subtle or incomplete, or only appears in segregating generations, it may go undetected. We suggest that tests for postmating isolation be more thorough to be certain it does not exist in some degree not easily revealed by the standard 0.0, 0.25, 0.50, and 1.00 scoring protocol. The lesson from studies of the genetics of sterility (Goulielmos and Zouros 1995; Hollocher and Wu 1996; Lamnissou et al. 1996) may apply to other rules and patterns as well.

Amid numerous and varied observations and specific results of experimentation, evolutionary biologists seek to impose order on the evolutionary process by proposing generalized mechanisms and patterns. The exhaustive surveys and insightful syntheses of Coyne and Orr (1989a,b) provide an invaluable framework for studies of speciation. Constructive analysis and experimentation have generated both support for and exceptions to these proposed rules, the relative weight of which will dictate their future acceptance, modification, or elimination. We hope that this brief analysis of the *D. arizonae/D. mojavensis* model system will inspire additional inquiry.

Acknowledgments   We thank Sean Murphy, Sue Bertram, Sophia Cleland, and Marsha St. Louis for assistance counting eggs and collecting flies. Our work is supported by NSF Grants INT 94 0261 and DEB 95 10645. GDH is an NSF predoctoral fellow.

References

Ashburner, M. 1989. Drosophila: A laboratory handbook. Cold Spring Harbor Press, New York.

Baker, W. K. 1947. A study of the isolating mechanisms found in *Drosophila arizonensis* and *Drosophila mojavensis*. University of Texas Publications 4752:126–136.

Begun, D. 1996. Origin and evolution of a new gene descended from alcohol dehydrogenase in *Drosophila*. Genetics 145: 375–382.

Brazner, J. C. 1983. The influence of rearing environment on sexual isolation between populations of *Drosophila mojavensis*: An alternative to the character displacement hypothesis. M.S. thesis, Syracuse University, New York.

Butlin, R. 1989. Reinforcement of premating isolation. In D. Otte and J. A. Endler (eds.), Speciation and Its Consequences. Sinauer, Sunderland, Mass., pp. 158–179.

Charlesworth, B., J. A. Coyne, and N. H. Barton. 1987. The relative rates of evolution of sex chromosomes and autosomes. American Naturalist 130:113–146.

Cleland, S., G. D. Hocutt, C. M. Breitmeyer, T. A. Markow, and E. Pfeiler. 1996. A new alcohol dehydrogenase polymorphism in barrel cactus populations of *Drosophila mojavensis*. Genetica 98:115–117.

Coyne, J. A., and H. A. Orr. 1989a. Patterns of speciation in Drosophila. Evolution 43:362–381.

Coyne, J. A., and H. A. Orr. 1989b. Two rules of speciation. In D. Otte and J. A. Endler (eds.), Speciation and Its Consequences. Sinauer, Sunderland, Mass., pp. 180–207.

Coyne, J. A., and H. A. Orr. 1997. Patterns of speciation in *Drosophila* revisited. Evolution 51:295–303.

Etges, W. J. 1990. Direction of life history evolution in *Drosophila mojavensis*. In J. S. F. Barker, W. T. Starmer, and R. J. MacIntyre (eds.), Ecological and Evolutionary Genetics of *Drosophila*. Plenum Press, New York, pp. 37–56.

Etges, W. J. 1992. Premating isolation is determined by larval substrates in cactophilic *Drosophila mojavensis*. Evolution 46:1945–1950.

Etges, W. J. 1993. Genetics of host-cactus response and life-history evolution among ancestral and derived populations of cactophilic *Drosophila mojavensis*. Evolution 47:750–767.

Etges, W. J., and W. B. Heed. 1987. Sensitivity to larval density in populations of *Drosophila mojavensis*: influences of host plant variation on components of fitness. Oecologia 71:375–381.

Goulielmos, G., and E. Zouros. 1995. Incompatibility analysis of male hybrid sterility in two *Drosophila* species: lack of evidence for maternal, cytoplasmic, or transposable elements effects. American Naturalist 145:1006–1014.

Grant, B. S. 1966. Fitness component analysis of the interspecific hybrids of *Drosophila arizonensis* and *Drosophila mojavensis*. M. S. thesis, North Carolina State University.

Haldane, J. B. S. 1922. Sex ratio and unisexual sterility in hybrid animals. Journal of Genetics 12:101–109.

Heed, W. B. 1978. Ecology and genetics of sonoran desert *Drosophila*. In P. F. Brussard (ed.), Ecological Genetics: The Interface, Springer-Verlag, New York, pp. 109–126.

Hollocher, H., and C.-I. Wu. 1996. The genetics of reproductive isolation in the *Drosophila simulans* clade: X vs. autosomal effects on male vs. female effects. Genetics 143:1243–1255.

Howard, D. J. 1993. Small populations, inbreeding, and speciation. In N. W. Thornhill (ed.), The Natural History of Inbreeding and Outbreeding: Theoretical and Empirical Perspectives, University of Chicago Press, Chicago, pp. 118–142.

Kence, A., and E. H. Bryant. 1978. A model of mating behavior in flies. American Naturalist 112:1047–1062.

Koepfer, H. R. 1987a. Selection for sexual isolation between geographic forms of *Drosophila mojavensis*. I. Interactions between the selected forms. Evolution 41:37–48.

Koepfer, H. R. 1987b. Selection for sexual isolation between geographic forms of *Drosophila mojavensis* II. Effects of selection on mating preference and propensity. Evolution 41:1409–1412.

Krebs, R. A. 1989. Courtship behavior, sexual selection, and genetics of sexual isolation in *Drosophila mojavensis*. Ph.D. dissertion, Arizona State University.

Krebs, R. A. 1990. Courtship behavior and control of reproductive isolation in *Drosophila mojavensis*: analysis of population hybrids. Behavioral Genetics 20:535–543.

Krebs, R. A. 1991. Variation among *Drosophila mojavensis* populations for locomotor activity does not cause sexual isolation. Evolutionary Theory 10:101–107.

Krebs, R. A., and T. A. Markow. 1989. Courtship behavior and control of reproductive isolation in *Drosophila mojavensis*. Evolution 43:908–912.

Lamnissou, K., M. Loukas, and E. Zouros. 1996. Incompatibilities between Y chromosome and autosomes are responsible for male hybrid sterility in crosses between *Drosophila virilis* and *Drosophila texana*. Heredity 76:603–609.

Liou, L. W., and T. D. Price. 1994. Speciation by reinforcement of premating isolation. Evolution 48:1451–1459.

Markow, T. A. 1981. Courtship behavior and control of reproductive isolation between *Drosophila mojavensis* and *Drosophila arizonensis*. Evolution 35:1022–1026.

Markow, T. A. 1982. Mating systems of cactophilic *Drosophila*. In J. T. S. Barker and W. T. Starmer (eds.), Ecological Genetics and Evolution: The Cactus-Yeast-Drosphila Model System, Academic Press, Sydney, pp. 273–287.

Markow, T. A. 1991. Sexual isolation among populations of *Drosophila mojavensis*. Evolution 45:1525–1529.

Markow, T. A. 1996. Evolution of *Drosophila* mating systems. Evolutionary Biology 29:73–106.

Markow, T. A. 1997. Assortative fertilization in *Drosophila*. Proceedings of the National Academy of Sciences 94:7756–7760.

Markow, T. A., and E. C. Toolson. 1990. Temperature effects on epicuticular hydrocarbons and sexual isolation in *Drosophila mojavensis*. In J. F. S. Barker, W. T. Starmer, and R. J. MacIntyre (eds.), Ecological and Evolutionary Genetics of Drosophila, New York, Plenum Press, pp. 315–331.

Maynard Smith, J. 1966. Sympatric speciation. American Naturalist 100:637–650.

Mettler, L. E. 1963. *D. mojavensis baja*, a new form in the mulleri complex. Drosophila Information Services 38:57–58.

Mills, L. E., P. Batterham, J. Alegre, W. T. Starmer, and D. T. Sullivan. 1986. Molecular genetic characterization of a locus that contains duplicate ADH genes in *Drosophila mojavensis* and related species. Genetics 112:295–310.

Moore, W. S. 1979. A single locus mass-action model of assortative mating with comments on the process of speciation. Heredity 42:173–186.

Moore, W. S. 1981. Assortative mating genes selected along a gradient. Heredity 46:191–195.

Nagle, J. J., and L. E. Mettler. 1969. Relative fitness of introgressed and parental populations of *Drosophila mojavensis* and *D. arizonensis*. Evolution 23:519–524.

Orr, H. A. 1993. A mathematical model of Haldane's Rule. Evolution 47:1606–1611.

Pantazidis, A. C., and E. Zouros. 1988. Location of an autosomal factor causing sterility in *Drosophila mojavensis* males carrying the *Drosophila arizonensis* Y chromosome. Heredity 60:299–304.

Park, L. K. 1989. Evolution in the repleta group of *Drosophila*: a phylogenetic analysis using mitochondrial DNA. Ph.D. dissertation, Washington University, Saint Louis.

Patterson, J. T. 1947a. Sexual isolation in the mulleri sub-group. University of Texas Publications 4752:32–40.

Patterson, J. T. 1947b. The insemination reaction and its bearing on the problem of speciation in the mulleri subgroup. University of Texas Publications 4752:41–77.

Patterson, J. T., and J. F. Crow. 1940. Hybridization in the mulleri group of *Drosophila*. University of Texas Publications 4032:251–256.

Patterson, J. T., and M. R. Wheeler. 1942. Description of new species of the subgenera *Hirtodrosophila* and *Drosophila*. University of Texas Publications 4213:67–109.

Pitnick, S., T. A. Markow, and G. Spicer. 1995. Delayed male maturity is a cost of producing large sperm in *Drosophila*. Proceedings of the National Academy of Sciences USA 92:10614–10618.

Ruiz, A., and W. B. Heed. 1978. Host plant specificity in the cactophilic *Drosophila mulleri* species complex. Journal of Animinal Ecology 57:237–249.

Ruiz, A., W. B. Heed, and M. Wasserman. 1990. Evolution of the mojavensis cluster of cactophilic *Drosophila*, with descriptions of two new species. Journal of Heredity 81: 30–42.

Spencer, H. G., B. H. McArdle, and D. M. Lambert. 1986. A theoretical investigation of speciation by reinforcement. American Naturalist 128:241–262.

Starmer, W. T., W. B. Heed, and E. S. Rockwood-Sluss. 1977. Extension of longevity in *Drosophila mojavensis* by environmental ethanol: differences between subraces. Proceedings of the National Academy of Sciences USA 74:387–391.

Turelli, M., and H. A. Orr. 1995. The dominance theory of Haldane's Rule. Genetics 140:389–402.

Turner, R. M., J. E. Bowers, and T. L. Burgess. 1995. Sonoran desert plants: an ecological atlas. University of Arizona Press, Tuscon.

Van den Berg, M. J., G. Thomas, H. Hendricks, and W. Van Delden. 1984. A reexamination of the negative assortative mating phenomenon and its underlying mechanisms in *Drosophila melanogaster*. Behavioral Genetics 14:45–61.

Vigneault, G., and E. Zouros. 1986. The genetics of asymmetrical male sterility in *D. mojavensis* and *D. arizonensis* hybrids: interactions between the Y-chromosome and autosomes. Evolution 40:1160–1170.

Wasserman, M. 1982. Cytological evolution in the *Drosophila repleta* species group. In J. S. F. Barker and W. T. Starmer (eds.), Ecological Genetics and Evolution: The Cactus-Yeast-*Drosophila* Model System, Academic Press, Sydney, pp. 49–64.

Wasserman, M., and H. R. Koepfer. 1977. Character displacement for sexual isolation between *Drosophila mojavensis* and *Drosophila arizonensis*. Evolution 31:812–823.

West-Eberhard, M. J. 1983. Sexual selection, social competition, and speciation. Quarterly Review of Biology 58: 155–183.

Wu, C.-I., and A. W. Davis. 1993. Evolution of postmating reproductive isolation: the composite nature of Haldane's Rule and its genetic bases. American Naturalist 142:187–212.

Zouros, E. 1973. Genic differentiation associated with the early stages of speciation in the mulleri subgroup of *Drosophila*. Evolution, 27:601–621.

Zouros, E. 1981a. An autosome–Y chromosome combination that causes sterility in *D. mojavensis* hybrids. Drosophila Information Service 56:167–168.

Zouros, E. 1981b. The chromosomal basis of sexual isolation in two sibling species of *Drosophila*: *D. arizonensis* and *D. mojavensis*. Genetics 97:703–718.

Zouros, E. 1981c. The chromosomal basis of viability in interspecific hybrids between *Drosophila arizonensis* and *Drosophila mojavensis*. Canadian Journal of Genetics and Cytology 23:67–72.

Zouros, E., and C. J. d'Entremont. 1980. Sexual isolation among populations of *Drosophila mojavensis*: response to pressure from a related species. Evolution 34:421–430.

Zouros, E., K. Lofdahl, and P. Martin. 1988. Male hybrid sterility in *Drosophila*: interactions between autosomes and sex chromosomes of *D. mojavensis* and *D. arizonensis*. Evolution 42:1321–1331.

18

Wolbachia and Speciation

John H. Werren

Bacteria in the genus *Wolbachia* are cytoplasmically inherited rickettsia found in the reproductive tissues (ovaries and testes) of invertebrates (reviewed in Werren 1997). This widespread and common group of bacteria cause a number of reproductive alterations in their hosts, including induction of cytoplasmic incompatibility, parthenogenesis, and feminization. *Wolbachia* are of particular interest because of their potential role as a mechanism for speciation (Laven 1959, 1967; Breeuwer and Werren 1990; Coyne 1992).

Cytoplasmic incompatibility (CI) is an incompatibility between sperm and egg induced by *Wolbachia* that causes either zygotic death (in diploid species) or increased male production (in haplodiploid species). CI can occur in crosses between strains, populations, or closely related species and can be either unidirectional (one cross is compatible and the reciprocal cross is incompatible) or bidirectional (crosses in both directions are incompatible). *Wolbachia*-induced CI has been documented in a wide range of arthropods (Werren et al. 1995a; Werren and O'Neill 1997).

CI has obvious potential implications for speciation. If *Wolbachia*-induced CI causes complete or nearly complete reproductive isolation between populations, then the populations can potentially diverge genetically and evolve into new species. Of particular interest is whether acquisition of *Wolbachia* can promote the rapid development of reproductive isolation, therefore promoting speciation.

Parthenogenesis-inducing (PI) *Wolbachia* cause parthenogenesis in hosts by manipulating chromosome behavior in unfertilized eggs (Stouthamer and Kazmer 1994; Stouthamer 1997). By causing the rapid development of parthenogenesis within populations, PI *Wolbachia* may promote the evolution of parthenogenetic "species." Feminizing *Wolbachia* have so far only been described in isopods (Rousset et al. 1992a). The direct role of these microorganisms in speciation is less apparent. However, a good case has been made for feminizing microorganisms causing rapid evolutionary changes in sex determination mechanisms (Rigaud and Juchault 1993), which could in turn lead to reproductive isolation between diverging populations.

Below, I explore the possible role of *Wolbachia* in speciation, focusing on the current evidence and potential directions for future research.

Distribution and Phylogeny of *Wolbachia*

Wolbachia are intracellular bacteria belonging to the Alpha subdivision of Proteobacteria (O'Neill et al. 1992). The closest relatives of *Wolbachia* are intracellular rickettsia in the genera *Ehrlichia, Cowdria, Anaplasma*, and *Rickettsia*. Many members of these latter genera are vectored by arthropods and cause disease in vertebrates. In contrast, *Wolbachia* are not known to cause vertebrate disease, but are typically found in reproductive (and sometimes other) tissues of invertebrates.

Phylogenetic studies using both 16S rDNA (O'Neill et al. 1992; Breeuwer et al. 1992; Rousset et al. 1992a) and protein coding sequences (Werren et al. 1995b) reveal two major subdivisions of *Wolbachia*, designated A and B (figure 18.1). These are estimated to have diverged approximately 50 million years ago, based on synonymous substitution rates in the protein-coding *fts*Z gene (Werren et al. 1995b). CI appears to be the common phenotype for *Wolbachia* and commonly occurs in both the A and B subdivisions (figure 18.1). Parthenogenesis induction has so far been found only in parasitic Hymenoptera, although it occurs within many different genera (Stouthamer 1997). *Wolbachia* have also been found in parthenogenetic beetles (Werren et al. 1995b), although curing experiments to establish a causative role of the bacteria have not been conducted in the beetles. Parthenogenesis induction also occurs in both of the major subdivisions of *Wolbachia* (figure 18.1). Feminizing B-group *Wolbachia* have been described in terrestrial isopods, most notably *Armadillidium vulgare* (Rousset et al. 1992a).

Within species, *Wolbachia* are typically transmitted through the egg cytoplasm, and therefore are inherited vertically from infected females to their progeny. However, phylogenetic studies (O'Neill et al. 1992; Rousset et al. 1992a; Werren et al. 1995b) reveal considerable

245

Figure 18.1. Phylogenetic tree of *Wolbachia* based on sequences of the *fts*Z gene (redrawn from Werren et al. 1995b). Name of the host arthropod species is followed by the strain designation. Parthenogenesis-associated bacteria are shown in boldface. The tree was generated by neighbor joining using the p-distance including insertions/deletions. Numbers next to nodes indicate the number of replicates confirming the node out of 100 (numbers less than 50 are excluded from the figure). Note the low sequence divergence of group Adm *Wolbachia* from different orders of insects, indicating intertaxon transfer of the bacteria.

horizontal (intertaxon) transfer of these bacteria between host species. For example, closely related *Wolbachia* (based on sequence information) are found in different orders of insects. One strain of *Wolbachia* in particular (designated Adm) appears to have undergone extensive "recent" horizontal transmission (figure 18.1).

Mechanisms of intertaxon transmission are unknown, although there is indirect evidence for exchange between parasitoids and their host insects (Werren et al. 1995b). Microinjection experiments show that CI *Wolbachia* can be moved between different species (e.g., *Aedes albopictus* and *Drosophila simulans*) and still cause CI (Braig et al. 1994). The ability of these bacteria to survive and function in different host cellular environments is no doubt important in their widespread distribution. However, we know little about whether strains of *Wolbachia* differ in their host range tolerances.

Horizontal transmission probably explains the common occurrence of double infections (Rousset and Solignac 1995; Werren et al. 1995a,b; Sinkins et al. 1995a; Perrot-Minnot et al. 1996). In some species, individual insects have been found to harbor infections with two different strains of *Wolbachia*. In a survey of neotropical insects, double infections with both A and B division *Wolbachia* were found in 35% of infected species. Double infections also have been found in *Aedes albopictus* (Sinkins et al. 1995a), three species of *Nasonia* wasps (Breeuwer et al. 1992), and *Drosophila* (Rousset and Solignac 1995). Multiple infections are possibly relevant to the role of *Wolbachia* in speciation and are discussed further below.

Wolbachia are both widespread and abundant. For example, over 16% of neotropical insect species examined in a survey were found to be infected, including each of the major insect orders (Werren et al. 1995a). Similar frequencies were found in a North American survey (Werren and Windsor, unpublished). Extrapolating to the global insect fauna gives estimates of 1.5–5.0 million species infected with these bacteria, making them among the most abundant parasitic bacteria on the planet. *Wolbachia* are also found in isopods (Rousset et al. 1992a), mites (Johaniwicz and Hoy 1996; Breeuwer and Jacobs 1996; Breeuwer 1997), and recently in a nematode (Sironi et al. 1995). The limits of distribution for these bacteria have not yet been determined. The observed widespread distribution and abundance of *Wolbachia* is a prerequisite for their potential importance as a speciation mechanism in invertebrates.

Cytoplasmic Incompatibility *Wolbachia*

Basic Biology of Cytoplasmic Incompatibility

CI is a bacterially induced reproductive incompatibility between sperm and egg that occurs following fertilization. Cytologically, CI involves a disruption of early mitoses in fertilized eggs, often manifested as improper condensation and/or loss of paternal chromosomes (O'Neill and Karr 1990; Reed and Werren 1995; Lassy and Karr 1996). In diploid species this typically results in zygotic death, whereas in haplodiploids it usually causes haploidization of the zygote and therefore male development. The biochemical mechanisms are still unknown. However, the pattern of CI is consistent with a modification-rescue system. *Wolbachia* in the testes modify the sperm (possibly by alteration of chromatin-binding proteins) and *Wolbachia* within the egg must rescue the modification, or incompatibility will occur. The bacteria are commonly transmitted in eggs, but only rarely through sperm (Hoffmann and Turelli 1988).

Unidirectional incompatibility typically occurs when the sperm from a *Wolbachia*-infected male fertilizes an uninfected egg (figure 18.2). In this cross, the bacteria have modified the sperm but are not present in the egg to effect rescue. The reciprocal cross (uninfected male and infected female) is compatible. Bidirectional incompatibility (bdCI) typically occurs when a male and a female harbor different strains of *Wolbachia* that are mutually incompatible (Clancy and Hoffmann 1996). BdCI strains apparently utilize different modification-rescue mechanisms and are unable to rescue the sperm modification of the reciprocal strain (figure 18.2).

CI was first described by Ghelelovitch (1952) and Laven (1951, 1959), who discovered that certain intraspecific crosses within *Culex* mosquitos were incompatible and that the incompatibility factor had a cytoplasmic inheritance pattern (i.e., was inherited through females but not through males). Laven (1967) subsequently uncovered a complex set of incompatibility patterns between strains of *Culex pipiens* from different geographic regions, including both unidirectional and bidirectional incompatibility. However, he was not aware of the causative agent of CI. Yen and Barr (1971) showed that CI is associated with presence of intracellular rickettsia (*Wolbachia*) within the reproductive tissues of the mosquitos, and established a causal relationship by antibiotic curing experiments. Subsequent studies have uncovered CI in a number of other species, including flour moths (Brower 1976), jewel wasps (Saul 1961), planthoppers (Noda 1984), flour beetles (Wade and Stevens 1985), and fruitflies (Hoffmann et al. 1986). However, it was not until the advent of the polymerase chain reaction and methods for molecular phylogenetic characterization of bacteria that CI was shown in these diverse host organisms to be associated with a closely related group of alpha bacteria (O'Neill et al. 1992; Breeuwer et al. 1992; Rousset et al. 1992a).

Bidirectional Incompatibility and Double Infections

Bidirectional incompatibility has been described in several systems, most notably between geographic strains in *Culex pipiens* (Laven 1967) and *Drosophila simulans*

UNIDIRECTIONAL
INCOMPATIBILITY

BIDIRECTIONAL
INCOMPATIBILITY

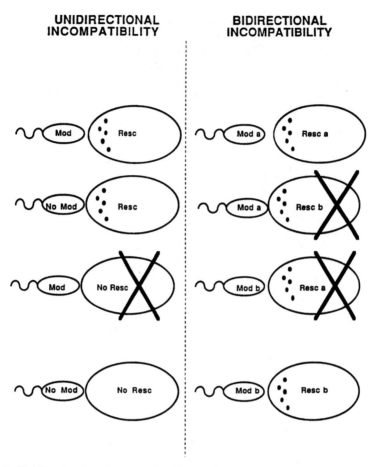

Figure 18.2. Unidirectional and bidirectional incompatibility. Unidirectional incompatibility occurs when an egg without *Wolbachia* (No Resc) is fertilized by a sperm from an infected male, that was modified by *Wolbachia* (indicated by Mod within the sperm). All other crosses are compatible. Bidirectional incompatibility can occur between strains with different modification-rescue mechanisms (a or b).

(O'Neill and Karr 1990) and between closely related species in the wasp genus *Nasonia* (Breeuwer and Werren 1990). Current evidence indicates that bdCI occurs in crosses between hosts infected with different strains of CI *Wolbachia*. It is also possible that bidirectional incompatibility can result when the same *Wolbachia* strain occurs in different host genetic backgrounds. However, this has not yet been demonstrated.

How many compatibility types exist? How readily do new compatibility types evolve? Can host genetic effects cause bidirectional incompatibility? These are important questions for assessing the potential role of *Wolbachia* in speciation. If there is a multitude of bdCI *Wolbachia* strains and significant rates of horizontal transfer, then the potential role of *Wolbachia* in reproductive isolation is increased. The implications of *Wolbachia* in speciation would be enhanced if new compatibility types frequently evolve within species, or if bidirectional incom-

patibility is often due to evolutionarily dynamical interactions with the host genome.

Experimental studies of double infections indicate a diversity of modification rescue systems. In *Aedes albopictus* (Sinkins et al. 1995a), *Drosophila simulans* (Mercot et al. 1995), and *Nasonia vitripennis* (Perrot-Minnot et al. 1996) it has been shown that double-infected strains are unidirectionally incompatible with single-infected strains (doubly infected males are incompatible with singly infected females; the reciprocal is compatible). Additionally, in *D. simulans* and *N. vitripennis*, segregation of the two bacterial types can result in single-infected strains that are bidirectionally incompatible. These findings indicate that the different compatibility types are "layered" upon each other and act somewhat autonomously. Therefore, double infections effectively create "new" compatibility types, providing a further source of compatibility variation. The implications of double infections with regard to speciation are discussed further below.

Population Dynamics of CI-Wolbachia

The population dynamics of CI have been extensively modeled (Caspari and Watson 1959; Fine 1978; Turelli 1994). The selective advantage to *Wolbachia* of causing CI is best explained by first evaluating unidirectional incompatibility. Consider a population with mostly uninfected females and rarely infected females. Under what circumstances will the infection increase in frequency? It is important to recall that the bacteria are usually cytoplasmically inherited through eggs, but not through sperm. Thus, CI-*Wolbachia* occurring in males are normally at an "evolutionary deadend"; they will not be transmitted to future generations. Nevertheless, *Wolbachia* within males indirectly increase frequency of the infection in the population by causing CI, because CI reduces the fitness of uninfected females. The modification-rescue effect of CI means that infected eggs are compatible with the sperm of both infected and uninfected males, whereas uninfected eggs are incompatible with the sperm from infected males (figure 18.2). Thus, the frequency of surviving uninfected eggs in the population is reduced because of fertilization by infected sperm. In other words, CI *Wolbachia* increase in frequency by decreasing the fitness of females not infected with them (Hurst 1991; Werren and O'Neill 1997).

Wolbachia readily increase in a population under a variety of conditions; however, if there is a fertility cost to the infection, a threshold frequency must be exceeded for *Wolbachia* to spread in a population (Turelli 1994). Turelli and Hoffmann (1991) have documented the spread of CI *Wolbachia* through uninfected North American populations of *Drosophila simulans*. Incomplete transmission of the bacteria to eggs will result in a polymorphic equilibrium of infection, and examples of polymorphisms have been documented (e.g., *D. melanogaster*, Hoffmann et al. 1994). However, in other systems, *Wolbachia* appear to be at or near fixation.

Prout (1994) pointed out that the benefits of CI (reduction in fitness of uninfecteds) would also apply to non-CI-inducing *Wolbachia* in the population, as long as they were able to rescue CI-modified sperm. But if initial establishment of infections is due to occasional intertaxon transfer events, then it is unlikely that both CI and non-CI bacteria would enter the population simultaneously. Once CI-*Wolbachia* are near fixation, the benefits to non-CI types are less pronounced. The long-term dynamics and stability of CI and non-CI *Wolbachia* infections are unclear (Hurst and McVean 1996).

What are the dynamics of bidirectional incompatibility? This topic has been less explored theoretically. BdCI within a population is unstable (Rousset et al. 1991; Turelli 1994). Generally, the uncommon *Wolbachia* variant that is bidirectionally incompatible with the common

Wolbachia will be rapidly eliminated from the population. However, if two populations are fixed for different bidirectionally incompatible *Wolbachia*, each will be highly stable to invasion by the alternative *Wolbachia* type, except under very high rates of migration. The extent of host gene flow between the populations will depend on the level of CI and other factors preventing gene flow (e.g., premating isolation, inviability, and sterility).

Empirical Studies of CI and Speciation

If *Wolbachia* are involved in speciation, then the bacteria should be associated with reproductive isolation between closely related species. We also expect to find earlier stages in the process where *Wolbachia* are involved in reproductive incompatibility between populations (e.g., races) that have not yet differentiated into separate species. Only a few CI systems have been studied in detail. Nevertheless, the patterns are suggestive of a possible role of *Wolbachia* in reproductive isolation. Below, I describe studies of CI relevant to the issue of speciation.

Nasonia Species Complex

Wasps in the genus *Nasonia* represent an interesting potential case of *Wolbachia*-induced speciation. *Nasonia* wasps parasitize the pupae of dipterans, primarily blowflies and fleshflies. There is a complex of three species, *Nasonia vitripennis* (a cosmopolitan species that parasitizes a variety of flies), *N. giraulti* (occurring in eastern North America and specializing on blowflies [*Protocalliphora*] that are found in bird nests), and *N. longicornis* (occurring in western North America and also specializing on birdnest blowflies) (Darling and Werren 1990). Molecular phylogenetic studies indicate that the species are young; the two sister species *N. giraulti* and *N. longicornis* are estimated to have diverged 250,000 years ago and their common ancestor diverged from *N. vitripennis* around 250,000 years earlier (Campbell et al. 1993). Both *N. giraulti* and *N. longicornis* occur microsympatrically with *N. vitripennis* over much of their range, often emerging from the same bird nests and even the same parasitized hosts (Darling and Werren 1990).

All three species of *Nasonia* harbor *Wolbachia* (Breeuwer and Werren 1990; Breeuwer et al. 1992), and individuals of each species typically have a double infection with species-specific variants of A group and B group *Wolbachia* (Breeuwer et al. 1992; Werren et al. 1995b; Perrot-Minnot et al. 1996). Crosses between *N. vitripennis* and *N. giraulti* typically fail to produce hybrids in either direction; however, when the wasps are antibiotically cured of their *Wolbachia* infections, fertile hybrids are produced (Breeuwer and Werren 1990). Significant levels of F2 hybrid breakdown are observed, indicating that the species have genetically diverged sufficiently for negative epistatic interactions to occur between their genomes (Breeuwer and Werren 1995). Similar *Wolbachia*-induced reproductive incompatibility is observed between *N. giraulti* and *N. longicornis*, and between *N. vitripennis* and *N. longicornis*, although incompatibility is partial in the latter case (Werren, unpublished). In addition, by introgressing the *vitripennis* nuclear genome into the *N. giraulti* cytoplasm, it has been shown that bidirectional incompatibility is due to differences in the *Wolbachia* strains, not to an interaction between *Wolbachia* and the host species genomes (Breeuwer and Werren 1993).

Were *Wolbachia* the primary cause of reproductive isolation and divergence of these species? It is difficult to say, since speciation has proceeded to the point that other isolating mechanisms are also present, such as F2 hybrid breakdown and variable levels of premating isolation (Breeuwer and Werren 1995). Thus, we do not know whether *Wolbachia*-induced incompatibility evolved first or following reproductive isolation by other causes.

Do *Wolbachia* maintain reproductive isolation between these sympatric species? Results show that under laboratory conditions *Wolbachia* do prevent hybridization between *N. vitripennis* and *N. giraulti*, but it is unknown whether the bacteria actively prevent hybridization in natural populations. Variation exists in the level of premating isolation, suggesting active evolution of this character. Unfortunately, interspecies incompatibility levels have been determined for only a few strains. As a result, the spectrum of incompatibility relationships in natural populations is not known.

Nasonia overwinter as diapausing larvae. In the laboratory, diapausing larvae can be kept alive under refrigeration (4°C) for 1–2 years. Perrot-Minnot et al. (1996) found that bacterial densities decline in larvae experiencing prolonged diapause and following diapause can be stochastically lost in some lineages. Using this method, uninfected, single A infected, single B infected, and double-infected (A+B) sublines were generated from a double-infected strain of *N. vitripennis*. Following stochastic segregation of the bacteria, single A and single B lines are bidirectionally incompatible with each other, even though they have the same nuclear genome. High levels of bidirectional incompatibility are found in these strains, showing rapid formation of "reproductive isolation" due to stochastic segregation of *Wolbachia* types.

Culex Pipiens

Laven (1951, 1959) and Ghelelovitch (1952) first characterized cytoplasmic incompatibility in the mosquito *Culex pipiens*, although they were unaware of the causative agent. Laven (1967) subsequently uncovered a complex system of uni- and bidirectional incompatibility between strains from different geographic regions, detecting a total of 17 different compatibility types. A general caution must be exercised in interpreting the older data on compatibility in *C. pipiens*. In only a few cases

was cytoplasmic inheritance of compatibility type firmly established for compatibility differences between strains. Therefore, some compatibility relationships could be genic rather than cytoplasmic (Rousset et al. 1991).

The diversity of compatibility types could result from either intertaxon transfer of different *Wolbachia* types from other species, or from rapid evolution of compatibility types within *Culex* (Barr 1982). In addition, diverse compatibility types may arise from a complex of single and double infections with different *Wolbachia* strains (Clancy and Hoffmann 1996; Hoffmann and Turelli 1997). To investigate *Wolbachia* diversity in *C. pipiens*, Guillemaud et al. (1997) sequenced the *Wolbachia ftsZ* gene from five mosquito strains with four different compatibility types and found no sequence variation. The result suggests rapid evolution of compatibility types following spread of a single *Wolbachia* infection within *C. pipiens* populations. Consistent with this scenario, *C. pipiens* shows a paucity of mitochondrial variation, with divergence time estimated to be 100,000 years or less (Guillemaud et al. 1997). An alternative interpretation is that (at least some) compatibility diversity in *Culex* is caused by *Wolbachia*–host genotype interactions.

The taxonomic status of *Culex pipiens* is complicated. It is composed of either a complex of sibling species or a complex of subspecies, depending upon the authority (Miles 1976; Barr 1982). The nomenclatural diversity makes it difficult to interpret patterns described in the literature. Nevertheless, is there any evidence that geographic races or subspecies of *Culex pipiens* are reproductively incompatible due to *Wolbachia*? Laven (1967) presented evidence, albeit based on small sample sizes, of geographic separation of CI types. In contrast, Irving-Bell (1983) found no association between compatibility type and subspecies types in Australia, Magnin et al. (1987) found six different CI types in southern France, and Barr (1980) detected three CI types in southern California. These findings appear to counter the view that geographic populations are isolated by *Wolbachia*-induced incompatibilty.

It is difficult to explain the long-term maintenance of such "within-population" polymorphisms in bidirectional incompatibility (Rousset et al. 1991). One possibility is that high migration rates between populations with different compatibility types maintain polymorphisms. There is good evidence for long-distance gene flow in *C. pipiens* (Raymond et al. 1991). *Culex* is widely distributed and is likely to have recently dispersed to many regions of the world as a result of human activity. Some populations, such as those in Australia, Europe, and North America, could be derived from relatively recent colonization events and therefore may not be most informative for understanding the origins of CI variability.

It remains to be seen what role, if any, CI and *Wolbachia* play in divergence between different geographic populations and subspecies of *Culex*. What is needed now is an extensive survey of the distribution of *Wolbachia*, compatibility types, and genetic differentiation of both bacteria and mosquitos in the *Culex pipiens* complex, with special attention to the patterns found in "indigenous" populations.

Drosophila simulans

CI has been studied extensively in *Drosophila simulans*. *D. simulans* most probably originated in Africa and has undergone a relatively recent worldwide expansion due to human activity. Several different compatibility types occur in worldwide populations, including bdCI, non-CI-expressing, double-infected, and uninfected strains (O'Neill and Karr 1990; Montchamp-Moreau et al. 1991; Clancy and Hoffmann 1996). So far, four different *Wolbachia* strains have been identified based upon sequence and compatibility relationships (*w*Ri, *w*Ha, *w*No, and *w*Ma).

CI may be relatively new in *D. simulans* because infections are tightly associated with particular mitochondrial haplotypes (Turelli et al. 1992; Rousset et al. 1992b; Clancy and Hoffmann 1996). In addition, Turelli and Hoffmann (1991) have documented a recent spread of CI *Wolbachia* (*w*Ri, Riverside strain) in previously uninfected populations in California. The *w*Ri strain appears to be spreading to fixation throughout North American populations, whereas initial surveys indicate that the *w*Ha strain is near fixation in a Hawaiian population and may be common in Pacific oceanic islands (Turelli and Hoffmann 1995). However, the worldwide distribution of these different compatibility types is still unclear.

Thus, the stage may be set for future divergence in geographic populations of *D. simulans*, partly mediated by CI *Wolbachia*. For example, given that *w*Ri *Wolbachia* are near fixation in North American populations, then stable bdCI with geographic populations fixed for other CI *Wolbachia* (e.g., *w*Ha in Hawaii) will result. *Wolbachia* would be responsible for the rapid formation of near-complete isolation between these populations, even though the populations have not yet diverged genetically. Subsequent genetic divergence would complete the process. BdCI can be nearly complete between different compatibility types (O'Neill and Karr 1990; Turelli and Hoffmann 1995). However, it is known in *D. simulans* that male aging reduces the level of uniCI, and that incompatibility levels from field-collected insects are lower than those found in laboratory strains (Turelli and Hoffmann 1995; Hoffmann and Turelli 1997). If such effects also apply to bdCI, then the level of reproductive isolation would be reduced, thus permitting gene flow between the incipient species. Nevertheless, it seems likely that *Wolbachia* can accelerate the process of speciation. Simply put, the genetic divergence necessary to evolve from 85% isolation (due to *Wolbachia*) to 100% isolation should occur more readily than that needed to go from 0% isolation (in the absence of *Wolbachia*) to 100%. Perhaps evolutionary biologists of future millennia will be able to document this process.

Other Systems

Both unidirectional and bidirectional incompatibilities are common in crosses between species of *Aedes* mosquitoes (Taylor and Craig 1985; Dev 1986; Trpis et al. 1981), as is the occurrence of *Wolbachia* in reproductive tissues (Wright and Barr 1980). For example, crosses between *A. polynesiensis* females and *A. kesseli* males are incompatible (the reciprocal being compatible); antibiotic and heat treatment restore compatibility (Trpis et al. 1981).

The alfalfa weevil (*Hypera postica*) is an introduced pest in North America, and colonization in eastern and western North America appears to have occurred from different source populations (Hsiao and Hsiao 1985). Western populations harbor *Wolbachia*, whereas eastern populations do not. As expected, unidirectional incompatibility occurs in crosses between these strains (Hsiao and Hsiao 1985). In the absence of strong negative epistatic interactions between genomes of the two races, models predict spread of the infected cytoplasm through the eastern population and coalescence of the two genomes. It will be interesting to see whether this occurs.

Recently, Breeuwer (1997) has established that *Wolbachia* cause CI in two spider mite species, *Tetranychus urticae* and *T. turkestani*. He further hypothesizes that *Wolbachia* could be playing a role in reproductive incompatibilities known to occur between populations and host races in spider mites (Gotoh et al. 1993, 1995). The enormous diversity of spider mites potentially makes them particularly promising for studying the role of *Wolbachia* in reproductive isolation and genetic divergence between populations.

In general, the distribution of *Wolbachia* within host genera is sporadic; some species have the bacteria whereas other closely related species are uninfected (Werren and Jaenike 1995). However, recent surveys have uncovered clusters of infected species within some genera. Examples include *Cissia* moths, *Nasonia* wasps and cassidine tortoise beetles, (Werren et al. 1995b; Werren, unpublished). The pattern could reflect a tendency of certain genera to acquire *Wolbachia* by horizontal transfer. Alternatively, the *Wolbachia* could have been acquired by a common ancestor and then coradiated with the hosts. Phylogenetic studies of the bacteria and hosts will reveal which scenario is correct. Systems with coradiation of *Wolbachia* and hosts are particularly interesting for studying the evolution of new compatibility types and the role of *Wolbachia* in reproductive isolation between species.

Could CI-*Wolbachia* Act as a Speciation Mechanism?

Laven (1959, 1967) first developed the idea that CI could be a speciation mechanism. The basic concept is that cytoplasmic incompatibility, by preventing or severely reducing gene flow between populations, could enhance the probability that populations diverge into separate species. This idea has been controversial (Caspari and Watson 1959, Mayr 1963). However, the discovery of bdCI between closely related species (Breeuwer and Werren 1990) and the widespread occurrence of *Wolbachia* (Werren et al. 1995a) argue for more serious exploration of the possibility.

There are two general ways that CI *Wolbachia* could be directly involved in speciation. First, bidirectional CI could be a primary cause of reproductive isolation between populations, with subsequent genetic divergence leading to speciation. Second, CI (either unidirectional or bidirectional) could be a contributing factor in reproductive isolation between diverging populations. Each of these scenarios is considered below.

Bidirectional CI as the Primary Cause of Reproductive Isolation

Could bidirectional incompatibility alone be the primary cause of reproductive isolation between incipient species? As described above, bdCI has been found between populations and closely related species, so the scenario is not completely unreasonable. In addition, there appears to be a diversity of incompatibility types, further increasing the possibility of bdCI arising between incipient species due to horizontal transmission of different *Wolbachia*.

Consider the following situation. Two allopatric populations of a species each acquire a different strain of CI *Wolbachia*. When these populations come into sympatry, the associated *Wolbachia* cause bidirectional CI, thus preventing or significantly reducing gene flow between the incipient species. As a result, coalescence of the two incipient species does not occur and they continue to diverge into separate species.

There are several important questions arising from this scenario. First, how do the populations acquire different *Wolbachia*, and how likely is this to occur? Second, what levels of combined bdCI and genetic divergence of the incipient species is necessary to prevent coalescence? Third, can bdCI actually help maintain stable coexistence of two incipient species, thus "providing" sufficient time for secondary divergence to occur?

1. How Would Populations Acquire Different *Wolbachia, and How Likely is This to Occur?* Allopatric populations could become bidirectionally incompatible by either (a) horizontal acquisition of different *Wolbachia* strains, (b) evolutionary changes in the resident *Wolbachia*, (c) host genome changes effecting compatibility type, or (d) a combination of the above. We currently do not know the relative importance of these different processes in causing bdCI.

Horizontal (intertaxon) transfer of CI-*Wolbachia* is now well documented (Werren et al. 1995b). However, to be important in speciation, the rates of intertaxon trans-

fer of different CI types would have to be sufficiently high to reciprocally occur in recently diverged allopatric populations. If recent allopatric populations already had a resident CI *Wolbachia*, then horizontal transfer of a different *Wolbachia* typically would result in double infections. Empirical studies indicate that double infections are unidirectionally incompatible with the resident single infection, and therefore could spread to (or near) fixation in the population due to unidirectional incompatibility (Perrot-Minnot et al. 1996; Sinkins et al. 1995a). However, without acquisition of a different compatibility type in the other population, the two incipient species would only be unidirectionally incompatible upon sympatry. Thus, two horizontal transfers are still required for bdCI to cause reproductive isolation.

New compatibility types may also evolve within species. Under this scenario, the evolution of bdCI still requires a two-step process (figure 18.3). Consider a population with a resident CI *Wolbachia* (designated W) at or near fixation. A newly arising *Wolbachia* mutant that is bidirectionally incompatible with the resident *Wolbachia* will quickly be eliminated by CI. In contrast, a mutant (W1) that is unidirectionally incompatible (i.e., mutant-bearing females are compatible with resident-bearing males, but the reciprocal is incompatible) will spread through the population under the same general conditions that favor increase of *Wolbachia* in an uninfected population (Turelli 1994). After replacement of W with W1, the population would be unidirectionally incompatible with its allopatric sister population that retained the original W. Upon sympatry, the new compatibility type would spread through the sister population (due to unidirectional CI), while nuclear genes from the sister population would move in the opposite direction. Coalescence of the two populations would likely result. For bdCI to occur, a second replacement event (e.g., with W2) is required (see figure 18.3). If the second replacement went to fixation in the allopatric sister population, then reproductive isolation between the populations would occur if W1 and W2 were bidirectionally incompatible (figure 18.3). If both replacements occurred in the same population, then reproductive isolation would result when W2 is bidirectionally incompatible with the original W. It should be emphasized that bdCI is not itself selected for, but is a by-product of repeated selective sweeps of unidirectional CI strains.

In *Nasonia* wasps, a phylogenetic analysis suggests that both intertaxon transfer and evolution of compatibility within the species complex has occurred. It appears that the B-*Wolbachia* of *N. longicornis* and *N. giraulti* evolved from a common ancestor (figure 18.1), possibly acquired prior to divergence of the two species, whereas the *N. vitripennis* B bacteria were acquired by horizontal transfer from a different source (Werren et al. 1995b). In contrast, *N. vitripennis* and *N. longicornis* A-*Wolbachia* appear to be more closely related (based on a single shared

EVOLUTION OF BIDIRECTIONAL INCOMPATIBILITY
IN ALLOPATRIC POPULATIONS

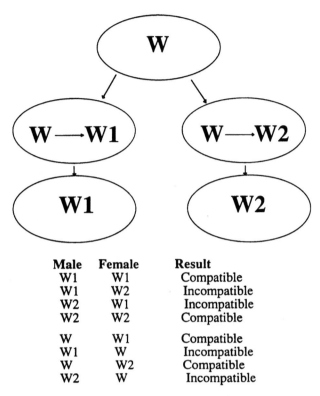

Male	Female	Result
W1	W1	Compatible
W1	W2	Incompatible
W2	W1	Incompatible
W2	W2	Compatible
W	W1	Compatible
W1	W	Incompatible
W	W2	Compatible
W2	W	Incompatible

Figure 18.3. Evolution of bidirectional incompatibility in allopatric populations. A population fixed for a CI *Wolbachia* strain (W) becomes divided into two allopatric populations. A new CI strain (W1) arises and sweeps through one population due to unidirectionally incompatibility with the resident W. A different CI type (W2) sweeps through the other population, also due to unidirectional CI. When these populations come into sympatry, W1 and W2 are bidirectionally incompatible. The evolution of bidirectional incompatibility results from successive sweeps of unidirectionally incompatible strains.

substitution in the *ftsZ* gene) than either are to the *N. giraulti* A-*Wolbachia*. The A-*Wolbachia* group show extensive horizontal transfer between insect orders.

In some systems, bdCI may be due to the effects of different host genomes on CI, rather than to differences in *Wolbachia*. This has been shown not to be the case in *Nasonia* (Breeuwer and Werren 1990). However, if genic *Wolbachia* interactions occur, then bdCI could arise between allopatric populations as a result of genetic changes in host populations.

Segregation of *Wolbachia* in double-infected species provides an interesting possible mechanism for development of rapid reproductive isolation. However, in large populations, those cytoplasmic lineages that have lost one or both *Wolbachia* types will tend to be eliminated because of incompatibility with double-infected lineages (Perrot-Minnot et al. 1996). Fixation of single infections from a double-infected source population is more likely to occur in peripheral or island populations during founding events. Still, stochastic loss of one or both bacterial types would result in unidirectional CI with the source population. Subsequent genetic divergence (e.g., in courtship) would be necessary to complete the isolation. As observed experimentally in *Nasonia*, nearly complete bdCI can occur between founder populations that become fixed for the alternative *Wolbachia* strains, even in the absence of any other genetic changes (Perrot-Minnot et al. 1996).

2. What Level of Bidirectional Incompatibility is Necessary to Prevent Coalescence? This question has not been extensively explored theoretically. In simple terms, if bdCI is complete (i.e., no hybrids occur between the two populations), then there will be no gene flow and the populations will be free to diverge.

In *D. simulans*, bdCI between strains (e.g., Hawaii and Riverside) is nearly complete (O'Neill and Karr 1990). However, sperm from older males shows higher levels of compatibility, apparently due to declining bacterial densities in testes (Bressac and Rousset 1993). This level of incompatibility would certainly not be sufficient to prevent gene flow between populations. In *Culex pipiens*, bdCI between geographic populations is variable, but can be nearly complete (Laven 1967). Similarly, interspecies crosses between laboratory strains of the parasitic wasps *Nasonia vitripennis* and *N. giraulti* give virtually complete incompatibility (Breeuwer and Werren 1990), as do crosses between *N. vitripennis* strains singly infected with A versus B *Wolbachia* (Perrot-Minnot et al. 1996). In *Tribolium*, natural antibiotics and other environmental factors can effect expression of CI (Stevens 1989; Stevens and Wicklow 1992).

The level and stability of bdCI are crucial issues, particularly in the absence of other genetic changes promoting isolation. In this situation, occasional hybrids escaping bdCI will result in gene flow and thus will homogenize the populations for neutral and near-neutral alleles. However, only moderate levels of hybrid fitness reduction (e.g., due to a breaking up of coadapted gene complexes) may be necessary for divergence of the populations to occur at loci subject to strong diverging selective pressures in the two populations. Formal modeling efforts are needed to determine what level of bdCI and negative epistasis in hybrids is needed to allow for divergence and speciation. The basic question is, For any given level of selection against hybrids, does the presence of bdCI *Wolbachia* significantly increase the probability of speciation?

3. Does Bidirectional Incompatibility Help Maintain Stable Coexistence of Two Incipient Species?

Coalescence is one possible outcome when two incipient species come into sympatry. A second possible outcome is that one type replaces the other by competitive exclusion. Even without significant levels of genetic divergence, it seems likely that bdCI could dramatically prolong the maintenance of (cytoplasmic) hybrid zones between two populations (Turelli 1994). The situation is analogous to maintenance of hybrid zones between populations with different chromosomal inversions that cause severe hybrid fitness reduction (Barton 1979; Barton and Hewitt 1985). Such zones are expected to stabilize in regions of low density and at barriers to migration. If bdCI can maintain stable hybrid zones between populations that differ in compatibility types, then the two populations could be stably maintained sufficiently long for subsequent genetic divergence to occur. Again, this is particularly likely if there are strong negative epistatic interactions between genes involved in adaptations in the two populations. These scenarios have yet to be explored in detail theoretically.

CI Wolbachia as a Contributing Factor to Reproductive Isolation

Many cases may exist where uniCI is an important contributing factor to reproductive isolation between sibling species. Imagine the following scenario: a species is divided into two allopatric populations. One population diverges in mating behavior (e.g., due to sexual selection). The other population acquires a *Wolbachia* that goes to fixation. When these populations come back into sympatry, hybrids are prevented in one direction by premating isolation and in the other direction by unidirectional CI. The same effect could occur for F1 sterility or inviability in one direction and CI in the other. Although CI would not be the only cause of reproductive isolation, it would be an essential factor in maintaining reproductive isolation between the incipient species.

For example, *Drosophila recens*, a North American drosophilid that develops on rotting mushrooms, harbors CI-inducing *Wolbachia* (Werren and Jaenike 1995). Jaenike (personal communication) now has evidence that CI may play a role in reproductive isolation between *D. recens* and its close relative *D. subquinaria*. Based so far on one strain of each species, CI causes a severe reduction in the viability of hybrids in crosses between *D. recens* males and *D. subquinaria* females, whereas there is significant premating isolation in the reciprocal direction.

Similarly, partial bdCI may be a contributing factor to reproductive isolation in many species. CI appears to be involved in reproductive isolation between African and European populations of the drosophilid pupal parasitoid *Trichopria drosophilae*. Based on examination of a few strains, European populations of *T. drosophilae* harbor an Adm strain of *Wolbachia*, whereas African populations harbor an Atc strain (van Alphen and Werren, unpublished). Adm and Atc are different subgroups of *Wolbachia* within the A subdivision (figure 18.1). Reciprocal crosses between an African strain and a European strain fail to produce hybrids. Crosses using antibiotically cured strains results in production of some hybrids, although other isolating factors are clearly also involved.

Such situations could be common. Given the widespread occurrence of *Wolbachia*, it is now necessary for researchers investigating the genetics of speciation to examine their species for presence of *Wolbachia* and possible involvement of these bacteria in reproductive isolation.

In some cases unidirectional CI will occur between two species in one direction but reproductive isolation may be insufficient in the reciprocal direction. Here, we can expect *Wolbachia* to "jump" the species barrier and spread through the previously uninfected species, bringing with it the mitochondrial type of the original infected species. Such a scenario may have occurred in the *D. simulans* complex (Rousset and Solignac 1995).

Sympatric Speciation and Reinforcement

All of the scenarios discussed above assume allopatric speciation. The importance of sympatric speciation has been debated for many years, and continues to be controversial (Rice and Hostert 1993; Berlocher 1997, ch. 8 this volume; Feder, this volume). In general, the conditions for *Wolbachia* to contribute to sympatric speciation are likely to be restrictive. Once established, *Wolbachia* are expected to spread rapidly through host populations (Turelli 1994). Thus, unless the barriers to gene flow are already complete, a *Wolbachia* infection acquired in one subpopulation is likely to readily sweep through the other subpopulation. For *Wolbachia* to play a role, we would have to assume a near simultaneous acquisition of different (bdCI) *Wolbachia* in the two populations and their increase to near fixation in each subpopulation prior to movement across the barrier. Given that the subpopulations are in sympatry, this scenario seems less probable than the allopatric case, where greater time scales for independent acquisition and fixation of bacteria would be available.

Parapatric speciation is another possibility. Again, one would have to assume near-simultaneous acquisition of different *Wolbachia* in different parts of the species range. Scilthuizen and Stouthamer (pers. communication) investigated the role of *Wolbachia* in promoting the evolution of premating isolation by reinforcement in parapatric populations. They concluded that the conditions were restrictive for the evolution of premating isolation in the contact zone.

Reinforcement is defined as "the evolution of prezygotic isolating barriers in zones of overlap or hybridization (or both) as a response to selection against hybridization" (Howard 1993, p. 46). Its role in speciation is controversial (Howard 1993; Butlin 1995; Liou and Price 1994). As with other postzygotic barriers, *Wolbachia*-induced CI may also select for reinforcement of premating isolation.

Wolbachia-*Induced Changes in Host Genomes*

Wolbachia may promote speciation, not directly by causing CI, but indirectly by causing evolutionary changes in host genomes. There are three basic scenarios.

1. Compensatory Changes in the Host Genome. *Wolbachia* reside within reproductive tissues and alter mitosis and chromosome behavior within host cells. Compensatory changes by the host are therefore expected to adapt to presence of *Wolbachia*, particularly in features relating to oogenesis, spermatogenesis, and mitosis. For example, Turelli (1994) has proposed that, when *Wolbachia* transmission is incomplete, the host will be selected to suppress modification of sperm and mimic rescue in eggs. Such changes may sufficiently alter the genetic architecture of reproduction in allopatric populations to cause postmating incompatibilities, independent of the CI effects of *Wolbachia*.

2. Hitchhiking of Mitochondria. During the initial stage of infecting a new species, *Wolbachia* are likely by chance to be associated with particular mitochondrial haplotypes. As the *Wolbachia* sweeps through the host population, the mitochondrial haplotype will spread with it by genetic hitchhiking (because they are associated cytoplasmic genomes) (Rousset et al. 1992b; Rousset and Solignac 1995). Any mutations within the mitochondrial genome will also spread to fixation. This rapid genetic change could then select for compensatory mutations in nuclear genes that interact with the mitochondrial product. Such changes could cause nuclear/cytoplasmic (i.e., mitochondrial) gene incompatibilities in subsequent hybrids. Consistent with this scenario, an incompatibility exists between an *N. vitripennis* nuclear gene(s) and an *N. giraulti* cytoplasmic (presumably mitochondrial) gene(s) (Breeuwer and Warren 1995). Both these species appear to have experienced successive *Wolbachia* sweeps.

3. Chromosome Rearrangements. CI *Wolbachia* induce paternal chromosome fragmentation. One manifestation of this is the formation of chromatin bridges during mitosis and generation of heritable centric fragments (Ryan et al. 1987; Reed and Werren 1995). Thus, CI can be a source for generating chromosomal rearrangements that, if fixed, could contribute to postzygotic incompatibilities between incipient species.

Parthenogenesis-Inducing Wolbachia and Speciation

Wolbachia are also known to induce parthenogenesis (Stouthamer et al. 1993). So far, PI by *Wolbachia* has been demonstrated only in the Hymenoptera, where there is evidence for bacterial-induced parthenogenesis in over 32 species (reviewed in Stouthamer 1997). In well-studied systems, *Wolbachia* have been detected by cytogenetic observation, PCR amplification, and sequencing of bacterial DNA. Antibiotic curing experiments have shown that elimination of the bacteria results in production of males, and in some species, these males can reproduce sexually with females. *Wolbachia* have also been found in parthenogenetic weevils, although curing experiments have not been performed (Werren et al. 1995b). In *Trichogramma* wasps, *Wolbachia* induce parthenogenesis (thelytoky) by causing endoduplication of the haploid egg in the first mitosis (Stouthamer and Kazmer 1994). Similar mechanisms appear to be operating in other systems (Stouthamer 1997). Genetically, this results in complete homozygosity of the asexual females.

We do not know how frequently PI *Wolbachia* evolve within a host species versus entering by horizontal transfer. Phylogenetic studies show that PI *Wolbachia* occur in both the A and B divisions. The pattern is consistent with PI evolving from CI *Wolbachia* multiple times independently (Werren et al. 1995b). However, it is also possible that the genetic machinery for parthenogenesis was transferred to different *Wolbachia* via plasmids, viruses, or some other mechanism. A third, although less likely, possibility is that PI is simply an artifact of CI *Wolbachia* expression in different host environments.

PI-*Wolbachia* and Speciation

Basic models tell us that PI-*Wolbachia* will increase in frequency in host populations as long as an infected female on average produces more infected daughters than an average female produces daughters in the population (Werren 1987; Stouthamer 1997). If there is complete transmission of the *Wolbachia* to eggs, then the infection is expected to drive to fixation in the population. Mixed populations of sexuals and asexuals are common in *Trichogramma* species, and may be due, in part, to a frequency dependent fertility cost of the infection (Stouthamer and Luck 1993). In addition, genetic exchange occurs between sexual and asexual forms, thus preventing genetic divergence of the two types (Stouthamer and Kazmer 1994) However, in other species, PI *Wolbachia* apparently have gone to fixation, and based on phenotypic divergence with related sexual species, these parthenogenetic "species" may have persisted for some time (Zchori-Fein et al. 1992).

How might parthenogenesis *Wolbachia* be involved in speciation? Clearly, if the infection goes to fixation within a population and persists sufficiently long for genetic divergence to occur, then an "asexual species" originates. Given sufficient time, we expect irreversible evolution to asexuality due to the accumulation of deleterious mutations in the genes involved in sexual reproduction. Examples of such genes include those involved in male sexual functions such as spermatogenesis and sperm performance, reproductive anatomy, courtship and copulation behavior, and female sexual functions such as pheromone production, mate acceptance, anatomical and physiological features involved in sperm transfer and storage, and egg fertilization. Given that the population is reproducing asexually, selection does not actively maintain such genes. Mutations causing loss of function will not be eliminated. Such mutations will inevitably increase to fixation by genetic drift, or could be actively selected for if they improve asexual performance of females.

There is good evidence that this process is occurring. In several species, curing of parthenogenetic females results in male production; however, either males are not functional or parthenogenetic females do not accept males. In *Aphytis lignanensis*, copulations occur, but sperm fails to fertilize the eggs (Zchori-Fein et al. 1995). In the asexual species *Encarsia formosa*, copulations do not occur (Zchori-Fein et al. 1992). In *Apoanagyrus*

diversicornis (which has both sexual and asexual strains), asexual females do not mate with either sexual males or males from cured asexual females, but sexual females mate with males of either form (Pijls et al. 1996). Once the process of mutation degeneration in sex functions has gone far enough, such "species" are no longer capable of reverting to sexuality, even if *Wolbachia* are lost.

It is generally believed that most parthenogenetic species cannot persist over long evolutionary time scales, although there are some possible exceptions such as bdelloid rotifers (Judson and Normark 1996). Consistent with this view, within the Hymenoptera, parthenogenetic species occur within genera that also contain sexual forms. One cost of asexuality is the general accumulation of deleterious mutations (as distinct from the degeneration of sexual genes described above) due to a Muller's Ratchet (Muller 1964). The rate of Muller's Ratchet in asexual populations depends on population size, the magnitude of the harmful effect of deleterious mutations, and the mutation rate. However, Muller's Ratchet will be much less effective in species with *Wolbachia*-induced parthenogenesis. The reason is that such parthenogenetic individuals are homozygous at all loci and deleterious mutations, which are usually recessive or partially recessive, are therefore more likely to be purged by selection before drifting to fixation (Charlesworth et al. 1993). As a result, we may expect *Wolbachia*-induced species to persist longer than diploid parthenogens, and for large populations such species could possibly resist the action of Muller's Ratchet almost indefinitely (Charlesworth, personal communication).

Parthenogenetic species also are believed to have difficulties adapting to new environments for two reasons: (1) an absence of recombination that brings together adaptive mutations occurring in different lineages (Maynard Smith 1978) and (2) background selection (Charlesworth et al. 1993), which eliminates many adaptive mutations in nonrecombining genomes because such mutations often occur in lineages with maladaptive deleterious mutations (Peck 1994). *Wolbachia*-induced parthenogenetic species will be less prone to the negative effects of background selection because deleterious mutations will be immediately expressed homozygously and rapidly purged from the population, thus maintaining larger effective population sizes. Another advantage of *Wolbachia*-induced parthenogenesis is that beneficial mutations that are recessive or partially recessive will be immediately made homozygous and exposed to positive selection. Thus, such mutations are more likely to become established and increase, possibly accelerating the rate of adaptive evolution in *Wolbachia*-induced parthenogenetic species relative to those with genetic mechanisms of parthenogenesis that maintain heterozygosity (i.e., apomixis).

A lot of work remains to be done to understand the evolution of parthenogenetic species. Systems with *Wolbachia*-induced parthenogenesis could be useful for investigating these processes. A particularly useful feature is that parthenogenetic forms can be "cured," with reversion to sexuality, thus facilitating studies of deleterious mutation accumulation, irreversible evolution, and adaptive mutation in parthenogenetic forms.

Models of *Wolbachia*-Associated Speciation

Based on the discussions above, we can summarize the different models for *Wolbachia*-induced speciation into the following general categories.

1. Independent Acquisition. Two allopatric populations acquire different *Wolbachia* strains by horizontal transfer. When the populations come back into sympatry, the level of reproductive incompatibility is sufficient to allow continued genetic divergence. This process may or may not require additional (nuclear) divergence to have occurred between the populations prior to sympatry, since a stable cytoplasmic "tension zone" can result between the populations (Turelli 1994), followed by genetic divergence or selection for reproductive character displacement.

2. Reciprocal Causes. Acquisition of *Wolbachia* occurs in one population; nuclear changes (e.g., mate discrimination) occur in the other. Bidirectional incompatibility results when the populations comes into sympatry, allowing continued genetic divergence.

3. Wolbachia Segregation. A double infected source population produces peripheral isolates which stochastically lose different *Wolbachia*. These isolates will be bidirectionally incompatible when they come back into sympatry. The problem with this model is that the isolates will be unidirectionally incompatible with the double infected source population. Therefore other changes seem necessary for *Wolbachia* segregation to promote speciation (e.g., see 1 and 2 above).

4. Nuclear Accommodation. Presence of *Wolbachia* results in evolutionary changes in the host, particularly in aspects of male and female gametogenesis due to the activity of *Wolbachia* in these tissues. These genetic changes result in nuclear incompatibilities between populations.

5. Sexual Degradation. In populations fixed for parthenogenesis-inducing *Wolbachia*, degradation of genes involved in sexual function will occur over time (due to mutation and/or selection). Such populations will become irrevocably parthenogenetic "species" due to this degradation.

Conclusions

Wolbachia may play an important role in speciation, at least in arthropods and perhaps in other phyla as well.

Arguments for this possibility include the widespread occurrence and abundance of *Wolbachia*, the phenotypic effects they have on hosts (i.e., CI and PI) that can contribute to reproductive isolation, and the association of *Wolbachia* with reproductive isolation in some sexual and parthenogenetic species. The primary argument against *Wolbachia* as an important speciation mechanism is that CI will usually be partial (due to incomplete transmission and expression of CI) and therefore insufficient to permit genetic divergence between populations. Both theoretical and empirical studies are needed to investigate the importance of these bacteria in speciation.

To empirically assess the importance of *Wolbachia* in speciation, we will need to determine (1) how often *Wolbachia* are associated with reproductive isolation between species, either as the primary or a contributing factor; (2) to what extent CI occurs between populations or races within a species, that is, whether early stages in the process occur where *Wolbachia* cause reproductive isolation prior to significant levels of genetic divergence; and (3) whether clades with *Wolbachia* infections show higher speciation rates than related clades without *Wolbachia*. We will also need to determine what levels of CI occur in natural populations, how often new bdCI types evolve within species or enter by lateral transfer of different *Wolbachia* strains from other hosts, and whether *Wolbachia* induce evolutionary changes in host genomes that contribute to reproductive isolation.

Important theoretical questions primarily concern what levels of CI (in combination with other isolating mechanisms) are necessary to prevent coalescence and allow divergence of incipient species. However, it should be kept in mind that theoretical treatments are, of necessity, simplifications of nature. Empirical studies are the ultimate arbiter of this or any other scientific question.

Acknowledgments   Thanks are extended to J. Jaenike for providing unpublished information, to S. O'Neill and F. Rousset for discussion of CI in mosquitoes, and to B. Charlesworth for discussion on mutation accumulation. In addition, S. Bordenstein, M. Drapeau, J. Jaenike, C. Perez, S. Perlman, D. Shoemaker, and R. Weston are thanked for comments on the manuscript.

References

Barr, A. R. (1980). Cytoplasmic incompatibility in natural populations of a mosquito, *Culex pipiens* L. Nature 283: 71–72.

Barr, A. R. (1982). The *Culex pipiens* complex. Champaign, Il., Stipes.

Barton, N. H. (1979). The dynamics of hybrid zones. Heredity 43:341–359.

Barton, N. H., and Hewitt, G. M. (1985). Analysis of hybrid zones. Annu. Rev. Ecol. Syst. 16:113–148.

Braig, H. R., Guzman, H., Tesh, R. B., and O'Neill, S. L. (1994). Replacement of the natural *Wolbachia* symbiont of *Drosophila simulans* with a mosquito counterpart. Nature 367:453–455.

Breeuwer, J. A. J. (1997). *Wolbachia* and cytoplasmic incompatibilty in the spider mite *Tetranychus urticae* and *T. turkestani*. Heredity. 79:41–47.

Breeuwer, J. A. J., and Jacobs, G. (1996). *Wolbachia*: intracellular manipulators of mite reproduction. Exp. Appl. Acarol. 20:421–434.

Breeuwer, J. A. J., and Werren, J. H. (1990). Microorganisms associated with chromosome destruction and reproductive isolation between two insect species. Nature 346: 558–560.

Breeuwer, J. A. J., and J. H. Werren. (1993). The effect of genotype on cytoplasmic incompatibility between two species of *Nasonia*. Heredity 70:428–436.

Breeuwer, J. A. J., and Werren, J. H. (1995). Hybrid breakdown between two haplodiploid species: the role of nuclear and cytoplasmic genes. Evolution 49:705–717.

Breeuwer, J. A. J., Stouthamer, R., Barns, S. M., Pelletier, D. A., Weisburg, W. G., and Werren, J. H. (1992). Phylogeny of cytoplasmic incompatibility microorganisms in the parasitoid wasp genus *Nasonia* (Hymenoptera: Pteromalidae) based on 16S ribosomal DNA sequences. Insect Mol. Biol. 1:25–36.

Bressac, C., and Rousset, F. (1993). The reproductive incompatibility system in *Drosophila* simulans, DAPI-staining analysis of the *Wolbachia* symbionts in sperm cysts. J. Invert. Pathol. 61:226–230.

Brower, J. H. (1976). Cytoplasmic incompatibility: occurrence in a stored product pest *Ephestia cautella*. Ann. Entomol. Soc. Am. 69:1011–1015.

Butlin, R. K. (1995). Reinforcement: an idea evolving. Trends Ecol. Evol. 10:432–434.

Campbell, B. C., Steffen-Campbell, J. D., and Werren, J. H. (1993). Phylogeny of the *Nasonia* species complex (Hymenoptera: Pteromalidae) inferred from an rDNA internal transcribed spacer (ITS2). Insect Mol. Biol. 2:255–237.

Caspari, E., and Watson, G. S. (1959). On the evolutionary importance of cytoplasmic sterility in mosquitoes. Evolution 13:568–570.

Charlesworth, B., Morgan, M. T., and Charlesworth, D. (1993). The effect of deleterious mutations on neutral molecular variation. Genetics 134:1289–1303.

Clancy, D. J., and Hoffmann, A. A. (1996). Cytoplasmic incompatibility in *Drosophila* simulans: evolving complexity. Trends Ecol. Syst. 11:145–146.

Coyne, J. A. (1992). Genetics and speciation. Nature 355:511–515.

Darling, D. C., and Werren, J. H. (1990). Biosystematics of two new species of *Nasonia* (Hymenoptera: Pteromalidae) reared from birds' nests in North America. Ann. Entomol. Soc. Am. 83:352–370.

Dev, V. (1986). Non-reciprocal fertility among species of the *Aedes (Stegomyia) scutellaris* group (Diptera: Culicidae). Experientia 42:803–806.

Fine, P. E. M. (1978). On the dynamics of symbiote-dependent cytoplasmic incompatibility in culicine mosquitoes. J. Invert. Pathol. 30:10–18.

Ghelelovitch, S. (1952). Sur le determinisme genetique de la sterilite dans les croisements entre differentes souches de *Culex autogenicus* Roubaud. CR Acad. Sci., Paris 234: 2386–2388.

Gotoh, T., Bruin, J., Sabelis, M. W., and Menken, S. B. J. (1993). Host race formation in *Tetranychus urticae*: genetic differentiation, host plant preference, and mate choice in a tomato and a cucumber strain. Entomol. Exp. Appl. 68: 171–178.

Gotoh, T., Oku, H., Moriya, K., and Odawara, M. (1995). Nucleus-cytoplasm interactions causing reroductive incompatiblity between two populations of *Tetranychus quercivorous* Ehara et Gotoh (Acari: Tetranychidae). Heredity 74:405–414.

Guillemaud, T., Pasteur, N., and Rousset, F. (1997). Contrasting levels of variability between between cytoplasmic genomes and incompatibility types in the mosquito *Culex pipiens*. Proc. Roy. Soc. Lond. B. 264:245–251.

Hoffmann, A. A., and Turelli, M. (1988). Unidirectional incompatibility in *Drosophila simulans*: inheritance, geographic variation and fitness effects. Genetics 119:435–444.

Hoffmann, A. A., and Turelli, M. (1997). Cytoplasmic incompatibility in insects. In S. L. O'Neill, A. A. Hoffmann, and J. H. Werren (eds). Influential Passengers: Inherited Microorganisms and Arthropod Reproduction. Oxford: Oxford University Press.

Hoffmann, A. A., Turelli, M., and Simmons, G. M. (1986). Unidirectional incompatibility between populations of *Drosophila simulans*. Evolution 40:692–701.

Hoffmann, A. A., Turelli, M., and Harshman, L. G. (1990). Factors affecting the distribution of cytoplasmic incompatibility in *Drosophila simulans*. Genetics 126:933–948.

Hoffmann, A. A., Clancy, D. J., and Merton, E. (1994). Cytoplasmic incompatibility in Australian populations of *Drosophila melanogaster*. Genetics 136:993–999.

Howard, D. J. (1993). Reinforcement: origin, dynamics and fate of anevolutionary hypothesis. In R.G. Harrison (ed.). Hybrid zones and the evolutionary process. New York: Oxford University Press.

Hsiao, C., and Hsiao, T. H. (1985). Rickettsia as the cause of cytoplasmic incompatibility in the alfalfa weevil, *Hypera postica*. J. Invert. Pathol. 45:244–246.

Hurst, L. D. (1991). The evolution of intra-populational cytoplasmic incompatibility or when spite can be successful. J. Theor. Biol. 148:69–77.

Hurst, L. D., and McVean, G. T. (1996). Clade selection, reversible evolution and the persistence of selfish elements: the evolutionary dynamics of cytoplasmic incompatibility. Proc. Roy. Soc. Lond. B 262:97–104

Irving-Bell, R. J. (1983). Cytoplasmic incompatibility within and between *Culex molestus* and *Culex quinquefasciatus* (Diptera Culicidae). J. Med. Entomol. 20:44–48.

Johanowicz, D. L., and Hoy, M. A. (1996). *Wolbachia* in a predator-prey system: 16S ribosomal DNA analysis of two phytoseiids (Acari: Phytoseiidae) and their prey (Acari: Tetranychidae). Ann. Entomol. Soc. Am. 89: 435–441.

Judson, O. P., and Normark, B. B. (1996). Ancient asexual scandals. Trends Ecol. Evol. 11:41–46.

Lassy, C. W., and Karr, T. L. (1996). Cytological analysis of fertilization and early embryonic development in incompatible crosses of *Drosophila simulans*. Mech. Dev. 57: 47–58.

Laven, H. (1951). Crossing experiments with *Culex* strains. Evolution 5:370–375.

Laven, H. (1959). Speciation by cytoplasmic isolation in the *Culex pipiens* complex. Cold Spring Harbor Symp. Quant. Biol. 24:166–173.

Laven, H. (1967). Speciation and evolution in *Culex pipiens*. In J.W. Wright and R. Pal (eds.). Genetics of insect vectors of disease. Amsterdam: Elsevier.

Liou, L. W., and T. D. Price. (1994). Speciation by reinforcement of premating isolation. Evolution 48:1451–1459.

Magnin, M., Pasteur, N., and Raymond, M. (1987). Multiple incompatibilities within populations of *Culex pipiens* L. in southern France. Genetica 74:125–30.

Maynard Smith, J. (1978). The evolution of sex. Cambridge: Cambridge University Press.

Mayr, E. (1963). Animal species and evolution. Cambridge: Belknap Press.

Mercot, H., Llorente, B., Jacques, M., Atlan, A., and Montchamp-Moreau, C. (1995). Variability within the Seychelles cytoplasmic incompatibility system in *Drosophila simulans*. Genetics 141:1015–1023.

Montchamp-Moreau, C., Ferveur, J.-F., and Jacques, M. (1991). Geographic distribution and inheritance of three cytoplasmic incompatibility types in *Drosophila simulans*. Genetics 129:399–407.

Miles, S. J. (1976). Taxonomic significance of assortative mating in a mixed field population of *Culex pipiens australicus*, *C. p. quinquefasciatus* and *C. globocoxitus*. Syst. Entomol. 1:262–270.

Muller, H. J. (1964). The relation of recombination to mutational advance. Mut. Res. 1:2–9.

Noda, H. (1984). Cytoplasmic incompatibility in a rice planthopper. J. Hered. 75:345–348.

O'Neill, S. L., and Karr, T. L. (1990). Bidirectional incompatibility between conspecific populations of *Drosophila simulans*. Nature 348:178–180.

O'Neill, S. L., Giordano, R., Colbert, A. M. E., Karr, T. L., and Robertson, H. M. (1992). 16S rRNA phylogenetic analysis of the bacterial endosymbionts associated with cytoplasmic incompatibility in insects. Proc. Natl. Acad. Sci. USA 89:2699–2702.

Peck, J. (1994). A ruby in the rubbish: beneficial mutations, deleterious mutations, and the evolution of sex. Genetics 137:597–606.

Perrot-Minnot, M.-J., Guo, L. R., and Werren, J. H. (1996). Single and double infections with *Wolbachia* in the parasitic wasp *Nasonia vitripennis*: effects on compatibility. Genetics 143:961–972.

Pijls, J., van Steenbergen, H. J., and van Alphen, J. J. M. (1996). Asexuality cured: the relations and differences between sexual and asexual *Apoanagyrus diversicornis*. Heredity 76:506–513.

Prout, T. (1994). Some evolutionary possibilities for a microbe that causes incompatibility in its host. Evolution 48:909– 911.

Raymond, M., Callaghan, A., Fort, P., and Pasteur, N. (1991). Worldwide migration of amplified insecticide resistance genes in mosquitoes. Nature 350:151–153.

Reed, K. M., and Werren, J. H. (1995). Induction of paternal genome loss by the paternal-sex-ratio chromosome and cytoplasmic incompatibility bacteria (*Wolbachia*): a comparative study of early embryonic events. Mol. Reprod. Dev. 40:408–418.

Rice, W. R., and Hostert, E. E. (1993). Laboratory experiments on speciation—what have we learned in 40 years? Evolution 47:1637–1653.

Rigaud, T., and Juchault, P. (1993). Conflict between feminizing sex ratio distorters and an autosomal masculinizing gene in the terrestrial isopod *Armadillidium vulgare* Latr. Genetics 133:247–252.

Rousset, F., and Solignac, M. (1995). Evolution of single and double *Wolbachia* symbioses during speciation in the *Drosophila simulans* complex. Proc. Natl. Acad. Sci. USA 92:6389–6393.

Rousset, F., Raymond, M., and Kjellberg, F. (1991). Cytoplasmic incompatibilities in the mosquito *Culex pipiens*: how to explain a cytotype polymorphism? J. Evol. Biol. 4: 69–81.

Rousset, F., Bouchon, D., Pintureau, B., Juchault, P., and Solignac, M. (1992a). *Wolbachia* endosymbionts responsible for various alterations of sexuality in arthropods. Proc. Roy. Soc. Lond. B 250:91–98.

Rousset, F., Vautrin, D., and Solignac, M. (1992b). Molecular identification of *Wolbachia*, the agent of cytoplasmic incompatibility in *Drosophila simulans*, and variability in relation with host mitochondrial types. Proc. Roy. Soc. Lond. 247:163–168.

Ryan, S. L., Saul, G. B., and Conner, G. W. (1987). Aberrant segregation of R-locus genes in male progeny from incompatible crosses in *Mormoniella*. J. Hered. 78: 21–26.

Saul, G. B. (1961). An analysis of non reciprocal cross incompatability in *Mormoniella vitripennis* (Walker). Zeitschr. Vererbungsl. 92:28–33.

Sinkins, S. P., Braig, H. R., and O'Neill, S. L. (1995a). *Wolbachia* superinfections and the expression of cytoplasmic incompatibility. Proc. Roy. Soc. Lond. 261:325– 330.

Sinkins, S. P., Braig, H. R., and O'Neill, S. L. (1995b). *Wolbachia pipientis*: bacterial density and unidirectional cytoplasmic incompatibility between infected populations of *Aedes albopictus*. Exp. Parasitol. 81:284–291.

Sironi, M., Bandi, C., Sacchi, L., Di Sacco, B., Damiani, G., and Genchi, C. (1995). Molecular evidence for a close relative of the arthropod endosymbiont *Wolbachia* in a filarial worm. Mol. Biochem. Parasitol. 74:223–227.

Stevens, L. (1989). Environmental factors affecting reproductive incompatibility in flour beetles, genus *Tribolium*. J. Invert. Pathol. 53:78–84.

Stevens, L., and Wicklow, D. T. (1992). Multispecies interactions affect cytoplasmic incompatibility in *Tribolium* flour beetles. Am. Nat. 140:642–653.

Stouthamer, R. (1997). *Wolbachia* induced parthenogenesis. In S.L. O'Neill, A.A. Hoffmann, and J.H. Werren (eds). Influential passengers: inherited microorganisms and arthropod reproduction. Oxford: Oxford University Press.

Stouthamer, R., and Kazmer, D. J. (1994). Cytogenetics of microbe-associated parthenogenesis and its consequence for gene flow in *Trichogramma* wasps. Heredity 73:317–327.

Stouthamer, R., and Luck, R. F. (1993). Influence of microbe-associated parthenogenesis on the fecundity of *Trichogramma deion* and *T. pretiosum*. Entomol. Exp. Appl. 67:183–192.

Stouthamer, R., Breeuwer, J. A. J., Luck, R. F., and Werren, J. H. (1993). Molecular identification of microorganisms associated with parthenogenesis. Nature 361:66–68.

Taylor, D. B., and Craig, G. B. (1985). Unidirectional reproductive incompatibility between *Aedes* (*Protomacleava*) *brelandi* and *A.* (*P.*) *hendersoni* (Diptera: Culicidae). Ann. Entom. Soc. Am. 78:769–774.

Trpis, M, Perrone, J. B., Reissig, M., and Parker, K. L. (1981). Control of cytoplasmic incompatibility in the *Aedes scutellaris* complex. J. Hered. 72:313–317.

Turelli, M. (1994). Evolution of incompatibility-inducing microbes and their hosts. Evolution 48:1500–1513.

Turelli, M., and Hoffmann, A.A. (1991). Rapid spread of an inherited incompatibility factor in California *Drosophila*. Nature 353:440–442.

Turelli, M., and Hoffmann, A. A. (1995). Cytoplasmic incompatibility in *Drosophila* simulans: dynamics and parameter estimates from natural populations. Genetics 140:1319–1338.

Turelli, M., Hoffmann, A. A., and McKechnie, S. W. (1992). Dynamics of cytoplasmic incompatibility and mtDNA variation in natural *Drosophila simulans* populations. Genetics 132:713–723.

Wade, M. J., and Stevens, L. (1985). Microorganism mediated reproductive isolation in flour beetles (genus *Tribolium*). Science 227:527–528.

Werren, J. H. (1987). The coevolution of autosomal and cytoplasmic sex ratio factors. J. Theor. Biol. 124:317–334.

Werren, J. H. (1997). Biology of *Wolbachia*. Annu. Rev. Entomol. 42:587–609.

Werren, J. H., and Jaenike, J. (1995). *Wolbachia* and cytoplasmic incompatibility in mycophagous *Drosophila* and their relatives. Heredity 75:320–326.

Werren, J. H., and O'Neill, S. L. (1997). The evolution of heritable symbionts. In S. L. O'Neill, A. A. Hoffmann, and J. H. Werren (eds.). Influential passengers: inherited microorganisms and arthropod reproduction. Oxford: Oxford University Press.

Werren, J. H., Windsor, D., and Guo, L. R. (1995a). Distribution of *Wolbachia* among neotropical arthropods. Proc. Roy. Soc. Lond. B 262:197–204.

Werren, J. H., Zhang, W., and Rong Guo, L. (1995b). Evolution and phylogeny of *Wolbachia*: reproductive parasites of arthropods. Proc. Roy. Soc. Lond. B 261:55–71.

Wright, J. D., and Barr, R. A. (1980). The ultrastructure and symbiotic relationships of *Wolbachia* of mosquitoes of the *Aedes scutellaris* group. J. Ultrastruct. Res. 72:52–64.

Yen, J. H., and Barr, A. R. (1971). New hypothesis of the cause of cytoplasmic incompatibility in *Culex pipiens*. Nature 232:657–558.

Zchori-Fein, E., Rousch, R. T., and Hunter, M. S. (1992). Male production induced by antibiotic treatment in *Encarsia formosa* (Hymenoptera: Aphelinidae), in asexual species. Experientia 48:173–178.

Zchori-Fein, E., Faktor, O., Zeidan, M., Gottlieb, Y., Czosnek, H., and Rosen, D. (1995). Parthenogenesis-inducing microorganisms in *Aphytis*. Insect Mol. Biol. 4:173–178.

19

Intergenomic Conflict, Interlocus Antagonistic Coevolution, and the Evolution of Reproductive Isolation

William R. Rice

Laboratory experiments have demonstrated that allopatric populations, and those connected by low levels of gene flow, can diverge genetically and that reproductive isolation frequently accrues as an incidental by-product (reviewed in Rice and Hostert 1993). In nature, it is still unclear which particular processes, if any, are primarily responsible for generating the genetic divergence that specifically leads to reproductive isolation. Here I develop the hypothesis that genetic conflict is frequently the primary factor driving the genetic divergence that leads to reproductive isolation.

To begin, I categorize the major forms of genetic conflict (figure 19.1). Conflict between a species and its enemies (parasites, pathogens, predators, and competitors) has been well documented. This led Van Valen (1973) to conclude that there is perpetual antagonistic coevolution between a species and its enemies, and that this is the dominant factor driving evolutionary change within a lineage. Van Valen's "Red Queen" process undoubtedly plays an important role in promoting genetic divergence among populations, but it is my evaluation that most reproductive isolation derives more from conflict that occurs within rather than between species.

Most theory and research concerning intraspecific conflict has focused on intragenomic conflict (e.g., meiotic drive genes, transposable elements, cytoplasmic versus nuclear genes). Here the antagonism occurs within a single genome (individual) and is directed between either allelic or nonallelic genes. While there is evidence that intragenomic conflict can lead to genetic change that produces reproductive isolation (e.g., Hurst 1992; Zeh and Zeh 1994; Werren, this volume), I think that it is intergenomic rather than intragenomic conflict that generates most of the reproductive isolation that accrues between populations.

Intergenomic conflict occurs between genes residing in different individuals of the same species. It can occur between genes at the same locus (intralocus conflict, e.g., sexually antagonistic alleles; reviewed in Rice 1996a), but it is conflict between nonallelic genes (interlocus conflict, e.g., between alleles at loci controlling male versus female behavior, male versus female genitalia, sperm versus egg, seminal fluid proteins versus other seminal fluid proteins or the female's reproductive tract and physiology) that sets the stage for the chronic antagonistic interlocus coevolution that I hypothesize drives most of the genetic change leading to reproductive isolation. Intergenomic conflict between alleles residing at different loci can, in theory, drive antagonistic coevolution between loci in a manner that is analogous to the Red Queen process between species.

Cross-fertilization creates unusually high opportunity for interlocus antagonistic coevolution, especially when there is internal fertilization and hence an infusion of chemicals from males into females. Internal fertilization generates intimate chemical interaction between a male and female, and between different males mated to the same female. With strict genetic monogamy (i.e., one male mates with one female and vice versa), the reproductive interests of male and female are identical and there is little opportunity for conflict between the sexes. But any deviation from monogamy creates multifarious opportunity for conflict.

As an example, consider the seminal fluid proteins of *Drosophila melanogaster* that are transferred along with sperm during copulation. These proteins act as pheromones that reduce the female's propensity to remate and induce her to ovulate at substantially higher rate (reviewed in Chen 1984). Recent work by Clark et al. (1995) demonstrates that seminal fluid proteins also play an important role in sperm competition. Everything that has been measured about seminal fluid proteins is beneficial to males.

But recent work demonstrates that seminal fluid is toxic to female *D. melanogaster*; the more they receive,

Figure 19.1. An overview of the major forms of evolutionary conflict.

the faster they die (Fowler and Partridge 1989; Chapman et al. 1995). This toxic effect has also been observed in the nematode *Caenorhabditis elegans* (Gems and Riddle 1996). To the extent that seminal fluid proteins are coded by genes that differ from those controlling female adaptations that reduce the toxic effects of seminal fluid, and/or reduce any deleterious impact of seminal pheromones, there is opportunity for interlocus antagonistic coevolution driven by intergenomic conflict.

When females mate with more than one male, internal fertilization produces direct, multifarious chemical communication between different males (via seminal fluids) creating a second opportunity for interlocus antagonistic coevolution. Consider a female who mates with two males. The seminal fluid of the first male is selected for "defense," that is, the capacity to pheromonally reduce remating in the female and, if this fails, to prevent his sperm from being displaced by sperm of the second male. The second male is selected for "offense" (I have borrowed the terms "offense" and "defense" from earlier conversations with Tim Prout), that is, the capacity to persuade the female to remate with him and to then displace the resident sperm. To the extent that the loci coding for offense and defense phenotypes are not identical, there is opportunity for interlocus antagonistic coevolution.

In summary, internal fertilization generates a high level of direct chemical interaction between individuals. This provides a large opportunity for antagonistic gene-product/gene-product interactions that can lead to interlocus antagonistic coevolution. Such gene product coevolution is expected to be a perpetual "arms race" since the gene products can change in a mutation/countermutation fashion. The fraction of the genome involved in such antagonistic coevolution is likely to be quite small. But these genes are expected to coevolve continually and therefore diverge between populations far faster than the rest of the genome. Because these antagonistically coevolving genes will code, in large part, for mating behavior and the male and female reproductive tracts, they may be primarily responsible for the genetic change that produces reproductive isolation between allopatric populations or populations connected by low levels of gene flow.

A major prediction of the interlocus antagonistic coevolution hypothesis is that genes controlling mating behavior and reproductive tracts should evolve far faster than the rest of the genome. Part of this prediction was tested by Thomas and Singh (1992) by comparing the rate of protein divergence of various organs among the four members of the *melanogaster* group of *Drosophila* (*D. melanogaster, D. simulans, D. sachellia*, and *D. mauritiana*). They used two-dimension gel electrophoresis (2DE) of proteins to quantify the proportion of diverged proteins (hundreds of different proteins are measured simultaneously) in three adult tissues: brain (as an example of typical somatic tissue), testes, and the male accessory gland (which produces most of the seminal fluid proteins). The brain proteins diverged the slowest, the male accessory proteins diverged almost twice as rapidly as those from the brain, and the divergence of testicular proteins was intermediate but still nearly 40% faster than brain. This very rapid evolution of male reproductive tract proteins was corroborated by Aguade et al. (1992), who found very rapid divergence in the gene sequence of two seminal fluid proteins in the *melanogaster* group and evidence that this change was driven by selection rather than drift.

If the observed rapid divergence of reproductive tract proteins contributes to reproductive isolation, then incompatibility of the male and female reproductive tracts and physiology should be observed during the very early stages of the evolution of reproductive isolation. Recent work with incipient species that still produce viable and fertile hybrid offspring supports this association (Gregory and Howard 1993, 1994, Howard et al., this volume). This research compared the capacity for heterospecific and conspecific males to induce ovulation in female crickets. Heterospecific mating induced a lower rate of ovulation (30% reduction), indicating a breakdown in the compatibility of male and female reproductive physiologies early during the evolution of reproductive isolation. They also found that while heterospecific males were effective at fertilizing females of the other species when they alone fertilized her, heterospecific males fertilized almost none of her eggs when she was also mated to a conspecific male (no matter what the mating order). This indicates a rapid breakdown in the effectiveness of the seminal fluid factors mediating sperm competition during the speciation process. Similar results were found by Wade et al. (1994) with two closely related flour beetle species.

Additional work between two more distantly related species, *D. melanogaster* and *D. simulans*, also demonstrates how incompatibility between reproductive systems can lead to reproductive isolation (Jamart et al. 1995). This work used lines of the two species that were previously selected for increased levels of interspecific mating. When these lines were mixed in a population cage, they found many cases where *D. melanogaster* females mated *D. simulans* males followed by *D. melanogaster* males but never in the reverse order. Thus, females mated by a heterospecific male readily remated (displacing the heterospecific sperm), while females mated to a con-

specific male did not remate to a heterospecific male. Heterospecific matings were therefore unable to pheromonally prevent rapid remating with a conspecific male but conspecific matings prevented remating with heterospecific males.

I am aware of only two studies that looked for incipient incompatibility, between male and female reproductive tracts, among strains of *D. melanogaster* maintained independently in the laboratory for many generations as separate stocks. Rates of sperm migration to the sperm storage organs was slower in homotypic crosses (among four wild-type strains; Yanders 1963), suggesting antagonistic male–female coevolution. Additional work by DeVries (1964), on two of the four *D. melanogaster* stocks studied by Yanders (1963), also indicated a slower rate of sperm storage in homotypic crosses. This later study also indicated that sperm survival, among sperm successfully stored, was higher in homotypic crosses. So, the very limited data on the evolution of males and females in isolated lines of *Drosophila* suggest antagonism with regard to sperm delivery, but mutual benefit concerning the retention of successfully stored sperm.

A second prediction of the interlocus antagonistic coevolution hypothesis is that hybrid infertility should evolve faster than hybrid inviability since loci engaged in interlocus antagonistic coevolution are expected to impact male and female reproductive tracts. Studies surveyed by Eberhardt (1985) indicate the external genitalia diverge between species far faster than the rest of the external phenotype. Recent work (reviewed in Wu and Davis 1993; True et al. 1996) also support the prediction of a faster divergence of reproductive tract organs. They transgressed small chromosomal regions between species of the *melanogaster* group of *Drosophila* and then looked for postzygotic isolation via hybrid infertility versus inviability. They found a tenfold excess of hybrid infertility introgressions despite the fact that mutation studies suggest a tenfold excess of in the opposite direction; that is, there are ten times more loci capable of mutating to produce inviability versus infertility. This work supports the conclusion that reproductive tracts evolve far more rapidly than the genome as a whole.

Studies by True et al. (1996) and Hollocher and Wu (1996) found that of the introgressions that cause infertility, there is a 5–10-fold excess of male infertility compared to female infertility. This result suggests that offense–defense interlocus antagonistic coevolution may be the major factor driving the evolution of reproductive isolation. Additional work by Civetta and Singh (1995) with *Drosophila*, however, measured the rate of protein divergence (via 2DE) of testes and ovaries in the *melanogaster* group and also among close and distant relatives within the *virilis* group. Virtually identical levels of protein divergence were found for testes and ovaries, indicating that (a) male-biased levels of hybrid infertility may not indicate a faster evolution of male compared to female reproductive tracts, and (b) spermatogenesis may be more sensitive to developmental breakdown, in hybrids, than oogenesis.

A Three-Way Tug-of-War

The available data are still fragmentary but they are consistent with the hypothesis that there is a three-way antagonism between offense, defense, and the female reproductive tract and physiology (figure 19.2). This three-way conflict generates many opportunities for both two-way and three-way interlocus antagonistic coevolution.

Thus far I have not considered interlocus antagonistic coevolution between the structural genes of sperm and egg. In *Drosophila melanogaster* a combination of (a) the restricted region of the egg (micropyle) where sperm can enter, (b) the limited number of sperm available at the mouth of the primary sperm storage organ (seminal receptacle), and (c) the path by which the egg engages with the sperm storage organ, all seem to restrict the opportunity for polyspermy (Lefevre and Jonsson 1962; Fowler 1973). In other species, however, the opportunity for polyspermy may cause the sperm and egg to be locked in an evolutionary arms race.

In those species where the number of sperm simultaneously encountering an egg is large, an important part of sperm competition will be mediated by the rate sperm transgress the cortex surrounding the egg's plasma membrane. Sperm are selected to bore through the outer cortex of the egg more rapidly than their competitors (i.e., sperm from other males). Eggs (i.e., the cortex surrounding the egg's plasma membrane) are selected to resist rapid penetration by sperm so that there is sufficient time, after the first sperm penetrates the plasma membrane, to recruit one or more blocks to polyspermy before additional sperm can enter the egg. Even when there are cytoplasmic mechanisms that prevent more than one sperm pronucleus from fertilizing the egg pronucleus, there may be selection to prevent multiple sperm from penetrating the egg; for example, multiple sperm penetration would increase the opportunity for pathogen entry.

In abalone (e.g., Lee et al. 1995) and *Tegula* snails (M. Hellberg, personal communication) the acrosomal protein lysin, mediating the process of a sperm boring through the vitelline envelope (i.e., the extracytoplasmic cortex that surrounds the egg), is evolving at an extremely rapid rate. This unusually rapid evolution may be in response to pathogen coevolution (i.e., the Red Queen process; Lee et al. 1995), but alternatively it may be in response to coevolution between sperm and egg that is driven by polyspermy or other detriments associated with multiple sperm entry into the egg.

As an example of a three-way tug-of-war, consider the remating behavior of a virgin female *Drosophila* who has just mated (figure 19.3). We can define, at least in principle, a remating threshold as some unidimensional measure of her propensity to remate, which typically will be with a different male. The optimum remating threshold

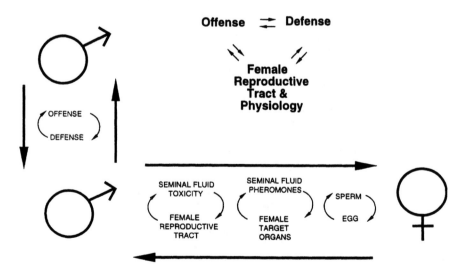

Figure 19.2. Internal fertilization generates evolutionary conflict between the genes coding for male offense, male defense, and female behavior, physiology, and reproductive tract. The conflict can generate interlocus antagonistic coevolution among the three sets of genes.

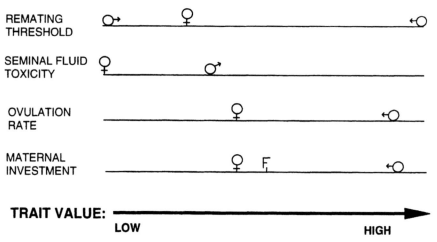

Figure 19.3. Contrasting optima for male offense (male symbol pointing horizontally to the right), male defense (male symbol pointing horizontally to the left), and females (female symbol). The standard male symbol is used to indicate the male optima without distinguishing between offense and defense, and the F_1 symbol represents the optima for the female's offspring.

from the perspective of genes controlling offense is zero, while that for genes coding for defense is a large threshold value. The optimal threshold for the female is likely to be intermediate (about 40% of *D. melanogaster* virgin females remate in the 6-hour period following their first mating, despite the fact that they usually get sufficient sperm from the first mating; Scott and Williams 1993; Van Vianen and Bijlsma 1993). There are many reasons why a female may gain by mating more than once, for example, infertility or low sperm levels of some males, the ability to gain access to resources without male harassment, and so on. The difference in the remating threshold optima of the three sets of genes controlling offense, defense, and female behavior can potentially drive interlocus antagonistic coevolution.

Since the physiological mechanisms mediating a female's remating threshold are still unresolved we can only make educated guesses as to how offense, defense, and the female phenotype may actually coevolve. Suppose, for example, that defense pheromones increase a female's remating threshold by reducing the signal transduction of certain sensory functions (e.g., olfaction), thereby making courting males less apparent. Genes controlling a female's remating threshold can respond by evolving increased sensory efficiency, reduced sensitivity to the male pheromone, and/or the use of alternative sensory input to trigger acceptance of a courting male. Offense genes can respond by increasing the vigor of courtship (see Markow and Hocutt, this volume), exploiting superstimulatory pathways in the female perception system, and so on. This simple, hypothetical example illustrates the multifarious ways by which the intergenomic conflict can lead to perpetual interlocus antagonistic coevolution.

In addition to the above example for remating rate, differences in the optima for many other aspects of offense, defense, and the female's reproductive tract and physiology are feasible (figure 19.3). Not all conflict will involve all three factors. For example, ovulation rate is impacted by seminal fluid proteins, and since a female may remate frequently in nature, genes mediating defense will tend to maximize short-term ovulation rate while those within the female will favor a rate that maximizes her lifetime reproductive rate. As a last example consider the toxicity of seminal fluid to females (Fowler and Partridge 1989; Chapman et al. 1995; Gems and Riddle 1996). The optimum for females is clearly zero, as it would be for males were there not some negative association between offense and defense performance and female survival. We know too little at present to rank order the optimal toxicity for offense versus defense, but the observed toxicity of seminal fluid suggests that one or both is substantially greater than zero. Genes reducing the toxic effect of seminal fluid will be favored in females. If these accumulate and reduce offense and defense performance, a counterresponse by the genes controlling offense and defense is expected. Similarly, genes coding for offense and defense may evolve in response to each other and additional toxicity of seminal fluid may be produced as a correlated character.

As a very speculative final point (i.e., meant to be provocative while totally unsubstantiated), an obvious limitation of seminal fluid is that it is a transient pulse of chemical input from male to female, analogous to a single application of a drug to a patient. A greater level of influence could be achieved if there were intermittent or continuous chemical transfer from male to female. This potentially can be achieved in several ways: (a) by a male repeatedly mating his mate, far in excess of the level needed to supply adequate sperm to fertilize her eggs, (b) by transferring sperm or spermlike cells that fuse with female reproductive tract cells and induce them to secrete male-benefit chemicals, and (c) by transferring living secretory cells that could continually release male-benefit chemicals into the female. In the latter case, females would be expected to respond by eliminating nonself tissue (i.e., internal fertilization may select for an elaborate nonself recognition process). Because females are not expected reject the sperm itself, any such male secretory cells placed in females may evolve to mimic sperm, or other female-benefit tissue such as placenta. Multiple sperm morphs have been found in many species yet their function, if any, is still unknown.

Viviparous species provide an unusual opportunity for a male to continually influence his mate, especially when tissue from the offspring (e.g., placenta) influences the rate of nutrient delivery between mother and offspring (see, for review, Haig 1993). Consider the level of maternal investment in a progeny from a polyandrous, outbred species (figure 19.3). The optima for the female and the offspring are likely to be most similar, owing to their 50% level relatedness, while the optima for the male (and the sperm haplotype) is likely to be larger, since he is unrelated to the mother and subsequent offspring produced by the female are likely to be sired by different males. The difference in the three optima can potentially drive perpetual interlocus antagonistic coevolution. Most such conflict will be male–female rather than offspring–mother owing to the greater dissimilarity of optima between male versus female compared to offspring versus mother.

Reproductive Isolation as an Incidental By-product of Interlocus Antagonistic Coevolution

The previous section outlines only a few of many possible ways by which offense, defense, and a female reproductive tract and physiology can antagonistically coevolve, causing the associated loci to evolutionarily diverge between populations at a rate far faster than the genome as a whole. Such coevolution can generate both pre- and postzygotic isolation. Prezygotic isolation will be produced by changes in the signals and sensory perception used by males and females during courtship. Postinsemination/prezygotic isolation will accrue from a breakdown in the compatibility of the reproductive tracts and secretions of males and females taken from different populations. Postzygotic isolation via hybrid infertility will occur due to changes in the structural characteristics of eggs, to a lesser extent, and especially sperm, which may be direct targets of many seminal fluid proteins that mediate offense and defense. As the structural genes for sperm and egg diverge, the fertility of hybrids will diminish owing to incompatibility of the divergent sperm and egg developmental programs. Postzygotic isolation via inviability should evolve far slower, at least on average, since genes involved in interlocus antagonistic coevolution are less likely to involve loci impacting viability.

Experiments to Detect Male–Female Antagonistic Coevolution

All of the above evidence for an substantial role of interlocus antagonistic coevolution in the evolution of reproductive isolation is correlational. As a first step in providing direct experimental evidence, I set out to induce rapid antagonistic coevolution between males and females (Rice 1996b). A major difficulty in observing coevolution between males and females can be expressed by analogy to an antagonistic, dancing couple. As each member moves there is a countermove by the partner. As a consequence, the relative position of the two is little changed despite their substantial movement across the dance floor. The adaptation-counteradpatation of the sexes would cause little conflict to be apparent at any single point in time despite an extensive history of antagonistic coevolution between them.

One could obtain evidence that each sex is antagonistically adapting to the other by holding one sex still and letting the other evolve in response to an evolutionary stationary partner. If the coevolution is antagonistic, a reduction in the fitness of the stationary partner should be observed. In my experiments, I evolutionarily held one partner still and looked for antagonistic evolution in response to the "tethered" partner.

To introduce the experimental protocol consider the natural history context of a unisexual, matriclinus species of fish of the genus *Poeciliopsis* (figure 19.4; see summary in Schultz 1961). First concentrate on the wild-type female at the bottom left of figure 19.4. She has two sets of homologous chromosomes (here after referred to as genomic haplotypes), one set she inherited from her father (collectively depicted by the single open rectangle) and one set from her mother (solid ellipse). When she produces eggs, the maternally and paternally derived chromosomes neither pair nor recombine, and only the maternally inherited genomic haplotype is transmitted through the egg. The female next mates with a male from a different bisexual species of the same genus, her daughter expresses both the genomic haplotype obtained from her mother and that from her father (i.e., at those loci that have been studied, females express a diploid rather than a haploid genotype), but then she only transmits, clonally, the maternally derived genomic haplotype to her eggs. In this way the unisexual female fish species has a clonal,

Clonal, female-limited genomic haplotypes in Poeciliopsis fish:

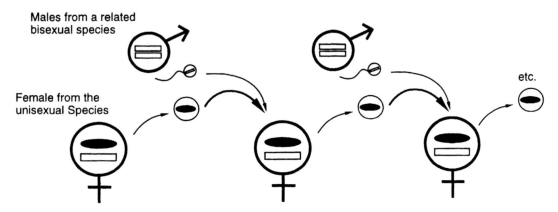

Males from a related
bisexual species

etc.

Female from the
unisexual Species

Figure 19.4. A schematic of the transmission genetics for a unisexual, matriclinus fish of the genus *Poeciliopsis*. This form of reproduction is termed hybridogenesis. The solid ellipse represents the set of chromosomes a female inherits from her mother (maternal genomic haplotype), and the open rectangle represents the set of chromosomes that a female inherits from her father (paternal genomic haplotype).

female-limited genomic haplotype. The paternally derived haplotypes are used for one generation and then discarded. The female-limited genomic haplotype can adapt to the males of the bisexual species and vice versa, but the two genomes are not homogenized by recombination each generation.

I was able to construct flies in a *Drosophila melanogaster* model system that recapitulate the unisexual *Poeciliopsis* genetic system with the exception that the clonal, genomic haplotypes were passed from father to son rather than from mother to daughter, and females could not evolve counteradaptations (figure 19.5).

The major "engine" behind the experimental design was the clone generator females. Wild-type, diploid male and female karyotypes of *D. melanogaster* are depicted at the top left of figure 19.5, with the sex chromosomes to the left and autosomes II and III to the right (these three chromosomes make up 99% of the genome; the "dot" fourth chromosome was not manipulated in these experiments and is not shown in the figure). Clone generator females have their two X-chromosomes attached via a common centromere (attached-X; both X-chromosomes cosegregate to the same pole during meiosis), they carry a Y-chromosome (which does not masculinize them), and they have their two major autosomal chromosomes translocated such that, among the surviving offspring of a heterozygous male, they segregate as if they were a giant single chromosome.

When wild-type males are mated to clone generator females, their sons inherit one genomic haplotype from their father (X, II, III; note that there is no molecular re-

combination in male *D. melanogaster*) and one from their mother (Y, Translocation [II, III], figure 19.5, top). When these sons are mated to new clone generator females, genetic markers on the translocation permit those sons that obtain a full set of genes (ignoring the "dot" chromosome IV) from their father to be identified and retained. All other offspring are removed from the breeding population.

By repeatedly crossing these males to clone generator females (figure 19.5, bottom) that are taken anew from a large stock population, the male genomic haplotypes can adapt to the female phenotype but the females cannot counterevolve in response to any evolutionary advance on the part of the male-limited genomic haplotypes. Thus, an asymmetry is created where the male-limited genomes can evolve in response to the females but not vice versa. There were a total of two experimental lines (EA and EB), cultured as described above, and two control lines (CA and CB), which were cultured side by side with the experimental lines, with a very similar protocol, effective population size, and opportunity for sexual selection, but with normal male and female karyotypes. Further details of the experimental protocol and the results reviewed below can be found in Rice (1996b).

After 30 generations of opportunity for the male-limited genomic haplotypes to evolve to the clone generator females, controls and experimental males were compared. I measured, when they were mated to clone generator females, their net fitness (adult sons produced per adult male) and a measure of offense (capacity to

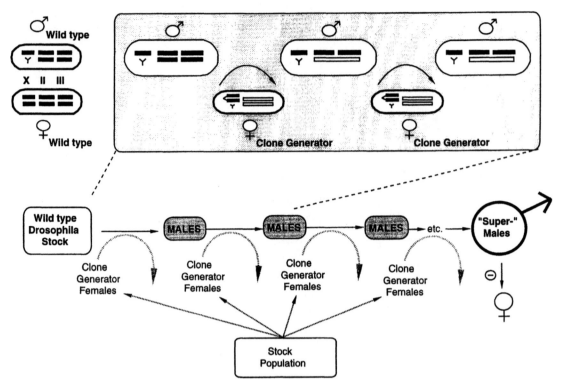

Figure 19.5. A schematic of the way *Drosophila melanogaster* was used to construct a male ana-
log of the *Poeciliopsis* unisexual reproductive system (hybridogenesis). See text for details. Four
percent of the male genomic haplotypes were recombined each generation (not illustrated in the figure)
to prevent the operation of Muller's Ratchet and to speed the rate of beneficial mutation accumulation. A complete description of the experimental protocol can be found in Rice (1996b).

remate nonvirgin females) and defense (capacity of a male
to prevent his sperm from being displaced when his mate
is exposed to another male). In all measures the experimental males outperformed the controls by at least 20%.
Thus, the experimental males (i.e., genomic haplotypes)
had clearly adapted to the clone generator females. This
indicates that there is ample genetic variation for males
to rapidly adapt to females.

To look for evidence for sexually antagonistic coevolution, the impact of the males on their mates was also
measured. Females mated to experimental and control
males had similar fecundity, but females mated to experimental males had substantially higher mortality. This
increased mortality was due, in both the EA and EB lines,
to experimental males remating females at a higher rate
(recall that seminal fluid is toxic to females). In the case
of line EB, there was additional mortality of their mates
owing to the seminal fluid becoming more toxic to females; that is, females that mated to EB males only once
had elevated mortality. Thus, there was evidence for the
fitness gain of the experimental males resulting in a fit-

ness decline in their mates, as predicted by the interlocus
antagonistic coevolution model.

The Relative Importance of Reproductive Isolation via Interlocus Antagonistic Coevolution

Interlocus antagonistic coevolution should be an important factor generating genetic divergence, and consequent
reproductive isolation, in all sexual populations; hence, it
has nearly universal application. Nonetheless, it is not
expected to always be the principal factor generating reproductive isolation during the speciation process. For
example, consider polar bears (*Thalarctos maritimus*) and
grizzly bears (*Ursus horribilis*). These species are fully
interfertile in forced sympatry (zoos), yet they are good
biological species in nature, apparently owing in part to
differences in their spatiotemporal habitat preference.
Adaptation of polar bears to a radically new environment
has generated strong prezygotic reproductive isolation.

Such reproductive isolation as an incidental by-product of divergent adaptation to new environments is commonly observed in laboratory experiments (see, for review, Rice and Hostert 1993). In those contexts where divergent ecological adaptation is substantial, interlocus antagonistic coevolution may play a minor role in the speciation process, although it may still be important in (a) generating unconditional postzygotic isolation via infertility, and/or (b) completing the speciation process when reproductive isolation via spatiotemporal habitat preference and/or hybrid inviability is incomplete. But in those cases (which I suspect are far more common) where divergent selection among populations is less extreme, the perpetual nature of interlocus antagonistic coevolution, in combination with the fact that the traits involved are directly relevant to reproductive compatibility, suggests to me that interlocus antagonistic coevolution may be the dominant factor generating reproductive isolation.

Summary and Conclusions

The interlocus antagonistic coevolution hypothesis states that (1) intergenomic conflict leads to perpetual antagonistic coevolution between sets of genes associated with male offense, male defense, and female reproductive tract, physiology, and behavior; and (2) the resulting genetic change is typically the predominant factor leading to the evolution of reproductive isolation between populations. Evidence for the hypothesis is still far from complete, but (1) data from the *melanogaster* group of *Drosophila* support the idea that reproductive tract organs are evolving a relatively rapid rate; (2) data from crickets and *Tribolium* beetles, in the very early stages of species formation, indicate a rapid breakdown in compatibility between male offense, male defense, and female reproductive tracts and physiology; (3) data from *Drosophila* species that have evolved more complete reproductive isolation indicate that hybrid infertility evolves far more rapidly than hybrid inviability and that both male and female reproductive tracts are evolving at a fast rate; and (4) recent experimental work indicates a high potential for antagonistic coevolution between the sexes. These four lines of evidence supporting the interlocus antagonistic coevolution hypothesis are fragmentary and need verification with a broader spectrum of organisms. Nonetheless, the preliminary evidence is that intergenomic conflict generates a mutation/countermutation genetical chain reaction that may prove to be a major catalyst in the speciation process.

References

Aguade, M., N. Miyashita, and Langhley, C. H. 1992. Polymorphism and divergence of the mst 355 male accessory gland gene region. *Genetics* 132:755–770.

Chapman, T., Lindsay, F., Liddle, F., Kalb, J. M., Wolfner, M. F., and Partridge, L. 1995. Cost of mating in *Droso-*

phila melanogaster females is mediated by male accessory gland products. *Nature* 373:241–244.

Chen, P. S. 1984. The functional morphology and biochemistry of insect male accessory glands and their secretions. *Annual Review of Biochemistry* 29:233–255.

Civetta, A., and Singh, R. S. 1995. High divergence of reproductive tract proteins and their association with postzygotic reproductive isolation in *Drosophila melanogaster* and *Drosophila virilis* group species. *Journal of Molecular Evolution* 41:1085–1095.

Clark, A., Agoude, G. M., Prout, T., Harshman, L., and Langley, C. H. 1995. Variation in sperm displacement and its association with accessory gland protein loci in *Drosophila melanogaster*. *Genetics* 139:189–201.

DeVries, J. K. 1964. Insemination and sperm storage in *Drosophila melanogaster*. *Evolution* 18:271–282.

Eberhard, W. G. 1985. *Sexual Selection and Animal Genitalia*. Harvard University Press, Cambridge, Mass.

Fowler, G. L. 1973. Some aspects of the reproductive biology of *Drosophila*: sperm transfer, sperm storage, and sperm utilization. *Advances in Genetics* 17:293–360.

Fowler, G. L., and Partridge, L. 1989. A cost of mating in female fruit flies. *Nature* 338:760–761.

Gems, D., and Riddle, D. L. 1996. Longevity in *Caenorhabditis elegans* reduced by mating but not gamete production. *Nature* 379:723–725.

Gregory, P. G., and Howard, D. J. 1993. Laboratory hybridization studies of *Allonemobius fasciatus* and *A. socius* (Orthoptera: Gryllidae). *Annals of the Entomological Society of America* 86:694–701.

Gregory, P. G., and Howard, D. J. 1994. A postinsemination barrier to fertilization isolates two closely related groups of crickets. *Evolution* 48:705–710.

Haig, D. Genetic conflicts in human pregnancy. 1993. *Quarterly Review of Biology* 68:495–531.

Hollocher, H., and Wu, C.-I. 1996. The genetics of reproductive isolation in the *Drosophila simulans* clade: X versus autosomal effects and male versus female effects. *Genetics* 143:1243–1255.

Hurst, L. D. 1992. Intragenomic conflict as an evolutionary force. *Proceedings of the Royal Society of London Series B* 247:189–194.

Jamart, J. A., Casares, P., Carracedo, M. C., and Pineiro, R. 1995. Consequences of homo- and heterospecific rapid remating on the fitness of *Drosophila melanogaster* females. *Journal of Insect Physiology* 41:1019–1026.

Lee, Y.-H., Tatsuya, O., and Vacquier, V. D. 1995. Positive selection is a general phenomena in the evolution of abalone sperm lysis. *Molecular Biology and Evolution* 12:231–238.

Lefevre, G., and Jonsson, U. B. 1962. Sperm transfer, storage, displacement, and utilization in *Drosophila melanogaster*. *Genetics* 47:1719–1736.

Rice, W. R. 1996a. Evolution of the Y sex chromosome in animals. *BioScience* 46:331–343.

Rice, W. R. 1996b. Sexually antagonistic male adaptation triggered by experimental arrest of female evolution. *Nature* 361:232–234.

Rice, W. R., and Hostert, E. E. 1993. Laboratory studies on speciation: what have we learned in 40 years? *Evolution* 47:1637–1653.

Schultz, R. J. 1961. Reproductive mechanisms of unisexual and bisexual strains of the viviparous fish *Poeciliopsis*. Evolution 15:302–325.

Scott, D., and Williams, E. 1993. Sperm displacement after remating in *Drosophila melanogaster*. *Journal of Insect Physiology* 39:201–206.

Thomas, S., and Singh, R. S. 1992. A comprehensive study of genetic variation in natural populations of *Drosophila melanogaster*. VII. Varying rates of genetic divergence as revealed by two-dimensional electrophoresis. *Molecular Biology and Evolution* 9:507–525.

True, J. R., Weir, B. S., and Laurie, C. C. 1996. A genome-wide survey of hybrid incompatability factors by the introgression of marked segments of *Drosophila mauritiana* chromosomes into *Drosophila simulans*. *Genetics* 142:819–837.

Van Valen, L. 1973. A new evolutionary law. *Evolutionary Theory* 1:1–30.

Van Vianen, A., and Bijlsma, R. 1993. The adult component of selection in *Drosophila melanogaster*: some aspects of early remating activity of females. *Heredity* 71:269–276.

Wade, M. J., Patterson, H., Chang, N. W., and Johnson, N. A. 1994. Postcopulatory, prezygotic isolation in flour beetles. *Heredity* 72:163–167.

Wu, C.-I., and Davis, A. W. 1993. Evolution of postmating reproductive isolation: the composite nature of Haldane's Rule and its genetic basis. *American Naturalist* 142:187–212.

Yanders, A. F. 1963. The rate of *D. melanogaster* sperm migration in inter- and intra-strain matings. *Drosophila Information Service* 38: 33–34.

Zeh, D. W., and Zeh, J. A. 1994. When morphology misleads: interpopulation uniformity in sexual selection masks genetic divergence in harlequin beetle-riding pseudoscorpion populations. *Evolution* 48:1168–1182.

20

Species Formation and the Evolution of Gamete Recognition Loci

Stephen R. Palumbi

The Ecological Context of Species Formation

The way different researchers look at speciation and the way they define species often are heavily influenced by the kinds of organisms they happen to study. Mayr, Carson, and Bush were influenced heavily by the ecology and natural history of birds, insular flies, and host-specific phytophagous insects, respectively, and their views of speciation (allopatry, founder-flush, and host-race formation) reflect processes that are particularly apparent in these groups. This diversity of approaches is a valuable part of speciation research because it emphasizes the varied ecologies and life histories of different taxa. Patterns and processes of speciation could differ between such taxa, and these differences expand our view of the action of evolutionary forces.

Among ecological and life history traits, the mobility of an individual plays a strong role in our perception of likely modes of speciation. Bush (1975) characterized different modes of speciation as being most appropriate for certain ranges of dispersal ability. For example, allopatric speciation was considered most likely for organisms with low movement. Species with high vagility require more stringent conditions for standard allopatric mechanisms to operate.

Many marine species have a phase in their life history that promotes long-distance dispersal. Most marine fish have planktonic eggs or larvae that develop for weeks in drifting currents (Hourigan and Reese, 1987). Speciose marine taxa like echinoderms, molluscs, and polychaete annelids have large numbers of species with larvae that show high dispersal potential (R. Strathmann, 1978; M. Strathmann, 1987). Other marine species have highly mobile adults that either drift with open ocean currents or actively swim thousands of kilometers. Although there are many examples of marine species with poor dispersal abilities, and although high dispersal *potential* does not necessarily result in high dispersal in nature (Palumbi, 1995), there are many cases of high dispersing species in marine environments.

Coupled with this high dispersal is the fact that the world's oceans are not extensively subdivided. As a result, there are very few absolute barriers to dispersal or gene flow in the sea (see Lessios, this volume, for one of the rare major exceptions, the Isthmus of Panama). The combination of high dispersal and incomplete geographic barriers often leads to huge ranges for many marine species. Particularly in the tropical Pacific (see compilations for *Conus* by Kohn and Perron, 1994; sea urchins by Emlet, 1995; corals by Veron, 1993), species ranges often span over 10,000 km. Across this range, cumulative population size can be very large, and there may be substantial populations structure (e.g., some whales, urchins, billfish; Baker et al., 1993; Graves and McDowell, 1995; Palumbi, 1996a), or very weak structure (some pelagic fish; Ward et al., 1994).

Large population sizes of taxa with high vagility living in an environment with few absolute geographic barriers might be expected to diverge slowly. The fossil record suggests that marine species with high dispersal form more slowly than do similar species with low dispersal (Hansen, 1980, 1983; Jablonski, 1986). Yet, Knowlton (1993) showed that virtually all marine invertebrate taxa show large numbers of phenotypically similar, cryptic species that have low genetic divergence. Thus, a classic allopatric view suggests speciation should be slow in the sea, but the high diversity of phenotypically similar species with low genetic divergences suggests that there are many recently diverged marine species. Understanding how large populations of highly dispersive individuals speciate rapidly in an environment with few absolute barriers to gene flow is a challenge for any conceptualization of speciation, and may greatly illuminate the process of genetic differentiation before and after species formation.

Evolution of Reproductive Isolation

Among different species in different habitats, there are many different types of reproductive isolation. Examining the genetic, morphological, physiological, and ecological bases for reproductive isolation has been a critically important part of understanding the speciation process. Allopatric or peripatric speciation begins with isolation imposed by the environment, either through a complete barrier or through isolation by distance (Slatkin, 1982). In this case genetic changes after isolation accumulate to limit interbreeding if the extrinsic, geographic barrier should ever break down (Bush, 1975). Sympatric speciation is similar in that it requires the accumulation of enough differences that assortative mating evolves in concert with genetic changes at other loci. However, these concerted genetic changes must take place within populations rather than between them (Rice, 1987).

Two Kinds of Premating Isolation

These ideas point out important distinctions in types of premating reproductive isolation: opportunity isolation and recognition isolation. Opportunity isolation can be defined as a decreased chance of encounter with a potential mate or gamete, and occurs when individuals in a population have unequal opportunities to mate with one another. Populations separated by geographic barriers are an obvious example. In addition, in heterogeneous habitats there may be differences in habitat selection such that individuals only come in contact with conspecifics that choose the same habitat type (Bush, 1975). Parasites are a good potential example of this type of isolation: only those individuals using a particular host have an opportunity to mate with one another (Johnson et al., 1996). Sedentary, free-living species also experience habitat variation that may limit reproductive interactions. Shallow-water marine organisms experience a wide array of environments due to wave exposure, intertidal position, temperature gradients, and so on. Like plants, these organisms probably mate most frequently with their nearest neighbors (Grosberg, 1987), and these neighbors are most likely to be in similar microhabitats. In these cases, assortative mating may occur with no variation in recognition between individuals (Rice, 1987).

The second category of isolation, recognition isolation, can be defined as a decreased chance of acceptance of a potential mate or gamete once encountered and comes from the ability of organisms to recognize and choose mates. Between species, recognition mechanisms are often ascribed to selection against gamete wastage and reinforcement (see Butlin, 1989). Within species, differential mate choice is often studied in the context of sexual selection (Kirkpatrick, 1987). The signals by which mate choice operates are not well known for most taxa, and may include visual perception of general body size or ornamentation in vertebrates (Kirkpatrick and Ryan, 1991), acoustic signals in insects and amphibia (Dagley et al., 1994), pheromone recognition in aquatic species, or kin recognition in vertebrates (Potts and Wakeland, 1993), invertebrates (Grosberg, 1988), and plants (Cornish et al., 1988).

Recognition without Brains

Although most attention has been focused on how behaviorally complex animals recognize one another before mating, there is another level at which prezygotic recognition can occur. Successful fertilization requires that male and female gametes bind together and fuse. This attachment and fusion requires interaction of gamete surfaces in such a way that gametes from the same species recognize one another. In hermaphroditic plants, products of the S-locus are used to discriminate among various pollen (Cornish et al., 1988) such that pollen from closely related individuals are less likely to successfully fertilize. Among animals, sperm–egg interactions are basic to almost all fertilization systems, although such systems differ widely between taxonomic groups.

The recognition requirement is especially strong for marine species that shed eggs and sperm into the water. In some cases, multiple species spawn at the same time (Babcock et al., 1986; Pennington, 1985), and in these cases, gamete recognition may be one of the major mechanisms by which reproductive isolation occurs (Palumbi, 1992). Although gamete recognition is probably most important in animals and plants that broadcast sperm or pollen, evidence from insects suggests that there may be a role for gamete recognition even for internally fertilizing species (see Howard et al., this volume).

Mechanisms by which gamete recognition evolves may provide a clue about the evolutionary forces shaping reproductive isolation. Moreover, mechanisms that control cell surface recognition by gametes are probably simpler than those controlling visual or acoustic recognition by adults, and it may be possible to obtain a clearer picture of the role of selection and drift in the evolution of recognition isolation.

Evolution of Gamete Interactions in Pacific Sea Urchins

In the tropical Indo-West Pacific, the sea urchin genus *Echinometra* is composed of at least four morphologically similar but genetically distinguishable species whose eggs and sperm show strong recognition isolation. Slight morphological differences in gonad and tube foot spicules were first reported by Uehara and colleagues in Okinawa (Uehara and Shingaki, 1985; Uehara et al., 1986). Allozymes, mitochondrial DNA and single-copy nuclear DNA renaturation (Matsuoka and Hatanaka,

1991; Palumbi and Metz, 1991) show these slight morphological variants are also different genetically, even when they occur sympatrically. Comparisons of genetic differences among these Pacific species suggest that the central and western Pacific *Echinometra* diverged within the last 1–3 million years (Palumbi, 1996b). Thus, they represent the most closely related suite of sea urchin species known.

Spawning in sea urchins occurs by release of gametes into the water, where fertilization takes place. In most cases, two to four species are found on the same reefs (Palumbi, 1996b). There is slight habitat differentiation between some species [e.g., *E. oblonga* (*sensu* Edmondson) dominates on exposed reef crests with high wave action (Russo, 1977)], but overall the four species tend to occur in similar microhabitats (Nishihira et al., 1991; Tsuchiya and Nishihira, 1984). Thus, there is little spatial segregation between spawning adults.

Although some urchins release gametes according to a lunar cycle (e.g., Lessios, 1984), *Echinometra* spawns seasonally or all year round, depending on the locality (Kelso, 1970), and sympatric species are ready to spawn at the same time. No natural spawning events have been observed, so it is possible that there are minor differences in spawning time like those reported for some corals during mass spawning events. But, to date, there is no evidence of any temporal segregation of spawning in *Echinometra*.

Instead, reproductive isolation in these closely related species strongly depends on gamete recognition. Eggs and sperm from different species do not fertilize well: generally fewer than 5% of eggs are fertilized even under conditions of excess sperm. Sperm can attach to heterospecific eggs but the numbers of sperm attached to eggs are lower in such experiments and sperm/egg fusion is inhibited (Metz et al., 1994). As a result, species of *Echinometra* in the Pacific show little opportunity isolation but strong recognition isolation. Understanding the rapid evolution of this recognition isolation may provide insight into how species such as these form in open marine environments, and may provide important information about the evolution of gamete interactions in other taxa.

Evolution of Gamete Recognition Genes

We have incomplete information on the mechanisms by which gametes interact, but recent advances have begun to identify the genes for proteins that are involved in sperm attachment and fusion. At least three evolutionarily important generalizations appear to be emerging from these studies. First, there is no universal system by which gametes recognize one another, and the proteins involved are not homologous in different phyla. For example, the proteins thought to be involved in gamete interaction in mammals are controversial (Snell, 1990), but none are

related to those known to be active in echinoderms or molluscs. Moreover, the basic mechanisms of gamete interaction are different. In mammals, enzymatic activity by a sperm protein acting on carbohydrates attached to the egg surface proteins has been implicated in gamete fusion (Miller et al., 1992). By contrast, in echinoderms, attachment of a sperm protein to an egg–surface receptor mediates both sperm binding and fusion (Glabe and Vacquier, 1977; Vacquier and Moy, 1977; Vacquier et al., 1995). In gastropods like abalone, the sperm burrows through the egg's chorion layer mediated by a protein called lysin whose activity is not yet well understood (Vacquier et al., 1990; Vacquier and Lee, 1993). The recruitment of fundamentally different mechanisms to the task of fertilization (Snell, 1990) is in stark contrast to the conservation of other fundamental cellular processes like glucose metabolism (Hochatchka and Somero 1973) or cell adhesion (Müller, 1995). This diversity suggests that stabilizing selection, which might be expected for cellular processes so closely related to fitness, is not always strong at gamete interaction loci.

The second generalization suggested by current research is that gamete recognition proteins appear to evolve quickly between closely related species. In sea urchins, both sperm attachment and fusion to eggs are mediated by a protein called bindin that coats the outside of the acrosome-reacted sperm (Vacquier and Moy, 1977). The gene sequence for bindin was reported by Gao et al. (1984) and Minor et al. (1990) and we have used this information to design PCR primers that amplify the entire coding region of the "mature" bindin protein from *Echinometra* species (Metz and Palumbi, 1996). Comparisons among these closely related species show that proteins involved in gamete recognition are evolving at a high rate and appear to be under positive selection for amino acid variation. There are more amino acid substitutions than silent substitutions, and more radical amino acid replacements than expected by chance alone (Metz and Palumbi, 1996).

Positive selection is seen only at the 5' end of the bindin gene. In this region there are up to 17% amino acid substitutions between closely related *Echinometra* species (Metz and Palumbi, 1996) but only 3–5% silent substitutions between species. By contrast, silent substitutions outnumber replacement substitutions in other regions of the protein.

Positive selection results from increased fitness conferred by proteins with novel amino acid sequences and has been observed in other recognition proteins, especially those involved in disease resistance. In these cases, selection appears to act strongly on those amino acid positions involved in protein–protein interactions. A similar pattern of positive selection has been described for the protein lysin (Vacquier et al., 1990). Amino acid differences in the lysin gene outstrip silent substitutions severalfold in closely related species (Lee et al., 1995), and amino acid changes are concentrated on the exterior of the folded protein (Shaw et al., 1993). The results for bindin and lysin suggest that gamete recognition is under selection for diversification and that closely related species can have very different cell surface recognition molecules.

Positive selection for amino acid substitution generates only a part of the variation in the bindin protein. Insertions and deletions of repeated blocks of amino acids occur frequently within and between species. Glycine chains are seldom the same from one allele to the other, and particular amino acid motifs have been duplicated in some alleles and deleted in others. In the 5' half of the bindin coding region, the combination of high amino acid substitution and insertions and deletions leads to great protein variation between species. Of the 128 amino acid positions in this region, only 59 are conserved among the three species we have examined to date.

In addition, recombination between alleles occurs within the bindin coding region. Using a very conservative "four-gamete test," we found at least three recombination events among 19 sequences within *E. mathaei*. These events shuffle unique amino acid sequences between different alleles and give rise to new proteins with potentially new properties. We identified these recombinations through the presence/absence of insertions and deletions, and so it is unlikely that parallel mutations are responsible for this pattern unless such mutations produce identical length changes.

A third evolutionary lesson derived from the study of gamete recognition proteins is that such genes can be highly polymorphic within species. In *E. mathaei*, we have sequenced 19 bindin alleles (after cloning PCR products into phagemid vectors; see Metz and Palumbi, 1996), and most have had different sequences. In some cases, the coding regions have been identical except for the number of amino acids in glycine chains. However, in most cases there is extensive amino acid variation in the 5' end of the gene. Up to 20 amino acid positions (out of 128) are different between some alleles in *E. mathaei*, and 19 are different in our current sample of alleles from *E. oblonga* (Metz and Palumbi, 1996). Although phylogenetic reconstruction of alleles within species is much more complicated because of recombination, bindin alleles coalesce within species, and there are species-specific amino acid substitutions.

Polymorphism in bindin sequences is also known for other urchin species (S. R. Palumbi, unpublished; C. Biermann, personal communication). To date, no polymorphism in lysin is known (Lee et al., 1995), but there has yet to be extensive sampling of abalone populations for lysin variation. Assuming that further sampling confirms lysin is less polymorphic than bindin, we must conclude that the forces allowing polymorphism within gamete recognition loci may not be as general as the forces generating amino acid variation between species.

Evolution of Assortative Mating within and between Species

Although some marine species show strong opportunity isolation (e.g., differences in habitat or spawning time; Palumbi, 1994), the above results suggest that differentiation of gamete recognition molecules occurs between closely related species whose major system of species recognition may reside in gamete surface interactions.

If divergence between species in gamete recognition proteins leads to reproductive isolation, what are the functional consequences of polymorphism of gamete recognition alleles within species? It is possible that intraspecific polymorphisms are neutral and that only interspecific differences have a selective basis. However, patterns of synonymous and replacement substitution in bindins are similar whether we compare alleles within species or those between species. This result suggests that selection is acting at both levels (Metz and Palumbi, 1996), although we have not yet identified the nature of these selective forces.

If gamete recognition polymorphisms have different functional effects, then it may be possible to observe these effects by examining the relationship between gamete recognition alleles and fertilization. More broadly, the action of recognition polymorphisms within populations should lead to different probabilities of mating between individuals based on their genotypes. Individuals in the population that share recognition alleles might be in the same random mating pool, whereas individuals with other alleles might be in other pools. These pools, or *mating guilds*, would be fully capable of interbreeding, but the probability of interguild matings would be less than within-guild matings.

Such differences in mating probability are commonly seen in systems in which female choice of mates operates (Kirkpatrick, 1987; Kirkpatrick and Ryan, 1991). Intraspecific variation in sender/receiver interactions is well documented for some types of acoustic signaling, where it has been shown to affect mate choice between populations within a species (Claridge and Morgan, 1993; Ryan and Wilczynski, 1988). The variation we observe in sperm attachment proteins suggests similar mating variation may occur in marine broadcast spawners, but to date no concerted attempts have been made to test for this variation.

Models of the evolution of loci involved in intraspecific recognition suggest that mating guilds may arise within populations. We have written a two-locus, two-allele model of gamete recognition that takes into account simple rules for the recognition of gametes by cell surface interactions (Palumbi and Asmussen, unpubl.). The simplest models show that linkage disequilibrum quickly develops between recognition loci, and that allele frequencies can change dramatically depending on starting conditions. In some conditions, polymorphisms at a sperm protein locus are stable if they are matched by polymorphisms at an egg locus. Wu (1985) modeled a two-locus, multi-allele system in which individuals with the ith allele at locus A mated best with individuals with the ith allele at locus B. Alleles mutated in a stepwise fashion, and a population that was initially monomorphic soon accumulated a large numbers of alleles at both loci. In many cases distinct mating guilds developed within populations, leading in some simulations to nearly complete reproductive isolation between guilds. This is a haploid model, and it is not clear how diploid transmission genetics would affect the results. Nevertheless, the model shows that polymorphism for functional recognition isolation can arise and persist in populations.

Mating Guilds and Bindin Polymorphisms

The above data and models show that there is abundant polymorphism for sperm proteins in sea urchins, and that such polymorphism could be involved in the evolution of reproductive isolation if certain conditions are met. The primary condition is that the polymorphisms are functional and result in individuals with unequal mating success. The prediction of the models is that such variants will fall into mating guilds that depend on the bindin genotype and the genotype at the receptor locus. Does such variation in mating (i.e., fertilization) exist in populations? Is there a genetic basis to such variation?

There are few published data sets that can be used to answer such questions. Fertilization of marine invertebrates is easily done in the laboratory, and in general sperm from every male can fertilize eggs of every female if the sperm concentration is high enough. As a result, the search for fertilization guilds in sea urchins must be done at limiting sperm concentrations in which less than 100% of the eggs are fertilized. Under these conditions, fertilization is highly dependent on environmental affects, and experiments must be done with exhaustive controls for gamete vitality, concentration, and age. Nevertheless, preliminary experiments with *Echinometra* suggest the widespread occurrence of fertilization guilds within species.

As an example, we have crossed eggs from a series of females with sperm at limiting concentrations from a series of males. To control for dilution errors, these experiments are done by diluting sperm into a large amount of seawater and using the same dilution to fertilize eggs from a number of different females. In general, sperm from a male tend to work better on some eggs than on others. In figure 20.1, sperm from male 31 (top panel) are most effective at fertilization of eggs from female 38 and least effective with eggs from female 48. Such variation in fertilization could be due to a number of nongenetic factors like egg condition or size. To try to test for such egg condition effects, we repeated the experiments with sperm from different males to see if similar results are obtained. Sperm from male 43 show a pattern

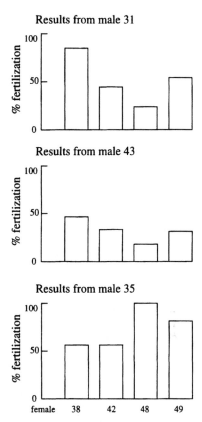

Figure 20.1.   Variation in fertilization in a cross of three males with each of four females of the species *Echinometra mathaei*. Under identical (but limiting) sperm concentrations, fertilization among eggs from different females varies almost threefold. Males 31 and 43 show similar patterns of fertilization, with eggs from female 38 being most easily fertilized and those of female 48 least easily fertilized. Sperm from male 35 show the opposite pattern. This variation can be due to a number of genetic and nongenetic causes. However, in these preliminary experiments, sequences of bindin from males showing similar fertilization patterns have the most similar bindin allele sequences.

similar to sperm from male 31 (middle panel). By contrast, sperm from male 35 show an opposite pattern: they fertilize eggs of female 48 best, and female 38 least (bottom panel).

These preliminary results suggest two important things. First, variation in fertilization ability among free spawners seems to exist and seems to result in large differences in fertilization when sperm are limiting. Second, differences in the rank order of sperm fertilization ability with different females suggest that the variation we see is due to interaction between eggs and sperm during fertilization. Are the interaction differences mediated by polymorphic gamete recognition alleles? We do not know the answer to this question, but preliminary results are encouraging.

We sequenced the bindin alleles of males 31, 35, and 43 used in the above experiments to see if, as predicted, males 31 and 43 had more similar alleles than did male 35. We found five different alleles in these three males. The alleles in male 31 differed from those in male 43 by 7–9 amino acid differences (at the 128 positions of the 5' gene region). Alleles in male 35 were much different, showing 8–12 amino acid differences to male 31 and 17–22 differences to male 43. These results are preliminary but suggest it may be possible to relate fertilization differences to the genetics of gamete recognition loci.

Mating as a Collaboration or a Conflict?

The combination of molecular data, fertilization variation, and modeling results summarized above suggests that mating interactions may be capable of evolving quickly even in behaviorally simple organisms like sea urchins. An important aspect of these interactions is that male and female gametes must combine, and this will usually depend on the interaction of separate loci expressed in males versus females. Such interactions have been considered before in the context of loci involved in mate signaling (e.g., the *period* locus in *Drosophila*; Wheeler et al., 1991) or sexual selection (Kirkpatrick, 1982; Wu, 1985), but the number of loci involved in complex behavioral signals is unknown. Moreover, signaling loci are probably expressed both in males and females, and probably have pleiotropic effects. As a result their evolution may be constrained. By contrast, gamete recognition probably involves fewer loci, including many that are expressed only in eggs or sperm. Thus, there is an increased opportunity to study the evolutionary dynamics of loci involved in mate signaling.

Interacting loci expressed only in males versus females may co-evolve, and this co-evolution might greatly increase the tempo of change of mate signals. To date, two types of co-evolutionary interactions have been suggested: collaboration, in which the male and female loci must match for successful reproduction (Wu, 1985), and conflict, in which male and female loci code for traits that reduce the fitness of the mating partner (see Rice, this volume). Rapid evolution in collaborative schemes depend on an intrinsically high mutation rate in at least one locus (Wu, 1985). In conflict schemes, there is a race between loci akin to the evolutionary arms race between predator and prey that drives rapid functional differentiation (Rice, this volume).

For systems with external fertilization, conflict between egg and sperm might reside in the basic asymmetry of the fate of gametes in the open ocean. A sperm is likely to encounter only a single egg: if it does not successfully fertilize that one egg, it dies without issue. Thus, selection will be for rapid egg entry once an egg is en-

countered. There will be little or no selection for sperm to give up attempts to fuse with an egg (even if the egg is already fertilized) unless there is a high probability that the sperm that was successful came from the same male. For free-spawning animals in which many adults spawn at the same time (e.g., Pennington, 1985; Babcock et al., 1986), this probablility may be low.

By contrast, an egg may encounter many sperm and, because polyspermy usually leads to the death of the zygote, mechanisms by which eggs can slow sperm entry might evolve. There also may be selection to slow sperm entry such that eggs can somehow choose among sperm. These ideas are speculative, but many marine invertebrates have elaborate, multiple blocks to polyspermy, suggesting that the problems of multiple sperm entry represent a potent evolutionary force. This asymmetry between egg and sperm may be strong enough to lead to the co-evolutionary race envisioned by Rice (this volume).

The Ecology of Species Formation: The View from the Egg

These arguments suggest that there is a link between the ecology of fertilization and the evolution of gamete interactions. Species whose eggs are fertilized in the presence of abundant sperm may be subject to different evolutionary pressures than are those whose eggs are sperm limited. One prediction is that taxa with highly evolved blocks to polyspermy (in which eggs presumably see many sperm) will have gamete recognition loci subject to positive selection but that taxa without such blocks will not. Future research may reveal more details about the interactions between fertilization ecology and gamete evolution.

These ideas focus on the role that recognition isolation may play in the evolution of reproductive isolation. The recognition process is a complex one in most organisms, may be subject to a wide variety of selective pressures, and may be controlled by a wide variety of different genetic mechanisms. For organisms that use protein-protein interactions on cell surfaces to recognize potential mates (e.g., mating type proteins in single celled ciliates [Miceli et al., 1992] or gamete recognition in animals [Hardy and Garbers, 1994; Rosati, 1995; Snell, 1990]), it may be possible to visualize some of the evolutionary forces acting on recognition isolation by examining particular protein-coding genes in detail. These studies will complement those on opportunity isolation or those that focus on the genetic control of postmating isolation.

Acknowledgments   I thank D. Howard, H. Lessios, E. Metz, and V. Vacquier for illuminating discussions about fertilization. H. Lessios and A. Martin provided many fine suggestions about how to improve the manuscript. This work was supported by grants from the NSF.

References

Babcock, R. C., G. D. Bull, P. L. Harrison, A. J. Heyward, J. K. Oliver, C. C. Wallace, and B. I. Willis. 1986. Synchronous spawning of 105 scleractinian coral species on the Great Barrier Reef. Mar. Biol. 90:379–394.

Baker, C. S., A. Perry, J. L. Bannister, M. T. Weinrich, R. B. Abernethy, J. Calambokidis, J. Lien, R. H. Lambertsen, J. Urban-Ramirez, O. Vasquez, P. J. Clapham, A. Alling, S. J. O'Brien, and S. R. Palumbi. 1993. Abundant mitochondrial DNA variation and world-wide population structure in humpback whales. Proc. Natl. Acad. Sci. (USA) 90:8239–8243.

Bush, G. L. 1975. Modes of animal speciation. Annu. Rev. Ecol. Syst. 6:339–364.

Butlin, R. 1989. Reinforcement of premating isolation, pp. 158–179. In D. Otte and J. A. Endler (eds.), Speciation and Its Consequences. Sinauer, Sunderland, Mass.

Claridge, M. F., and J. C. Morgan. 1993. Geographic variation in acoustic signals of the planthopper, Nilaparvata bakeri (Muir), in Asia: specie recognition and sexual selection. Biol. J. Linn Soc. Lond. 48:267–281.

Cornish, E. C., M. A. Anderson, and A. E. Clark. 1988. Molecular aspects of fertilization in flowering plants. Annu. Rev. Cell. Biol. 4:209–228.

Dagley, J. R., R. K. Butlin, and G. M. Hewitt. 1994. Divergence in morphology and mating signals, and assortative mating among populations of Chorthippus parallelus (Orthoptera: Acrididae). Evolution 48:1202–1210.

Emlet, R. B. 1995. Developmental mode and species geographic range in regular sea urchins (Echinodermata: Echinoidea). Evolution 49:476–489.

Gao, B., L. E. Klein, R. J. Britten, and E. H. Davidson. 1986. Sequence of mRNA coding for bindin, a species-specific sea urchin sperm protein required for fertilization. Proc. Natl. Acad. Sci. (USA) 33:8634–8638.

Glabe, C. G., and V. C. Vacquier. 1977. Species-specific agglutination of eggs by bindin isolated from sea urchin sperm. Nature 267:836–838.

Graves, J. E., and J. R. McDowell. 1995. Inter-ocean genetic distance of istiophorid billfishes. Mar. Biol. 122:93–203.

Grosberg, R. 1988. The evolution of allorecognition specificity in clonal invertebrates. Q. Rev. Biol. 63:377–412.

Grosberg, R. K. 1987. Limited dispersal and proximity-dependent mating success in the sessile colonial ascidian Botryllus schlosseri. Evolution 41:372–384.

Hansen, T. A. 1980. Influence of larval dispersal and geographic distribution on species longevity in neogastropods. Paleobiology 6:193–207.

Hansen, T. A. 1983. Modes of larval development and rates of speciation in early tertiary neogastropods. Science 220:501–502.

Hardy, D. M., and D. L. Garbers. 1994. Species-specific binding of sperm proteins to the extracellular matrix (zona pellucida) of the egg. J. Biol. Chem. 269:19000–19004.

Hochachka, P. W., and G. N. Somero. 1973. Strategies of Biochemical Adaptation. W. B. Saunders, N.Y.

Hourigan, T. F., and E. S. Reese. 1987. Mid-ocean isolation and the evolution of Hawaiian reef fishes. TREE 2:187–191.

Jablonski, D. 1986. Larval ecology and macroevolution in marine invertebrates. Bull. Mar. Sci. 39:565–587.

Johnson, P. A., F. C. Hoppensteadt, J. A. Smith, and G. L. Bush. 1996. Conditions for sympatric speciation: a diploid model incorporating habitat fidelity and non-habitat assortative mating. Evol. Ecol. 10:187–205.

Kelso, D. 1970. A Comparative Morphological and Ecological Study of Two Species of Sea Urchins, Genus Echinometra in Hawaii. Unpublished PhD dissertation, Dept. of Zoology, University of Hawaii.

Kirkpatrick, M. 1982. Sexual selection and the evolution of female choice. Evolution 36:1–12.

Kirkpatrick, M. 1987. Sexual selection by female choice in polygynous animals. Annu. Rev. Ecol. Syst. 18:43–70.

Kirkpatrick, M., and M. Ryan. 1991. The evolution of mating preferences and the paradox of the lek. Nature 350:33–38.

Knowlton, N. 1993. Sibling species in the sea. Annu. Rev. Ecol. Syst. 24:189–216.

Kohn, A. J., and F. E. Perron. 1994. Life History and Biogeography: Patterns in Conus. Clarendon Press, Oxford.

Lee, Y.-H., T. Ota, and V. D. Vacquier. 1995. Positive selection is a general phenomenon in the evolution of abalone sperm lysin. Mol. Biol. Evol. 12:231–238.

Lessios, H. A. 1984. Possible pre-zygotic reproductive isolation in sea urchins separated by the Isthmus of Panama. Evolution 38:1144–1148.

Matsuoka, N., and T. Hatanaka. 1991. Molecular evidence for the existence of four sibling species within the sea-urchin, Echinometra mathaei in Japanese waters and their evolutionary relationships. Zool. Sci. 8:121–133.

Metz, E. C., R. E. Kane, H. Yanagimachi, and S. R. Palumbi. 1994. Specificity of gamete binding and early stages of fusion in closely related sea urchins (genus Echinometra). Biol. Bull. 187:23–34.

Metz, E. C., and S. R. Palumbi. 1996. Positive selection and sequence rearrangements generate extensive polymorphism in the gamete recognition protein bindin. Mol. Biol. Evol. 13:391–406.

Miceli, C., A. LaTerza, R. A. Bradshaw, and P. Luporini. 1992. Identification and structural characterization of cDNA clone encoding a membrane-bound form of the polypeptide pheromone Er-1 in the ciliate protozoan Euplotes raikovi. Proc. Natl. Acad. Sci. (USA) 89:1988–1992.

Miller, D. J., M. B. Macek, and B. D. Shur. 1992. Complementarity between sperm surface ß-1,4-galactosyltransferase and egg-coat ZP3 mediates sperm-egg binding. Nature 357:589–593.

Minor, J. E., D. R. Fromson, R. J. Britten, and E. H. Davidson. 1990. Comparison of the bindin proteins of Strongylocentrotus fransiscanus, S. purpuratus, and Lytechinus variegatus: sequences involved in the species specificity of fertilization. Mol. Biol. Evol. 8:781–795.

Müller, W. E. G. 1995. Molecular phylogeny of metazoa [animals]: monophyletic origin. Naturwiss. 82:321–329.

Nishihira, M., Y. Sato, Y. Arakaki, and M. Tsuchiya. 1991. Ecological distribution and habitat preference of four types of the sea urchin Echinometra mathaei on the Okinawan coral reefs, pp. 91–104. In T. Yanagisawa, I. Yasumasu, C. Oguro, N. Suzuki, and T. Motokawa (eds.), Biology of Echinodermata. Balkema Press, Rotterdam.

Palumbi, S. R. 1992. Marine speciation on a small planet. Trends Evol. Ecol. 7:114–118.

Palumbi, S. R. 1994. Reproductive isolation, genetic divergence, and speciation in the sea. Annu. Rev. Ecol. Syst. 25:547–572.

Palumbi, S. R. 1995. Using genetics as an indirect estimator of larval dispersal, pp. 369–387. In L. McEdward (eds.), Ecology of Marine Invertebrate Larvae. CRC Press, Boca Raton, Fla.

Palumbi, S. R. 1996a. Macrospatial genetic structure and speciation in marine taxa with high dispersal abilities, pp. 101–117. In J. Ferraris and S. R. Palumbi (eds.), Molecular Zoology: Advances, Strategies and Protocols. Wiley, New York.

Palumbi, S. R. 1996b. What can molecular genetics contribute to marine biogeography? An urchin's tale. J. Exp. Mar. Biol. Ecol. 203:75–92.

Palumbi, S. R., and E. Metz. 1991. Strong reproductive isolation between closely related tropical sea urchins (genus Echinometra). Mol. Biol. Evol. 8:227–239.

Pennington, J. T. 1985. The ecology of fertilization of echinoid eggs: the consequences of sperm dilution, adult aggregation, and synchronous spawning. Biol. Bull. 169:417–430.

Potts, W., and E. K. Wakeland. 1993. The evolution of MHC genetic diversity: a tale of incest, pestilence and sexual preference. Trends Genet. 9:408–412.

Rice, W. R. 1987. Speciation via habitat specialization: the evolution of reproductive isolation as a correlated character. Evol. Ecol. 1:301–314.

Rosati, F. 1995. Sperm-egg interactions during fertilization in invertebrates. Bull. Zool. 62:323–334.

Russo, A. R. 1977. Water flow and the distribution and abundance of echinoids (genus Echinometra) on an Hawaiian reef. Austr. J. Mar. Freshwater Res. 28:693–702.

Ryan, M. J., and W. Wilczynski. 1988. Coevolution of sender and reciever: effect on local mate preference in cricket frogs. Science 240:1786–1789.

Shaw, A., D. E. McRee, V. D. Vacquier, and C. D. Stout. 1993. The crystal structure of lysin, a fertilization protein. Science 262:1864–1867.

Slatkin, M. 1982. Pleiotropy and parapatric speciation. Evolution 36:263–270.

Snell, W. J. 1990. Adhesion and signalling during fertilization in multicellular and unicellular organisms. Curr. Biol. 2:821–832.

Strathmann, M. F. 1987. Reproduction and Development of Marine Invertebrates of the Northern Pacific Coast. University of Washington Press, Seattle.

Strathmann, R. R. 1978. The length of pelagic period in echinoderms with feeding larvae from the northeast Pacific. J. Exp. Mar. Biol. Ecol. 34:23–27.

Tsuchiya, M., and M. Nishihira. 1984. Ecological distribution of two types of the sea urchin, Echinometra mathaei (Blainville), on Okinawa Reef Flat. Galaxea 3:131–143.

Uehara, T., and M. Shingaki. 1985. Taxonomic studies in the four types of the sea urchin, Echinometra mathaei, from Okinawa, Japan. Zool. Sci. 2:1009.

Uehara, T., M. Shingaki, and K. Taira. 1986. Taxonomic studies in the sea urchin, genus Echinometra, from Okinawa and Hawaii. Zool. Sci. 3:1114.

Vacquier, V. D., K. R. Carner, and C. D. Stout. 1990. Species specific sequences of abalone lysin, the sperm protein that creates a hole in the egg envelope. Proc. Natl. Acad. Sci. (USA) 87:5792–5796.

Vacquier, V. D., and Y.-H. Lee. 1993. Abalone sperm lysin: an unusual mode of evolution of a gamete recognition protein. Zygote 1:181–196.

Vacquier, V. D., and G. W. Moy. 1977. Isolation of bindin: the protein responsible for adhesion of sperm to sea urchin eggs. Proc. Natl. Acad. Sci. (USA) 74:2456–2460.

Vacquier, V. D., W. J. Swanson, and M. Hellberg. 1995. What have we learned about sea urchin sperm bindin? Dev. Growth Differ. 37:1–10.

Veron, J. E. N. 1993. Monograph Series: A Biogeographic Database of Hermatypic Corals. Australian Institute of Marine Science, Townsville.

Ward, R. D., N. G. Elliot, P. M. Grewe, and A. J. Smolenski. 1994. Allozyme and mitochondrial DNA variation in yellowfin tuna (Thunnus albacares) from the Pacific Ocean. Mar. Biol. 118:531–539.

Wheeler, D. A., C. P. Kyriacou, M. L. Greenacre, Q. Yu, J. E. Rutila, M. Rosbash, and J. C. Hall. 1991. Molecular transfer of a species-specific behavior from Drosophila simulans to Drosophila melanogaster. Science 251:1082–1085.

Wu, C.-I. 1985. A stochastic simulation study on speciation by sexual selection. Evolution 39:66–82.

21

The Evolution of Barriers to Fertilization between Closely Related Organisms

Daniel J. Howard
Marta Reece
Pamela G. Gregory
Jiming Chu
Michael L. Cain

Reproductive barriers between closely related species represent the here and now of speciation. They are an aspect of species and speciation that are not shrouded in the uncertainty of history, and evolutionists have long been interested in their characterization. Early in the Modern Synthesis, interest in isolating barriers was fueled by the perception that they were among the most important biological properties of species (Dobzhansky 1937, 1951; Mayr 1942, 1963) and that their description could aid in delimiting species and provide insight into the process and genetics of speciation.

Enthusiasm for studies of isolating barriers waned during the 1960s and 1970s as interest in describing new species declined and the conclusion emerged that species were typically isolated by multiple barriers (Mayr 1963; Dobzhansky 1970; Grant 1985). The complexity of reproductive isolation made its study seem rather futile—what trait should one concentrate on if multiple traits isolated species? Moreover, behavioral traits seemed to be especially important in isolating closely related animals (Mayr 1963), and such traits were perceived as intransigent to genetic analysis. Many evolutionists turned their attention to molecular tools, such as protein electrophoresis, in the hope that new insights into speciation could emerge from studies of the genetic differences between closely related species. This hope soon foundered on the bewildering diversity of results reported by various laboratories (Johnson and Selander 1971; Ayala et al. 1974; Johnson et al. 1977; Shaklee and Tamaru 1977; Johnson 1978; Zimmerman et al. 1978; Craddock and Johnson 1979; Ryman et al. 1979) and by the growing conviction that the genetics of speciation is not simply some measure of genetic differentiation between closely related species; rather, the genetics of speciation is the genetics of traits responsible for reproductive isolation between closely related species (Templeton 1981; Bush and Howard 1986; Howard 1993). As a consequence of this conviction, evolutionists have now redoubled their efforts to develop model systems of two or more closely related taxa in which the traits responsible for reproductive isolation can be identified and studied (Dawley 1986; Faden 1983; Heady and Denno 1991; Landolt and Heath 1987; Liebhold and Volney 1984; Macior 1983; Markow et al. 1983; Markow 1991; Morrison et al. 1994).

The development of model systems has been spurred on by the development of new techniques to detect molecular variation, most notably restriction fragment length polymorphism (RFLP) and random amplification of polymorphic DNA (RAPD). To resolve the presence of genes affecting a particular trait, one must work with taxa having many mapped markers, so that one can follow the segregation of chromosome regions and phenotypes in linkage studies. Until the advent of RFLP and RAPD techniques, such markers existed in only a few groups of organisms, such as *Drosophila*. However, RFLP and RAPD markers can be rapidly generated and mapped in virtually any group of organisms. Moreover, new statistical methods have considerably simplified and sped up the identification and mapping of loci affecting quantitative traits, including behavioral traits (Lander and Botstein 1989). These developments mean that the genetic basis of most traits in most organisms can be understood and that a true genetics of reproductive isolation, one extending beyond the genetics of sterility in *Drosophila* (e.g., Coyne 1984, 1985; Coyne and Kreitman 1986; Naveira and Fontdevila 1986, 1991; Orr 1989, 1992; Naveira 1992; Palopoli and Wu 1994), can come about.

It is important to note that the identification of traits responsible for reproductive isolation serves as more than a prelude to genetic studies. An understanding of these traits provides insight into the role of geographic isolation (or lack of it) in speciation as well as into the forces (natural selection, sexual selection, random drift) that drive the divergence of populations.

Barriers to Fertilization

Perhaps the biggest surprise to emerge from recent studies of reproductive barriers among terrestrial organisms has been evidence suggesting that postinsemination or postpollination barriers to fertilization play an important role in isolating closely related taxa (Howard and Gregory 1993; Arnold et al. 1993). Such barriers were not unknown to earlier evolutionists (Dobzhansky 1951; Mayr 1963), but they were considered to be of importance primarily in marine animals that do not interact behaviorally prior to gamete release. Behavioral differences that give rise to premating barriers were seen as more important in terrestrial animals (Mayr 1963), and differences in habitat and pollinators that give rise to prepollination barriers were regarded as more important in plants (Stebbins 1950).

These views began to shift when a series of sperm and pollen competition experiments between closely related terrestrial taxa uncovered evidence of conspecific sperm or pollen precedence (Hewitt et al. 1989; Bella et al. 1992; Howard and Gregory 1993; Arnold et al. 1993; Gregory and Howard 1994; Wade et al. 1994; Rieseberg et al. 1995). The results suggested that interactions between pollen and stigma, sperm and the female reproductive tract, or sperm and eggs can evolve as quickly as behavioral or ecological interactions, and may give rise to the primary barrier isolating closely related taxa. The importance of such barriers went unnoticed for so long because their operation was masked by the design of laboratory hybridization experiments (Howard and Gregory 1993; Gregory and Howard 1994) and because evolutionists failed to conduct interspecific sperm and pollen competition experiments.

Although the results from the recent gamete competition experiments are compelling, it is important to keep in mind that the number of such investigations is small and few have yet progressed beyond the relatively straightforward demonstration that conspecific sperm outcompetes heterospecific sperm for fertilizations. Thus, much remains to be done before evolutionists reach an understanding of the general significance of barriers to fertilization in isolating closely related species.

In the remainder of this chapter, we describe work in our laboratories that led us to suspect that the ground crickets *Allonemobius fasciatus* and *A. socius* are reproductively isolated by a postinsemination barrier to fertilization, and we report the outcomes of experimental tests of the isolating potential of conspecific sperm precedence. The results demonstrate that conspecific sperm precedence can severely limit gene flow between closely related species, even when one species is much less abundant than the other. Finally, we close with an examination of the factors that may account for the rapid evolution of barriers to fertilization.

Allonemobius fasciatus and *A. socius*

The ground crickets *Allonemobius fasciatus* and *A. socius* are small, ground-dwelling, morphologically indistinguishable, sister species in the subfamily Nemobiinae. They inhabit short grassland areas of eastern North America, particularly low-lying pastures and the edges of ponds and streams. *Allonemobius fasciatus* occurs from southeastern Canada to the northeastern and north-central United States, whereas populations of *A. socius* are found in the southeastern and south-central United States (Alexander and Thomas 1959; Howard 1983; Howard and Furth 1986). *Allonemobius fasciatus* and *A. socius* meet and hybridize, to a limited extent, in a contact zone of varying width that stretches from New Jersey at least as far west as Illinois (Howard and Waring 1991). Within mixed populations (populations containing members of both species), pure species individuals usually predominate and individuals classified as hybrid typically possess genotypes characteristic of backcrosses (Howard 1986; Howard and Waring 1991). The percentage of individuals with genotypes characteristic of F_1 hybrids is typically 1–7%. Thus, reproductive isolation between the two species in areas of contact appears quite strong, although not complete.

The barrier to gene flow responsible for this isolation eluded our best efforts at identification for a number of years. The two species are not isolated by habitat differences or phenological differences (Howard et al. 1993). They occur in the same low grassland areas in the zone of overlap, and their life histories are very similar. They both overwinter in the egg stage, and adults are abundant in the field from late July through September. Females mate multiply both in the field and in the laboratory, and females from single species populations typically produce progeny sired by more than one male (Gregory and Howard 1996). Slight calling-song differences exist (Benedix and Howard 1991; Veech et al. 1996), but females from sympatric and allopatric populations do not exhibit preferential movement toward conspecific songs (Doherty and Howard 1996). Postzygotic barriers are weak, as well. Individuals with genotypes indicative of mixed ancestry are quite viable in the field (Howard et al. 1993), and laboratory hybridization (with multiple mating allowed) results in the formation of many viable, fertile offspring (Gregory and Howard 1993).

The first break in solving the puzzle presented by these crickets emerged from sperm competition experiments

(Howard and Gregory 1993; Gregory and Howard 1994). In the experiments, an individual female of *A. fasciatus* or *A. socius* was mated once to each of two males: two conspecifics, a conspecific followed by a heterospecific, a heterospecific followed by a conspecific, or two heterospecifics. The results were quite clear. When females were mated to two conspecifics, the sperm of both males fertilized eggs. However, when one of the males was heterospecific, the conspecific male fertilized the vast majority of the eggs. This was true regardless of the order of matings and regardless of whether the female was from a pure population or a mixed population.

The fertilization advantage of conspecific males could not be attributed to lack of sperm transfer in heterospecific matings. The mean time from the beginning of copulation to spermatophore removal by the female did not differ significantly between conspecific and heterospecific matings. Moreover, dissections of *A. fasciatus* and *A. socius* females subsequent to a single heterospecific mating revealed the presence of numerous sperm in all. However, sperm appeared to be less motile in the spermathecae of heterospecific females than in the spermathecae of conspecific females (Gregory and Howard 1994).

Thus, a postinsemination barrier to fertilization exists between *A. fasciatus* and *A. socius*. The barrier eluded early detection because we did not incorporate sperm competition experiments into our hybridization studies, and hence could not perceive the fertilization advantage enjoyed by conspecific sperm.

A postinsemination barrier to fertilization should lead to positive assortment even between taxa that mate at random. But can such a barrier acting alone explain the genetic isolation between *A. fasciatus* and *A. socius* and other closely related species? This is the question that has guided much of our recent work.

Multiple Mating in *Allonemobius*

Strong conspecific sperm precedence would serve most effectively as a reproductive barrier when two species occur with equal abundance (Arnold et al. 1993) and when females mate repeatedly over a relatively short time period (Gregory and Howard 1994). Under such circumstances, a female is likely to mate at least once with a conspecific prior to laying most of her eggs, and therefore the majority of her progeny will have a conspecific father. As the abundance of a species decreases, so too does the chance that a female of that species will encounter and mate with a conspecific male. At some threshold of relative abundance, conspecific sperm precedence, by itself, will not serve as an effective barrier to gene flow between two taxa. The threshold level will depend on the strength of the conspecific sperm precedence, the intensity of positive assortative mating exhibited by females and males, and the number of times males and females

mate. Thus, an assessment of the isolating potential of conspecific sperm precedence calls for, among other things, information on multiple mating.

A recent study of the genotypes of offspring produced by field collected females demonstrated that females mate more than once in the field (Gregory and Howard 1996), but the level of resolution of the study was too low to provide insight into the number of times a female mates. Because the dense grassland habitats of *A. fasciatus* and *A. socius* preclude field studies of mating frequency, we further examined this aspect of life history in the laboratory by adapting the techniques of Burpee and Sakaluk (1993). Ten-day post-eclosion males and females were randomly paired (one pair per cage) in transparent plastic cages. Mating behavior was monitored continuously over a one-week period via time-lapse video photography. Tables 21.1 and 21.2 show the results of this multiple mating study. On average, pairs of *A. socius* mated 14.3 times, and pairs of *A. fasciatus* mated 10.9 times. Clearly, females of *A. fasciatus* and *A. socius* mate frequently under laboratory conditions, a finding that enhances the possibility that a barrier to fertilization can serve as a significant barrier to gene flow.

Patterns of Fertilization

Understanding patterns of sperm utilization by females is critical to understanding the isolating potential of conspecific sperm precedence. Obviously, conspecific sperm precedence would most effectively isolate two species if a single conspecific insemination ensured conspecific fertilization, regardless of the number of times a female mated with heterospecific males. When females of *A. fasciatus* and *A. socius* are mated twice, once to a conspecific and once to a heterospecific, conspecific sperm fertilizes the vast majority of eggs, regardless of the order of matings (Howard and Gregory 1993; Gregory and Howard 1994). What happens under other mating regimes? In particular, what are the patterns of fertilization when a female is mated once to a conspecific and more

Table 21.1. Number of matings over the course of a week in four cages, each containing one male and one female of *A. socius*.

Date	Pair 1	Pair 2	Pair 3	Pair 4	Mean
7/26/95	21	16	8	18	15.8
8/10/95	16	26	18	15	18.8
8/18/95	19	21	16	6	15.5
8/28/95	12	10	15	12	12.2
9/12/95	15	12	15	13	13.8
9/25/95	12	15	14	15	14.0
9/29/95	12	9	7	13	10.2
				Overall	14.3

Each row represents four different pairs.

Table 21.2. Number of matings over the course of a week in four cages, each containing one male and one female of *A. fasciatus*.

Date	Pair 1	Pair 2	Pair 3	Pair 4	Mean
7/27/95	13	14	20	20	16.8
8/29/95	14	18	12	17	15.2
9/7/95	12	14	10	12	12.0
9/15/95	10	2	5	6	5.8
9/20/95	1	5	8	8	5.5
10/6/95	7	12	11	—	10.0
				Overall	10.9

Each row represents four different pairs.

than once to a heterospecific?

We have been exploring these questions in the laboratory through sperm competition experiments. In the experiments, field-collected nymphs of each species were reared to adulthood in the laboratory, and two virgin males were mated to a virgin female in one of the following sequences: (1) one conspecific mating followed by two heterospecific matings; (2) one conspecific mating followed by three heterospecific matings; (3) one conspecific mating followed by four heterospecific matings, and so on. We have also run this protocol in reverse, allowing the heterospecific male to mate multiply with the female before mating her with a conspecific. However, few of the progeny of these latter matings have been analyzed thus far. Each series of matings took place over the course of 5 days and the maximum interval between two matings was 72 hours.

Tables 21.3 and 21.4 exhibit the proportion of conspecific versus heterospecific offspring produced by females involved in the various mating sequences. There was considerable variation in the results, but in general, it appears that a single conspecific insemination ensures conspecific fertilization over a wide range of multiple heterospecific inseminations. For example, when an *A. fasciatus* female was mated first to a conspecific male and then five times to a heterospecific male, 93% of the eggs were fertilized by the conspecific male. More re-

Table 21.3. Patterns of fertilization when females of *A. socius* were mated once to a conspecific male and subsequently more than once to a heterospecific male.

Number of Matings			Mean Proportions of Progeny	
1st Male	2nd Male	N	Conspecific	Hybrid
1	2	9	0.89	0.11
1	3	6	0.91	0.09
1	4	3	0.99	0.01
1	5	2	1.00	0.00
1	6	1	0.88	0.12

Table 21.4. Patterns of fertilization when females of *A. fasciatus* were mated once to a conspecific male and subsequently more than once to a heterospecific male.

Number of Matings			Mean Proportions of Progeny	
1st Male	2nd Male	N	Conspecific	Hybrid
1	2	3	0.96	0.04
1	3	2	0.98	0.02
1	4	5	0.66	0.34
1	5	3	0.93	0.07

markably, five *A. socius* females mated once to a conspecific male and then four or five times to a heterospecific male produced fewer than 1% hybrid offspring. It appears that conspecific sperm outcompetes heterospecific sperm for fertilizations even when heterospecific sperm are much more abundant in the female reproductive tracts of *A. fasciatus* and *A. socius*.

A Laboratory Test of the Isolating Potential of a Barrier to Fertilization

Because the operation of a postinsemination barrier to fertilization should not depend on environmental interactions or an ecological context, the isolating potential of the barrier can be tested directly in the laboratory. Howard et al. (1998) did so in a series of three population cage experiments.

In the first set of experiments, four replicate populations were established, each with 10 virgin males and 10 virgin females of both species (40 individuals total per cage). Individuals were marked so that observers could distinguish between the species. We allowed crickets in each population to mate and oviposit freely for three weeks. Cages were observed several times each day, and the nature of the matings taking place (conspecific or heterospecific) was noted. Because *A. fasciatus* and *A. socius* were in equal frequency in these cages, females of each species were likely to mate often with conspecifics. Hence, conditions in this experiment were optimal for conspecific sperm precedence to serve as a barrier to hybridization.

The second and third sets of experiments examined the effectiveness of the fertilization barrier when one of the species was in the minority. In four replicates, *A. fasciatus* was the minority species (four males and four females versus 16 males and 16 females of *A. socius*) and in the other four replicates *A. socius* was in the minority. Because females of the minority species were likely to mate infrequently with members of their own species, these experiments increased the chance that conspecific sperm precedence could be overwhelmed by repeated matings to heterospecifics. Once again, crickets were allowed to mate and oviposit freely for 3 weeks and the cages were observed several times each day to note the types of matings taking place.

The results of the mating observations are shown in table 21.5. When the two species were in equal abundance, conspecific matings were more frequent than heterospecific matings, but the deviation from random mating expectations did not quite achieve significance. Clearly, a great deal of interspecific mating occurred, and the potential for the production of hybrid offspring was high. When *A. socius* was in the majority, there were significant deviations from random mating expectations. In general, there were more matings between *A. fasciatus* males and females than expected. This is an intriguing result and suggests that females of *A. fasciatus* may more actively seek out conspecific males when they are in the minority in a population. However, interspecific matings were frequent, and in the absence of a postmating reproductive barrier, one would expect females in this cage to produce many hybrid progeny. Finally, when *A. fasciatus* was in the majority, there were no significant deviations from random mating expectations, and again, one would expect to find a large number of hybrids among the progeny of females.

The species composition (determined via protein electrophoresis; Howard 1986; Howard and Waring 1991) of the progeny of each cage is summarized in table 21.6. Hybrids were rare in the F_1 generation in 11 of the 12

population cages—ranging from 0 to 6%. These levels are consistent with the percentage of F_1 hybrids found in natural populations. We pooled data from the three experiments and found that the frequency of the parental species did not affect the number of hybrid offspring (ANOVA, $F_{2,9} = 0.74$, p = 0.50); thus, the effectiveness of conspecific sperm precedence as a barrier to hybridization did not break down when *A. fasciatus* or *A. socius* were rare. The one exception to this pattern was cage C1, a cage with *A. socius* in the minority: 19% of the progeny in this cage were hybrids. We have no ready explanation for this result other than to suggest that one or more of the *A. socius* females did not mate with any *A. socius* males and hence produced a large number of hybrid offspring. Another general pattern observed in these experiments was the greater abundance of *A. socius* among the progeny than *A. fasciatus*. This result can be attributed to the greater fecundity and survivorship of *A. socius* females (Gregory and Howard 1993). Overall, the results in table 21.6 provide strong evidence that conspecific sperm precedence can serve as an effective barrier to hybridization between *A. fasciatus* and *A. socius*.

Why Do Barriers to Fertilization Evolve Rapidly?

It is now clear that conspecific sperm precedence is a major—probably the major—factor responsible for repro-

Table 21.5. Mating patterns in the mixed-species cage experiments.

Cage	*fas* Male *fas* Fem	*fas* Male *soc* Fem	*soc* Male *soc* Fem	*soc* Male *fas* Fem	χ^2	Significance
A1	10	10	15	10	1.67	P>0.50
A2	15	9	10	10	2.00	P>0.25
A3	16	4	12	8	8.00	P<0.05
A4	11	8	10	6	1.68	P>0.60
Total	52	31	47	34		
					Pooled 7.74	P>0.05
B1	5	4	15	0	20.86	P<0.001
B2	3	3	24	5	3.21	P>0.30
B3	3	6	11	3	7.23	P>0.05
B4	0	3	11	3	0.74	P>0.80
Total	11	16	61	11		
					Pooled 14.08	P<0.005
C1	18	4	1	3	0.44	P>0.90
C2	21	1	2	1	7.06	P>0.05
C3	14	3	3	5	4.75	P>0.10
C4	20	4	0	8	3.16	P>0.30
Total	73	12	6	17		
					Pooled 2.48	P>0.40

In cages A1–A4, the two species were in equal abundance. In cages B1–B4, *A. socius* was in the majority (80% of the population). In cages C1–C4, *A. fasciatus* was in the majority (80% of the population).

Table 21.6. Species composition of the progeny of the 12 population cages.

Cage	Parental Composition	N	Proportion of Progeny		
			A. socius	*A. fasciatus*	Hybrid
A1	Equal abund.	247	0.652	0.287	0.061
A2	Equal abund.	196	0.714	0.260	0.025
A3	Equal abund.	157	0.790	0.197	0.013
A4	Equal abund.	164	0.817	0.146	0.036
B1	0.80 *socius*	131	0.901	0.084	0.015
B2	0.80 *socius*	175	0.834	0.154	0.011
B3	0.80 *socius*	177	0.938	0.028	0.033
B4	0.80 *socius*	144	0.958	0.028	0.014
C1	0.80 *fas*	168	0.262	0.548	0.190
C2	0.80 *fas*	146	0.432	0.527	0.041
C3	0.80 *fas*	126	0.206	0.754	0.040
C4	0.80 *fas*	90	0.411	0.589	0.000

ductive isolation between *A. fasciatus* and *A. socius*. This finding, while interesting in and of itself, is made more interesting when combined with other evidence suggesting that barriers to fertilization evolve rapidly between diverging taxa. The most direct evidence comes from studies of sperm or pollen competition between closely related species, which almost invariably report conspecific sperm or pollen precedence (Nakano 1985; Katakura 1986; Katakura and Sobu 1986; Hewitt et al. 1989; Bella et al. 1992; Arnold et al. 1993; Wade et al. 1994; Rieseberg et al. 1995). Extremely rapid evolution is also indicated by molecular studies of fertilization proteins, particularly in broadcast spawning marine invertebrates (Minor et al. 1991; Lee and Vacquier 1992; Palumbi 1992; Shaw et al. 1993; Swanson and Vacquier 1995; Metz and Palumbi 1996). Based on the ratio of nonsynonymous to synonymous substitutions in their DNA sequences, these proteins show some of the highest evolutionary rates known. One of the best characterized of all gamete recognition proteins is lysin, the acrosomal sperm protein of abalone that binds to and breaks down the vitelline envelope surrounding eggs. For lysin, the ratio of nonsynonymous to synonymous substitutions for a given pair of species can be as high as 3.4 (Lee and Vacquier 1992). An even more intriguing property of lysin is its propensity to acquire nonconservative substitutions, that is, substitutions that change the class of the amino acid, more rapidly than would be expected by chance (Vacquier and Lee 1993). It appears that selection for novelty is taking place—a puzzling result in a protein whose major functions include binding to a receptor.

On the surface, the rapid divergence of proteins and traits related to fertilization seems paradoxical. After all, doesn't fertilization involve coadaptation between males and females, and shouldn't the traits responsible for

fertilization be subject to strong stabilizing selection? These rhetorical questions reflect an assumption common among evolutionists that the interests of males and females coincide (when it comes to fertilization). Since this assumption leads to a paradox, it seems worthwhile to investigate alternative possibilities.

At least four basic hypotheses, all seemingly capable of explaining the rapid evolution both of reproductive barriers and of proteins and traits related to fertilization, can be formulated: (1) response to pathogens; (2) sexual selection by male–male competition; (3) female choice; and (4) avoidance of polyspermy (fertilization of an egg by more than one sperm). Focusing on the first hypothesis, response to pathogens is known to lead to rapid evolution (Tanaka and Nei 1989; Hughes et al. 1990; Hughes and Nei 1988, 1989), and the surface of the egg as well as the female reproductive tract appears to offer an easily accessible port of entry for disease organisms. The hostility of the female reproductive tract to sperm, as well as the composition of envelopes surrounding eggs of broadcast spawning invertebrates, may have evolved, at least in part, in response to pathogens (Sheldon 1993). The question is whether response to pathogens is the major cause of rapid evolution. While the possibility cannot be ruled out, such a scenario appears unlikely, at least according to some of the available molecular genetic evidence. For instance, lysin is monomorphic within species and highly divergent between species (Lee et al. 1995), suggesting that its receptor is also monomorphic within species. This pattern of variation stands in contrast to that of proteins known to evolve in response to pathogens, such as the MHC complex. These proteins are exceedingly diverse within species (Figueroa et al. 1988; Lawlor et al. 1988).

The second hypothesis, sexual selection by male–male competition with the female serving as a passive, relatively unchanging playing field, appears to be an unlikely explanation for the rapid evolution of reproductive barriers, as well. Male–male interactions may provide an opportunity for interlocus antagonistic coevolution (see Rice, this volume, for a discussion) and hence may explain the rapid evolution of proteins found in the reproductive tissue (testis and accessory gland) of *Drosophila* (Thomas and Singh 1992). However, such changes are unlikely to affect the ability of the sperm of a single male to move through the reproductive tract of a particular female and attach to an egg. If the female reproductive tract or egg surface remain relatively stable over evolutionary time, a locally optimal solution to transport and attachment would be expected to evolve in semen and sperm. Thus, for the keen competition between males, which no doubt exists, to produce a continuous stream of novelties and the evolution of reproductive barriers, the conditions under which the competition occurs would need to change.

The simplest formulation of the female choice hypothesis is the two-locus, two alleles per locus, "sexy sperm" model investigated theoretically by Curtsinger (1991). Curtsinger demonstrated that a more competitive sperm allele sweeps through a population very quickly, while a choosy female allele responds much more slowly and loses any advantage following a sweep. Should there be any cost to selection, the choosy allele will be rapidly lost from the population after a sweep. This form of the female choice hypothesis, therefore, does not appear capable of explaining the rapid evolution of proteins and traits related to fertilization.

However, other more complex scenarios involving female choice may be constructed. Vigorous, highly motile sperm with large amounts of certain proteins such as lysin may be indicative of the general health of the male. General health has been shown to be both sufficiently variable and heritable to underlie sexual selection (Charlesworth 1987). According to this scenario, females evolve a hostile reproductive tract and erect protective armor around their eggs to ensure that their eggs are fertilized by a vigorous spermatozoon, while males adapt to the female reproductive tract and compete with one another for fertilizations. The end result is an antagonistic coevolution between males and females. One of the major problems with this scenario is that the connection between healthy, well-provisioned sperm and the fitness of the resulting offspring has yet to be established.

The polyspermy avoidance hypothesis also assumes a difference between male and female interests. Males are assumed to be under pressure to fertilize, while females, even though they need to have each of their eggs fertilized, must ensure that each is fertilized only once. The potentially lethal effects associated with polyspermy (Wilson 1928; Jaffe and Gould 1985; Brawley 1987; Payne et al. 1994) are presumed to lead to the evolution of hostile female reproductive tracts and the evolution of envelopes around eggs that slow sperm down enough that

only a single spermatozoon makes initial contact with the plasma membrane of an egg. The importance of avoiding polyspermy is evidenced by the fast and slow blocks to polyspermy that arise in most animals after the fusion of egg and sperm plasma membranes (Miller et al. 1993; Goudeau et al. 1994; Kobayashi and Yamamoto 1994).

According to the avoidance of polyspermy hypothesis, an equilibrium solution to the conflict between males and females is difficult to achieve. Males are assumed to be under constant pressure to adapt to the female reproductive tract and eggs, and females are assumed to be under constant pressure to adapt to semen and sperm. Any mutation in males that provides a competitive advantage with regard to fertilization would likely lead to an increase in polyspermy and hence pressure on females to respond. The result is endless coevolution between males and females, and divergence between isolated populations in traits related to fertilization because the coevolutionary process is likely to take off in different directions in different populations.

Distinguishing between the avoidance of polyspermy and the female choice hypotheses will be difficult. First of all, both factors may act in concert. For example, as males increase both the quality and quantity of fertilization proteins, they may sometimes threaten females with polyspermy. Females, in turn, will become effectively more choosy as they protect themselves against this threat.

Second, the predictions of the two hypotheses are much the same and seemingly paradoxical—that conspecific males will outcompete heterospecific males for fertilization because they have had an opportunity to adapt to the reproductive tract of conspecific females or (and this is a less obvious prediction) that heterospecific males will be superior competitors because females have not had an opportunity to adapt to them. Although the two predictions appear contradictory, it is important to note that in the latter case the expectation is not fertilization proceeding normally; instead, the expectation is polyspermy and perhaps other problems. In fact, in the one instance that we know of in which heterospecific sperm appears to outcompete conspecific sperm for fertilizations, the result is polyspermy and abnormal development of the eggs (Gomez and Cabada 1994).

One area where the predictions of the female choice hypothesis and the avoidance of polyspermy hypothesis differ is with regard to monogamous species. If gene expression in sperm is generally lacking, female choice based on genetic differences between sperm can operate only if a female mates with more than one male. The female choice hypothesis would therefore predict slower evolution of barriers to fertilization in monogamous species. Polyspermy avoidance, on the other hand, would remain as important in monogamous as in polyandrous species. Studies of conspecific sperm precedence among closely related species characterized by monogamous mating systems, such as termites, may therefore have the potential to falsify one of the hypotheses. Unfortunately for this proposed test of the hypotheses, recent evidence indicates that at least some genes are expressed in sperm (Eddy et al. 1993; Eddy 1995). If this phenomenon is widespread, in terms of both the taxa and the suite of genes expressed, then female choice can operate even in monogamous species and this proposed test of the hypotheses collapses.

In closing, we should note that certain observations appear, at least on the surface, to be inconsistent with either the polyspermy avoidance hypothesis or the female choice hypothesis. Polyspermy appears to be common in birds (Birkhead 1995), an observation that seems quite damaging to the polyspermy avoidance hypothesis. However, even in birds, only one of the sperm that penetrates the inner perivitelline layer fuses with the female pronucleus. Thus, birds have evolved mechanisms to prevent multiple fusions; it is just that some of these mechanisms operate at a later stage of fertilization than those of other organisms.

A problematic observation for the female choice hypothesis is that utilization of sperm by females in many species is largely determined by mating order (Gwynne 1984; Birkhead 1995). For example, in many spiders, first male sperm precedence occurs (Austad 1984; Eberhard et al. 1994). This means that female choice of sperm may be too constrained in many organisms to serve as a force driving the rapid divergence of traits related to fertilization. Again, a caveat—this observation represents less of a problem for the female choice hypothesis if gene expression is common in sperm.

The importance of barriers to fertilization in the reproductive isolation of closely related organisms has only recently come to the attention of evolutionists, and much remains to be learned about the genetic control of these barriers, their mode of action, and the forces that lead to their evolution. A complete understanding will require concerted effort and the combined expertise of evolutionists, molecular geneticists, and fertilization biologists. But the reward will be great: new insights into that most fascinating of evolutionary processes—speciation.

Acknowledgments We are grateful to Steve Palumbi and Bill Rice for their comments on the manuscript. The work described in this chapter was largely supported by grants from the National Science Foundation, most recently NSF Grant DEB-9407229 to DJH and MLC.

References

Alexander, R. D., and Thomas, E. S. 1959. Systematic and behavioral studies on the crickets of the *Nemobius fasciatus* group (Orthoptera: Gryllidae: Nemobiinae). Ann. Entomol. Soc. Am. 52:591–605.

Arnold, M. L., Hamrick, J. L., and Bennett, B. D. 1993. Interspecific pollen competition and reproductive isolation in *Iris*. J. Hered. 84:13–16.

Austad, A. N. 1984. Evolution of sperm priority patterns in spiders. In R. L. Smith (ed.). Sperm Competition and Evolution of Animal Mating Systems. Orlando, Fla.: Academic Press, pp. 223–249.

Ayala, F. J., Tracey, M. L., Hedgecock, D., and Richmond, R. C. 1974. Genetic differentiation during the speciation process in *Drosophila*. Evolution 28:576–592.

Bella, J. L., Butlin, R. K., Ferris, C., and Hewitt, G. M. 1992. Asymmetrical homogamy and unequal sex ratio from reciprocal mating-order crosses between *Chorthippus parallelus* subspecies. Heredity 68:345–352.

Benedix, J. H., Jr., and Howard, D. J. 1991. Calling song displacement in a zone of overlap and hybridization. Evolution 45:1751–1759.

Birkhead, T. R. 1995. Sperm competition: evolutionary causes and consequences. Reprod. Fertil. Dev. 7:755–775.

Brawley, S. H. 1987. A sodium-dependent, fast block to polyspermy occurs in eggs of fucoid algae. Dev. Biol. 124: 390–397.

Burpee, D. M., and Sakaluk, S. K. 1993. Repeated matings offset costs of reproduction in female crickets. Evol. Ecol. 7:240–250.

Bush, G. L., and Howard, D. J. 1986. Allopatric and non-allopatric speciation; assumptions and evidence. In S. Karlin and E. Nevo (eds.). Evolutionary Processes and Theory. Orlando, Fla.: Academic Press, pp. 411–438.

Charlesworth, B. 1987. The heritability of fitness. In J. W. Bradbury and M. B. Andersson (eds.). Sexual Selection: Testing the Alternatives. New York: Wiley, pp. 21–40.

Coyne, J. A. 1984. Genetic basis of male sterility in hybrids between two closely related species of *Drosophila*. Proc. Natl. Acad. Sci. U.S.A. 81:4444–4447.

Coyne, J. A. 1985. Genetic studies of three sibling species of *Drosophila* with relationship to theories of speciation. Genet. Res. Camb. 46:169–192.

Coyne, J. A., and Kreitman, M. 1986. The evolutionary genetics of two sibling species of *Drosophila*. Evolution 40: 673–691.

Craddock, E. M., and Johnson, W. E. 1979. Genetic variation in Hawaiian *Drosophila*. V. Chromosomal and allozymic diversity in *Drosophila silvestris* and its homosequential species. Evolution 33:137–155.

Curtsinger, J. W. 1991. Sperm competition and the evolution of multiple mating. Am. Nat. 138:93–102.

Dawley, E. M. 1986. Behavioral isolating mechanisms in sympatric terrestrial salamanders. Herpetologica 42:156–164.

Dobzhansky, T. 1937. Genetics and the Origin of Species (1st ed.). New York: Columbia University Press.

Dobzhansky, T. 1951. Genetics and the Origin of Species (3rd ed.). New York: Columbia University Press.

Dobzhansky, T. 1970. Genetics and the Evolutionary Process. New York: Columbia University Press.

Doherty, J. A., and Howard, D. J. 1996. Lack of preference for conspecific calling songs in female crickets. Anim. Behav. 51:981–990.

Eberhard, W. G., Guzmanogomez, S., and Catley, K. M. 1994. Connection between spermathecal morphology and mating systems in spiders. Biol. J. Linn. Soc. 50:197–209.

Eddy, E. M. 1995. 'Chauvinist genes' of male germ cells: gene expression during mouse spermatogenesis. Reprod. Fertil. Dev. 7:695–704.

Eddy, E. M., Welch, J. E., and O'Brien, D. A. 1993. Gene expression during spermatogenesis. In D. de Kretser (ed.). Molecular Biology of the Male Reproductive System. San Diego: Academic Press, pp. 181–232.

Faden, R. B. 1983. Isolating mechanisms among five sympatric species of Aneilema R. Br. (Commelinaceae) in Kenya. Bothalia 14:997–1002.

Figueroa, F., Gunther, E., and Klein, J. 1988. MHC polymorphism pre-dating speciation. Nature 335:167–170.

Gomez, M. I., and Cabada, M. O. 1994. Amphibian cross-fertilization and polyspermy. J. Exp. Zool. 269:560–565.

Goudeau, H., Depresle, Y., Rosa, A., and Goudeau, M. 1994. Evidence by a voltage clamp study of an electrically mediated block to polyspermy in the egg of the ascidian Phallusia mammillata. Dev. Biol. 166:489–501.

Grant, V. 1985. The Evolutionary Process: A Critical Review of Evolutionary Theory. New York: Columbia University Press.

Gregory, P. G., and Howard, D. J. 1993. Laboratory hybridization studies of Allonemobius fasciatus and A. socius (Orthoptera: Gryllidae). Ann. Entomol. Soc. Am. 86:694–701.

Gregory, P. G., and Howard, D. J. 1994. A post-insemination barrier to fertilization isolates two closely related ground crickets. Evolution 48:705–710.

Gregory, P. G., and Howard, D. J. 1996. Multiple mating in natural populations of ground crickets. Entomol. Exp. Appl. 78:353–356.

Gwynne, D. T. 1984. Male mating effort, confidence of paternity, and insect sperm competition. In R. L. Smith (ed.). Competition and the Evolution of Animal Mating Systems. Orlando, Fla.: Academic Press, pp. 117–149.

Heady, S. E., and Denno, R. F. 1991. Reproductive isolation in Prokelisia planthoppers (Homoptera: Delphacidae): acoustic differentiation and hybridization failure. J. Insect Behav. 4:367–390.

Hewitt, G. M., Mason, P., and Nichols, R. A. 1989. Sperm precedence and homogamy across a hybrid zone in the alpine grasshopper Podisma pedestris. Heredity 62:343–353.

Howard, D. J. 1983. Electrophoretic survey of eastern North American Allonemobius (Orthoptera: Gryllidae): evolutionary relationships and the discovery of three new species. Ann. Entomol. Soc. Am. 76:1014–1021.

Howard, D. J. 1986. A zone of overlap and hybridization between two ground cricket species. Evolution 40:34–43.

Howard, D. J. 1993. Small populations, inbreeding, and speciation. In N. W. Thornhill (ed.). The Natural History of Inbreeding and Outbreeding: Theoretical and Empirical Perspectives. Chicago: The University of Chicago Press, pp. 118–142.

Howard, D. J., and Furth, D. G. 1986. Review of the Allonemobius fasciatus (Orthoptera: Gryllidae) complex with the description of two new species separated by electrophoresis, songs, and morphometrics. Ann. Entomol. Soc. Am. 79:472–481.

Howard, D. J., and Gregory, P. G. 1993. Post-insemination signalling systems and reinforcement. Philos. Trans. Roy. Soc. Lond. B 340:231–236.

Howard, D. J., and Waring, G. L. 1991. Topographic diversity, zone width, and the strength of reproductive isolation in a zone of overlap and hybridization. Evolution 45:1120–1135.

Howard, D. J., Waring, G. L., Tibbets, C. A., and Gregory, P. G. 1993. Survival of hybrids in a mosaic hybrid zone. Evolution 47:789–800.

Howard, D. J., Gregory, P. G., Chu, J., and Cain, M. L. 1998. Conspecific sperm precedence is an effective barrier to hybridization between closely related species. Evolution 52:506–511.

Hughes, A. L., and Nei, M. 1988. Pattern of nucleotide substitution at major histocompatibility complex class I loci reveals overdominant selection. Nature 335:167–170.

Hughes, A. L., and Nei, M. 1989. Nucleotide substitution at major histocompatibility complex class II loci: evidence for overdominant selection. Proc. Natl. Acad. Sci. U.S.A. 86:958–962.

Hughes, A. L., Ota, T., and Nei, M. 1990. Positive Darwinian selection promotes charge profile diversity in the antigen-binding cleft of class I major-histocompatibility-complex molecules. Mol. Biol. Evol. 7:515–524.

Jaffe, L. A., and Gould, M. 1985. Polyspermy-preventing mechanisms. In C. B. Metz and A. Monroy (eds.). Biology of Fertilization. New York: Academic Press, pp. 223–250.

Johnson, M. S. 1978. Founder effects and geographic variation in the land snail Theba pisana. Heredity 61:133–142.

Johnson, M. S., Clarke, B., and Murray, J. 1977. Genetic variation and reproductive isolation in Partula. Evolution 31:116–126.

Johnson, W. E., and Selander, R. K. 1971. Protein variation and systematics in kangaroo rats (genus Dipodomys). Syst. Zool. 20:377–405.

Katakura, H. 1986. A further study on the effect of interspecific mating on the fitness in a pair of sympatric phytophagous ladybirds. Kontyu 54:235–242.

Katakura, H., and Sobu, Y. 1986. Cause of low hatchability by the interspecific mating in a pair of sympatric ladybirds (Insecta, Coleoptera, Coccinellidae): incapacitation of alien sperm and death of hybrid embryos. Zool. Sci. 3:315–322.

Kobayashi, W., and Yamamoto, T. S. 1994. Fertilization of the lamprey (Lampetra japonica) eggs: implication of the presence of fast and permanent blocks against polyspermy. J. Exp. Zool. 269:166–176.

Lander, E. S., and Botstein, D. 1989. Mapping Mendelian factors underlying quantitative traits using RFLP linkage maps. Genetics 121:185–199.

Landolt, P. J., and Heath, R. R. 1987. Role of female-produced sex pheromone in behavioral reproductive isolation between Trichoplusia ni (Hubner) and Pseudoplusia includens (Walker) (Lepidoptera: Noctuidae, Plusiinae). J. Chem. Ecol. 13:1005–1018.

Lawlor, D. A., Ward, F. E., Ennis, P. D., Jackson, A. P., and Parham, P. 1988. HLA-A and B polymorphisms predate the divergence of humans and chimpanzees. Nature 335: 268–271.

Lee, Y., and Vacquier, V. D. 1992. The divergence of species-specific abalone sperm lysins is promoted by positive Darwinian selection. Biol. Bull. 182:97–104.

Lee, Y.-H., Ota, T., and Vacquier, V. D. 1995. Positive selection is a general phenomenon in the evolution of abalone sperm protein. Mol. Biol. Evol. 12:231–238.

Liebhold, A. M., and Volney, W. J. A. 1984. Effect of temporal factors on reproductive isolation between *Choristoneura occidentalis* and *C. retiniana* (Lepidoptera: Tortricidae). Can. Entomol. 116:991–1005.

Macior, L. W. 1983. The pollination dynamics of sympatric species of *Pedicularis* (Scrophulariaceae). Am. J. Bot. 70:844–853.

Markow, T. A. 1991. Sexual isolation among populations of *Drosophila mojavensis*. Evolution 45:1525–1529.

Markow, T. A., Fogelman, J. C., and Heed, W. B. 1983. Reproductive isolation in Sonoran Desert *Drosophila*. Evolution 37:649–652.

Mayr, E. 1942. Systematics and the Origin of Species. New York: Columbia University Press.

Mayr, E. 1963. Animal Species and Evolution. Cambridge, Mass.: Belknap Press.

Metz, E. C., and Palumbi, S. R. 1996. Positive selection and sequence rearrangements generate extensive polymorphism in the gamete recognition protein bindin. Mol. Biol. Evol. 13:397–406.

Miller, D. J., Gong, X., Decker, G., and Shur, B. D. 1993. Egg cortical granule N-acetylglucosaminidase is required for the mouse zona block to polyspermy. J. Cell Biol. 123: 1431–1440.

Minor, J. E., Fromson, D. R., Britten, R. J., and Davidson, E. H. 1991. Comparison of the bindin proteins of *Strongylocentrotus franciscanus*, *S. purpuratus*, and *Lytechinus variegatus*: sequences involved in the species specificity of fertilization. Mol. Biol. Evol. 8:781–795.

Morrison, D. A., McDonald, M., Bankoff, P., Quirico, P., and Mackay, D. 1994. Reproductive isolation mechanisms among four closely-related species of *Conospermum* (Proteaceae). Bot. J. Linn. Soc. 116:13–31.

Nakano, S. 1985. Sperm displacement in *Henosepilachna pustulosa* (Coleoptera, Coccinellidae). Kontyu 53:516–519.

Naveira, H. F. 1992. Location of X-linked polygenic effects causing sterility in male hybrids of *Drosophila simulans* and *D. mauritiana*. Heredity 68:211–217.

Naveira, H., and Fontdevila, A. 1986. The evolutionary history of *Drosophila buzzatii*. XII. The genetic basis of sterility in hybrids between *D. buzzatii* and its sibling *D. serido* from Argentina. Genetics 114:841–857.

Naveira, H., and Fontdevila, A. 1991. The evolutionary history of *Drosophila buzzatii*. XXI. Cumulative action of multiple sterility factors on spermatogenesis in hybrids of *D. buzzatii* and *D. koepferae*. Heredity 67:57–72.

Orr, H. A. 1989. Genetics of sterility in hybrids between two subspecies of *Drosophila*. Evolution 43:180–189.

Orr, H. A. 1992. Mapping and characterization of a 'speciation gene' in *Drosophila*. Genet. Res. Camb. 59:73–80.

Palopoli, M. F., and Wu, C.-I. 1994. Genetics of hybrid male sterility between Drosophila sibling species: a complex web of epistasis is revealed in interspecific studies. Genetics 138:329–341.

Palumbi, S. R. 1992 Marine speciation on a small planet. Trends Ecol. Evol. 7:114–117.

Payne, D., Warnes, G. M., Flaherty, S. P., and Matthews, C. D. 1994. Local experience with zona drilling, zona cutting, and sperm microinjection. Reprod. Fertil. Dev. 6:45–50.

Rieseberg, L. H., Desrochers, A. M., and Youn, S. J. 1995. Interspecific pollen competition as a reproductive barrier between sympatric species of *Helianthus* (Asteraceae). Am. J. Bot. 82:515–519.

Ryman, N., Allendorf, F. W., and Stahl, G. 1979. Reproductive isolation with little genetic divergence in sympatric populations of brown trout (*Salmo trutta*). Genetics 92: 247–262.

Shaklee, J. B., and Tamaru, C. S. 1977. Biochemical and morphological evidence of sibling species of bonefish, "*Albula vulpes*." Am. Zool. 17:973.

Shaw, A., McRee, D. E., Vacquier, V. D., and Stout, C. D. 1993. The crystal structure of lysin, a fertilization protein. Science 262:1864–1867.

Sheldon, B. C. 1993. Sexually transmitted disease in birds—occurrence and evolutionary significance. Philos. Trans. Roy. Soc. Lond. B 339:491–497.

Stebbins, G. L. 1950. Variation and Evolution in Plants. New York: Columbia University Press.

Swanson, W. J., and Vacquier, V. D. 1995. Extraordinary divergence and positive Darwinian selection in a fusagenic protein coating the acrosomal process of abalone spermatozoa. Proc. Natl. Acad. Sci. U.S.A. 92:4957–4961.

Tanaka, T., and Nei, M. 1989. Positive Darwinian selection observed at the variable-region genes of immunoglobulins. Mol. Biol. Evol. 6:447–459.

Templeton, A. R. 1981. Mechanisms of speciation; a population genetic approach. Annu. Rev. Ecol. Syst. 12: 23–48.

Thomas, S., and Singh, R. S. 1992. A comprehensive study of genic variation in natural populations of *Drosophila melanogaster*: VII. Varying rates of genic divergence as revealed by two-dimensional electrophoresis. Mol. Biol. Evol. 9:507–525.

Vacquier, V. D., and Lee, Y.-H. 1993. Abalone sperm lysin: unusual mode of evolution of a gamete recognition protein. Zygote 1:181–196.

Veech, J. A., Benedix, J. H., Jr., and Howard, D. J. 1996. Lack of calling song displacement between two closely related ground crickets. Evolution 50:1982–1989.

Wade, M. J., Patterson, H., Chang, N. W., and Johnson, N. A. 1994. Postcopulatory, prezygotic isolation in flour beetles. Heredity 72:163–167.

Wilson, E. B. 1928. The Cell in Development and Heredity (3rd ed.). New York: Macmillan.

Zimmerman, E. G., Kilpatrick, C. W., and Hart, B. J. 1978. The genetics of speciation in the rodent genus *Peromyscus*. Evolution 32:565–579.

Part V

The Genetics of Speciation

22

The Genetics of Sexual Isolation

Michael G. Ritchie
Stephen D. F. Phillips

For many, the term *speciation* is synonymous with the evolution of reproductive isolation between diverging taxa (Mayr 1963; Dobzhansky 1951). Even if one accepts a less demanding species definition, for example, involving levels of clustering or occupancy of adaptive peaks, the achievement of reproductive isolation is a major evolutionary event as it "fixes" genetic divergence, vastly reducing the potential for subsequent reintegration. The determinants of reproductive isolation are usually classified into premating or postmating factors, after Dobzhansky, depending on whether they come into play before or after sexual partners mate. It is increasingly apparent that isolating factors can act after mating but before fertilization (Howard et al., this volume), blurring the distinction somewhat, but the major classes of premating isolation involve ecological factors (breeding period, season, or location) and mating behaviors contributing to sexual isolation between species, while those of postmating isolation involve hybrid inviability and sterility.

Coyne and Orr (1989a) have carried out the only detailed comparison of the rate of appearance of pre- versus postmating isolation and concluded that both evolved at a similar rate in *Drosophila*, though premating isolation appeared most quickly between sympatric (partially overlapping) species. Against this, there are a number of more anecdotal studies suggesting that many animal species are isolated only by mating behaviors, possibly as a result of sexual selection generating rapid evolution (reviewed in Butlin and Ritchie 1994), though counterexamples can be found. In Dobzhansky's and others' highly influential early studies of North American *Drosophila*, postmating isolation was used in practice as the main criterion for defining species (possibly because it was easier to measure in the laboratory). This could have inflated the apparent importance of postmating factors in this group. Perhaps as a consequence, much of the early literature about the genetics of speciation concerns identifying genetic causes of hybrid dysfunction. Another

considerable stimulant behind much of this research is Haldane's Rule, the pattern whereby the heterogametic sex shows most dysfunction in the F_1 of crosses between species (Coyne and Orr 1989b). This is consistent for animals with nonconventional sex chromosome systems such as birds, Lepidoptera, and Orthoptera, and usually holds for both hybrid inviability and infertility. Haldane's Rule has multiple causes (Hollocher and Wu 1996; Wu et al. 1996). In the same species pair, hybrid inviability and sterility can be caused by different genes (Orr 1993). One consistent finding in early studies of the genetics of Haldane's Rule was that epistatically acting sex-linked genes were commonly involved (Coyne and Orr 1989b). This would be expected if the X accumulates differences between developing species more quickly than autosomes (Charlesworth et al. 1987). However, recent studies suggest that sex linkage may simply appear to be involved because of differential expression due to the hemizygosity, rather than accumulation, of X-linked genes in *Drosophila* (Hollocher and Wu 1996; True et al. 1996). Thus there may be no "special effect" of the X-chromosome in postmating isolation (at least in *Drosophila*) in the sense that sex-linked genes are not inherently more likely to effect reproductive isolation, though their divergence is more likely to be seen due to their greater probability of expression. The recent renaissance of the "Dominance Theory" (Turelli and Orr 1995) may provide the most general explanation for Haldane's Rule (Davies and Pomiankowski 1995; but see Hollocher and Wu 1996).

In comparison to postmating isolation, little concerted effort has been spent on analyzing either the genetic causes of, or identifying general rules about, premating isolation. Here we ask if studies of the genetics of sexual isolation show any evidence of a disproportionate affect of sex-linked genes. We also ask to what extent is there evidence that "major genes" may influence premating isolation, that is, that single genes may play a significant role in this aspect of speciation.

Behavior, Sex Linkage, and Major Genes

Following his studies of *Drosophila* courtship song, Ewing (1969) argued that X-linked genes were more likely to be involved in behavioral differences between closely related species because favorable partial recessives would be immediately exposed to selection, allowing more rapid evolution. This clearly could apply to genes with favorable effects on any trait expressed in males (Ewing supposed that male and female genes affecting sexual isolation might show "genetic coupling"). Charlesworth et al. (1987) suggest this could favor more rapid divergence of the X-chromosome. If there is any tendency for genes to accumulate on the X-chromosome, this could encourage further differentiation of any affecting sexual isolation because linkage disequilibrium can facilitate rapid coevolution through reduced recombination between male and female components of sexual communication. The fact that many traits involved in sexual isolation are sex-limited in expression must also favor sex linkage (Rice 1984).

One possible reason why sex-linkage may be less likely is if behaviors are polygenic. It has been the position of some authorities that the conventional allopatric model of speciation (and, indeed, a neodarwinian view of evolution) requires that most adaptive changes are polygenic and as a consequence divergence is piecemeal and gradual. However, recent reviews do not strongly support this position (Orr and Coyne 1992), and it seems likely that the increasing application of Quantitative Trait Loci (QTL) marker methodologies to evolutionary studies will provide further examples of major genes for traits previously thought to be polygenic (e.g., Bradshaw et al. 1995; Mitchell-Olds 1995; Liu et al. 1996). Single genes may be particularly important in rapid differentiation, especially involving selection (and changes between "adaptive peaks"). Behavior may seem particularly likely to be under polygenic control. Behavior genetics is traditionally studied using the methods of quantitative genetics, but mutant studies are finding many examples of major genes, such as *period* (Kyriacou and Hall 1980), which influence behavioral traits. Henry (1994) has argued that bisexual courtship communication may favor the appearance of major genes because coordinated communication systems are resistant to minor disruption, and Bakker and Pomiankowski (1995) found that 8 of 36 studies of the genetics of female mating preferences (not necessarily involved in premating isolation) were suggestive of or compatible with major gene effects. Asymmetrical selection in communication systems might allow mutations of large effect to persist in either signals or receivers (see below; see also Löfstedt 1993; Butlin and Trickett, 1996). We therefore might expect the involvement of major genes in sexual isolation not to be exceptional. However, other authors have argued that coevolving sexual selection may lead to a polygenic architecture (Coyne and Orr 1989a; Coyne et al. 1994). Given the lack of a clear prediction in the background literature, it seems appropriate to consider also evidence for the involvement of major genes in premating isolation.

Scope of the Review

We have reviewed research examining the genetics of assortative mating and behaviors implicated in this—primarily male mating signals and female preferences. We have excluded host plant choice or seasonality unless these are strongly implicated in premating isolation. Methods include conventional crosses between species with examination of F_1s and sometimes the first or second segregating generations. These can identify sex linkage fairly clearly, and sometimes implicate major genes. With *Drosophila,* the use of marked chromosomes in recombinant individuals allows better identification of potential major genes. We only consider a trait polygenic if every marked chromosome contributes to interspecific differences. Of course, under these circumstances major genes could still exist, especially as many studies only involve one or two marker genes per chromosome. The distinction between polygenic and "major gene" systems is necessarily somewhat arbitrary, there will be a continuum from a few genes up to so many that separate analysis is unlikely or unfeasible. Other techniques include biometrical methods which assume polygeny, though allow identification of linkage and nonadditive effects. Mutant studies are fairly common in *D. melanogaster,* but only the *period* example provides a clear connection with interspecific differences, and the mutant recovery techniques are usually biased toward finding sex-linked genes. We have therefore excluded mutant studies. Finally, we have concentrated on interspecific differences but have freely included studies between well-differentiated forms. In some ways, intraspecific studies may be better models of changes contributing to the process of speciation, because many differences between species may have accumulated following the speciation event and therefore be incidental to the process. Polymorphisms within populations are sometimes included (e.g., where a selection experiment has been carried out to mimic interspecific differences), but we have tried not to overemphasize such studies as they may not be representative of speciation. We do not imagine that our survey is exhaustive.

Background

Detailed studies of natural populations are contradicting the once common notion that variation in assortative mating or in "species-specific" mating behaviors is low within species (Paterson 1993; Henderson and Lambert 1982). Numerous studies testify to the continuous nature of speciation, with races of species showing partial isolation, numerous hybrid zones (Hewitt 1989) and considerable variability in signals and preferences (Butlin [1995] found 69 cases of variation in signals or preferences within recognized species versus 18 cases without such variation). It is therefore not surprising that arguments persist regarding species definitions. Recently Pomiankowski and Møller (1995) have shown that traits involved in courtship may be characterized by particularly high levels of genetic variation within populations. A few studies have documented intraspecific variation in mating preferences (Bakker and Pomiankowski 1995), and there are even a handful of studies that have measured genetic covariances between trait and preference. It is therefore surprising to find how few studies really contribute detail to our review. A minority were capable of detecting major genes. Although this is a general issue in quantitative genetics (Barton and Turelli 1989), the major reason for the paucity of good data is probably the frustrating scarcity of crossable species (Wu 1996), especially those with easily analyzable behaviors.

An important question is what constitutes significant sex linkage. Any difference between reciprocal F_1s is evidence of a role of the sex chromosomes (ignoring for the moment maternal effects). Evidence of a disproportionate role of the sex chromosomes requires this to be greater than expected were genes distributed randomly throughout the karyotype. For a sex-limited male trait, the expectation under the standard quantitative genetic model (all additive effects, (+)alleles in one species, (−)alleles in another, assuming further the X-chromosome is of approximately average size) is that the difference between reciprocals as a proportion of the parental difference should be $1/(2n − 1)$. So, for example, the expected magnitude is 14% in D. melanogaster (where effectively $n = 3$). This prediction is complicated by unequal chromosome sizes, dosage compensation, and asymmetrical gene action. In D. melanogaster, sex-linked genes are doubled in expression, which results in an expectation of simply $1/n$, that is, 33% of the difference (nearer 25% accounting for chromosome size). Plainly, one would expect to find a substantial role for X-linked genes in organisms with low chromosome number even in the absence of differential accumulation of these genes. Many studies do not give the haploid number of the organisms involved.

A substantial literature exists regarding the significance of finding major gene effects, usually in the context of QTL, but also for biometrical methods. In practice most of the studies here have either looked for segregation following crosses or cosegregation with marked chromosomes. Sometimes we have reanalyzed these data using appropriate statistical models to examine the significance of major gene effects where these were not given by the authors.

Case Studies

Table 22.1 summarizes the major components of the database. Below we describe major studies and issues they raise, taking different groups of organisms in turn.

Drosophila

Understandably the most detailed studies available concern Drosophila. Traits examined include assortative mating, species-specific acoustic and pheromonal components of courtship, and female preferences (for specific traits or via assortative mating).

Assortative mating (albeit in the laboratory) is probably the most direct method of ascertaining premating isolation. Only six studies are available. Tan (1946) examined assortative mating between recombinant females of crosses between D. pseudoobscura and D. persimilis (from 50% to 100% D. pseudoobscura, effective n = 4) when combined with different combinations of wild-type males and females. The best-studied crosses involve recombinant females with wild-type D. persimilis males and D. pseudoobscura females. The tendency to mate "homogametically" was significantly associated with only the marker for chromosome II with a subsidiary positive interaction between the X and IV (Tan defined homogametic matings here as the tendency for the D. persimilis males to mate with the recombinant females—this is positive assortment for D. persimilis genes). When the alternative wild-type female in the trio was D. persimilis rather than D. pseudoobscura, heterogametic matings (once more, matings with the recombinant females) were also mainly associated with chromosome II, though IV played a role. "Homogametic" matings (with the wild-type D. persimilis female) were influenced by the origin of the X in the recombinant females. Tan (1946) interpreted this last pattern as effectively male discrimination against sex-linked D. pseudoobscura genes, but recent work (Noor 1996) using a more conventional design has shown a lack of species discrimination in males of these species, implying the patterns of assortment seen in Tan's rather complex experimental design is probably due to female preference for unidentified male traits. Using recombinant males in a similar experiment does not reveal an X-effect (Noor, 1997).

Zouros's (1981) study of assortative mating between D. mojavensis and D. arizonae used a more conventional design involving recombinants (F₁ backcrossed into D. mojavensis) of both sexes. Male behaviors were assessed by scoring mating success of recombinant males with wild-type D. arizonae males and females. There was a strong effect of both sex chromosomes, with only marginal effects among other chromosomes (and interactions). Female behaviors (pure D. mojavensis males with females from backcrosses to D. arizonae) also showed clear evidence of major gene effects, with only two auto-

Table 22.1. Major studies of the genetics of sexual isolation in *Drosophila*, Orthoptera, and Lepidoptera.

Organism	Traits	Sex Linkage?	Major Genes?	Authors	Comments
Drosophila pseudoobscura and *D. persimilis*	Assortative mating	Possibly	Yes	Tan 1946	Chromosome II, plus X-linked effect
D. simulans and *D. sechellia*	Assortative mating (BX females to (+)males)	No	Yes	Coyne 1992	Few markers / No dominance
D. auraria	Song (ipi)	No	No	Tomaru and Oguma 1994	3 markers
D. paulistorum	Assortative mating	No	No	Ehrman 1961	Assortment weak in one pair
D. mojavensis and *D. arizonensis*	Assortative mating / BX females / BX males	No / Yes	Yes, 2 of 5 markers / Yes, 1 or 2 of 4	Zouros 1981	
D. simulans and *D. mauritiana*	Assortative mating	No	No	Coyne 1989, 1992	Dominance in F_1 marker arms additive
	Song	No	No	Pugh and Ritchie 1996	Additive
D. melanogaster	Song	No	No	Ritchie et al. 1994 / Ritchie and Kyriacou 1996	Intraspecific
D. pseudoobscura and *D. persimilis*	Song "types"	Yes	—	Ewing 1969	Quantitative aspects autosomal
D. virilis and *D. littoralis*	Song traits	Yes	Yes	Hoikkala and Lumme 1987	Disproportionate role of X between *virilis* and *montana* phylads
D. virilis and *D. lummei*	Song	No	No	Hoikkala and Lumme 1984	Within *virilis* phylad
D. simulans and *D. sechellia*	7-T/7-11HD pheromones	No	Yes	Coyne et al. 1994	Maps to chromosome 3
D. melanogaster	7-T/7-P pheromone races	No	Yes	Ferveur and Jallon 1996	
	Female preference for pheromone races	No	Yes	Scott 1994	
D. melanogaster	Assortative mating	No	—	Wu et al. 1995	Intraspecific
Chorthippus parallelus and *C. erythropus*	Various song traits	Possible, also maternal effects	No	Butlin and Hewitt 1988	Intraspecific
Laupala paranigra and *L. kohalensis*	Pulse rate	Yes, but proportional	No; Ne = n	Shaw 1996	Trait importance? / Fits polygenic model well, though a genetic background effect
Acheta spp.	Song	Yes	—	Bigelow 1960	
Poecilimon veluchianus subspp.	Body size (under sexual selection)	Yes	—	Reinhold 1994	Testis size Y-linked (in XO system!)
C. biguttulus and *C. mollis*	Song	Yes	—	von Helversen and von Helversen 1975a,b	male and female songs maternal? Behaviorally uncoupled

Species pair	Trait			Reference	
Ephippiger ephippiger races	Song and preference	Yes	No	Ritchie 1996	Intraspecific
Teleogryllus oceanicus / T. commodus	Song;				
	Intertrill interval	Yes	Yes	Bentley and Hoy 1972	Probably polygenic, but quite large differences between F₁s and n = 29
	Pulses per trill, etc.	No		Bentley and Hoy 1972	
Gryllus spp.	Preference	Yes	Yes	Hoy et al. 1977	Sometimes intermediate, but highly variable
	Song			Bentley and Hoy 1972	
Heliothis virescens / H. subflexa	Hairpencil glands	Yes	Yes	Teal and Oostendorp 1995	H. virescens males normally have hair pencil glands; H. subflexa males lack hair pencil glands
Ostrinia nubilalis E/Z pheromone races	Female pheromone production	No	Yes	Roelofs et al. 1987	Intraspecific
	Detection of pheromones by males	No	Possibly	Roelofs et al. 1987	
	Male behavioral contact with female—hair pencil display	No	Possibly (could be linked to the female autosomal factor for determining pheromone components)	Roelofs et al. 1987	
Colias philodice/C. eurytheme	Male ultraviolet wing color	Yes	Yes	Silberglied and Taylor 1978	
	Pheromone composition in males	Yes	Yes	Grula and Taylor 1980	
	Mate selection by females	Yes	Yes	Silberglied and Taylor 1978	
Choristneura fumiferana / C. pinus	Pheromone composition	Yes	Yes	Sanders et al. 1977	
	Female calling time	Yes	Yes	Sanders et al. 1977	
Y. ponomeuta padellus / Y. malinellus	Pheromone composition of females	No	—	Hendrikse 1988	
Papilio glaucus / P. canadensis	Oviposition preference	Yes	—	Scriber 1992	
Papilio machaon / P. zelicaon	Oviposition preference	Yes	—	Thompson 1988	

Sex linkage is entered as "No" if any effect detected was not disproportionate, and "—" indicates no result reported or not tested.

somes accounting for variation in insemination rates. The X had no effect, clearly showing a lack of any common genetic basis to the sexual behaviors of males and females.

Against those examples, Ehrman's (1961) study of crosses between two pairs of subspecies of *D. paulistorum* showed that every chromosome carries genes contributing to assortment, with no clear major gene effects, and Coyne's (1992) study of isolation in *D. simulans/D. sechellia* showed every autosome contributing, apart from the X (providing a striking contrast to his examination of postmating isolation between this pair). Wu et al. (1995) obtained similar results in a study of assortative mating between the differentiated Zimbabwe and control strains of *D. melanogaster*, and Welbergen et al.'s (1992) biometrical study of male sexual isolation between *D. simulans* and *D. melanogaster* (a diallel analysis of strains of *D. melanogaster* differing in frequency of interspecific mating with *D. simulans*) showed no sex-linkage effect (and directional dominance rather than additivity). Genetic variation for premating isolation between these species is fairly easily found (e.g., Jamart et al. 1993). The Hawaiian species *D. heteroneura* and *D. silvestris* also show substantial genetic polymorphism for premating isolation, and there is some evidence for epistatic interactions (Ahearn and Templeton 1989).

Turning to traits implicated in premating isolation, the most detailed studies concern cuticular hydrocarbons and male courtship song. Cuticular hydrocarbons are known to have a strong influence on premating isolation in the melanogaster group of *Drosophila*. The 7,11-Heptacosadiene (7,11-HD)/7-tricosene (7-T) polymorphism plays an important role in male recognition of females (Cobb and Jallon 1990; Coyne and Oyama 1995). Male *D. simulans* (7-T) will not court *D. sechellia* females (who carry the other pheromone compound, 7,11-HD). If 7,11-HD is transferred from *D. sechellia* to *D. simulans* females, males will cease to court their conspecific females. Similarly, *D. simulans* males will court *D. sechellia* females (even dead ones) who have acquired 7-T. Analysis of the pheromone blend of backcross females showed that most of the variation was dependent only on the origin of chromosome III (figure 22.1). A pheromone blend polymorphism exists in *D. melanogaster* that influences female mating speed. Females of the common form (with 7-T as the main component) mate more slowly with males carrying 7-pentacosene (7-P) (found in the Tai-Y strain, of Afrotropical origin). The mating discrimination also maps to chromosome III. As Tai-Y females mate randomly, this could be a gene that blocks discrimination against 7-P (Scott 1994). The difference in production of this blend maps to chromosome II (Ferveur 1991). Recent intrachromosomal studies have suggested that both the 7-T/7-P polymorphism and 7-T/7,11-HD species difference are produced by more than one linked gene (Ferveur and Jallon 1996; Coyne 1996). Pheromonal differences between *D. persimilis* and *D. pseudoobscura*

are determined by X- and second chromosome-linked genes (Noor and Coyne 1996).

The courtship song of *Drosophila* has been studied for many years, especially in *D. melanogaster*. The X-linked *period* (*per*) gene has been suggested as a possible example of a "speciation gene." It was identified by mutant studies but has been shown to influence species specific differences in courtship song (Wheeler et al. 1991) and activity cycles (Petersen et al. 1988). The song difference influences female mating speed in playback experiments (but the *period* gene itself does not determine the effect in females; Greenacre et al. 1993). However, the total contribution of *per* encoded traits to premating isolation is unclear as Ritchie and Kyriacou (1994a) found *D. melanogaster* females could not be selected to distinguish between males transformed with either the *D. melanogaster* or *D. simulans* allele. Other song traits, notably interpulse interval (IPI), are also likely to play a role. IPI in *D. melanogaster* has an unusual pattern of variability. Mean IPI varies little between and within populations, has low heritability, and shows considerable environmental variation in the laboratory (Ritchie et al. 1994; Ritchie and Kyriacou 1994b). Where variation between populations has been found it has been shown to be additive autosomal (Cowling 1980; Ritchie et al. 1994). Genetic variation can be uncovered by artificial selection experiments. Ritchie and Kyriacou (1996) managed to take the IPI of one line of *D. melanogaster* outside the range of the species within six generations. The response fitted expectations for a polygenic quantitative trait—about 25% of the variation was due to sex-linked genes, which fits expectations for *D. melanogaster* very well. Differences in IPI between *D. simulans* and *D. mauritiana* show a similar architecture (figure 22.2; see also Kyriacou and Hall 1986; Cowling and Burnet 1981).

Other *Drosophila* species have been less well studied. Hoikkala and colleagues have extensive information on the virilis group; for example, Hoikkala and Lumme (1987) carried out a diallel analysis of interspecific song differences. There is evidence of substantial nonadditive effects, directional dominance for characteristics of the pulse thought to be under strongest sexual selection and ambidirectional dominance (indicating possible stabilising selection) for pulse length. In contrast, crosses between species of the virilis subgroup suggested an additive autosomal architecture to species differences, whereas crosses between *D. virilis* and *D. littoralis* (from different subgroups) indicated strong, disproportionate, sex linkage (figure 22.3). Hoikkala and Lumme (1987) speculate that the change in song pattern associated with the X of the Montana group was a key event in the evolution of song pattern within this group, facilitating many structural differences. Ewing (1969) similarly found in crosses between *D. subobscura* and *D. persimilis* that, while quantitative differences in song were additive autosomal, qualitative differences (notably the presence or

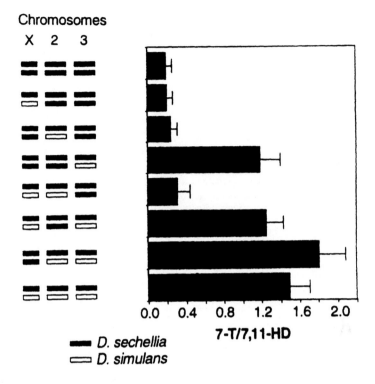

Figure 22.1. The ratio of the pheromones 7-T and 7,11-HD present in backcrosses between *D. simulans* and *D. sechellia* is almost solely determined by the origin of the third chromosome (when heterozygous, more 7,11-HD was produced than 7-T). Shading indicates the species of origin of each chromosome. Reprinted with permission from Coyne et al. (1994). Copyright 1994 American Association for the Advancement of Science.

absence of song types) were sex linked. Tomaru and Oguma (1994) found song to be polygenic in *D. auraria*.

Perhaps the technique with the greatest power to resolve the number of genes influencing a continuous trait is the use of molecular markers for quantitative trait loci. A relevant example is Liu et al.'s (1996) recent study of the shape of genitalia, potentially a character influencing sexual isolation, in crosses between *D. simulans* and *D. mauritiana*. Using 15 markers, they found that nine showed significant linkage effects with the trait and concluded this is sufficiently few to reject the classic "polygenic" model. Exactly where a boundary lies between a trait influenced by major genes and polygenic determination is not clearly defined. Fewer than 10 genes each explaining a significant proportion of the variance of the trait is probably sufficiently few to contemplate independent characterization and study of each locus, which may be as good a definition of a "major gene trait" as any.

Lepidoptera

The Lepidoptera are unusual in that sex linkage of traits that differ among species seems to be common, despite a relatively high chromosome number in this group

(Sperling 1994). Prowell (this volume) discusses this phenomenon in some detail and makes some suggestions why this might occur. Here we only briefly summarize some major studies as these relate to sexual isolation.

The method of chromosomal sex determination within the Lepidoptera causes females to be the heterogametic sex (ZZ for male and ZW for female). The traits involved in assortative mating include visual (e.g., *Heliconius*), pheromonal (e.g., *Ostrinia*), and phenological (e.g., many noctuid moths) differences. Of these, the most amenable to study is pheromonal communication. An important aspect of pheromonal communication is that the sex roles are usually also reversed; hence, the highly investing females are the signalers (emitting the pheromone) and males the responders (though further mate assessment might occur once in close proximity). This sex role reversal can have important implications for the ease with which shifts in signal might occur (see discussion below).

At least four Lepidopteran species groups or strains have been clearly shown to demonstrate sex-linked differences in factors affecting premating isolation. The two pierid butterflies *Colias philodice* and *C. eurytheme* are genetically very similar with offspring from F_1, F_2, and

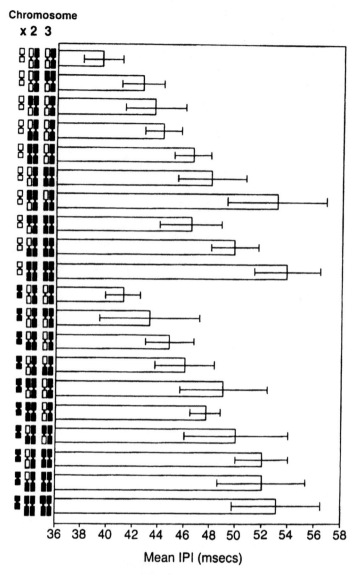

Figure 22.2. The interpulse interval (± 1 standard error) of backcross genotypes generated by cross-ing F₁ *D. simulans* × *D. mauritiana* females into *D. simulans*. Shading of chromosomes indicates the species of origin of each chromosome arm (open = *D. mauritiana*; solid = *D. simulans*, ignoring recombination in the F₁). Note that the genotypes are first arranged according to the origin of the X-chromosome, but no step is apparent in the IPI data. This would be expected if sex-linked genes made a major contribution to the differences between species. Reprinted from Pugh and Ritchie (1996) with permission of Blackwell Science Ltd.

backcross matings being at least partially viable and fer-tile (Ae 1979). There is significant Z-linkage for a dra-matic number of interspecific differences between these two species. At least three of these contribute to premating isolation: mate selection by females, via female response to species-specific male visual and olfactory cues (Sil-berglied and Taylor 1978); composition of male phero-mones (Grula and Taylor 1980); and male ultraviolet

wing color, an important visual signal (Silberglied and Taylor 1978).

The two pheromone races of the European corn borer moth *Ostrinia nubilalis* Hübner (Kochansky et al. 1975) are not considered biological species but have been ex-tensively studied and probably represent an early stage of speciation (Cardé et al. 1978). They show different female calling times, voltinism, and host-plant specific-

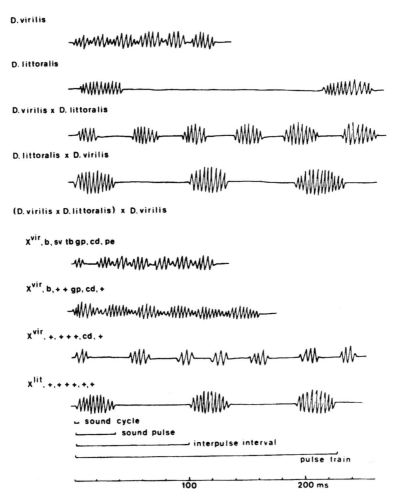

Figure 22.3. The inheritence of courtship song pattern in crosses between *D. virilis* and *D. littoralis*. Clear differences can be seen between the shape and patterning of pulses of the reciprocal hybrids (female × male). The backcrosses confirm a major involvement of the X-chromosome (the bottom two differ in the origin of the X and a single autosomal marker). Reprinted from Hoikkala and Lumme (1987) with permission of *Evolution*.

ity leading to considerable premating isolation (Liebherr and Roelofs 1975). Hybrids can easily be produced in the laboratory. Voltinism (Showers 1981; Glover et al. 1992) and male behavioral response to phero-mones (Roelofs et al. 1987) have been shown to be Z-linked. However pheromone composition (Klun and Maini 1979) and the physiological basis of pheromone detection by males (Roelofs et al. 1987; Löfstedt et al. 1989) are autosomally controlled.

Two species of tortricid moth, *Choristoneura fumi-ferana* and *C. pinus,* are sympatric over much of their ranges yet hybridization is unknown in the wild (Smith 1954). Premating isolation is due to at least three factors, female calling time (Sanders et al. 1977), larval-pupal development rate leading to phenological separation

(Smith 1953; Campbell 1962), and female pheromone composition (Sanders et al. 1977). All of these have been shown to be significantly Z-linked.

Papilio machaon and *P. zelicaon* are sympatric across much of western North America (Sperling 1987). The major interspecific differences include adult and larval color pattern (Clarke and Sheppard 1955; Sperling 1987), voltinism (Sperling 1994), and larval host-plant (Thompson 1988). The extent of premating isolation in the wild is primarily related to the spatial geography of the different larval host plants. Oviposition preference is partially Z-linked in these species and may thus result in some Z-linked control of premating isolation.

As a counterexample, the most closely related species pair of the *Yponomeuta* complex of ermine moths (consisting of several sympatric cryptic species) are *Yponomeuta padellus* complex and *Y. malinellus*. Premating isolation is controlled by species specific pheromones. F_1 hybrid females generally demonstrate intermediate pheromone composition, suggesting autosomal inheritance (Hendrikse 1988).

Orthoptera

The major component of the mating system of Orthoptera is female phonotaxis to relatively easily analyzed male song, making this group an excellent model for the study of sexual communication. Male songs and female phonotaxis must provide a major component of premating isolation in many sympatric cricket species, as they enable receptive males and females to find each other. Moreover, male songs can provide cues upon which females discriminate among conspecific males, so these songs are also interesting from a sexual selection viewpoint. However, Harrison and Bogdanowicz (1995) recently showed that major structural details of *Gryllus* chirp patterns need not change during speciation (see also Bigelow 1965), and Howard et al. (this volume) have shown how two hybridizing species of cricket can have high assortative fertilization without overt behavioral differentiation. The importance of the song preference system to sexual isolation needs to be considered carefully for individual studies of species pairs.

Two major early studies of orthopteran mating systems gave rise to a substantial literature on the genetics of premating isolation, because they stimulated debate about the "Genetic Coupling" hypothesis (Alexander 1962). Studies of F_1 hybrids between *Teleogryllus oceanicus* and *T. commodus* showed that, although some song traits were intermediate, reciprocal males strongly differed in song pattern. This suggests significant sex-linkage despite a high chromosome number of 29 (Bentley 1971; Bentley and Hoy 1972). Hoy et al. (1977) studied the phonotactic preferences of reciprocal female hybrids and found they preferred song of the appropriate reciprocal hybrid type. Thus, despite the fact that females are heterozygous for sex-linked genes, they still showed a behavioral preference for males from the same reciprocal cross as themselves. In contrast, von Helversen and von Helversen (1975a,b; see also Elsner and Popov 1978) studied F_1 hybrids between the acridid grasshoppers *Chorthippus mollis* and *C. biggutulus* (using female receptive stridulation rather than phonotaxis). Males and females produced songs that could be either intermediate or parental-like in structure, and females could express parental-like patterns of preference (though, e.g., a hybrid female could sing with mollis-like song but respond preferentially to biguttulus-like song!). It has been argued that the reciprocal specificity of preference and trait within the gryllid hybrids suggests a common inheritance, possibly even pleiotropy (hence the traits are genetically "coupled"), whereas the acridid hybrids suggest independent inheritance. In truth, neither of these analyses is sophisticated enough to distinguish these alternatives (Elsner and Popov 1978; Butlin and Ritchie 1989). The ability of the F_1 grasshoppers to behave like the parental species is intriguing and might occur if each haploid complement of genes is sufficient to produce almost' parental behaviors. An F_1 female possesses one complete haploid complement from each parental species. Females of crosses between *D. melanogaster* and *D. simulans* show preferences for song that has a combination of traits appropriate for each species; hence, they also show an ability to behave like either parental species (though they also prefer exactly intermediate song; see Kyriacou and Hall 1986; see also figure 22.4). Hybrid males lack a complete haploid complement from the paternal species—this could be a substantial proportion of the relevant genes if there is strong sex linkage or few chromosomes. This does not seem to result in difficulties in expressing parental-like male behaviours in the hybrid grasshoppers. In contrast, in the *Drosophila* example, hybrid males cannot produce the most preferred intermediate song as one of the traits is sex-linked. In the dominance theory of Haldane's Rule, the breakdown of fertility in male hybrids could involve the lack of the sex-linked component of one haploid complement of the paternal genome (Muller 1940; Turelli and Orr 1995; Davies and Pomiankowski 1995).

There is a suggestion in the research we have reviewed that the inheritance of calls and preferences in Orthoptera is unusual. Consider *Teleogryllus* again. Females would be expected to prefer intermediate songs if the genes acted additively (whether or not the relevant genes are sex-linked). Intermediate songs were not presented to these females as the males did not produce them. Hoy et al. (1977) suggested the preference for appropriate reciprocals seen in the homogametic sex might occur if the genes for female preference were sex-linked and a system of dosage compensation involving preferential inactivation of the paternal X-chromosome occurred. Another alternative is maternal (cytoplasmic) inheritance. Ritchie (1996) has found evidence for reciprocal differences for both song and preference in the tettigoniid bush-cricket *Ephippiger ephippiger*, as in *Teleogryllus*. Maternal effects were also sometimes seen in the *Chorthippus biguttulus/mollis* studies, and some song traits within the *C. parallelus* group show inheritance patterns more compatible with maternal than nuclear inheritance (Butlin and Hewitt 1988). Reinhold (1994) found X or maternal effects and a curious paternal effect in sexually selected traits in the tettigoniid *Poecilimon*. Bigelow (1960) found that hybrids between *Acheta* species were intermediate in pulse rate, but that the chirp structure was determined by sex-linked genes (not maternal). Ross (1992) found evidence for linkage or maternal effects in male mate choice in the cockroach *Blatella* (though morphological differences were confirmed to be X rather than maternal). Against these examples, Shaw (1996) recently

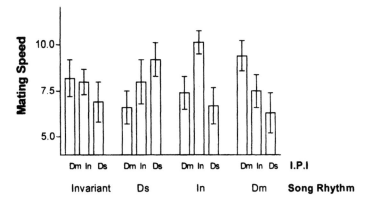

Figure 22.4. Responses of F_1 females produced by crossing *D. melanogaster* females with *D. simulans* males to synthetic song. The synthetic song varies both IPI (which is around 34 ms in *D. melanogaster*, 48 ms in *D. simulans*) and rhythm length (55 s in *D. melanogaster*, 35 s in *D. simulans*). Hybrid males produce an intermediate IPI but their rhythm is determined by the sex-linked *period* gene. Hybrid females can behave like either parental species in that, when presented with song with a normal rhythm length, they prefer the appropriate parental IPI (second and fourth group of histograms). But, when presented with an intermediate rhythm length, they prefer intermediate IPI (third group of histograms). The first group of histograms is for song lacking a rhythm. Reprinted with permission from Kyriacou and Hall (1986). Copyright 1986 American Association for the Advancement of Science.

completed a detailed study of song in crosses between *Laupala* species. She expected to find a simple inheritance pattern (possibly involving single genes) because this is a very recent speciation event and the behavioral basis of the song pattern difference seemed simple (a change in pulse rate). In fact, she found on average about 11% of the difference was sex-linked (though there were surprising differences in this between reciprocals). This corresponds very well with polygenic expectations given n = 8 in these species. Similarly, the number of "effective factors" was found to be 8.4. So, in contrast to the expectations, this study probably provides the best example of a quantitative, additive, polygenic inheritance in the Orthoptera.

Although few of the studies described above are sufficiently detailed to provide more than tentative evidence there is a clear suggestion of sex linkage or maternal effects in several. There have been few hypotheses proposed that could explain maternal effects. One intriguing possibility arises from sexual selection. Hastings (1994) suggested that ornate male traits were more likely to evolve in birds and butterflies because the sex determining system causes females to be the heterogametic sex. If the preference gene is on the sex-determining chromosome, it will therefore only be passed on to daughters. This will facilitate viability indicator processes of sexual selection because the preference gene avoids any association with the negative viability selection imposed on males, who carry the expensive indicator trait. Currently this simple idea lacks supporting empirical evidence, but a pattern of cytoplasmic inheritance would have the same

effect. Hoy et al.'s (1977) suggestion of paternal X-chromosome inactivation is not supported by the evidence available suggesting conventional random inactivation in Orthoptera, though this does rely on a single study of *Gryllotalpa* (Rao and Padmaja 1992). At any rate, we should probably be cautious in concluding unusual patterns exist within Orthoptera because the most detailed study available, Shaw (1996) with *Laupala*, shows conventional inheritance.

Other Groups

Few groups have been studied in as much detail as *Drosophila*, Lepidoptera, and Orthoptera. Henry's (1985) studies of neuropteran lacewing *Chrysoperla* seem to show conventional inheritance of song pattern (but see Henry 1994). Wells and Henry (1994) showed that some F_1 song parameters show reciprocal differences. F_1 females prefer hybrid song over one parental, but not the other (curiously, the one that seems most different from an intermediate pattern is not discriminated against). In the hemipteran *Nilaparvata* initial crosses between host races (probably isolated species) implied that differences in vibrational signal are polygenic (Claridge et al. 1985), though subsequent work shows that the number of loci is less than the recombination index, and could be quite low (Butlin 1996).

There are virtually no detailed studies of the genetics of sexual isolation outside of the insects. Important behaviours and morphological traits in *Schizocosa* spiders show apparently simple inheritance patterns (Stratton and

Uetz 1986). Doherty and Gerhardt (1983, 1984) have shown intermediate traits and preferences in crosses between species of tree frog, and a curious example is that of Beiles et al. (1984), who found evidence of a major gene effect for female mating preferences in subterranean mole rats.

Discussion

Surveying the literature reinforces the impression that our knowledge of the genetics of sexual isolation lags behind that of postmating isolation. Limited data are available from a range of species, but only a few have been studied in sufficient depth to give real confidence about the genetic architecture of the species differences. We are therefore cautious about inferring general trends from this survey, but there are some conclusions.

Major Genes

The frequency with which major genes are implicated in the control of sexual isolation differs depending on the trait involved in the sexual signaling system. Song differences are usually polygenic in origin, even between very closely related species with superficially "simple" differences, and in selection experiments. This contradicts Henry's (1994) recent suggestion that major genes may be common in such systems. Polygenic determination may be common in acoustic signals because such behaviors involve the coordinated action of multiple morphological traits and neurons. In contrast, changes in pheromone components or blends are much more commonly dictated by major genes. Again, one can imagine why this might be the case—pheromone synthesis is a complex multistage pathway, but there is a clearer potential for blockage at specific points, producing large changes in the final chemical product. Similarly, it may also be easier to envisage a single gene influencing female preference for a pheromone detection system as opposed to a neural signal processing system. For example, a single protein may have a large effect on the sensitivity or permeability of a surface membrane of an antennal receptor to a specific pheromone. Preferences for pheromones show numerous examples of major genes. Similar results are not found in studies of acoustic preferences, though there are many fewer such studies, and one could imagine the detection of frequency differences having a simple genetic basis (e.g., involving "clock" genes that are known to be expressed in neural tissues). Assortative mating between *Drosophila* species can show major gene effects, and pheromones are probably the most important trait determining the initiation of courtship (via mate recognition). Subterranean mole rat recognition is presumably pheromonal.

Another trend is that major genes are more likely to be found for qualitative than quantitative trait differences,

even for traits normally subject to polygenic determination (e.g., the presence and absence of song types in *D. pseudoobscura* and *D. persimilis* is possibly subject to major genes whereas differences in IPI are not; Ewing 1969). Again, this could reflect the possibility of single genes "blocking" the expression of a polygenically controlled behavioral trait rather than a single gene encoding a complex behavior.

Sex Linkage

There are phylogenetic differences in the likelihood of finding sex linkage of genes influencing sexual isolation. There is little convincing evidence that sex-linked genes commonly provide a disproportionate effect except in the Lepidoptera and perhaps the Orthoptera. Moths primarily communicate by pheromones and so possibly will be predisposed to major gene effects. Sex linkage in Lepidoptera is also evident in morphological and other behavioral traits, and may only be part of the phenomenon whereby most traits differing between closely related species are sex-linked (Sperling 1994; Prowell, this volume). The pattern within the Orthoptera is more ambiguous and somewhat confusing. In the absence of further detailed study, we should probably consider sex linkage of genes influencing sexual isolation unproven as a general pattern in the Orthoptera.

Speculation

The trends described above support the idea that the method of chromosomal sex determination and sex roles influences the evolution of communication systems (e.g., Hastings 1994; Butlin and Trickett 1996). In theory, the principal selection pressures in sexual signaling systems act against novel preferences and signals, resulting in strong stabilizing selection for maintenance of the coordinated status quo. Yet rapid divergence occurs during speciation. Certain permutations of signaling systems may be more prone to changes in signals and preferences by mutation than others. In particular, there are three parameters influencing the probability of shifts in sexual signaling systems, which must also influence the expected incidence of sex linkage: nonequal selection pressures acting on the two sexes, nonequal selection pressures acting on rare/mutant signals and preferences, and autosomal versus sex-linked loci.

Nonequal Selection Pressures Acting on the Two Sexes

Trivers (1972) showed that the asymmetry in parental investment in offspring between males and females leads to increased sexual selection acting on males. The fitness cost to a male with a variant sexual trait is likely to be very high due to greatly reduced or lack of mating success. The fitness cost to females with mutant sexual traits is probably lower. De Jong and Sabelis (1991) describe this as the "wallflower" effect where mutant females are likely to be mated eventually but do suffer the cost of increased predation risk prior to reproduction, and the reduced survival chances of offspring produced late in the season.

Nonequal Selection Pressures Acting on Rare/Mutant Signals and Preferences

The "response window" for many species is often wider than its corresponding range of signals (Basolo 1990; Ryan and Rand 1995; Collins and Cardé 1989; Löfstedt 1993; Butlin 1993), due either to arbitrary sensory biases or differences in sexual selection. The specieswide response window could result from either a range of narrow response windows of different centres, or each individual having a wide response window (Butlin and Trickett 1996). Differences in the width of the distribution of signal and preference would influence the selection against mutants affecting either trait (Löfstedt 1993). In figure 22.5 the mean and most preferred signal match (at X), but the preference function has a much greater width. A shift of signal from X to $(x - y)$ causes the new signal to cross the preference curve at $(l - m)$. The corresponding decrease in preference (m) is small. A similar shift of preference from X to $(x - y)$ causes the preference function to cut the signal distribution at $(l - n)$. The corresponding decrease in signal frequency (n) is much larger than (m). A shifted signal therefore remains relatively closer to the preference optimum than a shifted preference does to the peak of the signal distribution. Hence, in the absence of sex differences, the strength of selection against small shifts in signal will be weaker than that for preference. This may affect the expected genetic architecture of signal and preference traits. Mutations of large effect in signals will be more likely to be "tolerated" with the reestablishment of coordination coming through a more gradual polygenic retuning of preference genes.

Autosomal versus Sex-Linked Loci

Recessive mutations at autosomal loci are not initially expressed. For expression to occur it is necessary for the allele to increase via drift to a sufficiently high frequency that homozygotes occur. Dominant mutations will be expressed immediately, but most mutations are recessive. Recessive mutations at sex-linked loci will be expressed immediately and thus, if favored, have a reduced likelihood of being lost stochastically than favored but unexpressed autosomal recessives. This may be particularly true for sex-limited traits that are not available to selection in the homogametic sex.

Putting these three parameters together allows crude predictions to be made about the incidence of finding disproportionate contributions of sex-linked genes in sexual communication systems (figure 22.6). The most likely scenario for a shift is in an organism where the female is the signaler and the heterogametic sex, and where the signal locus is sex-linked. This fits the observation that the Lepidoptera show linkage most clearly. However, apart from the Lepidoptera, our survey provides little evidence of the above patterns. This may be for several reasons. Many examples of homogametic sex-limited characters were found, contrary to predictions, to be sex-linked (e.g., male ultraviolet wing color in *Colias*).

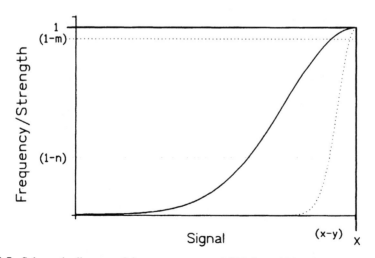

Figure 22.5. Schematic diagram of the consequences of differing widths of signal and preference distributions for the selection against a mutant. Horizontal axis is the value of a signal; vertical axis is the frequency of that signal value in the population (for trait; dotted curve) or strength of the preference for that signal value in the population (for preference; solid curve). For simplicity, only the left sides of the distributions are shown. See text for further details.

| | Female * | Female * | Male | Male |
	Homo	Hetero	Homo	Hetero
Signaller *	= **	x ***	= *	x **
Receiver	=*	x **	=	x *

Figure 22.6. Possible predictions for the likelihood of linkage in sexual isolation. =, alleles affecting premating isolation are equally likely to be present on any chromosome; x, alleles affecting premating isolation are more likely to be X-linked. The number of asterisks correlates with the ease of a shift in the signaling behaviors (e.g., easier for females, for signalers, and with linkage).

In Lepidoptera possible confounding factors are mechanisms of correctly expressing and developing sex-limited sexually dimorphic traits (Sperling 1994). In fact, most species-specific differences in Lepidoptera seem to be sex-linked, not just those associated with sexual isolation (Prowell, this volume). Reasons for this are obscure, but clearly Lepidoptera represent an important resource for understanding the genetics of speciation.

The generalizations above do allow some predictions to be made. In birds, sex-linked genes may be expected for preferences rather than male traits (perhaps even W-linked; Hastings 1994). Wax moths and a few other Lepidopterans that have secondarily evolved "conventional" sex roles (usually involving males singing or producing pheromones) might have a similar architecture due to the other two effects described above. We would also predict pheromone differences in *Drosophila* to be sex-linked (not apparent in the few detailed studies currently available).

Why No Rule?

Haldane's Rule provides a general framework for studies of the genetics of postmating isolation. The phenomenon, heterogametic sex-limited sterility or inviability, remains constant across taxa, even though the underlying explanations remain elusive and are probably heterogeneous. There does not seem to be any analogous cross-taxa generalization in sexual isolation. Why not? This may be due to an empirical bias in the few studies currently available, but is more likely to be due to fundamental differences in the selective forces that act on pre- and postmating isolation. The genes responsible for post mating isolation are not expressed except upon rare hybridizations. On the whole, they are therefore not subject to natural selection because of their affects on this trait (except for rare cases such as selection for the amelioration of hybrid unfitness in hybrid zones). In contrast, natural and sexual selection must constantly be acting on behaviors influencing mating behavior and, directly or more likely indirectly, sexual isolation. Shifts in signals and preferences seem to occur commonly and rapidly (there are numerous examples of substantial variability despite probable stabilizing selection on communication systems). Reproductive character displacement, abiotic environmental factors, or sexual selection can produce strong selection. Postmating isolation is only a pleiotropic consequence of divergence which occurs in another context. This must produce a more heterogeneous set of genetic systems underlying changes in sexual isolation, militating against common "rules" (Turelli and Orr 1995). Aberrant courtship behaviors seem likely to show more compelling analogies with Haldane's Rule, and there is some evidence that this is indeed the case (Davies et al. 1997; Noor 1997).

Conclusions

Detailed studies of the genetics of sexual isolation are rare. In contrast to studies of postmating isolation, sexual isolation does not have major patterns that are consistent across taxa. Some generalizations do seem to be present, however. Lepidoptera show a preponderance of sex-linked genes involved in sexual isolation, and probably general divergence between closely related species. Reasons for this are not clear, though the combination of role-reversed sexual communication and chromosomal sex determination is probably a factor. Another generalization is that the likelihood of major genes (and hence possibly of rapid speciation) varies with the mode of signaling—pheromonal systems showing many more examples of major gene effects than other systems such as acoustic systems, even where the differences superficially appear simple. The same is probably true of female signal detection systems, though more studies are needed of these. Moths and *Drosophila* provide the most promising systems for finding and characterizing major genes involved in sexual isolation.

Acknowledgments   We thank Dan Howard and Mohamed Noor for extensive and helpful comments on the manuscript. Ian Hastings, Roger Butlin, Jenny Gleason, Adrian Pugh, Chung-I Wu, and an anonymous referee also provided useful ideas, discussion, and feedback. The authors are funded by the N.E.R.C. (U.K.).

References

Ae, S. A. 1979. The phylogeny of some *Papilo* species based on interspecific hybridization data. Syst. Entomol. 4:1–16.

Ahearn, J. N., and Templeton, A. R. 1989. Interspecific hybrids of *Drosophila heteroneura* and *D. silvestris* I: courtship success. Evolution 43:347–361.

Alexander, R. D. 1962. Evolutionary change in cricket acoustical communication. Evolution 16:433–467.

Bakker, T. C. M., and Pomiankowski, A. 1995. The genetic basis of female mate preferences. J. Evol. Biol. 8:129–171.

Barton, N. H., and Turelli, M. 1989. Evolutionary quantitative genetics—how little do we know? Annu. Rev. Genet. 23: 337–370.

Basolo, A. L. 1990. Female preference predates the evolution of the sword in swordtail fish. Science 250:808–810.

Beiles, A., Heth, G., and Nevo, E. 1984. Origin and evolution of assortative mating in actively speciating mole rats. Theor. Pop. Biol. 26:265–270.

Bentley, D. R. 1971. Genetic control of an insect neural network. Science 174:1139–1141.

Bentley, D. R., and Hoy, R. R. 1972. Genetic control of the neuronal network generating cricket song patterns. Anim. Behav. 20:478–492.

Bigelow, R. S. 1960. Interspecific hybrids and speciation in the genus *Acheta* (Orthoptera, Gryllidae). Can. J. Zool. 38: 509–514.

Bigelow, R. S. 1965. Hybrid zones and reproductive isolation. Evolution 19:449–454.

Bradshaw, H. D., Wilbert, S. M., Otto, K. G., and Schemske, D. W. 1995. Genetic-mapping of floral traits associated with reproductive isolation in monkey flowers (*Mimulus*). Nature 376:762–765.

Butlin, R. K. 1993. The variability of mating signals and preferences in the brown planthopper, *Nilaparvata lugens* (Homoptera: Delphacidae). J. Insect Behav. 6:125–140.

Butlin, R. K. 1995. Genetic variation in mating signals and responses. Pages 327–366 in *Speciation and the Recognition Concept: Theory and Application*. J. Masters, D. M. Lambert, and H. Spencer (eds.). Baltimore: Johns Hopkins University Press.

Butlin, R. K. 1996. Co-ordination of the sexual signalling system and the genetic basis of differentiation between populations in the brown planthopper, *Nilaparvata lugens*. Heredity 77:369–377.

Butlin, R. K., and Hewitt, G. M. 1988. Genetics of behavioural and morphological differences between parapatric subspecies of *Chorthippus parallelus* (Orthoptera: Acrididae). Biol. J. Linn. Soc. 33:233–248.

Butlin, R. K., and Ritchie, M. G. 1989. Genetic coupling in mate recognition systems: what is the evidence? Biol. J. Linn. Soc. 37:237–246.

Butlin, R. K., and Ritchie, M. G. 1994. Mating behaviour and speciation. Pages 43–79 in *Behaviour and Evolution*. P. J. B. Slater and T. R. Halliday (eds.). Cambridge: Cambridge University Press.

Butlin, R. K., and Trickett, A. 1996. Can population genetic simulations help to interpret pheromone evolution? Pages 548–562. In *Pheromone Research: New Directions*. R. T. Cardé and A. K. Minks (eds.). New York: Chapman and Hall.

Campbell, I. M. 1962. Reproductive capacity in the genus *Chorisoneura* Led. (Lepidoptera: Tortricidae). II. Quantitative ineritance and genes as controllers of rates. Can. J. Genet. Cytol. 4:272–288.

Cardé, R. T., Roelofs, W. L., Harrison, R. G., Vawter, A. T., Brussard, P. F., Mutuura, A., and Munroe, E. 1978. European corn borer: pheromone polymorphism or sibling species? Science 199:555–556.

Charlesworth, B., Coyne, J. A., and Barton, N. H. 1987. The relative rates of evolution of sex chromosomes and autosomes. Am. Natur. 130:113–146.

Claridge, M. F., Den Hollander, J., and Morgan, J. C. 1985. Variation in courtship signals and hybridisation between geographically definable populations of the rice Brown planthopper, *Nilaparvata lugens* (Stål). Biol. J. Linn. Soc. 24:35–49.

Clarke, C. A., and Sheppard, P. M. 1955. A preliminary report on the genetics of the machaon group of swallowtail butterflies. Evolution 9:385–399.

Cobb, M., and Jallon, J.-M. 1990. Pheromones, mate recognition and courtship stimulation in the *Drosophila melanogaster* species sub-group. Anim. Behav. 39:1058–1067.

Collins, R. D., and Cardé, R. T. 1989. Selection for altered pheromone-component ratios in the pink bollworm moth, *Pectinophora gossypiella* (Lepidoptera: Gelechiidae). J. Chem. Ecol. 15:2647–2659.

Cowling, D. E. 1980. The genetics of *Drosophila melanogaster* courtship song—diallel analysis. Heredity 45:401–403.

Cowling, D. E., and Burnet, B. 1981. Courtship song and genetic control of their acoustic characteristics in sibling species of the *Drosophila melanogaster* subgroup. Anim. Behav. 29:924–935.

Coyne, J. A. 1992. Genetics of sexual isolation in females of the *Drosophila simulans* complex. Genet. Res. Camb. 60:25–31.

Coyne, J. A. 1996. Genetics of differences in pheromonal hydrocarbons between *Drosophila melanogaster* and *D. simulans*. Genetics 143:353–364.

Coyne, J. A., and Orr, H. A. 1989a. Patterns of speciation in *Drosophila*. Evolution 43:362–381.

Coyne, J. A., and Orr, H. A. 1989b. Two rules of speciation. In *Speciation and Its Consequences*. Pages 180–207 in D. Otte and J. Endler (eds.). Sunderland, Mass.: Sinauer.

Coyne, J. A., and Oyama, R. 1995. Localization of pheromonal sexual dimorphism in *D. melanogaster* and its effect on sexual isolation. Proc. Natl. Acad. Sci. USA 92:9505–9509.

Coyne, J. A., Crittenden, A. P., and Mah, K. 1994. Genetics of a pheromonal difference contributing to reproductive isolation in *Drosophila*. Science 265:1461–1464.

Davies, N., and Pomiankowski, A. 1995. Haldane's Rule: old theories are best. Trends Ecol. Evol. 10:350–351.

Davies, N., Aiello, A., Mallet, J., Pomiankowski, A., and Silberglied, R. E. 1997. Speciation in two neotropical butterflies: extending Haldane's Rule. Proc. Roy. Soc. London B. 264:845–851.

De Jong, M. C. M., and Sabelis, M. W. 1991. Limits to runaway sexual selection: the wallflower paradox. J. Evol. Biol. 4:637–656.

Dobzhansky, T. 1951. Genetics and the origin of species (3rd ed.). New York: Columbia University Press.

Doherty, J. A., and Gerhardt, H. C. 1983. Hybrid tree frogs: vocalisations of males and selective phonotaxis of females. Science 220:1078–1080.

Doherty, J. A., and Gerhardt, H. C. 1984. Acoustic communication in hybrid tree frogs: sound production by males and selective phonotaxis by females. J. Comp. Physiol. A 154: 319–330.

Ehrman, L. 1961. The genetics of sexual isolation in Drosophila paulistorum. Genetics 46:1025–1038.

Elsner, N., and Popov, A. V. 1978. Neuroethology of acoustic communication. Adv. Insect Physiol. 13:229–335.

Ewing, A. W. 1969. The genetic basis of sound production in Drosophila pseudoobscura and D. persimilis. Anim. Behav. 17:555–560.

Ferveur, J.-F. 1991. Genetic control of pheromones in D. simulans. I. Ngbo, a locus on the second chromosome. Genetics 128:293–301.

Ferveur, J.-F., and Jallon, J.-M. 1996. Genetic control of male cuticular hydrocarbons in Drosophila melanogaster. Genet. Res. Camb. 67:211–218.

Glover, T., Campbell, M., Robbins, P. S., Eckenrode, C. J., and Roelofs, W. L. 1992. Genetic control of voltinism characteristics in European corn borer races assessed with a marker gene. Arch. Insect Biochem. Physiol. 20: 107–117.

Greenacre, M., Ritchie, M. G., Byrne, B. C., and Kyriacou, C. P. 1993. Female song preference and the period gene of Drosophila melanogaster. Behav. Genet. 23:85–90.

Grula, J. W., and Taylor, O. R. 1980. Some characteristics of hybrids derived from the sulphur butterflies, Colias eurytheme and C. philodice: phenotypic effects of the X-chromosome. Evolution 34:673–687.

Harrison, R. G., and Bogdanowicz, S. M. 1995. Mitochondrial DNA phylogeny of North American field crickets: perspectives on the evolution of life cycles, songs, and habitat associations. J. Evol. Biol. 8:209–232.

Hastings, I. M. 1994. Manifestations of sexual selection may depend on the genetic basis of sex determination. Proc. Roy. Soc. Lond. B 258:83–87.

Henderson, N. R., and Lambert, D. M. 1982. No significant deviation from random mating of worldwide populations of Drosophila melanogaster. Nature 300:437–440.

Hendrikse, A. 1988. Hybridization and sex pheromone responses among members of the Yponomeuta padellus-complex. Entomol. Exp. Appl. 48:213–233.

Henry, C. S. 1985. Sibling species, call differences, and speciation in green lacewings (Neuroptera, Chrysopidae, Chrysoperla). Evolution 39:965–984.

Henry, C. S. 1994. Singing and cryptic speciation in insects. Trends Ecol. Evol. 9:388–392.

Hewitt, G. M. 1989. The subdivision of species by hybrid zones. Pages 85–110 in Speciation and Its Consequences. D. Otte and J. A. Endler (eds.). Sunderland, Mass.: Sinauer.

Hoikkala, A., and Lumme, J. 1984. Genetic control of the difference in male courtship sound between Drosophila virilis and D. lummei. Behav. Genet. 14:257–268.

Hoikkala, A., and Lumme, J. 1987. The genetic basis of the evolution of the male courtship sounds of the Drosophila virilis group. Evolution 41:827–845.

Hollocher, H., and Wu, C.-I. 1996. The genetics of reproductive isolation in the Drosophila simulans clade: X vs. autosomal effects and male vs. female effects. Genetics 143:1243–1255.

Hoy, R. R., Hahn, J., and Paul, R. C. 1977. Hybrid cricket auditory behavior: evidence for genetic coupling in animal communication. Science 195:82–84.

Jamart, J. A., Carracedo, M. C., and Casares, P. 1993. Sexual isolation between Drosophila melanogaster females and D. simulans males. Male mating propensities versus success in hybridisation. Experientia 49:596–598.

Klun, J. A., and Maini, S. 1979. Genetic basis of an insect chemical communication system: the European corn borer (Ostrinia nubilalis). Environ. Entomol. 8:423–426.

Kochansky, J., Cardé, R. T., Liebherr, J., and Roelofs, W. 1975. Sex pheromone of the European corn borer, in New York. J. Chem. Ecol. 1:225–231.

Kyriacou, C. P., and Hall, J. C. 1980. Circadian rhythm mutations in Drosophila melanogaster affect short-term fluctuations in the male's song. Proc. Nat. Acad. Sci. USA 77:6729–6733.

Kyriacou, C. P., and Hall, J. C. 1986. Interspecific genetic control of courtship song production and reception in Drosophila. Science 232:494–497.

Liebherr, J., and Roelofs, W. 1975. Laboratory hybridization and mating period studies using two pheromone strains of Ostrinia nubilalis. Ann. Entomol. Soc. Am. 68:305–309.

Liu, J., Mercer, J. M., Stam, L. F., Gibson, G. C., Zeng, Z.-B., and Laurie, C. C. 1996. Genetic analysis of a morphological shape difference in the male genitalia of Drosophila simulans and D. mauritiana. Genetics 142:1129–1145.

Löfstedt, C. 1993. Moth pheromone genetics and evolution. Philos. Trans. R. Soc. Lond. B 340:167–177.

Löfstedt, C., Hansson, B. S., Roelofs, W., and Bengtsson, B. O. 1989. No linkage between genes controlling female pheromone production and male pheromone response in the European corn borer, Ostrinia nubilalis Hübner (Lepidoptera; Pyralidae). Genetics 123:553–556.

Mayr, E. 1963. Animal Species and Evolution. Cambridge, Mass.: Harvard University Press.

Mitchell-Olds, T. 1995. The molecular basis of quantitative genetic variation in natural populations. Trends Ecol. Evol. 10:324–328.

Muller, H. J. 1940. Bearing of the Drosophila work on systematics. Pages 185–268 in The New Systematics. J. Huxley (ed.). Oxford: Clarendon Press.

Noor, M. A. F. 1996. Absence of species discrimination in Drosophila pseudoobscura and D. persimilis males. Anim. Behav. 52:1205–1210.

Noor, M. A. F. 1997. Genetics of sexual isolation and courtship dysfunction in male hybrids of Drosophila pseudoobscura and D. persimilis. Evolution 51:809–815.

Noor, M. A. F., and Coyne, J. A. 1996. Genetics of a difference in cuticular hydrocarbons between Drosophila pseudoobscura and D. persimilis. Genet. Res. Camb. 68:117–123.

Orr, H. A. 1993. Haldane's rule has multiple genetic causes. Nature 361:532–533.

Orr, H. A., and Coyne, J. A. 1992. The genetics of adaptation—a reassessment. Am. Nat. 140:725–742.

Paterson, H. E. H. 1993. *Evolution and the Recognition Concept of Species*. Baltimore: The Johns Hopkins University Press.

Petersen, G., Hall, J. C., and Rosbash, M. 1988. The *period* gene of *Drosophila* carries species specific behavioural instructions. EMBO J. 7:3939–3947.

Pomiankowski, A., and Møller, A. P. 1995. A resolution of the lek paradox. Proc. Roy. Soc. Lond. B 260:21–29.

Pugh, A. R. G., and Ritchie, M. G. 1996. Polygenic control of a mating signal in *Drosophila*. Heredity 77:378–382.

Rao, S. R. V., and Padmaja, M. 1992. Mammalian-like dosage compensation mechanism in an insect—*Gryllotalpa fosser* (Scudder) Orthoptera. J. Biosci. 17:253–273.

Reinhold, K. 1994. Inheritance of body and testis size in the bushcricket *Poecilimon veluchianus* Ramme (Orthoptera; Tettigoniidae) examined by means of subspecies hybrids. Biol. J. Linn. Soc. 52:305–316.

Rice, W. R. 1984. Sex chromosomes and the evolution of sexual dimorphism. Evolution 38:735–742.

Ritchie, M. G. 1996. The shape of female mating preferences. Proc. Nat. Acad. Sci. USA 93:14628–14631.

Ritchie, M. G., and Kyriacou, C. P. 1994a. Reproductive isolation and the *period* gene of *Drosophila*. Mol. Ecol. 3: 595–599.

Ritchie, M. G., and Kyriacou, C. P. 1994b. The genetic variability of courtship song in a population of *Drosophila melanogaster*. Anim. Behav. 45:425–434.

Ritchie, M. G., and Kyriacou, C. P. 1996. Artificial selection for a courtship trait in *Drosophila melanogaster*. Anim. Behav. 52:603–611.

Ritchie, M. G., Yate, V. H., and Kyriacou, C. P. 1994. Genetic variability of the interpulse interval of courtship song among some European populations of *Drosophila melanogaster*. Heredity 72:459–464.

Roelofs, W., Glover, T., Tang, X., Sreng, I., Robbins, P., Eckenrode, C., Lofstedt, and C., Ennis, J. 1987. Sex pheromone responses of *Choristoneura* spp. and their hybrids (Lepidoptera: Torticidae). Can. Entomol. 109:1203–1220.

Ross, M. H. 1992. Hybridisation studies of *Blattela germanica* and *Blattela asahinai* (Dictyoptera: Blattelidae): dependence of a morphological and a behavioral trait on the species of the X chromosome. Ann. Entomol. Soc. Am. 85:348–354.

Ryan, M. J., and Rand, A. S. 1995. Female responses to ancestral advertisement calls in Túngara frogs. Science 269: 390–392.

Sanders, C. J., Daterman, G. E., and Ennis, T. J. 1977. Sex pheromone responses of *Choristoneura* spp. and their hybrids (Lepidoptera: Torticidae). Can. Entomol. 109: 1203–1220.

Scott, D. 1994. Genetic variation for female mate discrimination in *Drosophila melanogaster*. Evolution 48:112–121.

Scriber, J. M. 1992. Latitudinal clines in oviposition preferences: ecological and genetic influences. Pages 212–214 in Proceedings of the 8th International Symposium on Insect-Plant Relationships. S. B. J. Menken, J. H. Visser, and P. Harrewjin (eds.). Dordecht: Kluwer.

Shaw, K. L. 1996. Polygenic inheritence of a behavioral phenotype—interspecific genetics of song in the Hawaiian cricket genus *Laupala*. Evolution 50: 256–266.

Showers, W. B. 1981. Geographic variation of diapause response in the European corn borer. Pages 97–111 in Insect Life History Patterns: Habitat and Geographic Variations. R. F. Denno and H. Dingle (eds.). New York: Springer.

Silberglied, R. E., and Taylor, O. R. 1978. Ultraviolet reflection and its behavioural role in the courtship of the sulphur butterflies, *Colias eurytheme* and *C. philodice* (Lepidoptera, Piridae). Behav. Ecol. Sociobiol. 3:203–243.

Smith, S. G. 1953. Reproductive isolation and integrity of two sympatric species of *Choristoneura* (Lepidoptera: Tortricidae). Can. Entomol. 85:141–151.

Smith, S. G. 1954. A partial breakdown of temporal and ecological isolation between *Choristoneura* species (Lepidoptera: Tortricidae). Evolution 8:206–224.

Sperling, F. A. H. 1987. Evolution of the *Papilo machaon* species group in western Canada. Q. Entomol. 23:198–315.

Sperling, F. A. H. 1994. Sex-linked genes and species differences in Lepidoptera. Can. Entomol. 126:807–818.

Stratton, G. E., and Uetz, G. W. 1986. The inheritence of courtship behavior and its role as a reproductive isolating mechanism in two species of *Schizocosa* wolf spiders (Araneae: Lycosidae). Evolution 40:129–141.

Tan, C. C. 1946. Genetics of sexual isolation between *Drosophila pseudoobscura* and *D. persimilis*. Genetics 31:558–573.

Teal, P. E. A., and Oostendorp, A. 1995 Production of pheromone by hairpencil glands of males obtained from interspecific hybridization between *Heliothis virescens* and *Heliothis subflexa* (Lepidoptera, nouctuidae). J. Chem. Ecol. 21:59–67.

Thompson, J. N. 1988. Evolutionary genetics of oviposition preference in swallowtail butterflies. Evolution 42:1223–1235.

Tomaru, M., and Oguma, Y. 1994. Differences in courtship song in the species of the *Drosophila auraria* complex. Anim. Behav. 47:133–140.

Trivers, R. L. 1972. Parental investment and sexual selection. Pages 136–179 in *Sexual Selection and the Descent of Man*. B. Campbell (ed.). Chicago: Aldine.

True, J. R., Weir, B. S., and Laurie, C. C. 1996. A genome-wide survey of hybrid incompatibility factors by the introgression of marked segments of *Drosophila mauritiana* chromosomes into *Drosophila simulans*. Genetics 142:819–837.

Turelli, M., and Orr, H. A. 1995. The dominance theory of Haldane's Rule. Genetics 140:389–402.

von Helversen, D., and von Helversen, O. 1975a. Verhaltengenetische untersuchungen am akustischen kommunikationssystem der feldheuschrecken (Orthoptera: Acrididae). I. Der gesang von artbastarden zwischen *Chorthippusbiguttulus* und *Ch. mollis*. J. Comp. Physiol. 104:273–299.

von Helversen, D., and von Helversen, O. 1975b. Verhaltengenetische untersuchungen am akustischen kommunikationssystem der feldheuschrecken (Orthoptera: Acrididae). II. Das lautschema von artbastarden zwischen *Chorthippus biguttulus* und *Ch. mollis*. J. Comp. Physiol. 104:301–323.

Welbergen, P., Spruijt, B. M., and van Dijken, F. R. 1992. Mating speed and the interplay between female and male courtship responses in *Drosophila melanogaster* (Diptera: Drosophilidae). J. Insect Behav. 5:229–244.

Wells, M. M., and Henry, C. S. 1994. Behavioral responses of hybrid lacewings (Neuroptera: Chrysopidae) to courtship song. J. Insect Behav. 7:649–662.

Wheeler, D. A., Kyriacou, C. P., Greenacre, M. L., Yu, Q., Rutila, J. E., Rosbash, M., and Hall, J. C. 1991. Molecular transfer of a species-specific behaviour from *Drosophila simulans* to *Drosophila melanogaster*. Science 251: 1082–1085.

Wu, C.-I. 1996. Now blows the east wind. Nature 380:105–107.

Wu, C.-I., Hollocher, H., Begun, D. J., Aquadro, C. F., Xu, Y., and Wu, M.-L. 1995. Sexual isolation in *Drosophila melanogaster*: a possible case of incipient speciation. Proc. Nat. Acad. Sci. USA 92:2519–2523.

Wu, C.-I., Johnson, N. A., and Palopoli, M. F. 1996. Haldane's rule and its legacy: why are there so many sterile males? Trends Ecol. Evol. 11:281–284.

Zouros, E. 1981. The chromosomal basis of sexual isolation in two sibling species of *Drosophila*: *D. mojavensis* and *D. arizonensis*. Genetics 97:703–718.

23

Sex Linkage and Speciation in Lepidoptera

Dorothy Pashley Prowell

Lepidoptera have undergone extensive radiation. They are the second largest insect order and contain over 112,000 described species (Arnett 1985). An unusual feature of speciation in Lepidoptera is a disproportionate association between traits that distinguish closely related species and the X-chromosome (Sperling 1994). In a survey of the six best-documented examples of linkage relationships of traits that differ between such taxa, more than half of the morphological, behavioral, and physiological genes were traced to the X-chromosome. Because lepidopterans have large numbers of small chromosomes (an average of 31; Robinson 1971), the bias of traits on the X-chromosome is considerably disproportionate to its relative abundance in the genome.

A relationship between X-linkage and speciation has been reported previously (Charlesworth et al. 1987; Coyne and Orr 1989). The importance of the X-chromosome was first emphasized by Haldane (1922) when he proposed that the hemizygous sex (XY) would be the sterile, inviable, or rare sex in interracial crosses (now referred to as Haldane's Rule). Empirical evidence supporting Haldane's Rule is extensive and includes cases within Lepidoptera, Diptera, Hemiptera, Anoplura, Cladocera, salamanders, birds, and mammals (Haldane 1922; Coyne and Orr 1989). Although the evolutionary phenomenon is widely accepted, the underlying mechanism has been controversial (Virdee 1993; Davies and Pomiankowski 1995). Mechanisms proposed have included an X–Y interaction (Haldane 1932; Coyne 1985), an X-autosome imbalance (Muller 1940; Charlesworth et al. 1987), a dominance effect (Turelli and Orr 1995), and meiotic drive or sex-ratio distorter genes (Frank 1991; Hurst and Pomiankowski 1991; Pomiankowski and Hurst 1993). Still others have suggested a composite phenomenon with separate mechanisms or effects for different sterility and inviability genes (Wu and Davis 1993; Orr 1993a; Hollocher and Wu 1996; Wu et al. 1996).

In addition to Haldane's Rule, Coyne and Orr (1989) proposed an "X-effect" to account for the observation that most of the genes affecting sterility and inviability are on the X-chromosome. Recently, the X-effect for sterility genes has been challenged (Hollocher and Wu 1996), but it is still valid for traits without multiple genetic pathways and causes (i.e., there may be over a hundred genes acting independently throughout the genome that cause sterility in *Drosophila*; Wu et al. 1996; Wu and Hollocher, this volume).

Much of the debate about Haldane's Rule and the X-effect has centered around postreproductive traits, almost exclusively inviability and sterility in *Drosophila*. A question of greater relevance to speciation is whether they apply to the evolution of prereproductive isolating mechanisms. These types of genes are more likely to initiate or reinforce differentiation. Although an X-effect in morphological and behavioral traits has not been observed in *Drosophila* (Wu and Hollocher, this volume), it has been observed within Lepidoptera and crickets (Charlesworth et al. 1987). In a review of Lepidoptera, Sperling (1994) reported numerous cases of prereproductive trait differences (Ritchie and Phillips, this volume).

Since Haldane (1922), several others have dealt theoretically with the importance of the hemizygous condition and evolutionary rates (Hartl 1971, 1972; Lester and Selander 1979; Avery 1984; Charlesworth et al. 1987; Orr 1993b). In general, X-linked loci evolve more rapidly than autosomal loci under natural selection if mutations are favorable and partially or fully recessive. Recessive alleles are shielded from selection by the heterozygous condition but are exposed to selection when hemizygous (Haldane 1924). Faster rates of evolution of X-linked genes is particularly relevant to genes coding for sexual dimorphism, many of which are sex limited in expression, because they are predicted to be disproportionately on the X-chromosome (Rice 1984a). The best empirical evidence of this is in *Colias* butterflies, where sexual dimorphism in body color, pheromones, and mating behavior are primarily sex limited in expression and are X-linked (Grula and Taylor 1980a,b).

In this chapter, I review speciation in Lepidoptera in light of these X-associated phenomena. First, I provide an overview of speciation in one well-studied species containing two host strains, the fall armyworm. This is followed by an update on Sperling (1994) and a review of the literature on traits that distinguish lepidopteran species and strains. Next, results are evaluated in the context of current knowledge on X-linkage, the gender of the heterogametic sex, and sex-limited expression. In the final section, I address the following questions: (1) Do traits that distinguish closely related lepidopterans exhibit an X-chromosome bias? (2) If so, what are the possible explanations for the bias? (3) Are speciation rates influenced by it? (4) Is the bias unique to Lepidoptera?

Speciation-Related Genes and Their Location in Lepidoptera

An Example from Fall Armyworm

The fall armyworm (*Spodoptera frugiperda* J.E. Smith) is a noctuid moth that has received considerable attention over the past century because of its significance as a pest of corn, rice, and forage grasses. Over 1,300 accounts have been published on its biology or control (Ashley et al. 1989). Despite this attention, the presence of two genetically differentiated forms went undetected until an allozyme study was conducted across its geographic range (Pashley et al. 1985). A difference was detected at an esterase locus that was coincident with host use. One set of genotypes was found feeding on corn (referred to as the corn strain), and another set of genotypes was found feeding on rice and forage grasses (referred to as the rice strain). Cross-rearing studies demonstrated that differences were not caused by divergent selection in the two host environments, but were more likely due to reduced gene flow between strains (Pashley 1986).

Subsequent to this discovery, numerous behavioral and physiological differences were detected (Pashley 1988; Pashley et al. 1995; Veenstra et al. 1995). Despite these differences, genetic markers are the only reliable means of distinguishing the two strains. The esterase allozyme is nearly diagnostic. Different mitochondrial DNA (mtDNA) haplotypes characterize each strain (Pashley 1989; Pashley and Ke 1992) and a rice-strain-specific DNA repeat has been reported (Lu et al. 1994).

An array of prereproductive isolating mechanisms exist that probably function in concert to minimize gene flow between the strains as there is no evidence of postzygotic isolation (Pashley et al. 1992). These include seasonal differences, minor female pheromone differences, and assortative mating in the laboratory. The most striking isolating mechanism between the strains is nightly mating time. Coupling in the corn strain occurs in the first half of the night, whereas mating in the rice

strain occurs in the second half. Despite these diverse mechanisms, none are completely fixed and asymmetries exist with respect to the sexes (e.g., corn strain males show no assortative mating whereas corn females do).

Results from a genetic analysis based on mtDNA and allozymes suggest differential host use by the two strains and interbreeding in the field (table 23.1). Considering mtDNA patterns, there are almost no corn mtDNA genotypes (2%) in the rice strain habitat. In contrast, rice strain mtDNA occurs at a frequency of 18% in the corn habitat. These results are probably due to differential ovipositional specificity in the two strains.

Composite genotypes at esterase and mtDNA indicate potential interstrain hybridization. Approximately 11% of the individuals that have an allozyme marker of one strain have an mtDNA genotype of the other (table 23.1). If "hybrid" allozyme genotypes are included (middle CR column in table 23.1), 16% of the individuals have conflicting genotypes.

To establish if hybridization is occurring in nature, diagnostic nuclear markers have been sought. Three approaches have been taken but have not produced diagnostic differences. Over 80 RAPD (random amplified polymorphic DNA) primers have been screened and indicate high within-strain variability but no strain-specific bands (D. Prowell, unpublished). The ITS1 gene has been completely sequenced for both strains. The approximately 350–base pair (bp) region contains numerous point mutations and some small indels, but, like RAPDs, the variation is not strain specific. A more extensive study of a sodium channel intron (190 bp) exhibited the same pattern (Adamcyzk et al. 1996).

Most of the traits that distinguish fall armyworm strains exhibit an unusual linkage association or inheritance pattern. Of nine traits for which inheritance patterns have been established (table 23.2), two are X-linked, two are maternally inherited or influenced, one is Y-linked, and four are autosomal. Given approximately 30 chromosomes for fall armyworm, this represents a strong bias (33%) of strain-divergent traits on the sex chromosomes. Furthermore, the four autosomal traits are allozymes that exhibit limited divergence (10–50%) relative to esterase (90%).

Evolution in fall armyworm appears to be accelerated for sex-linked genes and maternal traits. Indeed, the difficulty in identifying morphological and autosomal nuclear markers may be due to slower rates of evolution for those types of traits. If strains are not interbreeding, faster rates of evolution for mtDNA and sex-linked genes are implied. If hybridization is occurring, faster rates may also apply but the maintenance of differences could be due to differential rates of introgression of genes across the strain boundary (Harrison 1989). Genes linked to traits associated with fitness in each habitat may remain differentiated, whereas other genes may move freely between taxa. If the X-chromosome contains habitat fitness genes, portions of the chromosome surrounding those

Table 23.1. Frequency of cytonuclear (mtDNA and allozyme) genotypes in fall armyworm individuals collected in corn and grass (including rice) fields at various sites and dates.

| Habitat | MtDNA Genotype | Esterase Genotype | | | |
		CC♂/C♀	CR♂	RR♂/R♀	Total
Corn	C	0.73	0.05	0.04	0.82
(N = 340)	R	0.12	0.00	0.06	0.18
	Total	0.85	0.05	0.10	1.00
Grass	C	0.01	0.00	0.01	0.02
(N = 244)	R	0.02	0.07	0.89	0.98
	Total	0.03	0.07	0.90	1.00

C and R refer to genotypes characterizing the corn and rice strains, respectively.

genes could exhibit limited gene flow between the strains. Different population genetic processes are required to explain differentiation in mtDNA. Whether traits that differ between fall armyworm strains were instrumental in facilitating divergence initially or, instead, reinforced differences after the strains diverged in allopatry will be difficult to determine.

Summary of Trait Divergence in Lepidoptera

To assess the types of traits that evolve during the speciation process, results are summarized for 11 well-studied lepidopteran taxon pairs (table 23.2). Taxa included contain data for at least two loci from different classes of traits. Eight moth pairs from four lepidopteran families and three butterflies from two families were included. Five are races or strains and six are species. Six of the taxon pairs were reviewed and discussed in detail by Sperling (1994). New information has been added to two pairs.

Overall, there is a strong bias of X-linkage of traits. Out of a total of 77 traits, 39% are X-linked, 10% are maternal or Y-linked, and 51% are autosomally inherited. This bias is not uniformly distributed among taxon pairs. In *Colias* butterflies, all traits are X-linked, whereas in *Yponomeuta* moths, all are autosomal except mtDNA. Eight of 11 taxon pairs are distinguished by some X-linked traits. In all taxon pairs, the chromosomal number is around 30 with the possible exception of *Ostrinia* (pyralid chromosome numbers vary from 10 to 41; Robinson 1971). The proportion of traits on the X-chromosome is predicted to be 1/30th rather than the 1/3 that is observed.

Summarization of traits by functional category provides insight into the X-chromosome bias (table 23.3). Traits recognized to be important for speciation are X-linked in the majority of cases. These types of traits include host preference, performance on the host, and mate choice (Bush 1975). Inheritance patterns of traits in these three categories indicate a 67% X-linkage for mating behavior and ovipositional preference genes, and 60% for performance traits. In contrast, morphological traits exhibit a 38% X-linked bias. Allozymes exhibit only 24% X-linkage. In addition, the average frequency difference between taxa for X-linked allozymes is 83% compared to only 50% for autosomal loci. In four of five cases where the list includes both X-linked and autosomal allozyme loci, the X-linked loci exhibit greater frequency divergence. These data indicate an X-chromosome bias in all traits, with the greatest bias in traits relevant to speciation, including mating behavior and habitat choice.

Almost a third (31%) of the traits that distinguish taxa are sex limited in their expression (table 23.3). Of these traits, 71% are female limited and 29% are male limited. An association is apparent between sexual expression and chromosomal location of traits with the majority of sex-limited traits (63%) occurring on the sex chromosomes. All mating behavior traits and host preference are sex limited in expression. Sex-limited expression is much lower for the morphological (38%) and physiological (20%) traits.

Finally, mtDNA has been examined in five taxon pairs and is diagnostic in all of them. These findings are consistent with numerous other studies supporting a fast rate of evolution for this genome (Avise 1986; Harrison 1989).

Role of Female Heterogamety and X-Linkage in Speciation

The X-Effect in Lepidoptera

The X-chromosome bias of traits in Lepidoptera is consistent with a faster rate of evolution for X-linked traits. Perhaps the strongest evidence for this explanation comes

Table 23.2. Genetic traits that differ between species or strains of Lepidoptera.

Taxon Pair (Family)	Trait	Difference[1]	Inheritance	Expression	Reference
Heliothis virescens/H. subflexa (Noctuidae)	Oviposition preference	Significant	Autosomal	Females	Sheck and Gould 1995a
	Larval growth	Significant	Autosomal	Both sexes	Sheck and Gould 1995b
Spodoptera frugiperda host strains (Noctuidae)	*Est3* allozyme	90%	X-linked	Both sexes	Pashley 1986; D. Heckel (unpublished)
	Female calling time	Diagnostic	X-linked	Females	E. Lima, J. McNeil (unpublished)
	Hbdh allozyme	50%	Autosomal	Both sexes	Pashley 1986
	PepF allozyme	25%	Autosomal	Both sexes	Pashley 1986
	Pgm allozyme	15%	Autosomal	Both sexes	Pashley 1986
	Gpi allozyme	10%	Autosomal	Both sexes	Pashley 1986
	DNA repeat	Diagnostic	Y-linked	Females	Lu et al. 1994
	Egg color	Diagnostic?	Maternal	Females	Pashley and K. Veenstra (unpublished)
	Mitochondrial DNA	Diagnostic	Maternal	Both sexes	Pashley 1989
Papilio glaucus/P. canadensis (Papilionidae)	Pupal diapause	Significant	X-linked	Both sexes	Rockey et al. 1987
	Oviposition preference	Significant	X-linked	Females	Scriber 1992
	Mimetic morph suppression	Significant	X-linked	Both sexes	Hagen et al. 1991
	Ldh allozyme	100%	X-linked	Both sexes	Hagen and Scriber 1989
	Pgd allozyme	90%	X-linked	Both sexes	Hagen and Scriber 1989
	Hk allozyme	95%	Autosomal	Both sexes	Hagen and Scriber 1991
	Larval feeding	Significant	Autosomal	Both sexes	Hagen et al. 1991
	Mimetic morph determination	Significant	Maternal	Females	Hagen et al. 1991
	Mitochondrial DNA	Diagnostic	Maternal	Both sexes	Sperling 1993a
Papilio machaon/P. zelicaon (Papilionidae)	Oviposition preference	Significant	X-linked	Females	Thompson 1988
	Larval development	Significant	Autosomal	Both sexes	Thompson et al. 1990
	Larval spot color	Significant	Autosomal	Both sexes	Sperling 1987
	Hindwing eyespot	Significant	Autosomal	Both sexes	Clarke and Sheppard 1955
	Forewing apex	Significant	Autosomal	Both sexes	Clarke and Sheppard 1955
	Submarginal spots	Significant	Autosomal	Both sexes	Clarke and Sheppard 1955
	Adult body color	Significant	Autosomal	Both sexes	Clarke and Sheppard 1955
	Est4 allozyme	85%	Autosomal	Both sexes	Sperling 1987
	G6pd allozyme	50%	Autosomal	Both sexes	Sperling 1987
	Mitochondrial DNA	Diagnostic	Maternal	Both sexes	Sperling and Harrison 1994
Colias philodice/C. eurytheme (Pieridae)	Development rate	Significant	X-linked	Both sexes	Grula and Taylor 1980a
	Wing pigmentation	Significant	X-linked	Both sexes?	Grula and Taylor 1980a
	Adult size	Significant	X-linked	Both sexes	Grula and Taylor 1980a
	Fertility/fecundity	Significant	X-linked	Females	Grula and Taylor 1980a
	Ultraviolet wing	Significant	X-linked	Males	Silberglied and Taylor 1973
	Mate selection	Significant	X-linked	Females	Silberglied and Taylor 1978
	Pheromone composition	Significant	X-linked	Males	Grula and Taylor 1978
	Mating vigor	Significant	X-linked	Females	Grula and Taylor 1980a

Taxon (Family)	Trait	Difference[1]	Inheritance	Expression	Reference
Ostrinia nubilalis pheromone strains (Pyralidae)	Voltinism	Diagnostic	X-linked	Both sexes	Glover et al. 1992
	Pheromone response	Diagnostic	X-linked	Males	Roelofs et al. 1987
	Tpi allozyme	100%	X-linked	Both sexes	Glover et al. 1991
	Got1 allozyme	5%	Autosomal	Both sexes	Harrison and Vawter 1977
	Got2 allozyme	5%	Autosomal	Both sexes	Harrison and Vawter 1977
	Pgm allozyme	5%	Autosomal	Both sexes	Harrison and Vawter 1977
	Pheromone composition	Diagnostic	Autosomal	Females	Klun and Maini 1979
	Pheromone detection	Diagnostic	Autosomal	Males	Roelofs et al. 1987
Choristoneura fumiferana/C. pinus (Tortricidae)	Egg weight	Significant	X-linked	Both sexes	Campbell 1958, 1962
	Pupal weight	Significant	X-linked	Both sexes	Campbell 1958
	Pupal hemolymph	Significant	X-linked	Sexually dimorphic	Stehr 1959
	Adult wing color	Significant	X-linked	Females	Stehr 1955, 1964
	Larval/pupal development	Significant	X-linked	Both sexes	Smith 1953; Campbell 1962
	Aat1 allozyme	95%	X-linked	Both sexes	Harvey 1996
	Pheromone composition	Diagnostic	X-linked	Females	Sanders et al. 1977
	Female calling time	Significant	X-linked	Females	Sanders et al. 1977
	Female fecundity and body size	Significant	X-linked	Females	Campbell 1962
	Est5 allozyme	35%	Autosomal	Both sexes	Harvey 1996
	Pgi1 allozyme	35%	Autosomal	Both sexes	Harvey 1996
	Larval diapause	Significant	Autosomal	Both sexes	Harvey 1967
	Body weight available for egg production	Significant	Autosomal	Females	Campbell 1958
	Mitochondrial DNA	Diagnostic	Maternal	Both sexes	Sperling and Hickey 1995
Ctenopseustis obliquana pheromone races (Tortricidae)	Pheromone response	Significant	X-linked	Males	Hansson et al. 1989
	28S RNA	Diagnostic	Autosomal?	Both sexes	Sin et al. 1995
	Hkl allozyme	100%	Autosomal	Both sexes	White and Lambert 1994
Planotortrix excessana pheromone races (Tortricidae)	28S RNA	Diagnostic	Autosomal?	Both sexes	Sin et al. 1995
	Hbdh3 allozyme	100%	Autosomal	Both sexes	White and Lambert 1994
Zeiraphera diniana larch/pine host races (Tortricidae)	Idh-S allozyme	20%	X-linked	Both sexes	Emelianov et al. 1995
	Adult color morph	Significant	Autosomal	Males	Priesner and Baltensweiler 1987
	Mdh-S allozyme	75%	Autosomal	Both sexes	Emelianov et al. 1995
	Pgm allozyme	15%	Autosomal	Both sexes	Emelianov et al. 1995
	Pheromone response	Diagnostic	Autosomal	Males	Priesner and Baltensweiler 1987
Yponomeuta padella/Y. malinellus (Yponomeutidae)	Coccoon webs	Significant	Autosomal	Both sexes	Hendrikse 1988
	Larval gustation	Significant	Autosomal	Both sexes?	van Drongelen and van Loon 1980
	Pheromone composition	Diagnostic	Autosomal	Females	Hendrikse 1988
	Est1 allozyme	95%	Autosomal	Both sexes	Arduino and Bullini 1985
	Cal allozyme	50%	Autosomal	Both sexes	Arduino and Bullini 1985
	Mpi allozyme	80%	Autosomal	Both sexes	Arduino and Bullini 1985
	Mitochondrial DNA	Diagnostic	Maternal	Both sexes	Sperling et al. 1995

[1]Allozyme differences are approximations rounded off to the nearest 5%.

Table 23.3. Summary of linkage associations and expression of traits from table 23.2.

Traits	Maternal/ Y-Linked	X-Linked	Autosomal	Female-Biased Expression	Male-Biased Expression
Performance/physiology	0	9	6	3	0
Mating behavior	0	8	4	7	5
Oviposition preference	0	2	1	3	0
Morphology	2	5	6	3	2
Cocoon webbing	0	0	1	0	0
Allozymes	0	6	19	0	0
mtDNA	5	0	0	0	0
Nuclear DNA	1	0	2	1	0
Totals	8	30	39	17	7

from the allozyme loci that are not causally related to speciation. Nearly a quarter of the loci are X-linked. In addition, when multiple allozyme loci are examined, genetic divergence is almost always much greater in X-linked loci.

A second factor enhancing the X-effect is mode of expression for traits involved in the divergence process. Traits important to speciation often involve mate and habitat choice, which tend to be sex limited in expression. Within the lepidopteran examples, all mating behavior and host choice traits have a sexually biased mode of expression (tables 23.2 and 23.3). Moreover, excluding the 33 allozyme and DNA loci, over half of all traits studied (55%) are sex biased in expression. Arguments have been made for the evolutionary placement of sexually dimorphic or sex-limited genes on the sex chromosomes (Rice 1984a). Within Lepidoptera, sex-limited traits tend to occur on sex chromosomes more often than traits expressed in both sexes. Of the 25 traits that are sex biased or dimorphic (table 23.3), 60% are X-linked, 12% are maternal, and 28% are autosomal. In contrast, of the 52 traits with dual expression, 29% are X-linked, 10% are maternal, and 61% are autosomal. If traits that diverge during speciation are biased in favor of sex-limited traits, and sex-limited traits tend to be on the X-chromosome, an X-effect is expected. If the hemizygous condition contributes to faster rates of evolution, female-limited expression of X-linked traits in combination with female heterogamety in Lepidoptera could elevate rates of evolution for these traits.

Habitat Differentiation Mediated by X-Linkage and Female Heterogamety

Under certain conditions, female heterogamety and X-linkage can cause decreased fitness of progeny from interhabitat matings. The first condition is that habitat choice must be determined by the female, a requirement generally met by ovipositing females in insects. Second, fitness genes associated with habitat use must be on the

X-chromosome. In Lepidoptera, many fitness traits are X-linked (tables 23.2 and 23.3). In this model, host choice (C_i) and larval performance (P_i) are on the X-chromosome (figure 23.1). The X-chromosome location of host choice is not critical for a single generation but is important in subsequent generations if hybrid progeny survive. The i subscript refers to alleles associated with the ith habitat. Each locus contains two alleles, associated with either habitat 1 or 2 (figure 23.1A). The ith habitat is characterized by individuals with oviposition preferences for that habitat. Larvae are more fit when their performance alleles match the habitat in which they occur.

If females are the heterogametic sex and interhabitat matings occur, male and female progeny will exhibit fitness disadvantages (figure 23.1B). Parental females that mate with males from the other habitat will lay eggs in their own habitat because they have the oviposition allele for that habitat. Female progeny will receive performance genes from their male parent that will not match the habitat in which they occur. Male progeny will be heterozygous for performance genes. Their fitness will depend on dominance relationships of alleles and will be reduced if there is codominance.

In species with male heterogamety, only female progeny are compromised in their performance (figure 23.1C). Male progeny receive their performance genes from their female parent and match the habitat in which they occur. Thus, male progeny fitness is equal to progeny produced by intrahabitat matings. Female progeny are heterozygous for performance genes and, as with male progeny in figure 23.1B, their fitness depends on dominance relationship between performance alleles.

This model is presented in its simplest form considering only a single generation with complete segregation of genes within each of the two habitats. The outcome is considerably more complex if there are differences in male and female movement between habitats, intrahabitat gene polymorphisms at each locus, different selection coefficients associated with larval genotypes, dominance

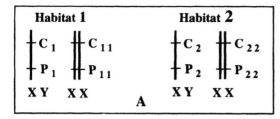

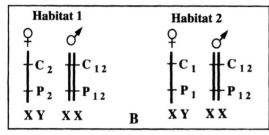

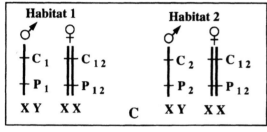

Figure 23.1. Effects of female versus male heterogamety on fitness of progeny from interhabitat matings. C_i = habitat choice, and P_i = larval performance, both X-linked, in the ith habitat. In habitat 1, fitness = $P_2 \leq P_{12} < P_1$, and in habitat 2, fitness = $P_1 \leq P_{12} < P_2$. A, genotypes of homogametic (XX) and heterogametic (XY) sexes in each habitat without interhabitat matings. B, progeny genotypes from interhabitat matings with female heterogamety. C, progeny genotypes from interhabitat matings with male heterogamety.

interactions, and so on. The integration of mate choice genes could also influence divergence likelihoods and rates. Moreover, linkage considerations become important with multiple generations (Rice 1984b, 1987; Diehl and Bush 1989).

Finally, the generality of this model also depends on which sex chooses habitats for progeny production. In the model, habitat use is determined by females. In other insects with female heterogamety such as certain Diptera and Trichoptera, this is likely. However, in birds and other vertebrates with female heterogamety, habitat use can be controlled by either or both sexes and predictions of this model might not apply. I present this model as a first step in understanding the role of female heterogamety in lepidopteran speciation. More extensive evaluations of the model are being developed elsewhere (Promislow, unpublished).

Haldane's Rule and Maternal Differentiation

X-linkage and female heterogamety have implications for the evolution of maternally inherited traits when Haldane's Rule is operational. If populations or species have evolved genetic differences, Haldane's Rule dictates that the heterogametic sex will suffer greater fitness disadvantages (Haldane 1922). In Lepidoptera, F_1 females will be preferentially eliminated in intertaxon matings. There is ample evidence of Haldane's Rule for viability and sterility genes in Lepidoptera (Haldane 1922). There is also evidence of prereproductive effects. F_1 females in butterflies (N. Davies, unpublished) and moths (Pashley and Martin 1987) refrained from mating whereas F_1 males readily mated.

Haldane's Rule can act to maintain distinctiveness of mtDNA among lepidopteran taxon pairs (Sperling 1993b; table 23.2) as well as Y-linked or maternal differences. If female hybrids are infertile or do not mate, but males disperse among habitats and successfully mate and transmit sperm, introgression of autosomal and X-linked traits between habitats would occur via males. In contrast, maternally transmitted traits such as mtDNA and Y-linked genes would remain restricted to a particular habitat or species. This process and behavior could explain maternal and Y-linked differences, but not X-linked trait differences.

Summary of Speciation in Lepidoptera and the Relevance of X-Linkage

At the beginning of this chapter, I posed the question of whether traits that distinguish closely related lepidopterans exhibit an X-chromosome bias. The answer is clearly yes. A more challenging question is why the bias exists. I have proposed a threefold explanation. First, traits on the X-chromosome evolve faster than autosomal ones and are more likely to differ among taxa. Autosomes contain the vast majority of genes, but they do not diverge as rapidly as X-linked genes and do not distinguish newly differentiated taxa. Second, many traits that are related to speciation tend to be sex limited in expression, and sex-limited traits tend to be on sex chromosomes. Finally, female-limited, X-linked traits may undergo fast rates of evolution relative to other traits when the female is the heterogametic sex and traits are advantageous.

The third question I posed was whether these X-chromosome-related phenomena enhance speciation rates. Undoubtedly, an herbivorous life style contributes substantially to speciation in Lepidoptera by providing opportunities for habitat shifts. However, an X-chromosome bias in combination with female heterogamety could facilitate speciation. Accelerated rates of evolution for traits on the X-chromosome such as mate choice could enhance the rate of development of assortative mating by shortening the time required for evolution of prereproductive isolation. In addition, X-linkage of performance genes associated with host use and mate choice could lead to decreased fitness in interhabitat matings, which would facilitate divergence. These processes can operate simultaneously to enhance the likelihood that taxa will develop new species if a shift in habitat use occurs. Other examples of rapid evolution of traits relevant to reproductive isolation that have effects similar to X-linkage include sexual selection (Lande 1982; West-Eberhart 1983), and gonad specific microorganisms (Werren, this volume).

The final question I posed in the introduction was whether the X-chromosome bias was unique to Lepidoptera. None of the explanations for the X-effect given above should apply only to Lepidoptera. Faster rates of evolution of sex-linked traits are expected in other organisms because the theoretical principles are general in nature (Charlesworth et al. 1987). Furthermore, when mating behavior is involved, as it commonly is in speciation, regardless of the organism, sex-biased expression is likely and sex linkage is predicted. An X-effect is not expected if (1) traits on autosomes and sex chromosomes evolve at the same rate; (2) speciation is not accompanied by a bias in divergence of sex-limited traits; and (3) sex-limited traits are not biased in their location on sex chromosomes. The universality of these genetic phenomena will be known when more linkage and inheritance data are available for a wide variety of organisms. It is possible that X-linkage is common in other closely related taxa including vertebrates, but has not been detected because linkage data are harder to gather for larger, long-lived animals.

If Lepidoptera are unique in this X bias, then the explanations given above for the X-effect are not sufficient and another interpretation must be considered. There are only a few intrinsic differences that stand out for Lepidoptera, and their relevance to the X bias is not obvious. For example, the numerous, small, holocentric chromosomes that characterize Lepidoptera are unusual and perhaps unique, but their relevance to X-linkage is not clear. Another unusual feature is female heterogamety. Among invertebrates, this trait is shared with the nearest ordinal relative to Lepidoptera, the Trichoptera, and a few dipteran groups (White 1973). Based on monophyly of these three insect orders (Liu and Beckenbach 1992; Pashley et al. 1993), it is likely that female heterogamety evolved once in the common ancestor to Trichoptera, Lepidoptera, and Diptera and was lost in certain Diptera. In vertebrates, heterogamety in females is nearly as common as in males. It occurs in all birds, some fishes, amphibians, and reptiles but not in mammals (Bull 1983). The primary difference between Lepidoptera and other groups with female heterogamety, with the exception of birds, is the lack of dosage compensation in males (Cock 1964; Johnson and Turner 1979). The effect of dosage compensation on evolutionary rates may depend on the degree of dominance of alleles, with conditions being somewhat stricter

without dosage compensation (Charlesworth et al. 1987). Further studies are required to fully understand the effects and prevalence of no dosage compensation in Lepidoptera.

High levels of speciation among diverse groups within the animal world are not likely to be explained by a single factor. Specific life history attributes, such as size, vagility, or feeding habits, play a major role by influencing habitat segregation and, consequently, speciation rates. However, broadly occurring, general mechanisms may supplement more basal speciation rates. In beetles and hummingbirds, the process may be sexual selection. In certain invertebrates, it may be microorganisms in the germ line. In other groups, including Lepidoptera, it may be X-linked traits and fortuitous evolutionary events such as the gender of the heterogametic sex.

Acknowledgments   Parts of this work were supported by USDA-NRICGP Grant #93-37302-9128. I am grateful to D. Promislow for first recognizing the importance of the gender of the heterogametic sex and for his input into the model, to F. Sperling for providing most of the information in table 23.2, and to J. Bossart, C. Carlton, D. Foltz, M. McMichael, M. Macnair, H. Naveira, M. Oard, and D. Promislow for providing constructive criticisms of the manuscript.

References

Adamcyzk, J. A., Prowell, D. P., and Silvain, J.-F. 1996. Intra- and interspecific DNA variation in a sodium channel intron in *Spodoptera* (Lepidoptera: Noctuidae). Ann. Entomol. Soc. Am. 89:812–821.

Arduino, P., and Bullini, L. 1985. Reproductive isolation and genetic divergence between the small ermine moths *Yponomeuta padellus* and *Y. malinellus* (Lepidoptera: Yponomeutidae). Atti della Accademia Nazionale dei Lincei Memorie, Serie 8, Sezione 3, 18:33–61.

Arnett, R. H. 1985. American Insects. New York: Van Nostrand Reinhold.

Ashley, T. R., Wiseman, B. R., Davis, F. M., and Andrews, K. L. 1989. The fall armyworm: a bibliography. Florida Entomol. 72:152–202.

Avery, P. J. 1984. The population genetics of haplo-diploids and X-linked genes. Genet. Res. 44:321–341.

Avise, J. C. 1986. Mitochondrial DNA and the evolutionary genetics of higher animals. Philos. Trans. R. Soc. Lond. B 312:325–342.

Bull, J. J. 1983. Evolution of Sex Determining Mechanisms. Menlo Park, Calif.: Benjamin/Cummings.

Bush, G. L. 1975. Modes of animal speciation. Annu. Rev. Ecol. Syst. 6:334–364.

Campbell, I. M. 1958. A genetic study of factors determining fecundity in the genus *Choristoneura* Led. (Lepidoptera: Tortricidae). PhD. dissertation, University of Toronto.

Campbell, I. M. 1962. Reproductive capacity in the genus *Choristoneura* Led. (Lepidoptera: Tortricidae). I. Quantitative inheritance and genes as controllers of rates. Can. J. Genet. Cytol. 4:272–288.

Charlesworth, B., Coyne, J. A., and Barton, N. H. 1987. The relative rates of evolution of sex chromosomes and autosomes. Am. Nat. 130:113–146.

Clarke, C. A., and Sheppard, P. M. 1955. A preliminary report on the genetics of the *machaon* group of swallowtail butterflies. Evolution 9:182–201.

Cock, A. G. 1964. Dosage compensation and sex-chromatin in non-mammals. Genet. Res. 5:354–365.

Coyne, J. A. 1985. The genetic basis of Haldane's Rule. Nature 314:736–738.

Coyne, J. A., and Orr, H. A. 1989. Two rules of speciation. In D. Otte and J. A. Endler (eds.). Speciation and Its Consequences. New York: Sinauer, pp. 180–207.

Davies, N., and Pomiankowski, A. 1995. Haldane's Rule: old theories are the best. Trends Ecol. Evol. 10:350–351.

Diehl, S. R., and Bush, G. L. 1989. The role of habitat preference in adaptation and speciation. In D. Otte and J. A. Endler (eds.). Speciation and Its Consequences. New York: Sinauer, pp. 345–365.

Emelianov, I., Mallet, J., and Baltensweiler, W. 1995. Genetic differentiation in *Zieraphera diniana* (Lepidoptera: Tortricidae, the larch budmoth): polymorphism, host races or sibling species. Heredity 75:416–424.

Frank, S. A. 1991. Haldane's Rule: a defense of the meiotic drive theory. Evolution 45:1714–1717.

Glover, T. J., Knodel, J. J., Robbins, P. S., Eckenrode, C. J., and Roelofs, W. L. 1991. Gene flow among three races of European corn borers (Lepidoptera: Pyralidae) in New York State. Environ. Entomol. 20:1356–1362.

Glover, T. J., Robbins, P. S., Eckenrode, C. J., and Roelofs, W. L. 1992. Genetic control of voltinism characteristics in European corn borer races assessed with a marker gene. Arch. Insect Biochem. Physiol. 20:107–117.

Grula, J. W., and Taylor, O. R. 1978. Genetics of mate-selection behavior in two species of *Colias* butterflies. Genetics 88:34–35.

Grula, J. W., and Taylor, O. R., Jr. 1980a. The effect of X-chromosome inheritance on mate-selection behavior in the sulfur butterflies, *Colias eurytheme* and *C. philodice*. Evolution 34:688–695.

Grula, J. W., and Taylor, O. R., Jr. 1980b. Some characteristerics of hybrids derived from the sulfur butterflies, *Colias eurytheme* and *C. philodice*: phenotypic effects of the X-chromosomes. Evolution 34:673–687.

Hagen, R. H., and Scriber, J. M. 1989. Sex-linked diapause, color, and allozyme loci in *Papilio glaucus*: linkage analysis and significance in a hybrid zone. J. Hered. 80:179–185.

Hagen, R. H., and Scriber, J. M. 1991. Systematics of the *Papilio glaucus* and *P. troilus* species groups (Lepidoptera: Papilionidae): inferences from allozymes. Ann. Entomol. Soc. Am. 84:380–395.

Hagen, R. H., Lederhouse, R. C., Bossart, J. L., and Scriber, J. M. 1991. *Papilio canadensis* and *P. glaucus* (Papilionidae) are distinct species. J. Lepidopt. Soc. 45:245–258.

Haldane, J. B. S. 1922. Sex ratio and unisexual sterility in hybrid animals. J. Genet. 12:101–109.

Haldane, J. B. S. 1924. A mathmatical theory of natural and artificial selection. Part I. Trans. Camb. Philos. Soc. 23:19–41.

Haldane, J. B. S. 1932. The Causes of Evolution. London: Longmans, Green.

Hansson, B. S., Lüfstedt, C., and Foster, S. P. 1989. Z-linked inheritance of male olfactory response to sex pheromone components in two species of tortricid moths, *Ctenopseustis obliquana* and *Ctenopseustis* sp. Entomol. Exp. Appl. 53:137–145.

Harrison, R. G. 1989. Animal mitochondrial DNA as a genetic marker in population and evolutionary biology. Trends Ecol. Evol. 4:6–11.

Harrison, R. G., and Vawter, A. T. 1977. Allozyme differentiation between pheromone strains of the European corn borer, *Ostrinia nubilalis*. Ann. Entomol. Soc. Am. 70:717–720.

Hartl, D. L. 1971. Some aspects of natural selection in arrhenotokous populations. Am. Zool. 11:309–325.

Hartl, D. L. 1972. A fundamental theorem of natural selection for sex linkage or arrhenotoky. Am. Nat. 106:516–524.

Harvey, G. T. 1967. On coniferophagous species of *Choristoneura* (Lepidoptera: Tortricidae) in North America. V. Second diapause as a species character. Can. Entomol. 99:486–503.

Harvey, G. T. 1996. Genetic relationships among *Choristoneura* species (Lepidoptera: Tortricidae) in North America as revealed by isozyme studies. Can. Entomol. 128:245–262.

Hendrikse, A. 1988. Hybridization and sex-pheromone responses among members of the *Yponomeuta padellus*-complex. Entomol. Exp. Appl. 48:213–233.

Hollocher, H., and Wu, C.-I. 1996. The genetics of reproductive isolation in the *Drosophila simulans* clade: X vs. autosomal effects and male vs. female effects. Genetics 143:1243–1255.

Hurst, L. D., and Pomiankowski, A. 1991. Causes of sex ratio bias may account for unisexuality in hybrids: a new explanation for Haldane's Rule and related phenomena. Genetics 128:841–858.

Johnson, M. S., and Turner, J. R. G. 1979. Absence of dosage compensation for a sex-linked enzyme in butterflies (*Heliconius*). Heredity 43:71–77.

Klun, J. A., and Maini, S. 1979. Genetic basis of an insect chemical communication system: the European corn borer. Environ. Entomol. 8:423–426.

Lande, R. 1982. Rapid origin of sexual isolation and character divergence in a cline. Evolution 36:213–223.

Lester, L. D., and Selander, R. K. 1979. Population genetics of haplodiploid insects. Genetics 92:1329–1345.

Liu, H., and Beckenbach, A. T. 1992. Evolution of the mitochondrial cytochrome oxidase II gene among 10 orders of insects. Mol. Phylogenet. Evol. 1:41–52.

Lu, Y.-L., Kochert, G. D., Isenhour, D. J., and Adang, M. J. 1994. Molecular characterization of a strain-specific repeated DNA sequence in the fall armyworm *Spodoptera frugiperda* (Lepidoptera: Noctuidae). Insect Mol. Biol. 3:123–130.

Muller, H. J. 1940. Bearing of the *Drosophila* work on systematics. In J. S. Huxley (ed.). The New Systematics. Oxford: Clarendon Press, pp. 185–268.

Orr, H. A. 1993a. Haldane's Rule has multiple genetic causes. Nature 361:532–533.

Orr, H. A. 1993b. A mathematical model of Haldane's Rule. Evolution 47:1606–1611.

Pashley, D. P. 1986. Host associated genetic differentiation in fall armyworm: a sibling species complex? Ann. Entomol. Soc. Am. 79:898–904.

Pashley, D. P. 1988. Quantitative genetics, development and physiological adaptation in sympatric host strains of fall armyworm. Evolution 42:93–102.

Pashley, D. P. 1989. Host-associated differentiation in armyworms: an allozymic and mtDNA perspective. In H. Loxdale and M. F. Claridge (eds.). Electrophoretic Studies on Agricultural Pests. London: Oxford University Press, pp. 103–114.

Pashley, D. P., and Ke, L. D. 1992. Sequence evolution in mitochondrial ribosomal and ND-1 genes in Lepidoptera: implications for phylogenetic analyses. Mol. Biol. Evol. 9:1061–1075.

Pashley, D. P., and Martin, J. A. 1987. Reproductive incompatibility between host strains of fall armyworm (Lepidoptera: Noctuidae). Ann. Entomol. Soc. Am. 80:731–733.

Pashley, D. P., Johnson, S. J., and Spark, A. N. 1985. Genetic population structure of migratory moths: the fall armyworm (Lepidoptera: Noctuidae). Ann. Entomol. Soc. Am. 78:756–762.

Pashley, D. P., Hammond, A. M., and Hardy, T. N. 1992. Reproductive isolating mechanisms in fall armyworm host strains (Lepidoptera: Noctuidae). Ann. Entomol. Soc. Am. 85:400–405.

Pashley, D. P., McPheron, B. A., and Zimmer, E. A. 1993. Systematics of holometabolous insect orders based on 18S ribosomal RNA. Mol. Phylogenet. Evol. 2:132–142.

Pashley, D. P., Hardy, T. N., and Hammond, A. M. 1995. Plant effects on developmental and reproductive traits in fall armyworm host strains (Lepidoptera: Noctuidae). Ann. Entomol. Soc. Am. 88:748–755.

Pomiankowski, A., and Hurst, L. D. 1993. Genomic conflicts underlying Haldane's Rule. Genetics 33:425–432.

Priesner, E., and Baltensweiler, W. 1987. Studien zum Pheromone-Polymorphismus von Zeiraphera diniana Gn. (Lep., Tortricidae). Zeitschrift angewandte Entomol. 104:433–448.

Rice, W. R. 1984a. Sex chromosomes and the evolution of sexual dimorphism. Evolution 38:735–742.

Rice, W. R. 1984b. Disruptive selection on habitat preference and the evolution of reproductive isolation: a simulation study. Evolution 38:1251–1260.

Rice, W. R. 1987. Speciation via habitat specialization: the evolution of reproductive isolation as a correlated character. Evol. Ecol. 1:301–314.

Robinson, R. 1971. Lepidoptera Genetics. New York: Pergamon Press.

Rockey, S. J., Hainze, J. H., and Scriber, J. M. 1987. Evidence of sex-linked diapause response in Papilio glaucus subspecies and their hybrids. Physiol. Entomol. 12:181–184.

Roelofs, W., Glover, T., Tang, X.-H., Sreng, I. , Robbins, P., Eckenrode, C., Lüfstedt, C., Hansson, B. S., and Bengtsson, B. O. 1987. Sex pheromone production and perception in European corn borer moths is determined by both autosomal and sex-linked genes. Proc. Nat. Acad. Sci. USA 84:7585–7589.

Sanders, C. J., Daterman, G. E., and Ennis, T. J. 1977. Sex pheromone responses of Choristoneura spp. and their hybrids (Lepidoptera: Tortricidae). Can. Entomol. 109:1203–1220.

Scriber, J. M. 1992. Latitudinal clines in oviposition preferences: ecological and genetic influences. Pp. 212–215 in Menken, S. B. J., Visser, J. H., and Harrewijn, P. (Eds.). Proceedings of the 8th International Symposium on Insect-Plant Relationships. Dordrecht: Kluwer.

Sheck, A. L., and Gould, F. 1995a. Genetic analysis of differences in oviposition preferences of Heliothis virescens and H. subflexa (Lepidoptera: Noctuidae). Environ. Entomol. 24:341–347.

Sheck, A. L., and Gould, F. 1995b. The genetic basis of host range in Heliothis virescens: larval survival and growth. Entomol. Exp. Appl. 69:157–172.

Silberglied, R. E., and Taylor, O. R. 1973. Ultraviolet differences between the sulphur butterflies, Colias eurytheme and C. philodice, and a possible isolating mechanism. Nature 241:406–408.

Silberglied, R. E., and Taylor, O. R. 1978. Ultraviolet reflection and its behavioral role in the courtship of the sulphur butterflies Colias eurytheme and C. philodice (Lepidoptera, Pieridae). Behav. Ecol. Sociobiol. 3:203–243.

Sin, F. Y. T., Suckling, D. M., and Marshall, J. W. 1995. Differentiation of the endemic New Zealand greenheaded and brownheaded leafroller moths by restriction fragment length variation in the ribosomal gene complex. Mol. Ecol. 4:253–256.

Smith, S. G. 1953. Reproductive isolation and the integrity of two sympatric species of Choristoneura (Lepidoptera: Tortricidae). Can. Entomol. 85:141–151.

Sperling, F. A. H. 1987. Evolution of the Papilio machaon species group in western Canada. Q. Entomol. 23:198–315.

Sperling, F. A. H. 1993a. Mitochondrial DNA phylogeny of the Papilio machaon species group (Lepidoptera: Papilionidae). Mem. Entomol. Soc. Can. 165:233–242.

Sperling, F. A. H. 1993b. Mitochondrial DNA variation and Haldane's Rule in the Papilio glaucus and P. troilus species groups. Heredity 71:227–233.

Sperling, F. A. H. 1994. Sex-linked genes and species differences in lepidoptera. Can. Entomol. 126:807–818.

Sperling, F. A. H., and Harrison, R. G. 1994. Mitochondrial DNA variation within and between species of the Papilio machaon group of swallowtail butterflies. Evolution 48:408–422.

Sperling, F. A. H., and Hickey, D. A. 1995. Amplified mitochondrial DNA as a diagnostic marker for species of conifer-feeding Choristoneura (Lepidoptera: Tortricidae). Can. Entomol. 127:277–288.

Sperling, F. A. H., Landry, J.-F., and Hickey, D. A. 1995. DNA-based identification of introduced ermine moth species in North America (Lepidoptera: Yponomeutidae). Ann. Entomol. Soc. Am. 88:155–162.

Stehr, G. 1955. Brown female—a sex-linked and sex-limited character. J. Hered. 46:263–266.

Stehr, G. 1959. Hemolymph polymorphism in a moth and the nature of sex-controlled inheritance. Evolution 13:537–560.

Stehr, G. 1964. The determination of sex and polymorphism in microevolution. Can. Entomol. 96:418–428.

Thompson, J. N. 1988. Evolutionary genetics of oviposition preference in swallowtail butterflies. Evolution 42:1223–1235.

Thompson, J. N., Wehling, W., and Podolsky, R. 1990. Evolutionary genetics of host use in swallowtail butterflies. Nature 344:148–150.

Turelli, M., and Orr, H. A. 1995. The dominance theory of Haldane's Rule. Genetics 140:389–402.

van Drongelen, W., and van Loon, J. 1980. Inheritance of gustatory sensitivity in F_1 progeny of crossses between *Yponomeuta cagnagellus* and *Y. malinellus* (Lepidoptera). Entomol. Exp. Appl. 28:199–203.

Veenstra, K., Pashley, D. P., and Ottea, J. A. 1995. Host plant adaptation in fall armyworm host strains: comparison of food consumption, utilization, and detoxification enzyme activities. Ann. Entomol. Soc. Am. 88:80–91.

Virdee, S. R. 1993. Unravelling Haldane's Rule. Trends Ecol. Evol. 8:385–386.

West-Eberhard, M. J. 1983. Sexual selection, social competition, and speciation. Q. Rev. Biol. 58:155–183.

White, C. S., and Lambert, D. M. 1994. Genetic differences among pheromonally distinct New Zealand leafroller moths. Biochem. Syst. Ecol. 22:329–339.

White, M. J. D. 1973. Animal Cytology and Evolution (3rd ed.). Cambridge: Cambridge University Press.

Wu, C.-I., and Davis, A. W. 1993. Evolution of postmating reproductive isolation: the composite nature of Haldane's Rule and its genetic basis. Am. Nat. 142:187–212.

Wu, C.-I., Johnson, N. A., and Palopoli, M. F. 1996. Haldane's Rule and its legacy: why are there so many sterile males? Trends Ecol. Evol. 11:281–284.

24

The Role of Chromosomal Change in Speciation

Franco Spirito

The frequent occurrence of karyotypical differences among related species in several groups of organisms suggests that karyotypic changes may occur frequently in association with the speciation process (White 1968, 1978; Bush et al. 1977; Bengtsson 1980). It is important from the outset to emphasize that, while some types of chromosome change may accompany speciation without playing a causal role in it (e.g., quantitative variation in the amount of heterochromatin), the major chromosomal rearrangements involving a change in the position of genes on chromosomes without a gain or a loss of euchromatin, that is, the main categories of inversions, reciprocal translocations, Robertsonian rearrangements (centric fusions and fissions), and tandem fusions, have the opportunity to play an active role (see the discussion in King 1987). In fact, these rearrangements are potentially negatively heterotic, insofar as unbalanced gametes with duplications and deficiencies may be produced as a result of meiotic malsegregation in the heterokaryotype. These types of chromosomal rearrangements (with some reservations in the case of inversions, for which efficient mechanisms for avoiding the production of unbalanced gametes are often present; see the discussion in King 1993), insofar as they reduce the fertility of the F_1 hybrids between two populations that are monomorphic for the two different chromosome types, belong by definition to the most classical category of postzygotic reproductive isolation (partial hybrid sterility) (Dobzhansky 1970). On the basis of this consideration and of the available evidence of association between chromosomal divergence and cladogenetic processes, White in 1968 put forth the most general and best-articulated hypothesis of chromosomal speciation (stasipatric speciation model), which was further developed in subsequent years (see White 1968, 1973, 1978). Some tenets of this hypothesis are nowadays considered invalid by most evolutionists (especially the geographic pattern of this type of speciation), and it is apparent that White exaggerated the importance of chromosomal speciation (Futuyma and Mayer 1980; Key 1981; Templeton 1981; Charlesworth et al. 1982;

Sites and Moritz 1987). However, the debate is still open about the core of his hypothesis, that is, the possibility that chromosomal rearrangements with partial heterozygote sterility can play a primary role in some processes of speciation. The prevailing attitude is rather skeptical, but different views are present in the scientific debate (for a brilliant exposition of the evidence in support of chromosomal speciation, see King's [1993] recent book; but see the criticisms by Butlin 1993; Coyne 1993; Sites 1995). My goal is to discuss the results obtained up to now, outlining the gaps in our empirical knowledge that prevent us from linking with certainty the findings of field studies and the results of population genetics models.

The main issues discussed are the following:

1. What is the most appropriate genetic model to describe the behavior of chromosomal rearrangements in populations?
2. Under what conditions can these rearrangements become established?
3. Under what conditions can several successful processes of chromosomal establishment occur at the same time in the same area?
4. What is the effectiveness of karyotypical divergence, due to the establishment of one or several chromosomal rearrangements, as a barrier to gene exchange?

The discussion focuses on speciation in bisexually reproducing animal organisms.

THE GENETIC MODEL

The Reduction in Fitness of the Heterozygote for a Single Chromosomal Rearrangement

The first question concerns the proportion of unbalanced gametes produced by the heterozygote for a single rearrangement, and their effect on fertility. This proportion depends on several features of the rearrangement and of the organism considered:

1. *The class to which a particular rearrangement belongs.* For example, heterozygous reciprocal translocations and tandem fusions typically cause semisterility due to the production of 50% of unbalanced gametes, while the corresponding proportion is lower in the case of inversions and Robertsonian translocations (0–30% is a rough estimate based on Lande [1979] and Searle [1993]).

2. *The organism considered, mainly with regard to the chromosomal behavior in meiosis.* For example, paracentric inversions do not give rise to unbalanced gametes in *Drosophila* because recombination is lacking in the male and the recombinant dicentric and acentric chromatids, which originate by a crossing over in the inversion loop, pass into the polar bodies in female meiosis. On the other hand, in organisms lacking these features unbalanced gametes may be produced (White 1973).

3. *The individual characteristics of the particular rearrangement considered.* For example, the different centric fusions found in natural populations of *Mus musculus domesticus* that have been introduced into laboratory strains cause different levels of unbalanced gametes in the heterozygote: the range observed in this case is 0.05–0.30 (see Cattanach and Moseley 1973; Ford and Evans 1973).

4. *The genetic background.* For example, the above mouse centric fusions, unlike the case of laboratory strains, often segregate normally in single heterozygotes found in hybrid zones between chromosomal races (see Winking 1986; Wallace et al. 1992).

5. *Sex.* It is not infrequent that the same rearrangement determines considerably different effects in the two sexes, as a consequence of the difference in the process of meiosis (and gametogenesis; see again the case of mouse centric fusions) (Gropp et al. 1982).

The knowledge of the frequency of unbalanced gametes is insufficient, however, to enable us to state with certainty the amount of selection against the heterokaryotype for two main reasons. First, some other negative effects on meiosis or gametogenesis, in addition to the production of unbalanced gametes, may contribute to lowering heterozygote fertility. For instance, in male mammals, germ cell death has been observed in several cases as a consequence of unpaired chromosomal regions at pachytene (Chandley 1984; DeBoer 1986). This primary infertility due to spermatogenic breakdown is especially important, however, when several rearrangements leading to complex meiotic configurations are present (see Searle 1993). Second, compensatory effects may be present, such as those envisaged by Bengtsson (1980), which may lead to a recovery of complete fertility, in spite of the production of unbalanced gametes.

In conclusion, the reduction in fitness due to the presence of a chromosomal rearrangement (especially in the case of inversions and Robertsonian rearrangements) is not foreseeable a priori solely on the basis of the nature of the structural rearrangement. The absence of definite rules means that it is necessary to experimentally analyze the level of selection against the heterozygote for each particular rearrangement of evolutionary interest.

The Possible Presence of a Systematic Advantage of the Rearrangement

Two different mechanisms have been proposed, which give rise to a systematic pressure in favor of the new rearrangement in populations: selective advantage of the homozygote for the rearrangement over the other homozygote and meiotic drive (White 1968, 1978). As for the possible presence of selection in favor of the new homokaryotype, it is rather improbable that such an effect is due to the chromosomal structure per se; an indirect advantage, as a consequence of the association of the rearrangement with a favorable gene complex, is more probable (see the discussion in Searle 1993). This indirect advantage may certainly play a role because recombination is often suppressed in the heterozygote in the chromosomal region close to the break-points of the rearrangement (suppression of recombination is especially important in the case of chromosomal inversions) (White 1973). However, homozygote advantage is generally considered comparatively unimportant in the most recent literature, even by those who emphasize the role of systematic factors in the evolutionary success of chromosomal rearrangements (see the discussion in King 1993). An important role has been attributed since White's first paper on chromosomal speciation to meiotic drive (White 1968, 1978; King 1993). Non-Mendelian segregation of the two chromosome types in the heterokaryotype (generally only in one of the two sexes) has been found in several cases (for a recent review, see King 1993). However, many successful chromosomal rearrangements segregate normally (e.g., Britton-Davidian et al. 1990; Viroux and Bauchau 1992), and therefore meiotic drive does not appear to be a general feature explaining the establishment of chromosomal rearrangements.

In light of what has been said, the model with strict underdominance (heterozygote disadvantage and no difference in fitness between the two homozygotes) and absence of meiotic drive is the most realistic in most cases and therefore deserves special attention.

The Reduction in Fitness of a Heterozygote for Several Rearrangements

If the different rearrangements are independent, the most likely expectation a priori is that fitnesses combine in a multiplicative way; that is, fitness of a multiple heterozygote is equal to the product of the fitnesses of each single heterozygote. This is especially true if selection against the heterokaryotype is almost exclusively the result of the production of unbalanced gametes. However, the situation may be more complicated, and there is some evidence of a synergistic effect of the individual rearrangements, which causes a higher than expected level of selection against the multiple heterozygotes (see King 1993; Searle 1993). A synergistic interaction between nonindependent rearrangements is far more important. For instance, two different mouse centric fusions may give rise to two metacentric chromosomes sharing a chromosomal arm (monobrachial homology); in this case, a heterozygous individual with both metacentrics may have a much more severe reduction in fertility than each single heterozygote as a consequence of both malsegregation and disruption of meiosis (Gropp et al. 1982).

The Establishment of a Chromosomal Mutant

The process of establishment of an underdominant chromosomal rearrangement has been theoretically studied by several authors in a variety of models. Here I limit myself to discussing only the main results found in discrete population models (results in continuous population models are not very different; see Barton and Rouhani 1991). The process can be separated into two sequential steps (see Slatkin 1981; Lande 1985): (1) the fixation of the rearrangement, or at least its establishment at a high stable frequency in the particular deme in which it originated; (2) the spread of the rearrangement and its complete fixation in a vaster region occupied by many demes. The first step is the most problematical, while the subsequent step does not present serious difficulties under certain conditions.

The Fixation Probability in a Single Deme

The probability of fixation of an underdominant rearrangement in a single local population, starting from a single mutant, has been studied quite completely (Wright 1941; Bengtsson and Bodmer 1976; Lande 1979; Hedrick 1981; Walsh 1982). The general solution reported herein was obtained by Walsh (1982) and Lande (1985) assuming that 1, $1 - s$, $1 + s'$ are the fitnesses, respectively, of the homozygote for the standard chromosomal arrangement, of the heterozygote, and of the homozygote for the new rearrangement; this formula gives the fixation probability (U) of an underdominant mutant as a function of s, s', and the actual and effective number of the population (N and N_e, respectively). It results, when $s \cdot N_e/N \ll 1$:

$$U = \frac{(2/N) \exp\{-N_e \cdot s^2/b\} \sqrt{N_e \cdot b/\pi}}{\mathrm{erf}\{(s + s') \sqrt{N_e/b}\} + \mathrm{erf}\{s\sqrt{N_e/b}\}}, \quad (24.1)$$

in which $b = s + s'/2$ and $\mathrm{erf}(x)$ is the error function evaluated at x. In the case of strict underdominance ($s' = 0$), the chance of fixation (U) is lower than in the case of a neutral mutant ($1/2N$) and decreases very rapidly with increasing effective population size (N_e), and with increasing selection against the heterokaryotype (s) (see table 24.1). This is due to the fact that fixation of an underdominant mutant is possible only when genetic drift causes an increase in the frequency of the mutant against the systematic pressure of selection (an underdominant mutant is selected against when in the minority, while it is selected for when in the majority). It is apparent that the relative probability of fixation compared to that of a neutral mutant (2NU) becomes negligible when $N_e \cdot s$ is rather large. A chromosomal mutant with a high s value (a reciprocal translocation with $s = 0.5$, e.g.,) has a probability of becoming established that is very close to zero in populations composed of only a few dozen individuals. Only a mutant with a very small s value can become established with a nonnegligible probability in a population of a few hundred individuals. These results hold for an isolated population. If there is a significant level of migration, the corresponding probability is considerably lowered (see Lande 1979).

The presence of an advantage of the homozygote for the rearrangement over the other homozygote ($s' > 0$) increases the chance of fixation of the underdominant rearrangement, because there is a lowering of the unstable equilibrium point below which the rearrangement is counterselected [from 0.5 to $s/(2s + s')$]. This increase is considerable for significant levels of s', but when population is large, the fixation probability relative to a neutral mutant becomes close to zero even if there is a significant level of selection in favor of the rearrangement homozygote (see table 24.1). A qualitatively similar pattern is found when a low level of meiotic drive for a strictly underdominant mutant is assumed. In fact, weak meiotic drive only lowers the unstable equilibrium point,

Table 24.1. The relative probability of fixation of an underdominant chromosomal rearrangement compared to that of a neutral mutant (2NU).

	$s = 0.02$	$s = 0.10$	$s = 0.50$
$N_e = 20$; $s' = 0$	0.761	0.226	2×10^{-4}
$N_e = 20$; $s' = s/2$	0.861	0.375	0.001
$N_e = 20$; $s' = s$	0.963	0.543	0.006
$N_e = 100$; $s' = 0$	0.226	2×10^{-4}	$<10^{-13}$
$N_e = 100$; $s' = s/2$	0.375	0.001	$<10^{-13}$
$N_e = 100$; $s' = s$	0.543	0.006	$<10^{-13}$
$N_e = 500$; $s' = 0$	2×10^{-4}	$<10^{-13}$	$<10^{-13}$
$N_e = 500$; $s' = s/2$	0.001	$<10^{-13}$	$<10^{-13}$
$N_e = 500$; $s' = s$	0.006	$<10^{-13}$	$<10^{-13}$

For each s value, the nine values shown refer to different pairs of N_e and s' values.

analogous to what happens in the case of homozygote advantage (the shape of the curve is different, however, and when similar unstable equilibria are present in the two models, a weaker selection against the rearrangement at low frequency is found in the model with meiotic drive; Hedrick 1981). In contrast, a different pattern occurs when meiotic drive is strong; in fact, if s is not too large, there exists a critical level of meiotic drive (depending on the amount of selection against the heterozygote) above which the unstable equilibrium disappears and there is therefore selection in favor of the rearrangement at all frequencies. If k (>0.5) is the proportion of gametes with the rearrangement produced by the heterozygote of the sex in which distorted segregation occurs, the critical k value is equal to $(1 + s)/2(1 - s)$ (see eq. 5 in Hedrick (1981). For k larger than (or equal to) this critical value, directional selection in favor of the rearrangement occurs, and therefore the fixation probability is similar to that of an advantageous mutant and can greatly exceed that of a neutral mutant. The solution in the case of meiotic drive can be found in Walsh (1982).

The Spread of the Rearrangement and Its Fixation in the Entire Population

The local fixation of the chromosomal mutant in a single deme is only the first step in the process of creation of a new chromosomal race. The rearrangement has then to become established in a vaster region. This may occur when individuals with the new karyotype colonize an area temporarily unoccupied, and/or migrate into other populations. Let us consider first the case of strict underdominance. Spreading by migration is nearly impossible for high levels of migration; instead, if the migration rate is very low, spreading is possible and occurs with an efficiency comparable to the spread of a neutral mutant; the probability of fixation in the entire population consisting of n demes, starting from initial fixation or establishment in one of the demes, is 1/n, as in the case of a neutral mutant (Slatkin 1981). The same result is obtained in models with local extinction and colonization instead of, or in addition to, migration (Lande 1979, 1985). In other words, the most difficult step is the fixation in the first deme, while under certain conditions, there are not subsequent difficulties in the spreading process. On the basis of the above finding, the fixation rate per generation (R) of a class of chromosomal rearrangements with a given s value in a multideme system is equal to the corresponding fixation rate in a single deme (Lande 1979, 1985). In fact, to obtain R, the fixation rate per generation in each single deme ($2N\mu U$, where μ is the spontaneous rate of mutation for a given class of rearrangements) must be multiplied by the number of demes (n), and by the probability of fixation in the entire population starting from fixation or establishment in one of the demes (1/n). This gives $R = 2N\mu U$, independent of n. Consequently, the expected waiting time between two successful fixation

processes (t_A), which is the reciprocal of the fixation rate ($t_A = 1/R$), is also obviously independent of n.

The presence of homozygote advantage, assuming "soft" selection (Wallace 1968) gives rise to the following pattern. When extinction and colonization are slow compared to the spread by migration, the probability of fixation in the entire population starting from fixation or establishment in one of the demes is equal to $1 - \exp(-2N_e \cdot s')$, for large n and $N_e \cdot s' > 1$ (see Slatkin 1981; Lande 1985). This value may be considerably higher than 1/n, and therefore the second step in the process, that is, the spread of the rearrangement, may be much easier than in the case of a netural mutant. However, because the first step is, as we have seen, more difficult, the overall result is that fixation in the entire population of a rearrangement starting from a single copy may be more or less probable than in the case of a neutral mutant, according to which of the two effects described is predominant. On the other hand, when the effect of local extinction and colonization is prevalent over that of migration, the spread of the rearrangement is as efficient as that of a neutral mutant (analogously to what happens in the absence of homozygote advantage).

The overall pattern in the presence of a low level of meiotic drive is expected to be similar to that described for the case of homozygote advantage. Conversely, when meiotic drive is strong and the rearrangement behaves as an advantageous mutant, both the first and the second steps will obviously be more probable than in the case of a neutral mutant.

In conclusion, the establishment of a chromosomal rearrangement in a vast region requires somewhat special conditions. In the absence of systematic factors of advantage demes must be very small and semiisolated. If population structure is different only a considerable homozygote advantage or meiotic drive in favor of the rearrangement can lead to its establishment.

Temporal Overlapping of Several Processes of Fixation of Chromosomal Rearrangements

Some insights on the chance of temporal overlapping of several successful fixation processes of chromosomal rearrangements can be gained by comparing the expected waiting time for the appearance of a rearrangement, which is destined to become fixed, with the time of duration of the process (see Barton and Rouhani 1991; Spirito et al. 1993). The expected waiting time between two successful events of fixation (t_A) for a class of rearrangements with given coefficients of selection is equal (see the preceding section) to the reciprocal of the fixation rate. The average time between the appearance by mutation of a single copy of the rearrangement and its fixation in the entire multideme system approximately corresponds (because the duration of the first step is negligible compared to that of the second step) to the time elapsing between the fixation or establishment at high frequency of the rearrangement in the first deme and its fixation in the entire population (t_F). When t_F is much smaller than t_A, several processes of fixation in the same area tend to occur one at a time. Vice versa, if t_F and t_A are comparable, or t_F is even greater than t_A, the next successful process of fixation is likely to start before the completion of the previous one. Thus, the pattern of karyotypic evolution of a species subdivided into a series of multideme isolates is crucially dependent on the ratio between t_F and t_A. Formulas for the ratio t_F/t_A (denoted as L) in some different models are shown in table 24.2. They have been obtained on the basis of Lande's solutions (Lande 1985). It is apparent that the value of L depends on population structure, on the mechanism of spreading (whether migration is prevalent over local extinction and colonization, or vice versa), and on the values of several parameters (mutation rate, effective deme size, number of demes, in addition to migration and extinction rates). Let us consider a strictly underdominant mutant in some extreme cases (circular stepping stone model, or island model; local extinction and colonization [at a rate λ per deme per generation] with migration absent [m = 0], or migration [at a rate m per generation] without local extinction [$\lambda = 0$]. When the rearrangements spread only by migration, the L ratio is independent in both models of population structure of the population size and of the amount of selection against the heterozygote, depending only on μ (the spontaneous rate of occurrence per generation of chromosomal rearrangements with given selection coefficients), m, and n. The value of L increases with increasing n and with increasing μ, while it decreases with increasing m. Figure 24.1 shows the values of the L ratio in the circular stepping stone model and in the island model for some parameter values. In particular, the value chosen is 10^{-4}, which is a realistic estimate based on empirical evidence (see Lande 1979); the m value is 0.01. It can be seen that already for a rather small number of demes the value of the L ratio becomes larger than one (especially in the circular stepping stone model). Thus, the chance that

several successful establishment processes occur one at a time is low in this case for a large number of demes. The situation is somewhat different when there is local extinction and colonization without migration (m = 0). In these models, the L value does not only depend on λ, μ and n, but also on N and U. Figure 24.2 shows the curves (for $\mu = 10^{-4}$ and $\lambda = 0.01$) in the two different population structure models for three different values of the relative probability of fixation compared to a neutral mutant: 2NU = 0.5, 0.1, 0.02. In this case (exclusive presence of local extinction and colonization), conditions that prevent temporal overlapping of several fixation processes (L << 1) occur more frequently.

The pattern found when there is selection in favor of the homozygote for the rearrangement is similar. Also in this case, the value of L is independent of deme size and of selective coefficients when local extinction does not occur (see table 24.2 and figure 24.1). The presence of homozygote advantage generally causes an increase in the value of L and therefore raises the chance of temporal overlapping of several processes (see table 24.2 and figures 24.1 and 24.2).

In conclusion, there exists a fairly wide range of conditions under which the overlapping of several fixation processes of underdominant rearrangements in the same area is more probable than the occurrence of several fixation processes one at a time. This finding can be useful in accounting for the numerous cases of parapatric chromosomal races differing for several chromosomal rearrangements (the most striking case of this type is perhaps represented by the chromosomal races of *Mus musculus domesticus* found in Europe; for a recent review, see Bauchau 1990).

The Effectiveness of Structural Heterozygosity as a Barrier to Gene Exchange

Another important issue to consider in the discussion of chromosomal speciation is to what extent the gene ex-

Table 24.2. Formulas for the L value (t_F/t_A) in the various models.

	Circular Stepping Stone Model	Island Model
Strict underdominance (s' = 0)		
m > 0; $\lambda = 0$	$\mu(n^2 - 1)/6m$	$2\mu(n - 1)/m$
m = 0; $\lambda > 0$	$\mu NU(n^2 - 1)/3\lambda$	$4\mu NU(n - 1)/\lambda$
Underdominance with homozygote advantage (s' > 0)		
m > 0; $\lambda = 0$	$\mu n^2/2m$	$\mu n \cdot \ln(n)/m$
m = 0; $\lambda > 0$	$\mu NU(n^2 - 1)/3\lambda$	$4\mu NU(n - 1)/\lambda$

The formulas for L in models without migration are the same for both s' > 0 and s' = 0. However, it must be considered that the U value (fixation probability in a single deme), which appears in the formulas, is different in the two cases (the U value is larger when s' > 0, all other things being equal).

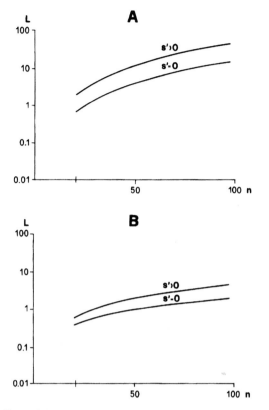

Figure 24.1. The value of the ratio L (t_F/t_A) for an underdominant chromosomal rearrangement in the absence of local extinction and colonization ($\lambda = 0$). It is assumed that m = 100μ. A, circular stepping stone model. B, island model. The two curves in each case refer to strict underdominance (s' = 0), and to homozygote advantage (s' > 0). Each curve gives the L values as a function of the number of demes (n), for n ≥ 20.

chromosomal rearrangement, with a small or moderate level of underdominance, there is no significant reduction in the genetic exchange between chromosomal races, except for genes that are very closely linked to the chromosomal rearrangement. This conclusion is valid for neutral genes and a fortiori for advantageous genes (see Barton 1979b). Some results obtained in a continent-island model will be described, by way of example. To measure the reduction in the genetic exchange the MRE (Migration Reduction Equivalent) index may be used (Spirito et al. 1983). The value of this index indicates to what reduction of migration the effect of the hybrid unfitness corresponds (e.g., an MRE value of 0.30 means that the reduction of the gene exchange produced by the presence of the rearrangement is the same as that caused by a reduction of the migration rate to 30%). In the continent-

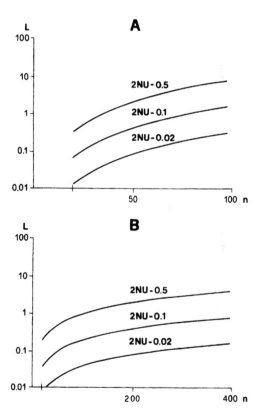

Figure 24.2. The value of the ratio L (t_F/t_A) for an underdominant chromosomal rearrangement in the absence of migration (m = 0). It is assumed that $\lambda = 100$μ. A, circular stepping stone model. B, island model. The three curves in each case refer to three different values of the relative fixation probability compared to a neutral mutant (2NU) and are suitable to represent both the case of strict underdominance and the case of homozygote advantage. Each curve gives the L values as a function of the number of demes (n), for n ≥ 20.

change between populations is reduced by the presence of karyotypic differences. Two populations of the same species can preserve their karyotypical identity in the presence of gene flow rather easily (Bazykin 1969; Barton 1979a, 1983; Spirito et al. 1991). Thus, a stable hybrid zone arises in the contact area between the two chromosomal races, and this hybrid zone ("tension zone") is maintained by a balance between migration and selection against hybrids (Barton and Hewitt 1985). The selection against hybridization will obviously cause a reduction in the amount of gene exchange between populations. The extent of this effect in the presence of one or several strictly underdominant rearrangements has been analyzed by several authors in a variety of models (Barton 1979b; Bengtsson 1985; Barton and Bengtsson 1986; Spirito et al. 1983, 1987). The general sense of this theoretical work is the same in all cases, apart from some quantitative differences among models. In the case of a single island model (and in a different two-population model; see Spirito et al. 1987), for small m values, the following limiting solution was found (Spirito 1986):

$$MRE = \frac{r(1 - s)}{r(1 - s) + s} , \qquad (24.2)$$

in which s is the selection against the heterozygote and r is the frequency of recombination between the gene and the rearrangement. Some numerical results computed using this formula are shown in table 24.3 and are in accordance with the above statement on the weak effectiveness of a single karyotypic difference as a reproductive isolating mechanism.

If two chromosomal races differ by several chromosomal rearrangements, the effect is obviously stronger for two reasons: first, the amount of selection against the hybrids is larger; second, the portion of the genome that is strictly linked to one rearrangement or other (and is therefore more strongly affected) becomes larger. For the sake of example, I will give the formula relative to K chromosomal rearrangements for low migration rates (it is assumed that the K rearrangements are not linked and are equivalent; that is, each single heterozygote has the same reduction in fitness):

$$MRE = \frac{r(1 - S)}{(1 - (1 - r)(1 - S)^{1/K}) \cdot (2 - (1 - S)^{1/K})^{K-1}} \qquad (24.3)$$

in which S is the selection against the F₁ hybrid, that is, the heterozygote for K rearrangements, and r is the frequency of recombination between the gene considered and the possibly linked rearrangement. This formula has been obtained using a procedure similar to that described in Spirito (1986). Some numerical results for this formula are shown in table 24.4. It is apparent that only if the fitness of the F₁ hybrid is considerably reduced, is genetic exchange lowered to a large extent for a significant portion of the genome.

The above solution was obtained assuming that fitnesses at the different rearrangements combine in a multiplicative way. It is possible, however, that the rearrangements interact in such a way as to determine a

Table 24.3. Values of MRE (Migration Reduction Equivalent) in the presence of a single chromosomal rearrangement in a continent-island model.

	s = 0.02	s = 0.1	s = 0.5
r = 0.5	0.961	0.818	0.333
r = 0.1	0.831	0.474	0.091
r = 0.02	0.495	0.153	0.020

s is the selection coefficient against the heterozygote; r is the frequency of recombination between the gene considered and the rearrangement.

Table 24.4. Values of MRE (Migration Reduction Equivalent) in the presence of K unlinked chromosomal rearrangements.

	K = 2	K = 5	K = 10
r = 0.5	0.669	0.367	0.134
r = 0.1	0.388	0.212	0.078
r = 0.02	0.125	0.068	0.025

Each single heterozygous rearrangement causes the same level of partial sterility (s = 0.1); fitnesses are multiplicative, and therefore selection against the heterozygote for all the K rearrangements (S) is equal to 0.190, 0.410, and 0.651, for K = 2, 5, and 10, respectively. Three different values of frequency of recombination between the gene considered and the rearrangement possibly linked are considered: r = 0.5, 0.1, and 0.02.

stronger reduction in fitness of a multiple heterozygote (see above). To get a general idea about the effect of the epistatic interaction among rearrangements on genetic exchange, let us consider the most extreme pattern of interaction: fitness of the heterozygote for all the K rearrangements is substantially reduced, while single heterozygotes have a negligible reduction in fitness. In this model, the limiting solution for low migration rates is

$$MRE = \frac{(1 - S) \cdot \left(1 - \frac{1-r}{2^{K-1}}\right)}{1 - \frac{(1 - r)(1 - S)}{2^{K-1}}} , \qquad (24.4)$$

in which S is the selection against the F₁ hybrid, that is, against the heterozygote for K rearrangements, and r is the frequency of recombination between the gene and the rearrangement to which it is possibly linked. In this model, the barrier to genetic exchange only slightly increases with a decrease in the frequency of recombination. In particular, when the number of rearrangements is large, MRE is nearly equal to 1 − S, which is the fitness of the heterozygote for K rearrangements, independently of r. A comparison between the effectiveness of the isolation in the two models of reproductive isolation for the same values of S and r is shown in figure 24.3: the value of the ratio (M) between the MRE value in the second model (computed using eq. 24.4) and the corresponding MRE value in the first model (computed using eq. 24.3) is shown for some parameter values. It is apparent that differences in results between the models are not very conspicuous for unlinked genes (unless S is very large). Instead, genes that are linked to one chromosomal rearrangement or the other behave in a very different manner in the two models. However, it must be emphasized that the second model considered represents the most extreme case of deviation from multiplicativity. In less extreme and more realistic situations, differences among models will be consid-

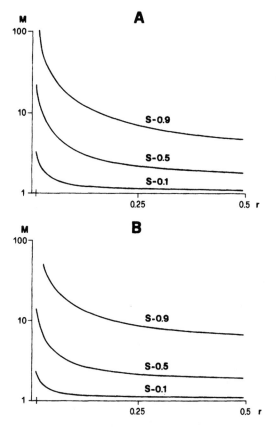

Figure 24.3. The ratio (M) between the MRE values in the two different models of interaction among re-arrangements (nonmultiplicative and multiplicative). A, 5 underdominant rearrangements (K = 5). B, 10 underdominant rearrangements (K = 10). In both cases the curves are shown for three different levels of selection against the heterozygote for all the K rearrangements (S = 0.1, S = 0.5, S = 0.9). Each curve gives the M values as a function of the frequency of recombination (r), for r ≥ 0.01.

erably smaller even for those genes that are linked to the rearrangements.

Concluding Remarks

Structural chromosomal heterozygosity due to underdominant chromosomal rearrangements represents a classical barrier of postzygotic reproductive isolation. The simplicity of the hereditary basis of this kind of reproductive isolation has caused this isolating mechanism to be an important object of research in the field of the genetics of speciation. The results obtained not only provide direct insights on the plausibility of chromosomal speciation, but clarify the more general debate on speciation. However, there is an important difference between karyotypic divergence and some other postzygotic isolating mechanisms. While polygenic models are hypothesizable and realistic, in which each mutant contributing to the reproductive isolation is neutral or advantageous per se, and can therefore become established without any difficulties (see the discussion by Dobzhansky 1970), the establishment of chromosomal rearrangements with a significant negative effect on fitness of the heterozygote is a process that can occur only under very special circumstances (see the above discussion). In spite of this, chromosomal rearrangements that are expected a priori to be negatively heterotic have become established in natural populations of several species, in the absence in most cases of other types of genetic divergence causing reproductive isolation (King 1993). The evolutionary role of this observed karyotypical divergence is unclear because of the wide variability in the effects on meiosis and gametogenesis even within a particular class of rearrangements. The pattern of geographic variation strongly suggests that in most cases underdominance is present to some extent; in fact, unlike electrophoretic variant and other variants that are certainly not underdominant, the two alternative chromosome types are generally segregating only in zones of primary or secondary contact between monomorphic chromosomal races (see the discussion by Searle 1993). However, the effects on fitness of each single heterozygous rearrangement, and the overall effect when several rearrangements are present, are not well known in most cases, and this makes any conclusion about their role in speciation uncertain. In the literature the following paradox (from the point of view of supporters of chromosomal speciation) has often been pointed out: chromosomal mutants with a very low level of selection against the heterokaryotype can become established rather easily, but are not likely to play an important role in speciation; on the other hand, chromosomal mutants with significant negative heterosis, which could be more important in cladogenetic processes, require very special conditions in order to become established (Sites and Moritz 1987). The feasibility of chromosomal speciation can be correctly asserted only if this difficulty is overcome. This can be done in one of the following three ways.

(1) A first possibility is that the special circumstances (the particular population structure required and/or the presence of systematic factors in favor of the rearrangement) under which the establishment of a chromosomal rearrangement with a significant level of heterozygote sterility is possible, occur rather frequently in some groups of organisms. This hypothesis is plausible and will not be discussed further. Here I limit myself to making some comments on the possible role of meiotic drive, which is the more effective force favoring the establishment of a rearrangement. If there is a high level of meiotic drive, the rearrangement will increase in frequency and replace the primitive chromosome type; therefore, such a rearrangement will not tend to exert a divisive

effect on gene pool and trigger speciation (for this argument, see Templeton 1981). To overcome this difficulty, an ad hoc hypothesis has been proposed (King 1993): rearrangements involved in chromosomal speciation could be driven only in the first period after their appearance, while subsequent changes in the genetic background would lead to more regular segregation. In my opinion, a more important cladogenetic role could be played by rearrangements with lower levels of meiotic drive; in fact, these rearrangements can become established more easily than strictly underdominant mutants with Mendelian segregation, but maintain the basic property of exerting a divisive effect on gene pool.

(2) A second possibility is that, although each rearrangement exerts only a slight effect on fertility (and can therefore become established without serious difficulties), the presence of several rearrangements can lead to a high degree of sterility. Indeed, there is some evidence in support of this hypothesis. The most recent work on chromosomal hybrid zones has shown that the reduction in fertility due to the presence of single heterozygous rearrangements is generally very slight (see the results of the analysis of chromosomal hybrid zones in mammals summarized by Searle 1993; see also the studies on the *Sceloporus grammicus* complex by Reed and Sites 1995; Reed et al. 1995a,b; Sites et al. 1995). On the other hand, there are several examples in which a strong reduction in fertility seems to be largely due to the presence of several heterozygous rearrangements (e.g., Tettenborn and Gropp 1970; Capanna et al. 1976), suggesting a synergism among the rearrangements. Several different mechanisms may be assumed to underlie this pattern. Some clear examples of the synergistic effect of several rearrangements on fertility have been described in the case of centric fusions giving rise to metacentric chromosomes with monobrachial homology; a special role in speciation by these rearrangements has been proposed (Capanna 1982; Baker and Bickham 1986).

(3) A third and less probable possibility requires that the degree of sterility due to chromosomal structural heterozygosity be to some extent under genetic control. Some cases have been described previously, in which this seems to occur. In these cases, it is feasible in principle that a chromosomal mutant that is only slightly underdominant at the time of its establishment increases its level of underdominance as a result of the genetic divergence between populations and of the consequent problems in pairing and segregation in the heterokaryotype. In such a case, difficulties in establishment are largely avoided, while the barrier to genetic exchange caused by several rearrangements can be very effective.

The incompleteness of our knowledge on these points does not allow us to formulate a final opinion concerning the feasibility of chromosomal speciation. The experimental work of the next few years will probably make a substantial contribution in this area, although there are good reasons for considering as overoptimistic the following sentence in the final chapter of White's (1978) book:

> This book is hence only the forerunner of the definitive work on modes of speciation that should be written, about the year 2,000, by someone with a much more extensive knowledge of the basic facts and with the ability to incorporate them in mathematical models. (p. 324)

Acknowledgments   Many thanks are due to Prof. M. Rizzoni for his critical reading of the manuscript. I also thank Mr. V. Salviati for his graphical work.

References

Baker, R. J., and Bickham, J. W. 1986. Speciation by monobrachial centric fusions. Proc. Natl. Acad. Sci. USA 83: 8245–8248.

Barton, N. H. 1979a. The dynamics of hybrid zones. Heredity 43:341–359.

Barton, N. H. 1979b. Gene flow past a cline. Heredity 43: 333–339.

Barton, N. H. 1983. Multilocus clines. Evoution 37:454–471.

Barton, N. H., and Bengtsson, B. O. 1986. The barrier to genetic exchange between hybridising populations. Heredity 56:357–376.

Barton, N. H., and Hewitt, G. M. 1985. Analysis of hybrid zones. Annu. Rev. Ecol. Syst. 16:113–148.

Barton, N. H., and Rouhani, S. 1991. The probability of fixation of a new karyotype in a continuous population. Evolution 45:499–517.

Bauchau, V. 1990. Phylogenetic analysis of the distribution of chromosomal races of *Mus musculus domesticus* Rutty in Europe. Biol. J. Linn. Soc. 41:171–192.

Bazykin, A. D. 1969. Hypothetical mechanism of speciation. Evolution 23:685–687.

Bengtsson, B. O. 1980. Rates of karyotype evolution in placental mammals. Hereditas 92:37–47.

Bengtsson, B. O. 1985. The flow of genes through a genetic barrier. In P. H. Harvey and M. Slatkin (eds.). Evolution: Essays in Honour of John Maynard Smith. Cambridge: Cambridge University Press, pp. 31–42.

Bengtsson, B. O., and Bodmer, W. F. 1976. On the increase of chromosome mutations under random mating. Theor. Popul. Biol. 9:260–281.

Britton-Davidian, J., Sonjaya, H., Catalan, J., and Cattaneo-Berrebi, G. 1990. Robertsonian heterozygosity in wild mice: fertility and transmission rates in Rb(16.17) translocation heterozygotes. Genetica 80:171–174.

Bush, G. L., Case, S. M., Wilson, A. C., and Patton, J. L. 1977. Rapid speciation and chromosomal evolution in mammals. Proc. Natl. Acad. Sci. USA 74:3942–3946.

Butlin, R. K. 1993. Barriers to gene flow. Nature 366:27–28.

Capanna, E. 1982. Robertsonian numerical variation in animal speciation: *Mus musculus*, an emblematic model. In C. Barigozzi (ed.). Mechanisms of Speciation. New York: Liss, pp. 155–177.

Capanna, E., Gropp, A., Winking, H., Noack, G., and Civitelli, M. V. 1976. Robertsonian metacentrics in the mouse. Chromosma (Berl.) 58:341–353.

Cattanach, B. M., and Moseley, H. 1973. Non-disjunction and reduced fertility caused by the tobacco mouse metacentric chromosomes. Cytogenet. Cell Genet. 12:264–287.

Chandley, A. C. 1984. Infertility and chromosome abnormality. Oxford Rev. Reprod. Biol. 8:1–46.

Charlesworth, B., Lande, R., and Slatkin, M. 1982. A neo-darwinian commentary on macroevolution. Evolution 36:474–498.

Coyne, J. A. 1993. Speciation by chromosomes. Trends Ecol. Evol. 8:76–77.

DeBoer, P. 1986. Chromosomal causes for fertility reduction in mammals. In F. J. de Serres (ed.). Chemical Mutagens (vol. 10). New York: Plenum Press, pp. 37–76.

Dobzhansky, T. 1970. Genetics of the Evolutionary Process. New York: Columbia University Press.

Ford, C. E., and Evans, E. P. 1973. Robertsonian translocations in mice: segregational irregolarities in male heterozygotes and zygotic unbalance. Chromosomes Today 4:387–397.

Futuyma, D. J., and Mayer, G. C. 1980. Non-allopatric speciation in animals. Syst. Zool. 29:254–271.

Gropp, A., Winking, H., and Redi, C. 1982. Consequences of Robertsonian heterozygosity: segregational impairment of fertility versus male-limited sterility. In P. G. Crosignani and B. L. Rubin (eds.). Genetic Control of Gamete Production and Function. Orlando, Fla: Grune and Stratton, pp. 115–134.

Hedrick, P. W. 1981. The establishment of chromosomal variants. Evolution 35:322–332.

Key, K. 1981. Species, parapatry and the morabine grasshoppers. Syst. Zool. 30:425–458.

King, M. 1987. Chromosomal rearrangements, speciation and the theoretical approach. Heredity 59:1–6.

King, M. 1993. Species Evolution: The Role of Chromosome change. Cambridge: Cambridge University Press.

Lande, R. 1979. Effective deme sizes during long-term evolution estimated from rates of chromosomal rearrangement. Evolution 33:234–251.

Lande, R. 1985. The fixation of chromosomal rearrangements in a subdivided population with local extinction and colonization. Heredity 54:323–332.

Reed, K. M., and Sites, J. W. 1995. Female fecundity in a hybrid zone between two chromosome races of the Sceloporus grammicus complex (Sauria, Phrynosomatidae). Evolution 49:61–69.

Reed, K. M., Greenbaum, I. F., and Sites, J. W. 1995a. Cytogenetic analysis of chromosomal intermediates from a hybrid zone between two chromosome races of the Sceloporus grammicus complex (Sauria, Phrynosomatidae). Evolution 49:37–47.

Reed, K. M., Greenbaum, I. F., and Sites, J. W. 1995b. Dynamics of a novel chromosomal polymorphism within a hybrid zone between two chromosome races of the Sceloporus grammicus complex (Sauria, Phrynosomatidae). Evolution 49:48–60.

Searle, J. B. 1993. Chromosomal hybrid zones in eutherian mammals. In R. G. Harrison (ed.). Hybrid Zones and the Evolutionary Process. Oxford: Oxford University Press, pp. 309–352.

Sites, J. W. 1995. Chromosomal speciation. Evolution 49:218–222.

Sites, J. W., and Moritz, C. 1987. Chromosomal evolution and speciation revisited. Syst. Zool. 36:153–174.

Sites, J. W., Barton, N. H., and Reed, K. M. 1995. The genetic structure of a hybrid zone between two chromosome races of the Sceloporus grammicus complex (Sauria, Phrynosomatidae) in central Mexico. Evolution 49:9–36.

Slatkin, M. 1981. Fixation probabilities and fixation times in a subdivided population. Evolution 35:477–488.

Spirito, F. 1986. Reduction in gene flow among populations due to a reproductive isolating mechanism based on a unifactorial or bifactorial heredity. Mem. Accad. Naz. Lincei Ser. 8 18, Sez. 3 (5):335–370.

Spirito, F., Rizzoni, M., and Rossi, C. 1983. Reduction of gene flow due to the partial sterility of heterozygotes for a chromosomal mutation. I. Studies on a "neutral" gene not linked to the chromosome mutation in a two population model. Evolution 37:785–797.

Spirito, F., Rizzoni, M., Lolli, E., and Rossi, C. 1987. Reduction of neutral gene flow due to the partial sterility of heterozygotes for a linked chromosome mutation. Theor. Popul. Biol. 31:323–338.

Spirito, F., Rizzoni, M., and Rossi, C. 1991. Populational interactions among underdominant chromosome rearrangements help them to persist in small demes. J. Evol. Biol. 4:501–512.

Spirito, F., Rizzoni, M., and Rossi, C. 1993. The establishment of underdominant chromosomal rearrangements in multideme systems with local extinction and colonization. Theor. Popul. Biol. 44:80–94.

Templeton, A. R. 1981. Mechanisms of speciation—a population genetic approach. Annu. Rev. Ecol. Syst. 12:23–48.

Tettenborn, U., and Gropp, A. 1970. Meiotic nondisjunction in mice and mouse hybrids. Cytogenetics 9:272–283.

Viroux, M. C., and Bauchau, V. 1992. Segregation and fertility in Mus musculus domesticus (wild mice) heterozygous for the Rb(4.12) translocation. Heredity 68:131–134.

Wallace, B. 1968. Polymorphism, population size, and genetic load. In R. C. Lewontin (ed.). Population Biology and Evolution. Syracuse, N.Y.: Syracuse University Press, pp. 87–108.

Wallace, B. M. N., Searle, J. B., and Everett, C. A. 1992. Male meiosis and gametogenesis in wild mice. (Mus musculus domesticus) from a chromosomal hybrid zone; a comparison between "simple" Robertsonian heterozygotes and homozygotes. Cytogenet. Cell Genet. 61:211–220.

Walsh, J. B. 1982. Rate of accumulation of reproductive isolation by chromosome rearrangements. Am. Nat. 120:510–532.

White, M. J. D. 1968. Models of speciation. Science 159:1065–1070.

White, M. J. D. 1978. Modes of speciation. San Francisco: Freeman.

White, M. J. D. 1973. Animal Citology and Evolution. Cambridge: Cambridge University Press.

Winking, K. 1986. Some aspects of Robertsonian karyotype variation in European wild mice. Curr. Top. Microbiol. Immunol. 127:68–74.

Wright, S. 1941. On the probability of fixation of reciprocal translocations. Am. Nat. 75:513–522.

25

The Genetics of Hybrid
Male Sterility in *Drosophila*

Horacio Fachal Naveira
Xulio Rodríguez Maside

Recent times have witnessed renewed interest in genetic analysis of hybrid male sterility in *Drosophila*, an extraordinarily frequent outcome of interspecific crosses in this genus (Bock, 1984). After pioneering works in the first half of this century (Dobzhansky, 1936; Muller and Pontecorvo, 1940; Sturtevant, 1920), this field of research was somewhat neglected, with a few exceptions (Dobzhansky, 1975; Ehrman, 1962; Schäfer, 1978). The turning point may be traced to the writings of Coyne in the mid 1980s (1984, 1985), centered on the analysis of reproductive isolation between species of the *simulans* clade (*D. simulans, D. mauritiana,* and *D. sechellia*). Since then, an impressive amount of work on the genetics of hybrid disharmonies in *Drosophila* has been reported (see, for reviews, Coyne and Orr, 1989a,b; Wu and Davis, 1993; Wu and Palopoli, 1994; Wu et al., 1996).

Among other achievements, rather convincing evidence was obtained that (1) hybrid male sterility is usually the first postzygotic isolating barrier to develop between species, and (2) although many genetic differences dispersed throughout the genome contribute to hybrid sterility, they appear to be concentrated on the X-chromosome. With statement (1) we simply mean that crosses between recently diverged species (arbitrarily, genetic distance (≤ 0.6) give rise to sterile males and fertile females much more frequently than to other forms of postzygotic isolation (e.g., the ratio of cases of hybrid male sterility to hybrid male inviability is 40:8, according to Wu and Davis, 1993). But we would not like to be engaged in the controversy of the relative evolutionary rates of hybrid sterility and inviability (Coyne and Orr, 1989a; Wu, 1992). That is a different problem (involving an important stochastic component), whose eventual resolution will not affect the conclusion stated in (1), which simply bespeaks the pervasiveness of hybrid male sterility. As regards the second conclusion, the disproportionate contribution of the X-chromosome, a similar pattern is observed for hybrid inviability and hybrid female sterility.

What makes hybrid male sterility of great current interest is the increasing evidence that the building blocks of this isolating barrier may be radically different from what we had come to believe (Wu and Palopoli, 1994). Most probably, they are not a more or less abundant collection of genes with individually detectable effects on hybrid fertility (multigenic basis), but a large number of interacting gene sets, made out of minor factors whose individual introgression has virtually no effect (polygenic basis). It is clear that a new paradigm is emerging, which will force us to, first, revise many conclusions of past studies that had gathered almost unanimous agreement, and, second, try a completely different experimental approach to unveil the genetic basis of hybrid male sterility.

In this chapter, we first review the evidence supporting the polygenic hypothesis; then we develop an estimator of the number of interacting factors of hybrid male sterility, and apply it to autosome introgressions from *D. koepferae* into *D.buzzatii*; and finally, we present our first results on the molecular characterization of polygenic factors of hybrid male sterility in the *melanogaster* complex.

Single Major Factors versus Interacting Gene Sets

While the question of the number of factors contributing to hybrid male sterility between species of *Drosophila* may be considered more or less a settled matter (there are "many" of them), the nature of these factors has given rise to much controversy. In principle, sterility factors might be either of two kinds: single genes, each one of them able to produce sterility by itself when introgressed into the recipient species, or interacting gene sets, made out of couples, trios, or higher order combinations of genes, whose individual introgression would be of no detectable effect, male sterility only coming about after the cointrogression of a full gene set.

That second idea was stated by Pontecorvo years ago (1943), but most contemporary authors have favored, at least implicitly, the first alternative and have tried to map sterility factors by the conventional method of recombination with species-specific chromosome markers (Coyne and Charlesworth, 1989; Orr, 1989a,b; Zouros et al., 1988). The main problem with this kind of study lies in the difficulty of obtaining a sufficiently large number of diagnostic chromosome markers, so that undetected recombination between the genomes of the two species can be reduced to a minimum.

This problem was first overcome by Naveira and Fontdevila (1986, 1991a,b), by using the asynapsis between homologous polytene chromosomes as a marker of introgressed material in backcross hybrids between *D. buzzatii* and *D. koepferae*. In this approach, any polytenized chromosome region in the salivary glands of third-instar larvae can be easily diagnosed as introgressed (in heterozygosis), according to its pairing pattern. Then, its contribution to the phenotypic character under analysis (male sterility, in this case) can be conveniently evaluated. The results of this analysis in the hybrids between *D. buzzatii* and *D. koepferae* were quite unexpected. Despite the high incidence of male sterility when autosomes were introgressed (through repeated backcrosses of hybrid females to either parental species), no single factor produced hybrid male sterility by itself when the contribution of these autosomes was fully dissected. However, a correlation between the length of individually introgressed segments and the fertility/sterility outcome was observed: long segments produced sterility, while short segments did not. If the maximum length observed in fertile introgressions ("maximum for fertility") of different regions from the same chromosome is compared with the the minimum length observed in sterile introgressions ("minimum for sterility"), it may be concluded that, on average, interspecific substitution of roughly up to 30% of any of the autosomes allows hybrid male fertility, whereas a 40% substitution leads to hybrid male sterility (table 25.1). In chromosome 3, we observed an overlap between the ranges of the minima for sterility and maxima for fertility, which is due to the heterogeneity of the records along this chromosome, with its central regions always producing lower scores than those bound by the telomere or the centromere. But the rule that an introgressed segment associated with male sterility is never contained in a segment that allows male fertility has absolutely no exception. In other words, for any particular chromosome region, maxima for fertility are always smaller than minima for sterility. In the best-documented cases, the transition from fertility to sterility as the length of the introgression is increased can be shown to be quite sharp, thus revealing the existence of genuine threshold length intervals (figure 25.1). What is more, the threshold can be crossed by recombining in the genome of a hybrid male two fertile introgressions from distant locations on the same chromosome, or even from different chromosomes (Naveira and Fontdevila, 1986, 1991b). Therefore, in *buzzatii-koepferae* hybrids, the question of whether a hybrid male will be sterile or fertile, after the introgression of an autosome, is equivalent to asking how long the introgressed chromosome segment received from his hybrid mother is.

In contrast to these findings on autosomes, both short and long segments of the X-chromosome from *D. koepferae* always produced sterile males when introgressed into *D. buzzatii*, no matter what region they came from (Naveira and Fontdevila, 1986). According to this evidence, one might be tempted to conclude that this chromosome contains a high number of factors whose single introgression is enough to produce male sterility. But asynapsis is known to interfere with crossing over, thus setting lower bounds to the length of introgressed segments. So, the genetic architecture of hybrid male sterility might be the same for the X and the autosomes, the difference being simply that the length that should be

Table 25.1. Average maximum length of introgressed chromosome segments in fertile backcross male hybrids between *D. buzzatii* and *D. koepferae,* and average minimum length in steriles, for autosomes 3, 4, and 5.

Autosome	Minima for Sterility	n_s	Maxima for Fertility	n_f
3	0.368 (0.0325)	10	0.277 (0.0249)	17
4	0.441 (0.0153)	9	0.277 (0.0070)	15
5	0.415 (0.0304)	2	0.333 (0.0361)	2

Length is expressed as a percentage of the total length of the involved chromosome. Average of different regions from the same chromosome and standard error (in parentheses) are given for each autosome. The number of different introgressions used for the calculations (n_s and n_f, for sterile and fertile introgressions, respectively) is also indicated.

Figure 25.1. Some examples of maxima for fertility (thin lines) and minima for sterility (thick lines) in the introgression of chromosome 4 from *D. koepferae* into *D. buzzatii*. Each line represents a separately introgressed segment, in heterozygosis. Segments marked by thin lines are the longest ones that have been found in fertile males, for the chromosome regions considered. Thick lines correspond to the shortest segments found in steriles, for those same regions.

introgressed to produce sterility is much shorter for the X than for the autosomes (and shorter than the operational limit imposed by the asynapsis technique).

To test this hypothesis, we decided to study a pair of more closely related species, *D. simulans* and *D. mauritiana*. Former results based on the analysis of recombination frequencies between the sterility character and species-specific markers of the X-chromosome (Coyne and Charlesworth, 1989) had led to the conclusion that one locus with large effects on hybrid male sterility was tightly linked to each one of the markers used in the experiment (*yellow-white*, *forked*, and *miniature*). Nevertheless, one of us was able to show that male sterility could be produced by recombining in the same genome two or three fertile introgressions, which apparently contained interacting hidden genetic determinants of hybrid male sterility (Naveira, 1992).

Our thesis is that the majority of hybrid male sterility factors reported in the literature (Coyne, 1984; Coyne and Charlesworth, 1989; Coyne and Kreitman, 1986; Dobzhansky, 1936; Orr, 1987, 1989a,b; Orr and Coyne, 1989; Pontecorvo, 1943; Zeng and Singh, 1995; Zouros et al., 1988) do not correspond to genes with large effects, but to the undetected cointrogression, in an intact segment from the donor species, of a certain number of interacting factors. The separate introgression of any of these factors would be expected to have no effect on hybrid male fertility.

This perspective has received very strong support by the high-resolution experiments of Wu and co-workers, who used DNA markers to follow X-chromosome introgressions beween species of the *D. simulans* clade (*D. simulans*, *D. mauritiana*, and *D. sechellia*), and carried out a dissection of chromosome effects never before reached (Cabot et al., 1994; Palopoli and Wu,

1994; Perez and Wu, 1995; Wu and Palopoli, 1994). They concluded that multilocus weak allele interactions must be a very common cause of hybrid male sterility. At least two factors in some cases, at least three in others (the actual number may be higher), are necessary to confer full sterility, while the introgression of each individual factor by itself has no effect. Similar results have recently been reported for autosomal introgressions (in homozygosis) in these same species (Hollocher and Wu, 1996). The important consequence of these findings is that, again, but this time at the level of DNA segments, the question of hybrid male sterility is set in terms of the size of introgressed segments: when an introgressed segment that produces sterility is partitioned by recombination into shorter segments, sterility vanishes.

A Generalized Polygenic Model

Several hypotheses have been advanced to explain all these results. The latest one, for the autosome-mediated sterility in the hybrids between *D. buzzatii* and *D. koepferae*, has recently been published (Marín, 1996). By overlapping the introgressions reported previously (Naveira and Fontdevila, 1986, 1991a,b), this author claims to have mapped four zones both in chromosome 3 and 4 of *D. koepferae* that, with few exceptions, produce sterility when any two of them are introgressed together. However, only a few of the possible pairwise introgressions have actually been obtained, and most of them are expected to be sterile because they exceed the threshold length for sterility that we have postulated. In those few cases where the threshold is not exceeded, either the predictions of Marín fail (i.e., no sterility is observed in spite of introgressing a couple of putative factors), or there are yet no data available to test his hypothesis. Most of the introgressions spanning only one of the mapped putative epistatic factors are expected to be fertile under our hypothesis, because they involve such short chromosome segments that the threshold will not be exceeded. Again, in those few cases where the threshold is exceeded, either the predictions of Marín fail (i.e., sterility is observed although a single putative factor is introgressed), or no data are yet available.

In the same report, Marín introduced an excellent method for estimating the number of interacting factors of hybrid male sterility. By assuming that the number of sterility factors per introgressed chromosome segment (k) follows a Poisson distribution, and that sterility is brought about whenever this number exceeds an arbitrary threshold (x), the probability that an introgression produces sterility can be expressed as a function of the average number of factors per segment (λ). As the length of the introgressed segments increases, so does the probability of exceeding the threshold, and thus of getting a sterile male. Let us call this probability $P(\lambda)$, which is given by:

$$P(\lambda) = 1 - \Sigma\, e^{-\lambda}\lambda^k/k! \qquad (25.1)$$

Marín did not use this equation. He chose instead to use $\Delta P(\lambda)$ (or $\Delta s/\Delta N$, in his notation) and obtained an expression for what actually is the point of inflection of this function, which is hardly informative of the number

of sterility factors per chromosome. We think we have made some improvements by working on $P(\lambda)$. We have represented this function in figure 25.2, for $x = 5$ (i.e., when six co-introgressed factors are enough to produce sterility). This is a strictly growing function, with two extreme asymptotic values, $P = 0$ (all introgressions are fertile) for short chromosome segments, and $P = 1$ (all introgressions are sterile) for relatively long chromosome segments. There is a most remarkable intervening length interval, where the transition between $P \approx 0$ and $P \approx 1$ takes place rather quickly, which corresponds to the threshold length between fertile and sterile introgressions. The same general result would be obtained if we represented this function for any other value of x, even $x = 0$ (sterility brought about by the introgression of a single factor), because it is a natural consequence of the underlying Poisson distribution of the sterility factors. However, as the value of x increases, the function is displaced toward ever higher values of λ, and, most interestingly, the threshold-length interval becomes ever narrower (and vice versa, when the value of x decreases). This result can be seen in table 25.2, where the chromosome lengths (in average number of sterility factors) that produce sterility with probabilities 0.1 ($L_{0.1}$), and 0.9 ($L_{0.9}$) are indicated for different values of x. The last column in table 25.2 shows the ratio between both chromosome lengths ($L_{0.9}/L_{0.1}$), which gives the suddenness of the transition from fertile to sterile introgressions (i.e., of the relative

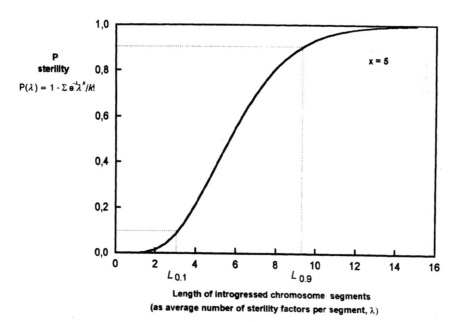

Figure 25.2. Probability that an introgressed chromosome segment produces sterility, as a function of its length, when sterility is brought about by the co-introgression of a minimum of six epistatic minor factors. The probability is given by equation 25.1, when the threshold (x) is 5. The length of the segments is expressed as the average number of factors they contain. A Poisson distribution of the factors along the chromosome is assumed.

Table 25.2. Threshold-length interval between fertile and sterile introgressions.

Threshold Number of Co-introgressed Factors (x)	Length of the Segments with Probability 0.1 of Exceeding the Threshold ($L_{0.1}$)	Length of the Segments with Probability 0.9 of Exceeding the Threshold ($L_{0.9}$)	$L_{0.9}/L_{0.1}$
1	0.54	3.89	7.20
2	1.1	5.3	4.82
6	3.9	10.6	2.72
10	7.1	15.4	2.17
15	11.1	21.3	1.92
20	15.4	27.1	1.76
25	19.7	32.7	1.66
40	33.1	49.4	1.49
60	51.2	71.2	1.39
100	88.5	114.0	1.29

Length of chromosome segments is given as the average number of epistatic sterility factors contained, assuming a Poisson distribution. Lengths with probabilities 0.1 ($L_{0.1}$) and 0.9 ($L_{0.9}$), respectively, of exceeding different thresholds for sterility (x) are indicated. The ratio between these two lengths ($L_{0.9}/L_{0.1}$) measures the width of the threshold-length interval between fertile and sterile introgressions.

width of the threshold-length interval). For example, when $x = 1$ (sterility produced by at least two co-introgressed epistatic factors), the chromosome lengths that usually give rise to sterility ($L_{0.9}$) are at least 7.2 times longer than the chromosome lengths that usually allow fertility ($L_{0.1}$). On the contrary, when $x = 100$ (sterility produced by at least 101 co-introgressed epistatic factors), they are only 1.29 times longer, and therefore the transition from fertility to sterility as the length of the introgressed segments increases would be seen as quite sudden by the observer. Thus, it should be possible to estimate the number of epistatic sterility factors from an analysis of the lengths of fertile and sterile introgressions.

By making use of the data in table 25.1 for the introgression from *D. koepferae* into *D. buzzatii*, we can obtain an underestimate of $L_{0.1}$, namely, the average maximum for fertility minus twice the standard error. Similarly, an overestimate of $L_{0.9}$ can be obtained from the average minimum for sterility plus twice the standard error. These estimates are shown in table 25.3, for chromosomes 3, 4, and 5, together with its ratio ($L_{0.9}/L_{0.1}$). The similarity among the three values of this ratio is striking, ranging from 1.8 (chromosome 4) to 1.9 (chromosome 3). This means that the genetic architecture of hybrid male sterility is probably the same in the three autosomes. Returning to table 25.2, we find that such values for $L_{0.9}/L_{0.1}$ are obtained for a threshold number of epistatic sterility factors (x) in between 15 and 20. Accordingly, the average number of sterility factors in $L_{0.1}$ segments may go from 11 to 15 (table 25.2), and since these segments represent approximately 25% of the whole chromosome (table 25.3), the total number of epistatic sterility factors in each autosome of *D. koepferae*

may be finally estimated as between 44 and 60. It must be stressed that this is most probably an underestimation of the real number.

Hybrid male sterility would, then, be produced by the cumulative effects of probably not less than 15 epistatic sterility factors introgressed from *D. koepferae* into *D. buzzatii*, sampled from a total of at least 60 per autosome. These factors are interchangeable, dispersed all over the chromosomes, and able to interact independently of the chromosome distance that separates them. The effect of their individual interspecific substitution is virtually undetectable, and, accordingly, they should be considered minor factors (polygenes), able to bring about sterility only when co-introgressed in a sufficient number (polygenic combination). It must be clearly understood that when we speak of "interacting" factors, we mean interaction in the final phenotypic outcome (ab-

Table 25.3. Estimates of $L_{0.1}$ and $L_{0.9}$ for autosome introgressions of *D. koepferae* into *D. buzzatii*.

Chromosome	$L_{0.1}$	$L_{0.9}$	$L_{0.9}/L_{0.1}$
3	0.227	0.433	1.907
4	0.263	0.472	1.795
5	0.260	0.476	1.831

The estimation of $L_{0.1}$ is obtained by subtracting twice the standard error from the average maximum for fertility given in table 25.2. The estimation of $L_{0.9}$, by adding twice the standard error to the average minimum for sterility given in that same table. See text for further explanation.

normalities in spermatogenesis), not in the immediate effects of each factor substitution. On the contrary, we postulate that these factors probably make an additive contribution to a still undetermined cellular balance. Interaction results from the cumulative contribution of both introgressions to this same balance, which might be established by an antagonism between two distinct groups of interacting elements, one for the X-chromosome and another for the autosomes. This is what Palopoli and Wu (1994) have called a generalized polygenic model.

A different polygenic model has been suggested by the same Wu and co-workers. It emphasizes the importance of linkage relationships by postulating (1) that interactions among closely linked conspecific genes introgressed in the recipient species are probably stronger and more frequent than among unlinked factors, and (2) that the interaction involves a specific gene set (Palopoli and Wu, 1994; Perez and Wu, 1995), rather than a large collection of promiscuous factors. Many combinations of fertile introgressions from different chromosome regions should be examined to discriminate between these two models. However, the available evidence from *buzzatii-koepferae* hybrids, which shows similar abnormalities in the spermatogenesis of hybrids for intact (noninterrupted) and nonintact (interrupted by recombination tracts) introgressed chromosome segments, or even for combinations of introgressions from different chromosomes (Naveira and Fontdevila, 1991b), militates against the general importance of linkage relationships in the determination of hybrid male sterility.

Walking on the Edge

Perhaps, no two alternative hypotheses can be more different than the single major gene versus the polygenic set as an underlying basis for a phenotypic character. However, it may not be easy to discriminate between them,

although the generalized polygenic model offers testable predictions against null hypotheses based on single factors with major effects (Maside and Naveira, 1996a,b). Polygenic effects are intrinsically difficult to dissect, because the average effect of each individual allelic substitution is obscured by environmental factors and genetic modifiers. However, when dealing with a threshold character, the genetic dissection may be more feasible. The idea is that the transition from hybrid male fertility to sterility may be so abrupt that if we could get near the threshold to sterility, without, so to say, crossing it, the effect of additional individual substitutions might be detected by a sudden increase in sterility frequencies.

To investigate this idea, we constructed a fixed hybrid strain, by introgressing into a multiple mutant stock of *D. simulans* three X-chromosome sections from *D. mauritiana*, one after the other, marked by the wild-type alleles of the loci *yellow* (*y*), *miniature* (*m*), and *forked* (*f*). Almost all males from this fixed hybrid strain ($y^+m^+f^+$) would be considered fertile according to the usual criterion, namely, the presence of motile sperm in their seminal vesicles, if examined when they are 3–4 days old (at 25°C). In these conditions, the sterility frequency in the fixed hybrid strain is in fact not different from that in the *D. simulans* stock ($\approx 1\%$). However, something in the spermatogenesis of these hybrid males is wrong. The first indication in this direction was the observation of a typical engrossment of the testis, just before the constriction that marks the entrance to the seminal vesicle. When examined at the microscope, this region appears to be filled with degenerating coiled spermatid cysts. A similar phenotype was observed in the hybrids between *D. buzzatii* and *D. koepferae*, and classified as semisterile (Naveira and Fontdevila, 1991b, figure A2b). In addition, there is a significant time lag, as compared to *D. simulans*, in the release of sperm to the seminal vesicle (table 25.4). In *D. simulans* the seminal vesicles of adult males get completely filled with sperm at 18–24

Table 25.4. Timing of sperm release into the seminal vesicle.

Fly Stock	Seminal Vesicle[1]	Time interval (hrs)							
		6–12	12–18	18–24	24–30	30–42	42–54	54–66	66–78
simulans	0	1	0.69	0	0.08	—	—	—	—
	1/2	0	0.19	0.12	0.08	—	—	—	—
	1	0	0.12	0.88	0.84	—	—	—	—
$y^+m^+f^+$	0	1	1	0.92	1	0.20	0.28	0.10	0.04
	½	0	0	0.08	0	0.70	0.67	0.65	0.46
	1	0	0	0	0	0.10	0.05	0.25	0.50

Data show relative frequency of observed vesicles in the indicated state, for each time interval after the ecdysis of adults, at 21°C. Sample size: 6–12 males.

[1]*0*, Seminal vesicle completely deprived of sperm; *1/2*, seminal vesicle less than 50% filled with sperm; *1*, seminal vesicle more than 50% filled with sperm.

hours after emergence, at 21 °C. In contrast, $y^+m^+f^+$ males take 30–42 hours to release some sperm, and 66–78 hours must elapse for 50% of the seminal vesicles to be completely filled. This delay is also manifest in the temporal pattern of change of the distribution of spermatid cyst classes at 25°C (table 25.5). Just emerged (0–6 hours old) D. simulans males are characterized by roughly equal frequencies of preindividualized and individualized spermatid cysts, and they usually have sperm in their seminal vesicles. On the contrary, hybrid $y^+m^+f^+$ males of the same age have nearly three times as many preindividualized as individualized cysts, and no sperm. It is later, at 48 hours, when sperm has generally already been released to their seminal vesicles, that the distribution is more like that of younger males of D. simulans. This probably means that the whole spermatogenesis process is slowed down in the fixed hybrid strain. Based on all this evidence, we think that males from this strain should be considered semisterile. All these properties are entirely lost when any two of the three chromosome segments from D. mauritiana are replaced by the corresponding D. simulans homologs. Under our hypothesis, the combination of the three chromosome segments introgressed in this strain ($y^+m^+f^+$) brings an indeterminate number of polygenes of hybrid male sterility, still not enough to cross the phenotypic threshold between fertility and sterility, but nearly so.

By P-element-mediated transformation of the fixed hybrid strain, we have produced insertions of a 6.7 kilobase pair (kbp) restriction fragment from the achaete-scute region of D. melanogaster, which includes the transcription unit of T4, one of the numerator elements of the X:A ratio that determines sex in Drosophila (Torres and Sánchez, 1989). Also, as a control, we produced insertions of a 1.9 kb adjacent fragment, which contains noncoding DNA. We obtained three different single insertion lines with the 6.7 fragment, and four with the 1.9 fragment. Table 25.6 shows the frequencies of adult males with few or no sperm in their seminal vesicles in these

different insertion lines, when males are aged for 3–4 days at 25°C. All three insertions of the 6.7 fragment led to a significant increase in the frequencies of males with no sperm (14–36%), as compared with the nontransformed fixed hybrid strain (1.7%). Few additional males (2–7%) were in the category of ≤25 motile sperm. On the other hand, in one of the four insertion lines of the 1.9 fragment, 23% of the males had no sperm at all, and the frequency raised to 43% when considering all males with ≤25 motile sperm (1.7% in controls). The three other insertions of noncoding DNA had only a minor effect (17% of males with ≤25 motile sperm, in line II), or no effect at all. When the different insertions of the 6.7 kb were tested on genetic backgrounds with less introgression from D. mauritiana, a significant decrease of sterility frequencies was observed, with all the three marked chromosome regions making a contribution to this effect (table 25.7). When tested on pure D. simulans background (i.e., without the three markers from D. mauritiana), all the males examined were fertile when aged for 3–4 days at 25°C (table 25.7). The analysis of the effect of these insertions in homozygosis has not been finished yet, but we can report that they have no effect at all on fertility when tested in the D. simulans background. On the contrary, when recombined into the genome of the fixed hybrid strain, sterility frequencies are generally higher than in heterozygotes. For example, in the line denoted I, for the 6.7 fragment, nearly 40% of the homozygous males aged for 7 days at 25°C have no sperm in their seminal vesicles (18% in heterozygotes aged for 3–4 days at this temperature; table 25.6). Moreover, an additional 10% of these males show extreme atrophy of one or both testes! For one or another reason, this particular line seems to have been brought to the verge of absolute sterility.

Therefore, it seems clear that this approach allows the detection of the effect on spermatogenesis of very small changes in the hybrid genome (at least those involving the addition of short DNA sequences), which is the first

Table 25.5. Temporal pattern of change in the distribution of classes of spermatid cysts.

Fly Stock	Time Interval (hrs)	Stage in Spermatogenesis			Presence of Sperm[2]
		(i)	(ii)	(iii)	
simulans	0–6	15.1	16.9	5.7	1
$y^+m^+f^+$	0–6	32.7	12.7	0.6	0
	48–54	23.6	18.1	12.3	0.9

Data show average number of cysts observed in each of the indicated stages, for males grown at 25°C, at different time intervals after their ecdysis. Sample size: 5 males.

[1](i) Preindividualization. (ii) Individualization. (iii) Coiling.

[2]Whether sperm has been released to the seminal vesicle (relative frequency of vesicles containing sperm).

Table 25.6. Relative frequencies (± standard deviation) of males with no motile sperm in their seminal vesicles ($n = 0$), or at most 25 motile sperm in both seminal vesicles ($n \leq 25$), in the nontransformed fixed hybrid strain ($y^+m^+f^+$), and in the different insertion lines obtained from it.

Strains	Sample Size	Motile Sperm in Seminal Vesicles	
		$n = 0$	$n \leq 25$
$y^+m^+f^+$	60	0.02 ±0.018	0.02 ±0.018
6.7–I	105	0.18 ±0.038	0.20 ±0.039
6.7–II	76	0.14 ±0.040	0.21 ±0.046
6.7–III	81	0.36 ±0.053	0.37 ±0.054
1.9–I	60	0.02 ±0.016	0.02 ±0.016
1.9–II	60	0.05 ±0.029	0.16 ±0.048
1.9–III	60	0.02 ±0.016	0.07 ±0.032
1.9–IV	65	0.23 ±0.052	0.43 ±0.061

The results correspond to $y^+m^+f^+$ adult hybrid males bearing the insertion in heterozygosis, dissected after being aged for 3–4 days at 25°C

condition for the molecular isolation of a sterility factor. On the other hand, the significant decrease in the penetrance of the insertions from *D. melanogaster* (sterility frequencies) when any of the three chromosome sections from *D. mauritiana* is eliminated, and the absolute normality of the spermatogenesis when tested individually in the pure *D. simulans* background, indicates that they behave exactly as expected for polygenes of hybrid male sterility. Finally, the effect detected after inserting noncoding DNA suggests that the coding potential of the introgressions from *D. mauritiana* might be also irrelevant for hybrid male fertility. It might be only a question of foreign DNA amount, in agreement with the thresholds observed in the chromosome substitution experiments described above. That may explain the generality of apparent epistasis among minor factors dispersed throughout the genome. This hypothesis does not exclude the existence of major genes of hybrid male sterility, but there are apparently very few, if any at all.

Concluding Remarks

Current evidence suggests that the genetic basis of hybrid male sterility in *Drosophila* is generally polygenic, and the total number of sterility factors must probably be numbered at least in the hundreds. The individual effect on fertility of any of these factors is virtually undetectable, but it can be accumulated to others. So, hybrid male sterility results from the co-introgression of a minimum number of randomly dispersed factors (polygenic combination). The different factors linked to the X, on the one hand, and to the autosomes, on the other, are interchangeable. All these properties can be incorporated into a generalized polygenic model, with testable predictions. The transition from fertility to sterility is first manifest by a slowdown of spermatogenesis, which may lead to a lag of several days in the release of sperm into the seminal vesicles. Recent experiments on the nature of these polygenes suggest that the coding potential of their DNA may be irrelevant.

Acknowledgments   Part of this work was supported by grants from Xunta de Galicia (XUGA 10305B95) and Ministerio de Educación y Ciencia (PB92–0386), Spain, awarded to HFN. We thank D. Howard, D. Prowell, and C.-I. Wu for their very helpful comments and criticism of the manuscript.

References

Bock, I. R. 1984. Interspecific hybridization in the genus *Drosophila*. Evol. Biol. 18:41–70.
Cabot, E. L., Davis, A. W., Johnson, N. A., and Wu, C.-I. 1994. Genetics of reproductive isolation in the *Drosophila*

Table 25.7. Relative frequencies of sterile males (± standard deviation) in the insertion lines produced with the fragment of 6.7 kbp from *D. melanogaster*, after the substitution of each of the three markers from *D. mauritiana*, or all of them, by their corresponding homologs in *D. simulans*.

Lost marker from *D. mauritiana*	Insertions of the 6.7 kbP Fragment		
	I	II	III
y^+	0.00 (28)	0.00 (38)	0.04 ±0.039 (25)
m^+	0.00 (20)	0.03 ±0.031 (31)	0.10 ±0.053 (31)
f^+	0.11 ±0.074 (18)	0.04 ±0.035 (28)	0.06 ±0.057 (17)
$y^+m^+f^+$	0.00 (32)	0.00 (15)	0.00 (30)

All the results correspond to adult males bearing the insertion in heterozygosis, after being aged for 3–4 days at 25°C. Sample size is indicated in parentheses.

simulans clade: complex epistasis underlying hybrid male sterility. Genetics 137:175–189.

Coyne, J. A. 1984. Genetic basis of male sterility in hybrids between two closely related species of *Drosophila*. Proc. Natl. Acad. Sci. U.S.A. 81:4444–4447.

Coyne, J. A. 1985. The genetic basis of Haldane's Rule. Nature 314:736–738.

Coyne, J. A., and Charlesworth, B. 1989. Genetic analysis of X-linked sterility in hybrids between three sibling species of *Drosophila*. Heredity 62:97–106.

Coyne, J. A., and Kreitman, M. 1986. Evolutionary genetics of two sibling species of *Drosophila*, *D. simulans* and *D. mauritiana*. Evolution 40:673–691.

Coyne, J. A., and Orr, H. A. 1989a. Patterns of speciation in *Drosophila*. Evolution 43:362–381.

Coyne, J. A., and Orr, H. A. 1989b. Two rules of speciation. In D. Otte and J. Endler (eds.). Speciation and Its Consequences. Sunderland: Sinauer, pp. 180–207.

Dobzhansky, T. 1936. Studies on hybrid sterility. II. Localization of sterility factors in *Drosophila pseudoobscura* hybrids. Genetics 21:113–135.

Dobzhansky, T. 1975. Analysis of reproductive isolation within a species of *Drosophila*. Proc. Natl. Acad. Sci. U.S.A. 72:3638–3641.

Ehrman, L. 1962. Hybrid sterility as an isolating mechanism in the genus *Drosophila*. Q. Rev. Biol. 37:279–302.

Hollocher, H., and Wu, C.-I. 1996. The genetics of reproductive isolation in the *Drosophila simulans* clade: X versus autosomal effects and male versus female effects. Genetics 143:1243–1255.

Marín, I. 1996. Genetic architecture of autosome-mediated hybrid male sterility in *Drosophila*. Genetics 142:1169–1180.

Maside, X. R., and Naveira, H. F. 1996a. On the difficulties of discriminating between major and minor hybrid male sterility factors in *Drosophila* by examining the segregation ratio of sterile and fertile sons in backcrossing experiments. Heredity 77:433–438.

Maside, X. R., and Naveira, H. F. 1996b. A polygenic basis of hybrid sterility may give rise to spurious localizations of major sterility factors. Heredity 77:488–492.

Muller, H. J., and Pontecorvo, G. 1940. Recombinants between *Drosophila* species the F₁ hybrids of which are sterile. Nature 146:199–200.

Naveira, H. F. 1992. Location of X-linked polygenic effects causing sterility in male hybrids of *Drosophila simulans* and *D. mauritiana*. Heredity 68:211–217.

Naveira, H. F., and Fontdevila, A. 1986. The evolutionary history of *Drosophila buzzatii*. XII. The genetic basis of sterility in hybrids between *D. buzzatii* and its sibling *D. serido* from Argentina. Genetics 144:841–857.

Naveira, H. F., and Fontdevila, A. 1991a. The evolutionary history of *Drosophila buzzatii*. XXII. Chromosomal and genic sterility in male hybrids of *D. buzzatii* and *D. koepferae*. Heredity 66:233–239.

Naveira, H. F., and Fontdevila, A. 1991b. The evolutionary history of *Drosophila buzzatii*. XXI. Cumulative action of multiple sterility factors on spermatogenesis in hybrids of *D. buzzatii* and *D. koepferae*. Heredity 67: 57–72.

Orr, H. A. 1987. Genetics of male and female sterility in hybrids of *Drosophila pseudoobscura* and *D. persimilis*. Genetics 116:555–563.

Orr, H. A. 1989a. Genetics of sterility in hybrids between two subspecies of *Drosophila*. Evolution 43:180–189.

Orr, H. A. 1989b. Localization of genes causing postzygotic isolation in two hybridizations involving *Drosophila pseudoobscura*. Heredity 63:231–237.

Orr, H. A., and Coyne, J. 1989. The genetics of postzygotic isolation in the *Drosophila virilis* group. Genetics 121: 527–537.

Palopoli, M. F., and Wu, C.-I. 1994. Genetics of hybrid male sterility between *Drosophila* sibling species: a complex web of epistasis is revealed in interspecific studies. Genetics 138:329–341.

Perez, D. E., and Wu, C.-I. 1995. Further characterization of the *Odysseus* locus of hybrid sterility in *Drosophila*: one gene is not enough. Genetics 140:201–206.

Pontecorvo, G. 1943. Hybrid sterility in artificially produced recombinants between *Drosophila melanogaster* and *D. simulans*. Proc. R. Soc. Edinb. Sect. B. 61:385–397.

Schäfer, U. 1978. Sterility in *Drosophila hydei* × *D. neohydei* hybrids. Genetica 49:205–214.

Sturtevant, A. H. 1920. Genetic studies on *Drosophila simulans*. I. Introduction. Hybrids with *Drosophila melanogaster*. Genetics 5:488–500.

Torres, M., and Sánchez, L. 1989. The *scute* (*T4*) gene acts as a numerator element of the X:A signal that determines the state of activity of *Sex-Lethal* in *Drosophila melanogaster*. EMBO J. 8:3079–3086.

Wu, C.-I. 1992. A note on Haldane's Rule: hybrid inviability vs. hybrid sterility. Evolution 46:1584–1587.

Wu, C.-I., and Davis, A. W. 1993. Evolution of postmating reproductive isolation: the composite nature of Haldane's Rule and its genetic bases. Am. Nat. 142:187–212.

Wu, C.-I., and Palopoli, M. F. 1994. Genetics of postmating reproductive isolation in animals. Annu. Rev. Genet. 28: 283–308.

Wu, C.-I., Johnson, N. A., and Palopoli, M. F. 1996. Haldane's Rule and its legacy: why are there so many sterile males? TREE 11:281–284.

Zeng, L.-W., and Singh, R. S. 1995. A general method for identifying major hybrid male sterility genes in *Drosophila*. Heredity 75:331–341.

Zouros, E., Lofdhal, K., and Martin, P. A. 1988. Male hybrid sterility in *Drosophila*: interactions between autosomes and sex chromosomes in crosses of *D. mojavensis* and *D. arizonensis*. Evolution 42:1321–1331.

26

Subtle Is Nature

The Genetics of Species Differentiation and Speciation

Chung-I Wu
Hope Hollocher

With all the recent advances that have been made in molecular genetics and developmental biology, it seem paradoxical what little impact those advances have had in bringing us closer to answering the simple question, How many genes does it take to make a new species? One impression from molecular evolution studies is that genetic differences between closely related species are generally very small (Wilson et al. 1974a,b; Nei 1975). The small difference between human and chimpanzee at the molecular level, currently estimated to be about 1.7% (Sibley and Ahlquist 1984; Koop et al. 1989), created a sensation when it was first reported (Goodman 1962; Wilson et al. 1974b). However, what does knowing that humans and chimpanzee have diverged by 1.7% really tell us about the genes involved in our own speciation? The relatively small difference of 1.7% represents about 50 million base pairs (bp) in a genome of 3×10^9 bp. How many base pairs have to be changed in order to convert a chimpanzee embryo into a human in all the morphological, physiological, and behavioral senses? Our ignorance about the genetics of species differences becomes quite obvious here. We cannot refute the view that 50 of the most critical base pair changes could convert a chimpanzee into a passable human being. Neither can we rule out the extreme opposite view that 500,000 bp, about 1% of the total difference, would be needed. Obviously, our ignorance about the genetics of species difference spans many orders of magnitude.

To address the question of species differentiation, we may study the underlying genetic basis of characters that define closely related species, such as morphological, physiological, or behavioral traits. In this report, we limit our discussion to the genetics of hybrid male sterility between three sibling species in the *Drosophila simulans* clade and briefly present the genetics of premating isolation between two *D. melanogaster* races. All in all, these species, as shown in figure 26.1, probably offer the best genetic tool kits among higher eukaryotes for asking about the genetics of species divergence. Hybrid males between each pair of the *D. simulans* clade are sterile, whereas all pairwise hybridizations produce fertile females. These species are so closely related that most molecular polymorphisms are still shared (e.g., Hey and Kliman 1993). The *D. simulans* clade is very well differentiated from its more familiar sibling species, *D. melanogaster*. In order to investigate the genetic differentiation between truly nascent species, we also study two races of this species (Begun and Aquadro 1993; Wu et al. 1995; Hollocher et al. 1997a,b). These

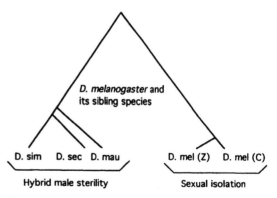

Figure 26.1. *D. simulans, D. mauritiana,* and *D. sechellia* are all sibling species of *D. melanogaster* and have been subjected to intensive analysis for the genetics of hybrid sterility. The Zimbabwe/southern Africa populations of *D. melanogaster* (Z-type) harbor genetic variants for sexual isolation from the cosmopolitan type (C-type). Recent studies of these two systems are reviewed in the text.

D. melanogaster races show strong premating isolation but little detectable postmating isolation between them (Wu et al. 1995) and serve as a good point of comparison with the *D. simulans* group for investigating how the genetic patterns of differentiation change as one moves between different levels of divergence.

Often a distinction is made between the "genetics of species differences" and the "genetics of speciation" (Templeton 1981, 1994), the reasoning being that if one truly wishes to understand speciation, then one should naturally concentrate on the changes that can be said to have directly contributed to speciation rather than changes that may have accumulated subsequent to speciation. Because hybrid male sterility is a component of reproductive isolation, these types of studies are often perceived as the studies of the "genetics of speciation." However, the genetic divergence underlying hybrid male sterility could not possibly have evolved for reasons of sterility. Postmating isolation is a pleiotropic by-product of selection that operates on something else, such as improving the competitiveness of sperm. Conceptually, it is probably simpler to accept studies of hybrid sterility as the studies of "species differences."

How, then, does the study of the genetics of species difference tell us anything about the genetics of speciation? In the strict allopatric mode of speciation, the two questions are in fact equivalent except perhaps in time scale as shown in figure 26.2a. If two diverging species have never had any contact, at least until they become truly good species, then the evolution in one lineage is quite independent of the evolution in the other lineage. In figure 26.2a, the divergence at T2 (the genetics of species difference) and the divergence at T1 (the genetics of speciation) are different only in scale but not necessarily in kind (see also Orr 1995; Palopoli et al. 1996). In strict allopatry, the forces driving the evolution along each lineage from T0 to T1 should not be any different than those driving the evolution from T1 to T2. In this view, hybrid male sterility can be considered the manifestation of functional divergence between genes of male reproduction. The fitness valley in the form of hybrid sterility or inviability is interesting mainly because it allows us to measure the differences between the two adjacent fitness peaks, as shown in figure 26.2b. Thus, studying hybrid sterility gives us a gauge as to how diverged two species have become and is equivalent to studying functional or morphological divergence between species. The intermediate form is gametogenic arrest and, naturally, hybrid sterility.

In short, the distinction between the genetics of "species differences" and "speciation" is a rather subtle one. If one considers mainly allopatric speciation, the difference is more in scale than in kind. It is of course desirable to study the genetic differences between nascent species (i.e., at T1 or earlier). The reason, however, is not necessarily because genetic divergence in the incipient stage is unique or special in kind, but rather because there are fewer genetic changes to deal with.

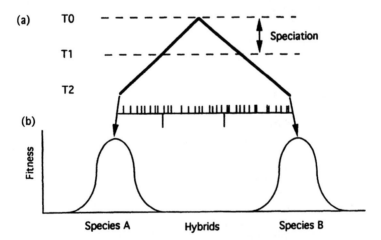

Figure 26.2. Species differentiation and speciation. a, speciation is completed at time T1, and the genetics of species divergence is assayed at time T2. In a strict allopatric mode, evolutionary forces between T0 and T1 and between T1 and T2 along each lineage should not be any different. Genetics of speciation only differ in scale, not in kind, from subsequent divergence. b, postmating reproductive isolation (such as hybrid male sterility) represents hybrid fitness reduction between the adaptive peaks of the two species. How many genetic changes separate the two adaptive peaks? Two possibilities are depicted: many small changes (ticks above the line) and a few large changes (below the line).

Hybrid Male Sterility and Haldane's Rule

Male sterility is highly prevalent in animal hybrids. In animals with heterogametic males, the first trait to suffer hybrid incompatibility is almost always male fertility (Wu 1992; Wu and Davis 1993). This is an important component of Haldane's Rule—that the heterogametic sex is more severely affected in species hybridization than the homogametic sex (Haldane 1922). The observation that male sterility evolves faster than male lethality is intriguing because the mutation rate to lethality is much greater than the mutation rate to male sterility. It has been suggested that male reproductive characters in general diverge rapidly (Eberhardt 1985), and hybrid male sterility (or, more precisely, divergence in spermatogenic development) may be another manifestation of this general phenomenon. If this view is correct, an implication is that sexual selection may be driving the evolution of postmating isolation in the form of hybrid male sterility (Wu et al. 1996).

Fine Structure of X-Chromosomal Regions Causing Hybrid Male Sterility

There are two approaches to analyzing the genetics of hybrid male sterility—by F1 crosses to obtain F2s or by repeated backcrosses to obtain introgressions. The former is more commonly used but it attempts to dissect a complex phenotype by examining a complex array of F2 genotypes. The results are often uninformative, and may even be misleading (Wu and Palopoli 1994). Instead, we have been analyzing introgressions as shown in figure 26.3; all four genotypes in that figure yield complete male sterility. It is natural to assume that, within the introgressed segment, a single gene causes incompatibility with the genetic background and results in male sterility. If there is indeed only one male sterility factor, it should be localizable to the same position from either flanking marker; furthermore, the recombinant genotypes should be either completely fertile or completely sterile. Perez et al. (1993) found that both criteria were met. DNA marker–assisted mapping also delineates the putative sterility factor, named Ods (for Odysseus), within a well-defined interval (figure 26.4). If this single-gene interpretation was true, it implied limited genetic divergence for the genome as a whole. Since hybrid male sterility evolves much more rapidly than other traits (Wu 1992; Wu and Davis 1993), the interpretation would likely be true for other aspects of species differentiation as well. This view is represented in the left panel of figure 26.5.

The simple interpretation presented above, however, has proven wrong when the analysis was extended. The Ods factor of figure 26.4 does not cause male sterility when introgressed alone. Another unidentified factor is

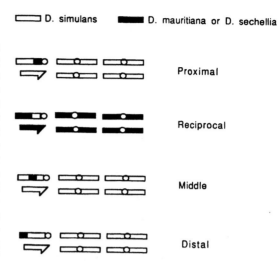

□ D. simulans   ■ D. mauritiana or D. sechellia

Proximal

Reciprocal

Middle

Distal

Figure 26.3. Each of the four X-linked male-sterile introgressions (proximal, its reciprocal, middle, and distal) can cause complete male sterility. Each introgression represents 2–5% of the total genome. Given here are X, Y, and the two major autosomes. Genetic analysis was made on each of these introgressions—proximal (Perez et al. 1993; Perez and Wu 1995), reciprocal (Palopoli and Wu 1994), middle (Davis and Wu 1996), and distal (Cabot et al. 1994).

also necessary, and hybrid male sterility can be observed only if both factors are present in the introgression (lower half of figure 26.4; from Perez and Wu 1995). Likewise, when the reciprocal introgression of figure 26.4 was carried out, four factors could be mapped in a slightly larger region, and none of these factors could singly cause sterility by itself (Palopoli and Wu 1994). As seen in figure 26.6, a single sterile introgression could be decomposed into two overlapping fertile introgressions, and if the two fertile introgressions are joined again, sterility is reconstituted. This pattern requires at least two factors to explain. It is worth noting that the factors mapped by the reciprocal introgression are different (see figure 26.7). Such an asymmetry was expected and have been observed (Wu and Beckenbach 1983; Zouros 1989).

The results from the series of high-resolution studies are summarized in figure 26.7 and table 26.1. Each sterility factor is defined when, in certain genetic backgrounds, its presence causes a transition from high fertility (often a penetrance of more than 60%) to complete or near sterility. A few of these factors were mapped because each causes a change in the spermatogenic phenotype, for example, from the production of motile sperm to a complete spermatogenic arrest. In this type of analysis, no factors that cause only a small additive reduction in fertility could be detected. In conjunction with the autosomal study discussed in the next section (Hollocher

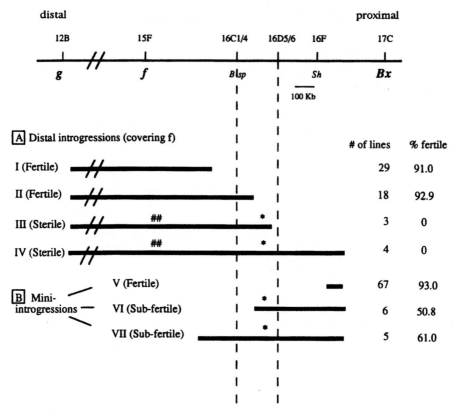

Figure 26.4. Detailed physical structure and the fertility penetrance of introgressions from *D. mauritiana* to *D. simulans* (Perez and Wu 1995). * denotes the putative sterility factor (Ods) mapped to the 16D region of the X-chromosome. ## represents an unidentified factor(s) that is also necessary for full sterility. Cytological locations of the visible and molecular markers are given above the top line. The Ods factor has now been cloned and determined to be a rapidly evolving homeobox gene (Ting et al. submitted).

and Wu 1996), we estimate the number of genes that have diverged in spermatogenic functions between *D. simulans* and *D. mauritiana* to be at least 120. This level of divergence is surprisingly high, especially in light of the results from mutagenic studies, which place the number of male sterility loci at around 1,000 in *Drosophila* (Lindsley and Tokuyasu 1980; Fuller 1993). One either accepts that more than 10% of spermatogenic loci have diverged in function between the two sibling species of *Drosophila* (a somewhat dubious proposition) or accepts that these hybrid sterility loci are not entirely a subset of male sterility mutations. This is a caution against the view that molecular characterization of male sterility mutations is a short cut to studying reproductive isolation at the molecular level. The overall results suggest extensive divergence in the genes of spermatogenesis between closely related species, as portrayed in the right panel of figure 26.5.

Autosomal (vis-à-vis X) Effects on Hybrid Male Sterility

It is evident from the fine-structure analysis of the X-chromosome that the underlying genetic architecture of hybrid male sterility involves many interacting loci. These X-chromosome analyses bring us much closer to answering our original question of how many genes it takes to make a new species. Compared to the X-chromosome, autosomes are harder to analyze but may tell us more about the genetics of reproductive isolation. For example, it is not possible to compare the effects of X-linked introgressions on female versus male sterility. The comparison between male and female effects will require autosomal studies (see the next section).

The main reason that the genetic analysis of hybrid male sterility has concentrated on the X-chromosome (Dobzhansky 1936; Bentley and Hoy 1979; Hennig 1977;

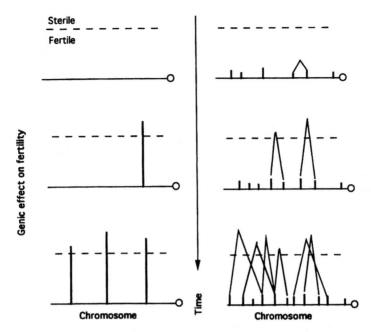

Figure 26.5. Two different views on the effect of hybrid sterility genes. The left panel portrays the simple view of "a few genes of large effect." Each locus acquires the capacity to cause strong sterility in one step. The right panel presents "weak allele–strong interaction" as the basis of hybrid sterility. When reproductive isolation is first attained as shown in the second stage, the right panel suggests extensive genetic divergence as opposed to the simpler case on the left.

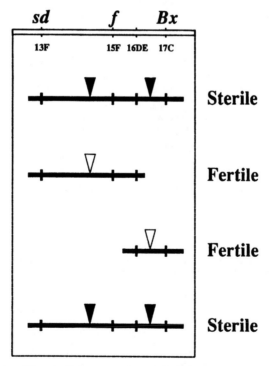

Figure 26.6. An example of "weak allele–strong interaction" that underlies hybrid male sterility (the "Reciprocal" in figure 26.3; Palopoli and Wu 1994). Both fertile introgressions show strong fertility penetrance (>95%) but joining them leads to complete sterility.

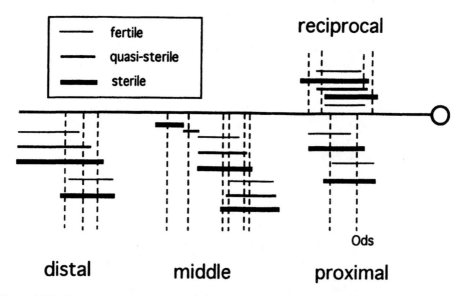

Figure 26.7. Summary of the genetic analysis of hybrid male sterility associated with the introgressions of figure 26.3. Dashed lines denote the locations of sterility factors. Note that these factors often interact epistatically as depicted in figure 26.6. The scale of the introgressions is only approximate.

Zouros 1981; Coyne and Orr 1989; Richler et al. 1989) with much less effort on the autosomes can be seen in the example of figure 26.8 (Hollocher and Wu 1996). It is at least possible in principle (but not necessarily so in practice) to analyze the full effects of X-linked genes on hybrid male sterility or inviability in F2, whereas autosomal effects can be fully analyzed only after several more generations of backcrosses. In F2, males may have their sole X-chromosome replaced by a foreign one; for autosomes, only one of the two copies can be replaced. Naturally, it is expected that the X-chromosome would appear to have a much stronger effect on hybrid fitness (fertility or viability, e.g.). This is one of the limitations of the "backcross F2" approach (Wu and Davis 1993).

This apparent involvement of the X-chromosome in reproductive isolation, termed the "large X-effect," was viewed to be as general a principle as Haldane's Rule and was referred to as the "second rule of speciation" (Coyne and Orr 1989). However, for reasons just discussed, the "large X-effect" is probably an observational bias rather than representing an intrinsically greater involvement of the X-chromosome in reproductive isolation. Proper analysis of autosomal effects requires regions of the autosomes to be made homozygous in order to detect recessive effects that would normally remain hidden in a typical F2 backcross analysis. When such an analysis of homozygous, autosomal effects is performed, as sketched in figure 26.8, the "large X-effect" disappears (Hollocher and Wu 1996) or is seen to be at most 50% greater than the effect of homozygous autosomes (True et al. 1996). The latter still leaves the "large X-effect" an order of magnitude less than the original backcross observations

Table 26.1. Estimating the number of genes for hybrid male sterility between *D. simulans* and *D. mauritiana*.

Chromosomal Regions	% Genome	No. of Genes	Reference
4A-5C	1.5	3	Cabot et al. (1994)
9A-11F	3	6	Davis and Wu (1996)
15F-16F	1	2	Perez and Wu (1995)
13–17	5	4	Palopoli and Wu (1994)
Total	10.5	15	

Based on these results on the X-chromosome as well as recent studies on autosomes (True et al. 1996; Hollocher and Wu 1996), we estimate the number of genes contributing to hybrid male sterility between these two sibling species to be greater than 120 (Wu et al. 1996).

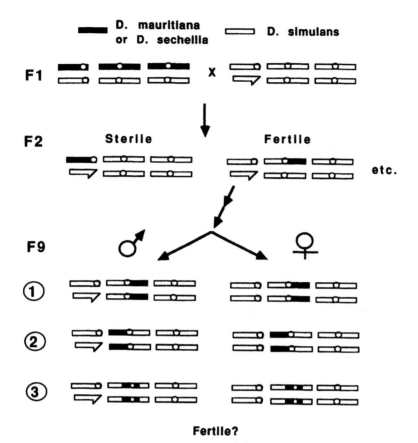

Figure 26.8. Analysis of autosomal introgressions for both male and female sterility. Note that in F2 males, X-introgressions are expected to have greater sterility effects than autosomal ones, because the latter replace only one of the two homologs. Autosomal introgressions are made homozygous in F9. Three regions of the second chromosome are analyzed and compared for their effects on males and females (Hollocher and Wu 1996).

had indicated. In short, the X-chromosome is not particularly unique in its capacity to cause hybrid sterility. (Interestingly, the X often contributes disproportionately less to premating isolation; see Hollocher et al. 1997b for discussion.)

Rapid Evolution of Hybrid Male Sterility vis-à-vis Hybrid Female Sterility

Autosomal studies have also provided a framework for making comparisons between hybrid male sterility and hybrid female sterility that cannot be easily addressed by X-chromosome analyses. Because males carrying X-chromosome introgressions are sterile, the crosses needed to make the introgressions homozygous in females simply cannot be performed. Davis et al. (1994) bypassed this problem by using the attached-X technique and suggested that there is much less genetic divergence for female sterility than for male sterility between sibling species of

Drosophila. Fortunately, in autosomal studies, the difficulty of making homozygous introgressions does not exist because most heterozygous introgressions are both male and female fertile and therefore can be crossed to produce homozygotes of both sexes, allowing a natural comparison of homozygous autosomal effects on hybrid male and female sterility (see figure 26.8). Such an analysis has shown that male sterility far outstrips female sterility even when controlling for hidden recessive effects (Hollocher and Wu 1996; True et al. 1996). Homozygous second chromosome regions introgressed from *D. sechellia* into *D. simulans* cause over 20 times more male sterility than female sterility; for introgressions from *D. mauritiana* into *D. simulans*, male sterility is four times greater (Hollocher and Wu 1996). An analysis by True et al. (1996) of introgressions from *D. mauritiana* into *D. simulans*, randomly covering the entire genome, showed male sterility to be about 10 times more prevalent than female sterility. It is evident that male sterility evolves much more quickly than female sterility at least when males are the heterogametic sex (Wu et al. 1996). It would be extremely interesting to see whether this pattern holds for analysis of species in which females are the heterogametic sex, such as butterflies and birds.

Y-Chromosome Effect on Male Sterility

In *Drosophila*, when hybrid males are sterile, the Y-chromosome is often part of the ensemble of genes that contribute to hybrid sterility (Johnson et al. 1993; Zeng and Singh 1993a; Pantazidis et al. 1993), but between very closely related species that produce sterile male hybrids the Y may not be involved (Johnson et al. 1992). Because of the difficulties in dissecting Y, we do not consider its effect further.

Genetics of Premating Isolation between Races

The main conclusions from the genetic studies of hybrid male sterility between sibling species (i.e., extensive divergence and strong epistasis) are open to two criticisms. First, hybrid incompatibility is the pleiotropic aspect of other unknown traits that may themselves be under selection. Since natural selection does not promote postmating isolation directly, the genetic basis of hybrid male sterility may not truly reflect the genetic basis of adaptive divergence. Second, the species being studied may have long passed the time of speciation, and hence, most of the genetic changes could have accumulated afterward (i.e., $T_2 >> T_1$ in figure 26.2). Although we have addressed these two issues in the introduction, it is desirable to analyze a different trait that characterizes the divergence of truly nascent species. Below, we shall highlight the analysis of sexual isolation between the Zimbabwe and the cosmopolitan race of *D. melanogaster*. A fuller account is presented elsewhere (Hollocher et al. 1997a,b; Ting et al. submitted).

D. melanogaster from Zimbabwe, Africa, was initially studied by Begun and Aquadro (1993) for their molecular divergence from the worldwide collection. A later study reveals strong asymmetric sexual isolation between many isofemale lines from Zimbabwe (Z-type) and all other lines of a cosmopolitan origin (referred to as the C-type in Wu et al. [1995] but now often referred to as the M-type). In multiple-choice experiments where both sexes of both Z-type and C-type are present, Z-females often mate exclusively with Z-males while C-males do somewhat better than Z-males in mating with C-females, as shown in figure 26.9a. Clearly, the genetics of both male and female behaviors have differentiated between these populations.

The Z versus C divergence is likely an event of nascent speciation for several reasons: (1) There exist extensive polymorphisms for both male and female mating behaviors in southern Africa (Hollocher et al. 1997a). For example, some Z-lines show intermediate Z-behaviors in both sexes (figure 26.9b). A collection of isofemale lines from Zambia and Botswana exhibit large variation in sexual preference ranging from the strong Z-type to the standard cosmopolitan type. (2) There is no detectable fitness reduction in F1s, F2s, or more complex genotypic combinations between the two types; all hybrids are apparently fully viable and fertile in laboratory culture. (3) There are no easily detectable morphological differences, even in male genitalia. These observations strongly suggest that the two types have not reached the point of "good species." On the other hand, the mate choice between the Z- and C-type can be strong and robust under

(a)

♀ \ ♂	Z	C
Z	38	0
C	16	47

(b)

♀ \ ♂	Z'	C
Z'	26	5
C	7	44

(c)

♀ \ ♂	CCZ	C
Z	31	0
C	19	38

(d)

♀ \ ♂	Z	CCZ
Z	30	11
C	25	34

(e)

♀ \ ♂	ZCC	C
Z	3	4
C	32	33

(f)

♀ \ ♂	CCZ	ZZC
CCZ	27	24
ZZC	28	28

Figure 26.9. A partial presentation of the genetic analysis of sexual isolation between the Zimbabwe (Z) and the cosmopolitan race (C) of *D. melanogaster* (from Hollocher et al. 1997b). In each experiment, about 120 pairs of virgin males and females, with equal representation from both types, are released into a mating cage, and copulating pairs are removed for type identification (see Wu et al. 1995). Numbers given are observed copulating pairs. CCZ denotes flies whose first and second chromosomes are of C-origin while the third chromosome is derived from a Z-source. (ZZC is similarly defined.) Z' denotes isofemale lines whose sexual behaviors are intermediate between Z and C.

a variety of environmental conditions, justifying the classification of race for the "mating types" (Hollocher et al. 1997b).

One advantage of studying reproductive isolation within the species of *D. melanogaster* is the convenience of genetic analysis. The first step is to construct "chromosome substitution lines" that may carry the first and second chromosomes from the C-type and the third from the Z-type, which is denoted CCZ. All combinations of substitutions between each of two C- and Z-lines have been constructed. These are whole chromosome substitutions with little genetic heterogeneity within lines and are maintained as stocks with no need of special care.

It can be seen that the third chromosome alone can make males sufficiently Z-like such that they monopolize the mating with Z-females (figure 26.9c). However, CCZ males are not full Z-males, as can be seen in their competition with pure Z-males; figure 26.9d suggests that either the first or the second chromosome (or both) must also be important for Z-maleness. Hollocher et al. (1997b) clearly show a role of the second chromosome. What is interesting is the absence of a detectable effect of the X-chromosome. Figure 26.9e shows that ZCC-males do not perform better than pure C-males. (The results, referred to earlier, corroborate the conclusion that the X-chromosome usually does not play a disproportionately larger role in reproductive isolation.) In a series of more refined analyses, an effect of the X-chromosome on Z-maleness could in fact be detected but the magnitude is rather small. Therefore, the contribution to Z-maleness can be summarized as III > II >> X $\geq$ 0, where III and II denote the two autosomes. Interestingly, when the chromosomal effect of Z-femaleness was measured (by analyzing female receptivity toward C-males), the same relationship III > II >> X > 0 was observed. Therefore, there is at least one genetic determinant for both male and female behavioral traits on each chromosome. Preliminary data from a finer genetic analysis of different regions of the third chromosome suggest multiple factors on the third chromosome (Ting and Wu, unpublished results).

Given the genetic determinants for both Z-femaleness and Z-maleness, we can then ask how male and female loci correspond to one another. In figure 26.9f, the two genotypes are maximally different (CCZ vs. ZZC). If female loci on the third chromosome correspond mostly to male loci on, say, the third chromosome, then strong assortative mating should be observed. The data indicate near random mating in this design. The implication is that female preference genes on both autosomes respond to male character genes on both autosomes equally well. Such a pattern also supports the view of extensive genetic divergence underlying sexual isolation between the two races of *D. melanogaster*.

Finally, the observation of strong sexual isolation in the absence of any forms of postmating isolation provides yet another example that reinforcement is not always necessary (Butlin 1989; Howard 1993; Noor 1995). The strength of this analysis is twofold: (1) The absence of postmating isolation is more thoroughly investigated than in many previous ones. There are many cases of postmating isolation in F2s where F1 hybrids are apparently normal. Between the Z- and C-types, F2s and other more complex genotypic combinations are all viable and fertile. (2) The number of genes contributing to sexual isolation in this case is large (>10). Each of these loci has evolved to affect certain aspects of mating choice, and each has done so in the absence (or near absence) of reinforcement. We return to this subject further below.

Discussion

The above observations of extensive genetic divergence between closely related species and between races often prompted the following question: How many genes are required for speciation and how many are in fact accumulated after speciation is completed? The distinction seems artificial in a strict allopatric mode of speciation because the time point that supposedly demarcates the periods before and after speciation (T1 of figure 26.2) has no biological significance. However, if nascent species come into contact during their initial divergence (i.e., in sympatry or parapatry), or if some evolutionists would want to know their species status by bringing them together to test for reproductive isolation from time to time (some very long-lived evolutionists), the questions are relevant, but need to be phrased differently. To the question, What is the minimal number of genes required for reproductive isolation?, the answer would be "two," or "one gene per species." As proposed by Dobzhansky (1936) and Muller (1940), the underlying model assumes that the ancestral population has the genotype A1B1, which evolved to A1B2 in one population and A2B1 in the other. The combination of A2B2 causes hybrid incompatibility. Such a simple model is so widely used in the literature that it could be referred to as the "standard model." The discovery of hybrid rescue mutations (Watanabe 1979; Hutter et al. 1990; Sawamura et al. 1993) indeed suggests that one gene per species could *in concept* be sufficient for reproductive isolation (see Wu and Palopoli [1994] for further interpretation).

The more appropriate question is, Does nature use the simple genetic system for reproductive isolation? In other words, how many genes actually contribute to postmating isolation when two diverging species first reach the status of good species? The two panels of figure 26.5 present two contrasting views. On the left is the simple genetic system that represents the shortest route to reproductive isolation (or the "standard model"). In our own analysis and the survey of animal genetic studies, however, there is no support for this scenario (Wu and Palopoli 1994). Instead, our studies support the view represented by the right panel of figure 26.5. When some genes have diverged sufficiently to cause postmating isolation, many other loci in the genome also have reached a comparable stage of reproductive incompatibility.

With respect to premating sexual isolation, the genetic divergence can also be quite extensive (table 26.2) even between races that show little sign of postmating isolation. In the case discussed here, the absence (or near absence) of postmating isolation can be demonstrated in F1s, F2s, and various chromosome substitutions (Hollocher et al. 1997b). In many earlier studies (Hennig 1977; Zouros 1981; Davis et al. 1994), these complex genotypes are often associated with easily detectable fitness reduction even though F1s are normal. This is certainly not true in this Z-C system. One may ask if some of these hybrid genotypes may experience fitness loss in a more realistic ecological context. While this cannot be ruled out, hybrid breakdown in most studies of reinforcement results from some sort of physiological incompatibilities that are manifested in all environments; for example, hybrid male sterility between D. simulans and D. mauritiana is complete in all environments. We can safely conclude that postmating isolation between the Z- and C-type, if existent at all, is probably much weaker than what is generally considered sufficient to drive reproductive character displacement.

Two more observations on the Z-C divergence are also relevant. First, the number of loci involved in the divergence of sexual behaviors is quite large (>10; Hollocher et al. 1997b; Ting et al. submitted). It seems unlikely that postmating isolation, too weak to detect in all hybrid genotypes, could drive the evolution at so many loci. Second, Hollocher et al.'s (1997a) survey of natural populations in southern Africa reveals a smooth gradation from the strong Z-type to pure C-populations. Such a smooth transition implies the absence of genes of postmating isolation. Therefore, to account for the Z-C divergence in mating behavior, sexual selection models that do not require reinforcement appear to suit the observations better. Runaway process (Fisher 1930; Lande 1981) is one such example.

The between-species comparison in table 26.2 sheds further light on the role of sexual selection in speciation. The contrast between the number of genes causing hybrid male sterility and those causing either inviability or female sterility has been extensively discussed (Wu et al. 1996). One explanation for why there are so many sterile males is analogous to the explanation for the rapid evolution of male genitalia (Eberhardt 1985). Male–male competition for mates can be so intense that male reproductive characters, be they genital morphology, male ejaculates (Clark et al. 1995; Howard et al., this volume), or spermatogenic properties (Lee and Vacquier 1992; Zeng and Singh 1993a; Metz and Palumbi 1996; Wu and Palopoli 1994; Tsaur and Wu 1997), may evolve at a very high rate. We hypothesize that sperm competition is intense and males are under constant selection pressure to outperform other males. Mutations at spermatogenic loci, as opposed to those at, say, oogenic or neurogenic loci, may have a greater chance of conferring differences (either positive or negative) in males' reproductive success. These mutations could affect sperm's maturation rate, their sizes, the mitochondrial density, receptors on sperm surface, and so on. The relatively large sperm in many Drosophila species may also be the consequence of such a process of sperm competition.

Both male–male competition and male–female coevolution (Rice 1996; Markow and Hocutt, this volume) could drive the evolution of sexual characters in speciation. With male–male competition, female reproductive traits are not expected to evolve as rapidly as male characters, but in male–female coevolution models, female reproductive traits should experience accelerated evolution as well. The studies of sexual characters have so far emphasized male traits (Rice 1996; Wu et al. 1996; Palumbi, this volume). Some analyses of female characters (e.g., Davis et al. 1994; True et al. 1996; Hollocher and Wu 1996; S. Palumbi, personal communication) have been carried out, but more will be needed to resolve the issue.

If the genetics of racial differentiation summarized in table 26.2 can serve as a guidepost, we may be led to the startling conclusion that the genetic differentiation is even greater for sexual behaviors than for male sterility. In other words, the genetic differences between D. simulans and D. mauritiana with respect to their sexual behaviors

Table 26.2. Summary of the genetic divergence underlying different aspects of reproductive isolation at two levels of species differentiation.

	Sexual Isolation	Hybrid Male Sterility	Hybrid Female Sterility or Inviability
Between races	>10[1]	0	0
Between species	?	> 120	< 20

Data are the estimated numbers of genes for each trait. "Between races" denotes the divergence between the Zimbabwe and worldwide populations of D. melanogaster (Wu et al. 1995), while "between species" denotes that between D. simulans and D. mauritiana.

[1]Including data of Hollocher et al. (1997b).

could be in the hundreds. There are reasons why such an extrapolation may not be valid (Orr 1995). It will thus be interesting to find out just how much genetic differentiation is associated with sexual isolation between good species. From published studies (e.g., Tan 1946; Zouros 1981; Coyne 1996), the number of loci is probably large but the resolution cannot yet provide a substitute for the question mark in table 26.2.

Conclusions

In the last five years, detailed genetic analyses of traits of reproductive isolation in *Drosophila* have revealed extensive genetic differentiation between sibling species as well as between races of the same species. For hybrid male sterility alone, the number of functionally divergent loci can be greater than 100 between species that are barely distinguishable by molecular means. This strongly supports the neodarwinian view of evolution (Fisher 1930; Mayr 1963; Dobzhansky 1970; Wright 1977) and is less in accord with the macroevolution argument of large and sudden changes (e.g., Raff and Kaufman 1983).

When different species are compared, strong epistasis between conspecific genes can often be observed. For traits of postmating isolation, "weak allele–strong interaction" is the rule rather than the exception (Wu and Palopoli 1994). Wright's (1977, 1982) idea of extensive epistasis between fitness-determining loci fits very well with these observations. The strong interaction has made it possible to clone a gene of reproductive isolation (Ting et al. submitted).

Sexual selection is a strong driving force of species differentiation, certainly with respect to premating isolation but likely for postmating isolation as well. The hypothesis will be testable when some of the genes are cloned (Perez and Wu 1995; Wu et al. 1996).

Genes underlying premating and postmating isolation are no more likely to be on the X than on an autosomal region of comparable size (Wu et al. 1995; True et al. 1996; Hollocher and Wu 1996; Hollocher et al. 1997b).

References

Begun, D. J., and Aquadro, C. F. 1993. African and North American populations of *Drosophila melanogaster* are very different at the DNA level. Nature 365:548–550.

Bentley, D. R., and Hoy, R. I. 1979. Genetic control of the neuronal network generating cricket (*Teleogryllus gryllus*) song. Anim. Behav. 20:478–492.

Butlin, R. 1989. Reinforcement of premating isolation. In D. Otte and J. Endler (eds.). Speciation and Its Consequences. Sunderland Mass.: Sinauer, pp. 158–179.

Cabot, E. L., Davis, A. W., Johnson, N. A., and Wu, C.-I. 1994. Genetics of reproductive isolation in the *Drosophila simulans* clade: complex epistasis underlying hybrid male sterility. Genetics 137:175–189.

Clark, A. G., Aguade, M., Prout, T., Harshman, L. G., and Langley, C. H. 1995. Variation in sperm displacement and its association with *Accessory gland protein* loci in *Drosophila melanogaster*. Genetics 139:189–201.

Coyne, J. A. 1996. Genetics of differences in pheromonal hydrocarbons between *Drosophila melanogaster* and *D. simulans*. Genetics 143:353–364.

Coyne, J. A., and Orr, H. A. 1989. Two rules of speciation. In D. Otte and J. Endler (eds.). Speciation and Its Consequences. Sunderland Mass.: Sinauer, pp. 180–207.

Davis, A. W., and Wu, C.-I. 1996. The broom of the sorcerer's apprentice: the fine structure of a chromosomal region causing reproductive isolation between two sibling species of *Drosophila*. Genetics 143:1287–1298.

Davis, A. W., Noonburg, E., and Wu, C.-I. 1994. Complex genic interactions between conspecific chromosomes underlying hybrid female sterility in the *Drosophila simulans* clade. Genetics 137:191–199.

Dobzhansky, T. 1936. Studies on hybrid sterility. II. Localization of sterility factors in *Drosophila pseudoobscura* hybrids. Genetics 21:113–135.

Dobzhansky, T. 1970. Genetics of the Evolutionary Process. New York: Columbia University Press.

Eberhardt, W. G. 1985. Sexual Selection and Animal Genitalia. Cambridge, Mass.: Harvard University Press.

Fisher, R. A. 1930. The genetical theory of natural selection. Clarendon Press, Oxford.

Fuller, M. T. 1993. Spermatogenesis. In M. Bate and A. Martinez-Arias (eds.). Development of *Drosophila*. Cold Spring Harbor, N.Y.: Cold Spring Harbor Press, pp. 71–147.

Goodman, M. 1962. Evolution of the immunologic species specificity of human serum proteins. Hum. Biol. 34:104–150.

Haldane, J. B. S. 1922. Sex ratio and unisexual sterility in hybrid animals. J. Genet. 12:101–109.

Hennig, W. 1977. Gene interactions in germ cell differentiation of Drosophila. In G. Weber (eds.). Advances in Enzyme Regulation. Oxford: Pergamon Press, pp. 363–371.

Hey, J., and Kliman, R. M. 1993. Population genetics and phylogenetics of DNA sequence variation at multiple loci within the *D. melanogaster* species complex. Mol. Biol. Evol. 10:804–822.

Hollocher, H., and Wu., C.-I. 1996. The genetics of reproductive isolation in the *Drosophila simulans* clade: X vs. autosomal effects and male vs. female effects. Genetics 143:1243–1255.

Hollocher, H., Ting, C. T., and Wu, C.-I. 1997a. Incipient speciation by sexual isolation in *Drosophila melanogaster*: variation in mating preference and correlation between sexes. Evolution 51:1175–1181.

Hollocher, H., Ting, C. T., Wu, M. L., and Wu, C.-I. 1997b. Incipient speciation by sexual isolation in *Drosophila melanogaster*: Extensive genetic divergence without reinforcement. Genetics 147:1191–1201.

Howard, D. J. 1993. Reinforcement: origin, dynamics, and fate of an evolutionary hypothesis. In R. G. Harrison (eds.). Hybrid Zones and the Evolutionary Process. New York: Oxford University Press, pp. 46–69.

Hutter, P., Roote, J., and Ashburner, M. 1990. A genetic basis for the inviability of hybrids between sibling species of *Drosophila*. Genetics 124:909–920.

Johnson, N. A., Perez, D. E., Cabot, E. L., Hollocher, H., and Wu, C.-I. 1992. A test of reciprocal X-Y interactions as a cause of hybrid sterility in *Drosophila*. Nature 358: 751–753.

Johnson, N. J., Hollocher, H., Noonberg, E., and Wu, C.-I. 1993. The effects of interspecific Y chromosome replacements on hybrid sterility within the *Drosophila simulans* clade. Genetics 135:443–453.

Koop, B. F., Tagle, D. A., Goodman, M., and Slightom, J. L. 1989. A molecular view of primate phylogeny and important systematic and evolutionary questions. Mol. Biol. Evol. 6:580–612.

Lande, R. 1981. Models of speciation by sexual selection on polygenic traits. Proc. Natl. Acad. Sci. USA 78:3721–3725.

Lee, Y.-H., and Vacquier, V. D. 1992. The divergence of species-specific abalone sperm lysins is promoted by positive Darwinian selection. Biol. Bull. 182:97–104.

Lindsley, D. L., and Tokuyasu, K. T. 1980. Spermatogenesis. In M. Ashbruner and T. R. F. Wright (eds.). The Genetics and Biology of *Drosophila*. New York: Academic Press, pp. 226–294.

Mayr, E. 1963. Animal Species and Evolution. Cambridge, Mass.: The Belknap Press.

Metz, E. C., and Palumbi, S. R. 1996. Positive selection and sequence rearrangements generate extensive polymorphism in the game recognition protein bindin. Mol. Biol. Evol. 13:397–406.

Muller, H. J. 1940. Bearing of the *Drosophila* work on systematics. In J. S. Huxley (ed.). The New Systematics. Oxford: Clarendon Press, pp. 185–268.

Nei, M. 1975. Molecular Population Genetics and Evolution. Amsterdam: North-Holland.

Noor, M. A. 1995. Speciation driven by natural selection in *Drosophila*. Nature 375:674–675.

Orr, H. A. 1995. The population genetics of speciation: the evolution of hybrid incompatibilities. Genetics 139:1805–1813.

Palopoli, M., and Wu, C.-I. 1994. Genetics of hybrid male sterility between *Drosophila* sibling species: a complex web of epistasis is revealed in interspecific studies. Genetics 138:329–341.

Palopoli, M. F., Davis, A. W., and Wu, C.-I. 1996. Discord between the phylogenies inferred from molecular versus functional data: uneven rates of functional evolution or low levels of gene flow. Genetics 144:1321–1328.

Pantazidis, A. C., Galanopoulos, V. K., and Zouros, E. 1993. An autosomal factor from *Drosophila arizonae* restores normal spermatogenesis in *D. mojavensis* males carrying the arizonae Y chromosome. Genetics 134:309–318.

Perez, D. E., and Wu, C.-I. 1995. Further characterization of the hybrid sterility gene, Odysseus (Ods), in the *Drosophila simulans* clade: one gene is not enough. Genetics 140:201–206.

Perez, D. E., Wu, C.-I., Johnson, N. A., and Wu, M.-L. 1993. Genetics of reproductive isolation in the *Drosophila simulans* clade: DNA-marker assisted mapping and characterization of a hybrid male sterility gene, Odysseus (Ods). Genetics 134:261–275.

Raff, R. A., and Kaufman, T. C. 1983. Embryos, Genes and Evolution. New York: Macmillan.

Rice, W. R. 1996. Sexually antagonistic male adaptation triggered by experimental arrest of female evolution. Nature 381:232–234.

Richler, C., Uliel, E. Rosenmann, A., and Wahriman, J. 1989. Chromosomally derived sterile mice have a fertile active XY chromatin conformation but no XY body. Chromosoma 97:465–474.

Sawamura, K., Yamamoto, M.-T., and Watanabe, T. K. 1993. Hybrid lethal systems in the *Drosophila melanogaster* species complex. II. The *zygotic hybrid rescue (zhr)* gene of *Drosophila simulans*. Genetics 133:307–313.

Sibley, C. G., and Ahlquist, J. E. 1984. The phylogeny of the hominoid primates, as indicated by DNA-DNA hybridization. J. Mol. Evol. 20:2–15.

Tan, C. C. 1946. Genetics of sexual isolation between *Drosophila pseudoobscura* and *Drosophila persimilis*. Genetics 31:558–573.

Templeton, A. R. 1981. Mechansims of speciation—a population genetic approach. Annu. Rev. Ecol. Syst. 12:23–48.

Templeton, A. R. 1994. The role of molecular genetics in speciation studies. In B. Shierwater, B. Streit, G. P. Wagner, and R. DeSalle (eds.). Molecular Ecology and Evolution: Approaches and Applications. Basel: Birkhauser, pp. 455–477.

Ting, C.-T., Tsaur, S. C., Wu, M. L., and Wu, C.-I. A rapidly evolving homeobox gene at the site of the Odysseus locus of reproductive isolation. Manuscript submitted for publication.

True, J. R., Weir, B. S., and Laurie, C. C. 1996. A genome-wide survey of hybrid incompatibility factors by the introgression of marked segments of *Drosophila mauritiana* chromosomes into *Drosophila simulans*. Genetics 142:819–837.

Tsaur, S. C., and Wu, C.-I. 1997. Positive selection and the molecular evolution of a gene of male reproduction, Acp26Aa of *Drosophila*. Mol. Biol. Evol. 14:544–549.

Watanabe, T. K. 1979. A gene that rescues the lethal hybrids between *Drosophila melanogaster* and *D. simulans*. Japan. J. Genet. 54:325–331.

Wilson, A. C., Maxon, L. R., and Sarich, V. M. 1974a. Two types of molecular evolution. Evidence from studies of interspecific hybridization. Proc. Natl. Acad. Sci. USA 71:2843–2847.

Wilson, A. C., Sarich, V. M., and Maxon, L. R. 1974b. The importance of gene rearrangement in evolution: evidence from studies of rates of chromosomal, protein and anatomical evolution. Proc. Natl. Acad. Sci. USA 71:3028–3030.

Wright, S. 1977. Evolution and the Genetics of Populations. Vol. 3. Experimental Results and Evolutionary Deductions. Chicago: University of Chicago Press.

Wright, S. 1982. Character change, speciation and the higher taxa. Evolution 36:427–443.

Wu, C.-I. 1992. A note on Haldane's Rule: hybrid inviability vs. hybrid sterility. Evolution 46:1584–1587.

Wu, C.-I., and Beckenbach, A. T. 1983. Evidence for extensive genetic differentiation between the sex-ratio and the standard arrangement of *Drosophila pseudoobscura* and *D. persimilis* and identification of hybrid sterility factors. Genetics 105:71–86.

Wu, C.-I., and Davis, A. W. 1993. Evolution of postmating reproductive isolation: the composite nature of Haldane's Rule and its genetic bases. Am. Nat. 142:187–212.

Wu, C.-I., and Palopoli, M. F. 1994. Genetics of postmating reproductive isolation in animals. Annu. Rev. Genet. 28: 283–308.

Wu, C.-I., Hollocher, H., Begun, D. J., Aquadro, C. F., Xu, Y., and Wu, M.-L. 1995. Sexual isolation in *Drosophila melanogaster*: a possible case of incipient speciation. Proc. Natl. Acad. Sci. USA 92:2519–2523.

Wu, C.-I., Johnson, N. A., and Palopoli, M. F. 1996. Haldane's Rule and its legacy: why are there so many sterile males? Trends Ecol. Evol. 11:281–284.

Zeng, L.-W., and Singh, R. S. 1993a. A combined classical genetic and high resolution two-dimensional electrophoretic approach to the assessment of the number of genes affecting hybrid male sterility in *Drosophila simulans* and *Drosophila sechellia*. Genetics 135:135–147.

Zeng, L.-W., and Singh, R. S. 1993b. The genetic basis of Haldane's Rule and the nature of asymmetric hybrid male sterility between *Drosophila simulans, D. mauritiana* and *D. sechellia*. Genetics 134:251–260.

Zouros, E. 1981. The chromosomal basis of sexual isolation in two sibling species of *Drosophila*: *D. arizonensis* and *D. mohavensis*. Genetics 97:703–718.

Zouros, E. 1989. Advances in the genetics of reproductive isolation in *Drosophila*. Genome 31:211–220.

27

The Genetics of Speciation

Promises and Prospects of
Quantitative Trait Locus Mapping

Sara Via
David J. Hawthorne

The processes that lead to speciation are of unquestionable importance to our understanding of organic evolution. It is thus surprising that we remain so ignorant about the type and magnitude of genetic changes that occur during species formation. In part, our understanding of the requirements of the process has been blurred by controversy over the definition o f the endpoint, that is, of what constitutes a species (Templeton, 1989; see also chapters by Templeton, Shaw, de Queiroz, and Harrison, this volume). Importantly, much of this controversy may be resolved by Harrison's (this volume) suggestion that different species concepts simply represent different points on a continuum of genetic divergence.

One of the most commonly accepted views of species among microevolutionists is the "biological species concept" (Mayr, 1963), in which reproductive isolation among groups is the core of speciation. This species concept is particularly useful for experimental studies of speciation because it permits the speciation process to be studied mechanistically as the evolution of intrinsic barriers to gene flow among groups, that is, as the evolution of reproductive isolation. According to this view, the genetic processes leading to the evolution of character differences that cause reproductive isolation between groups define the genetics of speciation (Templeton, 1981). Thus, "when we understand the origin of reproductive isolating factors, we understand the origin of species" (Coyne, 1992, p. 511). This chapter concerns how we might experimentally probe the genetic basis of characters that lead to reproductive isolation in order to answer crucial questions about the genetic mechanisms of speciation.

What kinds of genetic changes have actually led to reproductive isolation between groups? A wide range of population genetics models have been formulated to explore the kinds of genetic change that can produce speciation (e.g., Maynard Smith, 1966; Lande, 1980; Templeton, 1981; Felsenstein, 1981; Barton and Charlesworth, 1984; Carson and Templeton, 1984; Kondrashov 1983a,b; Kondrashov and Mina, 1986; Orr and Orr, 1996). However, with the notable exception of the detailed genetic work on postzygotic isolation in *Drosophila* (Coyne, 1984, 1985; Orr, 1989; reviewed in Wu and Palopoli, 1994), empirical work on speciation has focused more on whether reproductive isolation evolves under particular situations (reviewed in Rice and Hostert, 1993; Templeton, 1996) than on determination of the genetic changes that accompany the evolution of reproductive isolation.

Unfortunately, available studies on the genetics of species differences (e.g., examples in Gottlieb, 1984; Coyne and Lande, 1985; Orr and Coyne, 1994) cannot necessarily be used to understand the genetics of speciation. First, not all phenotypic differences between species influence reproductive isolation, even though they may be of great interest for understanding adaptation. Second, genetic differences continue to accumulate between species after they are reproductively isolated. Thus, only some of the observable genetic differences between species are relevant to the evolution of reproductive isolation (Templeton, 1981). Identifying the characters that lead to reproductive isolation is the undeniable first step in understanding the genetics of speciation. Then we must ask, What magnitude and type of genetic changes in these traits are required to effect sufficient reproductive isolation that speciation will occur?

Three decades of debate on the evolutionary mechanisms of speciation (e.g., Mayr, 1963; Bush, 1969, 1994; Templeton, 1981; Barton and Charlesworth, 1984; Carson and Templeton, 1984; Rice and Hostert, 1993) have revealed at least two general facts about the evolution of reproductive isolation.

First, it is clear that species can be formed in a number of different ways. No single description of either the ecology or the genetics of reproductive isolation will suffice for all groups. Therefore, experimental approaches that can be used in a wide array of systems are very desirable in order to develop a comparative viewpoint.

Second, the likely mode of speciation in a given group of plants or animals depends crucially on both its ecological situation and its genetic makeup. Detailed ecological studies are necessary to determine the characters that confer reproductive isolation. In addition, ecological factors affect the ways in which natural selection acts on incipient species; they determine the possibility for gene flow, and they influence the balance between genetic drift and selection through effects on population structure. Thus, the ecology of diverging populations affects parameters that are thought to be crucial in the evolution of characters that lead to reproductive isolation. In turn, the magnitude and structure of genetic variation in these (and correlated) characters influence the rate and direction of both the response to selection and of genetic drift (Lande, 1979). Therefore, we are unlikely to understand the evolution of reproductive isolation without continued efforts to learn more about both the ecological and genetic conditions that facilitate speciation. Although we focus here on the genetics of speciation, we stress that no amount of genetic work will reveal the mysteries of speciation if conducted outside of a solid ecological context.

Understanding the genetic architecture of speciation involves estimating the number of loci that affect characters that lead to different forms of reproductive isolation and determining the magnitude and types of the individual and joint effects of these loci on the relevant phenotypic traits. This information will not only permit crucial hypotheses about the mechanisms of speciation to be tested; it will also facilitate the development of more realistic mathematical models of the speciation process. Locating the genes that affect the relevant traits is useful in the analysis of the causes of genetic correlations among traits and is a prelude to fine-scale mapping and to future detailed molecular analyses of reproductive isolation.

Unfortunately, analyzing the genetic architecture of phenotypic characters is complicated by the fact that many of the characters involved in adaptation and reproductive isolation are continuously varying "quantitative" traits. Unlike characters in which phenotypic variants fall into discrete classes that correspond to different genotypes, the observed continuous phenotypic distribution of quantitative traits cannot be used to determine the number of loci involved or their individual effects.

Past studies of the genetic architecture of quantitative traits have relied on indirect statistical approaches for estimating the number of loci involved (e.g., Lande, 1981; Hartl and Clark, 1989; Zeng et al., 1990; Zeng, 1992, 1994) or for evaluating overall gene action (i.e., additivity, dominance, or epistasis; reviewed in Mather and Jinks, 1971; Lynch and Walsh, 1998). These methods are based on the means and variances of F_1, F_2, and backcross generations from a cross between phenotypically distinct parents. Although relatively easy to perform, these analyses are not very powerful, and they are essentially limited to determining the net effects of quantitative trait loci (QTLs) on a given trait (Zeng et al., 1990). It is important to note that they do not permit determination of the location, linkage relationships or individual effects of the QTLs.

Until recently, attempts to locate and enumerate the genes involved in phenotypic traits have been largely restricted to a few model organisms such as *Drosophila* spp. because it has been difficult to identify sufficient visible markers in nonmodel systems. However, even when model systems can be used, studies relying on visible markers only permit genes influencing quantitative traits to be mapped to the level of chromosomes or large chromosomal fragments (e.g., Dobzhansky, 1936; Coyne et al., 1994).

Analyses of the genetics of speciation have been similarly constrained. Our current knowledge of the genetic mechanisms of reproductive isolation in animals is largely limited to the genus *Drosophila* because of its convenience as a model organism for genetics (e.g., Orr, 1989; Wu and Palopoli; 1994; Wu and Hollocher, this volume; Naviera and Maside, this volume). However, such a strong reliance on any single group may not give a balanced view of the speciation process as it occurs in other organisms. For example, because postzygotic isolation through hybrid incompatibility seems strong in *Drosophila*, most of the work on the genetics of speciation in this group has focused on inviability and sterility of hybrids (e.g., Coyne, 1984, 1985; Orr, 1989; reviewed in Wu and Palopoli, 1994). We are only beginning to learn about the genetic architecture of other forms of reproductive isolation (Coyne et al., 1994; Wu and Hollocher, this volume; Ritchie and Phillips, this volume, Prowell, this volume).

This chapter examines the extent to which molecular techniques for the location and enumeration of loci involved in quantitative phenotypic characters may be a useful new tool in the study of speciation. These methods are collectively known as QTL mapping, and they are based on associating phenotypic values in recombinant progeny of crosses between phenotypically differentiated parents (which may or may not be in different species) with the presence of chromosome fragments inherited from one parent or the other (reviewed in Tanksley, 1993; Lander and Shork, 1994). Molecular markers that differ between the parents tag the chromosome fragments inherited from each parent. Given a linkage map of the parents in which the location of these markers is known, QTL mapping can be used to locate and enumerate the loci that influence quantitative characters, at least to the resolution of the linkage map. Even though the advent of techniques that make molecular markers readily available permits much greater resolution than in previous studies employing visible markers, the chromosomal fragments used in QTL mapping are still large enough on average to contain many genes. It is therefore appropriate to think of a QTL localized to a given fragment as a "genetic factor," which may involve more than one closely adjacent locus. The number of QTLs is thus a minumum number of loci contributing to the genetic differences between the parents.

Using QTL mapping, the number of genetic factors causing a difference between species in any phenotypic trait can potentially be evaluated, along with the magnitude of their individual phenotypic effects. These methods are just beginning to be used to address fundamental questions about the genetic basis of adaptation in organisms (reviewed in Mitchell-Olds, 1995). As Orr and Coyne (1994) suggest, detailed studies of gene numbers and effects will be crucial in understanding the genetic mechanisms of adaptation, including the extent to which adaptation is conferred by a few genes with major phenotypic effects rather than by more genes with smaller effects on the phenotype (see also Lande, 1981, 1983; Coyne and Lande, 1985).

To use QTL mapping to study the genetics of speciation, one would cross individuals from different sister species or incipient species (highly divergent and partially reproductively isolated populations). The recombinant (usually F_2) progeny from this cross would be scored for the value of characters affecting reproductive isolation and genotyped at the marker loci to determine which chromosomal fragments have been inherited from each parent and the contribution of each to the characters that produce reproductive isolation. In sum, when applied to characters with a known role in reproductive isolation, QTL mapping may permit us to determine the minimum number of loci involved in a given case of speciation, and to estimate the magnitude of effect and the type of gene action for each genetic factor. One of the great promises of QTL mapping for studies of speciation is that it can be applied to many types of organisms and used to study any form of reproductive isolation. Only by evaluating the genetic architecture of many kinds of reproductive isolation in a variety of systems will we understand the array of different ways in which speciation can occur.

Though having detailed information on the genetic architecture of speciation for even a single case study would be immensely valuable, QTL mapping will be most useful if it is used to validate assumptions and/or test hypotheses of different speciation models. We wish to motivate our discussion by outlining several important questions in the genetics of speciation that might be answered with appropriate use of QTL mapping techniques:

1. Are Fewer Genetic Changes Required for Sympatric than Allopatric Speciation?   Genetic models of sympatric speciation have generally involved very few loci

(Maynard Smith, 1966; Felsenstein, 1981; Diehl and Bush, 1989; Johnson et al., 1996; but for quantitative genetic models, see Kondrashov, 1983a,b; Kondrashov and Mina, 1986 and Kondrashov et al., this volume), and it is widely thought that sympatric speciation occurs rapidly and may involve few genetic changes (e.g., Schluter, this volume). In contrast, the accumulation of genetic differences in allopatry that is thought to lead to reproductive isolation potentially requires many genetic loci (Orr, 1995). Consistent with this, empirical work on hybrid sterility and inviability in *Drosophila* suggests that the loci involved in postzygotic isolation may number in the hundreds (review in Wu and Palopoli, 1994). Moreover, Coyne and Orr's (1989) review of data on *Drosophila* species suggests that sympatric species pairs are often less genetically distant than are allopatric ones. Although increased genetic distance does not necessarily require the involvement of more loci, their tantalizing survey should motivate further work. Comparative QTL mapping studies of a variety of species groups in which other evidence suggests either a sympatric or allopatric origin would be particularly useful in order to obtain replicated estimates of the numbers of loci involved in these two modes of speciation.

2. Does the Genetic Architecture of Characters that Lead to Premating Isolation (Pheromones, Mating Behavior, Calling Songs) Differ from that of Postmating Isolation (Hybrid Sterility or Inviability)?   Perhaps relatively few genes can disrupt mating, but more genetic changes are required to produce enough hybrid incompatibility to lead to reproductive isolation. Again, comparative studies over a wide range of systems would be very useful.

The continued accumulation of genetic differences between reproductively isolated species is problematic for this type of comparison. Current theory suggests that postzygotic isolation is due to the accumulation of alleles that are incompatible in the genetic background of other independently evolving populations or species (Orr, 1995; Orr and Orr, 1996). Because species continue to accumulate genetic differences after speciation, it will be difficult to determine which or how many genetic changes actually contributed to the initial reproductive isolation. This could cause many loci to contribute to contemporary analysis of postzygotic isolation even though few loci were initially involved in speciation (Coyne, 1992). This difficulty may not be so problematic with characters that confer premating isolation, in which the continuous accumulation of genetic changes may be curbed by stabilizing selection on the phenotype.

3. What Role do Genes of Large Effect Play in Different Kinds of Reproductive Isolation?   Recent reports, including those by Gottlieb (1984) and Orr and Coyne (1994) have brought new life to the old question of to what extent adaptation (and speciation) is produced by a small number of genes of large effect. The contrasting view, popular with neodarwinists, is that most phenotypic traits are influenced by a relatively large number of genes with individually small effects. This view forms the basis of the science of quantitative genetics as a tool in the study of phenotypic evolution (e.g., Lande, 1979). The use of QTL mapping to both enumerate and determine the individual effects of loci affecting quantitative traits is likely to provide important data that may help resolve this long-standing controversy. However, as we describe more fully below, these methods are biased toward finding QTLs with large effects, and also toward overestimating the effects of the QTLs that are identified (Beavis, 1994). Thus, very large samples of recombinant progeny (>500; Beavis, 1997) will be required if the issue of the magnitude of gene effects is to be addressed in any rigorous way.

4. What is the Role of Gene Interactions (Epistasis) in Speciation? Some models of speciation are predicated on extensive epistatic interactions among genes in the original base populations (e.g., "genetic transilience," Templeton, 1981, 1989, 1996; "founder-flush speciation," reviewed in Carson and Templeton, 1984). Indeed, the view that a "genetic revolution" might be required during speciation is based in the idea that gene pools are held together by epistatic interactions that resist gradual evolutionary change (reviewed in Carson and Templeton, 1984). Although robust methods for detecting epistasis using QTL mapping are still being developed (Cockerham and Zeng, 1996; Li et al., 1997), this approach may prove very useful for exploring the extent of epistasis in reproductive isolation, particularly when used with crosses between populations rather than between species.

5. Is the Genetic Architecture of Reproductive Isolation (Excluding Polyploidization) Different Between Plants and Animals? Gottlieb (1984) suggested that species differences, and perhaps speciation, are more often due to single genes of large effect in plants than in animals. His view was that the potentially deleterious pleiotropic effects associated with genes of large effect (Lande, 1983) may be smaller in plants than in animals because plants are modular, frequently have indeterminate growth, and have developmental systems that are not as tightly coordinated by hormones as are those of animals. Coyne and Lande (1985) took issue with this view and argued that much more empirical work is needed before the conventional view of multifactorial inheritance of most phenotypic traits should be discarded. Rigorous comparative studies of similar modes of reproductive isolation in animals and plants using the methods of QTL mapping may provide the crucial data required to test Gottlieb's hypothesis.

What Is QTL Mapping and Why Is It Potentially Useful in Studies of Speciation?

To answer the questions about the genetics of speciation that we have outlined above will require that we determine how many loci influence reproductive isolation, the magnitude of their effects on the phenotype, and the extent to which these loci have additive, dominant, or epistatic gene action. This is not a simple process, because many of the phenotypic characters that confer premating reproductive isolation are continuously distributed "quantitative" traits (e.g., Falconer, 1989), and postzygotic inviability and sterility are also likely to have a polygenic basis (Orr, 1995).

Even though the differences between polygenic characters and traits determined by just one or two loci may lie more in the relative magnitude of gene effects rather than in any fundamental difference in gene action, the population genetics of these two types of traits have been studied very differently. On one hand, the population genetics of loci causing discrete phenotypic variation can be studied directly because Mendelian segregation of the alleles leading to different discrete phenotypic classes allows gene frequencies to be calculated from observed trait distributions. In contrast, the number of loci and gene frequencies underlying quantitative genetic variation cannot be reliably detected from observed phenotypic distributions, both because many different genotypes may produce the same phenotype, and because of effects of environmental variation on the phenotype. Thus, hypotheses about genetic architecture in quantitative traits that are based on attempts to observe Mendelian ratios among progeny are inappropriate (Coyne and Lande, 1985; Falconer, 1989). For this reason, the population genetics of quantitative traits has focused on statistically partitioning the total phenotypic variation into its genetic and environmental causes (reviewed in Falconer, 1989), rather than on the determination of gene numbers and allele frequencies. Although statistical methods exist for the estimation of the minimum number of loci influencing a quantitative trait (e.g., Lande, 1981; Hartl and Clark, 1998), they are not very powerful (Zeng et al., 1990; Zeng, 1992), nor do they permit the study of individual gene effects.

The goal of QTL mapping is to move from a statistical description of quantitative genetic variation to a Mendelian description of the action of individual factors. In so doing, we can describe the genetic architecture (number of loci and their locations, the relative magnitude and scope of their effects) of traits of interest. These methods are important to evolutionary geneticists because they potentially permit the study of the genetic architecture of any measurable trait in virtually any organism. Although there are certain problematic limitations (including a need for very large sample sizes, outlined below) these advances signal a clear opportunity to consider the genetic architecture of reproductive isolation in a wide array of systems.

In broad outline, QTL mapping involves the association of genetically caused phenotypic variation with different chromosomal segments that are identified on a genetic linkage map and tagged with neutral markers. A linkage map quantifies the genetic (recombination) distance between pairs of markers on a chromosome. By associating the phenotypic values of a segregating generation of progeny with markers inherited from one or the other of a pair of phenotypically different parents, the chromosomal segments containing putative QTLs are localized on the linkage map. There are five basic steps involved in mapping the QTLs that influence a given trait: (1) developing a set of reliable polymorphic markers for the system of interest, (2) choosing the parents to be crossed, (3) making the initial linkage map and a framework map in which markers are evenly spaced, (4) phenotyping and genotyping a large number of progeny in a segregating generation, and (5) mapping the phenotypic variation onto the framework map. We review only the basics here, and refer the reader to an excellent primer (Lynch and Walsh, 1998) and recent reviews (e.g., Tanksley, 1993; Cheverud and Routman, 1993a,b; Routman and Cheverud, 1994; Lander and Shork, 1994) for more details.

In order to use QTL mapping to study speciation, an additional step must first be performed: determining which traits confer reproductive isolation between the groups in question. These may include characters involved in premating isolation such as habitat choice, pheromones, mating songs, and displays, or aspects of postmating isolation such as gametic incompatibility or hybrid inviability or sterility. If this step is not carefully completed, it is possible that a great deal of effort may be spent studying phenotypic traits that have not been centrally involved in the formation or maintenance of species.

Developing a Set of Reliable Molecular Markers

The recent development of methods that generate a large supply of neutral molecular markers has led to increased interest in constructing linkage maps and locating the genes that influence phenotypic traits. These markers come in several forms and excellent reviews of their attributes exist (Rafalski and Tingey, 1993; Routman and Cheverud, 1994). For our purposes, it is relevant to summarize them by noting that they fall into two broad classes.

The first class includes microsatellites and RFLPs (restriction fragment length polymorphisms). These markers are expensive to develop, but they are relatively powerful per marker because they are codominant and often multiallelic, making all three genotypes of an F_2 distinguishable and the contribution from each parental chromosome readily determined. The second class consists of dominant markers such as RAPDs (random amplified polymorphic DNAs) and AFLPs (amplified fragment length polymorphisms; Vos et al., 1995). These are inexpensive to develop but are less powerful per marker because heterozygotes are typically not distinguishable from the dominant homozygote. However, because markers in this second class are very abundant and the development cost per marker is much less, a much denser map can be constructed for the same cost as a sparse microsatellite or RFLP map. At economically realistic map densities for each marker type, the larger number of dominant markers obtained will more than compensate for the lower power per marker (Jiang and Zeng, 1996). It is worth noting that when codominant markers are desired for map alignment or other purposes, these dominant RAPD or AFLP markers can be transformed into codominant RFLPs or sequence tagged sites amenable to polymerase chain reaction (PCR) analysis (Grattapaglia and Sederoff, 1994; Cho et al., 1996). In this way, the investment in codominant markers can be limited to demonstrably informative markers at evenly spread locations throughout the genome.

Choosing the Parents for the Mapping Cross

Efficient mapping requires a cross between parents that are phenotypically divergent and who differ at a large number of neutral markers. In inbred species or populations, homozygosity within populations will be high and individuals from different populations are likely to be fixed for alternative alleles. Thus, for inbred lineages, the choice of individual parents for the cross is not a problem once the phenotypically divergent lines, populations, or species have been chosen.

For outcrossed species, however, in which genetic variability among individuals within populations is the rule, the choice of parents can be more difficult. For mapping it is not necessary to identify markers that have fixed differences between the parental groups (populations or species); markers must only differ between the particular parents used in the mapping cross. How, then, to select parents in order to construct a map that would be most useful for QTL mapping, given variation within populations in both the quantitative trait(s) and the marker genotypes? First, one would want to select parents that are as phenotypically different between groups as possible. Anderson et al. (1993) describe a method in which possible parents of equal phenotypic difference are first identified, then genotyped at marker loci. Those that are separated by the largest "genetic distance" at marker loci will contribute the most useful set of polymorphisms for mapping analyses. Thus, parents are chosen through joint criteria of phenotypic differences and marker variation that is established during the development of the markers.

In addition, parents from genetically variable populations are unlikely to be fixed for alternative alleles at all QTLs, and so the F_1 from these crosses will not uniformly be multiple heterozygotes (unlike F_1 from crosses between highly inbred lines). Thus, a strategy needs to be devised for forming the F_2 in this type of mapping study. If inbreeding depression is not severe, F_1 may be selfed (e.g., Bradshaw et al., 1995). Other methods include crossing two sets of phenotypically divergent parents, then crossing one F_1 from each set to form an F_2 (Schemske, personal communication), or simply crossing two different F_1 progeny genotypes from the same parental cross.

Making the Initial Linkage Map and Choosing Equally Spaced Markers for the Framework Map

The genetic distance between polymorphic markers is estimated by observing the recombination frequency between markers during meiosis. Typical mapping protocols involve the generation and genotyping of F_2 or backcross families (Tanksley, 1993). When all parents and offspring have been genotyped at all marker loci, computer software (e.g., MAPMAKER; Lander et al., 1987) is used to calculate the recombination frequencies between markers and to align the markers onto linkage groups.

How many markers are required? The reliability of a particular linkage map has two components: the accuracy of marker order along the chromosomes and the thoroughness of genome coverage. Lin and Ritland (1996) evaluated both aspects of reliability in a linkage map constructed for an interspecific cross of *Mimulus* using 101 RAPD markers. They found that more markers were required for reliable determination of marker order than for thorough genome coverage. In addition, although the basic linkage map that one makes first is likely to have many markers in clumps by chance, QTL mapping is much more powerful if evenly spaced markers are chosen. It is thus likely that many more markers may be required to construct the initial linkage map than will ultimately be used for the phenotypic mapping step. Some of the initial markers will be discarded, leaving a map with fewer but more evenly spaced markers (the "framework" map). The density of markers on the framework map influences the precision of the estimate of both QTL number and location. As we discuss below, markers at 5–20 cM intervals will probably be adequate for most QTL mapping studies, although it must be remembered that the coarser the map, the more genes may be subsumed within an interval identified as a single QTL. In sum, the total number of markers required depends on genome size, recombination rates, the extent to which markers are clumped, and the density of markers desired in the final framework map.

Phenotype the Progeny and Determine Their Genotypes at the Marker Loci

The same experimental cross that is used to make the linkage map can also be used to map QTLs. In such a case, the phenotypic characters of interest must be measured on each individual parent or offspring before determining its genotype at the marker loci (unless nondestructive sampling or cloning of the progeny is possible).

The resolution and power of a QTL map is determined by a combination of factors including: the heritability of the traits (the proportion of phenotypic variance due to genetic vs. environmental causes), the magnitude of gene effects, the experiment size (number of F_2 or backcross progeny), the density and information content of markers, and the magnitude of phenotypic difference between the parental groups (van Ooijen, 1992; Beavis, 1997). Under all conditions, however, large family sizes (>500 progeny genotypes; Beavis, 1997) are likely to be required to detect QTLs with small to moderate effects (i.e., explaining 5% of the phenotypic variance). If this many progeny cannot be obtained from a single cross and the parents cannot be cloned, then various experimental designs that pool data from multiple families are possible although they result in lower power (Lynch and Walsh, in press).

The resolution necessary for a QTL mapping effort depends on the goals of the project. For initial studies of genetic architecture of speciation, the goal is not to clone the QTLs (which requires an extremely fine-scale map; see Kruglyak and Lander, 1995) but rather to enumerate and localize them and learn something of their individual and joint effects. For this, the precision provided by 300–500 progeny genotypes will be adequate in most cases (Tanksley, 1993; Beavis, 1997). However, if it is logistically possible, more progeny are desirable (Beavis, 1997), and for rigorous testing of hypotheses concerning the size of gene effects on the phenotype, larger sample sizes are likely to be required.

Environmentally induced phenotypic variation and measurement error both act to reduce power. For this reason, recombinant progeny are often inbred to produce "recombinant inbred lines," allowing replicate phenotypic measurement of each progeny genotype. This replication can increase experimental power by reducing the error variance (Lander and Botstein, 1988, see Mitchell-Olds, 1995, for an example). An even better way to replicate, however, is to clone the progeny when possible (i.e., in plants or cyclical parthenogens). Cloning the progeny produced from a sexual cross permits replication of progeny genotypes without the loss of lines, inbreeding depression, and residual genetic variation within lines that can result when recombinant inbreds are made in a normally outcrossing species. The best balance between replication versus adding more progeny genotypes is currently under debate, and depends crucially on the extent to which the character of interest is susceptible to environmental variance. In most cases, it is unlikely that the benefits of replication increase fast enough after the second replicate to justify additional replication for in lieu of testing more progeny genotypes (Churchill, personal communication). However, if additional progeny are not available, replicate testing of the progeny available is preferred over unreplicated phenotyping. Access to replicated progeny may also permit the measurement of more characters on each genotype and/or allow each progeny genotype to be phenotyped in several environments. This feature facilitates the study of the genetic mechanisms of genetic correlations including those responsible for QTL × environment interactions (differences in phenotypic effects of QTLs in different environments).

We caution that the phenotyping step of a QTL mapping study should not be undertaken lightly. Some morphological traits might be easily measured, possibly even on dead specimens. However, many of the characters that confer reproductive isolation (i.e., habitat or mate choice, mating behavior, hybrid inviability, or sperm incompatability) will be much more difficult to measure. These traits must be measured on live organisms at a particular life stage, which can be very difficult in the large number of individuals required. Moreover, behavioral characters are notoriously prone to environmental variation, increasing the need for replication. The desire to measure multiple traits or traits expressed in several environments produces additional experimental challenges.

Mapping the Loci that Influence Quantitative Traits

QTL mapping involves measuring the degree to which variation in the phenotype cosegregates with chromosomal segments from divergent parents as defined by the markers on a linkage map. Thus, each progeny must be genotyped at each of the marker loci as well as phenotyped, so that the parent from which a particular chromosomal interval is inherited can be determined.

Tanksley (1993) provides a lucid description of the relationship between the ability to detect a QTL, the population size, and the effect of the locus. He describes a QTL analysis of characters in tomato in which 432 progeny were tested. He shows that many of the QTLs discovered contribute approximately 5% of the phenotypic variance (the limit of detection) and a few contribute more than 12%. He cautions that QTLs with smaller effects are unlikely to be detected in experiments of the size typically reported (100–300 progeny genotypes). In a simulation study, van Ooijen (1992) found that to have a reasonable chance (e.g., 30%) of detecting a QTL that accounts for 5% of the observed variation, at least 200 progeny must be scored. Reinforcing this observation, also with results from simulations, Beavis (1997) cautions against attempting to identify small effect QTLs with a small number of progeny (< 500).

Although more progeny always help, more markers will not always increase the power to detect a QTL. Darvasi et al. (1993) and Jiang and Zeng (1996) demonstrate that a framework map with evenly spaced markers an average of 5–10 cM apart provides as much power to detect a QTL as a much denser map, and that a 20 cM or even 50 cM map had only slightly less power than the 10 cM map. In sum, the probability of detecting a QTL affecting a given trait (or suite of traits) is quite low unless it is responsible for at least 5–10% of the observed phenotypic variation and the heritability of the trait is relatively high (Beavis, 1994, 1997). We can tell however, how much variation is left unaccounted for (and therefore attributable to undetected QTLs) by computing the percentage of the genetic variation between the parents that is explained by the identified factors. Disturbingly, it is all too possible both that QTLs of small effect can be missed and that the effect size of the detected QTLs can be overestimated (Beavis, 1994). This leads to a double bias toward finding few QTLs with large effect.

Because of this bias, experiments must be carefully designed so that hypotheses of interest can be effectively addressed. For example, it will be easier to test for the influence of a few major genes than to precisely quantify their effects or to identify a larger number of genes of small individual effect. Failing to find QTLs of small effect may not be very meaningful in experiments of small size. For this reason, it may be risky to conclude that only major genes are involved in a particular case of speciation or adaptation just because genes of small effect are not found, unless the experiment is of high enough power that it is clear that minor genes would be found if they were present. As a rule, comparing the number of genes involved in particular forms of reproductive isolation will probably be more productive than trying to estimate absolute numbers, as long as experiments on each type are of comparable power.

As with QTL detection, the accuracy of QTL location (measured by the width of the confidence interval of the QTL location) gets better with increased effect of the QTL and a larger sample size of progeny. Location of individual QTLs on a chromosome requires a denser set of markers than does the detection of that same QTL (Darvasi et al., 1993; Visscher et al., 1996). The resolvability of QTLs on a linkage map is important because it determines the ability to distinguish between hypotheses of pleiotropy and tight linkage (required to understand the genetic mechanisms of genetic correlations). In most cases, if QTLs for two traits map to the same chromosome fragment, it will be impossible to distinguish whether there is a single locus that affects both traits or two loci in the same interval without a considerable effort. In contrast, the hypothesis that two traits are influenced by pleiotropic effects of the same gene can be immediately eliminated if QTLs for two correlated traits map to different chromosome fragments.

Statistical Analysis of QTL Maps

Associating phenotypic variation with allelic variation at marker loci is fundamentally a statistical problem, and the analysis of QTL maps is an extremely active research area. The main statistical issue in QTL mapping is how to detect a significant association between the phenotype and the markers with the greatest power and least potential for error or bias. Because specific details of the best analysis methods are constantly changing, we present only the general outline of the analyses, leaving details to more technical sources (e.g., Kruglyak and Lander, 1995; Doerge and Rebai, 1996; Lynch and Walsh, 1998).

The simplest and earliest method of analysis computes association between phenotypes and single markers (Soller and Brody, 1976). By dividing the progeny population into those with and without the given marker, a simple F test reveals whether the group with the marker differs in average phenotype from the group without the marker. However, because a QTL with weak effect that is closely linked to the marker will not be distinguishable from one with a stronger effect that is more loosely linked, this method does not provide a means of estimating location or magnitude of a detected QTL's effect, and so is most useful when hypotheses are limited to QTL detection (Lander and Botstein, 1989).

Interval mapping involves associating the phenotype with a chromosomal region between two markers (Lander and Botstein, 1989). Because the genetic distance between the QTL and each of the two markers can be estimated in the population of recombinant progeny, the location and effect of the QTL in the interval can be separated with this method. Interval mapping can be done using either maximum likelihood analysis (e.g., with software such as MAPMAKER QTL [Lincoln et al., 1992], or QTL CARTOGRAPHER [Basten et al., 1996]), or using least squares methods (Haley and Knott, 1992; Haley et al., 1994).

Composite interval mapping allows the use of markers associated with other QTLs to be used as covariates in the analysis of additional QTLs (Zeng, 1994). This reduces the error variance and increases the power to detect multiple QTLs. The software QTL CARTOGRAPHER (Basten et al., 1996) implements this method.

Regardless of the methods used to detect and localize QTLs, several complex statistical issues haunt the analyses (Churchill and Doerge, 1994). First, by design, QTL analyses require many statistical tests, increasing the possibility of making a Type I error (assigning significance to an interval-phenotype association that is detected only by chance). Second, the distribution of phenotypes in a typical mapping population is a mixture of distributions, which further complicates the definition of appropriate test statistics that rely on the normal distribution. Finally, attributes of specific crosses, such as segregation distortion, can also make testing for significant associa-

tions tenuous. Churchill and Doerge (1994) have developed a very robust method of permutation testing to empirically determine the appropriate test statistics for a specific mapping experiment. Once putative QTLs are identified (using any method), the traits of individuals are shuffled randomly. If the trait-marker association is destroyed by the shuffling, the original association must have been real. If the association is not destroyed, it reflects nothing more than a random association of traits and markers and does not indicate a real QTL (see Doerge and Rebai, 1996, for additional methods).

Another clear challenge in QTL mapping is the detection of epistatic interactions (the nonadditive effect of alleles at different loci). The ability to detect epistasis among QTLs is a key element in the evaluation of genetic mechanisms of speciation (Templeton, 1981, 1996; Carson and Templeton, 1984; Wu and Palopoli, 1994). Initial approaches to detecting epistasis have been based on a factorial ANOVA in which two or more QTLs are the sources of variation and the phenotypes of the offspring are the dependent variables (Paterson et al., 1988; Tanksley, 1993). A significant QTL × QTL interaction indicates that the joint effect of two loci on a given character is not simply the sum of the individual effects. Although conceptually straightforward, this method of analysis has low power (Tanksley, 1993), and few epistatic interactions among QTLs affecting various agronomic traits have been found using this method even in rather large experiments (Paterson et al., 1988; Stuber et al., 1992). Recently, Cheverud and Routman (1993a), Cockerham and Zeng (1996), and Li et al. (1997) have described new methods for analysis of QTL data that may significantly improve our ability to detect and measure the epistatic action of QTLs. When Cockerham and Zeng (1996) reanalyzed the data of Stuber et al. (1992), they found several significant epistatic interactions among the QTLs. Similarly, Li et al. (1997) found many more significant epistatic effects in their own data on tomato than they did using the earlier method of analysis. In sum, the study of epistasis is a rapidly advancing statistical aspect of QTL mapping, and we expect even better methods to be developed for its estimation. This will increase the utility of QTL mapping as a tool for the study of speciation.

Case Studies

The vast majority of QTL mapping analyses performed to date have concerned species of agricultural importance (reviewed in Tanksley, 1993; Mitchell-Olds, 1995). We are aware of only a handful of studies in which molecular markers are being used to study characters associated with speciation in natural populations (though more may be in the works). Results from only two of these studies have been published to date.

First, Bradshaw et al. (1995) describe a QTL mapping analysis of floral traits that appear to be involved in reproductive isolation between *Mimulus lewisii and M. cardinalis* by affecting pollinator behavior (*M. lewisii* is pollinated by bumblebees, *M. cardinalis* by hummingbirds). Although these two species are completely interfertile in the lab, hybrids are not seen in the field. In the published work, 93 F_2 progeny were scored for eight floral characters (aspects of color, size, shape, and nectar reward). Phenotypes were then mapped onto a framework map of 153 evenly spaced RAPD markers. Bradshaw and colleagues found that for each of the eight characters, there were one to three QTLs that explained in excess of 20% of the phenotypic variance, leading them to suggest that major genes may play a large role when speciation involves the evolution of pollination systems. Importantly, this study is now being extended to include a much larger F_2 sample size and more markers (Schemske, personal communication), which will permit the detection of QTLs with smaller effects and will also provide a more accurate estimation of the effects of the QTLs already detected. In addition, field studies of progeny bearing different combinations of the floral characters are presently being conducted in order to determine their attractiveness to pollinators. These field observations will move this study well into the next phase of experimental evaluation of the genetic changes involved in speciation (see below).

Second, Reiseberg et al. (1996) have used molecular markers to map chromosomal segments retained and rearranged during hybrid speciation in sunflowers (*Helianthus* spp.). Although their published analysis has focused more on tracking parentage than on mapping QTLs, it represents a very important application of the use of molecular markers in speciation studies. Reiseberg and colleagues found that in a set of replicate crosses between the two parental species, the same set of genetic markers appeared in advanced hybrid progeny in each case. Furthermore, these markers denote the same gene combinations as found in the extant hybrid species, *H. anomalus*. This repeated evolution of the same gene combinations suggests an important role for selection favoring these combinations among all possible recombinants to produce the hybrid species. Although this is a recognized mechanism of hybrid speciation (Templeton, 1981), Reiseberg and colleagues' elegant study provides the first concrete evidence that selection among recombinants actually occurs during hybrid speciation.

We are aware of several additional mapping studies of QTLs affecting characters that lead to reproductive isolation. Howard and colleagues (reviewed in this volume) have determined that interspecific sperm incompatibility is a major cause of reproductive isolation between the crickets *Allonemobius socius* and *A. fasciatus*, which can be found sympatrically over part of their range. They have produced a linkage map using RAPD markers, and plan to map the QTLs for conspecific sperm precedence.

Pea aphids (*Acyrthosiphon pisum*) that are found sympatrically on alfalfa and red clover are highly genetically differentiated and almost completely reproductively isolated (Via, 1991, 1994), despite their continued taxonomic status as races of a single species (R. Blackman, personal communication). Most of the reproductive isolation is attributable to habitat choice that is highly correlated with their relative performance on the two hosts, although a demographic disadvantage of hybrids also contributes to reproductive isolation (Via, unpublished). We (Via and Hawthorne) have begun a QTL mapping study on these divergent populations of pea aphids. Our goals are to determine the number of genes influencing both habitat choice and demographic performance in the two habitats, and to probe the mechanism of the genetic correlation between habitat choice and performance.

We are certain that other QTL mapping projects involving characters that lead to reproductive isolation are currently planned or in early stages. We eagerly await their results.

What Is Required for QTL Mapping to Be a Successful Tool in the Genetics of Speciation?

We cannot overemphasize the importance of extensive ecological knowledge in a program to evaluate the genetic mechanisms of speciation. The first requirement for such a program is to determine which phenotypic characters or character combinations produce reproductive isolation. These are the ones in which genetic changes are most relevant to speciation.

Unfortunately, one cannot assume that because a given trait confers some reproductive isolation between groups, it is the only trait involved in a given case of speciation. There may be multiple sources of reproductive isolation, and so it is crucial to determine the proportion of reproductive isolation conferred by different kinds of barriers to gene flow. This is unlikely to be easy. For example, Howard and colleagues first found calling differences between the sympatric species of *Allonemobius* crickets (Benedix and Howard, 1991). Only later did they learn that reproductive isolation may be caused almost entirely by postmating conspecific sperm precedence (Howard et al., this volume). When several characters each act to produce partial reproductive isolation, the mapping situation is also complicated. However, if we want to understand speciation, we must acknowledge that reproductive isolation may not have only one cause.

Some systems lend themselves to analysis by QTL mapping better than others. The study organism must be amenable to controlled matings and rearing. Large family sizes are useful in order to generate the required number of progeny. The ability to clonally propagate the parents or the sexually produced progeny is also very helpful. Plants are particularly useful because the parents can sometimes be propagated clonally to increase the number of seeds generated by a mating and clonal propagation of the resulting progeny allows for replicated progeny phenotyping. The ability to store seeds or cryopreserve replicated progeny also facilitates progeny trials by permitting tests in different time blocks. Cyclical parthenogens like the pea aphid or *Daphnia* are particularly favorable animals for mapping, because clonally produced copies of the parents can be used to obtain as many F_2 or backcross progeny as can be managed. Moreover, in cyclical parthenogens, the recombinant progeny can be propagated clonally and tested in replicate without having to resort to the production and use of recombinant inbred lines.

Some phenotypic traits are easier to measure than others. This facilitates progeny trials when many individuals must be tested. For example, a character like flower morphology can be measured relatively easily, perhaps even on pressed specimens. Also, replicate flowers are produced on a single progeny, reducing measurement error. Other characters, like sperm precedence, demographic performance in different environments, habitat choice, or courtship song are much more difficult to measure on the large number of progeny that must be tested, and they must be measured on live specimens, often at a particular life stage.

In sum, a successful QTL mapping study depends on knowing the characters responsible for reproductive isolation, being able to find suitable markers for an accurate linkage map, and being able to accurately phenotype a large number of progeny from a cross between phenotypically extreme parents.

Limitations of QTL Mapping

QTL mapping will not solve all the remaining mysteries in the genetics of speciation. As with any experimental technique, there are limitations. Some of these can perhaps be overcome with improved experimental designs, but at present, we regard QTL mapping as most suitable for testing hypotheses or addressing questions that do not require extremely high precision. Comparative studies as outlined in the introduction may be particularly useful. Even with the following limitations, the development of QTL mapping technology is an important bridge to the next phase of study. Some limitations of QTL mapping include the following.

Bias toward detection of genes with large and additive effects. Genes with major additive effects are much more likely to be found than those that contribute only a small fraction of the phenotypic variance and/or those that have epistatic effects. Thus, QTL mapping

studies must acknowledge the extent to which they could identify minor genes or detect epistasis if present. Because of these biases, QTL mapping will be most useful when hypotheses are framed that do not depend on accurately detecting all genes of small effect.

Overestimation of gene effects. It has been suggested that not only are genes of large effect more likely to be detected, but the effect of the QTLs that are found may be overestimated (Beavis, 1994, 1997). Clearly, in order to test very precise hypotheses about the magnitude of gene effects, very large sample sizes and statistical improvement of current methods will be necessary. One test for bias might be to increase the sample size in the analysis and see if the estimated effects of the detected QTLs is altered.

Numbers and locations of QTLs are estimated with error. Like any experimental technique, QTL mapping leads to estimates of gene numbers, locations, and magnitudes of effects that have an associated error. This error can be reduced if the investigator is able to invest more resources to increase the sample size of recombinant progeny and, secondarily, the number of markers. In some cases, however, even a rough mapping effort will be useful.

Linkage and pleiotropy can be hard to distinguish. If two characters map to different intervals, then any observed genetic correlation between them must be due only to linkage disequilibrium, not to pleiotropy. For example, in the pea aphid system, if QTLs for habitat choice and performance on the same host map to different chromosomal fragments, then the possibility of strong genetic correlation (and consequent assortative mating) due to pleiotropy will have been eliminated, suggesting that selection is strong enough to maintain the genes for these different characters in linkage disequilibrium. However, if QTLs for two genetically correlated traits such as these map to the same interval, very fine-scale mapping is required to reveal whether the correlation is due to pleiotropic effects of a single gene or to linkage disequilibrium between two relatively tightly linked genes that both fall in the same interval.

Problems detecting epistasis. It is currently difficult to detect epistasis in QTL mapping studies, though this is an active area of statistical research (e.g., Cockerham and Zeng, 1996; Li et al., 1997).

The Next Step

The discovery of an association between the phenotype and a chromosomal interval is just the first step toward a detailed understanding of the genetic basis of quantitative traits. However, the molecular and statistical methods developed for QTL mapping are crucial preliminaries to the next phase of study. We suggest that the following areas will be important as we try to understand the genetics of speciation.

Marker-Assisted Introgression of QTLs

Finding a phenotype-marker association does not prove that there is a QTL in the interval that actually causes the trait in question. For this, it is necessary to use marker-assisted backcrosses to introgress chromosomal segments with putative QTLs into the alternate background. In the backcrossed progeny, only the markers need to be assessed, faciliating the selection of backcross progeny with the desired chromosomal segments. Introgression studies will be most useful if they can be done with both single and multiple segments, as Wu and Palopoli (1994) have done with *Drosophila*. Such introgressions isolate the effects of particular QTLs on the phenotype and permit individuals to be generated with phenotypic combinations that may not necessarily be seen in natural populations. This may be very useful for analyses of selection on putative early stages of a complex phenotype.

Field Studies of QTLs

It is crucial to take recombinant and introgressed lines to the field both to evaluate the extent to which they are reproductively isolated from ancestors, and to measure the selection on components of a complex phenotype. This is being done by Schemske and colleagues with "nearly isogenic lines" of *Mimulus lewisii* (the bumble-bee-pollinated species) that bear particular QTLs from *M. cardinalis* for floral attributes involved in humming-bird pollination (reverse introgressions are also being tested). Such studies can be used to test the fitness of various phenotypic attributes caused by known combinations of QTLs. Schemske and colleagues will observe pollinator visitation to these manipulated flowers in order to determine the magnitude of reproductive isolation attributable to particular genetic changes (or combinations of changes) in floral characters.

Fine-Scale Mapping

Identification of a chromosomal region with an important QTL may motivate a detailed fine-scale mapping effort focused on that region. Such fine mapping, which requires the identification of many additional markers within the original single interval, may eventually permit isolation and cloning of genes that influence quantitative traits (Churchill et al., 1993; Krugylak and Lander, 1995). It will also permit the hypothesis that only a single gene lies in the interval to be tested, and thus may be warranted when it is important to distinguish

linkage disequilibrium from pleiotropy as the cause of an observed genetic correlation between traits.

Conclusions

At a minimum, the prospect that genes influencing quantitive traits can be located and enumerated in a wide variety of organismal groups will invigorate the study of the genetics of speciation. Investigators will be able to evaluate whether certain types of speciation are associated with fewer genetic changes than other types, and will also be able to evaluate the role of major genes and epistasis in speciation, and to compare the genetic architecture of reproductive isolation among groups. We hope that this effort will be accompanied by a refinement of the crucial questions in the genetics of speciation that will lead to a better understanding of the dynamics of this important set of processes. In addition, the discovery of chromosomal segments that harbor genes affecting reproductive isolation will permit a type of detailed genetic analysis of the speciation process that could never have been attempted in the past. These studies will no doubt cause new and important questions to be asked about how species are formed.

Acknowledgments   We are grateful to Gary Churchill for help in developing our ideas about QTL mapping; to Chung-I Wu and Doug Schemske for valuable conversations during the Asilomar meeting; and to Marina Caillaud, Tim Carr, Ruth Hufbauer, Stasia Skillman, Anne Stork, Kelley Tilmon, Saskya van Nouhuys, and Andy Zink for discussion of the early stages of the manuscript. Doug Schemske, Toby Bradshaw, Scott Hodges, and Dan Howard provided very illuminating comments on a later draft. Our work on the genetics of population divergence and speciation is supported by the NSF.

References

Anderson, J. A., Churchill, G. A., Autrique, J. E., Tanksley, S. D., and Sorrells, M. E. 1993. Optimizing parental selection for genetic linkage maps. Genome 36:181–186.

Barton, N., and Charlesworth, B. 1984. Genetic revolutions, founder effects and speciation. Annu. Rev. Ecol. Syst. 15:133–164.

Basten, C. J., Weir, B. S., and Zeng, Z.-B. 1996. QTL CARTOGRAPHER (computer software). Program in Statistical Genetics, North Carolina State University.

Beavis, W. D. 1994. The power and deceit of QTL experiments: lessons from comparative QTL studies. 49th Annu. Corn and Sorghum Research Conf. 49:250–266.

Beavis, W. D. 1977. QTL analyses: power, precision and accuracy. *In* Molecular Analysis of Complex Traits, A. H. Paterson (Ed.). CRC Press, Cleveland, OH, pp. 145–162.

Benedix, J. H., Jr., and Howard, D. 1991. Calling song displacement in a zone of overlap and hybridization. Evolution 45:1751–1759.

Bradshaw, H. D., Wilbert, S. M., Otto, K. G., and Schemske, D. W. 1995. Genetic mapping of floral traits assoicated with reproductive isolation in monkey flowers (*Mimulus*) Nature 376:762–765.

Bush, G. L. 1969. Sympatric host race formation and speciation in frugivorous flies of the genus *Rhagoletis* (Diptera: Tephritidae). Evolution 23:237–251.

Bush, G. L. 1994. Sympatric speciation in animals: new wine in old bottles. Trends Ecol. Evol. 9:285–288.

Carson, H. L., and Templeton, A. R. 1984. Genetic revolutions in relation to speciation phenomena: the founding of new populations. Annu. Rev. Ecol. Syst. 15:97–131.

Cheverud, J. M., and Routman, E. J. 1993a. Epistasis and its contribution to genetic variance components. Genetics 139:1455–1461.

Cheverud, J. M., and Routman, E. J. 1993b. Quantitative trait loci: individual gene effects on quantitative characters. J. Evol. Biol. 6:463–480.

Cho, Y. G., Blair, M., Panaud, O., and McCouch, S. R. 1996. Cloning and mapping of variety-specific rice genomic DNA sequences: amplified fragment length polymorphisms (AFLP) from silver stained gels. Genome 39: 373–378.

Churchill, G. A., and Doerge, R. W. 1994. Empirical threshold values for quantitative trait mapping. Genetics 138: 963–971.

Churchill, G. A., Giovannoni, J. J., and Tanksley, S. D. 1993. Pooled-sampling makes high-resolution mapping practical with DNA markers. Proc. Natl. Acad. Sci. U.S.A. 90: 16–20.

Cockerham, C. C., and Zeng, Z. B. 1996. Design III with marker loci. Genetics 143:1437–1456.

Coyne, J. A. 1984. Genetic basis of male sterility in hybrids between two closely related species of *Drosophila*. Proc. Natl. Acad. Sci. U.S.A. 81:4444–4447.

Coyne, J. A. 1985. Genetic studies of three sibling species of *Drosophila* with relationship to theories of speciation. Genet. Res. 46:169–192.

Coyne, J. A. 1992. Genetics and speciation. Nature 355:511–515.

Coyne, J. A., and Lande, R. 1985. The genetic basis of species differences in plants. Am. Nat. 126:141–145.

Coyne, J. A., Crittenden, A. P., and Mah, K. 1994. Genetics of a pheromonal difference contributing to reproductive isolation in *Drosophila*. Science 265:1461–1464.

Darvasi, A., Weinreb, A., Minke, V., Weller, J. I., and Soller, M. 1993. Detecting marker-QTL linkage and estimating QTL gene effect and map location using a saturated genetic map. Genetics 134:943–951.

Diehl, S. R., and Bush, G. L. 1989. The role of habitat preference in adaptation and speciation. *In* Speciation and Its Consequences, D. Otte and J. A. Endler (Eds.) Sinauer Press, New York, pp. 345–365.

Dobzhansky, T. 1936. Studies on hybrid sterility. II. Localization of sterility factors in *Drosophila pseudoobscura* hybrids. Genetics 21:113–135.

Doerge, R. W., and Churchill, G. A. 1996. Permutation tests for multiple loci affecting a quantitative character. Genetics 142:285–294.

Doerge, R. W., and Rebai, A. 1996. Significance thresholds for QTL interval mapping tests. Heredity 76:459–464.

Felsenstein, J. 1981. Skepticism towards Santa Rosalia, or why are there so few kinds of animals? Evolution 35:398–409.

Gottlieb, L. D. 1984. Genetics and morphological evolution in plants. Am. Nat. 123:681–709.

Grattapaglia, D., and Sederoff, R. 1994. Genetic linkage maps of *Eucalyptus grandis* and *Eucalyptus urophylla* using a pseudo-testcross: mapping strategy and RAPD markers. Genetics 137:1121–1137.

Haley, C. S., and Knott, S. A. 1992. A simple regression method for mapping quantitative trait loci in line crosses using flanking markers. Heredity 69:315–324.

Haley, C. S., Knott, S. A., and Elsen, R. C. 1994. Mapping quantitative trait loci in crosses between outbred lines using least squares. Genetics 136:1195–1207.

Hartl, D. L., and Clark, A. G. 1997. Principles of Population Genetics. 3rd ed. Sinauer Press, Sunderland, Mass.

Jiang, C., and Zeng, Z.-B. 1996. Multiple trait analysis of genetic mapping for quantitative trait loci. Genetics 140: 1111–1127.

Johnson, P. A., Hoppensteadt, F. C., Smith, J. J., and Bush, G. L. 1996. Conditions for sympatric speciation: a diploid model incorporating habitat fidelity and nonhabitat assortative mating. Evol. Ecol. 10:187–205.

Kondrashov, A. S. 1983a. Multilocus model of sympatric speciation I. One character. Theor. Pop. Biol. 24:121–135.

Kondrashov, A. S. 1983b. Multilocus model of sympatric speciation II. Two characters. Theor. Pop. Biol. 24:136–144.

Kondrashov, A. S. 1986. Multilocus model of sympatric speciation III. Computer simulations. Theor. Pop. Biol. 29: 1–15.

Kondrashov, A. S., and Mina, M. V. 1986. Sympatric speciation: when is it possible? Biol. J. Linn. Soc. Lond. 27:201–223.

Kruglyak, L., and Lander, E. S. 1995. High-resolution genetic mapping of complex traits. Am. J. Hum. Genet. 56: 1212–1223.

Lande, R. 1979. Quantitative genetic analysis of multivariate evolution, applied to brain:body size allometry. Evolution 33:402–416.

Lande, R. 1980. Genetic variation in phenotypic evolution during allopatric speciation. Am. Nat. 116:463–479.

Lande, R. 1981. The minimum number of genes contributing to quantitative genetic variation between and within populations. Genetics 99:541–553.

Lande, R. 1983. The response to selection on major and minor mutations affecting a metrical trait. Heredity 50:47–65.

Lander, E. S., and Botstein, D. 1989. Mapping Mendelian factors underlying quantitative traits using RFLP linkage maps. Genetics 121:185–199.

Lander, E. S., and Shork, N. J. 1994. Genetic dissection of complex traits. Science 265:2037–2048.

Lander, E. S., Green, P., Abrahamson, J., Barlow, A., Daley, M., Lincoln, S., and Newburg, L. 1987. MAPMAKER, an interactive computer package for constructing primary genetic linkage maps of experimental and natural populations. Genomics 1:174–181.

Li, Z., Pinson, S. R. M., Park, W. D., Paterson, A. H., and Stansel, J. W. 1997. Epistasis for three grain yield components in rice (*Oryza sativa L.*). Genetics 145:453–465.

Lin, J.-Z., and Ritland, K. 1996. Construction of a genetic linkage map in the wild plant *Mimulus* using RAPD and isozyme markers. Genome 39:63–70.

Lincoln, S., Daly, M., and Lander, E. 1992. Mapping genes controlling quantitative traits with MAPMAKER/QTL 1.1. Whitehead Institute Technical Report, 2nd ed.

Lynch, M., and Walsh, J. B. 1998. Genetics and Analysis of Quantitative Traits. Sinauer Press, Sunderland, Mass.

Mather, K., and Jinks, J. L. 1971. Biometrical Genetics. (3rd ed.). Chapman and Hall, London.

Maynard Smith, J. 1966. Sympatric speciation. Am. Nat. 100:637–650.

Mayr, E. 1963. Animal Species and Evolution. Belknap Press, Cambridge, Mass.

Mitchell-Olds, T. 1995. The molecular basis of quantitative genetic variation in natural populations. Trends Ecol. Evol 10:324–328.

Orr, H. A. 1989. Genetics of sterility in hybrids between two subspecies of *Drosophila*. Evolution 43:180–189.

Orr, H. A. 1995. The population genetics of speciation: the evolution of hybrid incompatibilities. Genetics 139:1805–1813.

Orr, H. A., and Coyne, J. A. 1994. The genetics of adaptation: a reassessment. Am. Nat. 140:725–742.

Orr, H. A., and Orr, L. H. 1996. Waiting for speciation: the effect of population subdivision on the time to speciation. Evolution 50:1742–1749.

Paterson, A. H., Lander, E. S., Hewitt, J. D., Peterson, S., Lincoln, S. E., and Tanksley, S. D. 1988. Resolution of quantitative traits into Mendelian factors by using a complete linkage map of restriction fragment length polymorphisms. Nature 335:721–726.

Rafalski, J. A., and Tingey, S. V. 1993. Genetic diagnostics in plant breeding: RAPDs, microsatellites and machines. Trends Genet. 9:275–279.

Rice, W. R., and Hostert, E. E. 1993. Laboratory experiments on speciation: what have we learned in 40 years? Evolution 47:1637–1653.

Rieseberg, L. H., Sinervo, B., Linder, C. R., Ungerer, M. C., and Arias, D. M. 1996. Role of gene interactions in hybrid speciation: evidence from ancient and experimental hybrids. Science 272:741–745.

Routman, E., and Cheverud, J. M. 1994. Individual genes underlying quantitative traits: molecular and analyti-cal methods. *In* Molecular Ecology and Evolution: Approaches and Applications, B. Schierwater, B. Streit, G. P. Wagner, and R. DeSalle (Eds.). Birkhauser, Basel, pp. 593–606.

Soller, M., and Brody, T. 1976. On the power of experimental designs for the detection of linkage between marker loci and quantitative loci in crosses between inbred lines. Theor. Appl. Genet. 47:35–59.

Stuber, C. W., Lincoln, S. E., Wolff, D. W., Helentjaris, T., and Lander, E. S. 1992. Identification of genetic factors contributing to heterosis in a hybrid from two elite maize inbred lines using molecular markers. Genetics 132:823–839.

Tanksley, S. D. 1993. Mapping polygenes. Annu. Rev. Genet. 27:205–233.

Templeton, A. R. 1981. Mechanisms of speciation—a population genetic approach. Annu. Rev. Ecol. Syst. 12:23–48.

Templeton, A. R. 1989. The meaning of species and speciation: a genetic perspective. *In* Speciation and Its Consequences, D. Otte and J. A. Endler (Eds.). Sinauer Press, Sunderland, Mass., pp. 3–27.

Templeton, A. R. 1996. Experimental evidence for the genetic-transilience model of speciation. Evolution 50:909–915.

Van Ooijen, J. W. 1992. Accuracy of mapping quantitative trait loci in autogamous species. Theor. Appl. Genet. 84:803–811.

Via, S. 1991. The genetic structure of host plant adaptation in a spatial patchwork: demographic variability among reciprocally transplanted pea aphid clones. Evolution 45:827–852.

Via, S. 1994. Population structure and local adaptation in a clonal herbivore. *In* Ecological Genetics, L. Real (Ed.). Princeton University Press, Princeton, N.J., pp. 58–85.

Visscher, R. M., Thompson, R., and Haley, C. S. 1996. Confidence intervals in QTL mapping by bootstrapping. Genetics 143:1013–1020.

Vos, P., Hogers, R., Bleeker, M., Reians, M., van de Lee, T., Hornes, M., Frijters, A., Pot, J., Peleman, J., Kuiper, M., and Zabeau, M. 1995. AFLP: a new technique for DNA fingerprinting. Nucleic Acids Res. 23:4407–4414.

Wu, C.-I., and Palopoli, M. F. 1994. Genetics of postmating reproductive isolation. Annu. Rev. Genet. 27:283–308.

Zeng, Z.-B. 1992. Correcting the bias of Wright's estimates of the number of genes affecting a quantitative character: a further improved method. Genetics 131:987–1001.

Zeng, Z.-B. 1994. Precision mapping of quantitative trait loci. Genetics 136:1457–1468.

Zeng, Z.-B., Houle, D., and Cockerham, C. C. 1990. How informative is Wright's estimator of the number of genes affecting a quantitative character? Genetics 235:247.

Part VI

Hybrid Zones and Speciation

28

What Do Hybrid Zones in General, and the *Chorthippus parallelus* Zone in Particular, Tell Us about Speciation?

Roger Butlin

Fifteen thousand years ago, the Pyrenees were covered by permanent ice. In France to the north and in Spain to the south were wide expanses of tundra unsuitable for *Chorthippus* grasshoppers or for many other taxa found in the Pyrenees today. By about 8,000 years ago, all of this had changed. Deciduous forest had reinvaded much of France and Spain and had reached all but the highest altitudes in the Pyrenean mountains. With the forests came the grasshoppers, rapidly colonizing grassy clearings and, soon afterward (about 6,000 years ago), meadows created by man for grazing animals.

Today, the grasshopper species *Chorthippus parallelus* is very widely distributed in the Pyrenees up to an altitude of 2,000 m or a little higher. Elsewhere, this species occurs throughout Europe and across Russia, probably all the way to the Bering Straits. A very close relative, *C. curtipennis*, inhabits North America. Throughout northern Europe, *C. parallelus* is consistent in morphology and behavior, but populations in Spain are sufficiently distinct to be recognized as the subspecies *C. p. erythropus*. In the Pyrenees, there is an abrupt transition from one form to another wherever the two meet. Details of the transition vary among localities, but everywhere the region of mixed or intermediate characteristics is less than 50 km wide, a very short distance by comparison with the range of each subspecies, but a long distance in terms of grasshopper movement (estimated to be about 30 m per generation on average; Virdee and Hewitt 1990). Some idea of the nature of the transition can be gained from a comparison of the clines in a variety of characters that differ between the subspecies in a single locality (figure 28.1). For the majority of characters, clines are centered close together but vary in width from less than 1 km to greater than 20 km. Characters with significantly displaced cline centers are of particular interest and are discussed below.

The simplest explanation for this transition zone is that the northern and southern slopes of the Pyrenees were colonized independently, at the end of the last glaciation, by populations of grasshoppers that had diverged genetically in allopatry. The divergence was sufficient to generate a partial barrier to gene exchange such that, following renewed contact, the populations have not interbred to such an extent that the transition from one to the other is no longer discernible. Some form of selection, either against hybrids or due to adaptation to different environments, must be invoked to explain the narrow clines in some characters while wider clines may be consistent with neutral diffusion of alleles since contact. In the center of the hybrid zone, "parental" and "hybrid" individuals cannot be distinguished: all grasshoppers are intermediate in form. Such secondary hybrid zones are often proposed as "natural laboratories for evolutionary studies" (Hewitt 1988, p. 158) and especially for the study of speciation. So, what does research on *Chorthippus parallelus* tell us about speciation? I consider several questions about this specific example, and about other hybrid zones, that are relevant to general questions about the origin of species and demonstrate that hybrid zones do tell us something useful, although real problems remain.

Where Did the Interacting Populations Come From, and How Long Had They Been Diverging before Contact?

The *Chorthippus parallelus* subspecies' ranges probably met in the Pyrenees around 6–8,000 years ago when climate and vegetation became suitable (Hewitt 1993), but it is more difficult to estimate the length of time prior to that during which they had been evolving independently and accumulating genetic differences. The duration of the last ice age would provide a lower limit but it was preceded by several other cycles of glacial and interglacial conditions. During cold periods, the ancestors of the present *C. p. erythropus* almost certainly persisted in

Clines in the Col de la Quillane

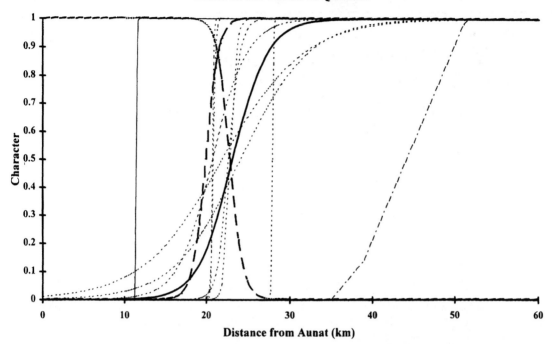

Figure 28.1. A compilation of clines for some of the characters that distinguish *Chorthippus parallelus parallelus* and *Chorthippus parallelus erythropus*. These are fitted tanh curves for transects across the Col de la Quillane in the eastern Pyrenees, with *C. p. parallelus* to the left side of the figure. The top of the Col is at about 24 km from Aunat. Each character has been scaled from 1 for *C. p. erythropus* to 0 for *C. p. parallelus* (except sterility, where fertile crosses to either female type have been scaled at 1, sterile at 0; see figure 28.3 for further details). The heavy black line is the cline for stridulatory peg number, used as a reference point because it is the morphological character that best separates the subspecies. The dashed black lines are the sterility clines (see figure 28.3). The black line displaced to the north is the cline for cuticular hydrocarbons. The steep cline displaced slightly to the south is for female preference. The dashed line displaced to the south is for a C-band on the X-chromosome (insufficient data are available to fit a cline in this case; the line connects sample points). The remaining clines are for morphology (male pronotum ratio, male hindwing length, male femur ratio, male peg row length, female ovipositor length) and song (echeme interval, syllable length). Note the close agreement in cline centers but variation in cline width. (Data are from Butlin 1989; Butlin and Ritchie 1991; Butlin et al. 1991; Ferris et al. 1993; Neems and Butlin 1994; Underwood 1994.)

refugia in southern or western Spain, where they can still be found at high altitudes. However, the location of the refuge from which France and the rest of northern Europe was colonized is more difficult to infer: it could have been in Italy, the Balkans, or farther east.

Both of these questions, the duration of divergence and the locality of refuges, can, in principle, be answered with molecular data. Allozyme analysis turns out not to be sufficiently sensitive: the two subspecies have a genetic distance of only about 0.05, which is in the range of interpopulation distances within subspecies (Butlin and Hewitt 1987; Hewitt, personal communication). An anonymous single-copy nuclear DNA sequence studied by Cooper et al. (1995) shows marked divergence across the Pyrenees (K_{ST} between Spain and France of 0.17–0.45). Again, there is substantial variation within Spain (K_{ST} between regions within Spain up to 0.124), but less within France (maximum 0.09). It is difficult to derive a time scale from these levels of divergence, but Cooper and Hewitt (1993) estimate that divergence between French and Spanish sequences began 200,000 to 2 million years ago. This does not conflict with estimates based on mitochondrial DNA sequence data (Lunt 1995; Lunt, Ibrahim, and Hewitt, unpublished), which indicate around 1% sequence divergence between subspecies, consistent with a divergence time of about 500,000 years ago. This is also consistent with the lack of allozyme divergence. If the figure is correct, and of course the errors on such estimates are large, it suggests that the subspecies have been diverging through four or five glaciations (Hewitt 1996).

The nuclear and mitochondrial sequence data also provide information on the similarities among grasshopper populations in other regions of Europe and thus suggest hypotheses for the locations of refuges. Refugial areas are expected to have high within-region diversity, while populations derived from a refuge will contain a sample of variants present in, and characteristic of, the refuge. Applying this logic, Cooper et al. (1995) infer that *C. parallelus parallelus* north of the Pyrenees originated from a refuge in the northern Balkans while *C. p. erythropus* originated in southern Spain. This implies very rapid range expansion across northern Europe, in the region of 0.5 to 1 km per year in order to reach Britain while a land bridge was still in place (figure 28.2).

A possible scenario for the original separation of the two subspecies (Hewitt 1996) is that a severe glaciation removed the species from southern Europe and then, in an interglacial warm period around 500,000 years ago, the whole region was recolonized, probably from Turkey.

Since then, the Spanish, Balkan, and Italian populations have been isolated for the majority of the time and have only had limited opportunities for gene exchange at hybrid zones during warm periods. The hybrid zone in the Pyrenees shows just how limited this exchange is, with each subspecies' alleles penetrating only a few kilometers into the range of the other subspecies (figure 28.1). At the onset of the next glaciation, these mixed genotypes will be among the first to perish because of their location, and any evolutionary modification within the zone, including reinforcement (see below), will thus be very limited in its long-term effect (cf. Hewitt 1989). The zone is a window on an intermediate stage in allopatric speciation, rather than an engine of parapatric speciation. For the study of speciation, it does not provide the chance to observe speciation in progress, but it does enable us to analyze the differences that have accumulated during a period of isolation and independent evolution, and their effects on gene exchange.

Whatever the geographic pattern, the molecular data imply that all genetic differentiation that we currently see across the Pyrenean hybrid zone has evolved in the relatively short interval of about 0.5 million years (= 0.5 million generations since these grasshoppers are univoltine).

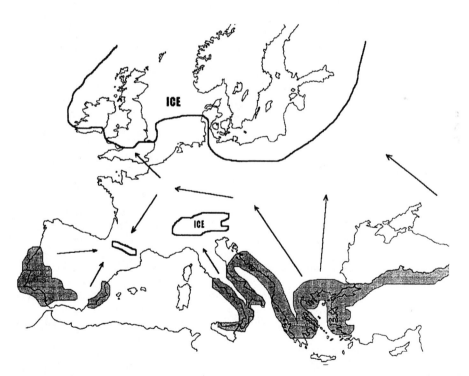

Figure 28.2. Expansion of *Chorthippus parallelus* from glacial refugia. The shaded area represents the approximate location of potential refugia based on the distribution of deciduous forest at the last glacial maximum. Arrows represent the colonization routes inferred from the data of Cooper et al. (1995).

Have the Subspecies Developed Postzygotic Isolation, What Is Its Genetic Basis, and What Happens to It in the Contact Zone?

F1 male hybrids between *C. p. parallelus* and *C. p. erythropus* are completely sterile, or very nearly so, with degenerate testes and severely disrupted meiosis (Hewitt et al. 1987). This does not appear to be due to chromosomal differences since the karyotypes of the subspecies differ only in heterochromatin and nucleolar organizer distribution (Gosalvez et al. 1988). The testes of backcross males are intermediate in condition: they produce some sperm and the males can fertilize females, but their likely fitness under field conditions is difficult to judge. F1 females have apparently normal fecundity.

A very influential report by Coyne and Orr (1989) considered the rate of evolution of reproductive isolation in *Drosophila*. Their survey confirmed Haldane's Rule that postzygotic isolation evolves more rapidly in the heterogametic sex, and the data provide an estimate for the mean time to appearance of male sterility or inviability of 1.25 million years (Nei's D = 0.25, with a wide range of <0.1 to 1.5, standard deviation 0.128). Other measures of the rate of appearance of reproductive isolation are rare. One example comes from sea urchin (*Echinometra*) taxa separated by the Isthmus of Panama 3–3.5 million years ago that now show 90% reduction in gamete compatibility in one direction of cross (Lessios and Cunningham 1990; Lessios, this volume), suggesting a slower rate of divergence than in *Drosophila*.

On this scale, the appearance of hybrid sterility in *Chorthippus* has clearly been rapid (almost 2 standard deviations below the mean for *Drosophila*). Rather few other hybrid zones can be added to this comparison. Two other cases of sterility of the heterogametic sex are at opposite ends of the scale: in pondskaters of the genus *Limnoporus*, Nei's D = 0.4 (Sperling and Spence 1991), while in flycatchers (*Ficedula*), D ≈ 0 (Tegelström and Gelter 1990). In one contact between pocket gopher species (*Thomomys bottae/umbrinus*), both sexes are apparently sterile despite a genetic distance of only 0.16 (expected D = 0.99, standard deviation 0.36; from Coyne and Orr) (Patton 1973). At the other end of the scale, there are hybrid zones that show no sterility of either sex in either direction of cross, with values of genetic distance ranging from zero up to at least 0.5 (*Bombina bombina/variegata*; Szymura 1993). These zones do, of course, show varying levels of selection against hybrids through inviability or loss of adaptation that provide barriers to gene exchange that can be as strong as those due to hybrid male sterility. However, these quantitative effects were not included in Coyne and Orr's comparisons. Perhaps not surprisingly, the consistency in the rate of evolution of postzygotic isolation observed within *Drosophila* does not readily extend to other taxa. However, it is quite surprising, given the observed range of genetic

distances between taxa that interact in hybrid zones (D = 0–2.6; Barton and Hewitt 1985), that male sterility is not encountered more frequently, and it remains true that the evolution of sterility in *Chorthippus* has been rapid.

The most plausible general prediction for the genetic basis of postzygotic isolation is the idea of "synthetic" or "complementary" lethals/steriles (Dobzhansky 1937). Debate continues over the role of sex chromosomes and the numbers of loci involved (Wu and Palopoli 1994; Wu and Hollocher, this volume) but laboratory crossing programs typically demonstrate the expected role of epistatic interactions among alleles derived from opposite species or races. A simple version of this system, where an ancestral aabb genotype has given rise to aaBB and AAbb populations independently and the A–B– combinations are sterile, has an interesting consequence with respect to secondary contact. Where only one sex is sterile, hybridization between the AAbb and aaBB genotypes can generate the ancestral aabb genotype in the F2 or subsequent generations. This genotype can cross with either of the derived taxa without producing sterile offspring. An analogous process can operate where more loci are involved.

The *Chorthippus* hybrid zone may provide evidence for this process. Although F1 males are sterile, no sterile males have been detected within the hybrid zone or in laboratory offspring from individuals collected in the zone center (Ritchie et al. 1992; Ritchie and Hewitt 1995). This surprising observation led us to investigate the clines for sterility alleles indirectly by crossing females from populations in two transects across the hybrid zone with males from either end of the transects (Virdee and Hewitt 1994; Underwood 1994). The expectation was that crosses spanning the cline for sterility alleles would produce sterile male offspring while crosses within subspecies would not. Females from the center of the sterility cline should produce partially sterile offspring, equivalent to laboratory backcrosses, when crossed with males of either subspecies. In fact, in both transects, we found that females from central populations produce fertile sons with both types of male. In effect, clines for sterility measured with the two types of male are displaced away from each other by about one cline width (figure 28.3).

The simplest explanation for this pattern is the reconstruction of an ancestral genotype within the zone as described above. Backcross distributions suggest that at least several loci are involved, possibly with some sex linkage (Virdee and Hewitt 1992), but without knowing much more about the genetics of sterility, this explanation cannot be tested in *Chorthippus*. However, a comparable example exists in the hybrid zone between the Oxford and Hermitage races of the common shrew, *Sorex araneus*. Here the divergence apparently responsible for selection against hybrids is chromosomal, and it is possible to compare the genotypes present in the center of the hybrid zone with the proposed ancestral genotype of the species. An all-acrocentric karyotype, like the proposed ancestral state,

Col de la Quillane

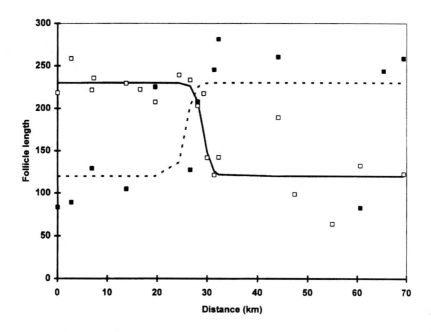

Col du Pourtalet

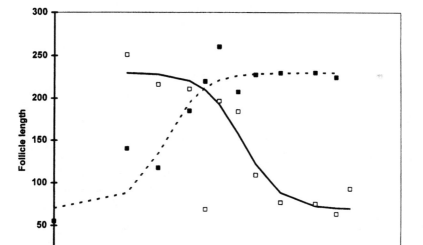

Figure 28.3. Clines in male sterility inferred from the testis follicle lengths of offspring from crosses between females from the localities indicated and males from localities outside the hybrid zones. a, Col de la Quillane transect, eastern Pyrenees. b, Col du Pourtalet transect, western Pyrenees. Open symbols and solid line, crosses to the *C. p. parallelus* male; solid symbols and broken line, crosses to the *C. p. erythropus* male. Points are means for the offspring of one to five females, lines are best fitting tanh curves, and distances are measured from the locality used to collect the *C. p. parallelus* males (i.e., the northern end of the transect). For further details, see Virdee and Hewitt (1994).

appears to have been recreated in the center of the hybrid zone because it is compatible with the metacentric chromosomes found in both races (Searle 1986).

In other hybrid zones, notably in *Podisma* and *Bombina*, it has been possible to estimate the numbers of loci responsible for hybrid inviability, and the relative contributions of epistasis versus underdominance, from the widths and shapes of clines, disequilibria among marker loci, and direct measures of fitness (Barton and Hewitt 1989; Szymura and Barton 1991; Barton and Gale 1993). The conclusions from two pairs of taxa with extremely different levels of genetic divergence (D ≈ 0 in *Podisma*, D = 0.5 in *Bombina*) are remarkably similar: many loci contribute to hybrid inviability, and epistatic interactions make a major contribution.

Thus, all of the (admittedly few) observations available from hybrid zones, which are free from the problems associated with laboratory crosses between species that may have accumulated differences after completion of speciation, are consistent with the view that postzygotic isolation is a result of complementary lethals/steriles and suggest that the accumulation of many small differences is involved. What is not clear from either type of data is the reason for the divergence, and this is unlikely to be discovered in *Chorthippus* while the only known phenotype of the loci involved is hybrid sterility. The rapid evolution of sterility might suggest that natural selection favored divergence, but one could also suggest that the history of range expansion and contraction gave ample opportunity for chance fixation of mutations. It may be possible to obtain some discrimination between these alternatives now that we have a reasonable idea of the positions of refugia and patterns of expansion. For example, if range expansion drives the substitutions that are responsible for hybrid sterility, crosses between Spanish and Balkan grasshoppers may produce fertile male offspring despite the sterility observed across the Pyrenees.

Have the Subspecies Developed Prezygotic Isolation, What Is Its Behavioral Basis, and Is It "Reinforced"?

Multiple-choice assortative mating tests between populations of *C. p. parallelus* and *C. p. erythropus* demonstrate that the subspecies have diverged sufficiently in characters associated with mate choice to produce an isolation index of 0.5 (0 equals random mating, 1 equals perfect assortative mating; Ritchie et al. 1989). This figure is derived from populations immediately on either side of the hybrid zone in the eastern Pyrenees, but we know that there can be significant assortment between populations within subspecies as well (Dagley et al. 1994). Therefore, we should be cautious about extrapolating this amount of prezygotic isolation to the level of a subspecific characteristic. Premating isolation is supplemented by a postmating, but still prezygotic, barrier to gene exchange: when mated with one homosubspecific and one heterosubspecific partner, a female grasshopper produces fewer hybrid offspring than would be expected from random fertilization (Bella et al. 1992; cf. Howard et al., this volume). This is true whatever the order of matings but the effect is stronger for *parallelus* females than for *erythropus* females.

Our initial interest in prezygotic isolation in the hybrid zone was driven by the idea of speciation by reinforcement. This theory suggests that selection will favor divergence in mating behavior, and increase in the strength of assortative mating, as a result of the production of hybrid genotypes of reduced fitness. There are many theoretical arguments which suggest that this selection pressure, which undoubtedly exists, is unlikely to be effective within a hybrid zone (Butlin 1989; Sanderson 1989). Some of these arguments have been disputed (Howard 1993), and recent models of reinforcement (Liou and Price 1994; Kelly and Noor 1996) suggest that there are circumstances in which it can occur with reasonable probability. However, the consensus that is emerging is that reinforcement can only complete the speciation process given secondary contact between populations that already show very substantial barriers to gene exchange: very low hybrid fitness and/or strong prezygotic isolation. Noor's (1995) observation of greater isolation between *Drosophila pseudoobscura* and *D. persimilis* in sympatry than in allopatry may fall into this category. Sympatry, or possibly multiple contacts in mosaic hybrid zones, may also facilitate reinforcement, because they allow more extensive contact between parental genotypes, but low hybrid fitness and/or high initial prezygotic isolation is still likely to be required (Liou and Price 1994).

The *Chorthippus* hybrid zone initially appeared to fit the requirements of reinforcement well: there was secondary contact between populations with F1 hybrid fitness reduced to 0.5 by male sterility and some initial prezygotic isolation. Selection against hybrids, possibly combined with adaptation to different environments, is enough of a barrier to impede gene flow across the hybrid zone, as evidenced by the narrow clines for some characters (figure 28.1). But the level of reproductive isolation is insufficient to prevent extensive mixing of genomes within the zone. As we now know, this mixing has permitted modification of the zone to reduce selection against hybrids: there is now apparently no selection against hybridization within the zone (at least in the form of male sterility). Therefore, while there may have been an opportunity for reinforcement following initial contact, there is no longer any known driving force.

Appropriate tests for reinforcement have been discussed elsewhere (Butlin 1989). The most common approach is to look for evidence that selection has altered mating signals and/or preferences in the area where the two divergent populations are in contact relative to allopatric populations. This may be achieved either by direct examination of signals or preferences, or by assortative mating tests. In the context of a hybrid zone, it predicts an "inverse cline," that is, an increased divergence close to the zone relative to the divergence farther away. An alternative approach is to ask whether the mating pattern within the zone is such that production of unfit hybrid genotypes is less frequent than would be expected under random mating. This is a more direct test of reinforcement because it examines the main predicted outcome. By contrast, patterns of variation in mating signals or preferences may be explained in many different ways.

We sought evidence for reinforcement in both of these ways: describing the pattern of change in mate choice and in mating signals across the hybrid zone, and testing the fitness consequences of mate choice. The pattern of mate choice across the zone was examined using two male–two female choice tests with females from zone populations and males from "reference" populations. The results were equivocal: a steep cline close to the center of the morphological zone (figure 28.1), or a broader cline with increased preference for *parallelus* males close to the zone on the *parallelus* (French) side, was indistinguishable statistically (Butlin and Ritchie 1991). In testing the fitness consequences of mate choice, we found no reduction in the fitness of offspring from females taken from the center of the hybrid zone when they were assigned mates at random as opposed to choosing their mates (Ritchie et al. 1992).

The mating signals used by these grasshoppers were initially considered to be almost exclusively acoustic. There are consistent differences in song characteristics across our eastern Pyrenean transect (Butlin and Hewitt 1985; Butlin 1989), but again, there is also variation among populations within subspecies (Dagley et al. 1994). Clines for song characters are variable in width (1.5–30 km; figure 28.1) but are all centered close to the center of the morphological transition, and none shows evidence for the excess divergence predicted by reinforcement.

Our observations of mating behavior during the initial song studies quickly showed a major difference in mate finding strategy between subspecies in this transect. Initially this showed up as a higher calling song output for *C. p. parallelus*, partly compensated by the courtship song produced only by *C. p. erythropus* (Butlin and Hewitt 1985). Later, we showed that this difference is present under field conditions and is associated with differences in movement: male *parallelus* remain stationary for longer bouts of song while male *erythropus* sing for short periods interspersed with movement through the habitat, apparently searching for females (Neems and Butlin 1993).

The difference in song output suggested that other signal channels might be used in addition to song. Ritchie's (1990) experiments with silenced or antennaless grasshoppers provided evidence that *parallelus* rely more on acoustic signals while *erythropus* use chemical or tactile signals in mate choice. These experiments, like most mate

choice experiments, could not distinguish between "male choice" and "female choice." Since only males produce acoustic signals, these must allow female *parallelus* to discriminate among males. However, chemical or tactile signals could be exchanged in either direction, and it may well be that female chemical signals are needed to initiate male courtship (see below). We have since shown that the subspecies differ in cuticular hydrocarbon blend in our eastern transect, and across other cols in the Pyrenees, although there is also substantial east–west variation within the mountains, and central Spanish populations are quite distinct (Neems and Butlin 1995). Once again, the subspecific difference appears to be complicated by within-subspecies variability.

Cuticular hydrocarbons have two known functions in insects: prevention of water loss, and communication (both sexual and social) (Howard and Blomquist 1982). In *Chorthippus* we have preliminary evidence that supports their role in sexual communication. Male grasshoppers will court freshly killed females, although the progression through the courtship sequence differs between subspecies as expected from the differences in mating strategy described above. When cuticular hydrocarbons are removed by washing with the organic solvent hexane, this courtship is largely abolished. It can be reestablished by "painting" the extract back onto the female and allowing the hexane to evaporate (thus excluding the possibility that the odor of hexane was responsible for the reduction in courtship) (table 28.1).

Table 28.1. Summary of responses of male *Chorthippus parallelus* to dead females placed in a lifelike position on the wall of a small cage.

Male behavior	Female		
	Dead	Washed	Painted
No courtship	52	34	21
Calling song	12	6	5
Courtship and mounting	11	0	2
Total	75	40	28
	$G_4 = 10.5$, $P < 0.05$		

"Dead" females were virgins aged 3 days or more from adult moult, killed by freezing at –40°C, and thawed completely before the experiment. "Washed" and "painted" females were treated as for "dead" females but the body was immersed in n-hexane for 1 hr after freezing and then the hexane was allowed to evaporate. For "painted" females, the hexane used in washing was allowed to evaporate to dryness, hydrocarbons were resuspended in 0.5 ml hexane, and this was applied to the female with a fine brush. Each male was observed with the test female for 30 min, and its behavior was categorized according to song production (calling vs. courtship) and mating attempts.

The major discriminating feature of hydrocarbon blend in our eastern transect is the proportion of a single compound or group of related compounds, detected in a gas chromatograph spectrum as peak 2, relative to the total hydrocarbons present. Variation among populations in this proportion fits a simple sigmoid cline with no excess divergence close to the center of the hybrid zone, thus showing no evidence of reinforcement. However, the cline is narrow (less than 1 km), indicating that the character is likely to be under selection, and it is displaced about 18 km north of the majority of other clines (figure 28.1; Neems and Butlin 1994). Such displacement is not a common feature of hybrid zones (Barton 1993) and requires explanation (see below).

The absence of evidence for reinforcement is, with the benefit of our current understanding of the hybrid zone, not at all surprising. If there was any tendency toward an increase in assortative mating when the subspecies initially came into contact, it made little progress before the source of the selection, hybrid male sterility, was removed. Such a race between reinforcement and other forms of modification may not be uncommon. Models involving "modifier" loci that reduce selection against hybrids are effectively equivalent to reinforcement models and so give no reason to expect modification to "win" this race (Ritchie and Hewitt 1995). However, if reconstruction of ancestral genotypes were a common possibility, it may provide a rapid route to increased hybrid fitness and greatly reduce the opportunity for reinforcement in hybrid zones.

Two questions remain to be answered about prezygotic isolation: what are the relative contributions of different signaling systems to the observed assortative mating, and how did the differences in these signaling systems evolve?

Ritchie's (1990) results suggest that acoustic signals are the dominant cause of assortment where the female is *C. p. parallelus*. This is despite the absence of strong phonotaxis by females to male song (Butlin and Hewitt 1986) and experiments by Dagley (1988) that failed to show a strong advantage to singing over silenced males where they competed for access to *parallelus* females. These conflicting results may reflect differences in the conditions used in the various experiments, but they are more likely to reflect genuine differences among the grasshopper populations used: from Gioux in central France (Butlin and Hewitt), the island of Jersey (Dagley), or the French Pyrenees (Ritchie). With regard to the characteristics of song, there is some indication (Butlin and Hewitt 1985) that syllable length is an important feature of male song, and this is consistent with interpopulation (Dagley et al. 1994) and interspecific comparisons (Bauer and von Helversen 1988), and data on other grasshoppers (von Helversen and von Helversen 1994; Butlin et al. 1985). However, syllable length clines are wide, indicating that it is not a strongly selected character. Clearly, demonstrating assortative mating plus differences in signal characters is not sufficient to infer that the signal divergence is responsible for the mating pattern. Nor can comparisons between pairs of populations be extended to the level of subspecies or species. Our current studies of interpopulation variation around Europe in relation to assortative mating may give more clues about important song variables.

Where the female was *C. p. erythropus*, Ritchie's (1990) experiments suggested that chemical cues were essential for assortment. Contact pheromones in insects typically function to elicit male courtship (e.g., Jallon 1984), rather than to provide cues on which female choice might operate. The "dead female" experiments mentioned above support the idea that chemical cues are needed to stimulate males to court. We are continuing this style of experiment in order to investigate subspecies specificity of this stimulation. The study of variation around Europe will relate assortative mating to cuticular hydrocarbon divergence as well as song variation.

With the cuticular hydrocarbons, the displaced cline in our eastern transect also provides an opportunity to study the function of the differentiation between subspecies. It is possible that the displacement is a chance result of colonization history, but this seems unlikely given the scale of the displacement relative to the close grouping of most other clines in this transect (figure 28.1). Otherwise, the displacement may be due to selection operating either through the waterproofing or the behavioral roles of the compounds involved. These explanations can be separated by a combination of behavioral experiments, detailed correlations between environmental parameters and hydrocarbon profiles, and comparison with other, environmentally different, transects.

As with the divergence that causes postzygotic isolation, the various changes described in this section must have evolved rapidly. Acoustic and chemical signals and mate finding strategy, plus some unknown aspect of postinsemination interactions, have all diverged within 0.5 million years. By comparison with Coyne and Orr's (1989) *Drosophila* data, an isolation index of 0.5 would be low for sympatric taxa at this level of genetic divergence (*Drosophila* mean 0.768, standard deviation 0.268) but greater than expected for allopatric taxa that had an average isolation of 0.21 (standard deviation 0.177) for all comparisons with $D \leq 0.5$ (up to 2.5 million years). As with postzygotic isolation, the major gap in our understanding is that we do not know the cause of divergence. In this case, the situation should be better since we know more about the characters involved and can therefore suggest ways in which selection might have influenced them. However, the range of possible modes of selection is considerable: adaptation to local signaling conditions, male–male competition, Fisherian or "good genes" sexual selection, sensory exploitation, reproductive character displacement, and so on. The divergence may be due to selection pressures that are not associated with signal function directly, such as selection on the waterproofing function of cuticular hydrocarbons, or it could be simply due to drift or to interactions between selection and drift such as in the Kaneshiro process (Kaneshiro 1989).

It is possible to tell an "adaptive story" about the evolutionary origin of the suite of differences observed between *parallelus* and *erythropus*. Silent, active searching with dependence on courtship song and contact chemoreception (*erythropus*) may be adaptive in high-density populations with many predators, whereas stationary singing in prominent sites, with female phonotaxis (*parallelus*), may be adaptive in lower density, less predator-rich habitats. Remating by females may be more common in the higher density habitats, leading to different strategies in relation to sperm competition that, as a by-product, produce the effect of assortative fertilization. Perhaps the conditions in the respective refuges differed in these ways, or perhaps the longer expansion route for *parallelus* has resulted in adaptations to the low-density populations that are likely to be a feature of repeated colonization of new habitat. It is difficult to see how one could substantiate such a scenario when so many other stories are possible. However, support may come from our current work on the distribution of signal variation and assortative mating in populations around Europe since we can assign these populations to categories such as refugial versus recently colonized, sympatric versus allopatric with close relatives, and marginal versus central ecologically. The variation among populations of *parallelus* noted in several contexts may be explicable in this way.

It seems likely that ecological conditions might be a major driving force in the origin of premating isolation in many systems. Schluter (this volume) describes how adaptive changes in size and body shape of sticklebacks might incidentally generate behavioral isolation, while Markow and Hocutt (this volume) attribute much of the divergence in mating strategy between *Drosophila arizonae* and *mojavensis* to the conditions imposed by their food plants. However, there are exceptions: in the brown planthopper, *Nilaparvata lugens*, signal divergence is apparently unrelated to host plant (Butlin 1997).

There are few comparable data from other hybrid zones. Divergence in mating signals and resulting assortative mating show little consistency in zones with a wide range of genetic distances: in *Colaptes* flickers (Moore 1987) where $D \approx 0$ there is substantial plumage divergence, but this does not generate assortment (at least near the center of the zone); in *Allonemobius* crickets (Howard 1986; Benedix and Howard 1991) where $D \approx 0.2$ slight song differences do not produce assortment, but assortative fertilization is a major barrier to gene exchange (Howard et al., this volume); in *Gryllus* crickets (Rand and Harrison 1989; Doherty and Storz 1992) where $D \approx 0.03$ song differences generate substantial assortment; in *Spalax* mole rats (Nevo 1991) where $D \approx 0.04$ there is significant assortment due to acoustic and chemical signal divergence; and in *Bombina* toads (Sanderson et al.

1992; MacCallum et al. 1995) where $D \approx 0.5$ there are major calling song differences, but assortment in the hybrid zone, in transects where it exists, is due mainly to habitat preference. Genetic divergence is, in this small sample, a poor predictor of signal divergence. Where signal divergence generates assortment between parentals, or would be expected to do so, this typically breaks down within the hybrid zone (*Colaptes, Bombina,* and *Chorthippus*). In *Bombina,* one can attribute at least some characteristics of the mating behavior to adaptation to different pond types, which is also responsible for many morphological and physiological differences between the taxa and contributes to selection against hybrids (MacCallum et al. 1995). *Allonemobius* provides a contrast: the slight differences in mating signals apparently contribute virtually nothing to reproductive isolation, which is due primarily to assortative fertilization. The reason for the divergence that underlies assortative fertilization in this case is not known.

What Are Species, Anyway?

Considerations about speciation are obviously influenced by one's view of the nature of species, and rightly so. It is clear that hybrid zones provide valuable lessons that must be accommodated by any general species concept. *Chorthippus parallelus* is not a single evolutionary unit: in addition to the distinct Spanish subspecies, groups of populations in Italy, Southern Greece, and Turkey are quite distinct genetically from the broadly distributed group of populations across northern Europe (as judged from an anonymous nuclear marker; Cooper et al. 1995). Recent data demonstrate that the Italian populations are distinct in other ways: they produce sterile hybrids with French grasshoppers and have distinct song characteristics (Flanagan 1998, personal communication). Where intermediates between these forms exist in the Pyrenees and in the Alps, they are confined to narrow zones of contact that represent a tiny fraction of the total range occupied by each form. These intermediates probably have little long-term evolutionary significance since they will be the first to disappear when climatic conditions next deteriorate. Within the *Chorthippus* contact zones that have been studied so far, all possible intermediate genotypes and morphologies exist with no hint of bimodality. Although there are barriers to gene exchange between these taxa, they are not sufficient to maintain the distinctness of the forms where they are in contact.

Within the ranges of the distinct genotypic groups, there is much less genetic differentiation. However, this lack of divergence is not a result of contemporary gene flow but reflects recent common ancestry before the last period of range expansion. Despite their genetic similarity (Cooper et al., 1995), populations in the United Kingdom, Brittany, and Finland, for example, must currently be evolving independently.

Chorthippus parallelus has two close relatives: *Chorthippus montanus,* with which it is sympatric in many areas from France to Siberia, and *C. curtipennis,* which occurs in North America. They are only slightly, if at all, more distinct genetically from northern European populations of *C. parallelus* than are the southern European populations (Butlin and Hewitt 1987). There is no contact between the ranges of *parallelus* and *curtipennis,* but *parallelus* and *montanus* occur together in wet meadow habitats, apparently without hybridization under field conditions, although they can easily be crossed in the laboratory.

This sort of complexity is common in well-studied taxa; indeed, it may well turn out to be the rule rather than the exception. It seems to me that the Biological Species Concept (BSC) copes with it well and that alternatives do not provide substantial improvement. This requires the proviso that the implication sometimes associated with the BSC that a species is a coherent evolutionary unit, held together by gene flow, be dropped. The common practice of recognizing subspecies as parapatric groups of populations with distinct characteristics already effectively acknowledges limitations on this coherence. Allopatric populations of a species, including populations separated by distance rather than by breaks in the distribution, are genetically similar and reproductively compatible because of their shared ancestry, rather than current gene exchange. The criterion for reaching species status is when the potential for gene exchange under natural conditions no longer exists, a point that has not been reached by the populations or subspecies of *C. parallelus* but has been reached in the divergence between *C. parallelus* and *C. montanus.*

Mallet's (1995) genotypic cluster definition would not differ from the BSC in its treatment of the *Chorthippus* populations described here. Where it would differ would be if a hybrid zone existed in which there was a deficiency of hybrid genotypes, but it is hard to see how much of a deficiency is required before the distinction is made, or why it should be considered critical. Populations either side of a hybrid zone exchange genes less freely than populations at a similar distance within the range of either taxon. The stronger the selection against hybrids, the greater the barrier and the longer the equivalent distance of uninterrupted territory. There is no qualitative difference unless gene exchange either is completely absent (the BSC) or is low enough, and is combined with sufficient ecological divergence, for the ranges of the interacting taxa to overlap rather than abut. It is this last category that provides the greatest difficulty in principle, although it does not often cause practical problems since hybrids are typically very rare in the areas of sympatry and have little impact on the distinctness of the parent taxa. Recognizing that hybrids occur between species is not a difficulty for the BSC for two reasons: because the occurrence of hybrids is not necessarily indicative of gene exchange, and because intermediate stages in the evolu-

tion of new species must occur. Intermediate stages must be treated pragmatically by taxonomists, perhaps by some sort of "genotypic cluster" criterion, but should not cause us to relinquish our species concept.

Acknowledgments   Many people have contributed to this work, especially Godfrey Hewitt and Mike Ritchie. I am particularly grateful to Godfrey Hewitt, Mike Ritchie, Tom Tregenza, and Stuart Buckley for recent discussions and comments on the manuscript.

References

Barton, N. H. 1993. Why species and subspecies? Curr. Biol. 3, 11–12.

Barton, N. H., and Gale, K. S. 1993. Genetic analysis of hybrid zones. In R. G. Harrison (ed.), Hybrid Zones and the Evolutionary Process. New York: Oxford University Press, pp. 13–45.

Barton, N. H., and Hewitt, G. M. 1985. Analysis of hybrid zones. Annu. Rev. Ecol. System. 16:113–148.

Barton, N. H., and Hewitt, G. M. 1989. Adaptation, speciation and hybrid zones. Nature 341, 497–503.

Bauer, M., and von Helversen, O. 1988. Separate localization of sound recognizing and sound producing neural mechanisms in a grasshopper. J. Comp. Physiol. A 161, 95–101.

Bella, J. L., Butlin, R. K., Ferris, C., and Hewitt, G. M. 1992. Asymmetrical homogamy and unequal sex ratio from reciprocal mating-order crosses between *Chorthippus parallelus* subspecies. Heredity 68, 345–352.

Benedix, J. H., Jr., and Howard, D. J. 1991. Calling song displacement in a zone of overlap and hybridisation. Evolution 45, 1751–1759.

Butlin, R. K. 1989. Reinforcement of premating isolation. In D. Otte and J. A. Endler (eds.), Speciation and Its Consequences. Sunderland, Mass.: Sinauer, pp. 158–179.

Butlin, R. K. 1997. Co-ordination of the sexual signalling system and the genetic basis of differentiation between populations in the brown planthopper, *Nilaparvata lugens.* Heredity 77, 369–377.

Butlin, R. K., and Hewitt, G. M. 1985. A hybrid zone between *Chorthippus parallelus parallelus* and *Chorthippus parallelus erythropus* (Orthoptera: Acrididae): behavioural characters. Biol. J. Linn. Soc. 26, 287–299.

Butlin, R. K., and Hewitt, G. M. 1986. The response of female grasshoppers to male song. Anim. Behav. 34, 1896–1899.

Butlin, R. K., and Hewitt, G. M. 1987. Genetic divergence in the *Chorthippus parallelus* species group (Orthoptera: Acrididae). Biol. J. Linn. Soc. 31, 301–310.

Butlin, R. K., and Ritchie, M. G. 1991. Variation in female mate preference across a grasshopper hybrid zone. J. Evol. Biol. 4, 227–240.

Butlin, R. K., Hewitt, G. M., and Webb, S. F. 1985. Sexual selection for intermediate optimum in *Chorthippus brunneus* (Orthoptera: Acrididae). Anim. Behav. 33, 1281–1292.

Butlin, R. K., Ritchie, M. G., and Hewitt, G. M. 1991. Comparisons among morphological characters and between localities in the *Chorthippus parallelus* hybrid zone (Orthoptera: Acrididae). Proc. R. Soc. Lond. B 334, 297–308.

Cooper, S. J. B., and Hewitt, G. M. 1993. Nuclear DNA sequence divergence between parapatric subspecies of the grasshopper *Chorthippus parallelus*. Insect Mol. Biol. 2, 185–194.

Cooper, S. J. B., Ibrahim, K. M., and Hewitt, G. M. 1995. Postglacial expansion and genome subdivision in the European grasshopper *Chorthippus parallelus*. Mol. Ecol. 4, 49–60.

Coyne, J. A., and Orr, H. A. 1989. Patterns of speciation in *Drosophila*. Evolution 43, 362–381.

Dagley, J. R. 1988. Population differentiation in the grasshopper, *Chorthippus parallelus* (Orthoptera: Acrididae): a study of the mate recognition system. Ph.D. thesis, University of East Anglia, Norwich, U.K.

Dagley, J. R., Butlin, R. K., and Hewitt, G. M. 1994. Divergence in morphology and mating signals, and assortative mating among populations of *Chorthippus parallelus* (Orthoptera: Acrididae). Evolution 48:1202–1210.

Dobzhansky, T. 1937. Genetics and the Origin of Species. New York: Columbia University Press.

Doherty, J. A., and Storz, M. M. 1992. Calling song and selective phonotaxis in the field crickets, *Gryllus firmus* and *G. pennsylvanicus* (Orthoptera: Gryllidae). J. Insect Behav. 5, 555–569.

Ferris, C., Rubio, J. M., Serrano, L., Gosalvez, J., and Hewitt, G. M. 1993. One way introgression of a subspecific sex chromosome marker in a hybrid zone. Heredity 71, 119–129.

Flanagan, N. S. 1998. Incipient speciation in the meadow grasshopper, *Chorthippus parallelus* (Orthopetera: Acrididae). Ph.D. thesis, University of East Anglia, Norwich, U.K.

Gosalvez, J., Lopez-Fernandez, C., Bella, L. J., Butlin, R. K., and Hewitt, G. M. 1988. A hybrid zone between *Chorthippus parallelus parallelus* and *Chorthippus parallelus erythropus* (Orthoptera: Acrididae): chromosomal differentiation. Genome 30, 656–663.

Hewitt, G. M. 1988. Hybrid zones—natural laboratories for evolutionary studies. Trends Ecol. Evol. 3, 158–167.

Hewitt, G. M. 1989. The subdivision of species by hybrid zones. In D. Otte and J. A. Endler (eds.), Speciation and Its Consequences. Sunderland, Mass.: Sinauer, pp. 85–110.

Hewitt, G. M. 1993. Postglacial distribution and species substructure: lessons from pollen, insects and hybrid zones. In D. R. Lees and D. Edwards (eds.), Evolutionary Patterns and Processes. London: Academic Press, pp. 97–123.

Hewitt, G. M. 1996. Some genetic consequences of ice ages, and their role in divergence and speciation. Biol. J. Linn. Soc. 58, 247–276.

Hewitt, G. M., Butlin, R. K., and East, T. M. 1987. Testicular dysfunction in hybrids between parapatric subspecies of the grasshopper, *Chorthippus parallelus*. Biol. J. Linn. Soc. 31, 25–34.

Howard, D. J. 1986. A zone of overlap and hybridisation between two ground cricket species. Evolution 40, 34–43.

Howard, D. J. 1993. Reinforcement: origin, dynamics and fate of an evolutionary hypothesis. In R. G. Harrison (ed.),

Hybrid Zones and the Evolutionary Process. New York: Oxford University Press, pp. 46–69.

Howard, R. W., and Blomquist, G. J. 1982. Chemical ecology and biochemistry of insect hydrocarbons. Annu. Rev. Entomol. 27, 149–172.

Jallon, J.-M. 1984. A few chemical words exchanged by *Drosophila* during courtship and mating. Behav. Genet. 14, 441–478.

Kaneshiro, K. Y. 1989. The dynamics of sexual selection and founder effects in species formation. In L. V. Giddings, K. Y. Kaneshiro, and W. W. Anderson (eds.), Genetics of Speciation and the Founder Principle. Oxford: Oxford University Press, pp. 279–296.

Kelly, J. K., and Noor, M. A. 1996. Speciation by reinforcement: a model derived from studies of *Drosophila*. Genetics 143, 1485–1497.

Lessios, H. A., and Cunningham, C. W. 1990. Gametic incompatibility between species of the sea urchin *Echinometra* on the two sides of the Isthmus of Panama. Evolution 44, 933–941.

Liou, L. W., and Price, T. D. 1994. Speciation by reinforcement of premating isolation. Evolution 48, 1451–1459.

Lunt, D. H. 1995. Mitochondrial DNA differentiation across Europe in the meadow grasshopper *Chorthippus parallelus* (Orthoptera: Acrididae). Ph.D. thesis, University of East Anglia, Norwich, U.K.

Lunt, D. H., Ibrahim, K. M., and Hewitt, G. M. In press. MtDNA phyleography and postglacial patterns of subdivision in the meadow grasshopper *Chorthippus parallelus*. Heredity.

MacCallum, C., Nurnberger, B., and Barton, N. H. 1995. Experimental evidence for habitat dependent selection in a *Bombina* hybrid zone. Proc. R. Soc. Lond. B 260, 257–264.

Mallet, J. 1995. A species definition for the Modern Synthesis. Trends Ecol. Evol. 10, 294–299.

Moore, W. S. 1987. Random mating in the northern flicker hybrid zone: implications for the evolution of bright and contrasting plumage patterns in birds. Evolution 41, 539–546.

Neems, R. M., and Butlin, R. K. 1993. Divergence in mate finding behaviour between two subspecies of the meadow grasshopper, *Chorthippus parallelus*. J. Insect. Behav. 6:421–430.

Neems, R. M., and Butlin, R. K. 1994. Variation in cuticular hydrocarbons across a hybrid zone in the grasshopper *Chorthippus parallelus*. Proc. R. Soc. Lond. B 257, 135–140.

Neems, R. M., and Butlin, R. K. 1995. Divergence in cuticular hydrocarbons between parapatric subspecies of the meadow grasshopper, *Chorthippus parallelus* (Orthoptera: Acrididae). Biol. J. Linn. Soc. 54:139–149.

Nevo, E. 1991. Evolutionary theory and processes of active speciation and adaptive radiation in subterranean mole rats, *Spalax ehrenbergi* superspecies, in Israel. Evol. Biol. 25, 1–125.

Noor, M. A. 1995. Speciation driven by natural selection in *Drosophila*. Nature 375, 674–675.

Patton, J. L. 1973. An analysis of natural hybridisation between the pocket gophers, *Thomomys bottae* and *Thomomys umbrinus*, in Arizona. J. Mammalogr. 54, 561–584.

Rand, D. M., and Harrison, R. G. 1989. Ecological genetics of a mosaic hybrid zone: mitochondrial, nuclear, and reproductive differentiation of crickets by soil type. Evolution 43, 432–449.

Ritchie, M. G. 1990. Does song contribute to assortative mating between subspecies of *Chorthippus parallelus* (Orthoptera: Acrididae)? Anim. Behav. 39, 685–691.

Ritchie, M. G., and Hewitt, G. M. 1995. Outcomes of negative heterosis. In D. M. Lambert and H. G. Spencer (eds.), Speciation and the Recognition Concept: Theory and Applications. Baltimore: Johns Hopkins University Press, pp. 157–174.

Ritchie, M. G., Butlin, R. K., and Hewitt, G. M. 1989. Assortative mating across a hybrid zone in *Chorthippus parallelus* (Orthoptera: Acrididae). J. Evol. Biol. 2, 339–352.

Ritchie, M. G., Butlin, R. K., and Hewitt, G. M. 1992. Fitness consequences of potential assortative mating within and outside a hybrid zone in *Chorthippus parallelus* (Orthoptera: Acrididae): implications for reinforcement and sexual selection theory. Biol. J. Linn. Soc. 45, 219–234.

Sanderson, N. 1989. Can gene flow prevent reinforcement? Evolution 43, 1223–1235.

Sanderson, N., Szymura, J. M., and Barton, N. H. 1992. Variation in mating call across the hybrid zone between the fire bellied toads, *Bombina bombina* and *B. variegata*. Evolution 46, 595–607.

Searle, J. B. 1986. Factors responsible for a karyotypic polymorphism in the common shrew, *Sorex araneus*. Proc. R. Soc. Lond. B 229, 277–298.

Sperling, F. A. H., and Spence, J. R. 1991. Structure of an asymmetric hybrid zone between two water strider species (Hemiptera: Gerridae: *Limnoporus*). Evolution 45, 1370–1383.

Szymura, J. M. 1993. Analysis of hybrid zones with *Bombina*. In R. G. Harrison (ed.), Hybrid Zones and the Evolutionary Process. New York: Oxford University Press, pp. 261–289.

Szymura, J. M., and Barton, N. H. 1991. Genetic analysis of a hybrid zone between the fire-bellied toads, *Bombina bombina* and *B. variegata*, near Cracow in southern Poland. Evolution 40, 1141–1159.

Tegelström, H., and Gelter, H. P. 1990. Haldane's Rule and sex biased gene flow between two hybridizing flycatcher species (*Ficedula albicollis* and *F. hypoleuca*, Aves: Muscicapidae). Evolution 44, 2012–2019.

Underwood, K. L. 1994. Mating pattern in a grasshopper hybrid zone. Ph.D. thesis, The University of Leeds, Leeds, U.K.

Virdee, S. R., and Hewitt, G. M. 1990. Ecological components of a hybrid zone in the grasshopper *Chorthippus parallelus* (Zetterstedt) (Orthoptera: Acrididae). Bol. San. Veg. (Fuenta de serie) 20, 299–309.

Virdee, S. R., and Hewitt, G. M. 1992. Postzygotic isolation and Haldane's Rule in a grasshopper. Heredity 69, 537–548.

Virdee, S. R., and Hewitt, G. M. 1994. Clines for hybrid dysfunction in a grasshopper hybrid zone. Evolution 48, 392–407.

von Helversen, O., and von Helversen, D. 1994. Forces driving co-evolution of song and song recognition in grasshoppers. In K. Schildberger and N. Elsner (eds.), The Neural Basis of Behavioural Adaptations. Stuttgart: Gustav Fischer Verlag, pp. 253–284.

Wu, C.-I., and Palopoli, M. F. 1994. Genetics of postmating reproductive isolation in animals. Annu. Rev. Genet. 27, 283–308.

29

Paradigm Lost

*Natural Hybridization and
Evolutionary Innovations*

Michael L. Arnold
Simon K. Emms

The evolution of a scientific concept can be followed in several interesting ways. One is to study the views of those people who shaped the discipline where the concept took root. The "Paradigm Lost" of our title is that natural hybridization can act as an important evolutionary stimulus, resulting in both adaptive evolution and evolutionary diversification (Anderson and Stebbins, 1954). We believe that this idea, which was held by many biologists until quite recently, but which fell into disfavor about 30 years ago, is now ripe for reassessment. Our goal in this chapter is to begin such a reassessment by asking two questions. First, can hybrids act as the founders of evolutionary lineages, even if early generations have much lower fitness than their parent species? This question is important because the views of Anderson and Stebbins were dismissed largely on the grounds that hybrids often had low viability and fecundity. Second, is it even true that hybrids typically have lower fitness than their parents? We address these questions by reviewing estimates of hybrid and parental fitness, associations between genotypes and habitats, patterns of introgression in contemporary populations and phylogenetic trees. Our conclusion is that hybrids frequently have relatively high fitness compared to parental genotypes, at least in some habitats, and that even when hybrids have low fitness, they have given rise to new evolutionary lineages in numerous plant and animal taxa.

History

In the first half of this century many plant scientists believed that natural hybridization had the potential to cause important evolutionary change (Lotsy, 1916, 1931; Anderson, 1948, 1949; Heiser, 1949; Stebbins, 1950, 1959; Grant, 1963). In contrast, with a few notable exceptions (e.g., Lewontin and Birch, 1966), evolutionary biologists working on animal systems believed the process to be evolutionarily unimportant. Indeed, the conceptual framework of the Modern Synthesis led many evolutionary biologists to see natural hybridization as an *impediment* to evolutionary diversification (e.g., Mayr, 1942). Others viewed it as a maladaptive process that, by causing selection against hybridizing individuals, could result in the reinforcement of barriers to reproduction (Dobzhansky, 1937, 1940; Mayr, 1942, 1963). Evolutionary zoologists in particular championed the view that hybridization was maladaptive because hybridizing individuals produced fewer and/or less fertile progeny (Darwin, 1872; Dobzhansky, 1937; Mayr, 1942, 1963). This viewpoint led to a preoccupation with natural hybridization as a violation of species integrity or species boundaries, where gene flow between bad species resulted in less fit mongrels (for a discussion of these views, see Paterson, 1985; Arnold, 1997).

The latter half of the century saw a shift in emphasis by plant scientists to an approach more typical of zoologists. An extreme view was expressed by Wagner (1969, 1970), who concluded that hybridization was merely "evolutionary noise." For example, he observed that although the phenomenon was extensive in pteridophytes, "Hybrids have made little or no contribution to the total diversity of the flora" (Wagner, 1969, p. 788). A more moderate stance was taken by Heiser (1973), who concluded that introgressive hybridization—the transfer of genetic material from one taxon to another through repeated backcrossing (Anderson and Hubricht, 1938)—"may play a very significant role: but it must be admitted, there is as yet no strong evidence to support such a claim" (p. 362). Recently, Mayr himself addressed the importance of natural hybridization in plants by reviewing morphological variation in the flora of Concord, Massachusetts. He concluded that hybridization did not play a major role in the evolution of species found in this region (Mayr, 1992). Indeed, he extended this conclusion to plants in general: "To be sure, the occasional production of an interspecific hybrid occurs frequently in plants. However, most of these hybrids seem to be sterile, or do not backcross with the parent species for other reasons" (p. 233). These last comments raise an issue that we address further below, the importance of rare events in determining the evolutionary effect of natural hybridization (Arnold and Hodges, 1995a; Arnold, 1997).

This brief history indicates that the paradigm of natural hybridization as a source of evolutionary innovations had fallen on hard times. The basis for renewed interest in the idea comes both from new cases (e.g., Grant and Grant, 1996) and reanalyses of classic examples of hybridization in nature (e.g., Rieseberg et al., 1988; Arnold et al., 1990a,b; Dowling and DeMarais, 1993; Soltis et al., 1995). A central conclusion of many of these studies is that natural hybridization often results in adaptive evolution and species diversification (reviewed in Arnold, 1997). As we show below, this conclusion differs greatly from that of the model most frequently used to explain and predict hybrid zone dynamics.

Models of Hybrid Zone Structure

If we adopt the definition of natural hybridization that "successful matings in nature between individuals from two populations, or groups of populations, which are distinguishable on the basis of one or more heritable characters" (Arnold, 1997, p. 4, modified from Harrison, 1990), then hybrid zones are "those instances in nature where two populations of individuals that are distinguishable on the basis of one or more heritable characters overlap spatially and temporally and cross to form viable and at least partially fertile offspring" (Arnold, 1997, p. 4). The most widely applied model of hybrid zone dynamics assumes selection against hybrids and dispersal of parental genotypes into the region of overlap. Such regions are called tension zones, defined as areas of parapatry where "there is no barrier to mating between the two forms, but [where] hybrids show reduced viability or fecundity" (Key, 1968, p. 19). Key's concept was adopted by Barton and Hewitt (1985), who concluded that "most of the phenomena referred to as hybrid zones are in fact clines maintained by a balance between dispersal and selection against hybrids" (p. 115). Arnold (1997) argued that the tension zone model was embraced by most evolutionary biologists interested in the genetic structure of hybrid zones. Moreover, he believed that the reason for its rapid acceptance was the assumption that hybrids were less fit than parental genotypes in all environments. Hybridization was viewed as inherently maladaptive and evolutionarily ephemeral, except when it caused the reinforcement of reproductive barriers (Dobzhansky, 1940;

Blair, 1955). The basis for such a view, of course, arose from the tenets of the Modern Synthesis.

Two other conceptual frameworks have been used to describe hybrid zones. These are the Bounded Hybrid Superiority model (Moore, 1977) and the Mosaic model (Howard, 1982, 1986; Harrison, 1986). Both models assume the existence of ecological gradients, along which the relative fitness of parental genotypes varies (Slatkin, 1973; May et al., 1975; Endler, 1977). The bounded hybrid superiority model (Moore, 1977) arose from the observation that hybrid zones often are associated with disturbed areas or ecotonal regions (Anderson, 1948; Remington, 1968), suggesting that parental taxa are less well adapted to these regions than certain hybrid genotypes. Hence, the model assumes that hybrids are fitter than both parents in some areas, but are less fit than parents in parental habitats (Anderson, 1949; Moore, 1977). Anderson (1948) stressed the importance of disturbance in creating new environments, but undisturbed habitats also may have unoccupied niches (Walker and Valentine, 1984; Futuyma, 1986).

The mosaic model of hybrid zone structure also is an ecological gradient model. It assumes that different parental genotypes are adapted to different habitats and that a patchy distribution of habitat types gives rise to a patchy distribution of parents (Howard, 1982, 1986; Harrison, 1986). Thus, it is consistent with the common observation that biogeographic regions contain numerous closely related species that have sympatric distributions but different habitat preferences (e.g., Ricklefs, 1976). The hybrid zone between two species of *Gryllus* crickets in New England is a good example (Harrison, 1986, 1990; Rand and Harrison, 1989). Parental genotypes are patchily distributed within the zone and are associated with different soil types: one species is found on sandy soils, the other on loam (Rand and Harrison, 1989). This association is thought to result from exogenous (positive or negative selection resulting from genotype × environment interactions) selection on parental genotypes (Rand and Harrison, 1989). The mosaic model differs from the bounded hybrid superiority model in assuming that hybrids are less fit than parental taxa in all habitats (Harrison, 1986; Rand and Harrison, 1989). However, the presence of open niches, the mosaic distribution of complex habitats, and the observation that hybrid classes may be as fit as, or fitter than, their parents argue against this assumption (for reviews, see Arnold and Hodges, 1995a,b; Arnold, 1997).

A New Model

The tension zone, bounded hybrid superiority, and mosaic models either assume that hybrids are less fit than parents in all habitats or are fitter than parents only in ecotonal regions. One of us (Arnold, 1997) recently has proposed a new model that relies heavily on earlier hypotheses of both botanists and zoologists (e.g., Anderson and Stebbins, 1954; Lewontin and Birch, 1966). Emphasizing the belief that reticulation often leads to the creation of novel evolutionary lineages founded by hybrid genotypes (see figure 29.1), the Evolutionary Novelty model makes the assumption that certain hybrid genotypes can be fitter than their parents in parental habitat as well as in unoccupied niches (figure 29.1; Arnold and Hodges, 1995a; Arnold, 1997). Consequently, it differs from both the tension zone and mosaic models in assuming the existence of relatively fit hybrid genotypes, and from the bounded hybrid superiority model in assuming that these genotypes are not necessarily restricted to ecotones. Another unique aspect of the evolutionary novelty model is the view that even if hybrid populations are established rarely, they can still give rise to novel, long-lived, evolutionary lineages (figure 29.2).

Maladaptive Hybridization and Lineage Founding in Animals

Natural hybridization has been viewed as evolutionarily ephemeral because selection should favor the elimination of those individuals that produce unfit (i.e., less viable or less fertile) hybrid offspring (e.g., Mayr, 1963). However, Arnold and Hodges (1995a) and Arnold (1997) provided several examples of the formation of hybrid lineages in which the earliest stages were marked by a preponderance of organisms with relatively low fitness. Thus, the observation that few hybrids are formed, or that most have reduced fertility, does not necessarily imply that they cannot form a new evolutionary lineage. This can be illustrated by findings from the animal genera *Caledia*, *Drosophila*, and *Gila*.

Caledia captiva

Two decades ago, Shaw (1976) identified an array of chromosomal races in the Australian grasshopper genus *Caledia*. The original description was followed by a series of experiments that led to a detailed understanding of the factors associated with reproductive isolation and genome organization (see Shaw et al., 1990, for a review). A major goal of this research was to describe the evolution of hybrid zones between two of the races, the Moreton and Torresian subspecies (Shaw et al., 1980). Both races have $2n = 24$ (females) or $2n = 23$ (males). Moreton individuals possess metacentric chromosomes with numerous interstitial heterochromatic regions. Torresian individuals have acrocentric chromosomes with little or no heterochromatin (Shaw et al., 1980). In one hybrid zone, the chromosomes changed from being largely Moreton to largely Torresian over a distance of only 200 m (Shaw et al., 1980), apparently because of hybrid breakdown. Although F_1 hybrids were as fertile

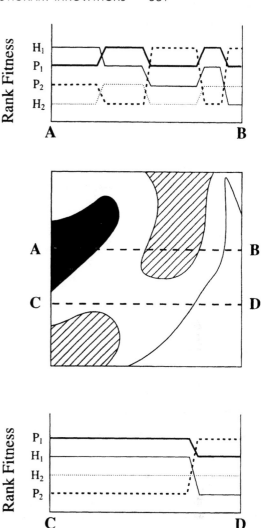

Figure 29.1. An illustration of the ecological component of the evolutionary novelty model. The environment in the central box consists of four habitats, indicated by the shading pattern. A transect across A–B would reveal a pattern of changing rank fitnesses among two parental species (P_1 and P_2) and two hybrid genotypes (H_1 and H_2), as shown in the upper graph. The two parental species and one of the hybrid genotypes (H_1) each has the highest rank fitness in at least one habitat. Hybrid genotype H_2 has low rank fitness in all habitats. A transect across C–D would sample only two of the habitats and would show that both hybrid genotypes had either intermediate fitness or lower fitness than both parental species, as shown in the lower graph. Many other patterns of changing rank fitness are, of course, possible.

and viable as intraspecific offspring, the F_2 generation was completely inviable, and only half of the B_1 offspring were viable (Shaw and Wilkinson, 1980). These patterns seemed to be caused by negative endogenous selection

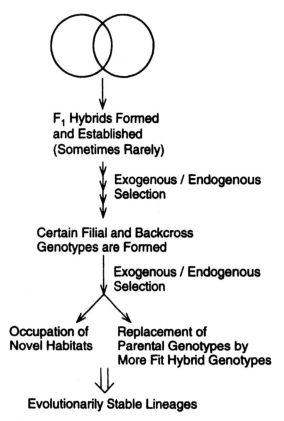

F₁ Hybrids Formed
and Established
(Sometimes Rarely)

Exogenous / Endogenous
Selection

Certain Filial and Backcross
Genotypes are Formed

Exogenous / Endogenous
Selection

Occupation of   Replacement of
Novel Habitats   Parental Genotypes by
                       More Fit Hybrid Genotypes

Evolutionarily Stable Lineages

Figure 29.2. The evolutionary novelty model of hybrid lineage formation.

(selection attributable entirely to genotypic constitution) due to (1) the disruption of gene complexes through abnormal chiasma positioning in F₁s and (2) the mixing of genetic elements from the two subspecies (Coates and Shaw, 1982; Shaw et al., 1982).

Shaw et al. (1990) summarized the findings from studies of chromosome, allozyme, mitochondrial DNA (mtDNA), ribosomal DNA (rDNA), and highly repeated DNA variation across the hybrid zone and in Moreton and Torresian allopatric populations. Allozyme, rDNA, and mtDNA markers characteristic of the Moreton taxon were found from 10 km to more than 16 km away from the present-day contact zone (Shaw et al. 1990), indicating the occurrence of contemporary introgression. However, further analyses discovered an area in which ancient hybridization had apparently established hybrid populations in an extensive geographical area. Moreton mtDNA, allozyme, and rDNA markers were found in Torresian populations located, respectively, 200, 300, and 400 km north of the present-day area of overlap (Shaw et al., 1990). This introgression was thought to have occurred at least 8,000 years ago, when the hybrid zone was located much farther north than it is today (Shaw et al., 1990). Incor-

poration of Moreton genetic elements was attributed to movement of the hybrid zone in response to climatic changes (Shaw et al., 1990). Thus, the high degree of hybrid breakdown in *Caledia* did not prevent the origin of long-standing hybrid lineages.

Drosophila

A number of studies suggest that the frequency of hybridization in *Drosophila* is rare (e.g., Bock, 1984). However, hybridization is not unknown in the genus, and its effects can be of evolutionary importance (Kaneshiro, 1990). One example is the species complex that includes *D. simulans*, *D. sechellia*, and *D. mauritiana* (Solignac and Monnerot, 1986; Aubert and Solignac, 1990). Phylogenetic analysis of this complex revealed that the mtDNA present in some *D. mauritiana* populations was identical to that found in some *D. simulans* samples. Solignac and Monnerot (1986) concluded that mtDNA had introgressed from *D. simulans* into *D. mauritiana*. However, experimental studies demonstrated strong barriers to gene flow between these two species. Offspring from crosses between female *D. mauritiana* and male *D. simulans* were extremely difficult to form (David et al., 1974; Robertson, 1983), and the reciprocal cross resulted in fertile females but sterile males (David et al., 1974; Robertson, 1983; Aubert and Solignac, 1990). Introgressive hybridization apparently has been a factor in the evolutionary history of this *Drosophila* species complex, in spite of strong barriers to interspecific crosses.

Gila

Cyprinid fishes are widely distributed in North America and show a relatively high level of natural hybridization in both the Pacific and Atlantic slope taxa (Hubbs, 1955). Dowling and DeMarais (1993) used nuclear and mitochondrial markers and a phylogenetic approach to examine five closely related and one distantly related species in the Pacific slope genus *Gila*. Earlier morphological studies had identified individuals that possibly were of hybrid origin (Dowling and DeMarais, 1993), and one variant (*G. seminuda*) was determined to be a stabilized, diploid, hybrid species (DeMarais et al., 1992).

Figure 29.3 shows the mtDNA and allozyme phylogenies for *Gila* species and populations (Dowling and DeMarais, 1993). Both trees clearly differ from the accepted taxonomic treatments. For example, *G. robusta* appears to be paraphyletic. Moreover, the two molecular trees are themselves nonconcordant. First, *G. elegans* and *G. seminuda* are placed in a single clade of the mtDNA phylogeny but in two different clades of the allozyme phylogeny. Second, *G. atraria* samples group into a single clade of the allozyme phylogeny, but are placed in two clades—with both *G. robusta* and *G. cypha*—of the mtDNA phylogeny. Dowling and DeMarais (1993)

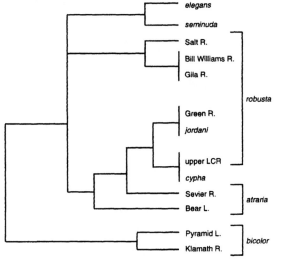

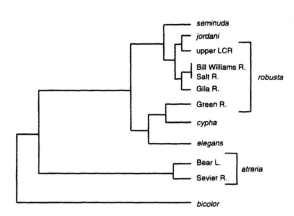

Figure 29.3. Phylogenetic trees for *Gila* based on mtDNA variation (top panel) and allozyme allele frequencies (bottom panel). Species names from the accepted taxonomic treatment are shown in italics; other populations are identified by capture location (from Dowling and DeMarais, 1993).

concluded that the shared sequence variation of *G. atraria* and *G. cypha* reflects relatively ancient introgression, whereas more recent hybridization has caused the patterns of mtDNA variation seen in other taxa (*G. robusta jordani*, *G. seminuda*, *G. cypha*, and *G. elegans*). Although the fitness of interspecific hybrids is not yet known for these species, parental taxa fail to spawn in hatcheries, possibly because of premating isolation mechanisms (Hamman, 1981). Despite this fact, natural hybridization has been, and continues to be, a major factor in the evolution of species complexes in the genus.

Maladaptive Hybridization and Lineage Founding in Plants

Helianthus

There are many instances of natural hybridization that result in the production of relatively few first generation hybrids, most of which have low or very low fertilities (Grant, 1963). The sunflower species *Helianthus annuus* and *H. petiolaris* provide an illuminating example from the plant world. Experimental crosses between these species gave rise to F_1 individuals with pollen fertilities ranging from 0 to 30%, with a mean of 14% (Heiser, 1947). In the next generation, only 1% of $F_1 \times F_1$ crosses and 2% of backcrosses produced seeds (Heiser et al., 1969). Despite the extremely low fitness of F_1, F_2, and backcross individuals, hybridization has given rise to at least three hybrid species (Rieseberg, 1991).

Iris

The ecology, taxonomy, and population genetics of Louisiana irises (series Hexagonae) have been studied for over 60 years (e.g., Viosca, 1935; Riley, 1938; Anderson, 1949; Randolph et al., 1967; Arnold et al., 1990a,b, 1991, 1993; Bennett and Grace, 1990; Arnold, 1992; Arnold and Bennett, 1993; Cruzan and Arnold, 1993, 1994; Carney et al., 1994, 1996; Hodges et al., 1996). These studies continue to be motivated by the observation of extreme morphological and genetic variation in natural populations. This variation was initially thought to reflect extensive taxonomic diversity (Small and Alexander, 1931). However, subsequent ecological and genetic analyses resulted in the recognition of only four species, with the remaining diversity being attributed to various hybrid classes (Viosca, 1935; Riley, 1938). The occurrence of numerous hybrid populations containing a wide array of hybrid phenotypes might suggest that the formation of hybrid populations is relatively easy. However, numerous studies have documented the existence of barriers to gene exchange between parental species. For example, *I. brevicaulis* differs from both *I. fulva* and *I. hexagona* in peak flowering time (Viosca, 1935; Cruzan and Arnold, 1993), causing hybridization to be limited by phenological barriers (Cruzan and Arnold, 1993). *Iris fulva* and *I. hexagona* have largely overlapping flowering periods, but in a mixed population of these two species less than 1% of the seeds examined were found to be hybrids (Arnold et al., 1993). Subsequent experiments documented a major effect of pollen competition on the formation of F_1 individuals (Arnold et al., 1993; Carney et al., 1994, 1996). Despite these barriers, hybrid populations are frequently encountered where the three species overlap (e.g., Randolph et al., 1967), and hybridization between *I. fulva*, *I. hexagona* and *I. brevicaulis* has resulted in a stabilized, diploid species, *I. nelsonii*

(Randolph, 1966; Arnold, 1993). Again, hybridization has led to apparent adaptive evolution (Cruzan and Arnold, 1993, 1994) and evolutionary diversification (Randolph, 1966) in the face of strong barriers to gene exchange between parental species.

The Fitness of Hybrid Genotypes

The evolutionary novelty model described earlier assumes that the rank fitness of parental and hybrid genotypes varies among environments (Arnold and Hodges, 1995a; Arnold, 1997). Some hybrids might be less fit because of the break up of coadapted gene complexes. The reduced fitness of such genotypes would then be due to negative endogenous selection. Other genotypes might be favored, or at least not disfavored, by endogenous selection and be favored by exogenous selection. A number of recent studies have tried to measure hybrid fitness (reviewed in Arnold and Hodges, 1995a; Arnold, 1997), but all have been to some extent inadequate. Problems have included (1) the estimation of individual fitness components, such as survival or fecundity, rather than lifetime fitness; (2) the grouping of individuals into broad classes, such as F_2s, rather than studying individual genotypes; and (3) the absence of estimates from more than one habitat. Despite these problems, estimates of hybrid fitness range from less fit to more fit than their parents (Arnold and Hodges, 1995a; Arnold, 1997).

One way to determine the effects of selection in animals is to perform population cage experiments. Using this approach, Hutter and Rand (1995) estimated fitness differences attributable to introgression of mtDNA haplotypes in *Drosophila persimilis* and *D. pseudoobscura*. Figure 29.4 illustrates the results of competition between the mtDNAs of these species against a background of either the conspecific or the heterospecific nuclear genome. In cages where all individuals contained the nuclear genome of *D. pseudoobscura*, the *D. pseudoobscura* mtDNA haplotype increased significantly with time (figure 29.4, top panel). This result supported the hypothesis that the nuclear and mtDNA genomes of this species are coadapted and that selection acts against hybrid genotypes (Hutter and Rand, 1995). However, of those cages where individuals possessed the *D. persimilis* nuclear genomes, one showed an increase in the *D. pseudoobscura* mtDNA, while the remaining three showed no significant change across generations (figure 29.4, bottom panel). Thus, individuals with *D. pseudoobscura* mtDNA appear to be at least as fit as those with the *D. persimilis* mtDNA haplotype in the *D. persimilis* nuclear background (Hutter and Rand, 1995). A similar conclusion was reached in an earlier experiment involving hybridization between *D. persimilis* and *D. pseudoobscura* (Van Valen, 1963). In all three population cages, *D. persimilis* genotypes were replaced by *D. pseudoobscura* when the flies were maintained at

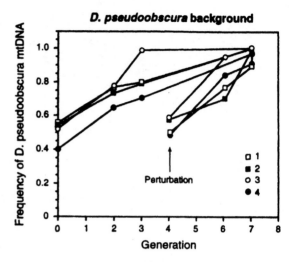

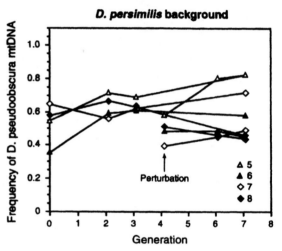

Figure 29.4. Frequencies of *Drosophila pseudoobscura* mtDNA haplotypes across eight generations in population cages. Haplotypes were crossed into either a *Drosophila pseudoobscura* (top panel) or a *D. persimilis* (bottom panel) nuclear background. Additional cages were started after four generations by perturbing haplotype frequencies back to their original values of approximately 0.5 (from Hutter and Rand, 1995).

16°C, but stable hybrid swarms formed at 25°C (Van Valen, 1963). Moreover, the genotypic constitution of these swarms differed among cages. Van Valen (1963) concluded, "The divergent courses followed by the three populations at 25° may well be due to the occurrence of different gene combinations in the three populations, with a resultant divergence in selective trends and available adaptive peaks" (p. 213).

These two studies illustrate the role of selection in the genetic structuring of experimental populations, but do not clearly distinguish between endogenous and exogenous selection. This is problematic, because there are likely to be environmental effects on fitness even at the earliest stages of development. However, some approaches do allow us to draw reasonably secure conclusions about the relative roles of gene interactions (the basis for endogenous selection) and environmental pressures (the basis for exogenous selection).

Endogenous Selection

A series of laboratory experiments with the *Caledia* grasshoppers discussed earlier revealed the existence of both positive and negative selection on various hybrid genotypes. In experimentally produced hybrids, some chromosomal genotypes occurred at a frequency greater than expected, whereas others were found rarely or never (Shaw et al., 1982). Furthermore, certain chromosomes were repeatedly involved in novel rearrangements in laboratory progeny (Shaw et al., 1983). The role of endogenous selection in controlling the viability of these genotypes was inferred from the fact that the same genotypes produced in laboratory crosses were also found in natural hybrid populations (Shaw et al., 1982, 1983). This argues against a purely environment-dependent role for selection. In *Helianthus* sunflowers endogenous selection may also affect the formation of diploid hybrid species. Rieseberg et al. (1995, 1996) mapped the genomes of *H. annuus* and *H. petiolaris*, their hybrid derivative, *H. anomalus*, and experimental hybrids resulting from crosses between the first two species. In experimentally produced backcross progeny very little genetic information (about 2.4%) was exchanged between regions possessing chromosomal structural changes, whereas 40% of the genome from collinear linkage groups introgressed (Rieseberg et al., 1995). The absence of recombination in the remaining 60% of collinear genomic regions was attributed to selection against these recombinants in hybrid individuals (Rieseberg et al., 1995). This conclusion was supported by high levels of concordance between the genomic composition of *H. anomalus* and the three experimental hybrid lineages (Rieseberg et al., 1996). However, there was also evidence for positive endogenous selection on novel combinations of *H. petiolaris* and *H. annuus* genes. Rieseberg et al. (1996) concluded that "The data presented here suggest that certain combinations of alien genes may in fact interact favorably" (p. 744).

Exogenous Selection

The effect of exogenous selection on the fitness of hybrid and parental genotypes has been demonstrated in field studies of Darwin's finches (*Geospiza*) and Louisiana irises. In *Geospiza*, fitness estimates derive from observations of the survivorship and reproductive success of natural hybrids. For the irises they derive from a reciprocal transplant experiment that monitored survivorship, leaf production, and clonal growth. The vegetative growth parameters are likely to be correlated with fitness in irises because larger plants generally produce more flowers and because clonal reproduction is important in the establishment of individual ramets in natural populations (Bennett, 1989; Arnold and Bennett, 1993).

Grant and Grant (1992, 1993, 1994, 1996) documented the effect of a major climatic perturbation (an El Niño event) on the fitness of hybrids between *Geospiza fuliginosa*, *G. fortis*, and *G. scandens*. Prior to the El Niño, hybrids were rarely formed and did not reproduce (Grant and Grant, 1993). After the event, F_1 hybrids (*G. fortis* × *G. fuliginosa* and *G. fortis* × *G. scandens*) and two classes of backcross (*G. fortis* × *fortis/fuliginosa* F_1 hybrids and *G. fortis* × *fortis/scandens* F_1 hybrids) had equal or higher survivorship, recruitment, and breeding success than the three parental species (Grant and Grant, 1992). Further studies indicated that the change in rank fitness of hybrid and parental classes was due to a change in the availability of certain types of seed (Grant and Grant, 1996). Individuals with novel beak morphologies (i.e., various hybrid genotypes) were better able to exploit the new seed types, resulting in more efficient feeding and higher levels of reproduction (Grant and Grant, 1996). Hybridization resulted in an enhancement of genetic variation (Grant and Grant, 1994) and gave rise to genotypes that were more fit in a new habitat. Another conclusion drawn from these analyses was that the likelihood of establishing a new evolutionary lineage may depend on (and be constrained by) the allometries of the hybridizing taxa. When they had similar allometries, hybridization led to strengthened genetic correlations and reduced the likelihood of novel evolutionary trajectories. In contrast, when they had "transposed allometries," genetic correlations could be weakened or eliminated by hybridization, and so lead to the founding of novel lineages (Grant and Grant, 1994).

The *Geospiza* studies support long-standing arguments that parental species may be replaced by hybrid individuals under novel environmental conditions (e.g., Anderson and Stebbins, 1954; Lewontin and Birch, 1966). Another potential consequence of hybridization, not generally considered, is the replacement of a parental species *in its own habitat* by hybrid recombinants (Arnold, 1997). Emms and Arnold (1997) used a reciprocal transplant experiment to estimate the relative fitness of hybrid and parental classes of Louisiana iris in both hybrid and parental habitats. Rhizomes of *I. fulva*, *I. hexagona*, and F_1 and F_2 hybrids were transplanted into four wild populations in southeastern Louisiana. These populations consisted of (1) pure *I. fulva* plants, (2) *I. fulva*-like plants with introgressed *I. hexagona* markers, (3) *I. hexagona*-like plants with introgressed *I. fulva* markers, and (4) pure *I. hexagona* plants. The four sites differed significantly in soil type, light availability, and associated vegetation (Emms and Arnold, 1997).

Figure 29.5 illustrates the pattern of growth revealed by this experiment. There were no significant differences in survivorship among plant classes at any site, and both F_1 and F_2 plants survived at least as well as both parental classes in all sites. Measures of leaf production showed that both hybrid classes performed significantly better than *I. fulva* everywhere except the *I. fulva* site, and that F_2 plants performed significantly better than *I. hexagona* at the *I. fulva* hybrid site (figure 29.5A). Even when differences were not significant, hybrids tended to perform better than parents. Finally, F_1 rhizomes gained as much, or significantly more, mass than either parent after one

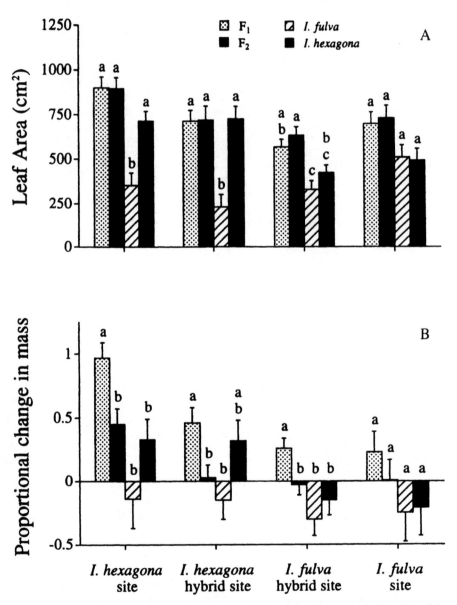

Figure 29.5. Leaf production (A) and the proportional change in rhizome mass (B) of *Iris fulva, I. hexagona*, and two classes of hybrid plants at four sites in southeastern Louisiana. Rhizomes were planted in October 1994; leaf production was measured in April-May 1995; rhizome mass change was measured in November 1995. The results are from analysis of covariance with initial rhizome mass as a covariate. Statistical analyses were performed on a site-by-site basis; plant classes sharing a letter do not differ significantly at $P < 0.05$ in a posteriori comparisons. Error bars are 1 SE of genotype means (from Emms and Arnold, 1997).

year of growth (figure 29.5B). There were no significant differences between F_2 rhizomes and either parent at any site (figure 29.5B). These findings suggest that the fitness of both F_1 and F_2 hybrids is equivalent to, or greater than, that of either parent in both parental and hybrid populations. If this is true, natural hybridization between these species could lead to the replacement of parental forms by hybrid individuals, eventually giving rise to the extensive introgression observed in southern Louisiana. Introgression has been well documented in many groups of organisms (see reviews by Arnold, 1992, 1997; Rieseberg and Wendel, 1993), suggesting that the replacement of parents by introgressed individuals may be a common process in the evolution of numerous lineages.

Paradigm Regained?

We have argued that natural hybridization should be seen as a process with evolutionarily significant consequences. What types of study would increase our knowledge of the process of natural hybridization and its outcomes? First, there is the need for theoretical exploration of more realistic models. Such models should incorporate mosaic habitat structures and allow selection coefficients to differ between hybrid classes and be greater than zero. Second, many more empirical studies are needed to determine the frequency and effects of natural hybridization. Third, phylogenetic analyses should be used to identify both historical and contemporary hybridization. In this regard, it is interesting to note that while botanists typically conclude that phylogenetic incongruence is due to reticulation, zoologists normally invoke incomplete lineage sorting. Arnold (1997) suggested that both groups should "evaluate their findings in light of the distributions and ecology of the species being studied, to determine whether lineage sorting or natural hybridization is the most likely explanation" (p. 183). Possibly the most important, yet least attempted, empirical studies are those that estimate the lifetime fitness of hybrid and parental genotypes in nature. Such studies are crucial if we are to assess the extent to which novel adaptations and new evolutionary lineages really do result from natural hybridization.

References

Anderson, E. 1948. Hybridization of the habitat. Evolution 2:1–9.

Anderson, E. 1949. Introgressive Hybridization. Wiley, New York.

Anderson, E., and Hubricht, L. 1938. Hybridization in *Tradescantia*. III. The evidence for introgressive hybridization. Am. J. Bot. 25:396–402.

Anderson, E., and Stebbins, G. L., Jr. 1954. Hybridization as an evolutionary stimulus. Evolution 8:378–388.

Arnold, M. L. 1992. Natural hybridization as an evolutionary process. Annu. Rev. Ecol. Syst. 23:237–261.

Arnold, M. L. 1993. *Iris nelsonii*: origin and genetic composition of a homoploid hybrid species. Am. J. Bot. 80: 577–583.

Arnold, M. L. 1997. Natural Hybridization and Evolution. Oxford University Press, Oxford.

Arnold, M. L., and Bennett, B. D. 1993. Natural hybridization in Louisiana irises: genetic variation and ecological determinants. In Hybrid Zones and the Evolutionary Process (ed. R. G. Harrison). pp. 115–139. Oxford University Press, Oxford.

Arnold, M. L., and Hodges, S. A. 1995a. Are natural hybrids fit or unfit relative to their parents? TREE 10:67–71.

Arnold, M. L., and Hodges, S. A. 1995b. The fitness of hybrids—a response to Day and Schluter. TREE 10:289.

Arnold, M. L., Bennett, B. D., and Zimmer, E. A. 1990a. Natural hybridization between *Iris fulva* and *I. hexagona*: pattern of ribosomal DNA variation. Evolution 44:1512–1521.

Arnold, M. L., Hamrick, J. L., and Bennett, B. D. 1990b. Allozyme variation in Louisiana irises: a test for introgression and hybrid speciation. Heredity 65:297–306.

Arnold, M. L., Buckner, C. M., and Robinson, J. J. 1991. Pollen mediated introgression and hybrid speciation in Louisiana irises. Proc. Natl. Acad. Sci. USA 88:1398–1402.

Arnold, M. L., Hamrick, J. L., and Bennett, B. D. 1993. Interspecific pollen competition and reproductive isolation in *Iris*. J. Hered. 84:13–16.

Aubert, J., and Solignac, M. 1990. Experimental evidence for mitochondrial DNA introgression between *Drosophila* species. Evolution 44:1272–1282.

Barton, N. H., and Hewitt, G. M. 1985. Analysis of hybrid zones. Annu. Rev. Ecol. Syst. 16:113–148.

Bennett, B. D. 1989. Habitat Differentiation of Iris fulva Ker Gawler, Iris hexagona Walter, and Their Hybrids. PhD thesis, Louisiana State University, Baton Rouge.

Bennett, B. D., and Grace, J. B. 1990. Shade tolerance and its effect on the segregation of two species of Louisiana iris and their hybrids. Am. J. Bot. 77:100–107.

Blair, W. F. 1955. Mating call and stage of speciation in the *Microhyla olivacea-M. carolinensis* complex. Evolution 9:469–480.

Bock, I. R. 1984. Interspecific hybridization in the genus *Drosophila*. Evol. Biol. 18:41–70.

Carney, S. E., Cruzan, M. B., and Arnold, M. L. 1994. Reproductive interactions between hybridizing irises: analyses of pollen tube growth and fertilization success. Am. J. Bot. 81:1169–1175.

Carney, S. E., Hodges, S. A., and Arnold, M. L. 1996. Effects of pollen-tube growth and ovule position on hybridization in the Louisiana irises. Evolution 50:1871–1878.

Coates, D. J., and Shaw, D. D. 1982. The chromosomal component of reproductive isolation in the grasshopper *Caledia captiva*. I. Meiotic analysis of chiasma distribution patterns in two chromosomal taxa and their F_1 hybrids. Chromosoma 86:509–531.

Cruzan, M. B., and Arnold, M. L. 1993. Ecological and genetic associations in an *Iris* hybrid zone. Evolution 47:1432–1445.

Cruzan, M. B., and Arnold, M. L. 1994. Assortative mating and natural selection in an *Iris* hybrid zone. Evolution 48:1946–1958.

Darwin, C. 1872. On the Origin of Species by Means of Natural Selection or the Preservation of Favored Races in the Struggle for Life. Prometheus Books, Buffalo, N.Y.

David, J., Lemeunier, F., Tsacas, L., and Bocquet, C. 1974. Hybridation d'une nouvelle espèce, *Drosophila mauritiana* avec *D. melanogaster* et *D. simulans*. Ann. Genet. 17:235–241.

DeMarais, B. D., Dowling, T. E., Douglas, M. E., Minckley, W. L., and Marsh, P. C. 1992. Origin of *Gila seminuda* (Teleostei: Cyprinidae) through introgressive hybridization: implications for evolution and conservation. Proc. Natl. Acad. Sci. USA 89:2747–2751.

Dobzhansky, T. 1937. Genetics and the Origin of Species. Columbia University Press, New York.

Dobzhansky, T. 1940. Speciation as a stage in evolutionary divergence. Am. Nat. 74:312–321.

Dowling, T. E., and DeMarais, B. D. 1993. Evolutionary significance of introgressive hybridization in cyprinid fishes. Nature 362:444–446.

Emms, S. K., and Arnold, M. L. 1997. The effect of habitat on parental and hybrid fitnesss: reciprocal transplant experiments with Louisiana irises. Evolution 51:1112–1119.

Endler, J. A. 1977. Geographic Variation, Speciation, and Clines. Princeton University Press, Princeton, N.J.

Futuyma, D. J. 1986. Evolutionary Biology. Sinauer, Sunderland, Mass.

Grant, B. R., and Grant, P. R. 1993. Evolution of Darwin's finches caused by a rare climatic event. Proc. Royal Soc. Lond. B 251:111–117.

Grant, B. R., and Grant, P. R. 1996. High survival of Darwin's finch hybrids: effects of beak morphology and diets. Ecology 77:500–509.

Grant, P. R., and Grant, B. R. 1992. Hybridization of bird species. Science 256:193–197.

Grant, P. R., and Grant, B. R. 1994. Phenotypic and genetic effects of hybridization in Darwin's finches. Evolution 48:297–316.

Grant, V. 1963. The Origin of Adaptations. Columbia University Press, New York.

Hamman, R. L. 1981. Hybridization of three species of chub in a hatchery. Progr. Fish-Cult. 43:140–141.

Harrison, R. G. 1986. Pattern and process in a narrow hybrid zone. Heredity 56:337–349.

Harrison, R. G. 1990. Hybrid zones: windows on evolutionary process. Oxford Surv. Evol. Biol. 7:69–128.

Heiser, C. B., Jr. 1947. Hybridization between the sunflower species *Helianthus annuus* and *H. petiolaris*. Evolution 1:249–262.

Heiser, C. B., Jr. 1949. Natural hybridization with particular reference to introgression. Bot. Rev. 15:645–687.

Heiser, C. B., Jr. 1973. Introgression re-examined. Bot. Rev. 39:347–366.

Heiser, C. B., Jr., Smith, D. M., Clevenger, S. B., and Martin, W. C., Jr. 1969. The North American sunflowers (*Helianthus*). Mem. Torrey Bot. Club 22:1–213.

Hodges, S. A., Burke, J., and Arnold, M. L. 1996. Natural formation of *Iris* hybrids: experimental evidence on the establishment of hybrid zones. Evolution 50:2504–2509.

Howard, D. J. 1982. Speciation and coexistence in a group of closely related ground crickets. Ph.D. Dissertation, Yale University, New Haven, Conn.

Howard, D. J. 1986. A zone of overlap and hybridization between two ground cricket species. Evolution 40:34–43.

Hubbs, C. L. 1955. Hybridization between fish species in nature. Syst. Zool. 4:1–20.

Hutter, C. M., and Rand, D. M. 1995. Competition between mitochondrial haplotypes in distinct nuclear genetic environments: *Drosophila pseudoobscura* vs. *D. persimilis*. Genetics 140:537–548.

Kaneshiro, K. Y. 1990. Natural hybridization in *Drosophila*, with special reference to species from Hawaii. Can. J. Zool. 68:1800–1805.

Key, K. H. L. 1968. The concept of stasipatric speciation. Syst. Zool. 17:14–22.

Lewontin, R. C., and Birch, L. C. 1966. Hybridization as a source of variation for adaptation to new environments. Evolution 20:315–336.

Lotsy, J. P. 1916. Evolution by Means of Hybridization. M. Nijhoff, The Hague, Netherlands.

Lotsy, J. P. 1931. On the species of the taxonomist in its relation to evolution. Genetica 13:1–16.

May, R. M., Endler, J. A., and McMurtrie, R. E. 1975. Gene frequency clines in the presence of selection opposed by gene flow. Am. Nat. 109:659–676.

Mayr, E. 1942. Systematics and the Origin of Species. Columbia University Press, New York.

Mayr, E. 1963. Animal Species and Evolution. Belknap Press, Cambridge, Mass.

Mayr, E. 1992. A local flora and the biological species concept. Am. J. Bot. 79:222–238.

Moore, W. S. 1977. An evaluation of narrow hybrid zones in vertebrates. Q. Rev. Biol. 52:263–277.

Paterson, H. E. H. 1985. The recognition concept of species. In Species and Speciation (ed. E.S. Vrba), pp. 21–29. Transvaal Museum Monograph No. 4, Transvaal Museum, Pretoria.

Rand, D. M., and Harrison, R. G. 1989. Ecological genetics of a mosaic hybrid zone: mitochondrial, nuclear, and reproductive differentiation of crickets by soil type. Evolution 43:432–449.

Randolph, L. F. 1966. *Iris nelsonii*, a new species of Louisiana iris of hybrid origin. Baileya 14:143–169.

Randolph, L. F., Nelson, I. S., and Plaisted, R. L. 1967. Negative evidence of introgression affecting the stability of Louisiana *Iris* species. Corn. Univ. Agric. Exp. Stat. Mem. 398:1–56.

Remington, C. L. 1968. Suture-zones of hybrid interaction between recently joined biotas. Evol. Biol. 2:321–428.

Ricklefs, R. E. 1976. The Economy of Nature. Chiron Press, Portland, Ore.

Rieseberg, L. H. 1991. Homoploid reticulate evolution in *Helianthus* (Asteraceae): evidence from ribosomal genes. Am. J. Bot. 78:1218–1237.

Rieseberg, L. H., and Wendel, J. F. 1993. Introgression and its consequences in plants. In Hybrid Zones and the Evolutionary Process (ed. R.G. Harrison), pp. 70–109. Oxford University Press, Oxford.

Rieseberg, L. H., Soltis, D. E., and Palmer, J. D. 1988. A molecular reexamination of introgression between *Helianthus annuus* and *H. bolanderi* (Compositae). Evolution 42: 227–238.

Rieseberg, L. H., Linder, C. R., and Seiler, G. J. 1995. Chromosomal and genic barriers to introgression in *Helianthus*. Genetics 141:1163–1171.

Rieseberg, L. H., Sinervo, B., Linder, C. R., Ungerer, M. C., and Arias, D. M. 1996. Role of gene interactions in hybrid speciation: evidence from ancient and experimental hybrids. Science 272:741–745.

Riley, H. P. 1938. A character analysis of colonies of *Iris fulva*, *Iris hexagona* var. *giganticaerulea* and natural hybrids. Am. J. Bot. 25:727–738.

Robertson, H. M. 1983. Mating behavior and the evolution of *Drosophila mauritiana*. Evolution 37:1283–1293.

Shaw, D. D. 1976. Population cytogenetics of the genus *Caledia* (Orthoptera: Acridinae) I. Inter- and intraspecific karyotype diversity. Chromosoma 54:221–243.

Shaw, D. D., and Wilkinson, P. 1980. Chromosome differentiation, hybrid breakdown and the maintenance of a narrow hybrid zone in *Caledia*. Chromosoma 80:1–31.

Shaw, D. D., Moran, C., and Wilkinson, P. 1980. Chromosomal reorganization, geographic differentiation and the mechanism of speciation in the genus *Caledia*. In Insect Cytogenetics (eds. R. L. Blackman, G. M. Hewitt, and M. Ashburner), pp. 171–194. Blackwell, Oxford.

Shaw, D. D., Wilkinson, P., and Coates, D. J. 1982. The chromosomal component of reproductive isolation in the grasshopper *Caledia captiva*. II. The relative viabilities of recombinant and non-recombinant chromosomes during embryogenesis. Chromosoma 86:533–549.

Shaw, D. D., Wilkinson, P., and Coates, D. J. 1983. Increased chromosomal mutation rate after hybridization between two subspecies of grasshoppers. Science 220:1165–1167.

Shaw, D. D., Marchant, A. D., Arnold, M. L., Contreras, N., and Kohlmann, B. 1990. The control of gene flow across a narrow hybrid zone: a selective role for chromosomal rearrangement? Genome 68:1761–1769.

Slatkin, M. 1973. Gene flow and selection in a cline. Genetics 75:733–756.

Small, J. K., and E. J. Alexander. 1931. Botanical interpretation of the iridaceous plants of the gulf states. Cont. New York Bot. Gard. 327:325–357.

Solignac, M., and Monnerot. M. 1986. Race formation, speciation, and introgression within *Drosophila simulans*, *D. mauritiana*, and *D. sechellia* inferred from mitochondrial DNA analysis. Evolution 40:531–539.

Soltis, P. S., Plunkett, G. M., Novak, S. J., and Soltis, D. E. 1995. Genetic variation in *Tragopogon* species: additional origins of the allotetraploids *T. mirus* and *T. miscellus* (Compositae). Am. J. Bot. 82:1329–1341.

Stebbins, G. L., Jr. 1950. Variation and Evolution in Plants. Columbia University Press, New York.

Stebbins, G. L., Jr. 1959. The role of hybridization in evolution. Proc. Am. Philos. Soc. 103:231–251.

Van Valen, L. 1963. Introgression in laboratory populations of *Drosophila persimilis* and *D. pseudoobscura*. Heredity 18:205–214.

Viosca, P., Jr. 1935. The irises of southeastern Louisiana: a taxonomic and ecological interpretation. Bull. Am. Iris Soc. 57:3–56.

Wagner, W. H., Jr. 1969. The role and taxonomic treatment of hybrids. BioScience 19:785–795.

Wagner, W. H., Jr. 1970. Biosystematics and evolutionary noise. Taxon 19:146–151.

Walker, T. D., and Valentine. J. W. 1984. Equilibrium models of evolutionary species diversity and the number of empty niches. Am. Nat. 124:887–899.

30

Mimicry and Warning Color at the Boundary between Races and Species

James Mallet
W. Owen McMillan
Chris D. Jiggins

Can Mimicry Cause Speciation?

Mimicry in butterflies is well known chiefly because it gives an easily understood visual example of adaptation. However, the major thrust of Bates's original paper "Contributions to an Insect Fauna of the Amazon Valley" (1862) concerned systematics and the origin of species, rather than merely mimicry. Bates felt that mimicry in ithomiine, heliconiine, and dismorphiine butterflies exemplified the continuum of geographic divergence and speciation by natural selection: "It is only by the study of variable species that we can obtain a clue to the explanation of the rest. But such species must be studied in nature, and with strict reference to the *geographic relations* of their varieties" (pp. 501–2). The geographic pattern Bates discovered was extraordinary: not only were there resemblances between species within any one area of the Amazon basin, but also the mimetic color patterns themselves changed every 100–200 miles. Repeated resemblances in different areas, using different color patterns, provided the highly convincing comparative evidence that clinched Bates's hypothesis of mimicry (see appendix 30.1 for discussion of Batesian vs. Müllerian mimicry). On top of this geographic divergence, closely related species within an area often belonged to different mimicry "rings" (see also Papageorgis 1975; Turner 1976; Mallet and Gilbert 1995). Bates's system had all the intermediate stages between local varieties, geographic races, and sympatric species.

Darwin (1863) wrote a glowing review of Bates's paper: "It is hardly an exaggeration to say, that whilst reading and reflecting on the various facts given in this memoir, we feel to be as near witnesses, as we can ever hope to be, of the creation of new species on this earth" (p. 92). Today, it may seem strange to propose that mimicry, normally viewed as an adaptation within species, should trigger speciation. On reflection, this idea is not so strange. "Isolating mechanisms" are divergently selected traits that must evolve initially within species, and reproductive isolation is usually, perhaps, a pleiotropic effect of environmental or genomic adaptation rather than directly selected. Selection for any ecological adaptation, such as mimicry, as well as any change directly promoting hybrid inviability, sterility, or mate choice, may instigate speciation.

Mimicry and warning color are in fact particularly good examples of traits that could maintain the separateness of species (i.e., "postmating isolating mechanisms"). Once a warning color pattern becomes abundant, the local predator community learns to avoid it. This favors the common pattern and causes frequency-dependent selection against rarer patterns. Rare hybrids and recombinants between divergent color patterns are not recognized as unpalatable and will form an adaptive trough between two adaptive peaks. A major difference between mimicry and traits more normally associated with speciation is of course that the agent of mimetic selection is ecological (predation) whereas postmating isolation is often genomic and independent of the environment. But, in terms of overall fitness, warning color selection and classical genomic incompatibility have much in common. Mimicry might therefore be an important cause of speciation in mimetic organisms.

We here explore selection at the boundary between geographic varieties and species in butterflies of the genus *Heliconius*. Armed with a criterion for species status, we ask which traits diverge before and after this criterion is reached. We use our data to test the idea put forward by Henry Walter Bates that mimicry and other ecological factors are important in speciation.

The Boundary between Races and Species

To study speciation, it helps to clarify what species are. Darwin and Wallace traveled widely and developed clear ideas of the continuum between geographic races and species: "independently of blending from intercrossing, the complete absence, in a well-investigated region, of varieties linking together any two closely-allied forms, is probably the most important of all the criterions of their specific distinctness" (Darwin 1871, pp. 214–5). Bates, like his traveling companion Wallace (1865), agreed with Darwin's view that species and geographic races formed a continuum, rather than conveniently delimiting themselves for us because of fundamental or essential differences: "distinct forms or species do not essentially differ from the undoubted varieties of the species cited" (Bates 1862, p. 501). A species differed from a race only in that it could coexist in sympatry without losing its integrity: "The new species cannot be proved to be established as such, unless it be found in company with a sister form which has had a similar origin, and maintaining itself perfectly distinct from it" (p. 530). By concentrating on the resultant distinctness rather than on mechanisms for its maintenance, Bates used a nonessentialist criterion of species (see appendix 30.2 for a discussion of essentialism in modern species concepts).

In practice, taxonomists still use something close to this operational criterion. This approach is often derided as the "morphological species concept" (e.g., Mayr 1982). But, after adding Mendelian genetics to this Darwinian cluster criterion, species can be seen as separate "genotypic clusters" that can coexist in sympatry (Mallet 1995a, 1996a,b; see also Feder, this volume). Genotypic clusters can be investigated in a local population sample (Darwin's "well-investigated region") characterized for a number of independent loci or independently inherited characters. If a single sample contains, at four recombining loci A, B, C, and D, one $AaBBCCDd$, 50 $AABBCCDD$, and 50 $aabbccdd$ genotypes, it is clear that there are two major genotypic clusters, plus a single individual intermediate between these clusters. Darwin's criterion implies two separate species. On the other hand, if a single population with the same overall allele frequencies gives mostly $AABbCcDD$, $AaBBCcDd$, $AABbCcDd$, and other intermediate forms, the evidence points to a single polymorphic species. Geographic variation across a species range will also, of course, give rise to different genotypic clusters in different areas, but the forms are viewed as geographic races if intermediates predominate in areas of overlap. A genotypic cluster definition makes reference neither to past or future evolution, nor to inferred phylogeny, nor to processes by which clusters are maintained (while not denying their existence); these are left open as interesting areas for investigation, since they form material for the study of speciation. Speciation then simply becomes the evolution of the tendency for genotypic clusters to coexist as recognizably distinct forms where they overlap. Recognizability might involve morphology, behavior, physiology, or biochemistry, as long as these are multiple, independently inherited traits. Some readers may have philosophical grounds for disagreeing with this and may attempt to define species using the "underlying essence" of species that Darwin, Wallace, and Bates were so keen to avoid (appendix 30.2; see also chapters by de Queiroz, Harrison, Shaw, and Templeton, this volume). However, the explicitness and simplicity of a genotypic cluster criterion allow clear questions to be posed about the causes of speciation.

Bates believed (and Darwin agreed) that there was evidence for causes of speciation in the continuum between varieties and species of mimetic butterflies. In addition to mimicry, adaptation to abiotic conditions played a role in geographic divergence: "The selecting agent, which acts in each locality by destroying the variations unsuitable to the locality, would not in these cases be the same as in *Leptalis* [i.e. mimicry]; it may act, for anything we know, on the larvae; in other respects, however, the same law of nature appears, namely, the selection of one or two distinct varieties by the elimination of intermediate gradations" (Bates 1862, p. 514). Bates also hypothesized that mate preferences were important in speciation:

> The process of the creation of a new species I believe to be accelerated in the *Ithomiae* and allied genera by the strong tendency of the insects, when pairing, to select none but their exact counterparts: this also enables a number of very closely allied ones to exist together, or the representative forms to live side by side on the confines of their areas, without amalgamating. (p. 501)

Bates had only scattered field evidence (and the systematics of these butterflies was then very crude) for his assertion about the causes of speciation, but we show in this chapter that he was largely correct for the *Heliconius* we have studied. As Bates suggested, assortative mating, coupled with divergence in warning color and ecology, triggers speciation.

Geographic Diversity of *Heliconius* Races

Heliconius are renowned not only for mimicry, but also for geographic diversity of mimetic patterns within species. In *Heliconius erato* and its Müllerian comimic *H. melpomene* this "subspeciation" has reached a feverish pitch, with 28 subspecies recognized in *erato* and 29 in *melpomene* across Central and South America (Brown 1979). Subspecies in *Heliconius* are usually monomorphic across their range and are separated from other subspecies by more or less narrow hybrid zones. A few hybrid zones are up to 200 km wide between races that differ only slightly in color pattern (e.g., Brown and Mielke 1972), and some Amazonian forms of *erato* and *mel-*

pomene have such wide bands of polymorphism that the "pure" races form little more than the ends of broad clines (Brown et al. 1974; Brown 1979). More normally, hybrid zones between forms that differ at one or two major pattern elements are about 20–80 km wide (Turner 1971a; Benson 1982; Mallet 1986); those between subspecies that differ at three or more major color pattern elements may be as little as 10 km wide (figure 30.1; Brown and Mielke 1972; Mallet et al. 1990).

Heliconius subspecies are unusual, perhaps, in that they are what they seem: they differ chiefly at the very traits we use for identification, their color pattern. For example, subspecies differ very little at allozyme loci, and Nei's genetic distances between races vary between D ≈ 0.01 and 0.05 (Turner et al. 1979; Jiggins et al. 1997a). Mitochondrial DNA (mtDNA) studies reveal two major clades within *H. erato* that diverged about 1.5–2 million years ago (Brower 1994a), but within each clade there is little patterning of the mtDNA genealogy with respect to color pattern races. Closely related mtDNA sequences are as likely to be found thousands of miles apart as in the same race (Brower 1994a, 1996a). Even major color pattern differences are inherited at a mere handful of loci (Sheppard et al. 1985; Mallet 1989; Jiggins et al. 1996; Jiggins and McMillan 1997). The ecologies of races are similar within each species; they are all denizens of sunny areas such as river edges, major tree falls, and man-made second growth. *Heliconius erato* lays eggs mainly on species of *Passiflora* in the subgenus *Plectostemma*, while its mimic *melpomene* specializes on subgenus *Granadilla* (Benson et al. 1976; Smiley 1978; Benson 1978). Some *erato* races do have minor differences in ecology or host plants (Benson 1978, 1982), but the differences are much greater between species.

It is not yet clear how geographic differentiation evolved in the face of stabilizing selection on warning color. One theory supposes that color patterns diverged in forest refuges formed during glacial maxima (Brown et al. 1974; Brown 1979, 1982, 1987; Turner 1971b, 1981, 1982), but geographic isolation would not be required if there were direct adaptation of particular color patterns to local mimetic (Turner 1982) or abiotic environments (Bates 1862; Benson 1982; Endler 1982) or if differentiation was initiated by genetic drift (Mallet 1986, 1993). These alternative geographic scenarios are the subject of much current argument (Mallet 1993; Brower 1996a; Turner and Mallet 1996; Mallet et al. 1996; Mallet and Turner 1998).

Selection in a Hybrid Zone between Races of *Heliconius erato*

More important for understanding the evolution of new species, hybrid zones between subspecies give information on the nature and strength of selection acting on geographically varying traits (Hewitt 1988; Barton and Hewitt 1989; Harrison 1990). The width, *w*, of a cline in allele frequency at a single locus (*w* = [maximum gradient of the cline]$^{-1}$) at equilibrium will be proportional to a ratio of migration and selection: $w \approx K\sigma/\sqrt{s}$, where σ is the dispersal distance, *s* is the selection acting on the cline, and *K* is a constant depending on the type of selection (Barton and Gale 1993). In *Heliconius*, frequency-dependent predation will remove rare color pattern variants on either side of the hybrid zone, giving $K \approx \sqrt{8}$ for a dominant gene and $K \approx \sqrt{12}$ for a codominant gene (Mallet and Barton 1989a).

To estimate selection, data were gathered from the hybrid zone between *Heliconius erato* races near Pongo de Cainarache, Peru. Across this zone, one codominant (D^{Ry}) and two dominant (Cr, Sd) unlinked loci determine the major color pattern differences, and we can therefore compare the theoretical prediction of cline width with field observations (figure 30.2; Mallet et al. 1990). The cline at

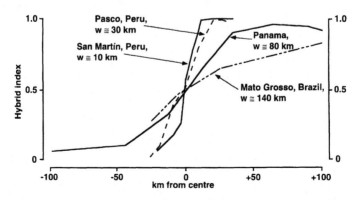

Figure 30.1. Hybrid zones between races of *Heliconius erato*. Each line represents an average "hybrid index" score based on one to four color pattern loci or characters between different races. Sources: Mato Grosso, Brazil (Brown and Mielke 1972); Panama (Mallet 1986); Pasco, Peru (Mallet, unpublished); and Pongo de Cainarache, San Martín, Peru (Mallet et al. 1990).

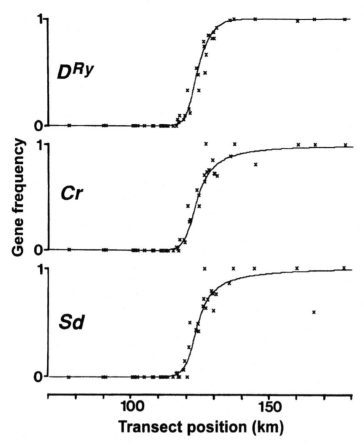

Figure 30.2. Clines making up the hybrid zone between *H. erato favorinus* (left) and *H. erato emma* (right) centered at Pongo de Cainarache, Peru. Top, symmetric cline at codominant locus D^{Ry}. The D^{Ry} allele adds rays and changes red band to yellow, and the cline shape fits the codominant model much better than the equivalent asymmetric recessive or dominant models ($P < 0.001$ in both cases). Middle, cline at dominant locus Cr, showing a long tail of introgression to the right of the hybrid zone, where the recessive allele escapes detection. In contrast, there is virtually no introgression of the highly visible dominant allele to the left of the hybrid zone. The recessive *cr* allele adds a yellow hindwing bar. Bottom, cline at dominant locus Sd. Again, there is a long tail of introgression to the right of the hybrid zone, and no tail to the left, as expected for a dominant gene. The Sd allele changes the shape of the forewing band from broad to narrow, and interacts with Cr to remove the full yellow hindwing bar. Both dominant loci fit asymmetric dominant clines significantly better than symmetric codominant clines ($P < 0.001$ and $P < 0.005$, respectively). The theoretically derived curves were fitted using likelihood to field data based on 1,572 individuals caught at 53 field sites (reprinted with permission from Mallet et al. 1990).

D^{Ry} is about 8.5 km wide, and the Cr and Sd clines are about 10.2 km wide (figure 30.2). This gives estimated ratios $\sigma/\sqrt{s_C} \approx 2.45$ km for the codominant D^{Ry}, and $\sigma/\sqrt{s_D} \approx 3.61$ km for the dominant Cr and Sd, so the ratio $s_C/s_D \approx 2.17$. In words, the codominant locus D^{Ry} is about twice as heavily selected (this is expected, since it has a greater phenotypic effect) as the two dominant loci Cr and Sd. If we knew the migration rate, σ, we would be able to estimate the absolute strength of selection, s, or vice versa.

The absolute strength of selection can be estimated indirectly, using gametic correlations (Barton and Gale 1993), or via direct experiments in the field. The first method is possible because migration across hybrid zones of a given width creates gametic correlations (also known as linkage disequilibria) between loci. Pairwise correlations between two genes A and B in the center of a hybrid zone are expected to be $R_{AB} \approx 4\sigma^2/c_{AB}w_Aw_B$ at equilibrium between selection and migration, where c_{AB} is the recombination fraction between loci (Barton and Gale 1993). In *H. erato* the correlations between color pattern loci ($R \approx 0.35$) and average cline width ($w \approx 9.6$ km) are known, and all three loci are unlinked ($c = 0.5$;

Mallet et al. 1990). Therefore, an estimate of dispersal is $\sigma \approx \sqrt{0.35 \times 0.5 \times 9.6 \times 9.6 \div 4} \approx 2.0$ km. Using the migration/dispersal ratios above gives per generation selection of $s_D \approx 0.31$, and $s_c \approx 0.67$. These high values of selection are somewhat dubious because the analytical theory depends on weak selection approximations. Simulations were used to circumvent the approximation problem, and suggest that the selection pressures were somewhat lower, $s \approx 0.23$ per locus on average, although still relatively intense (Mallet et al. 1990).

These estimates of selection can be compared with the results of a laborious field study using *H. erato*. Experimental butterflies were transported in both directions across the Pongo de Cainarache hybrid zone and released in four sites where the alternative color pattern was fixed. Controls with the local color pattern were collected from similar distances away from the release site and released alongside the experimentals, and their life expectancies were compared. Experimentals survived only 48% as long as controls (Mallet and Barton 1989b). Assuming multiplicative fitnesses and that $s_C/s_D \approx 2.17$ (above), we find a per locus selection of $s_C \approx 0.33$, with support limits (approximately equivalent to 95% confidence limits) of 0.14–0.51 for the codominant D^{Ry}, and $s_D \approx 0.15$ (0.07–0.23) for the dominant loci Cr and Sd. These field results therefore agree tolerably well with the analysis of gametic correlations.

It is conceivable that this selection maintaining the hybrid zone was caused by adaptation to abiotic or general ecological conditions, or by classical genomic incompatibility. However, there is compelling evidence that the strong selection was due mainly to warning color: (1) Beak marks of jacamars (*Galbula* spp., a bird that specializes on butterflies and other large insects) and other birds were found on the butterflies. More wing damage was found on experimentals than controls, showing that experimental butterflies were attacked, and suggesting (though not proving) a greater attack rate on foreign patterns. The strongest survival differences were found where jacamars were commonest (Mallet and Barton 1989b). (2) All three *erato* color pattern loci switch together in Pongo de Cainarache (figure 30.2). (3) The equivalent *melpomene* hybrid zone is also at an almost identical position, and disequilibrium-based estimates of selection ($s \approx 0.25$) are very similar to those in *erato*. The conjunction is expected since the two species have nearly identical patterns and should be under similar selection (Mallet et al. 1990). (4) Visual dominance of the pattern genes within *erato* correctly predicts the shapes of the clines. Genetic dominance should cause cline asymmetry because rare recessive alleles on the dominant side of the cline are mostly found as heterozygotes with the dominant phenotype, resulting in low selection and a long shallow "tail" of allele frequency on that side; rare dominants on the recessive side of the cline, on the other hand, are highly detectable, and will be strongly selected against, resulting in a steep gradient and no tail. As expected, a

theoretical codominant cline, which is symmetric, fits the visually codominant locus D^{Ry} highly significantly better than does a dominant cline, which is asymmetric. The visually dominant Cr and Sd show long tails on the dominant side of the cline, and are fitted by asymmetric dominant clines highly significantly better than by symmetric codominant clines (figure 30.2; Mallet et al. 1990). This is very strong evidence that visual selection acts directly on the color pattern phenotypes. (5) There are no allozyme or mtDNA differences across the hybrid zone (Nei's $D \approx 0.001$, Mallet, unpublished allozyme results; mtDNA results, Brower 1996a), suggesting a general lack of genomic differentiation. (6) Breeding experiments indicate full fertility and viability of hybrids, backcrosses, F2, and further hybrid generations (Mallet 1989). (7) The codominant locus D^{Ry} is at Hardy-Weinberg equilibrium, based on a powerful overall test of 883 individuals from 14 polymorphic populations, suggesting random mating and an absence of hybrid inviability in the field (see Mallet et al. 1990). Nor was there any evidence for assortative mating in crosses (Mallet 1989). (8) Ecology and host choice of the two forms of *erato* do not differ (Mallet and Barton 1989b). (9) The 52% reduction in longevity of experimental nonmimics agrees well with that found in other field studies of *Heliconius* mimicry: 22% in *erato* (Benson 1972) and 64% in *H. cydno* (Kapan 1998).

Heliconius Species That Hybridize

Hybrid zones like those within *erato* or *melpomene* have abundant hybrids, so the races have demonstrably not speciated. However, if hybrids are rare enough in areas of overlap, separate clusters of genotypes may be recognized. Using the existence of recognizable genotypic clusters as the criterion for the transition to species, comparisons between these two types of hybridization can lead to an understanding of speciation.

Natural interspecific hybrids are known between 25–28% of all *Heliconius* species (table 30.1). This is higher than the known fraction of hybridizing species among the world's birds (9% of species; see Grant and Grant 1992) or European butterflies (12%; see Guillaumin and Descimon 1976). However, American warblers (Parulinae) are known to hybridize almost as frequently (24% of species), perhaps because hybrids are readily noticed in this brightly colored subfamily (Curson et al. 1994). *Heliconius* are not only brightly colored; their hybrids are also highly prized by collectors.

Hybrid Zones between *H. erato* and *H. himera*

Although species that hybridize are common, hybrid individuals are very rare in comparison to their parent species. In *Heliconius*, hybrids between the sympatric

Table 30.1. Natural and laboratory hybridization between species of *Heliconius*.

Species 1	Species 2	Source[1]	Natural Hybrids[2]	Geographic Relationships	Laboratory Hybrids[2]	Assortative Mating[3]	Inviability or Sterility
numata	*melpomene*	2,6,9	+	Sympatric	−	(+)	?
ethilla	*melpomene*	2,9	+	Sympatric	−	(+)	?
ethilla	*numata*	1	+	Sympatric	−	(+)	?
ethilla	*heurippa*	1,9	+	Sympatric	−	(+)	?
ethilla	*besckei*	6	+	Sympatric	−	(+)	?
cydno	*melpomene*	2–6,9	+++	Sympatric	+++	++	Hybrid males are viable and fertile. F1 females are viable but sterile
heurippa	*melpomene*	4	+	Sympatric	−	(+)	?
pachinus	*cydno*	6,8	+	Parapatric	+++	+	All hybrids viable and fertile
ismenius	*cydno*	8	−	Sympatric	++	+++	Hybrids viable, F1 females sterile, F1 males have mechanical difficulties during mating
hecale	*melpomene*	8	−	Sympatric	+	+++	? (Only one laboratory hybrid known)
clysonymus	*hecalesia*	8,9	−	Sympatric	−	(+)	?
himera	*erato*	7,9	+++	Parapatric	+++	+	No detectable inviability/sterility; F1 males may have reduced mating propensity in one direction of cross

Hybrids between 11 species, or 28% of the 39 *Heliconius* species (*sensu* Brown 1979), have been collected. If we combine all allopatric *cydno* group species (including *pachinus, heurippa, timareta,* and *tristero*) into *cydno,* this reduces the number to 9 hybridizing out of 34 total species, or 25%.

[1]Sources: 1. Brown (1976); 2. Ackery and Smiles (1976); 3. Brown and Fernandez Yepez (1985); 4. Salazar (1993); 5. Posla-Fuentes (1993); 6. Holzinger and Holzinger (1994); 7. Jiggins et al. (1996); 8. L. E. Gilbert (personal communication); 9. W. Neukirchen (personal communicatiion).

[2]Frequency of hybrids: − = not known; + = 1–3 specimens known, ++ = 4–10 individuals known, +++ = over 10 individuals known.

[3]Assortive mating: +++ = extremely strongly assortative (<1% of trials when both sexes of both species kept together in laboratory); ++ = strongly assortive (~1% of trials), + = assortive (~10% of trials), (+) = assumed assortative; field hybrids are very rare.

species *H. cydno* and *H. melpomene* have been among the most regularly collected (table 30.1), but even they must occur at frequencies of ≤ 0.1% in nature compared with the pure forms. Hybrids between the parapatric *H. erato* and *H. himera* are much commoner, but contact zones are so narrow that they were discovered only within the last 20 years. Three *erato* races are known to abut with *himera*. In all three contacts, hybridization is infrequent but regular, and hybrids form about 10% of the population in an Ecuadorean overlap (figure 30.3; Descimon and Mast de Maeght 1984; Jiggins et al. 1996; Mallet et al. 1998). Because most butterflies in this hybrid zone population are parentals, even at cryptic allozyme loci (figure 30.3; Jiggins et al. 1997a), two genotypic clusters, species under our definition, are being maintained. All three contact zones correspond precisely with a transition from wet forest with *erato* to dry thorn scrub with *himera* (Mallet 1993; Jiggins et al. 1996). These hybrid zones are extremely narrow, $w \approx 5$ km, narrower than any known between *erato* races (figure 30.4), which implies very strong selection, $s \approx 1$ per locus (Jiggins et al. 1996).

Heliconius himera is similar to an ordinary geographic race of *erato* in many respects. It replaces *erato* in dry Andean valleys of southern Ecuador and northern Peru. Like *erato* races, *himera* differs from other *erato* in warning color pattern, and these pattern differences are controlled by genes homologous to those differing between races of *erato* (Jiggins et al. 1996; Jiggins and McMillan 1997). *Heliconius himera* feeds on the same *Passiflora* host plants as *erato* (Jiggins et al. 1996, 1997b), and differs in ecology only in its drier habitats and in lacking

close mimics; *erato* is typically found in wetter forest and is usually accompanied by its Müllerian comimic *melpomene* (Jiggins et al. 1996). Although *himera* wing and body shape is similar to that of *erato*, *himera* differs strongly from *erato* at allozymes: Nei's D ≈ 0.28 due to nearly fixed or strong frequency differences at 11 of 30 allozyme loci studied (Jiggins et al. 1997a). Their mtDNA sequences are also very different, suggesting about 1.5–2 million years of separation, although this level of differentiation is similar to that found across the Andes within *erato* (Brower 1994a, 1996a; Jiggins et al. 1997a).

These hybrid zones provide an ideal opportunity to investigate what maintains the distinctness of newly formed *Heliconius* species. We have performed extensive laboratory tests to determine the causes of the hybrid deficit in the contact zone between *H. erato cyrbia* and *H. himera* near Loja in southern Ecuador. The rarity of hybrids is partly explained by assortative mating: interspecific matings are only 11% (5–22%) as probable as intraspecific matings in laboratory choice experiments (McMillan et al. 1997); in the field, a similar value of 6% (0.3–27%) was found (Mallet et al. 1998). However, our laboratory crosses, which produced a total of 4,570 eggs and nearly 2,500 adult butterflies, showed no evidence for inviability or sterility of hybrids compared with parentals in terms of numbers of eggs laid per day, egg hatch, larval survival, developmental time, or sex ratio (table 30.2).

The narrowness of the hybrid zone, and the near-complete association between allozymes, mtDNA, envi-

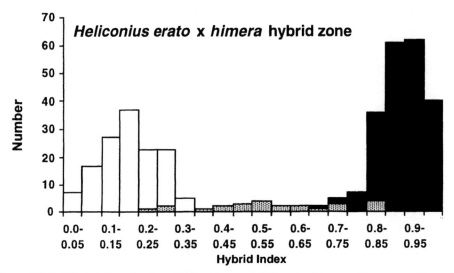

Figure 30.3. Distribution of scores of hybrid index of *Heliconius himera* × *H. erato cyrbia* in the centre of a hybrid zone near Guayquichuma, Ecuador. Individuals were classified using color pattern into three groups: *himera* (open), hybrids (stippled), and *erato* (solid). The hybrid index is calculated as the fraction of characteristic *erato* allozymes, based on 11 loci showing frequency differences between the species (Jiggins et al. 1997a).

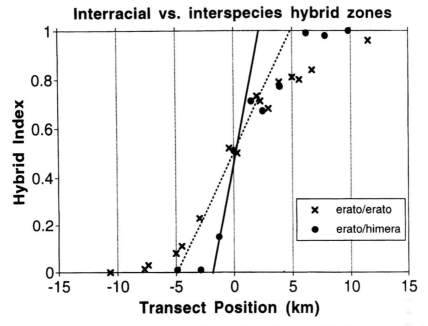

Figure 30.4. Comparison of the widths of an interracial hybrid zone within *H. erato*, and an interspecific hybrid zone between *H. erato* and *H. himera*. The interracial hybrid zone, near Pongo de Cainarache, Peru, is about 10 km wide (see also figure 30.2); the interspecific zone, near Guayquichuma, Ecuador, is about 5 km wide (Jiggins et al. 1996).

ronmental, and color pattern differences indicate that extremely strong selection maintains the hybrid zone and prevents amalgamation of the *himera* and *erato* genotypic clusters (figure 30.3). The supply of hybrids between *himera* and *erato* is undoubtedly limited by assortative mating, but 6% hybridization in each generation would quickly result in a hybrid swarm consisting largely of recombinant genotypes unless hybrids were removed by some form of selection, especially since pure forms mate as freely with F1 hybrids as with their own type (McMillan et al. 1997; Mallet et al. 1998). Classical hybrid inviability and sterility have been excluded, so this selection must be ecological.

There are two likely possibilities, and almost certainly both are involved. First, mimetic selection would take an interesting form in this hybrid zone compared with that

Table 30.2. Measures of viability of pure (E = *erato*, H = *himera*) and hybrid crosses (McMillan et al. 1997).

Cross Type	No. Broods	% Egg Hatch	% Larval Survival	Eggs/ Day	Development Time (d)	% Female
Pure						
E	12	90.3	57.5	1.9	29.8	48
H	14	92.7	66.7	2.6	28.5	49
F1 Hybrids						
E♀ × H♂	12	83.7	64.1	1.8	29.3	48
H♀ × E♂	8	89.2	53.8	2.4	29.2	45
Backcrosses						
F1 × E	7	90.0	62.3	2.2	28.9	53
F1 × H	10	85.7	53.8	2.0	28.1	44
Further crosses						
F2	7	92.0	84.5	1.5	30.2	51
F2 × E/H	6	99.3	83.4	2.5	28.1	50

in interracial hybrid zones. Because hybrids are rarer than the pure forms in the centre of the *himera/erato* zone, they should suffer more frequency-dependent predation than pure genotypes. If selection is as strong as that measured in interracial hybrid zones (survival of nonmimetic morphs ≈ 48% that of mimetic controls; see above), the hybrid inviability produced would be comparable to that caused by a major chromosomal rearrangement such as a reciprocal translocation, or by inviability or sterility of one sex of hybrids. Second, the abrupt ecological transition across all three hybrid zones between *erato* and *himera* suggests strong abiotic selection. If a hybrid (or pure form) were to fly across the hybrid zone, it could encounter temperature or humidity conditions unsuitable for survival or development. Studies of adult activity levels, fecundity and larval development rate all suggest differences in climatic adaptation between the species (McMillan et al. 1997; Davison et al. in press). Competition between sympatric *Heliconius* structures host-plant associations (Benson 1978), and competition may amplify the ecological differences such that slight advantages due to environmental adaptation are translated into strong competitive dominance on either side of the hybrid zone.

We are interested here more in the selective causes rather than the geographic context of speciation. However, correlations between allozymes, mtDNA, and morphology in hybrid zones have often been used as evidence for allopatric secondary contact. In this case, allozyme and mtDNA data show that the very strong genetic barrier produced by ecological selection must be as effective as a major geographic barrier in preventing the spread of approximately neutral alleles between the species (figure 30.3; Jiggins et al. 1997a). Strongly selected clines can accumulate other clines until extremely strong barriers result (Mallet 1993; Mallet and Turner 1998). Once a sufficiently strong barrier due to mimicry and ecological adaptation exists across a zone of this nature, novel genetic differences can accumulate faster than they are homogenized by gene flow. Thus, parapatric speciation remains difficult to rule out (Jiggins et al. 1996; see also Bush and Howard 1986; Bush 1993).

In conclusion, selection across this hybrid zone is strong. Its exact nature is obscure, but it must be mainly ecological rather than due to intrinsic hybrid inviability or sterility. Speciation between *himera* and *erato* has resulted from a combination of assortative mating and divergence of warning pattern and ecology.

Other *Heliconius* Species Pairs Related to *H. erato* and *H. himera*

Heliconius clysonymus and *H. telesiphe* are sister species native to the Andes (*clysonymus* is also found in the mountains of Central America), and are common in cloud forests at altitudes of between 800 and 2,000 m. Their closest relative, *erato*, is found in the lowlands at

0–1,500 m (Brown 1981; Brower 1994a). These species can be found overlapping between 800 and 1,500 m, but more usually only one of the three is present at any site, suggesting competitive exclusion. The larval host plants of *clysonymus* and *telesiphe* are, like those of *erato* and *himera*, in *Passiflora* subgenus *Plectostemma* (Benson 1978; Knapp and Mallet in press). These *Heliconius* are not known to hybridize either with *erato* or with each other, although hybrids have been found between *clysonymus* and another related species, *H. hecalesia*, in Costa Rica and Mexico (table 30.1). *Heliconius telesiphe*, *clysonymus*, and *erato* have clearly speciated under any definition but, like *erato* and *himera*, lack the ability to overlap extensively, presumably because they have failed to evolve suitable host-plant or microhabitat differences.

More distant relatives of *erato*, such as *H. hecalesia*, *H. charithonia*, and *H. demeter*, frequently coexist with *erato* and always differ in host-plant use. *Heliconius hecalesia* is found commonly in submontane and lowland wet forest in the company of *erato* relatives in central America and western South America; it is strongly associated with a cloud forest host plant, *Passiflora* (*Plectostemma*) *lancearia*, rarely used by the others (Benson et al. 1976; Mallet, personal observation). *H. charithonia* is found in a variety of lowland and submontane areas, where it often specializes on *Passiflora* ("*Tetrastylis*") *lobata* and *Passiflora* (*Granadilla*) *adenopoda*, both of which have hooked trichomes that kill other *Heliconius* larvae (Gilbert 1971); it also feeds on other *Passiflora* (*Plectostemma*) host plants more typical for the *erato* group (Benson et al. 1976; Jiggins et al. 1996; Jiggins and Davies 1998). Although its biology is poorly known, *demeter* is often sympatric with *erato* in lowland Amazonia, where it feeds especially on *Dilkea* and *Passiflora* (*Astrophea*) species (Brown and Benson 1975). Thus, closely related species in the *erato* group are often parapatric, but differ in aridity or altitudinal requirements. Where they become sympatric, they differ strongly in host-plant ecology (Jiggins et al. 1997b).

In the unrelated *melpomene* group of *Heliconius*, coexistence again seems to involve host-plant shifts. Related species in this group, such as *hecale*, *ismenius*, *melpomene*, *cydno*, *numata*, *elevatus*, *pardalinus*, and *ethilla*, partition species of *Passiflora* when they overlap (Benson et al. 1976; Benson 1978; Brown 1981; Gilbert and Smiley 1978; Gilbert 1991). The case of *H. cydno* and *H. melpomene* is particularly instructive. These are very closely related sister species: both mtDNA (Brower 1996a,b) and allozyme studies (Mallet, unpublished) suggest that *H. cydno* (together with semispecies *pachinus*, *heurippa*, *tristero*, and *timareta*) forms a monophyletic branch within a *melpomene-cydno* clade, making the species *melpomene* as a whole paraphyletic. *Heliconius melpomene* and *cydno* are sympatric throughout most of Central America and near the Andes (*melpomene* is also found east of the Andes in the Amazon basin, southern Brazil, and Argentina). *Heliconius melpomene* is typically red, black, and yellow and mimics *erato*, whereas *cydno* is typically blue-black and white or blue-black and yellow and usually mimics species in the *sapho/eleuchia* group of *Heliconius* (Linares 1996, 1997). (Three probable races or semispecies of *cydno* found on the eastern slopes of the Andes are exceptions to this rule and have some red coloration: *H. heurippa*, *H. tristero*, and *H. timareta*; see Brower 1996a,b). As already discussed, hybrids between *cydno* and *melpomene* are regularly found in the wild in Colombia, Venezuela, and Ecuador, although they are always exceedingly rare (table 30.1). Preliminary crosses show that female hybrids, although viable, are sterile, a "Haldane Rule" effect (Haldane 1922; Turelli and Orr 1995). Male hybrids between *cydno* and *melpomene*, on the other hand, are fully fertile and viable and can be used to transfer color pattern genes from one species to another in laboratory crosses (table 30.1; L. E. Gilbert, personal communication).

The ecology of *cydno* and *melpomene* has been well studied in Costa Rica. Whereas *melpomene* is a specialist on *Passiflora* (*Granadilla*) *menispermifolia* and *P.* (*G.*) *oerstedii* in open areas and young second growth, *cydno* oviposits and feeds on virtually every available species of *Passiflora* in Atlantic rainforest understory (Smiley 1978). This pattern of host and microhabitat partitioning is repeated in Pacific Costa Rica (Osa), where *pachinus* replaces *cydno* (Benson 1978; Gilbert 1991). However, the larvae of *melpomene* can use all the hosts of *cydno/pachinus*; host specialization in *melpomene* stems entirely from the oviposition behavior of the adult female (Smiley 1978).

The two species therefore differ in their mimetic allegiance and ecology (second growth specialist vs. forest understory generalist). F1 females are sterile, which is not found between *himera* and *erato*. These factors, together with extremely strong assortative mating, enable the coexistence, without fusing, of the two species. In this case it is not clear whether genomic incompatibilities (F1 female sterility) or ecological incompatibilities (host choice and mimicry) evolved first. But, although genomic selection against hybrids is strong (selection against F1s averaged across sexes is approximately 50%), selection against warning color intermediates (50% or more, judging from experiments with *erato*; see above) and on oviposition behavior (unmeasured, but presumably strong), is probably as or more intense. Even in this case, ecological selection and hybrid sterility are comparable in strength.

What Causes Speciation in *Heliconius*?

We can now summarize this information in the form of a sequence of events leading to speciation in *Heliconius* (figure 30.5), incorporating information from *erato* races, *himera/erato* hybrid zones, and the *cydno/melpomene* comparison. Warning color diverges most rapidly, in geographic races long before speciation. Related species

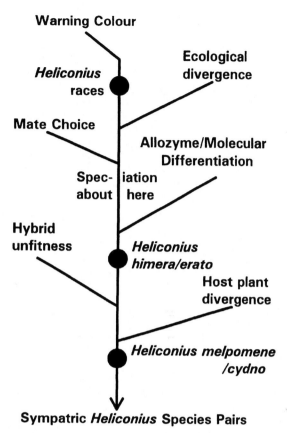

Figure 30.5. The sequence of divergence events leading from geographic race formation to speciation based on *Heliconius* butterflies. Events shown between studied divergent taxa (•) could have occurred in any order, and represent the approximate first occurrence of each event. For example, warning color evolution continues long after speciation but first occurs between races within species such as *erato*. Parapatric species pairs such as *erato* and *clysonymus* or *telesiphe* fall somewhere between *himera/erato* and *melpomene/cydno*, but hybrid compatibility status is unknown because of a lack of known hybrids.

usually differ in their mimicry rings (Turner 1976): there are nine pairs of sister *Heliconius* species inferred from a mtDNA phylogeny (Brower 1994b) and in only one pair, *H. sara* and *H. leucadia*, have the species not diverged strongly in their mimetic pattern. However, warning color evolution does not seem to trigger speciation on its own. Species such as *cydno* or *numata* may even become polymorphic within populations, as a result of mimicking several model species that vary in abundance in time and space (Brown and Benson 1974; Brown 1976; Linares 1997; Kapan 1998). More commonly, for Müllerian mimics, populations are monomorphic, and color patterns diverge geographically to form a diversity of color pattern races that show few signs of speciating.

Nonetheless, mimetic divergence will contribute strongly to speciation after assortative mating evolves because rare nonmimetic hybrids are strongly disfavored. Mimicry acts as a form of postmating isolation, as well as an interesting example of natural selection within species.

Another major correlate of speciation in *Heliconius* is a change of host or habitat use. The closest relatives of *Heliconius erato*, a lowland rainforest species, are all parapatric and occupy different biomes. *Heliconius clysonymus* and *telesiphe* replace *erato* in upland cloud forests, whereas *himera* replaces *erato* in rain-shadowed thorn scrub of Andean southern Ecuador and northern Peru. Although there are strong genetic differences in mtDNA, allozymes, and color pattern among these species, host-plant and microhabitat ecology, on the other hand, have diverged little. These observations strongly suggest that speciation occurred after climatic adaptation, but before any change in host-plant ecology in the *erato* group. In contrast, host-plant ecology or microhabitat could have driven speciation in other *Heliconius*, for example between *cydno* and *melpomene*. Regardless of whether host shifts and microhabitat changes are involved in speciation itself, they do appear necessary, in this genus, for the stable coexistence of sympatric sister species thereafter (Jiggins et al. 1997b).

Speciation in *Heliconius* always seems to involve a change in mating preference, as expected under the "recognition concept" of species (Paterson 1985), although we do not yet understand how this is achieved. In the *himera* and *erato* hybrid zone, it is unlikely that assortative mating is due to "reinforcement" (where selection against hybrids causes the evolution of mate choice; Dobzhansky 1940). Many experiments were done with *himera* collected >60 km from the hybrid zone, or *erato* collected >25 km from the hybrid zone, and these allopatric populations mated assortatively as strongly as those found in the area of overlap (McMillan et al. 1997). Of course, this is not strong evidence—conceivably, assortative mating evolved as a result of reinforcement inside the zone, and then spread to areas far away. However, for parapatric species such as *erato*, *himera*, *clysonymus*, and *telesiphe*, with narrow zones of overlap, reinforcement is unlikely on theoretical grounds (Butlin 1989). Mate choice is perhaps more likely a pleiotropic effect of some other evolutionary change. One candidate is a change of mimetic color pattern. Males, and sometimes females, of red, orange, and black species such as *erato* and *melpomene* can be attracted down from the forest canopy by means of red rags (Brown and Benson 1974; Mallet et al. 1990), but *cydno* are never attracted in this way. Black and white *cydno* can, however, be attracted by a white rag (Mallet and Kapan, unpublished). A profound mimetic switch may thus lead to a shift in long-range mating and social signals. However, color pattern change is not the only factor causing assortative mating: divergence in short-range pheromone signaling systems is almost certainly also involved. Whatever the signaling mode, coevolution between sexual traits and behavioral response may lead to rapid mating divergence (Lande 1982).

The compatibility of *himera* and *erato* demonstrates that hybrid unfitness induced by genomic, as opposed to ecological, incompatibilities may evolve some time after speciation. Even in *melpomene* and *cydno*, the fertility of male hybrids suggests that genes could flow relatively freely between the species. Pleiotropic changes leading to hybrid inviability and sterility, as in *cydno* and *melpomene*, may well evolve more readily after gene flow is effectively halted. Genomic incompatibility could therefore evolve as a consequence rather than a cause of speciation. Obviously, we cannot be sure that this is so in *cydno* and *melpomene* or more distantly related species pairs because hybrid incompatibilities and ecological differences are found together. The sequence of events may not always follow figure 30.5 exactly, and more comparative data are needed to test the generality of these ideas. Nonetheless, ecological factors—mimicry, and adaptation to particular biotopes—together with the evolution of assortative mating appear to be more important than genomic inviability and sterility in this group. Our results show that Henry Walter Bates's inferences were broadly correct. Mimicry, coupled with other ecological adaptations and the evolution of assortative mating, has caused speciation in heliconiine butterflies.

Appendix 30.1
Did Bates Discover Müllerian Mimicry?

Bates showed that palatable dismorphiines mimicked unpalatable ithomiine butterflies (Batesian mimicry). It is not as well known that Bates used exactly the same principle to explain why rare unpalatable forms such as *Napeogenes* and *Heliconius* mimic commoner, but also unpalatable ithomiine models. However, Bates did not fully understand why pairs of common unpalatable ithomiine species were also comimetic. He suggested "similar adaptation of all to the same local, probably inorganic, conditions." Mimicry between common unpalatable species was later explained by Müller (1879), who used one of the first explicit mathematical models in ecology to quantify the relative benefits to pairs of unpalatable species sharing a warning pattern. Mimicry where palatable species copy unpalatable species is now referred to as Batesian, whereas mimicry between any pair of unpalatable species is termed Müllerian.

Appendix 30.2
Essentialism in Modern Species Concepts

Reading Ernst Mayr (e.g., 1982), one obtains the impression that Darwin, Wallace, and other early evolutionists were confused about the nature of species. Our opinion is that the reverse is true (Mallet 1995a,b): Mayr himself confuses the causes of speciation (i.e., the way species are maintained) with species definitions. In our view, much of the current muddle over "species concepts" results from Mayr's adoption of an obviously essentialist view of species: that biological species, unlike lower or higher taxonomic categories, are "objectively real," defined by "isolating mechanisms" that "are a protective device for well-integrated genotypes"; that species are "evolutionary units" necessary for adaptive evolution, have "internal cohesion," and are "individuals," and so on. We fully realize, of course, that Mayr considers himself an enlightened *anti*-essentialist who frequently castigates other biologists, including Darwin, for adopting "typological" thought patterns (e.g., Mayr, 1982, p. 268). Mayr coined the derogatory term "typological" to imply a belief in the existence of a fixed "type," or Platonic "form" underlying a systematic category such as species; to Mayr, the "fixed type" was the essentialism he was fighting. Since it was Darwin's knowledge of natural variation that freed him from the essentialist view of species, it is odd that Mayr views the Darwinian "morphological species concept" as essentialist. The biological species concept can itself be seen as typological, in that Mayr, an admirer of of the great essentialist Hegel, seeks an underlying "essence" or "true meaning" for the term "species," rather than a useful and convenient criterion to delimit actual populations as species taxa; in fact, Mayr repeatedly emphasized that he was avoiding a taxonomically useful criterion in favor of a "concept" that embodied the biological meaning of species. Mayr decided that the true essence of species was reproductive isolation. Others reject this particular essence, but propose other essences which underlie "objectively real" species, for example ecology, mating behavior, "cohesion," or phylogeny/genealogy (see chapters by De Queiroz, Harrison, Shaw, and Templeton, this volume).

Darwin and Wallace both specifically rejected essentialist ideas of a "true meaning" of species. In Darwin's (1859) words:

> In short, we shall have to treat species in the same manner as those naturalists treat genera, who admit that genera are merely artificial combinations made for convenience. This may not be a cheering prospect; but we shall at least be freed from the vain search for the undiscovered and undiscoverable *essence* of the term species. (pp. 484–5, emphasis added).

Darwin and Wallace realized that species were often reproductively incompatible, but pointed out many difficulties with reproductive definitions of species (Mallet 1995a). Instead, they used a simple operational cluster criterion for species that was independent of the origin or maintenance of differences. The value of their species definition is manifest in the scientific revolution they achieved.

Acknowledgments   The Ecuador work was carried out under the auspices of Dr. Miguel Moreno and Germania Estévez ate the Museo de Ciencias Naturales, Quito; INEFAN provided collecting permits. María Arias Diaz, María del Carmen Avila, Roberto Carpio Ayala, José Carpio Ayala, Angus Davison, Sarah Dixon, and Ashleigh Griffin helped with the extensive rearing and fieldwork; Peter Wilson, Fausto Lopez, and Pablo Lozano at the Fundación Arcoiris, and Joy and Curtis Hoffman helped with logistics; Drs. Keith Brown, Andrew Brower, Henri Descimon, Larry Gilbert, Gerardo Lamas, John R. G. Turner provided discussion and advice. The work was funded by BBSRC. We are extremely grateful to all of these people and organizations, without whom the project would have been impossible. Finally, we acknowledge a strong influence of Guy Bush's "conceptually radical" ideas on our own work.

References

Ackery, P. R., and Smiles, R. L. 1976. An illustrated list of the type-specimens of the Heliconiinae (Lepidoptera: Nymphalidae) in the British Museum (Natural History). Bull. Brit. Mus. (Nat. Hist.) Entomol. 32:171–214.

Barton, N. H., and Gale, K. S. 1993. Genetic analysis of hybrid zones. In R. G. Harrison (ed.). Hybrid Zones and the Evolutionary Process. New York: Oxford University Press, pp. 13–45.

Barton, N. H., and Hewitt, G. M. 1989. Adaptation, speciation and hybrid zones. Nature 341:497–503.

Bates, H. W. 1862. Contributions to an insect fauna of the Amazon valley. Lepidoptera: Heliconidae. Trans. Linn. Soc. Lond. 23:495–566.

Benson, W. W. 1972. Natural selection for Müllerian mimicry in *Heliconius erato* in Costa Rica. Science 176:936–939.

Benson, W. W. 1978. Resource partitioning in passion vine butterflies. Evolution 32:493–518.

Benson, W. W. 1982. Alternative models for infrageneric diversification in the humid tropics: tests with passion vine butterflies. In G. T. Prance (ed.). Biological Diversification in the Tropics. New York: Columbia University Press, pp. 608–640.

Benson, W. W., Brown, K. S., and Gilbert, L. E. 1976. Coevolution of plants and herbivores: passion flower butterflies. Evolution 29:659–680.

Brower, A. V. Z. 1994a. Rapid morphological radiation and convergence among races of the butterfly *Heliconius erato* inferred from patterns of mitochondrial DNA evolution. Proc. Natl. Acad. Sci. USA 91:6491–6495.

Brower, A. V. Z. 1994b. Phylogeny of *Heliconius* butterflies inferred from mitochondrial DNA sequences (Lepidoptera: Nymphalinae). Mol. Phylog. Evol. 3:159–174.

Brower, A. V. Z. 1996a. Parallel race formation and the evolution of mimicry in *Heliconius* butterflies: a phylogenetic hypothesis from mitochondrial DNA sequences. Evolution 50:195–221.

Brower, A. V. Z. 1996b. A new mimetic species of *Heliconius* (Lepidoptera: Nymphalidae), from southeastern Colombia, revealed by cladistic analysis of mitochondrial DNA sequences. Zool. J. Linn. Soc. 116:317–332.

Brown, K. S. 1976. An illustrated key to the silvaniform *Heliconius* (Lepidoptera: Nymphalidae) with descriptions of new subspecies. Trans. Am. Entomol. Soc. 102:373–484.

Brown, K. S. 1979. Ecologia Geográfica e Evolução nas Florestas Neotropicais. Universidade Estadual de Campinas, Campinas, Brazil. Livre de Docencia.

Brown, K. S. 1981. The biology of *Heliconius* and related genera. Annu. Rev. Entomol. 26:427–456.

Brown, K. S. 1982. Historical and ecological factors in the biogeography of aposematic Neotropical butterflies. Am. Zool. 22:453–471.

Brown, K. S. 1987. Biogeography and evolution of neotropical butterflies. In T. C. Whitmore and G. T. Prance (eds.). Biogeography and Quaternary History in Tropical America. Oxford: Oxford University Press, pp. 66–104.

Brown, K. S., and Benson, W. W. 1974. Adaptive polymorphism associated with multiple Müllerian mimicry in *Heliconius numata* (Lepid.: Nymph.). Biotropica 6:205–228.

Brown, K. S., and Benson, W. W. 1975. The heliconians of Brazil (Lepidoptera: Nymphalidae). Part VI. Aspects of the biology and ecology of *Heliconius demeter* with description of four new subspecies. Bull. Allyn. Mus. 26:1–19.

Brown, K. S., and Fernandez Yepez, F. 1985. Los Heliconiini (Lepidoptera, Nymphalidae) de Venezuela. Bol. Entomol. Venez. N.S. 3:29–76.

Brown, K. S., and Mielke, O. H. H. 1972. The heliconians of Brazil (Lepidoptera: Nymphalidae). Part II. Introduction and general comments, with a supplementary revision of the tribe. Zoologica (N.Y.) 57:1–40.

Brown, K. S., Sheppard, P. M., and Turner, J. R. G. 1974. Quaternary refugia in tropical America: evidence from race formation in *Heliconius* butterflies. Proc. Roy. Soc. Lond. B 187:369–378.

Bush, G. L. 1993. A reaffirmation of Santa Rosalia, or why are there so many kinds of *small* animals? In D. R. Lees and D. Edwards (eds.). Evolutionary Patterns and Processes. London: Linnean Society of London/Academic Press, pp. 229–249.

Bush, G. L., and Howard, D. J. 1986. Allopatric and nonallopatric speciation; assumptions and evidence. In S. Karlin and E. Nevo (eds.). Evolutionary Processes and Theory. New York: Academic Press, pp. 411–438.

Butlin, R. 1989. Reinforcement of premating isolation. In D. Otte and J. A. Endler (eds.). Speciation and Its Consequences. Sunderland, Mass.: Sinauer, pp. 158–179.

Curson, J., Quinn, D., and Beadle, D. 1994. New World Warblers. Christopher Helm, London.

Darwin, C. 1859. On the Origin of Species by Means of Natural Selection, or the Preservation of Favoured Races in the Struggle for Life (1st ed.). John Murray, London.

Darwin, C. 1863. A review of H. W. Bates' paper on "mimetic butterflies." In P. H. Barrett (ed.). The Collected Papers of Charles Darwin, vol. 2. Chicago: University of Chicago Press, pp. 87–92. 1977.

Darwin, C. 1871. The Descent of Man, and Selection in Relation to Sex (2nd ed.). John Murray, London.

Davison, A., McMillan, W. O., Griffin, A. S., Jiggins, C. D., and Mallet, J. L. B. In press. Behavioural and physiologi-

cal adaptation between two parapatric *Heliconius* species (Lepidoptera: Nymphalidae). Biotropica.

Descimon, H., and Mast de Maeght, J. 1984. Semispecies relationships between *Heliconius erato cyrbia* Godt. and *H. himera* Hew. in southwestern Ecuador. J. Res. Lepid. 22:229–239.

Dobzhansky, T. 1940. Speciation as a stage in evolutionary divergence. Am. Nat. 74:312–321.

Endler, J. A. 1982. Pleistocene forest refuges: fact or fancy? In G. T. Prance (ed.). Biological Diversification in the Tropics. New York: Columbia University Press, pp. 641–657.

Gilbert, L. E. 1971. Butterfly-plant coevolution: has *Passiflora adenopoda* won the selectional race with heliconiine butterflies? Science 172:585–586.

Gilbert, L. E. 1991. Biodiversity of a Central American *Heliconius* community: pattern, process, and problems. In P. W. Price, T. M. Lewinsohn, T. W. Fernandes, and W. W. Benson (eds.). Plant-Animal Interactions: Evolutionary Ecology in Tropical and Temperate Regions. New York: Wiley, pp. 403–427.

Gilbert, L. E., and Smiley, J. T. 1978. Determinants of local diversity in phytophagous insects: host specialists in tropical environments. In L. A. Mound and N. Waloff (eds.). Symposia of the Royal Entomological Society of London, 9, Diversity of Insect Faunas. Oxford: Blackwell, pp. 89–104.

Grant, P. R., and Grant, B. R. 1992. Hybridization of bird species. Science 256:193–197.

Guillaumin, M., and Descimon, H. 1976. La notion d'espèce chez les lépidoptères. In C. Bocquet, J. Génermont, and M. Lamotte (eds.). Les Problèmes de l'Espèce dans le Règne Animal, vol. 1. Paris: Société zoologique de France, pp. 129–201.

Haldane, J. B. S. 1922. Sex ratio and unisexual sterility in hybrid animals. J. Genet. 12:101–109.

Harrison, R. G. 1990. Hybrid zones: windows on evolutionary process. In D. Futuyma and J. Antonovics (eds.). Oxford Surveys in Evolutionary Biology, vol. 7. Oxford: Oxford University Press, pp. 69–128.

Hewitt, G. M. 1988. Hybrid zones—natural laboratories for evolutionary studies. Trends Ecol. Evol. 3:158–167.

Holzinger, H., and Holzinger, R. 1994. *Heliconius* and Related Genera. Lepidoptera: Nymphalidae. The Genera *Eueides*, *Neruda* and *Heliconius*. Sciences Nat, Venette, France.

Jiggins, C. D., and Davies, N. 1998. Genetic evidence for a sibling species of *Heliconius* (Lepidoptera: Nymphalidae). Biol. J. Linn. Soc. 64:57–67.

Jiggins, C. D., and McMillan, W. O. 1997. The genetic basis of an adaptive radiation: warning color in two *Heliconius* species. Proc. Roy. Soc. Lond. B. 264:1167–1175.

Jiggins, C. D., McMillan, W. O., Neukirchen, W., and Mallet, J. 1996. What can hybrid zones tell us about speciation? The case of *Heliconius erato* and *H. himera* (Lepidoptera: Nymphalidae). Biol. J. Linn. Soc. 59:221–242.

Jiggins, C. D., McMillan, W. O., King, P., and Mallet, J. 1997a. The maintenance of species differences across a *Heliconius* hybrid zone. Heredity. 79:495–505.

Jiggins, C. D., McMillan, W. O., and Mallet, J. L. B. 1997b. Host plant adaptation has not played a role in the recent speciation of *Heliconius himera* and *Heliconius erato* (Lepidoptera: Nymphalidae). Ecol. Entomol. 22:361–365.

Kapan, D. 1998. Divergent natural selection and Müllerian mimicry in polymorphic *Heliconius cydno* (Lepidoptera: Nymphalidae). Ph.D. dissertation, University of British Columbia.

Knapp, S., and Mallet, J. In press. A new species of *Passiflora* (Passifloraceae) with notes on the natural history of its herbivore, *Heliconius* (Lepidoptera: Nymphalidae: Heliconiiti). Novon.

Lande, R. 1982. Rapid origin of sexual isolation and character divergence in a cline. Evolution 36:213–223.

Linares, M. 1996. The genetics of the mimetic coloration in the butterfly *Heliconius cydno weymeri*. J. Hered. 87:142–149.

Linares, M. 1997. The ghost of mimicry past: laboratory reconstitution of an extinct butterfly 'race.' Heredity 78:628–635.

Mallet, J. 1986. Hybrid zones in *Heliconius* butterflies in Panama, and the stability and movement of warning color clines. Heredity 56:191–202.

Mallet, J. 1989. The genetics of warning color in Peruvian hybrid zones of *Heliconius erato* and *H. melpomene*. Proc. Roy. Soc. Lond. B 236:163–185.

Mallet, J. 1993. Speciation, raciation, and color pattern evolution in *Heliconius* butterflies: evidence from hybrid zones. In R. G. Harrison (ed.). Hybrid Zones and the Evolutionary Process. New York: Oxford University Press, pp. 226–260.

Mallet, J. 1995a. A species definition for the Modern Synthesis. Trends Ecol. Evol. 10:294–299.

Mallet, J. 1995b. Reply to Dover and Gittenberger. Trends Ecol. Evol. 10:490–491.

Mallet, J. 1996a. The genetics of biological diversity: from varieties to species. In K. J. Gaston (ed.). Biodiversity: Biology of Numbers and Difference. Oxford: Blackwell, pp. 13–47.

Mallet, J. 1996b. What are 'good' species? Reply to Kerry Shaw. Trends Ecol. Evol. 11:174–175.

Mallet, J., and Barton, N. 1989a. Inference from clines stabilized by frequency-dependent selection. Genetics 122:967–976.

Mallet, J., and Barton, N. H. 1989b. Strong natural selection in a warning color hybrid zone. Evolution 43:421–431.

Mallet, J., and Gilbert, L. E. 1995. Why are there so many mimicry rings? Correlations between habitat, behaviour and mimicry in *Heliconius* butterflies. Biol. J. Linn. Soc. 55:159–180.

Mallet, J. L. B., and Turner, J. R. G. 1998. Biotic drift or the shifting balance—did forest islands drive the diversity of warningly coloured butterflies? In P. R. Grant and B. Clarke (eds.). Evolution on Islands. Oxford: Oxford University Press. pp. 262–280.

Mallet, J., Barton, N., Lamas, G., Santisteban, J., Muedas, M., and Eeley, H. 1990. Estimates of selection and gene flow from measures of cline width and linkage disequilibrium in *Heliconius* hybrid zones. Genetics 124:921–936.

Mallet, J., Jiggins, C. D., and McMillan, W. O. 1996. Mimicry meets the mitochondrion. Curr. Biol. 6:937–940.

Mallet, J., McMillan, W. O., and Jiggins, C. D. 1998. Mate choice between a pair of *Heliconius* species in the wild. Evolution. 52:503–510.

Mayr, E. 1982. The Growth of Biological Thought. Diversity, Evolution, and Inheritance. Belknap, Cambridge, Mass.

McMillan, W. O., Jiggins, C. D., and Mallet, J. 1997. What initiates speciation in passion-vine butterflies? Proc. Natl. Acad. Sci. USA. 94:8628–8633.

Müller, F. 1879. *Ituna* and *Thyridia*; a remarkable case of mimicry in butterflies. Trans. Entomol. Soc. Lond. 1879:xx–xxix.

Papageorgis, C. 1975. Mimicry in neotropical butterflies. Am. Sci. 63:522–532.

Paterson, H. E. H. 1985. The recognition concept of species. In E. S. Vrba (ed.). Transvaal Museum Monograph, 4, Species and Speciation. Pretoria: Transvaal Museum. pp. 21–29.

Posla-Fuentes, M. 1993. An unusual form of *Heliconius cydno* from Costa Rica. Trop. Lepid. 4:92.

Salazar, J. A. 1993. Notes on some populations of *Heliconius heurippa* in Colombia (Lepidoptera: Nymphalidae: Heliconiinae). Trop. Lepid. 4:119–121.

Sheppard, P. M., Turner, J. R. G., Brown, K. S., Benson, W. W., and Singer, M. C. 1985. Genetics and the evolution of muellerian mimicry in *Heliconius* butterflies. Philos. Trans. Roy. Soc. Lond. B 308:433–613.

Smiley, J. T. 1978. Plant chemistry and the evolution of host specificity: new evidence from *Heliconius* and *Passiflora*. Science 201:745–747.

Turelli, M., and Orr, H. A. 1995. The dominance theory of Haldane's Rule. Genetics 140:389–402.

Turner, J. R. G. 1971a. Two thousand generations of hybridization in a *Heliconius* butterfly. Evolution 25:471–482.

Turner, J. R. G. 1971b. Studies of Müllerian mimicry and its evolution in burnet moths and heliconid butterflies. In E. R. Creed (ed.). Ecological Genetics and Evolution. Oxford: Blackwell, pp. 224–260.

Turner, J. R. G. 1976. Adaptive radiation and convergence in subdivisions of the butterfly genus *Heliconius* (Lepidoptera: Nymphalidae). Zool. J. Linn. Soc. 58:297–308.

Turner, J. R. G. 1981. Adaptation and evolution in *Heliconius*: a defense of neo-Darwinism. Annu. Rev. Ecol. Syst. 12: 99–121.

Turner, J. R. G. 1982. How do refuges produce tropical diversity? Allopatry and parapatry, extinction and gene flow in mimetic butterflies. In G. T. Prance (ed.). Biological Diversification in the Tropics. New York: Columbia University Press, pp. 309–335.

Turner, J. R. G., and Mallet, J. L. B. 1996. Did forest islands drive the diversity of warningly coloured butterflies? Biotic drift and the shifting balance. Philos. Trans. Roy. Soc. Lond. B 351:835–845.

Turner, J. R. G., Johnson, M. S., and Eanes, W. F. 1979. Contrasted modes of evolution in the same genome: allozymes and adaptive change in *Heliconius*. Proc. Natl. Acad. Sci. USA 76:1924–1928.

Wallace, A. R. 1865. On the phenomena of variation and geographical distribution as illustrated by the Papilionidae of the Malayan region. Trans. Linn. Soc. Lond. 25:1–71.

31

Hybridization and Speciation in Darwin's Finches

The Role of Sexual Imprinting on a Culturally Transmitted Trait

B. Rosemary Grant
Peter R. Grant

Bush (1993, 1994) defines speciation as a process of divergence of lineages sufficiently genetically distinct from one another for each to follow independent evolutionary paths. He emphasizes two points: first, complete reproductive isolation is the end point of the speciation process and not the beginning; second, lineages are capable of hybridizing and exchanging genes for quite some time without losing their phenotypic identities.

The potential to hybridize is taxonomically widespread, being also prevalent in plants and vertebrates (Arnold 1992; Bush 1993, 1994; Howard 1993). For example, almost 10% of all birds have been known to hybridize (Grant and Grant 1992). Moreover, Prager and Wilson (1975) showed that birds can hybridize after an extraordinarily long time has elapsed since their divergence from a common ancestor. From an analysis of albumin and transferrin immunological distances between 36 pairs of hybridizing bird species, they calculated the length of the hybridizing period to extend for an average of 22 million years (Prager and Wilson 1975). This is more than one third of the time since the most common order of birds, the passerines, originated (e.g., see Boles 1995). The same analyses showed that amphibia retain a hybridization potential for a similar length of time, whereas for mammals it is much shorter (Prager and Wilson 1975).

In this chapter we investigate the effects of interbreeding on the speciation process, by studying directly the causes and consequences of hybridization in closely related sympatric populations of Darwin's finches on the Galápagos Islands. They are particularly suitable for this type of investigation for several reasons. First, the Galápagos archipelago holds its full complement of finch species (Grant 1986), in contrast to other archipelagos

such as Hawaii (James and Olsen 1991; Steadman 1995), the West Indies, New Guinea, and the western pacific islands, which, although rich in avian speciation, have lost many endemic species due to human disturbance (Steadman 1995). Second, the 13 species are closely related, having arisen relatively recently. They provide several examples of diverging lineages in secondary contact. Third, fieldwork during the last 20 years has established that at least half of the species hybridize, albeit rarely.

The outline of the chapter is as follows. We first examine the initial conditions of colonization of the archipelago, and a model of the speciation process. We then focus on Isla Daphne Major, where four species are resident, and give a brief summary of the species. Next, we examine the nature of the barrier to gene flow in sympatry and the degree to which it is permeable. We examine the fitness of hybrids and conditions under which hybrids survive long enough to reproduce. We then ask why, if interbreeding has been occurring sporadically, the species have not fused into one panmictic whole. Finally, we examine the evolutionary implications of introgressive hybridization for the speciation process, and find that under some circumstances it can play a creative role in facilitating further divergence.

Colonization and Radiation

Darwin's finches are related to emberizine finches in Central and South America, but the closest mainland relative to the ancestor of Darwin's finches is still not known. The initial colonization of the Galápagos archipelago by Darwin's finches is estimated to have occurred 2.8–3 million years ago (Grant 1994), based on a study of allozyme variation (Yang and Patton 1981) and an assumed molecular clock. At this time the Galápagos archipelago may have consisted of only five islands: San Cristóbal, Española and three others now submerged (P. R. Grant and Grant 1996a). The reconstruction of the geological history of the archipelago (P. R. Grant and Grant 1996a) assumes a constant southeasterly plate movement from a fixed hotspot beneath present-day Fernandina, and takes into account dates of submergence and radiometric dating of the islands (Christie et al. 1992).

The geological information, combined with revised estimates of the ages of finch species, reveals a relationship between the radiation of the finches and the origin of the islands (P. R. Grant and Grant 1996a). The starting point of the radiation occurred when a lineage that has led to the modern warbler finch, *Certhidea olivacea*, diverged from the ancestral stock (Yang and Patton 1981; Stern and Grant 1996). Subsequently, the number of species increased, as did the number of islands. Unless several species arose and went extinct, unrecorded, the number of species was always less than the number of islands. If this is correct, it implies that the increase in number of islands may have facilitated an increase in number of finch species (P. R. Grant and Grant 1996a). A similar pattern of finch radiation, with islands and species developing together, applies in the Hawaiian archipelago (Tarr and Fleischer 1995).

Within this dynamic geological context the adaptive radiation of the Darwin's finches can be explained in terms of the allopatric model of speciation (Stresemann 1936; Mayr 1942; Lack 1947; Bowman 1961; Grant 1981a, 1986; P. R. Grant and Grant 1996a). The archipelago was colonized by a moderately large number of founders (Vincek et al. 1997). The first stage of finch diversification is hypothesized to be associated with the development of one or more allopatric populations as a result of dispersal from the original island. Allopatric populations diverge under different selection regimes, and as a result of genetic drift. The second stage comes about when two related taxa come together and make secondary contact. During the first stage the degree of evolutionary change and genetic isolation between diverging lineages is a function of time since separation, whereas in the second stage it is additionally a function of the permeability of the barrier to gene flow. The speciation cycle is repeated several times.

At both stages of the speciation process ecological factors are believed to be of primary importance in causing adaptive change. These factors, primarily food factors, have been analyzed and discussed extensively (Lack 1947; Bowman 1961; Grant 1981a, 1986; Schluter and Grant 1984). Our concern here is not with the evolutionary origin of ecological isolation in the speciation process but with reproductive isolation. In order to reveal the factors retarding or preventing gene exchange between sympatric congeners, we carried out a detailed examination of reproductive interactions between closely related finch species on Isla Daphne Major. We give a brief description of the island and its resident population of finches before examining the nature of the barrier to interbreeding between the species.

Isla Daphne Major and Its Finches

Isla Daphne Major is a small (0.33 km^2) and low (120 m) tuff cone situated 8 km from the nearest, much larger, island of Santa Cruz (904 km^2). The age of Daphne is unknown, but is likely to be at least a million years (P. R. Grant and Grant 1996b). Daphne was alternately connected to Santa Cruz above sea level and separated from it as a result of fluctuations in sea level. Its most recent separation occurred approximately 15,000 years ago, and as sea level rose its area and elevation decreased and isolation increased (figure 31.1).

Four species of finches are resident on Daphne: *Geospiza fuliginosa* (small ground finch), *G. fortis* (medium ground finch), *G. magnirostris* (large ground finch), and *G. scandens* (cactus finch). Their average weights are 13, 17, 30, and 21 g, respectively. The same four species are present on Santa Cruz and may have been present on Daphne when it last became separated from Santa Cruz. However, it is doubtful if the two rarer species have occupied the island continuously. *Geospiza fuliginosa* persists on the island now as a result of repeated immigration, probably from Santa Cruz. *Geospiza magnirostris* colonized the island in 1983, probably also from Santa Cruz (Grant and Grant 1995). Both species were observed on the island, though rarely, earlier this century (Gifford 1919; Beebe 1924).

The study of breeding birds began in 1976. Breeding activity was followed throughout the breeding season every year until 1991, and from January to March in 1992 to 1995. Average breeding populations sizes from 1976 to 1995 were 270 *G. fortis*, 105 *G. scandens*, 14 *G. magnirostris*, and 7 *G. fuliginosa* individuals. After 1978 over 90% of all birds were banded, and this was gradually increased to 100% in 1992. Birds were banded as adults or as nestlings. In the latter case their parents were identified by observations of incubation (mother) and feeding of nestlings (both parents). The high heritabilities of morphological traits, the lack of parental sex bias in heritabilities, and the lack of observed extrapair copulations suggest that incorrect assignment of parents is rare (Grant and Grant 1995).

Some breeding took place in all years except the drought years of 1985, 1988, and 1989. Individuals bred almost exclusively with members of their own species; for diagnoses of the species, see Grant (1993). Hybridization occurred in almost every year but was always rare. *Geospiza fortis* hybridized with *G. fuliginosa* in 11 of the 13 years in which both species bred, and with *G. scandens* in six of the years. On average, 1.8% of breeding *G. fortis*

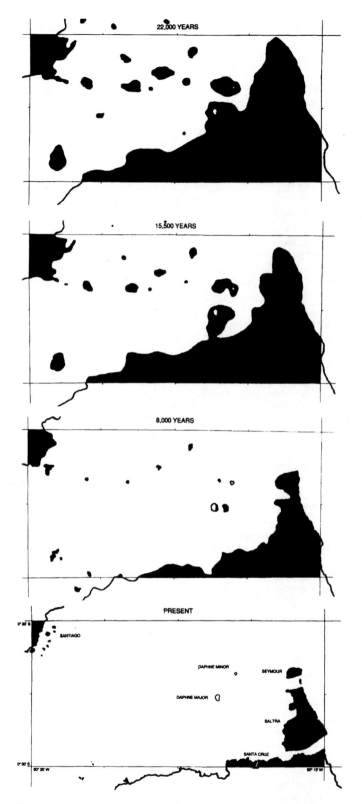

Figure 31.1. Changes in the center of the Galápagos archipelago in the last 22,000 years as a result of a lowering of the sea level. Reconstructed from data in Bard et al. (1990). Positions of Daphne Major and Daphne Minor are shown in white in all four maps. The above-water connection between Daphne Major and Santa Cruz was severed about 15,000 years ago.

individuals, 0.8% of breeding *G. scandens* individuals, but 73% of the very rare *G. fuliginosa* individuals hybridized. *Geospiza fuliginosa* may have hybridized frequently because it was difficult for them to find a mate of their own species (Grant and Grant 1997a). Although numbers of the fourth species, *G. magnirostris*, have increased since the island was colonized in 1983, only one individual has paired heterospecifically, and it failed to produce any nestlings (Grant and Grant 1995).

During the first six years of the study, from 1976 to 1982, none of the hybrids survived long enough to breed, whereas after 1983, under altered ecological conditions produced by an exceptionally severe El Niño event, hybrids did survive to breed. Throughout the following 12 years the F_1 hybrids produced were viable and fertile. Hybrids rarely bred with each other but backcrossed to the more numerous *G. fortis* or *G. scandens*. There was no indication of genetic incompatibilities in either sex, or sterility in the female (heterogametic sex; the Haldane effect), in any of the crosses (Grant and Grant 1992).

The four species, as in all members of the genus *Geospiza*, have identical plumages, nest structure, and courtship and reproductive habits. They differ in body size, bill size and shape, and song. Therefore, we examined morphology and song for evidence of a barrier to gene flow. An ability to discriminate between members of their own and other species on the basis of morphology and song would explain the almost exclusive breeding of finches with conspecifics even when there is no genetic incompatibility.

Nature of the Barrier to Gene Flow

Discrimination between Conspecifics and Heterospecifics on the Basis of Morphology and Song

Experiments using models of birds and song playback showed that *G. fortis* and *G. scandens* on Daphne Major have no difficulty in distinguishing between conspecific and heterospecific individuals on the basis of both size and shape of bill, and song (Ratcliffe and Grant 1983a,b, 1985). In contrast, *G. fortis* did not discriminate between conspecific models and songs and those of *G. fuliginosa*. These results are interesting because *G. fuliginosa* is more similar to *G. fortis* in song and morphology than is *G. scandens*, and hybridizes with it more frequently. Moreover, unlike the resident *G. scandens*, it is a repeated immigrant. Therefore, the situation on Daphne, with a resident species, *G. fortis*, encountering and occasionally hybridizing with an immigrant species, *G. fuliginosa* (speciation cycle, stage 2), is as close to the early stages of secondary contact of two taxa that have diverged in allopatry (stage 1) as we can find in the archipelago, especially as the Daphne form of *G. fortis* is unusually small.

These results are consistent with those of other experiments conducted on other islands, and can be generalized by saying that sympatric pairs of species show clear discrimination, but allopatric ones of similar morphology and song characteristics do not (Ratcliffe and Grant 1983a, 1983b, 1985).

The difference in behavior of sympatric and allopatric species can be interpreted in terms of sexual imprinting. Sexual imprinting is the phenomenon in which experience with adults, usually the parents, during a short sensitive period early in life influences the development of mating preferences (Immelmann 1975). Experience in first courtship encounters can modify the preferences (Bischoff and Clayton 1991). The experience component differs between sympatry and allopatry. Experience with similar-looking or -sounding heterospecifics in sympatry may narrow the range of potential phenotypes acceptable in a mate, thereby consolidating the effects of early learning (e.g., see Kruijt and Meeuwissen 1993). Since sexual imprinting on song and morphological traits of parents tends to restrict the mating of individuals to members of the same species (Ten Cate and Bateson 1988; Laland 1994), we investigated the transmission of song and morphology from one generation to the next, and the role of song and morphology in the choice of mates.

Transmission of Song between Generations

An individual male typically sings a single, structurally simple, song that remains unaltered throughout life; less than 1% of Daphne finches have a repertoire of two songs. Females do not sing. Songs of *G. fortis* and *G. scandens* differ to the human ear, sonagraphically (figures 31.2 and 31.3) and in quantitative characteristics (figure 31.4), whereas the songs of breeding individuals of *G. fuliginosa* resemble the songs of *G. fortis* (figures 31.3 and 31.5). Although there are exceptions, songs of sons typically resemble the songs their fathers sing (figure 31.4; see also Millington and Price 1985; Gibbs 1990) and in fine quantitative detail (figure 31.6). They also resemble the songs of their paternal grandfather but not the songs of their maternal grandfather (figure 31.6), indicating that song is culturally and not genetically inherited (B. R. Grant and Grant 1996a).

Bowman (1983) found that captive Darwin's finches have a short sensitive period for song learning early in life, extending from approximately day 10 to day 40. This period coincides with the last few days in the nest and extends through the stage when fledglings are being fed by their parents. Males sing vigorously on their territories before obtaining a mate. After males have paired their song rate declines, although males continue to sing near the nest, especially after feeding nestlings or fledglings (Millington and Price 1985; B. R. Grant and P. R. Grant, personal observations). Singing also occurs during interactions with intruders and neighbors (Ratcliffe 1981).

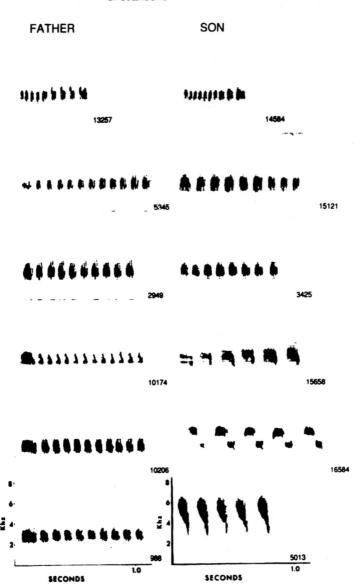

Figure 31.2. Sonagrams of the songs of selected *G. scandens* and their sons. Note the typical close resemblance of father and son songs in the top three examples, in contrast to the bottom three exceptions. Reprinted with permission from B. R. Grant and Grant (1996a).

Therefore, during their sensitive period young are exposed repeatedly to their father's song and to a lesser extent to the song of their natal neighbor and the song of other species holding overlapping or adjacent territories. In some bird species males learn a variety of songs during the sensitive stage and select a song similar to their neighbor at the time of breeding (Nelson and Marler 1993; Nelson et al. 1995), or copy the song from their breeding neighbor (Payne 1981; Baptista and Morton 1988). This is not the case with Darwin's finches. Instead, sons generally copy their father's song in fine detail, and even those that do not copy their fathers in precise detail learn to sing conspecific song, although it is not clear if they do so from the closest natal neighbor, a breeding neighbor, or another male (B. R. Grant and Grant 1996a).

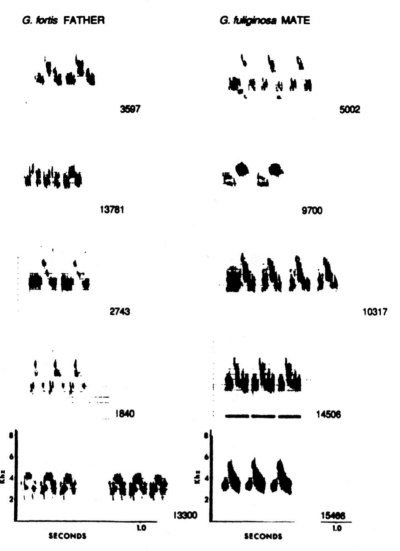

Figure 31.3. Sonagrams of the songs of *Geospiza fortis* and *G. fuliginosa* compared: songs of the fathers of *G. fortis* females are compared with the songs of the *G. fuliginosa* mates of those females.

Transmission of Morphology between Generations

In contrast to the cultural inheritance of song, morphology is genetically inherited. Three beak dimensions (length, depth, and width) display highly heritable variation and covariation in both *G. fortis* and *G. scandens*. Heritabilities, calculated from regressions of midoffspring values on midparent values, were in the range of 0.57–0.75 for *G. fortis* and 0.48–0.69 for *G. scandens* (Grant and Grant 1994). Body size is also heritable in both species. This is shown by the heritabilities of weight, wing, and tarsus length (Grant and Grant 1994), and the first component of a principal components analysis of all six morphological variables (Boag 1983). Genetic correlations between traits are positive and generally strong, especially between beak dimensions of *G. fortis* (Grant and Grant 1994). Thus, there is a strong genetic transmission from parents to offspring of factors that determine beak size and shape, and body size, in these species. *G. fuliginosa* has been too rare on the island for comparable analysis.

Mating Pattern of Females Influenced by Song

Females mate with males that sing a conspecific song, almost without exception (Grant and Grant 1997a). In a

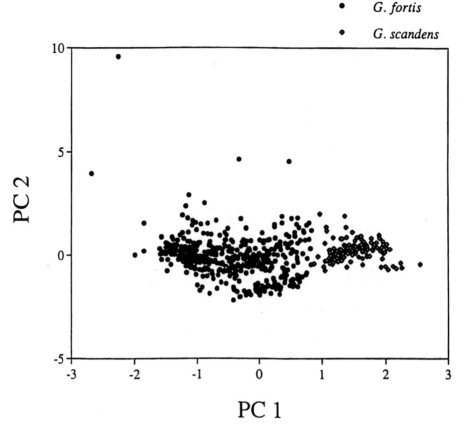

Figure 31.4. Multivariate separation of the songs of *G. fortis* and *G. scandens* on two principal components axes. Principal Components Analysis (PCA) was performed with measurements of the frequency and duration of individual notes. Reprinted with permission from B. R. Grant and Grant (1996a).

sample of 392 female *G. fortis*, in which both father's and mate's song were recorded, 378 paired with a *G. fortis* male singing a *G. fortis* song. Twelve of the 14 females that did not mate conspecifically mated with *G. fuliginosa* males that sang songs indistinguishable from *G. fortis* songs. Moreover, quantitative features of the songs of the mates and of the females' fathers were positively correlated (figure 31.7). The other two *G. fortis* females mated with *G. scandens* males that sang *G. scandens* songs, and thus all but two mated with males that sang a conspecific song. Of 90 *G. scandens* females for which the songs of both the father and mate were recorded, 86 mated with *G. scandens* males singing *G. scandens* songs. The remaining four females were daughters of a *G. scandens* male that sang a *G. fortis* song, and all four mated with *G. fortis* males. Therefore, paternal song apparently determined the mating pattern of all but two of 482 females of the two species combined, even though in 16 cases it led to mating with heterospecifics (B. R. Grant and Grant 1996a). These results imply that

the singing of heterospecific or similar song enhances the likelihood of hybridization occurring because females choose mates at least partly on the basis of song characteristics.

Mating Pattern of Females Influenced by Morphology

Although song is a strong factor determining mating patterns, size might also contribute to a choice of mates. We found no evidence of assortative mating among conspecifics on the basis of any of the measured morphological traits (Grant and Grant 1997a). Nevertheless three pieces of evidence indicate that hybridizing birds base their choice of mates partly on morphology. First, female *G. fortis* that mated with male *G. fuliginosa* (the smaller species) were significantly smaller in wing and tarsus length than those that mated only conspecifically (Grant and Grant 1997a). Second, male *G. scandens* that mated with *G. fortis* (the smaller species) were significantly

G. fuliginosa

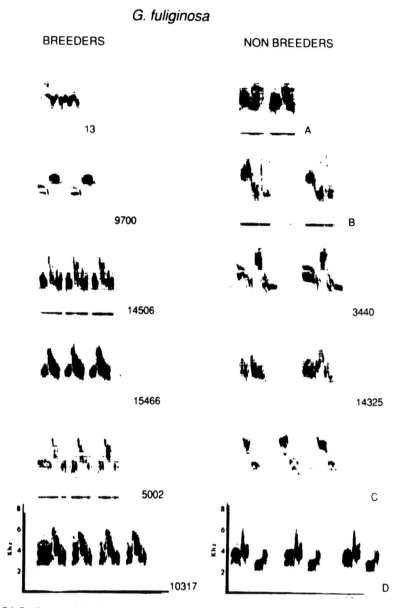

Figure 31.5. Songs of *G. fuliginosa* that bred compared with those that did not. Songs of most breeders were similar to the songs of *G. fortis* (see figure 31.3), and more similar than were the songs of those that did not breed. One father-son pair is shown; male 14506 is the father of 15466.

smaller than those *G. scandens* that mated conspecifically. Third, body size of the *G. scandens* mates was positively correlated with the body size of the mothers (but not the fathers) of the *G. fortis* females. Thus, hybridizing birds were unusual members of their populations in being morphologically similar to their mates, and the correlations suggest that hybridizing birds are influenced by the appearance of their parents, specifically mothers, in choosing a mate.

Imperfections in the Barrier to Gene Flow

Imprinting tends to constrain the mating of females to conspecifics, and thus plays an important role in species formation by acting as a barrier to gene flow on secondary contact. And yet the barrier is leaky. What is responsible for the leaks?

For the rarest species, *G. fuliginosa*, we need look no further for an answer than the scarcity of potential con-

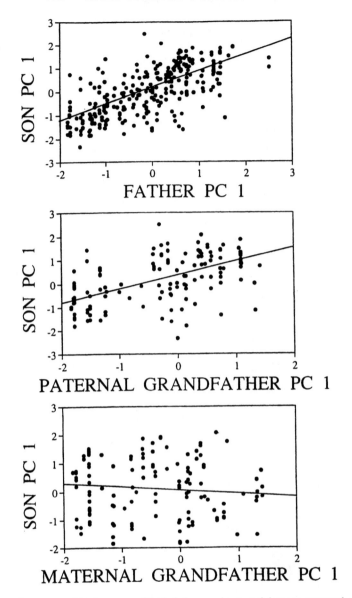

Figure 31.6. Songs of sons resemble the songs of their fathers and paternal, but not maternal, grandfathers. Reprinted with permission from B. R. Grant and Grant (1996a).

specific mates, and an abundance of *G. fortis* that are very similar to *G. fuliginosa* in morphology and song. Rarity of one species is well known as a predisposing factor in hybridization (Mayr 1963; Grant and Grant 1997a) and has been called the Hubbs principle in recognition of the work of Hubbs (1955) on the hybridization of fish species (Mayr, cited in Sibley 1961). A strongly male-biased sex ratio and hence a scarcity of conspecific females might also help to explain the fact that *G. scandens* males hybridized more frequently than females (Grant and Grant 1997a). Unanswered is why individuals of *G. fortis*, the common species, should choose to mate with a member of a rarer species. We know that they don't do so because they are poor reproductive competitors owing to inexperience or slow maturity, or because the sex ratio is skewed at the time that they secure mates (Grant and Grant 1997a). The most likely explanation is that they have become misimprinted; that is, they hybridize because they have been imprinted on the song and appearance of a heterospecific individual, or as a result of mating with a misimprinted male.

To learn a song a bird must hear it during the sensitive period early in life. If the father dies or for some reason does not sing frequently, his offspring are likely

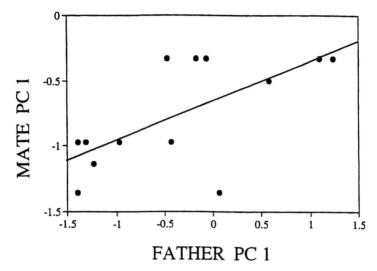

Figure 31.7. Similarity in the songs of the fathers of *G. fortis* females and the songs of their *G. fuliginosa* mates; see also figure 31.3. Reprinted from Grant and Grant (1997a) with permission of University of Chicago Press.

to imprint on the song they hear most frequently during their sensitive period. Note that as females do not sing, we have no means of knowing the identity of the models on which they imprint. For example, misimprinting on heterospecific models is a viable but untested hypothesis for the strange case of two *G. fortis* females from the same nest hybridizing with males of two different species! However misimprinting of males has been documented. In one case an exceptionally loud and vocal *G. magnirostris* nested between two adjacent *G. fortis* territories. All three nests were unusually close to each other. A male offspring from each of the *G. fortis* nests grew up to sing *G. magnirostris* song and not the *G. fortis* song of their fathers (figure 31.8). In this case the misimprinting did not lead to hybridization; the two males (and one female from one of the nests) mated conspecifically. Nor did it lead to hybridization in another instance of a related *G. fortis* male learning to sing a *G. scandens* song (figure 31.8).

Even though the direct link between misimprinting and hybridization is lacking, there is an association between the singing of heterospecific song and hybridization. Three of nine breeding *G. fortis* and *G. scandens* males that sang the other species' song hybridized, and they formed three of the total of eight pairs of *G. fortis* × *G. scandens*.

The Fitness Consequences of Hybridization

Hybridization will have no evolutionary consequence if hybrids do not survive long enough to breed, as was the case prior to 1983. However after 1983, although the frequency of hybridization did not increase, hybrids survived to reproduce and backcrossed to members of their parental species. Introgressive hybridization continued throughout the following 12 years (figure 31.9).

After 1983 F_1 hybrids of three cohorts (1983, 1987, 1991) survived as well as or better than individuals from the parental species born (hatched) at the same time (figure 31.10). The *G. fortis* × *G. scandens* F_1 hybrids from all three cohorts survived particularly well. The first and second backcross generations showed similar high survival in comparison with individuals from the parental species of the same cohorts.

Hybrids and backcrosses not only survived well but reproduced well. The number of clutches, production of eggs, and nestling and fledgling success of F_1 hybrids and first generation backcrosses did not differ significantly from the parental species in either of the first two cohorts (1983 or 1987). As a consequence of their high survival and reproduction rates, the most numerous cohorts of hybrids and backcrosses (1987) had more than replaced themselves by 1991 (Grant and Grant 1992), whereas the parental species cohorts born in the same year and experiencing the same environmental conditions did not do so until one or two years later.

The main conclusion we draw from these comparisons is that the hybridizing species are genetically compatible. There is no evidence of inviability or sterility in the female (heterogametic sex) as might be expected if some degree of postzygotic isolation had arisen in accordance with Haldane's Rule (Grant and Grant 1992). A second conclusion is that ecological circumstances determine whether or not hybridization leads to introgression of genes.

MISIMPRINTING

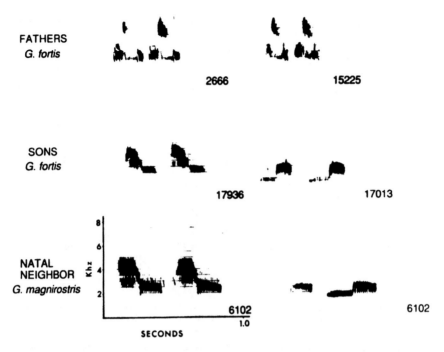

Figure 31.8. Miscopying of *G. magnirostris* songs by *G. fortis* as a result of an imprinting-like process. The two *G. fortis* sons copied the two songs sung by a neighboring *G. magnirostris* male and not the songs sung by their respective fathers. Their fathers were adjacent neighbors. Male 15225 was the son of 2666: a rare case of a son occupying a territory next to its father. The *G. magnirostris* male is also a rare example of an individual singing two songs. In neither case did the misimprinting lead to hybridization. Reprinted with permission from B. R. Grant and Grant (1996a).

Causes of the Change in Hybrid Fitness

Hybrids survived long enough to reproduce after 1983 as a result of altered ecological conditions produced by an exceptionally severe El Niño event (Gibbs and Grant 1987; Grant and Grant 1993). This brought a large amount of rain to the archipelago and altered the amount and composition of the plants and hence food supply of the finches, especially the dry season food supply. Prior to 1983 the seed supply of the finches had been dominated by the large and hard seeds of *Tribulus cistoides* and *Opuntia echios*. The abundant rainfall that year stimulated the production of many annuals, vines, and other plants, all of which produced small, soft seeds. *Tribulus cistoides*, a low-lying plant, was smothered by the fast-growing grasses, herbs, and vines. *Opuntia* also suffered. Many were covered by vines and were unable to photosynthesize, whereas others absorbed so much water they became top-heavy and were blown over in the high winds accompanying the rain storms. The altered plant composition has persisted to the present as the re-

sult of a second El Niño event in 1987 and a third extending from 1991 to 1993.

All finch populations benefited from the highly productive conditions in 1983, but when food supply and finch numbers subsequently declined the hybrids survived especially well. A strong determinant was bill morphology (B. R. Grant and Grant 1996b). Hybrids (and backcrosses) inherit bill traits from both parents and have on average intermediate bill sizes (Grant and Grant 1994). Bill shape and size covary with seed size in the diet of the parental species (Grant 1981b; Price 1987; Grant and Grant 1989); therefore, it is not surprising to find that hybrids with intermediate bill morphology have diets generally intermediate between the parental species (figure 31.11). A decrease in abundance of *Tribulus* and *Opuntia* seeds, the major food of large *G. fortis* and *G. scandens*, respectively, coincided with high mortality and a decline in numbers of these species (Gibbs and Grant 1987; B. R. Grant and Grant 1996b). Hybrids, in contrast may have been favored by a high biomass of small seeds (B. R. Grant and Grant 1996b). Hybrids of

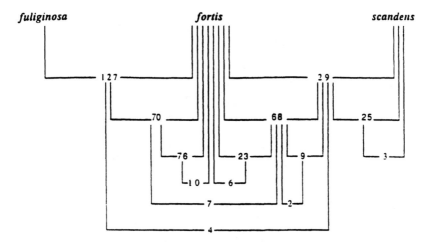

Figure 31.9. A network of interbreeding relationships of three species on Daphne Major island. Numbers refer to the total banded fledglings produced in 1976–92 by interspecific and various hybrid pairs. A small number of backcrosses produced by other types of crosses are not shown. From Grant (1993).

G. fortis × *G. fuliginosa* feed predominantly on small seeds, and are able to feed at least as efficiently and perhaps more efficiently on small seeds than *G. fortis* (B. R. Grant and Grant 1996b). The high survival of *G. fortis* × *G. scandens* hybrids can also partly be accounted for by an abundance of small seeds. They have a broader diet than *G. scandens*, and can exploit *Opuntia* seeds significantly more efficiently than *G. fortis* (B. R. Grant and Grant 1996b). Another possibility for the higher survival of hybrids is as yet untested: that they experienced hybrid vigor as a result of their heterozygous condition.

Population-Level Consequences of Introgressive Hybridization

If fluctuating environmental conditions have persisted in the past (P. R. Grant and B. R. Grant 1996b), and interbreeding without fitness loss has been occurring sporadically, the question arises as to why the species have not fused into a single panmictic population. There are two answers, one for the hybridization between *G. fortis* and *G. fuliginosa* and another for the hybridization between *G. fortis* and *G. scandens*.

First, although *G. fuliginosa* and *G. fortis* are in the process of fusing into a single population, with all F_1 hybrids backcrossing to *G. fortis*, repeated immigration of *G. fuliginosa* maintains this species on the island. Second, and in contrast, *G. fortis* and *G. scandens* maintain their specific distinctiveness in the face of gene exchange as a result of selective mating of the F_1 hybrids according to which song group they belong to (figure 31.12). All the F_1 hybrid females mated with males that sang the same species song as their fathers. Likewise, all

three F_1 hybrid males backcrossed to their paternal-song species, although in one year one of them, singing the song of *G. scandens*, paired with a *G. fortis* female. All of the backcrosses bred within their own song group. Thus imprinting on paternal song appears to be the most important factor in determining the mating pattern of F_1 hybrids and their offspring (Grant and Grant 1997b), as it is for the hybridizing species (Grant and Grant 1997a). Sexual imprinting on parental morphology appears to play an additional role (Grant and Grant 1997b). For example the one F_1 hybrid male that sang a *G. scandens* song but paired with *G. fortis* was the most like *G. fortis* in morphology. Although the morphological distinctiveness of *G. fortis* and *G. scandens* has become less sharp as a result of the backcrossing (figure 31.13), there is no evidence that the frequency of hybridization has been increased.

The pattern given above is illustrated in figure 31.13 with birds born during and after 1987. *Geospiza fortis* and *G. scandens* are discretely different in morphology and song (figure 31.13A). Backcrosses are morphologically intermediate between the two species (figure 31.13B). Nevertheless, backcrosses to *G. scandens* sing *G. scandens* songs and tend to be morphologically closer to the *G. scandens* line of allometry, and backcrosses to *G. fortis* sing a *G. fortis* song and are closer to the *G. fortis* line of allometry (figure 31.13B). Members of the second backcross generation (*FFS* × *F*) are morphologically intermediate between their parents and indistinguishable in morphology and song from *G. fortis* (figure 31.13C). Thus the close association of morphology with song that began to weaken in the first backcross generation is reestablished by the second generation, imprinting on song having constrained the phenotypic fusion of the two species but not the flow of genes from one population to the other.

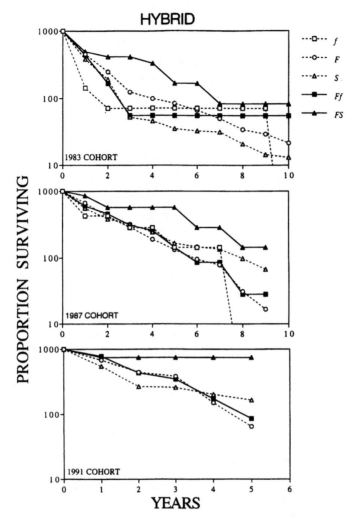

Figure 31.10. Survival of F_1 hybrids, parental species, and two generations of backcrosses; cohorts are complete and not sampled. Actual sizes of the 1983, 1987, and 1991 cohorts, respectively: *f* (*fuliginosa*), 14, 7, —; *F* (*fortis*), 1019, 955, 581; *S* (*scandens*), 763, 164, 108; *Ff* (*fortis* × *fuliginosa*), 18, 35, 23; *FS* (*fortis* × *scandens*), 12, 7, 4; *FFf* (first-generation backcross of *Ff* to *fortis*), —, 34, 31; *FFS* (first-generation backcross of *FS* to *fortis*), —, 43,—; *SSF* (first-generation backcross of *FS* to *scandens*), —, —, 19; *FFFf* (second-generation backcross of *Ff* to *fortis*), —, —, 65; *FFFS* (second-generation backcross of *FS* to *fortis*), —, —, 17.

Genetic Consequences of Hybridization

Even without a direct knowledge of the genes, we can infer genetic consequences of hybridization from the pattern of matings and from analyses of heritable phenotypic variation.

Gene Exchange

Mating patterns reveal an unequal exchange of genes between the hybridizing species (Grant and Grant 1997b).

Hybrids formed by the interbreeding of *G. fortis* and *G. fuliginosa* backcross solely to *G. fortis*. As a result both mitochondrial and nuclear genes flow unidirectionally from *G. fuliginosa* to *G. fortis*. Gene flow between *G. fortis* and *G. scandens* is bidirectional, but not equal. Breeding hybrids have been produced only by *G. scandens* males paired with *G. fortis* females. The genetic consequences have depended on which song the *G. scandens* male sang. When the song was the typical one for the species (*G. scandens*), all F_1 hybrids backcrossed to *G. scandens*, and mitochondrial genes and some nuclear genes should

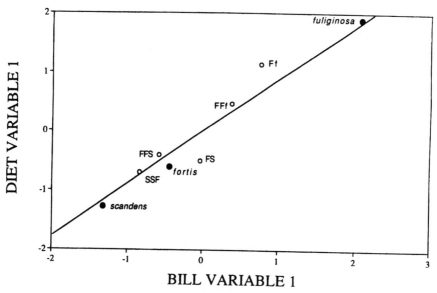

Figure 31.11. Canonical correlation of diets and morphology among F_1 hybrids, parental species, and the first-generation backcrosses. Based on the canonical correlation analysis reported in B. R. Grant and Grant (1996b), with the addition of SSF (the backcross class of *G. scandens* × *G. fortis* F_1 hybrids to *G. scandens*). The bill variable comprises bill length and width. The diet variable comprises the proportion of birds in each group feeding on *Opuntia* flowers, and the proportion of the seed-eating subset feeding on *Tribulus* seeds. *G. fuliginosa*, the smallest species, feeds almost entirely on small seeds, and *G. scandens* feeds largely on *Opuntia* flowers and seeds other than the large and hard *Tribulus* seeds. Hybrids and backcrosses are generally intermediate on both bill and diet axes (see also Fig. 31.13). See figure 31.10 for abbreviations.

HYBRIDIZATION

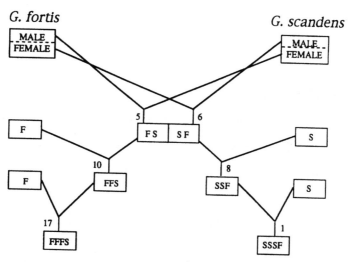

Figure 31.12. The mating pattern of hybrids and backcrosses. The pairing of F_1 hybrids produced by mixed *G. fortis* and *G. scandens* pairs, and all the backcrosses, is determined by paternal song. This applies to both sexes. Numbers refer to individuals in each category that produced offspring that bred. Other offspring that did not breed (not shown) include some that were still alive in 1996. Note the absence of any F_2 hybrids. For abbreviations, see figure 31.10.

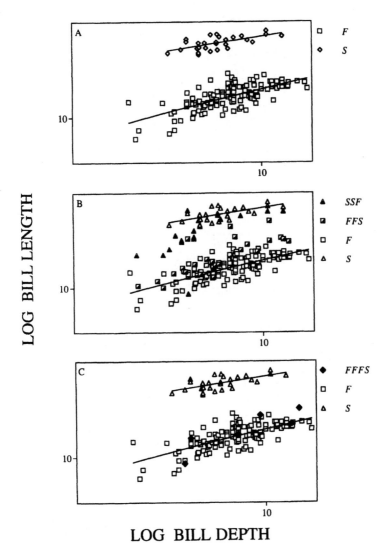

Figure 31.13. The morphological effect of adding first-generation backcrosses (*FFS* and *SSF*) and second-generation backcrosses (*FFFS*) to the parental species *G. fortis* (*F*) and *G. scandens* (*S*). All males and females produced in 1987 and after are included; their parents were known in all cases. Regression lines are drawn through the points for the species alone. A, *G. fortis* and *G. scandens* are discretely different in morphology (and song). B, first-generation backcrosses are morphologically intermediate between the two species, *SSF* being closer to the line of allometry of *G. scandens*, *FFS* to the line of allometry of *G. fortis* (*SSF* males sing *G. scandens* songs, *FFS* males sing *G. fortis* songs). C, as a result of all *FFS* breeding with *G. fortis*, the second-generation backcrosses are morphologically indistinguishable from *G. fortis* in morphology (and song).

have flowed across the species boundary, that is, from *G. fortis* to *G. scandens*. On the other hand, when hybridizing *G. scandens* males sang a *G. fortis* song, all backcrossing took place to *G. fortis*, and the *G. fortis* population would have gained nuclear genes from *G. scandens* but no mitochondrial genes. Unless *G. fortis* males successfully hybridize with *G. scandens* females under conditions not witnessed during our study, these results lead to a clear prediction: *G. scandens* should be distinctive in mitochondrial genes but not in selectively neutral nuclear genes.

Polygenic Effects

Gene mixing through hybridization can severely complicate attempts to establish phylogenetic relationships with molecular data. Our concern here is with the effects of introgression on speciation, rather than on phylogeny. Potentially those effects range from the destruction of sharp morphological boundaries between species to the creation of new taxa. Introgression of genes through hybridization produces novel combinations of alleles (Stebbins 1959; Lewontin and Birch 1966; Barton and Hewitt 1985; Woodruff 1989). The new combinations may be selectively superior, and provide the starting point of a new direction of evolutionary change that could lead to the formation of a new species.

By comparing samples of morphological measurements of the species with and without hybrids and backcrosses included, we calculated that the additive genetic variance introduced by hybridization was two to three orders of magnitude greater than that introduced by mutation alone (Grant and Grant 1994). Thus, one effect of hybridization of Darwin's finches is to increase the standing (polygenic) genetic variation and thereby enhance the population's evolutionary responsiveness to selection.

A second consequence of hybridization is to strengthen or weaken the genetic covariation of morphological traits, the effect depending on the nature of the allometries of the hybridizing species. The morphological analyses showed that introgressive hybridization of *G. fortis* and *G. fuliginosa* with similar allometries strengthens genetic correlations, thereby rendering evolution in a new direction away from the line of allometry less likely. In contrast to this, hybridization between *G. fortis* and *G. scandens* with different allometries weakens genetic correlations. For example, bill length becomes genetically more independent of bill depth and width as a result of the hybridization, and hence the potential for one trait to change under selection while another does not becomes enhanced. Thus, the opportunity arises with this pair of species for a new evolutionary trajectory to be followed, one in which bill shape differs from either of the ancestral forms (Grant and Grant 1994).

Discussion

Birds, especially the passerines (Raikow 1986; Baptista and Trail 1992), are rich in numbers of species partly because they have evolved a remarkable variety of plumage colors and patterns, songs, and courtship displays. General explanations for the phenotypic variety are currently being sought in terms of sexual selection arising from female preferences for bright and conspicuous plumage (Andersson 1994). The problem is one of explaining the origin of novel traits or the elaboration of existing traits in males, and why females should prefer the novelty or elaboration. Models explore the evolution of male traits by assuming a genetic covariation exists between the male traits and female preferences for them. Female preferences, or more generally mate preferences, are also responsible for the lack or scarcity of hybridiza-

tion. This is the connection between sexual selection and speciation (see also Ryan and Rand 1993). The connection is usually made explicit with models of the divergence of male signaling systems and female responses under a regime of reinforcement when mixed mating results in low hybrid fitness. Reinforcement is not necessary, however. The novel traits could arise in isolation and be subject to directional sexual selection that leads to speciation entirely in allopatry (Lande 1981).

Consideration of speciation in Darwin's finches leads us to emphasize three other processes. First, the evolutionary divergence of lineages takes place as a result of morphological adaptation of allopatric populations to different ecological conditions. Song, a culturally inherited trait, also diverges in isolation through an accumulation of small copying inaccuracies (Grant and Grant 1997c). Second, when secondary contact is made, the populations are not reproductively isolated from each other as a result of postzygotic isolating mechanisms that have arisen in allopatry. The populations are in fact reproductively compatible; hence, there is no scope for reinforcement. Third, they nonetheless usually do not interbreed because their choice of mates is governed by sexual imprinting on parental features of morphology and song. Although the first of these points concerning ecological divergence is well recognized and uncontroversial (e.g., Mayr 1942, 1963; Lack 1947; Grant 1986), the others are generally overlooked or underemphasized.

It may not be an exaggeration to suggest that imprinting syndromes are identical in closely related species derived from a common ancestor. The syndromes include the timing and duration of the sensitive period for learning, and receptivities to certain stimuli and not to others. Differences between the species reside in the particular stimuli that individuals experience during imprinting, and not in the imprinting mechanisms themselves. Independent of these influences during imprinting, there may be no inherent tendencies for members of either sex to prefer conspecific mates over heterospecific ones. If this is correct, male traits and female preferences for them are associated within species (and differ between species) not just because there is a close genetic coupling between them but because there is an environmental coupling through similar experience. Cross-fostering provides an experimental device for testing the null hypothesis (e.g., see Bischoff and Clayton 1991; Kruijt and Meeuwissen 1993).

These possibilities cast a different light on the scope for sexual selection and the evolution of premating isolation. They suggest that the place to look for salient genetic variation is in the details of the imprinting syndrome. Evolution of mate preferences may depend on genetic variation in the fine details of imprinting mechanisms, such as the length of the sensitive period and its neuronal and hormonal control (e.g., see Immelmann 1980; Nottebohm 1993). If genetic variation is limited, evolution will be slow.

Thus, the mating isolation of recently evolved species can involve an interesting combination of genetically and culturally inherited factors. It seems to us likely that the combined factors play a similar role in other groups where imprinting on parental, or even nonparental, features occurs. Furthermore, since the normal imprinting program can be disrupted or perturbed and lead to hybridization, we would expect hybridization to occur in those groups exhibiting sexual imprinting. Both imprinting (Ten Cate et al. 1993) and hybridization (Grant and Grant 1992) are widespread in birds, but no study has yet been undertaken to examine the functional connection between the two. Indications of a functional connection exist in the form of crude associations: between the importance of early learning of song or plumage in groups that are known to hybridize at a relatively high frequency, such as parrots, hummingbirds, ducks, and geese, and the lack or scarcity of hybridization in other groups that have genetically inherited and not learned song, such as doves and suboscine passerines.

Hybridization, like gene flow between conspecific populations, is often viewed as a process that retards or reverses differentiation. A counter view has long been held by many botanists. For example, in 1959 Stebbins wrote that through introgression (or polyploidy) the products of hybridization can form new evolutionary lines that are isolated from the ancestral type and are therefore free to evolve in new directions. Our study of hybridization in Darwin's finches suggests that the creative role of introgressive hybridization in the speciation process should not be thought of as confined to plants. Introgressive hybridization results in an enhancement of additive genetic variation. This facilitates directional evolutionary change, subject to constraints arising from genetic correlations between characters. But these constraints become weaker when the interbreeding species have different allometries; hence, evolutionary change in a new direction becomes easier (Grant and Grant 1994). Whether or not the evolutionary potential is translated into something novel will depend on population structure and ecological opportunity, on such factors as the persistence of hybridization, the strength of directional selection pressures, on whether the population goes through a bottleneck, and on the likelihood that new populations may be started through the colonization of nearby islands by a small number of individuals including hybrids and backcrosses (e.g., see figure 31.1).

Our suggestions on the evolutionary potential of introgressive hybridization in birds are general and not restricted to morphometric, continuously varying, traits. Consider the evolutionary origin of a new discontinuously varying trait. It is more difficult to explain than is the elaboration of an existing trait. A novel trait requires a genetic explanation independent of the sexual selection regime that might favor it. Single mutations are one possible answer, and Mayr's (1954) founder principle in small and peripherally isolated populations is another

(P. R. Grant and B. R. Grant 1996a, 1997c). A third possibility is hybridization. Hybrid ducks, for example, may exhibit plumage traits not exhibited in either of the parental species that produced them (Scherer and Hilsberg 1982).

We conclude by suggesting that when viewed over the long-term of millenia, introgressive hybridization may have contributed importantly to the generation of species diversity in birds.

Acknowledgments   We thank the Galápagos National Parks Service and the Charles Darwin Research Station for administrative and logistical support of our research on the Galápagos islands, and the National Science and Engineering Research Council of Canada and the National Science Foundation of the USA for financial support.

References

Andersson, M. 1994. Sexual Selection. Princeton, N.J.: Princeton University Press.

Arnold, M. L. 1992. Natural hybridization as an evolutionary process. Annu. Rev. Ecol. Syst. 23:237–261.

Baptista, L. F., and Morton, M. L. 1988. Song learning in montane white-crowned sparrows: from whom and when. Anim. Behav. 36:1753–1764.

Baptista, L. F., and Trail, P. W. 1992. The role of song in the evolution of passerine diversity. Syst. Biol. 41:242–247.

Bard, E., Hamelin, B., Fairbanks, R. G., and Zindler, A. 1990. Calibration of the ^{14}C timescale over the past 30,000 years using mass spectrometric U-Th ages from Barbados corals. Nature 345:405–410.

Barton, N. H., and Hewitt, G. M. 1985. Analysis of hybrid zones. Annu. Rev. Ecol. Syst. 15:133–164.

Beebe, W. 1924. Galápagos: World's End. Putnams, New York.

Bischoff, H.-J., and Clayton, N. 1991. Stabilization of sexual preferences by sexual experience in male zebra finches *Taeniopygia guttata castanotis*. Behaviour 118:144–155.

Boag, P. T. 1983. The heritability of external morphology in Darwin's ground finches (*Geospiza*) on Isla Daphne Major, Galápagos. Evolution 37:877–894.

Boles, W. E. 1995. The world's oldest songbird. Nature 374:21–22.

Bowman, R. I. 1961. Morphological differentiation and adaptation in the Galápagos finches. Univ. Calif. Publ. Zool. 58:1–302.

Bowman, R. I. 1983. The evolution of song in Darwin's finches. In R. I. Bowman, M. Berson, and A. E. Leviton (eds.). Patterns of Evolution in Galápagos Organisms. San Francisco: American Association for the Advancement of Science, pp. 237–537.

Bush, G. 1993. A reaffirmation of Santa Rosalia, or why are there so many kinds of small animals? In D. R. Lees and D. Edwards (eds.). Evolutionary Patterns and Processes. London: Linnean Society, pp. 229–245.

Bush, G. 1994. Sympatric speciation in animals: new wine in old bottles. Trends Ecol. Evol. 9:285–288.

Christie, D. M., Duncan, R. A., McBirney, A. R., Richards, M. A., White, W. M., Harpp, K. S., and Fox, C. B. 1992. Drowned islands downstream from the Galapagos hotspot imply extended speciation times. Nature 355:246–248.

Gibbs, H. L. 1990. Cultural evolution of male song types in Darwin's medium ground finches, *Geospiza fortis*. Anim. Behav. 39:253–263.

Gibbs, H. L., and Grant, P. R. 1987. Ecological consequences of an exceptionally strong El Niño event on Darwin's finches. Ecology 68:1735–1746.

Gifford, E. W. 1919. Field notes on the land birds of the Galápagos Islands and of Cocos Island, Costa Rica. Proc. Calif. Acad. Sci., ser. 4, 2:189–258.

Grant, B. R., and Grant, P. R. 1989. Evolutionary Dynamics of a Natural Population: The Large Cactus Finch of the Galápagos. University of Chicago Press, Chicago.

Grant, B. R., and Grant, P. R. 1993. Evolution of Darwin's finches caused by a rare climatic event. Proc. R. Soc. Lond. B 251:111–117.

Grant, B. R., and Grant, P. R. 1996a. Cultural inheritance of song and its role in the evolution of Darwin's finches. Evolution 50:2471–2487.

Grant, B. R., and Grant, P. R. 1996b. High survival of Darwin's finch hybrids: effects of beak morphology and diets. Ecology 77:500–509.

Grant, P. R. 1981a. Speciation and the adaptive radiation of Darwin's finches. Am. Sci. 69:653–663.

Grant, P. R. 1981b. The feeding of Darwin's finches on *Tribulus cistoides* (L.) seeds. Anim. Behav. 29:785–793.

Grant, P. R. 1986. Ecology and Evolution of Darwin's finches. Princeton, N.J.: Princeton University Press.

Grant, P. R. 1993. Hybridization of Darwin's finches on Isla Daphne Major, Galápagos. Philos. Trans. R. Soc. Lond. B 340:127–139.

Grant, P. R. 1994. Population variation and hybridization: comparison of finches from two archipelagos. Evol. Ecol. 8:598–617.

Grant, P. R., and Grant, B. R. 1992. Hybridization of bird species. Science 256:193–197.

Grant, P. R., and Grant, B. R. 1994. Phenotypic and genetic effects of hybridization in Darwin's finches. Evolution 48:297–316.

Grant, P. R., and Grant, B. R. 1995. The founding of a new population of Darwin's finches. Evolution 49:229–240.

Grant, P. R., and Grant, B. R. 1996a. Speciation and hybridization of island birds. Philos. Trans. R. Soc. Lond. B 351:765–772.

Grant, P. R., and Grant, B. R. 1996b. Finch communities in a climatically fluctuating environment. In M. L. Cody and J. A. Smallwood (eds.). Long-Term Studies of Vertebrate Communities. New York: Academic Press, pp. 343–390.

Grant, P. R., and Grant, B. R. 1997a. Hybridization, sexual imprinting and mate choice. Am. Nat. 149:1–28.

Grant, P. R., and Grant, B. R. 1997b. Mating patterns of Darwin's finch hybrids determined by song and morphology. Biol. J. Linn. Soc. 60:317–343.

Grant, P. R., and Grant, B. R. 1997c. Genetics and the origin of bird species. Proc. Nat. Acad. Sci. USA 94:7768–7775.

Howard, D. 1993. Reinforcement: origin, dynamics, and the fate of an evolutionary hypothesis. In R. G. Harrison (ed.). Hybrid Zones and the Evolutionary Process. Oxford: Oxford University Press, pp. 46–69.

Hubbs, C. L. 1955. Hybridization between fish species in nature. Syst. Zool. 4:1–20.

Immelmann, K. 1975. Ecological significance of imprinting and early experience. Annu. Rev. Ecol. Syst. 6:15–37.

Immelmann, K. 1980. Genetical constraints on early learning: A perspective from sexual imprinting in birds. In (J. R. Royce, ed.). Theoretical Advances in Behavior Genetics. Amsterdam: Van Nijhoff, pp. 121–133.

James, H. F., and Olson, S. L. 1991. Descriptions of thirty-two new species of birds from the Hawaiian islands: part II. Passeriformes. Ornithological Monographs No. 46. Washington, DC: American Ornithologists' Union.

Kruijt, J. P., and Meeuwissen, G. B. 1993. Consolidation and modification of sexual preferences in adult male zebra finches. Neth. J. Zool. 43:68–79.

Lack, D. 1947. Darwin's finches. Cambridge: Cambridge University Press.

Laland, K. N. 1994. On the evolutionary consequences of sexual imprinting. Evolution 48:477–489.

Lande, R. 1981. Models of speciation by sexual selection on polygenic traits. Proc. Nat. Acad. Sci. USA 78:3721–3725.

Lewontin, R. C., and Birch, L. C. 1966. Hybridization as a source of variation for adaptation to new environments. Evolution 20:315–336.

Mayr, E. 1942. Systematics and the Origin of Species. New York: Columbia University Press.

Mayr, E. 1954. Change of genetic environment and evolution. In J. Huxley, A. C. Hardy, and E. B. Ford (eds.). Evolution as a Process. London: Allen and Unwin, pp. 157–180.

Mayr, E. 1963. Animal Species and Evolution. Cambridge, Mass.: Belknap.

Millington, S., and Price, T. 1985. Song inheritance and mating patterns in Darwin's finches. Auk 102:342–346.

Nelson, D. A., and Marler, P. 1993. Innate recognition of song in white-crowned sparrows: a role in selective vocal learning? Anim. Behav. 46:806–808.

Nelson, D. A., Marler, P., and Palleroni, A.. 1995. A comparative approach to vocal learning: intraspecific variation in the learning process. Anim. Behav. 50:83–97.

Nottebohm, F. 1993. The search for neural mechanisms that define the sensitive period for song learning in birds. Neth. J. Zool. 43:193–234.

Payne, R. B. 1981. Song learning and social interaction in indigo buntings. Anim. Behav. 29:688–697.

Prager, E. R., and Wilson, A. C. 1975. Slow evolutionary loss of the potential for interspecific hybridization in birds: a manifestation of slow regulatory evolution. Proc. Nat. Acad. Sci. USA 72:200–204.

Price, T. D. 1987. Diet variation in a population of Darwin's finches. Ecology 68:1015–1028.

Raikow, R. J. 1986. Why are there so many kinds of passerine birds? Syst. Zool. 35:255–259.

Ratcliffe, L. M. 1981. Species recognition in Darwin's ground finches (*Geospiza*, Gould). Ph.D. thesis, McGill University.

Ratcliffe, L. M., and Grant, P. R. 1983a. Species recognition in Darwin's finches (*Geospiza*, Gould). I. Discrimination by morphological cues. Anim. Behav. 31:1139–1153.

Ratcliffe, L. M., and Grant, P. R. 1983b. Species recognition in Darwin's finches (*Geospiza*, Gould). II. Geographic variation in mate preference. Anim. Behav. 31:1154–1165.

Ratcliffe, L. M., and Grant, P. R. 1985. Species recognition in Darwin's finches (*Geospiza*, Gould). III. Male responses to playback of different song types, dialects and heterospecific songs. Anim. Behav. 33:290–307.

Ryan, M. J., and Rand, A. S. 1993. Species recognition and sexual selection as a unitary problem in animal communication. Evolution 47:647–657.

Scherer, S., and Hilsberg, T. 1982. Hybridisierung und Verwandtschaftsgrade innerhalb der Anatidae—eine systematische und evolutionstheoretische Betrachtung. J. Ornithol. 123:357–380.

Schluter, D., and Grant, P. R. 1984. Determinants of morphological patterns in communities of Darwin's finches. Am. Nat. 123:175–196.

Sibley, C. G. 1961. Hybridization and isolating mechanisms. In W. F. Blair (ed.). Vertebrate Speciation. Austin: University of Texas Press, pp. 69–88.

Steadman, D. W. 1995. Prehistoric extinctions of Pacific island birds: biodiversity meets zooarchaeology. Science 267:1123–1131.

Stebbins, G. L., Jr. 1959. The role of hybridization in evolution. Proc. Am. Philos. Soc. 103:231–251.

Stern, D. L., and Grant, P. R. 1996. A phylogenetic reanalysis of allozyme variation among populations of Galápagos finches. Zool. J. Linn. Soc. 118:119–134.

Stresemann, E. 1936. Zur Frage der Artbildung in der Gattung *Geospiza*. Orgaan der Club Van Nederlandische Vogelkunde 9:13–21.

Tarr, C. L., and Fleischer, R. C. 1995. Evolutionary relationships of the Hawaiian honeycreepers (Aves: Drepanidinae). In W. L. Wagner and V. A. Funk (eds.). Hawaiian Biogeography: Evolution on a Hot-Spot Archipelago. Washington, DC: Smithsonian Institution Press, pp. 147–159.

Ten Cate, C., and Bateson, P. P. G. 1988. Sexual selection: the evolution of conspicuous characteristics in birds by means of imprinting. Evolution 42:1355–1358.

Ten Cate, C., Vos, D. R., and Mann, N. 1993. Sexual imprinting and song learning: two of one kind? Neth. J. Zool. 43:34–45.

Vincek, V., O'Huigin, C., Satta, Y., Takahata, N., Boag, P. T., Grant, P. R., Grant, B. R., and Klein, J. 1997. How large was the founding population of Darwin's finches? Proc. R. Soc. Lond. B 264:111–118.

Woodruff, D. S. 1989. Genetic anomalies associated with *Cerion* hybrid zones: the origin and maintenance of new electromorph variants called hybrizymes. Biol. J. Linn. Soc. 36:281–294.

Yang, S.-Y., and Patton, J. L. 1981. Genic variability and differentiation in Galápagos finches. Auk 98:230–242.

Part VII

Perspectives

32

The Conceptual Radicalization
of an Evolutionary Biologist

Guy L. Bush

An Early Interest in Natural History

I can't recall when I first started thinking about evolution and how species evolve, but my interest in the biological world was kindled very early in my life. Like many evolutionary biologists, I seemed to have an inborn fascination with nature that my parents, Guy and Louise Bush, recognized and encouraged. They gave me free reign to pursue my interests in animals and plants. I was allowed to explore the countryside and do things on my own that I would be reluctant to let my own children do today. This freedom was probably the single most important factor contributing to the development of my interest in biology and evolution.

Living close to nature was also fairly easy in my early years because my parents always kept a summer cottage within commuting distance from our home town, first on Devils Lake in Baraboo, Wisconsin, and later at my aunt Clara Wilkinson's cattle ranch west of Denver near Evergreen, Colorado. I spent my summers at these cottages whose surroundings provided me with a never-ending inventory of things to collect and observe while I was fishing, swimming, or wandering through the woods. With the help of my aunt Edna Gibbs, a school teacher who often spent her summer months with us in Wisconsin and elsewhere, I made my first insect and leaf collections when I was six years old. From then on my room was home to all kinds of biological odds and ends both living and dead, including the occasional sunfish, frog, garter snake, injured bird, or rodent. When we moved to Denver, Colorado, in 1940, I began raising pigeons in a loft my father helped me build. I would spend hours observing their behavior, fascinated by their ability to find their way home from long distances, their strutting courtship displays, and their almost humanlike care of their young.

In the fall of 1943, with the outcome of World War II uncertain, my father, at the time a regional director of the U.S. Government's Agricultural Adjustment Administration, was appointed agricultural attaché to Brazil and our family moved to Rio de Janeiro. For a 13 year old with a biological bent, Rio was paradise gained. I was soon collecting almost everything I could lay my hands on. With four close school mates I explored and camped in the forests surrounding Rio and the islands in its vast and splendid Baia de Guanabara. Typical of these trips were the ones we took to the steep mountain slopes surrounding Rio. These trips were sometimes made on a train destined for the resort town of Petropolis. As the wooden-seated coaches were pulled slowly, one by one, up the mountain by a tiny, wood-burning, Swiss cog-railway engine, we would jump off and set off into the forest. There we camped within the crumbling, verdant walls of an old, early-nineteenth-century inn and stagecoach stop located on an abandoned cobblestone road. We would spend the days wandering, climbing high into the trees, stopping here and there to observe some new insect, plant, or bird. Sometimes we would follow a troop of marmosets. Often we just hung around the campsite talking and daydreaming. When our food supply was exhausted we hiked back to the cog-railroad tracks and caught one of the coaches coming down the mountain.

On these trips I collected orchids, bromeliads, and other plants for our yard, trapped birds for the aviary I installed outside my room on our home's third floor terrace, or collected insects that I kept in jars and jerry-built cages. Since our house was only a block from the Ipanema beach, I had easy access to the seashore. A lot of my time was spent swimming, surfing, fishing, and wandering along the beach searching for marine life. With help from my friends, I fashioned a crude diving helmet from a 20-liter cookie can, an old truck tire pump, and a garden hose, using homemade lead belt and shoes for ballast. It was lucky that my friends and I were not drowned trying to explore the shallow margins and rocky outcrops along Rio's Ipanema shoreline. For me, the five teenage years spent in Brazil was altogether one long, magnificent summer vacation. I seldom studied or took my schoolwork seriously. In retrospect, this period of my life was to have a far greater influence on my future than I appreciated at the time.

Experience in a Brazilian Yellow Fever Laboratory

The single most important event that sharpened my biological interests occurred on a road trip with my father to the state of Minas Gerais in the southern summer months of February and March, 1946. In those years there were few paved highways except between the major cities, so we traveled in a Jeep on loan from the U.S. Navy. My father had it fitted out with a 100-gallon oil drum filled with gasoline, and a judicious supply of spare parts. These trips commonly entailed bouncing down deeply rutted ox cart tracks and single-lane dirt roads to reach agricultural areas my father needed to visit. Whenever we entered a small village, local inhabitants would usually press around to inspect the Jeep and the gringos. On the day we reached Passos, a little village in western Minas Gerais, we were told that an American scientist and his wife lived there and had opened a research laboratory. In short order we were accompanied by half the village, with the local priest in the lead, to a small, newly constructed building and introduced to Dr. Ottis R. Causey and his wife.

The lab had been built by the Rockefeller Foundation, and Causey, a medical doctor, was there to assess the mosquito populations and level of yellow fever in the monkey population in anticipation of an outbreak in the local human inhabitants. I was fascinated and showed so much interest and excitement that Causey and his wife convinced my father to let me return to Passos for an extended stay in the lab on the completion of our trip. It was not long before I was back in Passos after a long train ride, freshly immunized against yellow fever and ready for anything.

During my stay in Passos I spent many sunrises and sunsets perched on small wooden platforms within the forest canopy some 10–30 m above the ground, allowing all mosquitoes except *Anopheles* to feed on my exposed arms. When a female mosquito had her fill of my blood, I would gently capture her in a shell vial and return her to the lab, where she was allowed to lay a raft or two of eggs. The objective was to rear males that could be used to accurately identify species and to allow Causey, who also handled the identifications and systematic descriptions, to associate males with females. I was taught how to prepare slides of male genitalia and to use taxonomic keys for species identification. I was hooked. I suddenly had visions of one day being ensconced as a curator of insects in a museum somewhere making field trips to exotic places to collect and study mosquitoes.

College Days

Toward the end of my high school days, I enrolled at Iowa State College at my parents insistence. Iowa was a place my parents called home because we still held part of the old family farm in central Iowa homesteaded by my great grandfather in 1856. But I had become more Brazilian than American, which, I later learned, greatly concerned my parents. So when the time came to leave Brazil in late August of 1948, I was put on a plane and with great reluctance returned to the United States.

Iowa State College was a small, first-rate agricultural institution that boasted a beautiful campus. At that time the college was surrounded by corn fields belonging to prosperous farmers whose ancestors in the not so distant past had expeditiously displaced the native inhabitants and converted the wild prairie into cash crops and livestock. Although I was born in Iowa, returning wasn't exactly like coming home. To acknowledge that my first Iowa winter was traumatic, after five years in the tropics, is an understatement.

But in my sophomore year I took a course in insect biology and systematics taught by two dedicated and superb teachers, James Slater and Jean Laffoon, that rekindled my flagging interest in a biological career. It was another turning point that excited me enough to take more courses and major in entomology. However, in those days being male and attending a land grant college meant compulsory military courses in ROTC for a minimum of two years, for which I unfailingly managed to receive failing grades, a calamity since the Korean war had started. In order to delay the inevitable call from Uncle Sam and complete my coursework, I somehow persuaded the ROTC to let me repeat the first two years while enrolled in the second two year series. I managed to pass both lower and upper division requirements and graduated a year late in 1953 with a B.S. in entomology and a reserve commission as a 2nd Lieutenant in the Signal Corps.

Army Days

I reported for active duty to Fort Monmouth, New Jersey, on August 15, 1953, a little over two weeks after the Korean war ended. Before completing signal school, I requested assignment to the Signal Corps's only remaining homing pigeon loft, but the waiting list was long. I was transferred instead to the Medical Service Corps and sent to Germany as an ambulance platoon leader assigned first to the 48th Armored Medical Battalion at the headquarters of the 2nd Armored Cavalry Regiment (Hell on Wheels) in Mainz. After a brief period of training I was reassigned to Hannau, where I served as an ambulance platoon leader in the 915th Medical Corps (Ambulance). I passed the remainder of my tour with the army doing very little of consequence, except to take every possible opportunity to travel. Although from my perspective the army was an unconscionable waste of time, it did make me aware that I had to return to school. As the months dragged on, I became more and more impatient to get back to my insects and biology. My goal was still to be an insect systematist working in a large museum like the Smithsonian.

Iowa State College, Mexico, and a Job with the U.S. Department of Agriculture

Shortly after my discharge I returned to Iowa State College where I worked during the summer of 1956 for Professor John Lilly as a research field assistant on soil insects, an experience that gave me much-needed training in data collection and experimental design. That summer my life took another fortuitous sharp turn when a research position with the USDA opened in Mexico. With Lilly's support, I landed the job. By early fall of 1956 I had joined the staff of a moderately large USDA research lab in Mexico City and began to study the host relationships of the Mexican fruit fly, *Anastrepha ludens*. This tephritid fly was a major pest of citrus fruits and a potential threat to the citrus industry of California and Texas.

My assignment was to establish the range of native and introduced host fruits infested by this pest. I traveled throughout Mexico on extended field trips in order to systematically collect and identify every fruit I could find infested by *Anastrepha* larvae. These fruits were returned to the lab and held for sufficient time to allow larvae to mature and pupate. Emerging adults were then used for accurate species identification. My early training with Causey and travels in the outback of Brazil with my father began to pay off. However, the whole rearing process proved time consuming and costly. To make things more difficult, fruits were infested by many more *Anastrepha* species than I expected. I needed a faster, less labor intensive way of identifying larvae so morphologically similar to one another that they often could not be distinguished.

At about this time I read that chromosomes were being used to identify *Drosophila* sibling species, and I had the rather vague and naive hope of using larval karyotypes to shorten the identification process. I spent a good deal of time reading everything I could find on the subject in the science and agriculture library of the Rockefeller Foundation located at their Mexico City headquarters. Two things happened at that library that profoundly altered my future. I discovered how to make chromosome preparations, and I met Donald Freeburn, an agricultural economist working with the Rockefeller Foundation. Don eventually introduced me to my future wife, Dolores Rose Alpisa, who was visiting her uncle, Dr. John Pino, then assistant director of the Rockefeller Foundation in Mexico.

It was not long before I was making chromosome preparations from larval brain squashes and exploring the

possibility of using karyotypes to identify species. Although the technique worked and the results of my comparative cytogenetic studies were eventually published (Bush 1962), these studies had a far more important spinoff. My cytogenetic research set me thinking about evolution and how *Anastrepha* and other phytophagous insects speciated, subjects about which I quickly realized I knew very little. But, during my hours in the library, I discovered a recently published edition of Michael J. D. White's book *Animal Cytology and Evolution* (1954), which proved to be a gold mine of information.

White's book opened a new window on the biological world and set me off on a fresh path of investigation that broadened over time to include other areas of evolution, a path from which I have never strayed. I can still feel the initial rush of excitement I experienced as I read White's book and suddenly began to understand and appreciate the scope and power that an evolutionary perspective brings to the analysis of biological problems. This revelation also made me aware that I had to return to graduate school if I wanted to work in the field of systematics and evolutionary biology. Although a return to school became my immediate goal, I did not see how this could be achieved.

I had not been the model undergraduate student at Iowa State. Not only had I flunked ROTC my freshman and sophomore years, but I had done poorly in several mandatory classes that did not interest me. These I either flunked and repeated or earned a barely passing grade. A contributing factor was my participation in Iowa State Players, a little theater group, and a campus modern dance troupe. If it came down to acting, dancing, serving as a stage hand, or hitting the books, the latter activity always lost out. Throughout my college days I also worked as a waiter in a sorority house, which provided a little extra cash, free food, and dates but cut further into my study time. In retrospect it is hard to imagine how I managed to graduate let alone get into graduate school—but I did.

A Fortuitous Meeting and Graduate School

Again it was one of those chance events that provided me with the opportunity to attend graduate school. As it so happened, the Entomological Society of America was meeting concurrently with the American Association for the Advancement of Science in New York City in the late fall of 1957 and I talked my boss, Alan Stone, and the USDA into sending me to the meeting. In an attempt to learn more about cytogenetic methods and interpretation, I selected three cytogeneticists and evolutionary biologists to visit during the trip who I perceived from my readings to be leaders in the field. I then wrote to them asking if I could visit their laboratories for counseling on how to resolve problems I was having interpreting results from my cytogenetic studies on *Anastrepha*. Two of the three, Theodosius Dobzhansky and Sally Hughes-

Schrader, were at Columbia University. The third, Wilson Stone, was head of the Department of Zoology at the University of Texas in Austin and I hoped to visit him on my way back to Mexico. All three agreed to see me, and to my surprise, each spent several hours showing me how to make better preparations, providing me with new information on chromosome evolution, and introducing me to other faculty members and graduate students. When I left Columbia University, Sally wrote a letter to Stone asking him to free me from some of my survey work so I could devote more time to my cytogenetic studies. From then on I spent more time on my chromosome studies.

A second, unforeseen event occurred at that AAAS meeting in New York City that eventually cleared my way for a return to graduate school the next fall. While having a beer in the hotel's tavern, I struck up a conversation with a fellow entomologist, G. Mallory Boush. Mal, I learned, was an associate professor working on crop pests at one of Virginia Polytechnic Institute's agricultural research stations located at Holland, Virginia. During the course of our conversation I expressed an interest in graduate school, and Mal offered to sponsor me and to help resolve the problems I anticipated over my poor academic record. To this day I am ever thankful for his offer. With Mal's support I was admitted to VPI on probation until I could prove myself academically. This restriction in no way deterred me from resigning from a secure position in the USDA and returning to school with financial support only from the GI Bill. I arrived in Blacksburg, Virginia, for the fall term of 1958 eager to start my studies and research with an energy and enthusiasm that from my current perspective seems inconceivable.

Insect Systematics, Cockroaches, and Peanuts at Virginia Polytechnic Institute

I had been reading about new biochemical methods for characterizing and applying blood protein variation to problems of insect systematics. A large and diverse collection of cockroach species, representing several genera, were used in various physiological and genetic studies underway in the entomology department. I decided to use them as guinea pigs to explore the suitability of using protein variation as a systematic tool. I minored in statistics and took on a heavy course schedule, determined to prove to myself that I could handle anything, and indeed managed to ace most of my courses. Meanwhile, to fulfill my commitments for a research assistantship, I spent the first summer at VPI working with Mal on the control and management of peanut insects at the research station in Holland, Virginia. In the fall, I returned to Blacksburg to finish my coursework and research. During a two-week period between the end of spring quarter and before I began my second tour of duty in Holland,

Dorie and I were married. We moved together to Holland, returning in the fall to VPI and Blacksburg, where Dorie taught in a local grade school while I worked on my thesis.

In the preceding spring of 1959 I had begun applying to Ph.D. programs in various entomology and zoology departments. Although my coursework and research had gone exceptionally well, I hesitated sending out applications because I was not sure any institution would take me once they viewed my undergraduate record. Mal again came to my rescue. Over my objections he wrote a letter to Harvard University. I was sure I would never be accepted at that institution. But I was wrong. After reviewing my record and asking me to explain my undergraduate performance (or, more accurately, lack thereof), they admitted me without qualification and offered me a teaching assistantship. The day I received the acceptance letter I was ecstatic.

Off to Harvard and a Ph.D.

In early December I completed writing my thesis on the systematics of blood proteins in cockroaches, passed my thesis exam, packed our belongings in an old trailer I purchased from a friend, and left with Dorie for a brief vacation. I began my studies at Harvard in January, 1960, at the beginning of the winter term.

Because I delayed selecting a specific professor to work with in the biology department until after I arrived, the paleoentomologist, Professor Frank M. Carpenter, with whom I had corresponded before coming to Harvard, took me on until I decided what I wanted to do. As it turned out, this was a lucky pairing. He remained my major professor throughout my stay at Harvard and gave me complete freedom to pursue my interests. I initially toyed with the idea of working on some aspect of cockroach evolution. Toward that end, during the first term I took a course on electron microscopy taught by George Chapman. For the required research paper, I carried out a thin section study of the symbiotic bacteria or bacteriodes inherited through the oocytes of all cockroaches (Bush and Chapman 1961). As fate would have it, George was denied tenure that spring, an experience common to most assistant professors at Harvard. This ended my foray into the evolution of cockroach endosymbionts, although my interest in the evolutionary role of symbiosis never left me. Then I took four courses that were to precipitate a major shift in my understanding of evolution and alter forever my views on how species arise in nature.

During my first summer at Harvard I audited two courses. The first, "Evolutionary Biology," was taught by then Associate Professor Edward O. Wilson. Ed gave me my first modern overview of the evolutionary process with an emphasis on population ecology and genetics as well as a large dose of examples documenting the evolution of behavior. I also took a companion course entitled "Animal Behavior" that Ed Wilson co-taught with Andrew J. Meyerriecks of the Massachusetts Audubon Society. Although I did not know it at the time, these courses were to provide me with essential clues on the evolution of behavior that, along with information gleaned from two subsequent courses and my own field observations, prompted me to develop my ideas on sympatric speciation.

At about this time I began sharing an office with another graduate student, Ellis MacLeod, who also worked with Frank Carpenter. Ellis and I were older than most other graduate students in the bio labs and we shared many interests, including a fondness for insects. Ellis, with a special interest in Neuroptera, was the best naturalist and teacher I have known. He, and another graduate student and close friend, Ira Rubinoff, who inhabited the office next door, served as nonjudgmental and thoughtful sounding boards for my ideas on speciation and evolution. Their open-minded probing and empathetic questioning served to focus and shape my thinking on speciation more than they probably realize.

An Evolution Course with Ernst Mayr and G. G. Simpson

That nonallopatric speciation occurs in many groups of animals now seems abundantly clear, but certainly this was not always so. As a new graduate student in the biology department at Harvard University in the fall of 1960, it did not take me long to be exposed to the doctrine that sympatric speciation in animals was, by all odds, an unlikely event. This point of view was emphasized repeatedly in a course co-taught by Ernst Mayr and George Gaylord Simpson, "Principles of the Evolutionary Process." For the required essay, I chose to write on "Sympatric Speciation: A Factor in Evolution?" In writing this paper I learned that *Rhagoletis pomonella* and other *Rhagoletis* species were regarded as likely examples of sympatric host race formation and speciation via a host shift (Brues 1924; Huxley 1942; Walsh 1864).

In the early 1860s this fly established a biologically distinct new host race on introduced apples (*Malus pumula*) from populations infesting North American hawthorns (*Crataegus* spp.) (Bush et al. 1989), apparently in the absence of geographic isolation. The sympatric origin of other species in this genus was also indicated by the fact that most *Rhagoletis* species groups are composed of sympatric, morphologically very similar if not indistinguishable species, each infesting a different host. Although these observations were suggestive, I rejected the *Rhagoletis* case as an example of sympatric speciation because the evidence was mostly anecdotal and conjectural. I concluded, "It now seems there is little doubt that some form of geographic isolation is needed before reproductive isolating mechanisms can become established." However, the *Rhagoletis* case firmly planted a seed of

uncertainty about the widely accepted view that all animal species arise only during periods of allopatric isolation. I therefore ended the paper with the qualification that "Conventional wisdom, however, has been proven wrong many times and it may well be that some day a good case of sympatric speciation will present itself, but I feel this is highly unlikely."

However, I realized *Rhagoletis* would provide an excellent group to study the process of host race formation and speciation while at the same time allowing me to fulfill my goal to become an insect systematist. Because *Rhagoletis* species are major economic pests, much had been written about their biology and distribution. A study on such a group would allow me to put to good use the experience I had acquired in 1956–57 while working on the host relationships of the related tephritid genus *Anastrepha*. When I discussed with Ernst Mayr the idea of undertaking a thorough systematic and cytogenetic study of the genus *Rhagoletis* for my Ph.D. thesis problem, he enthusiastically endorsed the idea, commenting that such a study would put that unresolved example of sympatric speciation to rest once and for all. So I embarked on a mission to demolish the claims that species of *Rhagoletis* had evolved by colonizing new hosts in the absence of geographic isolation.

As soon as my last class was over in the spring of 1961, my wife, Dorie, and I loaded up our old Chevy BelAir sedan and headed west to collect larvae and adult *Rhagoletis* species throughout North America and Mexico for my forthcoming morphological and cytogenetic analysis. Funds for the trip came from a small grant from Sigma Xi and from Harvard. After four months, mostly spent camping, we returned in late September with a complete set of chromosome slides and extensive collections of almost all previously described *Rhagoletis* species as well as a few new ones. That fall I had no coursework and began working over my field collections, preparing drawings, and photographing, comparing, and measuring karyotypes.

Sociobiology and the Origin of an Idea

Then, in the winter term of 1962, I took a new course from Ed Wilson that radicalized my thinking about how animals speciate, and ultimately altered forever the course of my research and, for that matter, my views on evolution. The course was entitled simply "Sociobiology." Ed explored a broad range of topics, not only on the evolution of social behavior, but also on the evolution of sexual behavior and mate recognition. Toward the end of the term I wrote the usual requisite term paper on "The Evolution of Epigamic Behavior in Diptera." In researching this paper, I discovered that mating in most primitive Diptera occurs in aerial swarms above a contrasting swarm marker, while species in the more highly derived families of the Acalypterates, which includes the most speciose families, mate almost exclusively on the plants they use for food or oviposition.

During the previous summer I had spent a great deal of time watching *Rhagoletis* males patrol their host fruit. These males would challenge and struggle to dislodge other males from the fruit, but would court females with wing-waving displays and stereotyped movements and posturing before attempting to mate. It was clear both male and female *Rhagoletis* were using their larval host fruits as rendezvous sites for courtship and mating. These observations came to mind while I was writing my term paper. Then the idea suddenly struck me. Contrary to my original views, if mate and host choice were tightly correlated, new host races and eventually distinct species of *Rhagoletis* could evolve sympatrically after all. Host-related mating behavior was the clue lacking in the original proposals of sympatric speciation in *Rhagoletis*. It was the clue that allowed the penny to drop.

With this insight, it was immediately clear to me that at least four biological attributes in *Rhagoletis*, acting in combination, contribute to the reduction in gene flow between populations adapting to different hosts and initiate the process of host race formation. First, these flies mate only on their hosts fruits. Mate recognition is therefore directly dependent on host preference. Second, it is reasonable to assume that individuals must possess a modicum of genetically based host fidelity if they are to establish a permanent population on a new host. Host preference must therefore have a genetic basis. Third, there must be genetically based trade-offs in fitness associated with colonizing and adapting to a new host. Alleles that are suitable on one host may be selected against on the other. These might be genes responsible for coping with nutritive or toxic qualities in the plants. A fourth factor directly related to the fitness trade-offs results from adaptation to host plants with different fruiting times. Such differences in fruiting times would favor alleles or genes that optimize eclosion time for oviposition and mating, and result in substantial allochronic isolation of insects on the new and old hosts. Allochronic isolation seems particularly important in *Rhagoletis* because host plant fruiting times of closely related sympatric species, although often overlapping, are usually significantly different. I reasoned that these genetic differences in host preference, fitness, and eclosion time, coupled with the fact that mating occurs only on the host, can provide strong ecological and allochronic isolation. In the absence of such host-associated genetic differences, host races are unlikely to evolve because mating would frequently take place between individuals infesting different host plants.

Sympatric Speciation in *Rhagoletis* and My Thesis Committee

Although to me it seemed self-evident and straightforward that the biological attributes of phytophagous insects like *Rhagoletis*, as well as many parasitoids, preadapted them to speciate sympatrically, it was not clear or acceptable to many others at Harvard. I was caught, as they say, between a rock and a hard place. Among the evolutionarily inclined cognoscenti housed within the ivy-clad walls of the Museum of Comparative Zoology, the Gray Herbarium, and the Department of Biology, sympatric speciation was viewed as little better than an heretical, unsupported fantasy. Fortunately, I knew through conversations that I would encounter little if any difficulty advocating the sympatric origin of *Rhagoletis* species from my thesis advisor, Frank Carpenter, or committee member Ed Wilson, but finding support for this view from the third member of my thesis committee, Ernst Mayr, was another matter.

Mayr was putting the final touches on *Animal Species and Evolution*, which was published in 1963. In this book he forcefully advocates the universality of allopatric speciation and rejects the possibility of sympatric speciation by way of host (p. 473) or seasonal races (p. 474). His views on the likelihood of sympatric speciation are encapsulated in his comments,

> One would think that it should no longer be necessary to devote much time to this topic, but past experience permits one to predict with confidence that the issue will be raised again at regular intervals. Sympatric speciation is like the Lernaean Hydra which grew two new heads whenever one of its old heads was cut off. There is only one way in which final agreement can be reached and that is to clarify the whole relevant complex of questions to such an extent that disagreement is no longer possible. (p. 451)

In regard to ecological races he concluded that "every race is simultaneously a geographic race and an ecological race. There is no evidence to support the notion of ecological races as a category distinct from that of geographical races" (p. 455).

In his class, lectures, and personal contacts, Mayr made it abundantly clear that he placed anyone advocating sympatric speciation in about the same crackpot category as other advocates of nonallopatric modes of speciation and hopeful monsters such as the saltationist Richard Goldschmidt (Bush 1982). Reading his rejection of all models of sympatric speciation, including one very similar to mine, in his book was a sobering and, for a short time, depressing experience. For several weeks after his book appeared, I found it very difficult to continue writing my dissertation.

For a graduate student to contest the views of Ernst Mayr, particularly when he was on your thesis committee, took a good deal more courage than I could muster at the time. I was, in fact, not absolutely sure I was right. But the more I worked on local *Rhagoletis* populations, the more I was convinced that speciation must often occur sympatrically via a host shift in these flies. As I worked on my thesis I saw less and less of Mayr. My solution was to propose both an allopatric and a sympatric mode of speciation in my thesis, thus leaving the door open for reconsideration later. Although I conceded that an allopatric scenario could be developed for all sympatric sibling species of *Rhagoletis* specializing on different host species, such scenarios required elaborate, unsupported assumptions about past and present host-plant distributions and local extinctions. A sympatric host shift was a much more parsimonious explanation.

I proposed instead that host race formation and speciation in *Rhagoletis* may occur as a result of allochronic isolation on sympatric host plants that fruited at different times of the summer and fall (Bush 1964). I speculated that allochronic isolation on plants fruiting at different times, coupled with the fact that courtship and mating is restricted to the host plant, would reduce the chance of hybridization and competition as well as the need to evolve strong isolating mechanisms associated with visual cues. In my thesis, I went on to outline a model of sympatric speciation promoted by a reduction of gene flow brought about by the colonization of a new, somewhat allochronically isolated host. I concluded that this mode of nonallopatric speciation is a definite possibility in *Rhagoletis* and best explains the origin of many morphologically indistinguishable sympatric sibling species in this genus (for details, see also Bush 1969a,b).

I did not complete a draft of my thesis until early one evening in late January, 1964, too late to deliver a copy to Mayr's office in the Museum of Comparative Zoology as I had promised. It had taken me much longer to put a clean copy together than I had anticipated. I knew Mayr was planning to leave on vacation, so that same evening I took a copy to his home a few blocks from the museum. I had no idea how he would view my proposal of sympatric speciation, but I was prepared to defend my conclusion. He had told me that if his porch light was out it would be too late for him to read my thesis and indeed his porch was dark. On the outside chance that he had forgotten to turn the light on, I rang his doorbell anyway. At the door he informed me that I was indeed too late, and suggested I get someone else to replace him. I left with mixed feelings. Inwardly I breathed a sigh of relief while at the same moment regretting he would not be on my committee or read my thesis. The following day I asked Howard Evans, curator of Hymenoptera at the museum and an authority on insect behavior for whom I had great respect, to take Mayr's place and he agreed.

To Australia and Back on a Postdoctoral Fellowship

Shortly after the defense of my thesis on February 7, Dorie and I left Harvard for the University of Melbourne in Australia. There I spent two productive and stimulating years on a National Institutes of Health postdoctoral fellowship with the cytogeneticist and evolutionary biologist Michael White, whose book had started me thinking about speciation while working in Mexico. Not only did I gain a great deal more experience in cytogenetics, but I also grew to appreciate the importance and role of chromosome rearrangements in speciation and evolution. I spent over six months in the field with Dorie, collecting, rearing, and studying the chromosomes of Australian Tephritidae. These studies only reinforced the views on sympatric speciation I had developed at Harvard. During these investigations I discovered that most if not all tephritids associated with the flowers, stems, and roots of the family Asterales are female heterogametic while those infesting fruits are normally male heterogametic (Bush 1966a). Michael White, at first skeptical about the role of sympatric speciation in my *Rhagoletis* fruit flies, began to develop his own model of nonallopatric divergence that he called stasipatric speciation to explain the pattern of chromosome evolution in some of his flightless grasshoppers (White 1968).

A few months before my fellowship came to an end I accepted an assistant professorship in the department of zoology at the University of Texas in Austin. Michael White had recommended me for the position and I was hired sight unseen. When it was time to leave Melbourne, we returned by way of Papua, New Guinea, and Guadalcanal Island, where we spent five weeks collecting tephritids and other insects for Harvard's Museum of Comparative Zoology. By late May, 1966, we were back at Harvard, where I had an interim fellowship to work on papers and prepare my thesis for publication before moving to Texas.

Sympatric Speciation and the Measles

Shortly after arriving in Cambridge, I attended the meetings of the Society for the Study of Evolution held jointly in late spring of 1966 in Washington, DC, with the AAAS. There I presented my first paper at a scientific meeting with an unexpected and disconcerting result, but one to which I eventually became accustomed. The title of my talk was "Sympatric Speciation in True Fruit Flies." When I arrived at the appointed room, I was surprised to see that the person chairing the session was none other than one of my idols, Theodosius Dobzhansky. Dobzhansky introduced me with a slightly drawn out emphasis on the "True" part of the title. I wasn't sure if he was poking fun at me or if it was just his Russian accent. I don't recall how I made it through the talk or what I said, but I do remember what happened when it ended: there was absolute silence. No questions were asked, so I concluded that either they were not interested in the topic or that no one was buying my idea of sympatric speciation in *Rhagoletis*. The latter conclusion was immediately reinforced when Dobzhansky politely thanked me and added, "That was a very interesting story," and after a pause, "but I don't believe it." He continued, "Sympatric speciation is like the measles; everyone gets it and we all get over it."

A Goal—Identify "Speciation" Genes

Following the society meeting, it was clear that if I wanted to convince evolutionary biologists that sympatric speciation was not only possible but likely to be common in animals such as phytophagous parasites, I would need to devote all my research efforts to the study of host race formation and speciation in *Rhagoletis* or some other suitable group. Because speciation involves events unique to each taxon, the only way to understand the process was to select a group undergoing active speciation and examine all aspects of adaptations associated with host shifts.

When I initiated my research program on host race formation and speciation at the University of Texas in 1966, my long-term goal was to determine the genetic basis for speciation. I wished to establish whether speciation required the accumulation of changes of small effect in many genes, as Mayr and others advocated, or whether it could occur as a result of alterations in only a few genes, as I suspected. This meant the identification of the genes responsible for what at that time was referred to as reproductive isolation. It was already clear from my preliminary field observations that host specificity played a major role in limiting gene flow between populations adapting to different host plants. Therefore, the most obvious kind of "reproductive isolating" genes to look for in *Rhagoletis* were the ones responsible for host recognition. There also appeared to be substantial allochronic isolation between the apple and hawthorn host races of *R. pomonella,* so I assumed that emergence time and diapause were under genetic control and contributed to allochronic isolation. Genes for fitness and conditioning (associative learning) could also provide additional isolation between diverging host races and warranted study. With these traits in mind, I developed a verbal genetic model of host race formation and speciation in *Rhagoletis* fruit flies (Bush 1969b). This model served as the basis for all my subsequent research.

The model was a simple one based on the following assumptions presented here in somewhat abbreviated form: (1) diapause and emergence times are under genetic control, (2) initial orientation to and selection of a host plant is in response to chemical cues (from the host plant), (3) host selection has a genetic basis, and (4) host-plant and mate selection are positively correlated. Individuals move preferentially to their respective host plants depending on their genotype, and mating occurs on the host plant.

I went on to conclude that, given these properties, the colonization of a new host with a fruiting time different from the original host would result in the selection of individuals with the appropriate genotypes for mating, oviposition, and larval survival on the new host. Follow-ing Levene (1953) and Maynard Smith (1966), I reasoned that if the trade-off in fitness, resulting in lowered heterozygote survival, was sufficiently strong, it would allow divergence even in the face of initial gene flow between the host races. The fact that there would be some gene flow, at least initially, also suggested that the resulting host races and sister species would differ primarily at loci responsible for adaptation to the new host. Divergence in mate recognition is the direct outcome of adaptations responsible for host preference, eclosion times, and fitness trade-offs in these flies. Postmating incompatibility may evolve later, but it is not an expected cause or outcome of sympatric speciation. Closely related sympatric sister species should, therefore, show little evidence of reproductive incompatibility.

The Reality: First Identify Key Traits Responsible for Speciation

However, before I could identify specific genes responsible for speciation, I needed a great deal more information on the specific ecological, behavioral, and physiological traits involved with host selection, diapause, and fitness. It was one thing to realize the importance of host selection in the speciation process. It was quite another to pinpoint the key trait or set of traits involved in the colonization of a new host plant. Only after I had a clear picture of the sequence of events and the cues used by *Rhagoletis* to locate their host could I begin a search for the genes involved.

Once specific traits were identified, establishing their genetic basis would require an accurate linkage map that, with the tools available in the late 1960s, was a very difficult and time-consuming undertaking. Not only was *Rhagoletis* poorly represented in Texas, but most species have one generation per year and are notoriously difficult to maintain in culture. Furthermore, no genetic traits had been identified. The task of locating such genes and establishing their mode of action was further hampered by the fact that details of the genetic code had only just been resolved and little was actually known about gene function or expression. At the time, allozymes were about the only means available for linkage studies on flies like *Rhagoletis,* but at best they provided only four to eight genetically variable markers. None of the allozymes we studied initially appeared to be remotely related to host race formation and speciation. We were to find much later, when more proteins were available for study, that the frequency of alleles at some allozyme loci differed significantly between the apple and haw races (Feder et al. 1988).

In hopes of developing a more easily manipulated model system, my first graduate student, Milton Huettel, and I started working on the genetics of host selection in *Procecidochares,* a genus of tephritid flies that galled the flowers and stems of various Asterales and whose sibling species were sympatric. The hosts of these multivoltine flies could be raised in the greenhouse throughout the year. They also had spectacular polytene chromosomes (Bush and Taylor 1969), which I hoped would provide a means for reconstructing phylogenetic relationships, provide information on the pattern of chromosome rearrangements, and simplify the construction of a linkage map. Although interspecific hybridization and host selection experiments demonstrated that host preference in *Procecidochares* was controlled primarily by a few major loci (Huettel and Bush 1972) , these flies proved more difficult to rear and maintain than expected and showed almost no evidence of chromosome evolution, so we returned to *Rhagoletis*.

Research on *Rhagoletis* in Wisconsin and Switzerland

There were two persons that contributed to this return. In the spring of 1970, Ronald J. Prokopy, a specialist on *Rhagoletis* ecology and behavior, joined my group in Austin. Ron and I decided to investigate the circumstances surrounding the recent colonization of introduced sour cherry, *Prunus cerasi*, by the apple maggot in Door County, Wisconsin. Reports indicated that the cherry race emerged in early summer well before the peak emergence of the apple and hawthorn races also present in the area. From 1970 to 1983, I spent most summers in Door County, first with Ron, then with graduate students, working on various aspects of the ecology, behavior, and population genetics of the cherry, apple, and hawthorn races. Our objective was to identify the key traits responsible for host and mate selection. During this period our research and that of my students (summarized in Bush 1993b) identified the cues and traits responsible for host-plant recognition and mating behavior.

The second person who contributed to my return to *Rhagoletis* was Ernst Boller, an entomologist at the Swiss Federal Research Station in Wädenswil, Switzerland, who was working on the biology and control of the European cherry fruit fly, *Rhagoletis cerasi*. After a summer with Ron in Door County, I spent 10 months on sabbatical leave with Ernst, who was developing a sterile male release program in order to reduce the level of pesticides necessary to control this fly on cherries. His broad grasp of *Rhagoletis* ecology and behavior and research experience provided me with the kind of knowhow I needed to develop my own research program on host race formation.

Working with Ernst also gave me the opportunity to investigate the biological status of a host race, morphologically indistinguishable from the population on cherries, which infested the fruits of honeysuckle, *Lonicera tartarica*, and related species. Our biological and genetic studies not only established that the cherry and honeysuckle races differed in habitat preference and emergence times (Boller and Bush 1974), but also revealed that the cherry race was composed of two large northern and southern parapatric populations between which there was strong unidirectional sterility (Boller et al. 1976). Our results suggested that population subdivision might be more widespread than we anticipated and could not only involve host shifts, but be promoted by factors such as endosymbionts. Studies on these races of *R. cerasi* have continued and the results have been summarized recently by Boller et al. (in press).

Biological Aspects of Host and Mate Preference in *Rhagoletis*

In the early years, we set out to learn how *Rhagoletis* species selected their host plants and fruits. We knew from Ron's earlier work on *R. pomonella* that flies do not perform normally in a laboratory environment, so most of our behavior studies were carried out in a natural field setting. In a series of experiments spanning several summers we used plywood models of trees of various shapes, sizes, and colors to establish which physical cues might be important in host plant recognition by *R. pomonella* as well as by other sympatric cherry-infesting *Rhagoletis*, such as *R. cingulata* and *R. fausta*.

These investigations revealed that each *Rhagoletis* species has very specific preferences for particular colors, shapes, and sizes of their host tree and host fruit (Moericke et al. 1975; Prokopy and Bush 1973a). In earlier experiments, we had already demonstrated that *R. pomonella* adult males and females selected dark red or black spheres over any other shape and began orienting to such spheres only after they were 6–7 days old (Prokopy and Bush 1973a). Physical cues, however, were not sufficiently specific to explain the ability of each *Rhagoletis* species to discriminate between host and nonhost. Such specificity in host recognition could only be attributed to specific odors or contact chemical cues emanating from the leaves or fruit. To test the efficacy of host odor to attract *Rhagoletis* to the host plant, we used our model trees and nonhost trees (Prokopy et al. 1973) and discovered that the odor emitted by the apple was responsible for host recognition.

The odors involved in host recognition were later identified by others to consist primarily of five esters (Fein et al. 1982). Later, Todd Bierbaum (Bierbaum and Bush 1987, 1988, 1990b) and Jürg Frey (Frey et al. 1992; Frey and Bush 1990) in my laboratory were able to show that the behavioral and electrophysiological response to host odors differed between *R. pomonella* and *R. mendax* as well as between the apple and hawthorn races. Host selection is thus mediated by a combination of physical and chemical cues, with the latter providing specific cues for host-plant specificity. Because both males and females of each *Rhagoletis* species or host race respond to the same chemical cues and all courtship and mate selection occurs on or near the host fruit (Prokopy 1975; Prokopy and Bush 1973b; Prokopy and Reissig 1976), these cues also serve as very specific, strong isolating mechanisms. Host selection alone is sufficient to inhibit hybridization between sympatric sister species such as *R. pomonella* and *R. mendax* (Feder and Bush 1989; Feder et al. 1994) even though these two species can be easily induced to form F₁ hybrids and backcrosses in the laboratory (Bierbaum and Bush 1990a).

Evolution and Modes of Speciation in Other Animals

My interest in speciation broadened in the mid-1970s when one of my students, Dorothy Prowell, began working on host race formation in the codling moth, *Laspyrisia pomonella* (Prowell 1980). This moth had established a new host race on walnuts in the recent past from a population infesting apples and pears in California. Dorothy has since gone on to study host race formation and speciation in other Lepidoptera (Powell, this volume). Her research confirmed my suspicion that sympatric host race formation and speciation were probably widespread among invertebrate groups that specialized on one or two host plants. On the basis of our research on *Rhagoletis* and the codling moth, I proposed a more refined, genetically based model of sympatric host race formation and speciation (Bush 1974, 1975b).

At about this same time, I also began to explore modes of speciation in animals other than insects. While writing an article for the *Annual Review of Ecology and Systematics* on "Modes of Animal Speciation" (Bush 1975a), I discovered an interesting correlation between karyotype and species diversity and the social structuring within certain vertebrate groups. I proposed that within those taxa whose social structuring promoted the establishment and maintenance of small demes, inbreeding might occasionally occur. This would facilitate and thus increase the rate of fixation of point and chromosome mutations and as a result increase the rate of speciation. I argued that social structuring could, under certain circumstances, result in nonallopatric isolation and speciation. With Alan Wilson and others I found a strong correlation between social structuring, rapid chromosome rearrangements, and speciation (Bush 1981; Bush et al. 1977; Wilson et al. 1975). As a result of this diversion into vertebrate speciation, I realized that there may be other ways habitat races might be established, and that nonallopatric speciation might be more widespread among animals, including vertebrates, than previously recognized. In a later treatment, Dan Howard and I (Bush and Howard 1986) provided further evidence of the likelihood of nonallopatric speciation in animals.

During this period I also initiated a survey of allozyme variation in the screwworm fly, *Cochliomyia hominivorax* (Diptera: Calliphoridae), at the request of the U.S. Department of Agriculture. This fly was being mass reared in a large facility in McAllen, Texas, for use in a sterile male release program. Apparently the mass-reared flies were experiencing difficulty in locating and competing with wild screwworm flies. In collaboration with one of my graduate students, Raymond Neck, and G. Barrie Kitto from the University of Texas chemistry department, we discovered that when each factory population was established, the rare α-GPD allele rapidly approached fixation. A kinetic study revealed that this allele functioned best at the abnormally high rearing temperatures maintained in the factory (Bush et al. 1976). By maximizing fly production, the factory managers had inadvertently selected for strains that could not fly at the low temperatures encountered by this fly in nature during the morning hours when most mating occurred. Elimination of the offending α-GPD allele restored competility and the fly was eventually eradicated from Texas and much of Mexico.

Michigan State University and a Change in Research Direction

It became clear by the late 1970s that in order to delve deeper into the genetics of speciation, I would need to employ molecular tools. I had also become interested in the role of microbial symbionts in the speciation process. However, my laboratory at the University of Texas was ill-equipped for such research. Then, in 1981, I was offered a Hannah Distinguished Professorship at Michigan State University. This chair would provide me with the facilities and support I needed for developing a research program in molecular evolution, as well as the opportunity to study the role of bacterial symbiosis in *Rhagoletis* speciation. I accepted the position and my family, which now included three children, moved to Michigan in the summer of 1981.

My interest in microbial symbionts grew out of reports that a gland or diverticulum in the head of *Rhagoletis* species, as well as pouchlike folds enclosing the ovipositor, were packed with a single species of pseudomonad bacteria. These bacteria were also reported to fill the larval gastric caecae. Several authors had concluded that *Rhagoletis* larvae could not survive without these bacteria and that the bacteria contribute to, if not determine, host specificity. When Dan Howard joined me at MSU as a postdoctoral fellow, we initiated a three-year study of symbiosis in *Rhagoletis*. Our research failed to substantiate these claims. Contrary to earlier reports, *Rhagoletis* harbored up to at least 10 bacterial species, none of which were pseudomonads, and larvae survived as well, if not better, without the bacteria than with them. Nor could we find any relationship between bacterial species or genetic strain and host preference (Howard and Bush 1989; Howard et al. 1985). It became clear these bacteria are not responsible for host specificity or involved with host shifts.

The Population Genetics of *Rhagoletis* Host Races and Species

During the 1980s one of my graduate students, Jeffery Feder, once again began using allozymes to examine the genetic structure of sympatric apple and hawthorn *Rhagoletis* populations in the field. I told him he was wasting his time as I was unable to find any differences using the same method in the early 1970s. I had concluded that allozymes were unlikely to resolve the fine differences between newly established host races and possibly even species. Jeff, however, ignored my pessimism. He added several new enzyme systems and, in a preliminary study, discovered that the sympatric apple and hawthorn populations in Michigan maintained significantly different allele frequencies in alleles at six loci over several generations (Feder et al. 1988). Similar results were obtained in the laboratory of one of my former students, Stewart Berlocher, by one of his own students, Bruce McPheron (McPheron et al. 1988). Another of Berlocher's students, D. Courtney Smith (1988), found that there were genetically based differences in the eclosion times of the apple and hawthorn populations. Further studies by Feder have determined that gene flow between sympatric Michigan populations of the two races is 6% or less and that there are strong negative genetic trade-offs associated with adaptations to different hosts (Feder et al. 1995, 1997a,b). These investigations have confirmed that host race formation can occur within a very short period of time in these flies. In many respects these races already have many of the characteristics of distinct species (Bush and Smith 1997).

The Phylogeny of *Rhagoletis* Host Races and Species

In order to establish the pattern and rate of host shifts and speciation among *Rhagoletis* species, it is necessary to determine their phylogenetic relationships. My first attempt to unravel evolutionary relationships among species in this genus was based on the morphological, cytological, and biogeographic evidence that emerged from my systematic analysis (Bush 1964, 1966b). In the early 1960s a few systematists were just beginning to explore a new numerical taxonomic approach to systematic analysis, but it was not clear if it was appropriate for phylogenetic reconstructions. I therefore endeavored to determine which characters were primitive and which were derived by comparing *Rhagoletis* species with species in genera recognized by contemporary systematists as close relatives. I was able to gain some insight into general patterns of divergence and biogeographic relationships, but it was impossible to determine actual directions of host shifts or to identify the original host used by the early progenitor of North American *Rhagoletis*.

Later Stewart Berlocher and I, using allozyme variation, established a better picture of relationships among *Rhagoletis* species (Berlocher 1976; Berlocher and Bush 1982). This approach, for the first time, revealed a poor correlation between the level of genetic and morphological divergence between species. For instance, the Nei unbiased genetic distances between species within the *R. pomonella* species group, which are morphologically almost indistinguishable, range from 0.034 to 0.604 (Berlocher and Bush 1982). Recently, James Smith and I have employed sequence differences in cytochrome oxidase II mitochondrial DNA to reconstruct phylogenetic relationships in *Rhagoletis* (Smith and Bush 1997). This analysis not only confirms the general pattern of phylogenetic relationships obtained with allozymes, but also provides a preliminary assessment of the sequence of host plant colonization and speciation. It seems that several sympatric *R. pomonella* group species are paraphyletic and have speciated very recently.

Genetic Models and the Simulation of Sympatric Speciation

My students and I have also explored the conditions promoting sympatric speciation and its likelihood by genetic simulation. Parameters used in these simulations are based on realistic biological conditions known to exist in phytophagous insects and on traits responsible for mate recognition and negative fitness trade-offs attendant with host-plant specialization. Scott Diehl and I found that when mate and host choice are coupled, as in many phytophagous insects, host races can rapidly develop and be maintained by realistic levels of selection acting on host-related polymorphisms (Bush and Diehl 1982; Diehl and Bush 1989). It was unclear, however, whether such host races could continue to diverge in sympatry until no gene flow occurred between them. In a subsequent multilocus simulation study with Paul Johnson (P. Johnson et al. 1996), we found that when variation in host-plant-related fitness and host preference were coupled with variation in non-host-related mate recognition, such as in a pheromone or visual cue, sympatric speciation could occur even under conditions of moderate levels of penetrance and selection.

Sympatric Speciation in Other Animals

There are a growing number of studies treated elsewhere (Bush 1993a, 1994; Bush and Smith 1997) and in this volume that suggest that other host-plant and habitat specialists have speciated sympatrically. Although insects and other invertebrates have provided the best examples of nonallopatric race formation and speciation, it is very likely that the same mode of speciation has occurred in some vertebrates, such as fish (J. C. Johnson et al. 1996; Schliewen et al. 1994; Schluter, this volume). Each case involves sister species or species flocks that occur in lakes of known geological age that lack physical barriers or rivers essential for geographic speciation. Sympatric divergence in other vertebrates may also have occurred, but in the absence of accurate knowledge of past geographical distributions it is difficult, if not impossible, to establish. However, the likelihood of sympatric speciation is supported by recent evidence that divergence in mate recognition between sympatric populations adapting to different habitats can arise in response to strong selection (Butlin 1995; Schluter 1994; Schluter and Nagel 1995; Schluter, this volume).

The take-home lesson from these investigations is that habitat and host-plant specialization can facilitate and promote the divergence of populations in animals that mate within a preferred habitat or that mate assortatively among individuals specializing in different habitats. This divergence can occur in the absence of spatial isolation simply in response to selection on intrinsic genetic variation already present in the parent species. Sympatric divergence may, in fact, be responsible for much of the incredible biodiversity in arthropods and other groups, such as nematodes.

Prospects for the Future

Before the advent of molecular biology, the methods for reconstructing phylogenetic relationships were based on little more than subjective criteria derived from morphological studies. Phylogenies based on morphology, although useful, sometimes provided inaccurate estimates of the rate and pattern of divergence because morphological traits seldom evolve in a clocklike way. Molecular evolution within a taxon fortunately is more clocklike. Sequence data, particularly when several unrelated portions of the genome are assessed, provide sufficiently accurate estimates of divergence to allow fairly robust phylogenetic reconstructions. These phylogenies allow a more critical test of certain hypotheses regarding modes of speciation, the history of habitat and host shifts, and actual levels of gene flow between populations, habitat races, and species.

Molecular biology also enables us to establish linkage maps sufficiently detailed to identify and explore the genetic basis of quantitative traits responsible for habitat and host preference, mate recognition, and other adaptive attributes directly involved with speciation. The long-sought goal of actually understanding the genetics of speciation is finally within our grasp. More important, the molecular tools are rapidly replacing speculation about the role of developmental changes and epigenetic factors in adaptation and speciation. When coupled with our increasingly more sophisticated understanding of the biological attributes and ecological interactions of animals in their natural settings, a more realistic understanding of the speciation process will inevitably emerge. It is clear that we are on the threshold of a new era in the study of adaptation and speciation. More advances in our understanding of the speciation process will likely occur in the first 35 years of the next millennium than were revealed during the preceding one thousand years. The prospect is exciting to contemplate. It is also apparent that some of our most cherished convictions regarding the way animals and plants speciate and evolve are undergoing or will face major revision. In the not too distant future these advances should be incorporated into a new, au courant synthesis. I only wish I could be given another 35 years of research on speciation.

Acknowledgments   Without the support and contributions of many outstanding students and associates, I would not have been able to make the advances over the past 35 years in our understanding of host race formation and speciation of *Rhagoletis* and other tephritids. Their ideas and dedicated research efforts as well as their fellowship, empathy, and willingness to spend long hours in the field and laboratory, sometimes under questionable conditions, made it possible to undertake widely different approaches to the study of speciation in *Rhagoletis* fruit flies. I have also benefited immeasurably from the support and encouragement of my wife, Dorie and our three children, Lisa, Guy, and Eliana. I thank them all.

References

Berlocher, S. H. 1976. The genetics of speciation in *Rhagoletis* (Diptera: Tephritidae). Ph.D. thesis, University of Texas.

Berlocher, S. H., and Bush, G. L. 1982. An electrophoretic analysis of *Rhagoletis* (Diptera: Tephritidae) phylogeny. Syst. Zool. 31:136–155.

Bierbaum, T. J., and Bush, G. L. 1987. A comparative study of host plant acceptance behaviors in *Rhagoletis* fruit flies. In: V. Labeyrie, G. Fabres, and D. Lachaise (eds.). Insect–Plants. Dordrecht, W. Junk, pp. 374–375.

Bierbaum, T. J., and Bush, G. L. 1988. Divergence in key host examining and acceptance behaviors of sibling species *Rhagoletis mendax* and *R. pomonella* (Diptera: Tephritidae). In: M. T. AliNiazee (ed.). Fruit Flies of Economic Importance: Bionomics, Ecology and Management. Agriculture Experiment Station, Oregon State University Special Report, pp. 26–55.

Bierbaum, T. J., and Bush, G. L. 1990a. Genetic differentiation in the viability of sibling species of *Rhagoletis* fruit flies on host plants, and the influence of reduced hybrid viability on reproductive isolation. Entomol. Exp. Appl. 55:105–118.

Bierbaum, T. J., and Bush, G. L. 1990b. Host fruit chemical stimuli eliciting distinct ovipositional response from sibling species of *Rhagoletis* fruit flies. Entomol. Exp. Appl. 56:165–177.

Boller, E. F., and Bush, G. L. 1974. Evidence for genetic variation in populations of the European cherry fruit fly, *Rhagoletis cerasi* (Diptera: Tephritidae) based on physiological parameters and hybridization experiments. Entomol. Exp. Appl. 17:279–293.

Boller, E. F., Russ, K., Vallo, V., and Bush, G. L. 1976. Incompatible races of European cherry fruit fly, *Rhagoletis cerasi* (Diptera: Tephritidae), their origin and potential use in biological control. Entomol. Exp. Appl. 20:237–247.

Boller, E. F., Katsoyannos, B. I., and Hippe, C. In press. Host races of *Rhagoletis cerasi* L. (Diptera: Tephritidae): effect of prior adult experience on oviposition site preference. J. Appl. Entomol.

Brues, C. T. 1924. The specificity of food-plants in the evolution of phytophagous insects. Am. Nat. 58:127–144.

Bush, G. L. 1962. The cytotaxonomy of the larvae of some Mexican fruit flies in the genus *Anastrepha* (Tephritidae, Diptera). Psyche 69:87–101.

Bush, G. L. 1964. A revision of the genus *Rhagoletis* in North America (Tephritidae, Diptera). Ph.D. thesis, Harvard University.

Bush, G. L. 1966a. Female heterogamety in the family Tephritidae (Acalyptratae, Diptera). Am. Nat. 100:119–126.

Bush, G. L. 1966b. The taxonomy, cytology, and evolution of the genus *Rhagoletis* in North America (Diptera, Tephritidae). Bull. Mus. Comp. Zool. 134:431–562.

Bush, G. L. 1969a. Mating behavior, host specificity, and the ecological significance of sibling species in frugivorous flies of the genus *Rhagoletis* (Diptera, Tephritidae). Am. Nat. 103:669–672.

Bush, G. L. 1969b. Sympatric host race formation and speciation in frugivorous flies of the genus *Rhagoletis* (Diptera, Tephritidae). Evolution 23:237–251.

Bush, G. L. 1974. The mechanism of sympatric host race formation of the true fruit flies (Tephritidae). In: M. J. D. White (ed.). Genetic Mechanisms of Speciation in Insects. Sydney, Australia and New Zealand Book Co., pp. 3–23.

Bush, G. L. 1975a. Modes of animal speciation. Annu. Rev. Ecol. Syst. 6:339–364.

Bush, G. L. 1975b. Sympatric speciation in phytophagous parasitic insects. In: P. W. Price (ed.). Evolutionary Strategies of Parasitic Insects and Mites. New York, Plenum, pp. 187–206.

Bush, G. L. 1981. Stasipatric speciation and rapid evolution in animals. In: W. R. Atchley and D. S. Woodruff (eds.). Evolution and Speciation: Essays in Honor of M.J.D. White. Cambridge, Cambridge University Press, pp. 201–218.

Bush, G. L. 1982. Goldschmidt's follies. Review of the material basis of evolution. R. Goldschmidt. Paleobiology 8:463–469.

Bush, G. L. 1993a. A reaffirmation of Santa Rosalia, or why are there so many kinds of small animals? In: D. R. Lees and D. Edwards (eds.). Evolutionary Patterns and Processes. London, Academic Press, pp. 229–249.

Bush, G. L. 1993b. Host race formation and sympatric speciation in *Rhagoletis* fruit flies (Diptera: Tephritidae). Psyche 99:335–357.

Bush, G. L. 1994. Sympatric speciation in animals: new wine in old bottles. Trends Ecol. Evol. 9:285–288.

Bush, G. L., and Chapman, G. B. 1961. Electron microscopy of symbiotic bacteria in developing oocytes of the American cockroach, *Periplaneta americana*. J. Bacteriol. 81:267–276.

Bush, G. L., and Diehl, S. R. 1982. Host shifts, genetic models of sympatric speciation and the origin of parasitic insect species. In: J. H. Visser and A. K. Minks (eds.). Proceedings 5th International Symposium on Insect-Plant Relationships. Wageningen, Netherlands, Pudoc, pp. 297–305.

Bush, G. L., and Howard, D. J. 1986. Allopatric and non-allopatric speciation: assumptions and evidence. In: S. Karlin and E. Nevo (eds.). Evolutionary Processes and Theory. Orlando, Academic Press, pp. 411–438.

Bush, G. L., and Smith, J. J. 1997. The sympatric origin of phytophagous insects. In: K. Dettner, G. Bauer, and W. Völkl (eds.). Vertical Food Web Interactions: Evolutionary Patterns and Driving Forces. Ecological Studies. Heidelberg, Springer-Verlag. 130:3–19.

Bush, G. L., and Taylor, S. C. 1969. The cytogenetics of *Procecidochares*. I. The mitotic and polytene chromosomes of the Pamakani fly, *P. utilis* (Tephritidae-Diptera). Caryologia 22:311–322.

Bush, G. L., Neck, R. W., and Kitto, G. B. 1976. Screwworm eradication: inadvertent selection for noncompetitive ecotypes during mass rearing. Science 193:491–493.

Bush, G. L., Case, S. M., Wilson, A. C., and Patton, J. L. 1977. Rapid speciation and chromosomal evolution in mammals. Proc. Natl. Acad. Sci. U.S.A. 74:3942–3946.

Bush, G. L., Feder, J. L., Berlocher, S. H., McPheron, B. A., Smith, D. C., and Chilcote, C. A. 1989. Sympatric origins of *R. pomonella*. Nature 339:346.

Butlin, R. K. 1995. Reinforcement: an idea evolving. Trends Ecol. Evol. 10:432–434.

Diehl, S. R., and Bush, G. L. 1989. The role of habitat preference in adaptation and speciation. In: D. Otte and J. Endler (eds.). Speciation and Its Consequences. Sunderland, Mass., Sinauer, pp. 345–365.

Feder, J. L., and Bush, G. L. 1989. A field test of differential host-plant usage between two sibling species of *Rhagoletis pomonella* fruit flies (Diptera: Tephritidae) and its consequences for sympatric speciation. Evolution 43:1813–1819.

Feder, J. L., Chilcote, C. A., and Bush, G. L. 1988. Genetic differentiation between sympatric host races of *Rhagoletis pomonella*. Nature 336:61–64.

Feder, J. L., Opp, S. B., Wlazlo, B., Reynolds, K., Go, W., and Spisak, S. 1994. Host fidelity is an effective pre-mating barrier between sympatric races of the apple maggot fly, *Rhagoletis pomonella*. Proc. Natl. Acad. Sci. U.S.A. 91:7990–7994.

Feder, J. L., Reynolds, K., Go, W., and Wang, E. C. 1995. Intra- and interspecific competition and host race formation in the apple maggot fly, *Rhagoletis pomonella* (Diptera: Tephritidae). Oecologia 101:416–425.

Feder, J. L., Roethele, J. B., Wlazlo, B., and Berlocher, S. H. 1997a. Selective maintenance of allozyme differences between sympatric host races of the apple maggot fly. Proc. Natl. Acad. Sci. U.S.A. 94:11417–11421.

Feder, J. L., U. Stolz, K. M. Lewis, W. M. Perry, J. B. Roethele, and A. Rogers. 1997b. The effects of winter length on the genetics of apple and hawthorn races of *Rhagoletis pomonella* (Diptera: Tephritidae). Evolution 51:1862–1876.

Fein, B. L., Reissig, W. H., and Roelofs, W. L. 1982. Identification of apple volatiles attractive to the apple maggot, *Rhagoletis pomonella*. J. Chem. Ecol. 8:1473–1487.

Frey, J. E., and Bush, G. L. 1990. *Rhagoletis* sibling species and host races differ in host odor recognition. Entomol. Exp. Appl. 57:123–131.

Frey, J. E., Bierbaum, T. J., and Bush, G. L. 1992. Differences among sibling species *Rhagoletis mendax* and *R. pomonella* (Diptera: Tephritidae) in their antennal sensitivity to host fruit compounds. J. Chem. Ecol. 18:2001–2023.

Howard, D. J., and Bush, G. L. 1989. The influence of bacteria on larval survival and development in *Rhagoletis* (Diptera: Tephritidae). Ann. Entomol. Soc. Am. 82:633–640.

Howard, D. J., Bush, G. L., and Breznak, J. A. 1985. The evolutionary significance of bacteria associated with *Rhagoletis*. Evolution 39:405–414.

Huettel, M. D., and Bush, G. L. 1972. The genetics of host selection and its bearing on sympatric speciation in *Procecidochares* (Diptera: Tephritidae). Entomol. Exp. Appl. 15:465–480.

Huxley, J. S. 1942. Evolution, the Modern Synthesis. London, Allen and Unwin.

Johnson, J. C., Scholtz, C. A., Talbot, M. R., Kelts, K., Ricketts, R. D., Ngobi, G., Beuning, K., Ssemmanda, I., and McGill, J. W. 1996. Late Pleistocene desiccation of Lake Victoria and rapid evolution of cichlid fishes. Science 273:1091–1093.

Johnson, P., Hoppensteadt, F., Smith, J., and Bush, G. L. 1996. Conditions for sympatric speciation: a diploid model incorporating habitat fidelity and non-habitat assortative mating. Evol. Ecol. 10:187–205.

Levene, H. 1953. Genetic equilibrium when more than one ecological niche is available. Am. Nat. 87:311–313.

Maynard Smith, J. 1966. Sympatric speciation. Am. Nat. 104:487–490.

Mayr, E. 1963. Animal Species and Evolution. Cambridge, Mass., Harvard University Press.

McPheron, B. A., Smith, D. C., and Berlocher, S. H. 1988. Genetic differences between host races of the apple maggot fly. Nature 336:64–66.

Moericke, V., Prokopy, R. J., Berlocher, S., and Bush, G. L. 1975. Visual stimuli eliciting attraction of *Rhagoletis pomonella* (Diptera: Tephritidae) flies to trees. Entomol. Exp. Appl. 18:497–507.

Prokopy, R. J. 1975. Mating behavior in *Rhagoletis pomonella*. V. Virgin female attraction to male odor. Can. Entomol. 107:905–908.

Prokopy, R. J., and Bush G. L. 1973a. Ovipositional responses to different sizes of artificial fruit by flies of *Rhagoletis pomonella* species group. Ann. Entomol. Soc. Am. 66:927–929.

Prokopy, R. J., and Bush, G. L. 1973b. Mating behavior in *Rhagoletis pomonella* (Diptera: Tephritidae). IV. Courtship. Can. Entomol. 105:873–891.

Prokopy, R. J., and Reissig, W. H. 1976. Marking pheromones deterring repeated oviposition in *Rhagoletis* flies. Entomol. Exp. Appl. 20:170–178.

Prokopy, R. J., Moericke, V., and Bush, G. L. 1973. Attraction of apple maggot flies to odor of apples. Environ. Entomol. 2:743–749.

Prowell, D. 1980. Genetic comparison between native and introduced populations of the codling moth, *Laspyrisia pomonella* (L.) (Tortricidae) and among other tortricid species. Ph.D. thesis, University of Texas.

Schliewen, U., Tautz, D., and Pääbo, S. 1994. Sympatric speciation suggested by monophyly of crater lake cichlids. Nature 368:629–632.

Schluter, D. 1994. Experimental evidence that competition promotes divergence in adaptive radiation. Science 266:798–801.

Schluter, D., and Nagel, L. M. 1995. Parallel speciation by natural selection. Am. Nat. 146:292–301.

Smith, D. C. 1988. Heritable divergence of *Rhagoletis pomonella* host races by seasonal asynchrony. Nature 336:66–68.

Smith, J. J., and Bush, G. L. 1997. Phylogeny of the genus *Rhagoletis* (Diptera: Tephritidae) inferred from DNA sequences of mitochondrial cytochrome oxidase II. Mol. Phylol. Evol. 7:33–43.

Walsh, B. D. 1864. On phytophagic varieties and phytophagic species. Proc. Entomol. Soc. Philadelphia 3:403–430.

White, M. J. D. 1954. Animal Cytology and Evolution, 2nd ed. Cambridge, Cambridge University Press.

White, M. J. D. 1968. Modes of speciation. Science 159:1065–1070.

Wilson, A. C., Bush, G. L., Case, S. M., and King, M. C. 1975. Social structuring of mammalian populations and the rate of chromosome evolution. Proc. Natl. Acad. Sci. U.S.A. 72:5061–5065.

33

Unanswered Questions and Future Directions in the Study of Speciation

Daniel J. Howard

Guy Bush ends his contribution to this volume with an optimistic view of the future of speciation studies—an optimism that many of us share. The era in which speciation could be regarded as a deep mystery beyond the reach of scientific study has come to a close. The chapters of this volume are a testament to the new understanding that can be imparted by process-oriented studies of speciation (studies that move beyond the description of pattern and attempt to directly test and explore process). At the same time, these chapters provide evidence of the fracturing of the field. Once primarily the province of systematics, speciation has increasingly attracted the attention of ecologists, geneticists, and behaviorists. Practitioners in each field feel, quite correctly, that they have something important to say about speciation. However, the emphases can be different enough that they are bewildering. What is it we are trying to understand about speciation? In this final chapter, I outline the major questions I believe we are trying to answer and some approaches for addressing the questions, drawing heavily on the thinking of the contributors to this volume. Along the way I highlight some particularly intriguing findings, but for the most part I do not attempt to provide answers to the general questions posed. Some tentative answers can be found in the chapters of this volume. More enduring answers await the outcome of future research.

Before beginning I digress briefly into the area of species definitions, not because I intend to provide insight into future directions for work in this area, but because these definitions have long influenced the questions biologists ask about speciation.

Species Definitions

Templeton (this volume) expresses the feelings of many when he argues that it is first necessary to have a definition of species before studying speciation, because a definition provides a theoretical and operational guide to

one's studies. Curiously, despite these sentiments, many biologists studying speciation, zoologists in particular, have stopped following the continuing controversy about species definitions. To some degree this state of affairs reflects the comfort many zoologists have with the most widely accepted definition of species, the so-called Biological Species Concept (BSC) associated with Dobzhansky (1937, 1951, 1970) and Mayr (1942, 1963). On another level, the lack of interest reflects discomfort with a debate that appears to have no end and seems to have become increasingly muddled. For those who have stopped following the controversy, the chapters 1–4 of this book represent good reasons to become reacquainted with the issues.

Rather than further roiling the waters, the authors attempt to clarify species definitions (Harrison and Templeton) and to find common ground among the definitions (Shaw and deQueiroz). What emerges from a consideration of the definitions and ideas contained in these chapters has a familiar ring—speciation is usually an extended process (the exceptions include speciation due to polyploidization) during which several significant events generally occur. The events may often, but do not necessarily, occur in the following order: a lineage splits and diverges, trait differences evolve between the diverging lineages, a trait difference evolves that reduces potential or actual gene flow between the lineages, separation of the lineages becomes irreversible, and exclusivity (see Shaw, this volume) of each lineage evolves. It is these events that provide the framework for questions about speciation.

Modes of Speciation

For most speciation researchers, "modes of speciation" refers to the geographic circumstances under which one lineage splits and diverges to form two new lineages. For many years, the only accepted mode of speciation was allopatric (as promulgated by Mayr, 1963). One of Bush's (1969, 1975) most important contributions to evolutionary biology was to advocate the idea that spatial separation of populations is not absolutely essential for genetic divergence and the formation of new species. Bush argued that individuals that shift onto a new host or into a new habitat because of a genetically encoded preference can give rise to a genetically isolated population if host/habitat fidelity is strong and mating occurs primarily within the preferred habitat. Bush's arguments provided the impetus for modeling (e.g., Bazykin 1973; Endler 1977; Udovic 1980; Kondrashov 1983a,b, 1984, 1986; Rice 1984; Kondrashov and Mina 1986; Diehl and Bush 1989; Johnson et al. 1996; Kondrashov et al., this volume) and empirical studies (Guttman et al. 1981; Wood and Guttman 1983; Macnair and Christie 1983; Barendse 1984; Feder et al. 1988, 1994, 1995; McPheron et al. 1988; Prokopy et al. 1988; Bierbaum and Bush 1990; Taylor and Bentzen 1993; Schliewen et al.1994; Theron and Combes 1995; Mina et al. 1996) that have now convinced the majority of biologists studying speciation that sympatric speciation is possible. The questions that currently confront us are whether sympatric and other, less extreme, forms of nonallopatric speciation have been important in generating animal and plant diversity and what conditions promote nonallopatric speciation.

Conditions that promote nonallopatric speciation can best be addressed by modeling studies and by intensive study of closely related organisms that are likely to have diverged while in spatial contact, such as host races in insects, edaphic endemics in plants, and species flocks in lakes. The insights that can be gained from such studies are on display in the chapters contributed by Kondrashov et al., Johnson and Gullberg, Feder, Menken and Roessingh, Macnair and Gardner, Schluter, and McCune and Lovejoy.

The question of the relative importance of nonallopatric modes of speciation will be answered by intensive studies of speciation in a variety of animal and plant groups and a clear understanding of phylogenetic relationships, habitats, and geographic distributions. The first step in any empirical investigation of speciation should be a detailed analysis of the systematics and natural history of the group under consideration. This analysis should include multiple populations of each species and every species likely to be closely related to the taxa under consideration. The resulting phylogeny will not only identify sister taxa that should be the object of more intensive study but will also provide insight into the geographic circumstances under which the sister taxa diverged (Patton and da Silva, this volume). For example, if the ranges of sister taxa do not overlap, it is likely they diverged in allopatry. On the other hand, if sister taxa exploit distinct habitats and have ranges that overlap extensively, sympatric divergence is a more likely scenario (Bush and Howard 1986; Berlocher, this volume). Unfortunately, even in well-studied groups, range maps

are often incomplete, habitat utilization patterns are poorly understood, and relevant taxa are not available for genetic and morphological analyses. Until we correct this situation, the relative importance of nonallopatric speciation in generating species diversity will remain highly conjectural (Berlocher, ch. 8, this volume).

Factors Affecting Divergence

Under the allopatric model of speciation as envisioned by Mayr (1963), the evolution of reproductive isolation was seen as a by-product of genetic divergence between geographically isolated populations. According to this model, reproductive isolation has a polygenic basis and evolves as populations accumulate genetic differences. The relative roles of random genetic drift and divergent environmental selection pressures in driving genetic divergence were not seen as particularly pressing issues—populations that were not in spatial contact would diverge, regardless of environmental differences. The one factor, besides geographic isolation, on which Mayr focused special attention was population structure. Impressed by the phenotypic divergence often seen in peripherally isolated populations, Mayr (1954) contended that great genetic divergence and the evolution of truly new lineages were likely to take place in small, peripherally isolated populations founded by one or a few individuals. He argued that in such populations there is a dramatic change in the genetic environment. Genetic variability decreases, and this change in genetic background may alter the selective values of alleles at many loci, setting off a genetic revolution. Although initially somewhat ignored, Mayr's ideas attracted a great deal of attention in later years, especially as Carson, Templeton, Kaneshiro, and others developed research programs centering on Hawaiian Drosophilidae, a species-rich group in which founder events were presumed to be frequent. Eventually, other models of founder-effect speciation, which differed considerably from Mayr's with regard to the genetic changes presumed to accompany a founder event, were developed by Carson (1968, 1971) and Templeton (1980).

The question of whether small founder populations play an important role in genetic divergence and speciation is still open, although there is probably less enthusiasm for the role of founder events in speciation now than existed a decade ago. Barton and Charlesworth have pressed the case that the probability of a shift from one adaptive peak to another during a founder event is low and that, should such a peak shift occur, it will not generate much reproductive isolation between a small founder population and its parental population (Barton and Charlesworth 1984; Barton 1989, 1996). Proponents of founder-effect speciation have mounted a vigorous defense, responding that the models investigated by Barton and Charlesworth fail to incorporate all aspects of the various models of founder-effect speciation and therefore are not good representations of the process (Carson and Templeton 1984; Hollocher 1996). A greater difficulty for founder-effect models, in the minds of many biologists, has arisen from laboratory tests of the models. In general, these tests have yielded rather equivocal results (Powell 1978; Galiana et al. 1989; Moya et al. 1995).

The fact that founder-effect models can be tested in the laboratory is one of their attractive features, and given the element of uncertainty inherent in any speciation event, more experimental work should be carried out. Natural populations should be a focus of attention, as well. The correlation between peripheral isolation and phenotypic divergence that motivated the development of founder-effect models still holds (Barton and Charlesworth 1984; Berry 1996; Hollocher 1996), and still tantalizes. Investigations of island species presumed to have arisen via a founder event should continue, but this work should be supplemented with studies of populations known (via historical records) to have been founded by one or a few individuals. The impact of a known founder event on levels of genetic variation has been investigated by a number of biologists (e.g., Taylor and Gorman 1975; Harrison et al. 1983; Baker and Moeed 1987), but up to now, there has been very little effort to ascertain whether a known founder event or population bottleneck has led to changes in mate recognition, patterns of mating, or other aspects of behavior and ecology that affect reproductive compatibility.

The transition from large population sizes to small population sizes is a critical aspect of founder-effect speciation models, and therefore, it is important to distinguish small founder populations from small populations per se. Nevertheless, whether a species broken into many small, isolated populations is more likely to give rise to new species than is a species with a few large, isolated populations is of both theoretical and practical importance. A recent modeling effort, albeit limited in scope, addresses this question and the results fail to support the importance of small populations in speciation. Orr and Orr (1996) demonstrated that when reproductive isolation involves "complementary genes" (genes that have no deleterious effect in their normal genetic background, but cause sterility or inviability when present together in hybrids), a species composed of many small populations is less likely to speciate than one composed of two large populations. Lessios (this volume) outlines another approach to examining the role of small population size in genetic divergence and speciation. He advocates analyzing the amount of divergence between populations with different population structures that are known to have been separated for the same period of time (e.g., by the rise of the Isthmus of Panama) to determine whether one sees more divergence between lineages characterized by small population sizes.

Turning back to ecology, the new respectability of sympatric speciation has brought with it a fresh emphasis on the role of the environment in speciation events. In particular, it has brought to the fore the question of whether resource-based divergent natural selection is responsible for much of the divergence between closely related species (Schluter, this volume). Clearly, divergence in sympatry is not possible without divergent natural selection playing some (typically major) role. The nature of the selection pressures can be resolved through detailed ecological and behavioral studies such as those that characterize the work on sticklebacks (Schluter 1993, 1994, 1995) and *Rhagoletis* (Prokopy 1968; Prokopy et al. 1971, 1972, 1988; Feder et al. 1993, 1994, 1995; Berlocher, ch. 8, this volume). Schluter (this volume) presses the case further, arguing that differing ecological circumstances often drive divergence and the evolution of reproductive isolation even in allopatric speciation events. He suggests that one can assess the role of divergent natural selection by comparing the amount of divergence between equal age pairs of sister populations that occur in similar and dissimilar environments. Although seemingly straightforward, the difficulty with such comparisons is providing convincing evidence that environments are truly similar or dissimilar.

Up to now, I have treated reproductive isolation as something that evolves after the genetic divergence of populations. Although this was once the dominant point of view among evolutionists, it has now given way to an awareness that reproductive isolation can arise before significant divergence of populations. Indeed, sympatric models of speciation based on host or habitat shifts postulate that the trait differences largely responsible for reproductive isolation—genetically based preferences for distinct hosts or habitats—are the first differences to arise between incipient species. Empirical evidence for a dissociation between genetic divergence and the acquisition of reproductive isolation has emerged from a number of studies; perhaps none more persuasive than allozyme studies carried out during the 1970s that failed to find a close association between genetic divergence and the acquisition of reproductive isolation (Johnson and Selander 1971; Carson et al. 1975; Avise and Smith 1977; Sene and Carson 1977; Avise 1978; Zimmerman et al. 1978; Ryman et al. 1979). As it has become evident that reproductive isolation is not necessarily coupled to amounts of genetic divergence, biologists interested in the evolution of reproductive isolation have turned their attention away from measurements of genetic divergence to a more direct focus on the trait differences responsible for reproductive isolation (Templeton 1981; Bush and Howard 1986).

Barriers to Reproduction

The emphasis on reproductive barriers has become strong enough over the past decade that Harrison (this volume) can claim that they are the focus of most investigations of speciation—and there is little doubt that he is correct. Although the evolution of trait differences that isolate two lineages represents but one aspect of speciation, it is seen as an especially important one by evolutionists—the event that irrevocably separates two lineages and assures their future independence. I have already touched on some of the questions that we would like to answer about the evolution of reproductive isolation, such as under what geographic circumstances reproductively isolating trait differences evolve, and what role divergent natural selection plays in the evolution of these trait differences. Additional questions include the role of sexual and social selection in the evolution of reproductive barriers, the role of reinforcement in strengthening barriers to reproduction, and the number and kind of genes that control reproductively isolating trait differences. A critical step in answering all of these questions is the identification of the trait differences actually responsible for reproductive isolation. The nature of the trait differences provide insight into the forces that drive divergence; for example, if the traits that isolate allow for efficient resource utilization, then divergent natural selection is implicated (Schluter, this volume). On the other hand, if the trait differences are associated with mate recognition, then sexual selection or some sort of antagonistic interaction between males and females is implicated (chapters by Rice, Palumbi, and Howard et al., this volume). Moreover, one cannot begin to study the genetics of reproductive isolation without some understanding of the trait differences that isolate two lineages.

In light of the clear importance of identifying reproductive barriers, there have been remarkably few in-depth assessments of the trait differences that isolate two closely related lineages (but see earlier references to the work on *Rhagoletis* of Bush, Prokopy, Feder, Berlocher, and their colleagues; see also Wells and Henry, this volume). Instead, biologists have tended to focus on differences presumed to be important in reproductive isolation, without clearly demonstrating their relevance. The peril of taking this shortcut is highlighted by the work of my laboratory on the ground crickets *Allonemobius fasciatus* and *A. socius*. Some of our early work on this group demonstrated that there were differences in the male calling songs of the two sister species (Howard and Furth 1986). Given the importance attributed to male calling song in mate recognition among singing Orthoptera (Walker 1957; Pollack and Hoy 1979, 1981; Doherty 1985; Doherty and Hoy 1985; Stout et al. 1988; Sergejeva and Popov 1994), we presumed that the song differences played an important role in reproductive isolation and engaged in a series of studies of the song differences, ranging from examinations of patterns of geographic variation (Benedix and Howard 1991; Veech et al. 1996) to an analysis of quantitative genetics (Mousseau and Howard in press). Only after this work had been initiated did it become clear from phonotactic studies that females do not prefer the songs of conspecific males and that song differences do

not serve as a premating reproductive barrier between the two species (Doherty and Howard 1996). A series of investigations, ranging from laboratory hybridization studies (Gregory and Howard 1993) and mate choice experiments (Gregory and Howard, unpublished) to field studies of phenology and habitat usage (Howard et al. 1993), finally determined that the two taxa are isolated by a postinsemination barrier to fertilization (Howard et al., this volume), which is the focus of much of our current research. The critical point here is that the presence of this barrier was completely unforeseen and emerged only through vigorous inquiry.

Intriguingly, postinsemination barriers to fertilization have now been described between a number of other closely related terrestrial animals, leading to the suggestion that such barriers arise quickly and may often be the first barrier to evolve between isolated populations (Howard and Gregory 1993; Gregory and Howard 1994; Rice, this volume). Other surprises to emerge from in-depth investigations of reproductive barriers are the potentially important roles played by endosymbiotic bacteria of the genus *Wolbachia* (Werren, this volume), and by differences in cuticular hydrocarbon blends (Butlin, this volume). The message from these studies is unmistakable—take nothing for granted when it comes to identifying barriers to gene exchange between closely related taxa.

Of the questions that need to be answered about the evolution of reproductive barriers, perhaps none are more difficult than resolving the force or forces that drive the evolution of isolating traits. In answering these questions, we confront the specter that haunts all too much our work on speciation—the difficulty of reconstructing processes that have taken place in the past and have left little or no interpretable record. Certain types of reproductive barriers implicate the action of particular forces (see above and Schluter, this volume), and a broad comparative approach (Howard et al., this volume) can help to distinguish between competing hypotheses, but some uncertainty will accompany all scenarios that purport to explain the evolution of trait differences that have arisen in the past. This uncertainty places a premium on studies of dividing lineages that are not yet reproductively isolated, such as host races of *Rhagoletis pomonella*. As demonstrated by the work of Feder (this volume), under such circumstances one can identify and measure the forces driving divergence and the final acquisition of reproductive isolation.

The Genetics of Reproductive Barriers

Ever since Mayr (1942) and Goldschmidt (1940) squared off over the role of macromutations in the evolution of reproductive isolation, the genetic basis of reproductive isolation has been one of the central problems of speciation. It is a problem that is not shrouded in the mists of history and hence is amenable to experimental investigation, as demonstrated by early work of Dobzhansky

(1933, 1936). However, a concerted assault on the problem did not begin until about a decade ago, when the question of interest was better framed; namely, how many and what kind of genes control the trait difference(s) responsible for reproductive isolation (Templeton 1981; Bush and Howard 1986), and the question again attracted the attention of *Drosophila* population geneticists, who could draw on a wealth of mapped markers to investigate the influence of different chromosome regions on reproductive isolation. Among the most notable *Drosophila* workers have been Coyne, Naveira, Orr, and Wu, who have concentrated, as did Dobzhansky, on the genetic basis of male hybrid sterility (Coyne 1984, 1985, 1989; Naveira and Fontdevila 1986, 1991; Orr 1989, 1992; Naveira 1992; Wu 1992; Wu and Davis 1993; Maside and Naveira 1996; Wu and Palopoli 1996).

The results of their studies, particularly the newest results from the laboratories of Wu and Naveira (see chapters by Wu and Hollocher, and Naveira and Maside, this volume) indicate that many genes spread throughout the genome contribute to hybrid male sterility, even between very closely related taxa, and that "weak allele–strong interaction" is the rule rather than the exception (Wu and Hollocher, this volume). Unfortunately, the narrow focus of the studies limits the generality of the results. We do not know if hybrid male sterility has a similar genetic basis in pairs of closely related taxa outside of *Drosophila*, nor do we have an adequate understanding of the genetic basis of other traits, such as signaling system traits, that often isolate closely related taxa (Ritchie and Phillips, this volume). The importance of studying other taxonomic groups is underscored by what has been learned about the genetics of trait differences in Lepidoptera. Here there is a disproportionate association between traits thought to be important in speciation and the X-chromosome (Prowell, this volume).

As alluded to earlier, to resolve the presence of genes affecting a particular trait, one must work with taxa having many mapped markers, so that one can follow the segregation of chromosome regions and phenotypes in linkage studies. Until recently, very few species other than *Drosophila* had such markers. However, the development of new techniques to detect molecular variation, most notably restriction fragment length polymorphisms (RFLPs) and random amplification of polymorphic DNA (RAPDs), have made it possible to quickly uncover such markers for virtually any group of organisms. Moreover, the development of new statistical methods has considerably simplified and sped up the identification and mapping of quantitative trait loci (QTL; Lander and Botstein 1989; Knott and Haley 1992; Jiang and Zeng 1995), such as those likely to be responsible for differences in reproductively isolating traits. Thus, genetic investigations of reproductive barriers can now be performed in virtually any taxonomic group.

Before the floodgates open and we are deluged with QTL studies of the genetic basis of reproductive isolation, a few cautionary notes are in order (for a fuller discussion of QTL studies, see Via and Hawthorne, this volume). First of all, it is important to recognize that species will continue to accumulate genetic differences after a speciation event and that these differences will influence traits responsible for reproductive isolation. If the critical question is how many genetic changes are necessary for the initial acquisition of reproductive isolation, then taxa that have very recently acquired, or are in the early stages of acquiring, reproductive isolation must be the focus of attention, so that the accumulation of genetic differences after the speciation event do not confound the results. Furthermore, QTL studies should not be initiated until the trait differences responsible for reproductive isolation between the groups of interest have been clearly identified. Shortcuts around this requirement will ultimately delay rather than accelerate progress in the field. Finally, we need to keep in mind that QTL studies will provide a fairly coarse first look at the genetic basis of reproductive barriers. Consequently, the results of these studies should be regarded as tentative and should be followed up with more detailed work, such as studies of introgression lines (see chapters by Naveira and Maside, and Wu and Hollocher, this volume).

No consideration of the genetic basis of reproductive isolation contained in a volume dedicated to Guy Bush can be considered complete without some mention of the potential role played by chromosomal change. The most forceful advocate of the importance of chromosomal change in speciation was Bush's postdoctoral advisor, M. J. D. White. In 1968, White proposed a stasipatric model of speciation that had as its centerpiece the contention that chromosomal rearrangements can serve as strong primary genetic barriers to gene exchange because of the diminished fecundity of heterozygotes. To explain how a rearrangement that decreased the fitness of heterozygotes could go from low to high frequency in a population, White (1978) pointed out that four factors alone or in combination could overcome negative heterotic selection: random genetic drift in small populations, inbreeding, meiotic drive, and selective advantage of the new rearrangement as a homozygote. Bush and his colleagues (Bush et al. 1977) added their voices to the controversy engendered by White's ideas by noting that social structuring could also accelerate chromosomal evolution and speciation by reducing effective population sizes, restricting gene flow, and increasing levels of inbreeding. A lack of empirical studies prevents final judgment on the merits of White's model and the modifications proposed by Bush et al. (Spirito, this volume); however, it is now clear that chromosomal rearrangements do not automatically have a negatively heterotic effect (Howard 1993a), a finding that weakens the arguments of White. The lack of definite rules means that the actual effect of a chromosomal variant on reproductive success will have to be evaluated for each group of organisms that comes under scrutiny, and understanding the role of chromosomal rearrangements in speciation will come about only after in-depth studies of a number of different closely related taxa.

Hybrid Zones and Speciation

Hybrid zones have been characterized as "windows on the evolutionary process" (Harrison 1990), and certainly much of the interest they have generated over the years can be attributed to the insights they might provide into the process of speciation. Early in the Modern Synthesis, Dobzhansky (1940) promulgated the idea that premating barriers to gene exchange between diverging lineages were unlikely to arise in allopatry and that selection against hybridization in zones of secondary contact (a process called reinforcement) drove the evolution of these barriers. Although reinforcement fell out of favor in the 1970s and 1980s, there has been a rebirth of interest in recent years based largely on new empirical evidence in favor of the process emanating from studies of hybrid zones and of zones of contact between closely related species (Coyne and Orr 1989; Howard 1993b; Noor 1995, 1997). Positive assortative mating seems to be a common feature of hybrid zones and reproductive character displacement (the predicted outcome of reinforcement) may be more widespread than previously thought (Howard 1993b). However, as Markow and Hocutt (this volume) make clear, patterns suggestive of reinforcement can be generated by other processes and a clear demonstration of reinforcement entails much effort. Consequently, reinforcement remains controversial (Butlin, this volume), and many more empirical studies of hybrid zones and of areas of overlap between closely related taxa are necessary to resolve the question of whether selection against hybridization often drives the evolution of prezygotic reproductive barriers.

Ironically, even as the controversy over reinforcement has been heating up, another group of biologists led by Arnold (1997; Arnold and Emms, this volume) and Grant and Grant (this volume) have been reviving other long-dormant questions that emphasize a more creative role for hybridization in the origin of new species; namely, can hybrids act as the founders of new evolutionary lineages, and can introgression play a creative role in the speciation process? With regard to the latter question, there is no doubt that introgression of genes from one species into another is a common occurrence in nature (e.g., Parsons et al. 1993; Bell 1996; Arnold 1997; Howard et al. 1997) and that this introgression increases the genetic variation of the recipient taxon, enhancing its ability to respond to selection (Grant and Grant 1992, 1994). However, whether introgression can provide the impetus for a population to move onto a separate evolutionary pathway is much less obvious. Detailed, long-term studies of the consequences of genetic exchange between closely related taxa, such as those being carried out by Grant and Grant and by Arnold, should help to resolve this issue.

The argument that hybrids of the sort found at the center of a hybrid zone can represent the starting point of new species is likely to encounter stiff resistance from biologists, even from those who acknowledge the importance of allopolyploidy in the evolution of plant diversity. Setting aside the question of how these genotypes are maintained in the face of gene flow (Arnold [1997] provides some possible answers), the notion that novel genotypes arising in a hybrid zone will be fit enough to compete with, and possibly replace, parental populations runs counter to a large hybrid zone literature that emphasizes hybrid unfitness. Foreseeing this response, Arnold (1997) has noted that estimates of hybrid fitness in the literature are largely inadequate, and he has enumerated many instances in which hybrids appear to be as fit or more fit than parentals. Moreover, hybrids may avoid competing with parentals by exploiting new habitats (Arnold 1997). Given the inadequacy of our understanding of hybrid zones and hybridization (Butlin, this volume), whether new species are frequently founded by hybrid individuals is likely to remain uncertain for some time. In the meantime, the controversy places new pressure on biologists to more carefully measure the fitness and habitat utilization patterns of hybrids.

Genetic analyses of hybrid zones using species-specific markers can reveal much information relevant to speciation, including the overall strength of selection against hybrids and the number of genes responsible for hybrid unfitness (Barton and Gale 1993). As mapped markers begin to be incorporated into these investigations, even more information can be gleaned from hybrid zones. We know that many species that remain distinct in a variety of ecologically and evolutionarily important traits exchange genes in and across hybrid zones (Mallet et al., this volume). Clearly, trait differences that are maintained in the face of gene flow cannot be under the control of regions of the genome that easily cross the species boundary. Thus, understanding how much and what part of the genome is *not* exchanged between hybridizing taxa by using mapped markers can provide insight into the genetic control of ecologically and evolutionarily important trait differences, such as those that reproductively isolate. I hasten to note that only those hybrid zones characterized by strong (but not complete) reproductive isolation between the interacting taxa can be exploited for the purpose of studying reproductive barriers—but such hybrid zones are fairly common (Howard 1993b). This sort of hybrid zone can also be used to test whether regions of the genome implicated in the control of reproductive isolation have the effect attributed to them. A region purported to affect reproductive isolation should rarely introgress from one taxon into another.

Conclusion

The questions that confront biologists about speciation are difficult but answerable. Most of the questions are framed in terms of "how important is," which means that the results emerging from studies on a single system, while meaningful, will provide only partial insight into speciation and the factors that promote it. A rich understanding demands investigation of a number of closely related taxa presumed to be at different stages of the process. Because very little can be taken for granted about any single speciation event, each investigation will require long-term commitment and a willingness on the part of the investigator to cross disciplinary lines. These investigations will also require considerable resources from private and federal funding agencies. The ultimate question, "How are new species formed?" is fundamental enough to biology and to human curiosity to deserve the time and money.

The questions are important from another perspective as well. The rise of conservation biology has fueled a new understanding of the importance of preserving species diversity. Thus far, our preservation efforts have concentrated on rescuing plants and animals skirting the edge of extinction, and in setting up preserves that are capable of maintaining viable, genetically diverse populations of plants and animals. But ultimately, efforts to preserve will not be successful without the generation of new diversity. It is in this area that what we learn about speciation may have its most profound impact.

Acknowledgments   I am grateful to Stewart Berlocher, Guy Bush, and Michael Cain for their insightful criticisms of an earlier draft of this chapter. My work on hybrid zones and speciation has been supported by the National Science Foundation for some time, most recently by NSF Grant DEB 9726502.

References

Arnold, M. L. 1997. Natural Hybridization and Evolution. Oxford University Press, New York.

Avise, J. C. 1978. Variances and frequency distributions of genetic distance in evolutionary phylads. Heredity 40: 225–237.

Avise, J. C., and Smith, M. H. 1977. Gene frequency comparisons between sunfish (Centrarchidae) populations at various stages of evolutionary divergence. Syst. Zool. 26:319–335.

Baker, A. J., and Moeed, A. 1987. Rapid genetic differentiation and founder effect in colonizing populations of common mynas (*Acridotheres tristis*). Evolution 41:525–538.

Barendse, W. 1984. Speciation in the genus *Crinia* (Anura: Myobatrachidae) in southern Australia: a phylogenetic analysis of allozyme data supporting endemic speciation in southwestern Australia. Evolution 38:1238–1250.

Barton, N. H. 1989. Founder effect speciation. In D. Otte and J. A. Endler (eds.), Speciation and Its Consequences. Sunderland, Mass.: Sinauer, pp. 229–256.

Barton, N. H. 1996. Natural selection and random genetic drift as causes of evolution on islands. Philos. Trans. R. Soc. Lond. B 351:785–795.

Barton, N. H., and Charlesworth, B. 1984. Genetic revolutions, founder effects, and speciation. Annu. Rev. Ecol. Syst. 15:133–164.

Barton, N. H., and K. S. Gale. 1993. Genetic analysis of hybrid zones. In R. G. Harrison (ed.), Hybrid Zones and the Evolutionary Process. Oxford: Oxford University Press, pp. 13–45.

Bazykin, A. D. 1973. Population genetic analysis of disruptive and stabilizing selection. Part II. Systems of adjacent populations and populations within a continuous area. Genetika 9:156–166.

Bell, D. A. 1996. Genetic differentiation, geographic variation and hybridization in gulls of the *Larus glaucescens-occidentalis* complex. Condor 98:527–546.

Benedix, J. H.., Jr., and Howard, D. J. 1991. Calling song displacement in a zone of overlap and hybridization. Evolution 45:1751–1759.

Berry, R. J. 1996. Small mammal differentiation on islands. Philos. Trans. R. Soc. Lond. B 351:753–764.

Bierbaum, T. J., and Bush, G. L. 1990. Genetic differentiation in the viability of sibling species of *Rhagoletis* fruit flies on host plants, and the influence of reduced hybrid viability on reproductive isolation. Entomol. Exp. Appl. 55:105–108.

Bush, G. L. 1969. Sympatric host race formation and speciation in frugivorous flies of the genus *Rhagoletis*. Evolution 23:237–251.

Bush, G. L. 1975. Modes of animal speciation. Annu. Rev. Ecol. Syst. 6:339–364.

Bush, G. L., Case, S. M., Wilson, A. C., and Patton, J. L. 1977. Rapid speciation and chromosomal evolution in animals. Proc. Natl. Acad. Sci. USA 74:3942–3946.

Bush, G. L., and Howard, D. J. 1986. Allopatric and nonallopatric speciation: assumptions and evidence. In S. Karlin and E. Nevo (eds.), Evolutionary Processes and Theory. Orlando: Academic Press, pp. 411–438.

Carson, H. L. 1968. The population flush and its genetic consequences. In R. C. Lewontin (ed.), Population Biology and Evolution. New York: Syracuse University Press, pp. 123–137.

Carson, H. L. 1971. Speciation and the founder principle. Stadler Genet. Symp. 3:51–70.

Carson, H. L., Johnson, W. E., Nair, P. S., and Sene, F. M. 1975. Allozymic and chromosomal similarity in two *Drosophila* species. Proc. Natl. Acad. Sci. USA 72:4521–4525.

Carson, H. L., and Templeton, A. R. 1984. Genetic revolutions in relation to speciation phenomena: the founding of new populations. Annu. Rev. Ecol. Syst. 15:97–131.

Coyne, J. A. 1984. Genetic basis of male sterility in hybrids between two closely related species of *Drosophila*. Proc. Natl. Acad. Sci. USA 81:4444–4447.

Coyne, J. A. 1985. Genetic studies of three sibling species of *Drosophila* with relationship to theories of speciation. Genet. Res. Camb. 46:169–192.

Coyne, J. A. 1989. Genetics of sexual isolation between two sibling species, *Drosophila simulans* and *Drosophila mauritiana*. Proc. Natl. Acad. Sci. USA 86:5464–5468.

Coyne, J. A., and Orr, H. A. 1989. Patterns of speciation in *Drosophila*. Evolution 43:362–381.

Diehl, S. R., and Bush, G. L. 1989. The role of habitat preference in adaptation and speciation. In D. Otte and J. A. Endler (eds.), Speciation and Its Consequences. Sunderland, Mass.: Sinauer, pp. 345–365.

Dobzhansky, T. 1933. On the sterility of the interracial hybrids in *Drosophila pseudoobscura*. Proc. Natl. Acad. Sci. USA 19:397–403.

Dobzhansky, T. 1936. Studies on hybrid sterility. II. Localization of sterility factors in *Drosophila pseudoobscura* hybrids. Genetics 21:113–135.

Dobzhansky, T. 1937. Genetics and the Origin of Species. Columbia University Press, New York.

Dobzhansky, T. 1940. Speciation as a stage in evolutionary divergence. Am. Nat. 74:312–321.

Dobzhansky, T. 1951. Genetics and the Origin of Species, 3rd ed., Columbia University Press, New York.

Dobzhansky, T. 1970. Genetics and the Evolutionary Process. Columbia University Press, New York.

Doherty, J. A. 1985. Trade-off phenomena in calling song recognition and phonotaxis in the cricket, *Gryllus bimaculatus* (Orthoptera, Gryllidae). J. Comp. Physiol. 156:787–801.

Doherty, J. A., and Hoy, R. R. 1985. Communication in insects. III. The auditory behavior of crickets: some views of genetic coupling, song recognition, and predator detection. Q. Rev. Biol. 60:457–472.

Doherty, J. A., and Howard, D. J. 1996. Lack of preference for conspecific calling songs in female crickets. Anim. Behav. 51:981–990.

Endler, J. A. 1977. Geographic Variation, Speciation, and Clines. Princeton University Press, Princeton, N.J.

Feder, J. L., Chilcote, C. A., and Bush, G. L. 1988. Genetic differentiation between sympatric host races of *Rhagoletis pomonella*. Nature 336:61–64.

Feder, J. L., Hunt, T. A., and Bush, G. L. 1993. The effects of climate, host plant phenology and host fidelity on the genetics of apple and hawthorn infesting races of *Rhagoletis pomonella*. Entomol. Exp. Appl. 69:117–135.

Feder, J. L., Opp, S., Wlazlo, B., Reynolds, K., Go, W., and Spisak, S. 1994. Host fidelity is an effective pre-mating barrier between sympatric races of the apple maggot fly. Proc. Natl. Acad. Sci. USA 91:7990–7994.

Feder, J. L., Reynolds, K., Go, W., and Wang, E. C. 1995. Intra- and interspecific competition and host race formation in the apple maggot fly, *Rhagoletis pomonella* (Diptera: Tephritidae). Oecologia 101:416–425.

Galiana, A., Ayala, F. J., and Moya, A. 1989. Flush-crash experiments in *Drosophila*. In A. Fontdevila (ed.), Evolutionary Biology of Transient Unstable Populations. Berlin: Springer-Verlag, pp. 58–73.

Goldschmidt, R. 1940. The Material Basis of Evolution. Yale University Press, New Haven, Conn.

Grant, P. R., and Grant, B. R. 1992. Hybridization and bird species. Science 256:193–197.

Grant, P. R., and Grant, B. R. . 1994. Phenotypic and genetic effects of hybridization in Darwin's finches. Evolution 48:297–316.

Gregory, P. G., and Howard, D. J. 1993. Laboratory hybridization studies of *Allonemobius fasciatus* and *A. socius* (Orthoptera: Gryllidae). Ann. Entomol. Soc. Am. 86:694–701.

Gregory, P. G., and Howard, D. J. 1994. A post-insemination barrier to fertilization isolates two closely related ground crickets. Evolution 48:705–710.

Guttman, S. I., Wood, T. K., and Karlin, A. A. 1981. Genetic differentiation along host plant line in the sympatric *Enchenopa binotata* Say complex (Homoptera: Membracidae). Evolution 33:205–217.

Harrison, R. G. 1990. Hybrid zones: windows on evolutionary processes. Oxford Surv. Evol. Biol 7:129–156

Harrison, R. G., Wintermeyer, S. F., and Odell, T. M. 1983. Patterns of genetic variation within and among gypsy moth, *Lymantria dispar* (Lepidoptera: Lymantriidae), populations. Ann. Entomol. Soc. Am. 76:652–656.

Hollocher, H. 1996. Island hopping in *Drosophila*: patterns and processes. Philos. Trans. R. Soc. Lond. B 351:735–743.

Howard, D. J. 1993a. Small populations, inbreeding, and speciation. In N. W. Thornhill (ed.), The Natural History of Inbreeding and Outbreeding: Theoretical and Empirical Perspectives. Chicago: University of Chicago Press, pp. 118–142.

Howard, D. J. 1993b. Reinforcement: origins, dynamics, and fate of an evolutionary hypothesis. In R. G. Harrison (ed.), Hybrid Zones and the Evolutionary Process. New York: Oxford University Press, pp. 46–69.

Howard, D. J., and Furth, D. G. 1986. Review of the *Allonemobius fasciatus* (Orthoptera: Gryllidae) complex with the description of two new species separated by electrophoresis, songs, and morphometrics. Ann. Entomol. Soc. Am. 79:472–481.

Howard, D. J., and Gregory, P. G.. 1993. Post-insemination signaling systems and reinforcement. Philos. Trans. Roy. Soc. Lond. B 340:231–236.

Howard, D. J., Waring, G. L., Tibbets, C. A., and Gregory, P. G. 1993. Survival of hybrids in a mosaic hybrid zone. Evolution 47:789–800.

Howard, D. J., Waring, G. L., Tibbets, C. A., and Gregory, P. G. 1993. Survival of hybrids in a mosaic hybrid zone. Evolution 47:789–800.

Howard, D. J., Preszler, R. W., Williams, J., Fenchel, S., and Boecklen, W. J. 1997. How discrete are oak species? Insights from a hybrid zone between *Quercus grisea* and *Q. gambelii*. Evolution 51:747–755.

Jiang, C., and Zeng, Z. B. 1995. Multiple trait analysis of genetic mapping for quantitative trait loci. Genetics 140: 1111–1127.

Johnson, W. E., and Selander, R. K. 1971. Protein variation and systematics in kangaroo rats (genus *Dipodomys*). Syst. Zool. 20:377–405.

Johnson, P. A., Hoppensteadt, F. C., Smith, J. J., and Bush, G. L. 1996. Conditions for sympatric speciation: a diploid model incorporating habitat fidelity and non-habitat assortative mating. Evol. Ecol. 10:187–205.

Knott, S. A., and Haley, C. S. 1992. Maximum likelihood mapping of quantitative trait loci using full-sib families. Genetics 132:1211–1222.

Kondrashov, A. S. 1983a. Multilocus model of sympatric speciation. I. One character. Theor. Pop. Biol. 24:121–135.

Kondrashov, A. S. 1983b. Multilocus model of sympatric speciation II. Two characters. Theor. Pop. Biol. 24:136–144.

Kondrashov, A. S. 1984. On the intensity of selection for reproductive isolation at the beginning of sympatric speciation. Genetika 20:408–415.

Kondrashov, A. S. 1986. Multilocus model of sympatric speciation. III. Computer simulations. Theor. Pop. Biol. 29:1–15.

Kondrashov, A. S., and Mina, M. V. 1986. Sympatric speciation: when is it possible? Biol. J. Linn. Soc. 27:201–223.

Lander, E. S., and Botstein, D. 1989. Mapping Mendelian factors underlying quantitative traits using RFLP linkage maps. Genetics 121:185–199.

Macnair, M. R., and Christie, P. 1983. Reproductive isolation as a pleiotropic effect of copper tolerance in *Mimulus guttatus*? Heredity 50:295–302.

Maside, X. R., and Naveira, H. F. 1996. A polygenic basis of hybrid sterility may give rise to spurious localizations of major sterility factors. Heredity 77:488–492.

Mayr, E. 1942. Systematics and the Origin of Species. Columbia University Press, New York.

Mayr, E. 1954. Change of genetic environment and evolution. In J. Huxley, A. C. Hardy, and E. B. Ford (eds.), Evolution as a Process. London: Allen and Unwin, pp. 157–180.

Mayr, E. 1963. Animal Species and Evolution. Belknap Press, Cambridge, Mass.

McPheron, B. A., Smith, D. C., and Berlocher, S. H. 1988. Genetic differences between *Rhagoletis pomonella* host races. Nature 336:64–66.

Mina, M. V., Mironovsky, A. N., and Dgebuadze, Y. 1996. Lake Tana large barbs: phenetics, growth and diversification. J. Fish Biol. 48:383–404.

Mousseau, T. A., and Howard, D. J. In press. Genetic variation in cricket calling song across a hybrid zone between two sibling species. Evolution.

Moya, A., Galiana, A., and Ayala, F. J. 1995. Founder-effect speciation theory: failure of experimental corroboration. Proc. Natl. Acad. Sci. USA 92:3983–3986.

Naveira, H. F. 1992. Location of X-linked polygenic effects causing sterility in male hybrids of *Drosophila simulans* and *D. mauritiana*. Heredity 68:211–217.

Naveira, H., and Fontdevila, A. 1986. The evolutionary history of *Drosophila buzzatii*. XII. The genetic basis of sterility in hybrids between *D. buzzatii* and its sibling *D. serido* from Argentina. Genetics 114:841–857.

Naveira, H., and Fontdevila, A. 1991. The evolutionary history of *Drosophila buzzatii*. XXI. Cumulative action of multiple sterility factors on spermatogenesis in hybrids of *D. buzzatii* and *D. koepferae*. Heredity 67:57–72.

Noor, M. 1995. Speciation driven by natural selection in *Drosophila*. Nature 375:674–675.

Noor, M. 1997. How often does sympatry affect sexual isolation in *Drosophila*? Am. Nat. 149:1156–1163.

Orr, H. A., and Orr, L. H. 1996. Waiting for speciation: the effect of population subdivision on the time to speciation. Evolution 50:1742–1749.

Orr, H. A. 1989. Genetics of sterility in hybrids between two subspecies of *Drosophila*. Evolution 43:180–189.

Orr, H. A. 1992. Mapping and characterization of a 'speciation gene' in *Drosophila*. Genet. Res. Camb. 59:73–80.

Parsons, T. J., Olson, S. L., and Braun, M. J. 1993. Unidirectional spread of secondary sexual plumage traits across an avian hybrid zone. Science 260:1643–1646.

Pollack, G., and Hoy, R. R. 1979. Temporal pattern as a cue for species-specific calling song recognition in crickets. Science 204:429–432.

Pollack, G., and Hoy, R. R. 1981. Phonotaxis to individual rhythmic components of a complex cricket calling song. J. Comp. Physiol. 144:367–373.

Powell, J. R. 1978. The founder-flush speciation theory: an experimental approach. Evolution 32:465–474.

Prokopy, R. J. 1968. The influence of photoperiod, temperature and food on the initiation of diapause in the apple maggot. Can. Entomol. 100:318–329.

Prokopy, R. J., Bennett, E. W., and Bush, G. L. 1971. Mating behavior in *Rhagoletis pomonella* (Diptera: Tephritidae). I. Site of assembly. Can. Entomol. 103:1405–1409.

Prokopy, R. J., Bennett, E. W., and Bush, G. L. 1972. Mating behavior in *Rhagoletis pomonella* (Diptera: Tephritidae). II. Temporal organization. Can. Entomol. 104:97–104.

Prokopy, R. J., Diehl, S. R., and Cooley, S. S. 1988. Behavioral evidence for host races in *Rhagoletis pomonella* flies. Oecologia 76:138–147.

Rice, W. R. 1984. Disruptive selection on habitat preference and the evolution of reproductive isolation: a simulation study. Evolution 38:1251–1260.

Ryman, N., Allendorf, F. W., and Stahl, G. 1979. Reproductive isolation with little genetic divergence in sympatric populations of brown trout (*Salmo trutta*). Genetics 92:247–262.

Schliewen, U. K., Tautz, D., and Paabo, S. 1994. Sympatric speciation suggested by monophyly of crater lake cichlids. Nature 383:613–616.

Schluter, D. 1993. Adaptive radiation in sticklebacks: size, shape, and habitat use efficiency. Ecology 74:699–709.

Schluter, D. 1994. Experimental evidence that competition promotes divergence in adaptive radiation. Science 266:798–801.

Schluter, D. 1995. Adaptive radiation in sticklebacks: trade-offs in feeding performance and growth. Ecology 76:82–90.

Sene, F. M., and Carson, H. L. 1977. Genetic variation in Hawaiian *Drosophila*. IV. Allozymic similarity between *D. silvestris* and *D. heteroneura* from the island of Hawaii. Genetics 86:187–198.

Sergejeva, M. V., and Popov, A. V. 1994. Ontogeny of positive phonotaxis in female crickets, *Gryllus bimaculatus* De Geer: dynamics of sensitivity, frequency-intensity domain, and selectivity to temporal pattern of the male calling song. J. Comp. Physiol A 174:381–389.

Stout, J. F., DeHaan, C. H., and McGhee, R. W. 1988. Attractiveness of the male *Acheta domesticus* calling song to females. I. Dependence on each of the calling song features. J. Comp. Physiol. 153:509–521.

Taylor, C. E., and Gorman, G. C. 1975. Population genetics of a "colonising" lizard: natural selection for allozyme morphs in *Anolis grahami*. Heredity 35:241–247.

Taylor, E. B., and Bentzen, P. 1993. Evidence for multiple origins and sympatric divergence of trophic ecotypes of smelt *Osmerus* in northeastern North America. Evolution 47:813–832.

Templeton, A. R. 1980. The theory of speciation via the founder principle. Genetics 94:1011–1038.

Templeton, A. R. 1981. Mechanisms of speciation—a population genetic approach. Annu. Rev. Ecol. Syst. 12:23–48.

Theron, A., and Combes, C. 1995. Asynchrony of infection timing, habitat preference, and sympatric speciation of shistosome parasites. Evolution 49:372–375.

Udovic, D. 1980. Frequency-dependent selection, disruptive selection, and the evolution of reproductive isolation. Am. Nat. 116:621–641.

Veech, J. A., Benedix, J. H., Jr., and Howard, D. J. 1996. Lack of calling song displacement between two closely related ground crickets. Evolution 50:1982–1989.

Walker, T. J. 1957. Specificity in the response of female tree crickets (Orthoptera, Gryllidae, Oecanthinae) to calling songs of the males. Ann. Entomol. Soc. Am. 50:626–636.

White, M. J. D. 1968. Models of speciation. Science 159:1065–1070.

White, M. J. D. 1978. Modes of Speciation. Freeman, San Francisco.

Wood, T. K., and Guttman, S. I. 1983. *Enchenopa binotata* complex: sympatric speciation? Science 220:310–312.

Wu, C.-I. 1992. A note on Haldane's Rule: hybrid inviability vs. hybrid sterility. Evolution 46:1584–1587.

Wu, C.-I., and Davis, A. W. 1993. Evolution of postmating reproductive isolation: the composite nature of Haldane's Rule and its genetic bases. Am. Nat. 147:187–212.

Wu, C.-I., and Palopoli, M. F. 1996. Haldane's rule and its legacy: why are there so many sterile males? Trends Ecol. Evol. 11:281–284.

Zimmerman, E. G., Kilpatrick, C. W., and Hart, B. J. 1978. The genetics of speciation in the rodent genus *Peromyscus*. Evolution 32:565–579.

Publications of Guy L. Bush

Chronological List

G. L. Bush. 1957. Some notes on the susceptibility of avocados in Mexico to attack by the Mexican fruit fly. *Journal of the Rio Grande Valley Horticulture Society* 11: 75–78.

G. L. Bush and G. B. Chapman. 1961. Electron microscopy of symbiotic bacteria in developing oocytes of the American cockroach, *Periplaneta americana. Journal of Bacteriology* 81:267–276.

G. L. Bush. 1962. The cytotaxonomy of the larvae of some Mexican fruit flies in the genus *Anastrepha* (Tephritidae, Diptera). *Psyche* 69:87–101.

T. H. Hamilton, I. Rubinoff, R. H. Barth, Jr., and G. L. Bush. 1963. Species abundance: natural regulation of insular variation. *Science* 142:1575–1577.

G. L. Bush. 1965. The genus *Zonosemata*, with notes on the cytology of two species (Diptera-Tephritidae). *Psyche* 72:307–323.

G. L. Bush. 1966. The comparative cytology of the Choristidae and Nannochoristidae (Mecoptera). *Year Book of the American Philosophical Society* 326–328.

G. L. Bush. 1966. Female heterogamety in the family Tephritidae (Acalyptratae, Diptera). *American Naturalist* 100:119–126.

G. L. Bush. 1966. The taxonomy, cytology, and evolution of the genus *Rhagoletis* in North America (Diptera, Tephritidae). *Bulletin of the Museum of Comparative Zoology* 134:431–562.

G. L. Bush. 1969. Mating behavior, host specificity, and the ecological significance of sibling species in frugivorous flies of the genus *Rhagoletis* (Diptera: Tephritidae). *American Naturalist* 103:669–672.

G. L. Bush. 1969. Sympatric host race formation and speciation in frugivorous flies of the genus *Rhagoletis* (Diptera, Tephritidae). Evolution 23:237–251.

G. L. Bush. 1969. Trail laying by larvae of *Chlosyne lacinia. Annals of the Entomological Society of America* 62:674–675.

G. L. Bush, and S. C. Taylor. 1969. The cytogenetics of *Procecidochares*. I. The mitotic and polytene chromosomes of the Pamakani fly, *P. utilis* (Tephritidae-Diptera). *Caryologia* 22:311–322.

G. L. Bush and M. D. Huettel. 1970. Cytogenetics and description of a new North American species of the Neotropical genus *Cecidocharella* (Diptera: Tephritidae). *Annals of the Entomological Society of America* 63:88–91.

B. A. Drummond, III, G. L. Bush, and T. C. Emmel. 1970. The biology and laboratory culture of *Chlosyne lacinia* Geyer (Nymphalidae). *Journal of the Lepidopterists' Society* 24:135–142.

R. W. Neck, G. L. Bush, and B. A. Drummond, III. 1971. Epistasis, associated lethals and brood effect in larval colour polymorphism of the patch butterfly, *Chlosyne lacinia. Heredity* 26:73–84.

R. J. Prokopy, E. W. Bennett, and G. L. Bush. 1971. Mating behavior in *Rhagoletis pomonella* (Diptera: Tephritidae). I. Site of assembly. *Canadian Entomologist* 103:1405–1409.

G. L. Bush. 1972. Mechanisms of sympatric host race formation in the true fruit flies (Tephritidae). *Abstracts of the 14th International Congress of Entomology* 16.

M. D. Huettel, and G. L. Bush. 1972. The genetics of host selection and its bearing on sympatric speciation in *Procecidochares* (Diptera: Tephritidae). *Entomologia Experimentalis et Applicata* 15:465–480.

R. J. Prokopy and G. L. Bush. 1972. Apple maggot infestation of pear. *Journal of Economic Entomology* 65:597.

R. J. Prokopy and G. L. Bush. 1972. Mating behavior in *Rhagoletis pomonella* (Diptera: Tephritidae). III. Male aggregation in response to an arrestant. *Canadian Entomologist* 104:275–283.

R. J. Prokopy, E. W. Bennett, and G. L. Bush. 1972. Mating behavior in *Rhagoletis pomonella* (Diptera: Tephritidae). II. Temporal organization. *Canadian Entomologist* 104:97–104.

R. J. Prokopy and G. L. Bush. 1973. Mating behavior in *Rhagoletis pomonella* (Diptera: Tephritidae). IV. Courtship. *Canadian Entomologist* 105:873–891.

R. J. Prokopy and G. L. Bush. 1973. Oviposition by grouped and isolated apple maggot flies. *Annals of the Entomological Society of America* 66:1197–2000.

R. J. Prokopy and G. L. Bush. 1973. Oviposition responses to different sizes of artificial fruit by flies of *Rhagoletis pomonella* species group. *Annals of the Entomological Society of America* 66:927–929.

R. J. Prokopy, V. Moericke, and G. L. Bush. 1973. Attraction of apple maggot flies to odor of apples. *Environmental Entomology* 2:743–749.

E. F. Boller and G. L. Bush. 1974. Evidence for genetic variation in populations of the European cherry fruit fly, *Rhagoletis cerasi* (Diptera: Tephritidae) based on physiological parameters and hybridization experiments. *Entomologia Experimentalis et Applicata* 17:279–293.

G. L. Bush. 1974. The mechanism of sympatric host race formation of the true fruit flies (Tephritidae). In: *Genetic Mechanisms of Speciation in Insects*. M. J. D. White, ed. Brookvale, N.S.W.: Australia and New Zealand Book Co., pp. 3–23.

M. D. Huettel and G. L. Bush. 1974. Enzyme polymorphisms and the differentiation of sibling species. *Second International Symposium on Biological Control* 145–150.

G. L. Bush. 1975. Genetic changes occurring in flight muscle enzymes of the screwworm fly during mass-rearing. *Journal of the New York Entomological Society* 83:275–276.

G. L. Bush. 1975. Genetic variation in natural insect populations and its bearing on mass-rearing programs. In: *Controlling Fruit Flies by the Sterile-Insect Technique*. Proceedings Series, International Atomic Energy Agency, Vienna, pp. 9–17.

G. L. Bush. 1975. Modes of animal speciation. *Annual Review of Ecology and Systematics* 6:339–364.

G. L. Bush. 1975. Sympatric speciation in phytophagous parasitic insects. In: *Evolutionary Strategies of Parasitic Insects and Mites*. P. W. Price, ed. Plenum, New York, pp. 187–206.

V. Moericke, R. J. Prokopy, S. Berlocher, and G. L. Bush. 1975. Visual stimuli eliciting attraction of *Rhagoletis pomonella* (Diptera: Tephritidae) flies to trees. *Entomologia Experimentalis et Applicata* 18:497–507.

A. C. Wilson, G. L. Bush, S. M. Case, and M. C. King. 1975. Social structuring of mammalian populations and rate of chromosomal evolution. *Proceedings of the National Academy of Sciences USA* 72:5061–5065.

E. F. Boller, K. Russ, V. Vallo, and G. L. Bush. 1976. Incompatible races of European cherry fruit fly, *Rhagoletis cerasi* (Diptera: Tephritidae), their origin and potential use in biological control. *Entomologia Experimentalis et Applicata* 20:237–247.

G. L. Bush. 1976. Sex, and the screwworm fly. *Discovery* 1:4–7.

G. L. Bush and M. D. Huettel. 1976. Population and ecological genetics of fruit flies. In: *Studies in Biological Control*. Volume 9. V. L. Delucchi, ed. International Biological Programs, Cambridge University Press, Cambridge, pp. 43–49.

G. L. Bush and R. W. Neck. 1976. Ecological genetics of the screwworm fly, *Cochliomyia hominivorax* (Diptera: Calliphoridae) and its bearing on the quality control of mass-related insects. *Environmental Entomology* 5:821–826.

G. L. Bush, R. W. Neck, and G. B. Kitto. 1976. Screwworm eradication: inadvertent selection for noncompetitive ecotypes during mass rearing. *Science* 193:491–493.

G. L. Bush. 1977. Planning a rational quality control program for the screwworm fly. In: *The Screwworm Problem: Evolution of Resistance to Biological Control*. R. Richardson, ed. University of Texas Press, Austin: pp. 37–47.

G. L. Bush. 1977. The use of gel electrophoresis to monitor genetic variation and maintain quality in mass reared fruit flies. In: *Quality Control: An Idea Book for Fruit Fly Workers*. E. F. Boller and D. L. Chambers, eds. International Organization for Biological Control Publication Series. WPRS Bulletin, pp. 143–145.

G. L. Bush and E. Boller. 1977. The chromosome morphology of the *Rhagoletis cerasi* species complex (Diptera: Tephritidae). *Annals of the Entomological Society of America* 70:316–318.

G. L. Bush S. M. Case, A. C. Wilson, and J. L. Patton. 1977. Rapid speciation and chromosomal evolution in mammals. *Proceedings of the National Academy of Sciences USA* 74:3942–3946.

G. L. Bush. 1978. Book review: *Geographic Variation, Speciation, and Clines*, John A. Endler. *Systematic Zoology* 27:482–483.

G. L. Bush and G. B. Kitto. 1978. Application of genetics to insect systematics and analysis of species differences. In: *Beltsville Symposia in Agricultural Research*. 2. Biosystematics in Agriculture. J. Rombergus, R. Gorte, L. Knutson, and P. Lent, eds. Allangeld, Osmern, Montclair, N.J., pp. 89–118.

G. L. Bush. 1979. Ecological genetics and quality control. In: *Genetics in Relation to Insect Management*. M. Hoy, C. Kochler, and J. McKelvey, eds. Rockefeller Foundation Special Report Series. Rockefeller Foundation, New York, pp. 145–152.

D. P. Pashley and G. L. Bush. 1979. The use of allozymes in studying insect movement with special reference to the codling moth, *Laspeyresia pomonella* (L.) (Olethreutidae). In: *Movement of Highly Mobile Insects: Concepts and Methodology in Research*. R. L. Rabb and G. G. Kennedy, eds. North Carolina State University, Raleigh, pp. 333–341.

G. L. Bush. 1980. The sympatric colonization of new hosts by parasites. *Second International Congress of Systematic and Evolutionary Biology* 40.

J. S. Morgante, A. Malavasi, and G. L. Bush. 1980. Biochemical systematics and evolutionary relationships of neotropical *Anastrepha*. *Annals of the Entomological Society of America* 73:622–630.

G. L. Bush. 1981. Book review: *Evolutionary Biology of Parasites*, Peter W. Price. *American Scientist* 69:95–96.

G. L. Bush. 1981. Stasipatric speciation and rapid evolution in animals. In: *Evolution and Speciation*. W. R. Atchley and R. S. Woodruff, eds. Cambridge University Press, Cambridge, pp. 201–218.

S. H. Berlocher, and G. L. Bush. 1982. An electrophoretic analysis of *Rhagoletis* (Diptera: Tephritidae) phylogeny. *Systematic Zoology* 31:136–155.

G. L. Bush. 1982. What do we really know about speciation? In: *Perspectives on Evolution*. R. Milkman, ed. Sinauer, Sunderland, Mass., pp. 119–131.

G. L. Bush, and S. R. Diehl. 1982. Host shifts, genetic models of sympatric speciation and the origin of parasitic insect species. In: *Proceedings 5th International Symposium on Insect-Plant Relationships*. J. H. Visser and A. K. Minks, eds. Pudoc, Wageningen, Netherlands, pp. 297–305.

G. L. Bush. 1982. Goldschmidt's follies. Review of *The Material Basis of Evolution*, R. Goldschmidt. *Paleobiology* 8:463–469.

S. R. Diehl and G. L. Bush. 1983. The use of trace element analysis for determining the larval host plant of adult *Rhagoletis* (Diptera: Tephritidae). *Proceedings 1st International Symposium on Fruit Flies of Economic Importance*, 276–284.

M. C. Rossiter, D. J. Howard, and G. L. Bush. 1983. Symbiotic bacteria of *Rhagoletis pomonella. Proceedings 1st International Symposium on Fruit Flies of Economic Importance*, 77–84.

G. L. Bush and M. A. Hoy. 1984. Evolutionary processes in insects. In: *Insect Ecology*. C. B. Huffaker and R. L. Rabb, eds. Wiley, New York, pp. 247–277.

S. R. Diehl, and G. L. Bush. 1984. An evolutionary and applied perspective of insect biotypes. *Annual Review of Entomology* 29:471–504.

H. Zwölfer and G. L. Bush. 1984. Sympatrische und parapatrische Artbildung. *Zeitschrift fur Zoologische Systematik und Evolutions forschung* 22:211–233.

D. J. Howard, G. L. Bush, and J. A. Breznak. 1985. The evolutionary significance of bacteria associated with *Rhagoletis. Evolution* 39:405–414.

G. L. Bush and D. J. Howard. 1986. Allopatric and nonallopatric speciation: assumptions and evidence. In: *Evolutionary Processes and Theory*. S. Karlin and E. Nevo, eds. Academic Press, New York, pp. 411–438.

T. J. Bierbaum and G. L. Bush. 1987. A comparative study of host plant acceptance behaviors in *Rhagoletis* fruit flies. In: *Insect-Plants*. V. Labeyrie, G. Fabres, and D. Lachaise, eds. W. Junk, Dorderecht, pp. 374–375.

E. Boller, R. Schöni, and G. L. Bush. 1987. Oviposition deterring pheromone in *Rhagoletis cerasi*: biological activity of a pure single compound verified in semi-field test. *Entomologia Experimentalis et Applicata* 45:17–22.

G. L. Bush. 1987. Evolutionary behavior genetics. In: *Evolutionary Genetics of Invertebrate Behavior*. M. Huettel, ed. Plenum Press, New York, pp. 1–5.

R. Sarma, G. B. Kitto, S. Berlocher, and G. L. Bush. 1987. Biochemical and immunological studies on an α-glycerophosphate dehydrogenase from the tephritid fly *Anastrepha suspensa. Archives of Insect Biochemistry and Physiology* 4:271–286.

T. J. Bierbaum and G. L. Bush. 1988. Divergence in key host examining and acceptance behaviors of sibling species *Rhagoletis mendax* and *R. pomonella* (Diptera: Tephritidae). In: *Fruit Flies of Economic Importance: Bionomics, Ecology and Management*. M. T. AliNiazee. ed. Oregon State University Special Report 830, pp. 26–55.

J. L. Feder, C. A. Chilcote, and G. L. Bush. 1988. Genetic differentiation between sympatric host races of *Rhagoletis pomonella. Nature* 336:61–64.

G. L. Bush, J. L. Feder, S. H. Berlocher, B. A. McPheron, D. C. Smith, and C. A Chilcote. 1989. Sympatric origins of *R. pomonella. Nature* 339:346.

S. Diehl and G. L. Bush. 1989. The role of habitat preference in adaptation and speciation. In: *Speciation and Its Consequences*. D. Otte and J. Endler, eds. Sinauer, Sunderland, Mass., pp. 345–365.

J. L. Feder and G. Bush. 1989. A field test of differential host-plant usage is an effective premating barrier between two sibling species of *Rhagoletis pomonella* fruit flies (Diptera: Tephritidae) and its consequences for sympatric models of speciation. *Evolution* 43:1813–1819.

J. L. Feder, C. A. Chilcote, and G. Bush. 1989. Are the apple maggot, *Rhagoletis pomonella* and the blueberry maggot. *R. mendax* (Diptera: Tephritidae) distinct species? Implications for sympatric speciation. *Entomologia Experimentalis et Applicata* 51:113–123.

J. L. Feder, C. A. Chilcote, and G. Bush. 1989. Gene frequency clines for host races of *Rhagoletis pomonella* in midwestern United States. *Heredity* 63:245–266.

D. J. Howard and G. L. Bush. 1989. The influence of bacteria on larval survival and development in *Rhagoletis* (Diptera: Tephritidae). *Annals of the Entomological Society of America* 82:633–640.

J. L. Feder, C. A. Chilcote, and G. Bush. 1989. Inheritance and linkage relationships of allozymes in the apple maggot fly. *Journal of Heredity* 80:277–283.

T. J. Bierbaum, and G. L. Bush. 1990. Genetic differentiation in the viability of sibling species of *Rhagoletis* fruit flies on host plants, and the influence of reduced hybrid viability on reproductive isolation. *Entomologia Experimentalis et Applicata* 55:105–118.

T. J. Bierbaum and G. L. Bush. 1990. Host fruit chemical stimuli eliciting distinct ovipositional responses from sibling species of *Rhagoletis* fruit flies. *Entomologia Experimentalis et Applicata* 56:165–177.

J. L. Feder, C. A. Chilcote, and G. Bush. 1990. Regional, local and microgeographic genetic variation between apple and hawthorn populations of *Rhagoletis pomonella* (Diptera: Tephritidae) in western Michigan. *Evolution* 44:595–608.

J. L. Feder, C. A. Chilcote, and G. Bush. 1990. The geographic pattern of genetic differentiation between host races of *Rhagoletis pomonella* (Diptera: Tephritidae) in eastern United States and Canada. *Evolution* 44:570–594.

J. E. Frey, and Guy L. Bush. 1990. *Rhagoletis* sibling species and host races differ in host odor recognition. *Entomologia Experimentalis et Applicata* 57:123–131.

G. L. Bush. 1991. Molecular genetics applied to systematics. In: *Entomology Serving Society: Emerging Technologies and Challenges*. S. B. Vinson and R. Metcalf, eds. Entomological Society American Centennial National Symposium. Entomological Society of America, Lanham, Md., pp. 86–97.

J. E. Frey, T. J. Bierbaum,, and G. L. Bush. 1992. Differences among sibling species *Rhagoletis mendax* and *R. pomonella* (Diptera: Tephritidae) in their antennal sensitivity to host fruit compounds. *Journal of Chemical Ecology* 18: 2001–2023.

S. H. Berlocher, B. McPheron, J. Feder, and G. Bush. 1993. Genetic differentiation at allozyme loci in the *Rhagoletis pomonella* (Diptera: Tephritidae) species complex. *Annals of the Entomological Society of America* 86:716–727.

G. L. Bush. 1993. A reaffirmation of Santa Rosalia, or why are there so many kinds of *small* animals? In: *Evolutionary Patterns and Process*. D. Edwards and D. R. Lees, eds. Academic Press, New York, pp. 229–249.

G. L. Bush. 1993. Host race formation and speciation in *Rhagoletis* fruit flies (Diptera: Tephritidae). *Psyche* 99:335–357.

Feder, J. L., T. A. Hunt, and G. L. Bush. 1993. The effects of climate, host phenology and host fidelity on the genetics of apple and hawthorn infesting races of *Rhagoletis pomonella*. *Entomologia Experimentalis et Applicata* 69:117–135.

R. J. Prokopy and G. L. Bush. 1993 Evolution in an orchard. *Natural History* 9:4–10.

G. L. Bush. 1994. Sympatric speciation: new wine in old bottles. *Trends in Ecology & Evolution* 9:285–288.

J. J. Smith, J. S. Scott-Craig, J. R. Leadbetter, G. L. Bush, D. L. Roberts,, and D. W. Fullbright. 1994. Characteriza-

tion of random amplified polymorphic DNA (RAPD) products from *Xanthomonus campestris* and some comments on the use of RAPD products in phylogenetic analysis. *Molecular Phylogenetics and Evolution* 3:135–145.

G. L. Bush. 1995. *Bemisia tabaci*: biotype or species complex? *Proc. 2nd Central American and Caribbean Workshop on White Flies* (Commission National de Moscas Blancas, Managua, Nicaragua) 25–30.

P. A. Johnson, F. C. Hoppensteadt, J. J. Smith, and G. L. Bush. 1996. Conditions for sympatric speciation: a diploid model incorporating habitat fidelity and non-habitat assortative mating. *Evolutionary Ecology* 10:187–205.

J. E. Frey and G. L. Bush. 1996. Impaired host odor perception in hybrids between the sibling species *Rhagoletis pomonella* and *R. mendax*. *Entomologia Experimentalis et Applicata* 80:163–165.

J. J. Smith and G. L. Bush. 1997. Phylogeny of the genus *Rhagoletis* (Diptera: Tephritidae) inferred from DNA sequences of mitochondrial cytochrome oxidase II. *Molecular Phylogeny and Evolution* 7:33–43.

G. L. Bush and J. J. Smith. 1997. The sympatric origin of phytophagous insects. In: *Vertical Food Web Interactions: Evolutionary Patterns and Driving Forces*. K. Dettner, G. Bauer, and W. Völkl, eds. Ecological Studies. Heidelberg, Springer-Verlag, 130:3–19.

G. L. Bush. 1998. The conceptual radicalization of an evolutionary biologist. This volume.

G. L. Bush and J. J. Smith. In press. The genetics and ecology of sympatric speciation: a case study. *Researches on Population Ecology*.

Index